D1532278

INTRODUCTORY ELECTRONICS FOR SCIENTISTS AND ENGINEERS

INTRODUCTORY ELECTRONICS FOR SCIENTISTS AND ENGINEERS

Robert E. Simpson
Associate Professor of Physics
University of New Hampshire

Allyn and Bacon, Inc.
Boston • London • Sydney • Toronto

Printed in the United States of America.

10 9 8 85 84 83 82 81 80

Library of Congress Catalog Card Number: 73-84908

ISBN: 0-205-03845-X

TO SUSAN AND JEFFREY

CONTENTS

Preface ix

chapter 1 / Direct Current Circuits 1

1.1 Electric Charge and Current 1
1.2 Voltage 3
1.3 Resistance 4
1.4 Ohm's Law 6
1.5 Batteries 11
1.6 Power 13
1.7 Kirchhoff's Laws 17
1.8 Voltage Dividers 22
1.9 Ideal Voltage and Current Sources 24
1.10 Thevenin's Theorem 26
1.11 Norton's Theorem 28

chapter 2 / Alternating Current Circuits 35

2.1 Periodic Waveforms 35
2.2 AC Power 39
2.3 Capacitance 41
2.4 Capacitive Reactance 46
2.5 Inductance 48
2.6 Mutual Inductance 52
2.7 Inductive Reactance 54
2.8 The Complex Voltage Plane 56
2.9 The *RC* High-Pass Filter 57
2.10 The *RC* Low-Pass Filter 61
2.11 *RLC* Circuits 64
2.12 Series and Parallel Resonance 70
2.13 "Q" (Quality Factor) 75

chapter 3 / Fourier Analysis and Pulses 83

3.1 Introduction 83
3.2 Description of a Pulse 83
3.3 Fourier Analysis 84
3.4 Integrating Circuit (Low-Pass Filter) 97
3.5 Differentiating Circuit 101
3.6 Compensated Voltage Divider 108

chapter 4 / Semiconductor Physics 115

4.1 Introduction 115
4.2 Energy Levels 115
4.3 Crystals 116
4.4 Energy Levels in a Crystal Lattice 118
4.5 The Pauli Exclusion Principle 119
4.6 Fermi-Dirac Statistics 120
4.7 Electron Energy Distribution 123
4.8 Conduction in Semiconductors 126
4.9 p–n Junctions 132
4.10 Diode Applications 147

chapter 5 / Transistor Action and Simplified Amplifier Design 157

5.1 Introduction 157
5.2 Transistor Construction 157
5.3 Biasing and Current Flow inside a Transistor 159
5.4 Amplification 163
5.5 Biasing and Graphical Treatment 165
5.6 Stability Factor 173
5.7 Common Emitter Amplifier Design 177

chapter 6 / *h* Parameter Equivalent Circuit and Amplifiers 187

6.1 Introduction 187
6.2 Transistor Equivalent Circuits 187
6.3 Common Emitter Configuration 192
6.3.1 Voltage Gain 6.3.2 Current Gain
6.3.3 Input Impedance 6.3.4 Output Impedance
6.4 Common Collector Configuration 206
6.5 Common Base Configuration 214

chapter 7 / The Field Effect Transistor (FET) 225

7.1 Introduction 225
7.2 FET Construction 225
7.3 FET Characteristic Curves 231
7.4 FET *y* Parameter Equivalent Circuit 233
7.5 FET Temperature Effects 238
7.6 Biasing and Practical Circuits 239
7.7 The Common Source FET Amplifier 243

7.8 The Common Drain FET Amplifier 247
7.9 The Common Gate FET Amplifier 250
7.10 The Metal Oxide Semiconductor Field Effect Transistor (MOSFET) 251

chapter 8 / Vacuum Tubes 259

8.1 Introduction 259
8.2 Thermionic Emission 260
8.3 The Vacuum Tube Diode 262
8.4 The Vacuum Tube Triode 264
8.5 The Vacuum Tube Pentode 270
8.6 Representative Vacuum Tube Circuits 274

chapter 9 / Feedback and Practical Amplifier Circuits 281

9.1 Introduction 281
9.2 Negative Feedback 282
9.3 Examples of Negative Feedback Amplifier Circuits 289
9.4 Operational Amplifiers 296
9.5 Positive Feedback 306
9.6 Practical Comments and Neutralization 317

chapter 10 / High Frequencies 325

10.1 Introduction 325
10.2 Electromagnetic Radiation 326
10.3 Series Inductance and Shunt Capacitance 331
10.4 High Frequency FET Equivalent Circuit and the Miller Effect 333
10.5 The Cascode Amplifier 336
10.6 The Gain–Bandwidth Product 338
10.7 Diffusion Time and Transit Time 340
10.8 Transmission Lines 342
10.9 Distributed Amplifiers 349
10.10 Tuned Amplifiers 350
10.11 Varactor Frequency Multipliers 352
10.12 Classes of Amplification 355
10.13 Microstrip and Strip Line Techniques 357
10.14 Waveguides 359
10.15 Microwave Oscillators 364

chapter 11 / Noise 371

11.1 Introduction 371
11.2 Interference 372
11.3 Thermal Noise or Johnson Noise 373
11.4 Shot Noise 378
11.5 Calculation of Amplifier Noise 381
11.6 Flicker Noise 390
11.7 Noise Temperature 393
11.8 Lock-in Detection 394
11.9 Signal Averaging Techniques 400
11.10 Correlation Techniques 402

chapter 12 / Pulse and Digital Circuits 407

12.1 Introduction 407
12.2 Logic Circuits 409
12.3 The OR Circuit 411
12.4 The AND Circuit 416
12.5 The NOT Circuit 419
12.6 The NOR Circuit 420
12.7 The NAND Circuit 421
12.8 The Exclusive OR Circuit 421
12.9 The Exclusive NOR Circuit 422
12.10 Binary Addition 423
12.11 The Pulse-Height Analyzer 427
12.12 The Bistable Multivibrator 431
12.13 The Monostable Multivibrator 441
12.14 The Astable Multivibrator 444
12.15 The Schmitt Trigger Circuit 445

chapter 13 / Integrated Circuits 451

13.1 Introduction 451
13.2 Integrated Circuit Construction 453
13.3 Special Design Features of Integrated Circuits 454
13.4 The μA 709 Operational Amplifier 457
13.5 I.C. RF Amplifier 460
13.6 An I.C. TTL NAND Gate 463
13.7 An I.C. TTL NOR Gate 464
13.8 An I.C. TTL Inverter 464
13.9 I.C. TTL Flip-Flops 465
13.10 An I.C. TTL Monostable Multivibrator 471
13.11 Design Examples Using I.C.'s 471
13.12 I.C. Regulated Power Supplies 477
13.13 Complementary MOSFET Circuts 483

Appendices 487

A Components: Resistors, Capacitors, Inductors, and Transformers 487
B Batteries 491
C Measuring Instruments: Voltmeters, Ammeters, Ohmmeters, VOM, Oscilloscope, and Bridges 495
D Cables and Connectors 507
E Complex Numbers 511
F Determinants 515
G Characteristic Curves and Parameters for Selected Semiconductor Devices 519
H Nuclear Radiation Detectors 529
I Suggested Laboratory Experiments 533

Index 567

PREFACE

This book was written for a course in experimental electronics for junior physics majors and junior and senior mechanical engineering majors. These students have previously had three semesters of college physics at the level of Weidner and Sells. They also have had four semesters of college math including differential and integral calculus in several dimensions and ordinary differential equations. I have assumed the student can readily differentiate and integrate simple functions such as sin (kt) and exp (kt), but not solve differential equations or do Fourier analysis, both of which topics are covered briefly in the text. A familiarity with complex numbers is assumed, and a short review of complex numbers as applied to ac circuits is contained in Appendix E. If a concept can be best explained by using a derivative or by appealing to the general idea of Fourier analysis, I will do so, but I will not spend several pages in deriving complicated formulae which are rarely used by practicing experimentalists and engineers.

My general philosophy of teaching electronics is to emphasize the practical rather than the theoretical. The course in which this text is used consists of two one-hour lectures and two three-hour laboratories each week. I have found that the two lectures are absolutely essential, and that it is often desirable to lecture for the first ten or fifteen minutes of some of the laboratory sessions. I also feel that a student should learn the approximate numerical value of the various voltages and resistances in a typical circuit, and to this end I have tried to include actual component and voltage values in most of the schematic diagrams.

The first two chapters are somewhat in the nature of a review of dc and ac circuits as most students have previously been exposed to this material. However, I have found by sad experience that students at the junior level do need a review of Kirchhoff's laws, capacitive and inductive reactance, and especially the practical use of the oscilloscope and the wiring of circuits.

Chapter 3 is devoted to the often neglected topics of pulses, Fourier analysis, and its application to modulation and timing.

Considerable time is spent in chapters 4 and 5 in giving the reader a physical feeling for the behavior of semiconductor material and a "seat-of-the-pants" design procedure for most common transistor amplifier circuits. Specific amplifier circuits are given with component values and *measured* circuit performance.

The concept of equivalent circuits is very carefully covered in chapter 6 with special attention given to the derivation of the final equivalent circuit. The concepts of input and output impedance are very carefully explained from both a physical and a mathematical viewpoint.

Chapter 7 is entirely devoted to field effect transistors (FET's) and specific FET circuits with component values given. Again, both a physical and a mathematical approach are used. The metal oxide semiconductor field effect transistor (MOSFET) is covered.

The essentials of vacuum tubes are treated in chapter 8 so that the student will have some familiarity with tube concepts if he encounters them in older or special-purpose equipment containing tube circuitry. The student should be familiar with the high plate voltages, the filament circuits, and the approximate component values of tube circuits.

Chapter 9 covers feedback with sample circuits using both positive and negative feedback, including operational amplifiers.

The elementary theory and experimental pitfalls of high frequency circuits are treated in chapter 10, including simple antennas, the gain–bandwidth product, the Miller effect, varactor frequency multipliers, microwave diode oscillators, and an introduction to strip lines.

Chapter 11 is entirely devoted to the topic of noise. This is one of the few modern introductory treatments of noise in electrical circuits in print today. A simplified physical derivation of the shot noise formula is given. The mysteries of transformer impedance matching to increase the signal-to-noise ratio are explained. Several modern low noise circuit techniques are treated such as lock-in detection and signal averaging.

The final two chapters cover pulse and digital circuits and integrated circuits. The similarities and differences between discrete component circuits and integrated circuits are outlined. The single channel and multichannel pulse height analyzer are discussed in some detail.

Finally, the appendices contain a wealth of practical information on circuit components, cables, and connectors. Actual measured characteristic curves and approximate prices are given for 16 inexpensive devices (diodes, bipolar transistors, and FET's), which will satisfy most circuit needs.

The last appendix contains a list of 32 suggested student laboratory experiments with a detailed description of the apparatus necessary, including manufacturer's names and approximate prices. Specific suggestions about organizing the laboratory are given and include the schematic for an inexpensive, short circuit protected, variable voltage power supply.

This text can be used for either a one or two semester course. In one semester of fifteen weeks a reasonable pace would cover chapters 1 through 6 or 7. I usually include an introduction to negative feedback from chapter 9 in the last two weeks so that this time can be spent on one project—the design, construction and debugging of a multistage negative feedback amplifier. This project serves to pull together most of the topics of the first semester and gives the student a real feeling of accomplishment.

The second semester should cover chapters 8 through 13. Integrated circuits, the subject of chapter 13, are replacing at a very rapid rate circuits made from individual components in most applications. Therefore, one might well ask why more emphasis was not given to integrated circuits in this text. The answer is simply that the text is aimed at students with little or no electronic experience, and such students should first learn the fundamental subjects of simple networks and single transistor circuits before graduating to the "black box" approach of integrated circuits.

In our laboratory at the University of New Hampshire, the students work in pairs or singly at a bench. Each pair of students has one good oscilloscope, a short-circuit protected 0–20V dc variable voltage power supply, a variable frequency sine wave generator, and a 5″ x 17″ "Vector" terminal board and several dozen spring clip terminals for breadboarding circuits. A plug-in socket for dual in-line integrated circuits is also available. The power supply and terminal board are mounted in a small table top relay rack on the bench. The circuits thus can be constructed without soldering in a vertical plane on the terminal board at eye level directly in front of the students. It is of immense educational value to have the circuit on the terminal board look exactly like the schematic diagram; e.g., if a resistance and a capacitance make a right angle in the schematic the corresponding resistor and capacitor should make a right angle on the terminal board. This feature is absolutely essential for students with little practical experience in wiring circuits. It is also important to stress that the circuits should be neat. Clip leads should be used sparingly; most wires should be cut to order and stripped by the student. A supply of wire, side cutters, and wire strippers as well as resistors, capacitors, transistors, etc., should be available in the laboratory. Several "goof-proof" VOM's and an impedance bridge are also included in the laboratory.

A word of caution should be given to the students before the vacuum tube experiment, as the voltages used are quite literally lethal. Working with transistor circuits and their relatively low voltages often develops sloppy safety habits.

In conclusion I would like to express my thanks to several people without whose cooperation and enthusiasm this book probably never would have been written. First, a word of thanks to Professor Robert E. Houston, Jr., who has served as chairman of the physics department here at New Hampshire and who has given me much needed encouragement. I would also like to thank Phil Isaacson of the UNH Space Science Center for numerous helpful consultations on modern solid-state devices and circuits. Going back a little further in time, I must mention Professor D. A. Bromley for his incredibly stimulating and lucid lectures at the University of Rochester when I was an undergraduate. My thesis advisor, Professor Francis M. Pipkin, deserves thanks for his enthusiasm and help during my graduate work. And finally, I would like to thank my wife Donna for inspiring me to solve several knotty problems.

Robert E. Simpson
University of New Hampshire

1

DIRECT CURRENT CIRCUITS

1.1 ELECTRIC CHARGE AND CURRENT

There are two kinds of electric charge, positive and negative. They are so named because they add and subtract like positive and negative numbers. All atoms contain charge; the usual picture of an atom is a small (10^{-13}cm diameter) positively charged nucleus around which negatively charged electrons move in orbits of the order of 10^{-8}cm diameter. Charge is measured in coulombs in mks units and in statcoulombs in the cgs system. The basic indivisible unit of charge is the charge on one electron which is -1.6×10^{-19} coulombs or -4.8×10^{-10} statcoulombs. All electric charges (positive and negative) are, in magnitude, integral multiples of the charge of the electron. However, in most electronic circuit problems the discrete nature of electric charge may be neglected, and charge may be considered to be a continuous variable. One of the most fundamental conservation laws of physics says that in any closed system the total net amount of electric charge is conserved or, in other words, is constant in time. For example, in a semiconductor if one electron is removed from a neutral atom then the atom minus the one electron has a net electric charge of $+1.6 \times 10^{-19}$ coulombs.

The flow of electric charge, either positive or negative, is called "current"; that is, the current I at a given point in a circuit is defined as the time rate of change of the amount of electric charge Q flowing past that point.

$$I \equiv \frac{dQ}{dt} \tag{1.1}$$

The direction of the current is taken by convention to be the direction of the flow of *positive* charge. If electrons are flowing from right to left in a wire, then this electron current is electrically equivalent to positive charge flowing from left to right; hence, we say the current is to the right. In a current carrying wire it is actually the electrons which flow, but nevertheless, the direction of current is taken as the direction of the equivalent positive charge flow, which is opposite to the direction of the electron flow.

Current is measured in "amperes"; one ampere of current is the flow of one coulomb of charge per second which is 6.25×10^{18} electrons per second. Other units of current are the milliampere (mA) which is 0.001 ampere, the microampere (μA) which is 10^{-6} ampere, the nanoampere (nA) which is 10^{-9} ampere, and the picoampere (pA) which is 10^{-12} ampere. Prefixes for various powers of ten are given in Table 1-1. Typical currents which flow in low power transistor electronic circuits, such as in small radio receivers and amplifiers, are of the order of 1 to 10 mA; typical currents in low power vacuum tube circuits are of the order of 10 to 100 mA. The largest current normally encountered in vacuum tube circuits is about 500 mA, but specially designed high current transistors are available which carry currents from 1 to 100 amperes.

TABLE 1-1. POWERS-OF-TEN PREFIXES

Prefix	*Symbol*	*Meaning*	*Example*
giga	G	10^9	1 gigahertz = 1 GHz = 10^9 hertz
mega	M	10^6	1 megohm = 1 MΩ = 10^6 ohms
kilo	k	10^3	1 kilovolt = 1 kV = 10^3 volts
milli	m	10^{-3}	1 milliampere = 1 mA = 10^{-3} amperes
micro	μ	10^{-6}	1 micro volt = 1 μV = 10^{-6} volts
nano	n	10^{-9}	1 nanoampere = 1 nA = 10^{-9} amperes
pico	p	10^{-12}	1 picofarad = 1 pF = 10^{-12} farads

There are two general kinds of current, direct current ("dc") and alternating current ("ac"). Direct current is a flow of charge in which the direction of flow is always the same. If the magnitude of the current varies from one instant of time to another, but the direction of flow remains the same, then this type of current is called "pulsating" direct current. If both the direction and the magnitude of the flow of charge are constant, then this current is called "pure direct current" or simply "direct current." If the charge flows back and forth

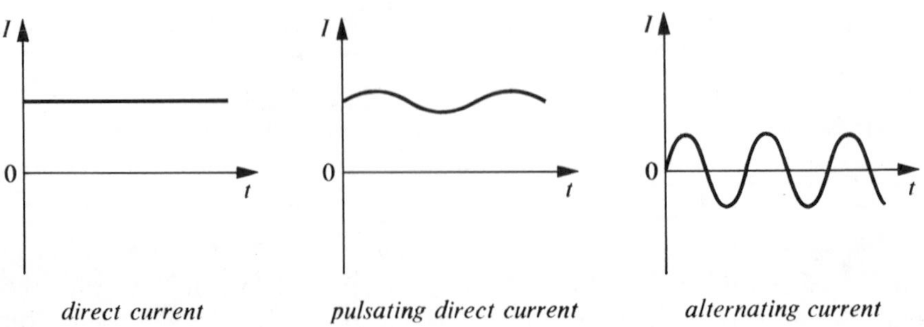

FIGURE 1.1 Types of current.

alternately, then this current is called "alternating current" (to be discussed further in Chapter 2). Graphs of these three kinds of currents are shown in Fig. 1.1.

It is sometimes useful to think of electric current in a wire as being similar to the flow of water in a pipe, with the water molecules being analogous to the electrons in the wire. A flow of water in grams/second is analogous to a flow of electric charge in coulomb/second or amperes.

1.2 VOLTAGE

The voltage, or electric potential, V at a given point in a circuit is defined as the potential energy W a positive charge Q would have at that point divided by the magnitude of the charge, $V \equiv W/Q$. The voltage V is measured relative to some other specified reference point in the circuit where we say the energy of all charges is zero. This reference point is usually called "ground" or "earth" in British literature. A good ground is a cold water pipe or sometimes just the metal chassis or box enclosing the circuit. The point is that the circuit "ground" has a constant, unvarying potential which we set equal to zero volts by convention. The earth has a constant potential or voltage simply because it is so large and is a reasonably good conductor. Any charge taken from or added to the earth through a circuit's ground wire will not appreciably change the total charge on the earth or the earth's voltage. The symbols used for ground in circuit notation are shown in Fig. 1.2.

FIGURE 1.2 Ground symbols.

The units of voltage are joules/coulomb; one volt (V) is defined as one joule/coulomb. Other units of voltage are the kilovolt (kV) which is 1000 V, the millivolt (mV) which is 0.001 V, the microvolt (μV) which is 10^{-6} V, and the nanovolt which is 10^{-9} V. Notice that the voltage at a given point has no *absolute* meaning, but only means the potential energy per unit charge *relative to ground.* This energy per unit charge definition of voltage is rather difficult to understand intuitively until one considers the direction the charges will tend to move. For, in all natural processes, things tend to move in such a way as to minimize the potential energy; thus, positive charges will move from points of higher voltage toward points of lower voltage, for example from a point with a voltage of +15 V toward a point of voltage +12 V. Similarly, negative charges will tend to move from points of lower voltage toward points of higher voltage, because this direction of charge flow will minimize their potential energy. For example, negative charge (electrons) will move from a point of voltage -12 V to a point of voltage -7 V. In terms of the analogy between current and water flow, the voltage is analogous to the water pressure, because water tends to flow from points of higher pressure toward points of lower pressure. It is also sometimes useful to think of the voltage as causing or forcing the flow of current, just as one thinks of the water pressure as causing or forcing the water flow. Thus, voltage is a "cause" and current an "effect."

1.3 RESISTANCE

If a current I flows through any two terminal circuit elements, then the static or dc resistance (usually referred to as simply the "resistance") of that circuit element is defined as the difference in voltage $V_2 - V_1$ between the two terminals divided by the current I (see Fig. 1.3). Strictly speaking, this definition applies

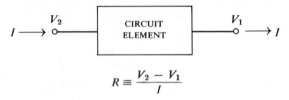

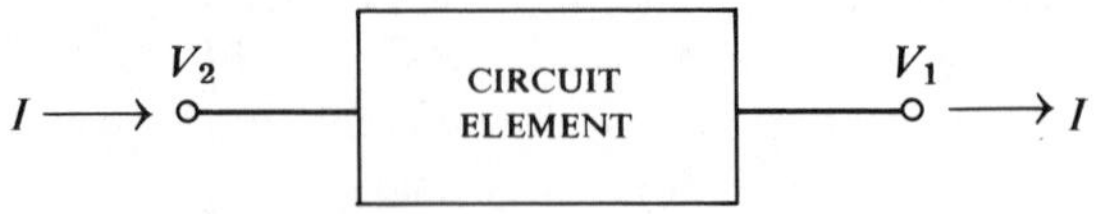

$I \longrightarrow$ V_2 ○—— CIRCUIT ELEMENT ——○ V_1 $\longrightarrow I$

$$R \equiv \frac{V_2 - V_1}{I}$$

FIGURE 1.3 Definition of resistance R.

only to a circuit element which converts electrical energy into heat, but this situation occurs in the overwhelming majority of cases in electronic circuitry.

$$R \equiv \frac{V_2 - V_1}{I} \tag{1.2}$$

The current into the circuit element must exactly equal the current leaving from the conservation of charge law. V_2 is the voltage at the terminal where the current enters the circuit element; V_1 is the voltage where the current leaves. If the circuit element is "passive", that is, if there is no energy given to the charge by the element, then the charge loses energy in the element. Thus, V_2 must be greater than V_1. Thus, for all passive circuit elements, the static or dc resistance is positive. The current–voltage graph for an ordinary positive resistance is shown in Fig. 1.4(a).

In a formal sense, a battery has a negative resistance, because the battery gives energy to the charge making V_1 greater than V_2, but this is because chemical energy is converted into electrical energy in a battery. Some circuit elements

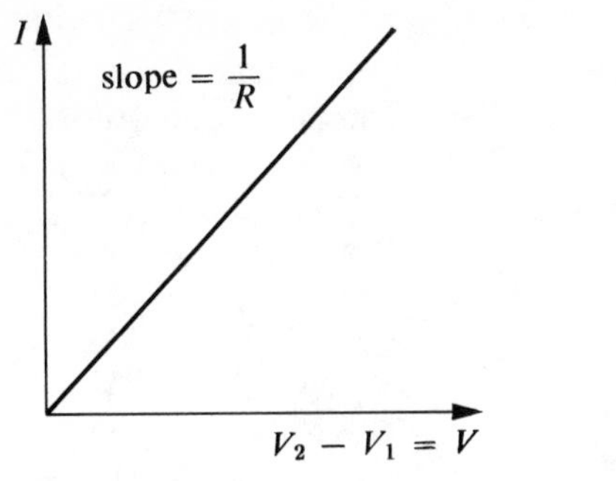

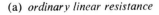

(a) *ordinary linear resistance*

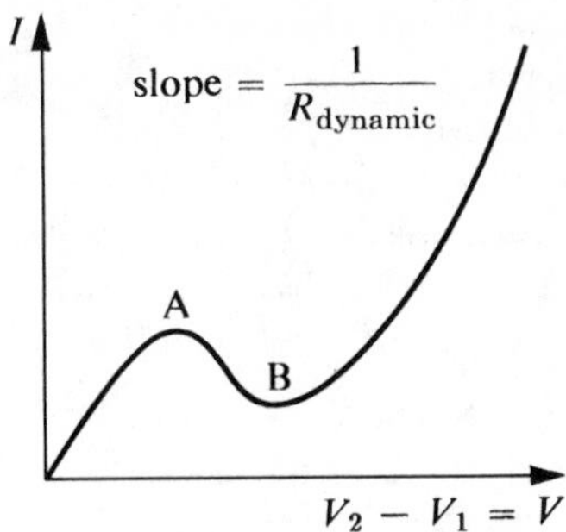

(b) *nonlinear resistance (tunnel diode)*

FIGURE 1.4 Current-voltage curves.

are said to have a negative resistance; but this always refers to *dynamic* resistance which is defined as the rate of change of voltage with respect to current:

$$R_{\text{dynamic}} \equiv dV/dI \tag{1.3}$$

where dV is the change in the voltage across the circuit element and dI is the change in the current through the element. In Fig. 1.4(b), which is the current-voltage characteristic for a tunnel diode, the static resistance is always positive for all values of V, but between A and B the dynamic resistance is negative because I decreases as V increases. Also note that both the static and the dynamic resistance vary with voltage in Fig. 1.4(b). A circuit element for which the current voltage curve is not a straight line is said to have a "nonlinear" resistance.

The units of resistance are volts/ampere; one ohm (Ω) is defined as one volt per ampere. Other units of resistance are the kilohm (kΩ sometimes just written k) which is 1000 ohms and the megohm (MΩ or just M) which is 10^6 ohms. A "resistor" is a two terminal circuit element specifically manufactured to have a constant resistance. Thus a 4.7 kΩ resistor has a resistance of 4,700 ohms; a 2.2 MΩ resistor has a resistance of 2,200,000 ohms.

Usually the resistance of a resistor is given by a "color code" (see Fig. 1.5). Bands of different colors specify the resistance, according to the following rule:

$R =$ (*first color number*) (*second color number*) × 10 (*raised to the third color number*)

The first digit of the resistance is given by the color band closest to the end of the resistor. The fourth color band gives the tolerance of the resistor. Silver means ±10% tolerance, gold means ±5% tolerance, and no fourth colored band means ±20% tolerance. A 2 kΩ 10% resistor will have a resistance somewhere between 1.8 kΩ and 2.2 kΩ.

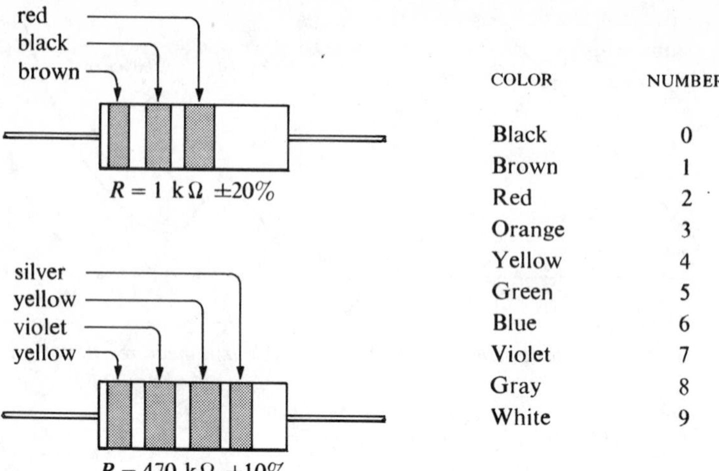

COLOR	NUMBER
Black	0
Brown	1
Red	2
Orange	3
Yellow	4
Green	5
Blue	6
Violet	7
Gray	8
White	9

FIGURE 1.5 Resistor color code.

If one has difficulty remembering the color code, then one can note that the colors follow the colors of the visible spectrum starting with red for the number two and going through violet for the number seven. Or, one can use the following scheme: "**B**lasphemous **B**oys **R**ape **O**ur **Y**oung **G**irls **B**ut **V**iolet **G**ives **W**illingly," the first letter of each word being the first letter of the colors for the numbers zero through nine. And it is extremely useful to note that if the third color is brown, the resistor is in the hundreds of ohms; red—thousands of ohms, orange—tens of thousands of ohms, yellow—hundreds of thousands of ohms, green—millions of ohms.

On certain precision metal film resistors, the resistance value is specified to three significant figures. These resistors have five colored bands; the first three bands represent the three significant figures, the fourth band the multiplier, and the fifth band the tolerance. Certain high reliability resistors are tested for failure rates under conditions of maximum power and voltage, and the results are expressed in percentage failure per thousand hours. The fifth colored band represents the failure rate per thousand hours according to the following scheme: brown 1%, red 0.1%, orange 0.01%, and yellow 0.001%.

1.4 OHM'S LAW

It is empirically true for many different kinds of circuit elements, that the resistance of the circuit element is constant if we keep the temperature and composition of the element fixed. This is true over an extremely large range of voltages and currents. That is, changing the voltage difference between the two terminals by any factor will cause the current to change by exactly the same factor, i.e., doubling the voltage difference $V_2 - V_1$ exactly doubles the current I.

Ohm's law is simply the statement that the resistance is constant. It can be written in three ways:

$$R = \frac{V_2 - V_1}{I} \qquad V_2 - V_1 = IR \qquad I = \frac{V_2 - V_1}{R}. \tag{1.4}$$

Any two terminal circuit element of constant resistance is shown in circuit diagrams as a zig-zag line [see Fig. 1.6(a)] and is called a resistor. The larger

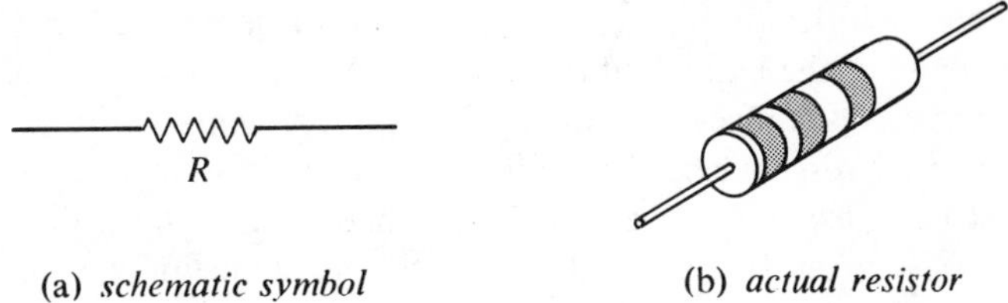

(a) *schematic symbol* (b) *actual resistor*

FIGURE 1.6 Resistance.

the voltage difference, the larger the current for a fixed resistance; the larger the resistance for a fixed voltage drop, the smaller the current. Thus, "resistance" is a very appropriate name; it literally means opposition to current flow.

These three forms of Ohm's law can be thought of in the following terms. $R = (V_2 - V_1)/I$ means that if there is a voltage difference $V_2 - V_1$ across a circuit element through which a current I is flowing, then the circuit element must have a resistance $(V_2 - V_1)/I$. $V_2 - V_1 = IR$ means that if a current I is forced through a resistance R, then a voltage difference IR will be developed between the two ends of the resistance. $I = (V_2 - V_1)/R$ means that if there is a voltage difference $V_2 - V_1$ across a resistance then a current $(V_2 - V_1)/R$ must be flowing through the resistance. Perhaps the most important thing to remember about Ohm's law is that it is only the *difference* $(V_2 - V_1)$ in voltage across a resistor which causes current to flow. Thus a $5\,\text{k}\Omega = 5000\,\Omega$ resistor with one end at 35 V and the other end at 25 V will pass a current of 2 mA because $I = (V_2 - V_1)/R = 10\,\text{V}/5000\,\Omega = 0.002\,\text{A} = 2\,\text{mA}$. A $5\,\text{k}\Omega$ resistor with one end at 1078 V and the other end at 1068 V will also pass a current of 2 mA, since the voltage difference is also 10 V. It is useful to remember the short cut that the voltage difference in volts divided by the resistance in kilohms equals current in milliamperes; in the previous example $I = 10\,\text{V}/5\,\text{k}\Omega = 2\,\text{mA}$.

If two resistors R_1 and R_2 are connected in "series" (see Fig. 1.7), that is if

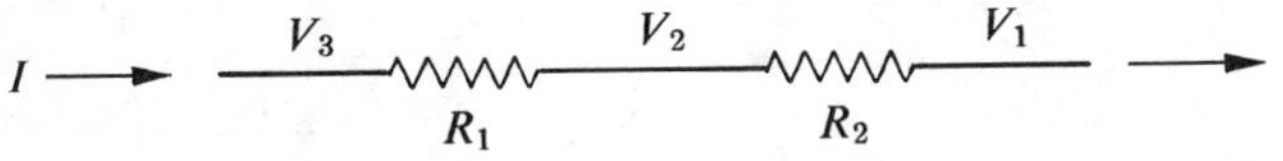

FIGURE 1.7 Two resistors in series.

they are connected "end-to-end" so that the same current flows through each of them, then the total effective resistance is simply the sum of the two individual resistances. This result follows from Ohm's law applied separately to R_1 and R_2.

$$R_{\text{total}} = \frac{V_3 - V_1}{I} = \frac{V_3 - V_2}{I} + \frac{V_2 - V_1}{I} = R_1 + R_2 \tag{1.5}$$

Thus, a 1 kΩ and a 3 kΩ resistor in series act like a single 4 kΩ resistor. This rule can be extended to N resistors in series, in which case the total effective resistance is equal to the sum of all the N individual resistances. $R_{\text{total}} = R_1 + R_2 + R_3 + \cdots + R_N$. Notice that the total resistance for a series connection always is greater than any of the individual resistances. Also notice that a straight line connecting the two resistances in a circuit diagram represents a wire or electrical connection of zero resistance. Thus all points of a straight line in a circuit diagram must be at exactly the same voltage; there is no voltage drop along a wire of zero resistance. In an actual circuit, the resistance of the wire used to connect various elements is usually negligibly small, e.g., a two-inch length of number 18 copper wire (0.04 in. diam) has a resistance of only 0.0011 Ω.

If two resistors are connected in "parallel" or "side by side" (see Fig. 1.8)

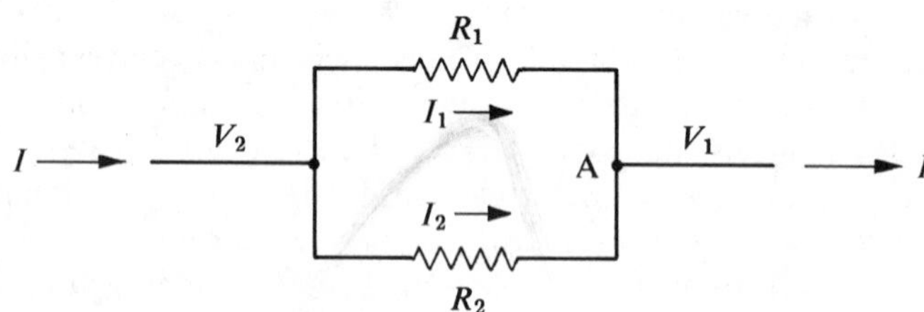

FIGURE 1.8 Two resistors in parallel.

so that the same voltage appears across each one, then the total effective resistance is given by

$$R_{\text{total}} = \frac{1}{\dfrac{1}{R_1} + \dfrac{1}{R_2}} = \frac{R_1 R_2}{R_1 + R_2}. \tag{1.6}$$

This result follows from Ohm's law and the conservation of current. At the point A, the current I entering splits up into two parts, $I = I_1 + I_2$.

$$R_{\text{total}} = \frac{V_2 - V_1}{I} = \frac{V_2 - V_1}{I_1 + I_2} = \frac{V_2 - V_1}{\dfrac{V_2 - V_1}{R_1} + \dfrac{V_2 - V_1}{R_2}} = \frac{1}{\dfrac{1}{R_1} + \dfrac{1}{R_2}} = \frac{R_1 R_2}{R_1 + R_2}$$

For example, a 6 kΩ and a 4 kΩ resistor in parallel act like a single 2.4 kΩ resistor.

If we have N resistors $R_1, R_2, R_3, \ldots, R_N$ all connected in parallel, the total effective resistance is given by:

$$R_{\text{total}} = \frac{1}{\dfrac{1}{R_1} + \dfrac{1}{R_2} + \dfrac{1}{R_3} + \cdots + \dfrac{1}{R_N}}. \tag{1.7}$$

Notice that the total resistance for a parallel connection is always less than any of the individual resistances and that the voltage drop is the same across all of the resistors in parallel. From Ohm's law applied to each resistor, the current divides among the various resistors in such a way that the most current flows through the smallest resistance, and vice versa. For example, in the circuit of Fig. 1.8:

If $R_1 = 10\text{k}\Omega$ and $R_2 = 2\text{k}\Omega$, then $R_{\text{total}} = (10\text{k})(2\text{k})/12\text{k} = 1.67\text{k}\Omega$.
If $V_2 - V_1 = 8\text{V}$, then $I = 8\text{V}/1.67\text{k}\Omega = 4.8\,\text{mA}$.
The current I_1 flowing through $R_1 = 10\text{k}\Omega$ is $I_1 = 8\text{V}/10\text{k}\Omega = 0.8\,\text{mA}$.
The current flowing through $R_2 = 2\text{k}\Omega$ is $I_2 = 8\text{V}/2\text{k}\Omega = 4\,\text{mA}$.

In general $I_1/I_2 = R_2/R_1$ which follows from the fact that R_1 and R_2 have the same voltage drop across them: $I_1R_1 = I_2R_2$.

Variable resistors, often called "potentiometers" or "pots," are also available. They come in many sizes and styles, and are usually adjusted by manually turning a shaft, as shown in Fig. 1.9. They have three terminals, one at each end of the resistor and one for the variable position of the tap. The total resistance R_T between the two end terminals A and B is always constant and equals the resistance value of the pot. The resistance R_1 between A and the tap and R_2 between B and the tap varies as the shaft is turned. Notice that if the shaft is turned fully clockwise the tap is electrically connected to terminal A and $R_2 = R_T$, $R_1 = 0$. The variation may be "linear taper" with shaft rotation as shown in Fig. 1.9(c), or "logarithmetic taper" as in Fig. 1.9(d). For example, a $100\,\text{k}\Omega$ pot has $R_T = R_1 + R_2 = 100\,\text{k}\Omega$ regardless of the shaft rotation, but R_1 and R_2 depend on the shaft position—always subject to the condition $R_1 + R_2 = 100\,\text{k}\Omega$. A logarithmetic taper is usually used in volume controls for audio equipment; a linear tuner is more commonly used in scientific apparatus.

It is worthwhile to note that Ohm's law must apply to a resistance even though it is connected with another circuit element which does not obey Ohm's law. For example, a "zener diode" is a solid-state device which has the non-linear property that the voltage drop across its two terminals is essentially constant, regardless of the current through it over a wide range of currents. Thus Ohm's law does not apply to the zener diode, or in other words the resistance of a zener diode varies with the current through the diode. Consider a resistance and a zener diode connected in series with a battery as shown in Fig. 1.10. The current through R and the zener diode must be equal as they are connected in series, and the voltage of the battery must equal the voltage across

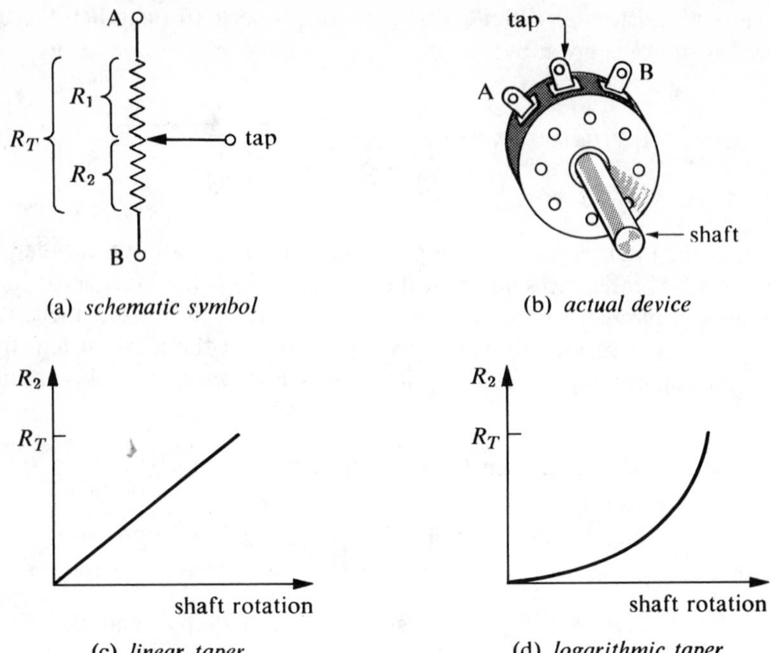

FIGURE 1.9 Variable resistor.

R, V_R, plus the voltage across the zener, V_z: $V_{bb} = V_R + V_z$. Ohm's law applied to the resistor alone implies $V_R = IR$. Thus $V_{bb} = IR + V_z$. If V_{bb} decreases, the constant zener voltage V_z implies that I must decrease so that the IR voltage drop across R always equals the difference between V_{bb} and V_z.

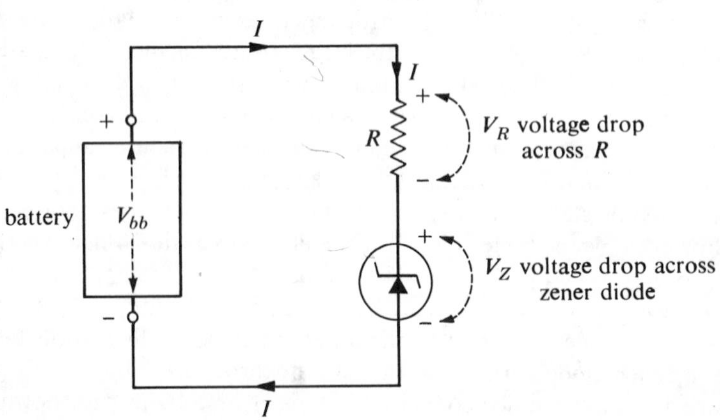

FIGURE 1.10 Zener diode circuit.

$$IR = V_{bb} - V_z \tag{1.8}$$

This circuit is often used to produce a constant output voltage which is taken from across the zener diode. If $V_{bb} = 12\,\text{V}$ and $V_z = 6.8\,\text{V}$, then $IR = 12\,\text{V} - 6.8\,\text{V} = 5.2\,\text{V}$. If $I = 10\,\text{mA}$, then $R = 5.2\,\text{V}/10\,\text{mA} = 520\,\Omega$ by Ohm's law. For $R = 520\,\Omega$, if V_{bb} falls to 10V, then I must decrease from 10mA to $(10\,\text{V} - 6.8\,\text{V})/520 = 3.2\,\text{V}/520 = 6.16\,\text{mA}$.

1.5 BATTERIES

A battery is a two terminal device in which chemical energy is converted into electrical energy, and a voltage difference is generated between the two battery terminals. A battery tends to spew positive charge out of the positive terminal and draw it into the negative terminal. A battery also tends to spew negative charge out of the negative terminal and draw it into the positive terminal. In an actual battery negatively charged electrons are spewed out the negative terminal and are drawn in the positive terminal, even though we may for convenience speak of positive charge flowing. A "dry" battery is one in which the chemicals are essentially dry or in a paste form, e.g., a flashlight battery. A "wet" battery is one containing liquids, e.g., an automobile battery that contains sulfuric acid solution. The circuit symbol for battery is a series of long and short parallel lines, as shown in Fig. 1.11—with the longer line on the

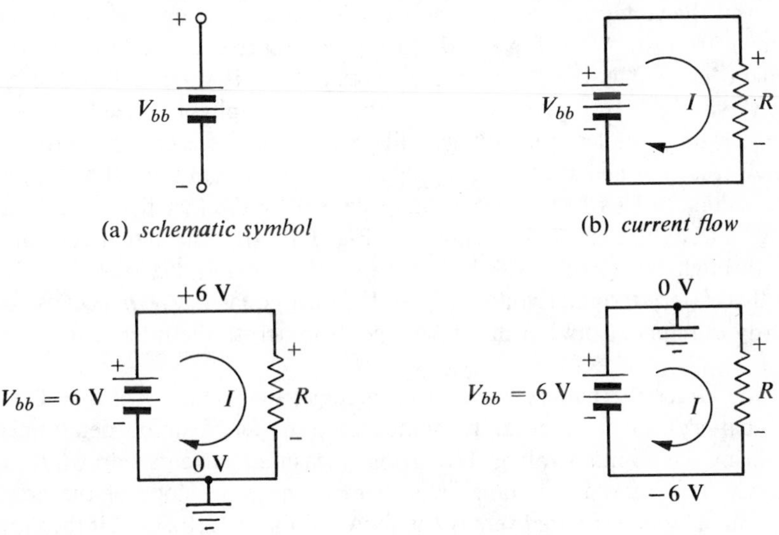

FIGURE 1.11 Batteries.

end representing the positive terminal and the shorter line on the other end representing the negative terminal. Thus, if a resistor R is connected between the two battery terminals as shown in Fig. 1.11(b), then positive current will flow out of the battery plus terminal, through the resistor, and back into the minus terminal of the battery. Notice that the (positive) current flows *out* of the battery "+" terminal and *into* the plus end of the resistor.

Notice again that the straight lines drawn in the circuit diagram represent wires with *zero resistance.* Thus there is no change in voltage along the wires represented by straight lines. The voltage at the positive battery terminal is exactly the same as the voltage at the top of the resistance in Fig. 1.11(b). In order to have a voltage difference between two points in a circuit, there must be some resistance between these two points from Ohm's law $V_2 - V_1 = IR$; that is, if $R = 0$, $V_2 - V_1 = 0$ even if $I \neq 0$. Notice also that the "+" and "−" signs represent the relative polarity of the voltages; that is, the top of the resistor is positive with respect to the bottom. Also notice that we may ground any *one* point in a circuit. Two such cases are shown in Fig. 1.11(c). In either case the voltage difference between the battery terminals is 6 V and the current is the same in each case. Notice that no current flows into the ground connection; the current flowing out of the positive terminal of the battery flows back into the negative terminal. The ground connection merely sets the zero voltage level.

A battery is rated in volts; its voltage rating V_{bb} indicates the difference in voltage which the battery will maintain between its two terminals. A perfect or ideal battery will always maintain the same voltage difference between its terminals regardless of how much current I it supplies to the rest of the circuit. However, the voltage of any real battery will decrease as more and more current is drawn from it. Thus for a real battery, V_{bb} represents the terminal voltage when no current is drawn from the battery. V_{bb} is often called the "open circuit" voltage. In general, the larger the battery's physical size (for the same voltage rating) the less the voltage will decrease as more current is drawn; in other words, a large battery can supply more current than a small battery at the same voltage. This behavior can be explained by the fact that a real battery has an internal resistance r as shown in Fig. 1.12(a). The actual current must flow through both r and R, which are in series, so $I = V_{bb}/(r + R)$. Notice that the battery terminal voltage $V_A - V_B$ must equal $V_{bb} - Ir$ because of the polarity difference between the Ir voltage drop across the internal resistance r and V_{bb}.

The larger the battery for a given voltage, the smaller r is ($r = 0$ for an ideal battery). The internal resistance of a battery can be determined by measuring the terminal voltage for various measured currents I drawn from the battery. The internal resistance r is then the negative slope of the graph of the terminal voltage plotted versus I as shown in Fig. 1.12(b). Or, if the terminal voltage is measured for two currents, then r is given by [see Fig. 1.12(b)] $r = (V_1 - V_2)/(I_2 - I_1)$.

When a battery goes "bad," its internal resistance r increases sharply.

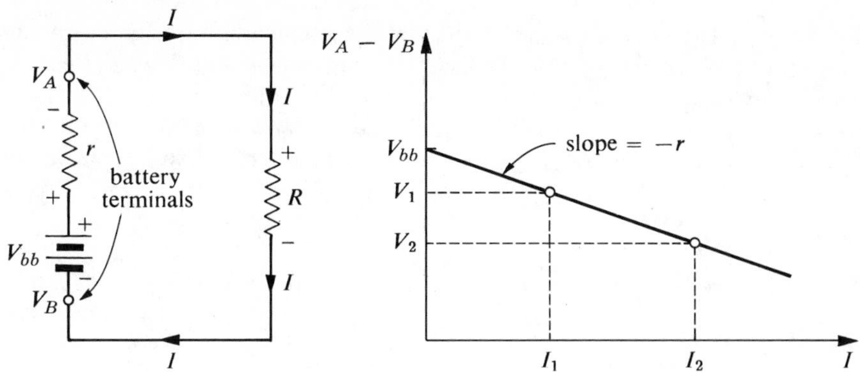

FIGURE 1.12 Internal resistance r of battery.

The typical good 12 V automobile battery has an internal resistance of about 0.03 ohm. A size D 1.5 V carbon–zinc dry cell such as is used in flashlights and in some portable transistor circuits has $r = 0.5$ ohm. A small $1'' \times \frac{1}{2}'' \times \frac{1}{4}''$ 9 V battery often used to power portable transistor radios has $r \cong 13\,\Omega$. Usually the internal resistance r is omitted from circuit diagrams, but this omission is valid only when r is much less than any other series resistances in the circuit.

In scientific circuitry, a battery which very gradually goes bad (that is, whose voltage slowly decreases) is a real disadvantage, because the circuit behavior may become erratic and difficult to diagnose as the battery slowly wears out. However, with a mercury battery which goes bad very abruptly (after a long life), the battery voltage is either all right or extremely low in which case the circuit will usually not function at all. Thus, in most scientific instruments mercury batteries are used, particularly if proper circuit behavior depends strongly upon a certain minimum battery voltage. The nickel-cadmium battery also goes bad abruptly after a long life, and it has the additional advantage of being rechargeable. It is, however, more expensive than the mercury battery. For a brief summary of the six different types of batteries, see Appendix B.

1.6 POWER

Power is defined as the time rate of doing work or the time rate of expending energy; that is, $P \equiv dW/dt$ where W is work or energy and t is time. The units of power are thus joules/sec; one "watt" is defined as one joule/sec. We will now show that a dc current I flowing through a resistor R develops a power of I^2R or VI or V^2/R, where V is the voltage drop across the resistor.

Recalling that voltage is electrical potential energy per unit charge, we see that a charge has less electrical potential energy when it leaves a resistor than it has when it enters because of the decrease in voltage, or voltage "drop" across the resistor. The time rate at which the flowing charge gives up electrical potential energy is thus the amount of charge flowing per second times the energy lost per unit charge, which is exactly equal to the current I times the voltage drop V. Thus, $P = IV$. But, from Ohm's law we know that $V = IR$; thus the power can also be expressed as $P = I^2R$ (see Fig. 1.13). And, again from

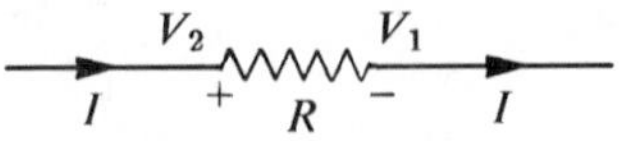

$$V \equiv V_2 - V_1$$

$$P = IV = I(IR) = I^2 R$$

$$P = IV = \left(\frac{V}{R}\right) V = \frac{V^2}{R}$$

FIGURE 1.13 Power dissipated in a resistor.

Ohm's law, $I = V/R$; thus another way of expressing the power is $P = V^2/R$. These three expressions for the power are equivalent and apply only to direct current flowing through a resistor. For alternating current the phase angle between the current and the voltage must be taken into account—more about this in Chapter 2.

The power developed in a resistor shows up as heating of the resistor. In other words, the loss of electrical potential energy (due to the IR voltage drop) of the charge flowing through the resistor is converted into random thermal motion of the molecules in the resistor. The kinetic energy of the flowing charges remains approximately constant everywhere in the circuit. Electrical potential energy is converted into heat energy in any dc circuit element across which there is a voltage drop and through which current flows. In the above three expressions for power, R represents the effective dc resistance of the circuit element, and V and I stand for the voltage drop across and the current through the circuit element, respectively.

If too many watts of power are converted into heat in a resistor or in any circuit element, the resistor may "burn up" in which case the resistor turns a brown or black carbonized color and may actually fragment into several pieces, thus breaking the electrical circuit. In other words the resistor is "open" or has an infinite resistance. Or, if the resistor is heated too much its resistance value may increase tremendously, thus changing the operation of the circuit drastically. For these reasons, resistors are rated by the manufacturer according to how much power they can safely dissipate without being damaged. Resistors are commonly available with wattage ratings of $\frac{1}{8}$ W, $\frac{1}{4}$ W (for low power transistor circuits), $\frac{1}{2}$ W, 1 W, 2 W, 5 W, 10 W, 20 W, 50 W, 100 W, and 200 W. The actual sizes of several commonly used resistors rated for various powers are shown in Fig. 1.14.

It is important to emphasize that the actual physical size does not depend upon the resistance, but only on the power rating; e.g., a $\frac{1}{2}$ W 2.2 kΩ resistor is the same size as a $\frac{1}{2}$ W 470 kΩ resistor. In designing circuits a resistor power

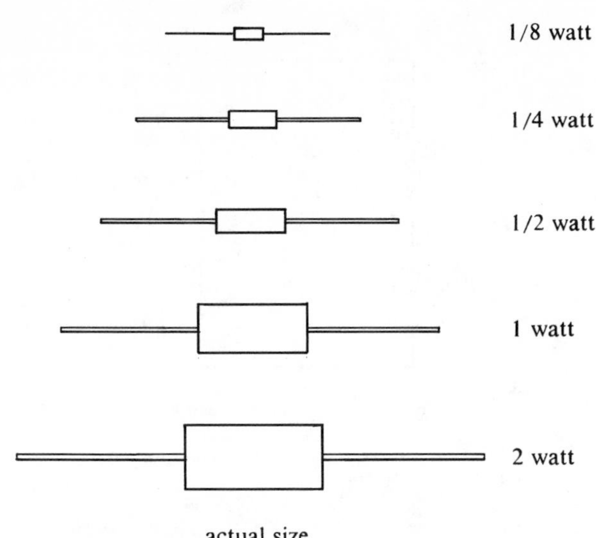

FIGURE 1.14 Resistor sizes for different power ratings.

rating of at least 3 or 4 times the expected power is usually chosen. For example, if a 1.5 kΩ resistor is to carry 20 mA of direct current, then the power dissipated as heat in the resistor will be $P = I^2R = (20 \times 10^{-3}\text{A})^2(1.5 \times 10^3\Omega) = 0.6\,\text{W}$. In the actual circuit, a 1.5 kΩ 2 W resistor would be used, or perhaps even a 1.5 kΩ 5 W resistor if the circuit were very sensitive to heat. If a large wattage is developed in a certain part of a circuit, care should be taken to provide an adequate vertical flow path for air around the hot element, so that the heat can be carried away by the resulting convection air currents. A resistor dissipating a large amount of power should never be placed in a closed chassis. Heat is an enemy of transistors as well as other circuit elements.

We will now derive an important theorem about power. If a battery has a certain fixed internal resistance r and we connect a load resistor R_L across its terminals, as shown in Fig. 1.15, how do we maximize the power $P_L = IV_L$ dissipated in R_L? If R_L is very small, I is large but $V_L = IR_L$ is small so the power in R_L is small. If R_L is very large, the voltage V_L across R_L is nearly equal to V_{bb}, but then the current is small so again the power in R_L is small. It seems reasonable that for some intermediate value of R_L the power dissipated in R_L is maximized. Let us set up an expression for P_L as a function of R_L, and maximize it.

$$P_L = IV_L \tag{1.9}$$

But

$$I = \frac{V_{bb}}{(r + R_L)} \qquad \text{and} \qquad V_L = IR_L = \left(\frac{V_{bb}}{r + R_L}\right)R_L.$$

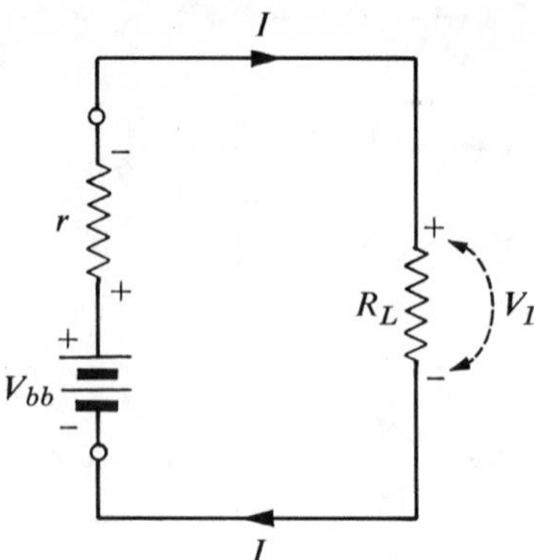

FIGURE 1.15 Circuit illustrating power transfer from source to load R_L.

Therefore,

$$P_L = \frac{V_{bb}}{(r + R_L)} \cdot \frac{V_{bb}}{(r + R_L)} \cdot R_L = \left(\frac{V_{bb}}{r + R_L}\right)^2 R_L \tag{1.10}$$

$$\frac{\partial P_L}{\partial R_L} = V_{bb}^2 \frac{(r + R_L)^2 - 2R_L(r + R_L)}{(r + R_L)^4}.$$

Set $\partial P_L/\partial R_L = 0$ to find the value of R_L for which P_L is an extremum.

$$\frac{\partial P_L}{\partial R_L} = 0 = \frac{V_{bb}^2}{(r + R_L)^4}[(r + R_L)^2 - 2R_L(r + R_L)]$$

$$(r + R_L)^2 = 2R_L(r + R_L)$$

$$R_L = r \tag{1.11}$$

It can be shown that P_L is a maximum not a minimum when $R_L = r$ by showing $\partial^2 P_L/\partial R_L^2$ is negative at $R_L = r$. In words, the maximum power is dissipated in the load resistance R_L when it equals the internal resistance of the battery. Under this condition, the voltage V_L across the load equals one half of V_{bb} which is the open circuit battery voltage. Also half of the total power is dissipated in r and half in R_L.

However, suppose we have a *fixed load resistance* R_L and V_{bb}, and variable r, and we ask the question: "How can we maximize the power dissipated in R_L?" The answer is *not* $r = R_L$, but $r = 0$! This answer is really obvious when we realize that any power dissipated in r is wasted as far as the load is concerned. It also follows mathematically from maximizing $P_L = V_{bb}^2 R_L/(r + R_L)^2$ with respect to r.

To sum up, if the load R_L is fixed, the smaller the internal resistance r the better. If we are presented with a fixed internal resistance r, then the load R_L gets maximum power when $R_L = r$. Also note that when r is fixed, the load *voltage* is very large when R_L is very large, and the load *current* is very large when R_L is very small.

1.7 KIRCHHOFF'S LAWS

The two basic laws of electricity that are most useful in analyzing circuits are Kirchhoff's current law and voltage law:

Kirchhoff's Current Law

At any junction of wires in a circuit, the sum of all the currents entering the junction exactly equals the sum of all the currents leaving the junction. In other words, electric charge is conserved.

Kirchhoff's Voltage Law

Around any closed loop or path in a circuit, the algebraic sum of all the voltage drops must equal zero. In other words energy is conserved.

The current law merely says that no electric charge is being created or destroyed at the junction in question; that is, the total current entering equals the total current leaving. The voltage law says that there is no net gain or loss in electrical potential energy for any charge making a round trip around any closed loop; that is the energy the charge gains (in passing through a battery) must be all lost (as heat, radiation, etc.) in the rest of the loop. (If a changing magnetic field is present, then an induced "emf" must be placed in series with the loop just as if a battery were actually present.)

Let us illustrate the use of Kirchhoff's laws by using them to solve for the currents and voltages in the circuit of Fig. 1.16, assuming that we know all the resistances and the battery voltage V_{bb}. First one draws the currents in all parts of the circuit using arrows, and then the polarities of the voltage drops, remembering that current always goes into the plus end and out of the minus end of a resistor. By current we mean the flow of positive charge. One usually can figure out the direction of the current quite easily, but once the current direction is drawn, then the voltage polarities are determined. If one has chosen the current direction incorrectly, the final numerical answer will come out negative, which merely means the current flows opposite to the direction of the arrow in Fig. 1.16.

The current law says that at junction B, $I = I_2 + I_3$ and that $I_2 + I_3 = I$ at junction C. The voltage law says that starting at G which is ground (0 volts) and going around the circuit clockwise,

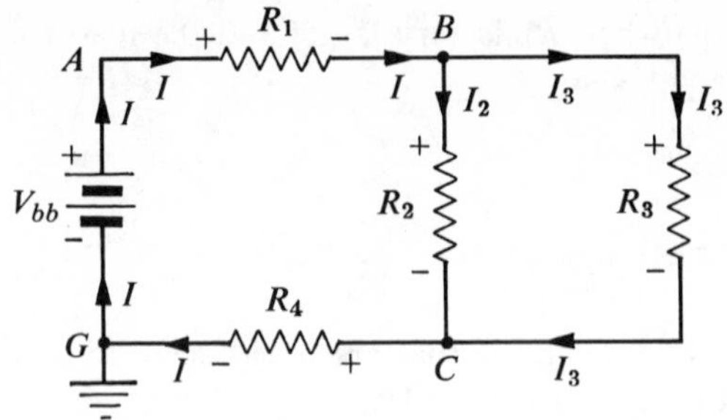

FIGURE 1.16 Circuit problem.

$$+V_{bb} - IR_1 - I_2R_2 - IR_4 = 0 \tag{1.12}$$

or

$$+V_{bb} - IR_1 - I_3R_3 - IR_4 = 0 \tag{1.13}$$

or

$$+V_{bb} - IR_1 - I(R_2 \parallel R_3) - IR_4 = 0 \tag{1.14}$$

where $R_2 \parallel R_3 = R_2R_3/(R_2 + R_3)$.

Equation (1.13) differs from (1.12) only in that I_2R_2 has been replaced by I_3R_3, because the same voltage drop must exist across R_2 and R_3. $R_2 \parallel R_3$ means the total effective resistance of R_2 in parallel with R_3. If we calculate $R_2 \parallel R_3$, then equation (1.14) can be solved for I. I_1 and I_2 can now be obtained from $I = I_1 + I_2$ and $I_1R_2 = I_2R_3$.

Suppose $V_{bb} = 24\,\text{V}$, $R_1 = 470\,\Omega = 0.47\,\text{k}\Omega$, $R_2 = 1\,\text{k}\Omega$, $R_3 = 2\,\text{k}\Omega$, $R_4 = 4.7\,\text{k}\Omega$. Then,

$$(R_2 \parallel R_3) = \frac{R_2R_3}{R_2 + R_3} = \frac{(1\,\text{k}\Omega)(2\,\text{k}\Omega)}{1\,\text{k}\Omega + 2\,\text{k}\Omega} = 0.67\,\text{k}\Omega = 670\,\Omega.$$

Solving equation (1.14) for I gives

$$I = \frac{V_{bb}}{R_1 + (R_2 \parallel R_3) + R_4} = \frac{24\,\text{V}}{0.47\,\text{k}\Omega + 0.67\,\text{k}\Omega + 4.7\,\text{k}\Omega} = 4.1\,\text{mA}.$$

Remembering that point G is at 0 V, we can calculate that point C is at $+IR_4$ or $+19.27\,\text{V}$, point B is at $IR_4 + I(R_2 \parallel R_3) = 19.27\,\text{V} + 2.75\,\text{V} = +22.02\,\text{V}$, and point A is at $+24\,\text{V}$ since it is connected to the positive terminal of the 24 V battery. Remember all these voltages are relative to ground (point G) which is 0 volts.

Another approach would be to write down the Kirchhoff voltage law equations using the minimum number (two) of unknown currents. In the example of Fig. 1.16, we would write going around the complete loop through the battery, R_1, R_3 and R_4:

$$V_{bb} - IR_1 - (I - I_2)R_3 - IR_4 = 0 \tag{1.15}$$

and

$$I_2R_2 - (I - I_2)R_3 = 0 \quad (\text{using } I_3 = I - I_2) \tag{1.16}$$

from going around the small loop containing R_2 and R_3. Thus, we have two equations that can be solved for the two unknown currents, I and I_2. There are many different ways to solve such problems, but they all eventually lead to the same solution. With a little experience, the reader will learn how to choose the most effective technique for a particular problem.

It can be shown that in circuit problems involving many loops, the simplest equations usually result if we draw "loop currents", that is, currents which flow in *closed* loops. The circuit problem of Fig. 1.17 illustrates this technique. We will solve for the currents and voltages assuming that we know the battery voltage and all the resistances. We have three current unknowns, so we need three equations in the currents. Note that $(I_1 - I_2)$ is the total current through R_2; $(I_1 - I_3)$ is the total current through R_5, and $(I_2 - I_3)$ is the total current through R_4.

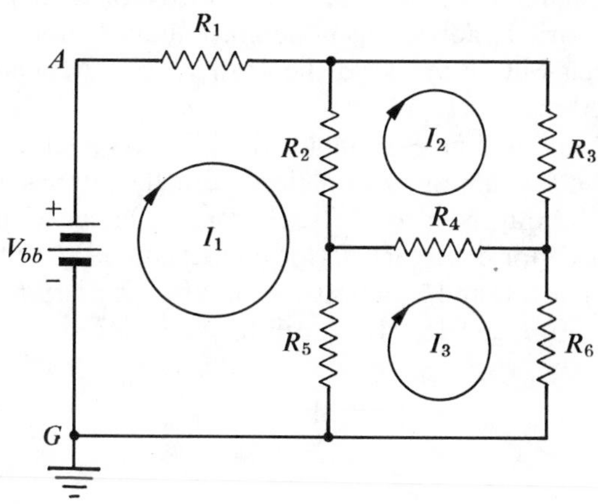

FIGURE 1.17 Circuit problem using loop currents.

By drawing loop currents, each of which flows *through* junctions, we automatically have satisfied the Kirchhoff current law, that the current entering each junction must exactly equal the current leaving. Thus, we need only to write down and solve the equations implied by the Kirchhoff voltage law.

The Kirchhoff voltage law applied to the three loops gives us: for the I_1 loop, $V_{bb} - I_1R_1 - (I_1 - I_2)R_2 - (I_1 - I_3)R_5 = 0$ remembering the sign convention for voltage drops across a resistor, as shown below,

for the I_2 loop, starting at point C and going clockwise,

$$(I_1 - I_2)R_2 - I_2R_3 - (I_2 - I_3)R_4 = 0; \tag{1.17}$$

for the I_3 loop, starting at point G and going clockwise,

$$(I_1 - I_3)R_5 - (I_3 - I_2)R_4 - I_3R_6 = 0. \tag{1.18}$$

Rewriting and grouping all the similar terms together in each equation gives us:

$$\begin{aligned} (R_1 + R_2 + R_5)I_1 - R_2I_2 - R_5I_3 &= V_{bb} \\ R_2I_1 - (R_2 + R_3 + R_4)I_2 + R_4I_3 &= 0 \\ R_5I_1 + R_4I_2 - (R_4 + R_5 + R_6)I_3 &= 0. \end{aligned}$$

These three equations can now be solved for the three loop currents I_1, I_2, and I_3 either by various adroit algebraic substitutions or by the method of determinants which will *always* yield the answer. See Appendix F for a discussion of the method of determinants.

We now provide a simple example which is very useful when designing transistor (and also vacuum tube) amplifiers. For the purpose of this example, we can consider the transistor as a "black box" or circuit element with three terminals, labeled C (for collector), E (for emitter), and B (for base). Suppose that the two terminals, C and E, a resistor R_L, and a battery V_{bb} are connected in series, as shown in Fig. 1.18. If we call the voltage between C and E of the

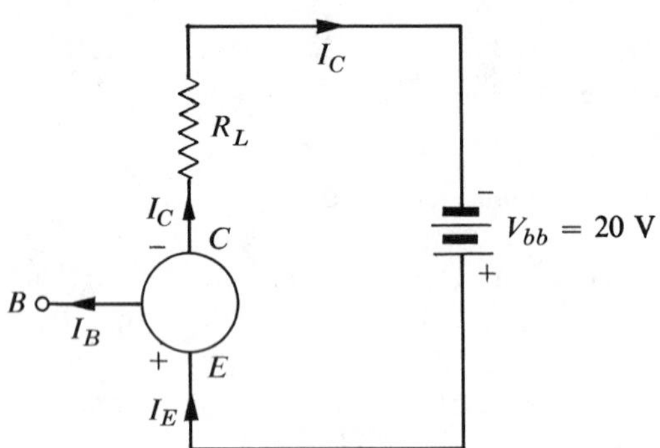

FIGURE 1.18 "Black box" transistor circuit.

transistor V_{CE} (V_{CE} is negative if C is negative with respect to E), then Kirchhoff's voltage law for the loop is:

$$-V_{CE} - I_CR_L + V_{bb} = 0. \tag{1.19}$$

From Kirchhoff's current law, regarding the transistor as a junction, we see that $I_E = I_B + I_C$, as the transistor can create no current. As we shall see later the base current I_B is usually about 50 or 100 times smaller than I_C, so $I_E \cong I_C$ is a good approximation. If we solve for the current I_C, we obtain

$$I_C = \frac{V_{bb}}{R_L} - \frac{1}{R_L} V_{CE}, \tag{1.20}$$

which is of the form $y = b + mx$ where m = slope and b = y intercept. This is called the load line equation and is usually graphed with V_{CE} as the independent variable and I_C as the dependent variable, as shown in Fig. 1.19.

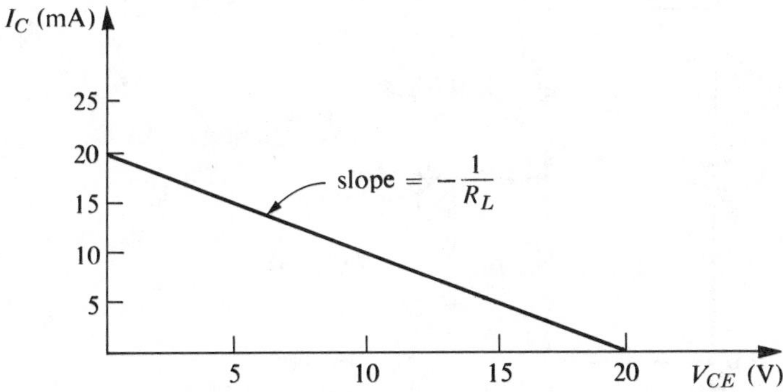

FIGURE 1.19 Transistor load line.

Notice that the graph is a straight line with a slope of $-1/R_L$, a vertical (current) intercept of V_{bb}/R_L, and a horizontal (voltage) intercept of V_{bb}. That is, the larger R_L, the flatter the line, and the more V_{CE} changes for a given change in I_C. Also notice that the V_{bb} intercept represents the condition when there is no current flowing through the transistor and the entire battery voltage appears across the transistor. In other words, the transistor has an infinite effective resistance or is "cut off." The V_{bb}/R_L intercept on the other hand represents the condition when the transistor has an effective resistance of 0 ohms or is fully conducting—"saturated" or "turned on." In this condition the only thing limiting the current flow is the resistance R_L. This argument has neglected any resistance presented to the current flow by the transistor itself; in a real transistor V_{CE} never falls quite to zero when it is turned "on."

This graph is usually called the load line and is completely determined by V_{bb} and R_L. Once the load line is determined, then if we know the current I_C, we can immediately find the voltage across the transistor V_{CE}. For example, referring to Fig. 1.19, if $I = 15\text{mA}$, then $V_{CE} = 5\text{V}$. And, the voltage drop across R_L must be $V_{bb} - V_{CE} = 20\text{V} - 5\text{V} = 15\text{V}$. The important thing to remember here is that regardless of what the transistor is doing its collector-emitter voltage V_{CE} and its current I must *always* lie on the load line because of Kirchhoff's voltage law and Ohm's law.

1.8 VOLTAGE DIVIDERS

Often one wishes to reduce a battery or power supply voltage; for example, in the previous circuit example of Fig. 1.18 suppose that a -4 V voltage is needed for some other part of the circuit. We have to devise some voltage divider circuit to get -4 V from the 20 V battery. One solution is shown in Fig. 1.20(a).

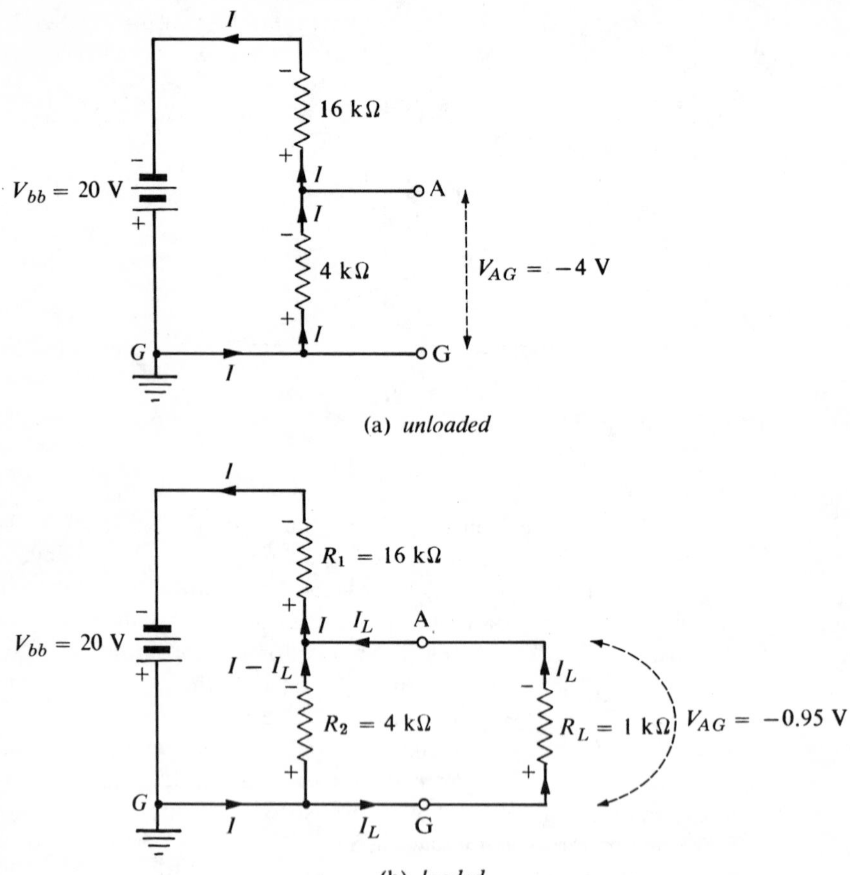

FIGURE 1.20 Voltage divider.

If no current is drawn from terminal A, then the voltage at terminal A will be -4 V, the total current I drawn from the battery by R_1 and R_2 will be

$V_{bb}/(R_1 + R_2) = 20\,\text{V}/(16\,\text{k}\Omega + 4\,\text{k}\Omega) = 1\,\text{mA}$; thus, the voltage at A will be $1\,\text{mA} \times 4\,\text{k}\Omega = 4\,\text{V}$. The terminal A is negative with respect to G because the current flows from G toward A. In general, the voltage at A will be $V_{bb} \times R_2/(R_1 + R_2)$ if no current is drawn from A. This can be derived as follows:

$$V_A = IR_2 = \frac{V_{bb}}{R_1 + R_2} R_2 = V_{bb} \frac{R_2}{R_1 + R_2}. \tag{1.21}$$

There are two things to keep in mind in connection with this type of voltage divider. First, one should keep the total current drawn from the battery low to ensure a reasonably long battery life. Second, one should realize that any current drawn from terminal A may greatly affect the voltage at A. For example, if a $1\,\text{k}\Omega$ load resistor R_L were connected between A and G as in Fig. 1.20(b) the voltage at A would fall to only

$$\begin{aligned} V_A &= I \times (R_2 \,||\, R_L) \\ &= \frac{V_{bb}}{R_1 + (R_2 \,||\, R_L)} (R_2 \,||\, R_L) = \frac{20\,\text{V}}{(16\,\text{k}\Omega + 0.8\,\text{k}\Omega)} (0.8\,\text{k}\Omega) = 0.95\,\text{V}. \end{aligned}$$

In this case, we say that the $1\,\text{k}\Omega$ resistor has "loaded down" the voltage divider. In a similar vein, any variation in the resistance R_L connected between A and G will cause the voltage at A to vary. This variation in V_A can be minimized by making R_2 much less than R_L, in other words by making the current I flowing through the R_1R_2 divider chain much greater than the current I_L drawn out terminal A through R_L. Then any change in R_L will affect the total current only slightly, and thus V_G will change only slightly. But, R_L then uses only a small fraction of the current drawn from the battery which is wasteful.

If one knows that R_L will remain constant, then a simpler circuit will do as shown in Fig. 1.21. In this circuit all the current drawn from the battery flows

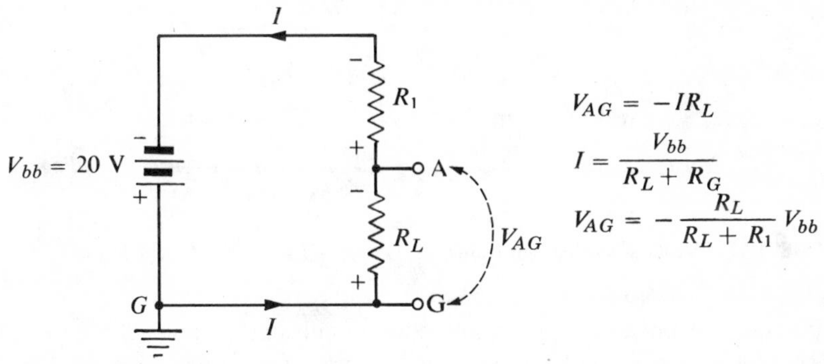

FIGURE 1.21 Resistance voltage divider.

through the load R_L which may be an important advantage if the load draws a sizeable current and the battery size is limited.

$$V_{AG} = \frac{R_L}{R_1 + R_L} 20\text{V} = 4\text{V} \qquad \text{if } R_L = 1\,\text{k}\Omega,\ R_1 = 4\,\text{k}\Omega \tag{1.22}$$

1.9 IDEAL VOLTAGE AND CURRENT SOURCES

It is often convenient to use the concept of an ideal or perfect voltage source in describing various circuits. An ideal voltage source simply means a source of voltage with *zero internal resistance*, in other words a perfect battery. The circuit symbol for an ideal voltage source is a circle as shown in Fig. 1.22(a).

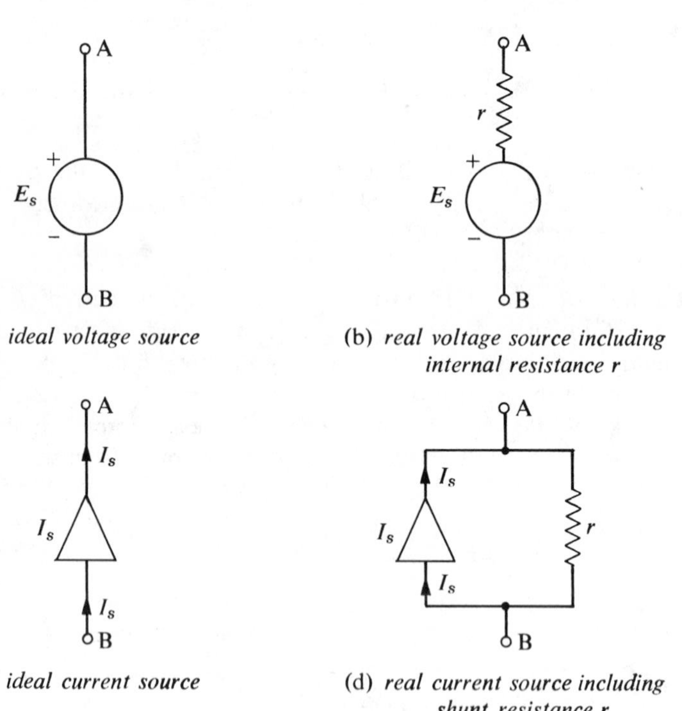

FIGURE 1.22 Voltage and current sources.

Any battery or power supply in an actual circuit can be considered ideal if its internal resistance is small compared to any other resistances connected in series with it. As mentioned in section 1.6, most batteries are considered to have zero internal resistance in circuits. Such a voltage source is called "ideal" because it keeps on supplying the same voltage regardless of how much current we draw from it. An ideal voltage source can be approximated by a good battery. A real voltage source or battery can be represented by an ideal battery or voltage source in series with a resistance r, as shown in Fig. 1.22(b). The terminal voltage of a real battery, of course, falls as we draw more and more current from it because of the voltage drop across the battery's internal resistance as explained in 1.6.

An ideal *current* source is a source which supplies a constant current regardless of what load it is connected to. Thus, an ideal current source has an *infinite internal resistance* so that changes in the load resistance will not affect the current supplied by the source. For example a 6 V battery with a 1 MΩ internal resistance will deliver essentially 6 μA current to a 1 kΩ or a 10 kΩ or a 100 kΩ resistor connected across it. The current remains almost constant so long as the load resistance is small compared to the internal resistance. The circuit symbol for an ideal current source is shown in Fig. 1.22(c); in this circuit a current I_s *always* flows through the triangle representing the ideal current source regardless of the rest of the circuit. Such a source can be approximated by a battery with a series resistance R_s much larger than any resistances in the load connected across the battery. Then, the battery will always supply a current of V_{bb}/R_s, regardless of any changes in the load resistances.

Any real current source can be represented by an ideal current source in parallel with a resistance r as shown in Fig. 1.22(d). If a load resistance R_L is then connected to the terminals A and B of the real current source, the current I_s is then split between r and R_L according to Ohm's and Kirchhoff's laws, the larger current flowing through the smaller resistance as is shown in Fig. 1.23.

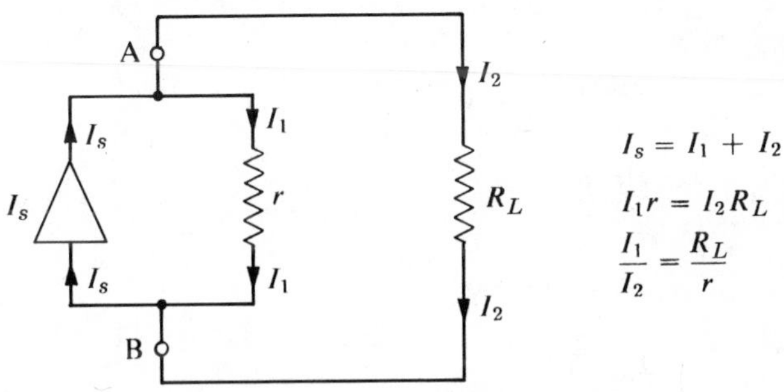

FIGURE 1.23 Current source with load R_L.

As we will see later, vacuum tubes can be conveniently represented by ideal voltage sources and transistors by ideal current sources. One can always convert from a current source to a voltage source and vice versa. The conversion is shown in Fig. 1.24. Notice that the resistance in series with the ideal voltage source is equal to the resistance across the ideal current source, and that $E_s = I_s r$. These results follow from Thevenin's theorem which is discussed in the next

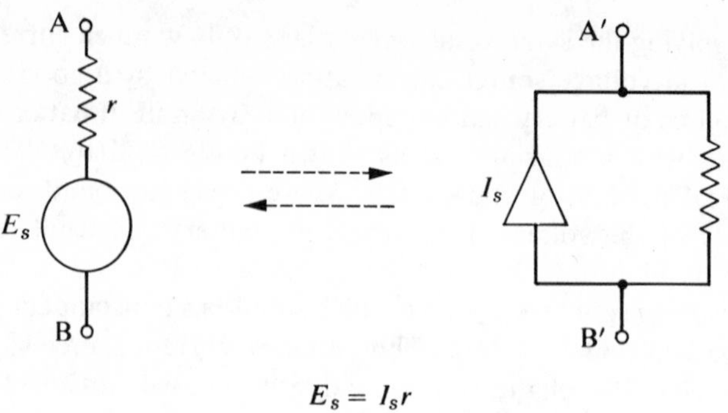

$$E_s = I_s r$$

FIGURE 1.24 Equivalence of current and voltage source.

section. The net result is that terminals A′B′ act electrically exactly like the terminals A and B. Which circuit you choose is a matter of convenience; one may yield an easier set of circuit equations to solve.

1.10 THEVENIN'S THEOREM

A theorem useful in analyzing many circuits is Thevenin's theorem which states that any combination of batteries and resistances with two terminals is electrically equivalent to one battery of voltage e in series with one resistance r as shown in Fig. 1.25. All the Kirchhoff voltage loop equations in the actual circuit

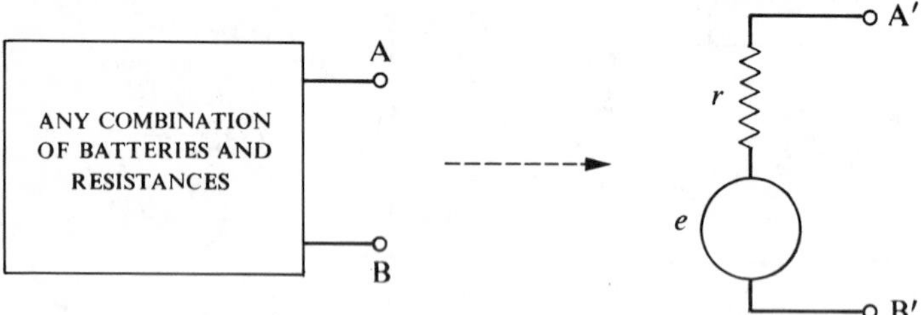

FIGURE 1.25 Thevenin equivalent circuit.

are linear; that is, the currents I and voltages V all occur raised to the first power. There are no terms containing products or quotients of I and V. The Kirchhoff equation for the Thevenin equivalent circuit is also linear. Thus, one way of looking at Thevenin's theorem is that it is impossible to make a nonlinear equation out of a set of linear equations. Any number of linear current-voltage equations in the actual circuit being analyzed is mathematically equivalent to *one* linear equation, which has as its equivalent circuit *one* battery in series with *one* resistance.

The voltage e of the equivalent battery is the open circuit voltage—the voltage V_{AB} measured when no current flows into or out of A and B. This measurement must be made by a device which draws negligible current from the circuit. Usually an oscilloscope will do nicely. The equivalent resistance r is the open circuit voltage V_{AB} divided by the short circuit current which is the current flowing from A and B when the terminals A and B are shorted together. In other words, terminals A′ and B′ of the Thevenin equivalent circuit will act electrically exactly like terminals A and B of the actual circuit. It is easy and safe to measure the open circuit voltage to determine e. However it is often disastrous (producing sparks, smoke, vile odors, destruction of the circuit, and embarrassment) actually to short terminals A and B to determine r. A better way to experimentally determine r is to measure the voltage V_{AB} between the two circuit terminals for a known load resistance R_L connected between A and B. Then, r is given by

$$r = \frac{e - V_{AB}}{I} = \frac{e - V_{AB}}{V_{AB}/R_L} = \frac{e - V_{AB}}{V_{AB}} R_L \tag{1.23}$$

where e is the open circuit voltage measured when $R_L = \infty$. Another useful way of calculating r is to mentally short out all the voltage sources in the circuit being analyzed. The resistance r is then the total net resistance between output terminals A and B.

In analyzing transistor amplifier circuits with a voltage divider, it is often useful to replace the actual divider with its Thevenin equivalent as shown in Fig. 1.26. Notice that both the actual circuit and the Thevenin equivalent produce the same open circuit voltage and the same short circuit current V_{bb}/R_1.

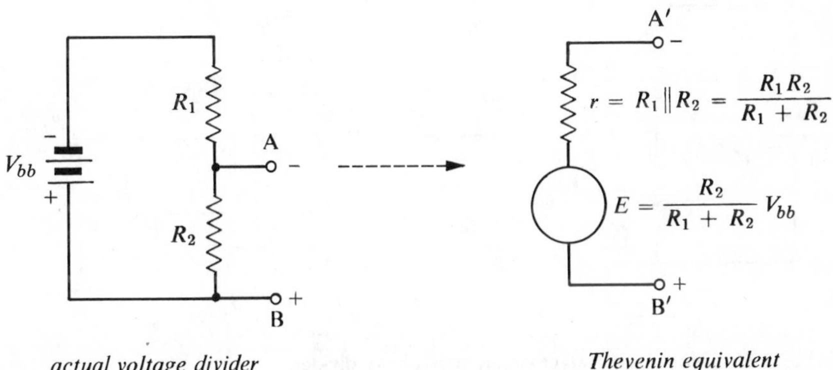

FIGURE 1.26 Thevenin equivalent circuit of a voltage divider.

1.11 NORTON'S THEOREM

Norton's circuit theorem simply states that any combination of batteries and resistances with two terminals is electrically equivalent to an ideal current source in parallel with a resistance as shown in Fig. 1.27. The equivalent resistance

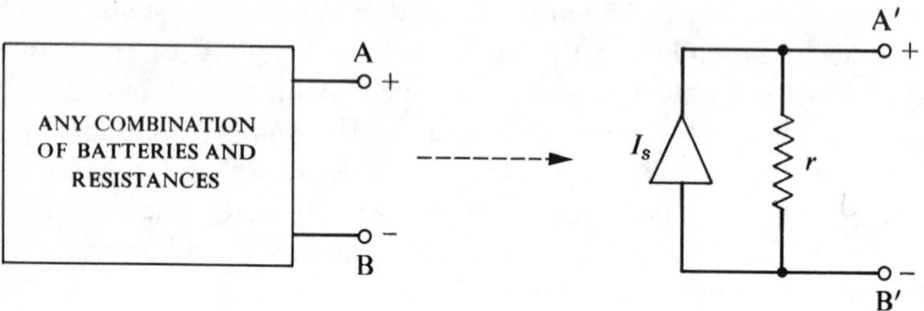

FIGURE 1.27 Norton equivalent circuit.

is the same as the equivalent resistance of the Thevenin equivalent circuit, and the ideal constant current source supplies a current of

$$I_s = \frac{V_{AB \text{ open circuit}}}{r}. \tag{1.24}$$

Thus, a voltage divider circuit can be replaced by a Norton or current equivalent circuit as shown in Fig. 1.28.

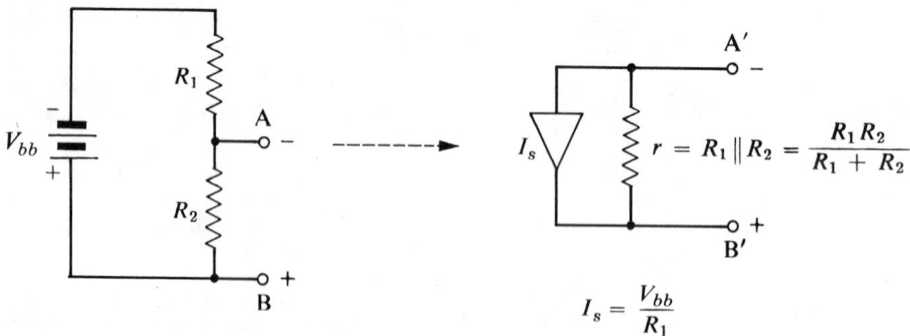

FIGURE 1.28 Norton equivalent circuit of a voltage divider.

As a final point it should be emphasized that the Thevenin equivalent is just as good as the Norton equivalent and vice versa. Each is electrically equivalent to the original circuit being analyzed and to the other. Which equivalent circuit one uses is a matter of taste and convenience, although there is a tendency to use the current equivalent circuit in describing transistor circuits and the voltage equivalent circuit for vacuum tube circuits.

Thevenin's and Norton's theorems can be extended to alternating current ("ac") circuits where they are used extensively in analyzing the ac behavior of transistor and vacuum tube circuits. When applied to ac circuits, the theorem states that any combination of ac circuit elements (such as resistors, capacitors, inductors, etc.) whose equations are *linear* in current and voltage can be replaced by a Thevenin or a Norton equivalent. Practically speaking, this means that almost any small signal circuit can be represented by a Thevenin or a Norton equivalent circuit. This important subject will be covered extensively in later chapters.

problems

1. What is the electric charge on one electron? On a He^+ ion? On a He^{++} ion? On an As^+ ion? Briefly state why the quantization of charge can be neglected in most electrical circuit problems.
2. Doubly ionized helium ions at a concentration of 10^{13} ions per cubic centimeter move with a velocity of 10^5 cm/sec. Calculate the current density in amperes/cm^2.
3. Briefly define current.
4. Calculate how many electrons flow per second past a fixed point in a wire carrying 10 mA of current. If the current moves from left to right, which way do the electrons move?
5. Briefly define electric potential or voltage. What are the mks units of voltage? What does "ground" mean?
6. Calculate the gain in kinetic energy for an electron which moves from a point of voltage 3 V to a point of voltage 5 V. Express your answer in joules and also in electron volts.
7. Briefly define resistance.
8. State Ohm's law, including the units of all the terms.
9. Calculate the voltage at points A, B, and C if:

   (a) A is grounded
   (b) B is grounded
   (c) C is grounded.

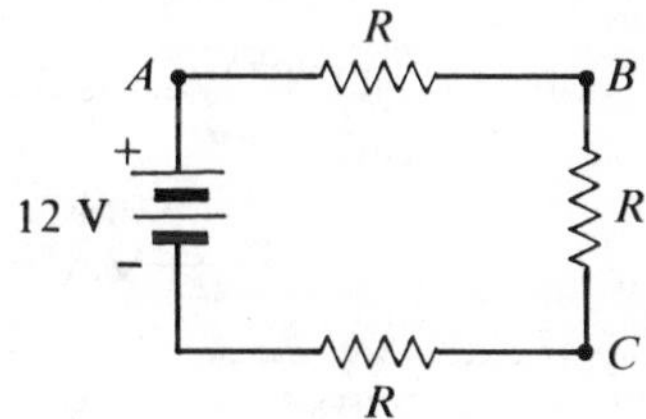

10. Calculate the voltage at points A, B, and C in the following circuit.

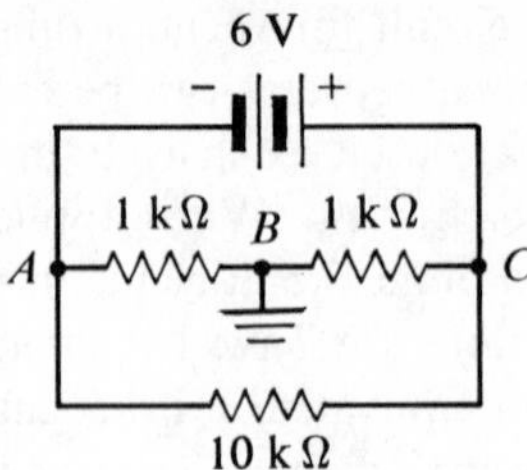

11. Calculate the resistance between terminals A and B.

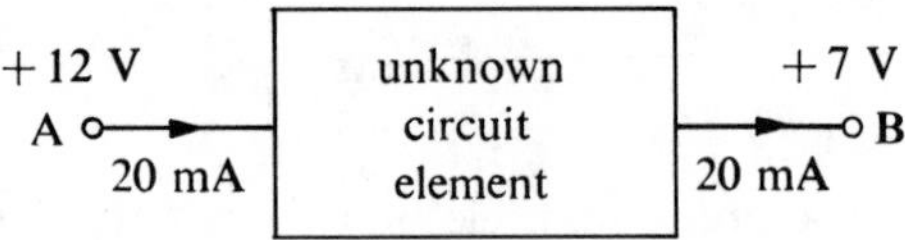

12. A gas discharge tube draws a current of 20 mA when a voltage difference of 800 V is maintained between its two ends. Calculate the effective dc resistance in ohms between the two ends of the tube.

13. Estimate quantitatively the static (dc) and the dynamic (ac) resistance at points A, B, and C of the device whose current–voltage curve is shown.

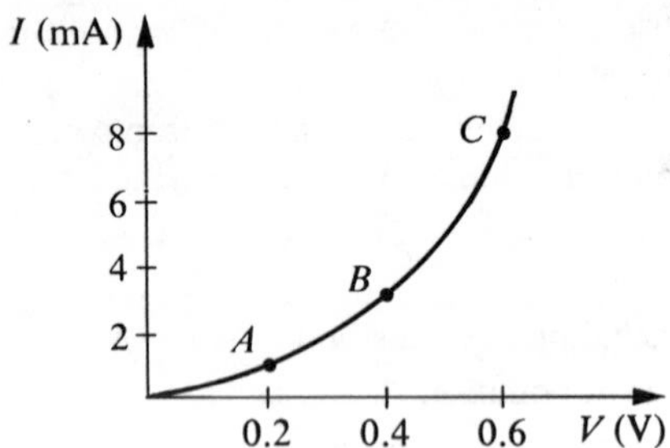

14. Calculate the voltage between A and B. What is the polarity of V_{AB}?

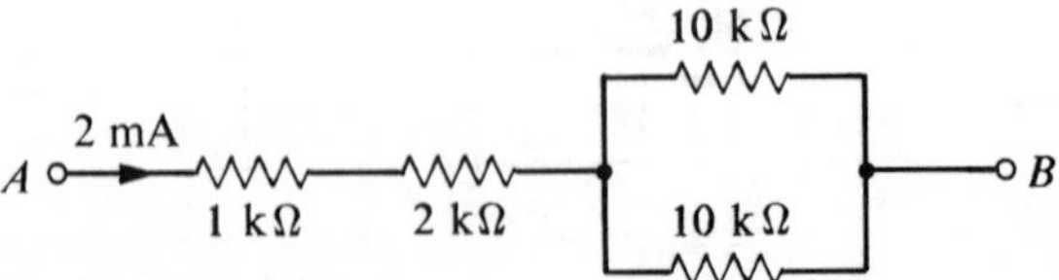

15. Calculate the current I.

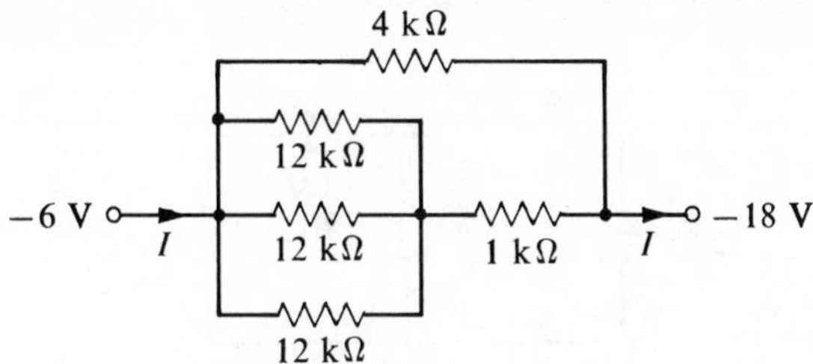

16. Briefly describe the energy conversion process which occurs in a battery. Does the negative battery terminal draw in electrons or spew electrons out? Does the current into one battery terminal always equal the current out of the other terminal?

17. Define power, including the units.

18. Assume that a "Durg" is a well-known physical quantity represented by the symbol D. What is a kilodurg? A megadurg? A gigadurg? A millidurg? A microdurg? A nanodurg? A picodurg?

19. How would you make a crude 10 watt heater–defroster for your car to run directly off the 12 V battery?

20. A one watt 1 kΩ carbon resistor carries a current of 30 mA. Calculate the power in watts dissipated as heat in the resistor. Would this situation be desirable in a circuit? Explain briefly.

21. Briefly describe, on a microscopic basis, what happens when electrical energy is converted into heat in a resistor. Is heat always generated when a current flows through a resistor?

22. An automobile battery has a terminal voltage of 12.8 V with no load. When the starter motor (which draws 90 A) is being turned over by the battery, the terminal voltage drops to 11 V. Calculate the internal resistance of the battery.

23. A 30 V dc power supply has an internal resistance of 2 ohms. Calculate the terminal voltage when a current of 500 mA is being drawn from the power supply.

24. How large should the heater resistance R_h be to draw maximum power from a 12 V battery with an internal resistance of 3 ohms? Calculate the power dissipated in the heater and in the battery under such conditions.

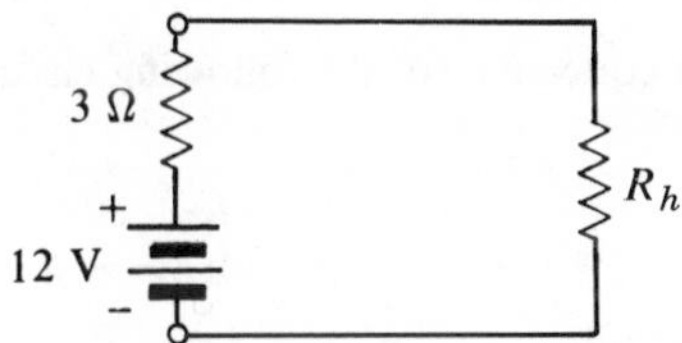

25. Briefly but clearly state Kirchhoff's two laws.

26. Calculate I_1 and I_2 in the following circuit.

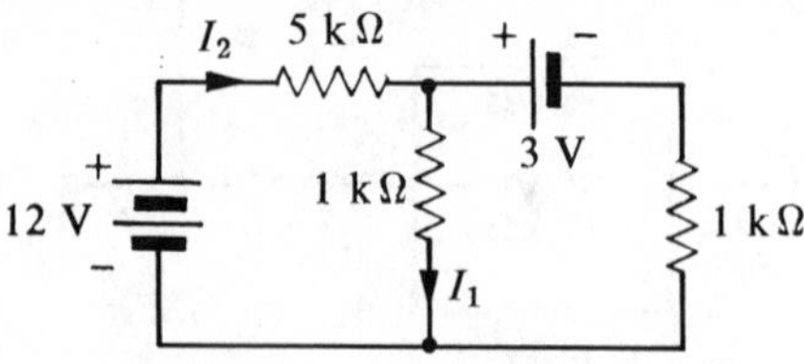

27. Calculate I_1, I_2, and I_3 in the following circuit.

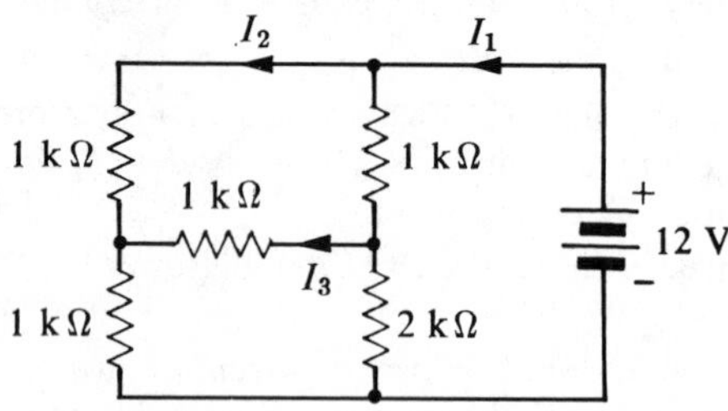

28. Briefly describe an ideal current source and an ideal voltage source.

29. Briefly state Thevenin's theorem.

30. Calculate the Thevenin equivalent of the following circuit.

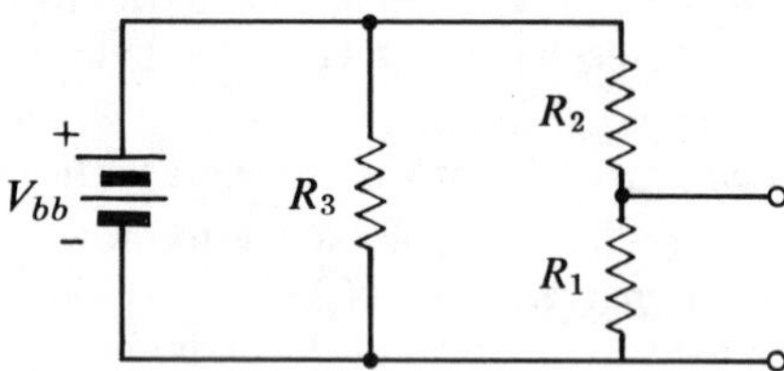

31. Calculate the Thevenin equivalent of the following circuit.

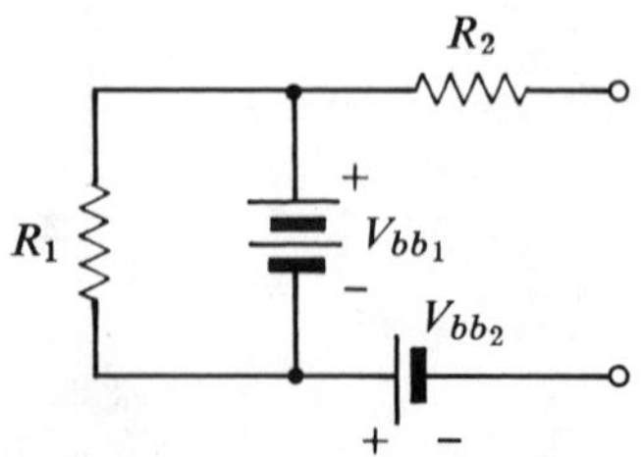

32. A "black box" with three terminals labeled E, B, and C is connected in the following circuit. (a) Calculate V_C and V_E. (b) If terminal B is 0.6 V more negative than terminal E, calculate R_1 and R_2 assuming that I_d is very large compared to the 40μA flowing in the "B" lead. (This black box is a silicon pnp transistor.)

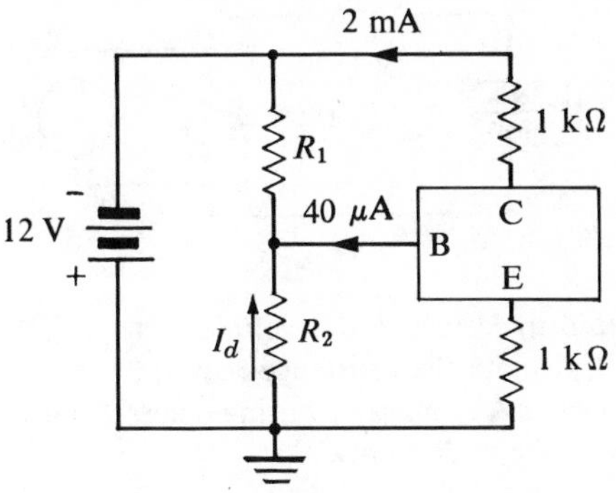

33. A "black box" with three terminals labeled K, G, and P is connected in the following circuit. The current flowing in the G lead is of the order of 10^{-8} A. Calculate V_P, V_G, and V_K. (This black box is a vacuum tube.)

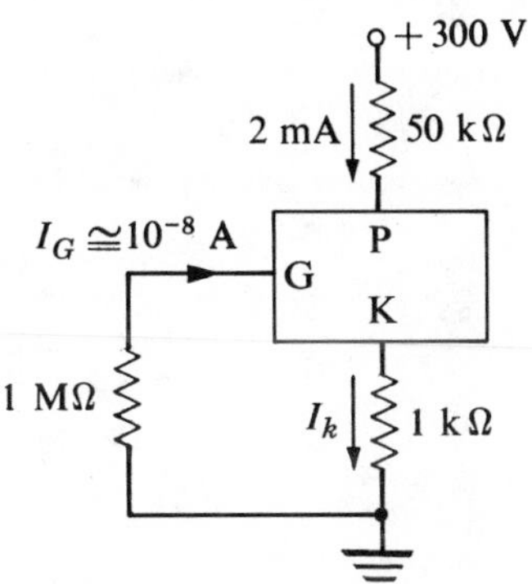

34. Calculate the voltmeter reading for (a) $R = 1\,\text{k}\Omega$ and (b) $R = 1$ megohm. You may assume the voltmeter is an oscilloscope with a 1 megohm input resistance.

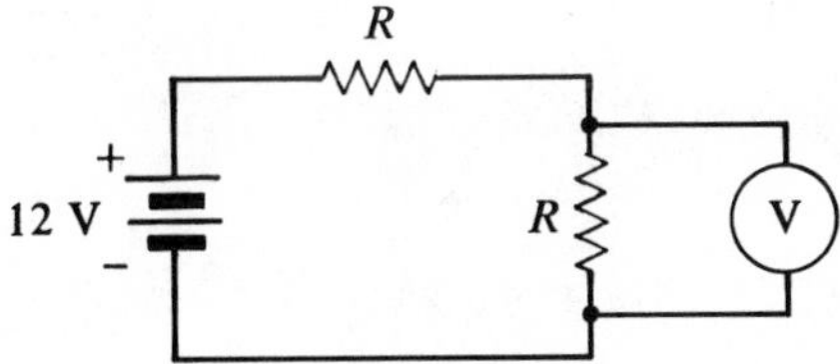

35. A voltmeter rated at 20,000 ohms per volt is connected as shown in the circuit at the left. Calculate the voltmeter reading if it is set (a) on the 0–5 V scale. (b) on the 0–50 V scale.

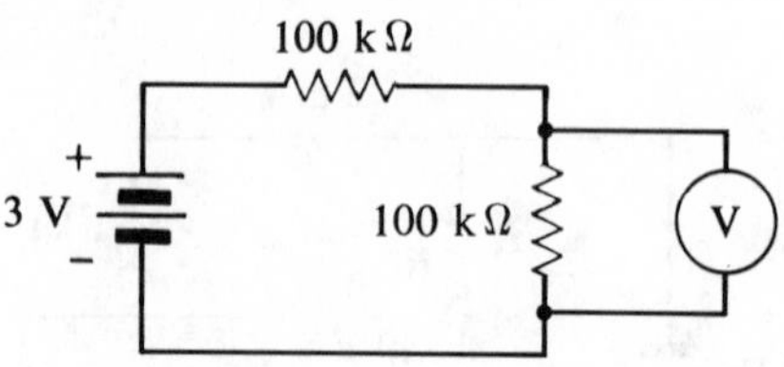

36. Show how you could make a 0–100 V voltmeter from a 0–50 μA microammeter and a resistance. Calculate the resistance required. How would you make this microammeter into a 0–5 A ammeter by using one resistance? The microammeter has an internal resistance of 20 ohms.

2

ALTERNATING CURRENT CIRCUITS

2.1 PERIODIC WAVEFORMS

In Chapter 1 we briefly mentioned alternating current (ac) and described it as current in which the electric charge "sloshed" back and forth along the wire. The most common type of ac is that in which the back and forth motion is sinusoidal; that is, if we plot a graph of the instantaneous value of the current $i(t)$ against time t, we get a sine wave as shown in Fig. 2.1(a). Voltage $v(t)$ can also be sinusoidal as shown in Fig. 2.1(b).

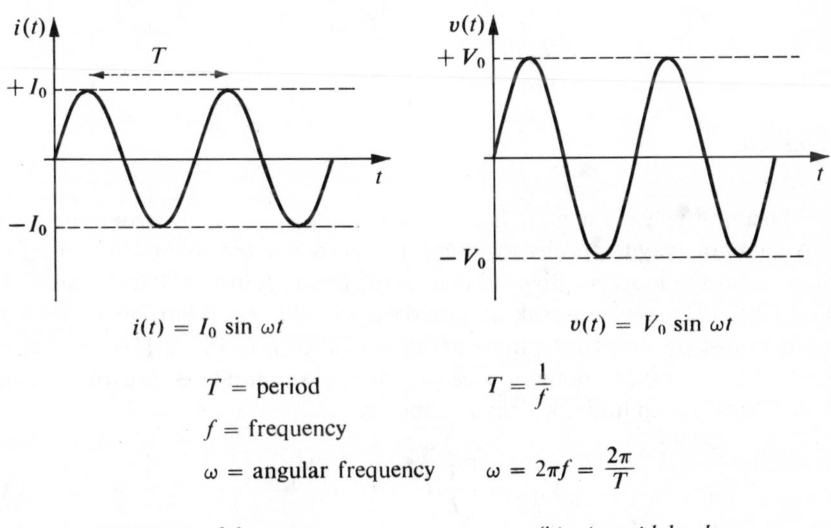

FIGURE 2.1 Sinusoidal waveforms.

Such a sine wave is completely determined by specifying three things: (1) the frequency f of the wave, (2) the amplitude of the wave, and (3) the phase of the wave. The frequency f of the wave is the number of complete cycles which occur in one second of time and is thus expressed in cycles per second, or cps. The "Hertz" (Hz) is a unit of frequency now widely used and named after Heinrich Hertz, who was the first person to demonstrate electromagnetic wave propagation. One Hz equals one cycle per second, 1 kilohertz (kHz) = 1000 Hz, 1 megahertz (MHz) = 10^6 Hz, 1 gigahertz (GHz) = 10^9 Hz. The period T of the wave is the time for one complete cycle, and therefore the period equals the reciprocal of the frequency: $T = 1/f$. For example, a wave of frequency 2000 cps or 2 kHz has a period of 0.0005 sec. The "60 cycle" (really 60 cycles per second) line voltage available in most laboratories has a period of 1/60 sec = 0.0167 sec. Often the angular frequency $\omega = 2\pi f = 2\pi/T$ is used; ω has units of radians per second. (2π radians = 360°; 1 radian = 57°.) Note that a radian is really a pure number, so radians per second = sec^{-1}. This follows from the definition of an angle θ (see Fig. 2.2) in radians: $\theta \equiv s/r$, where s is the arc length and r is the radius.

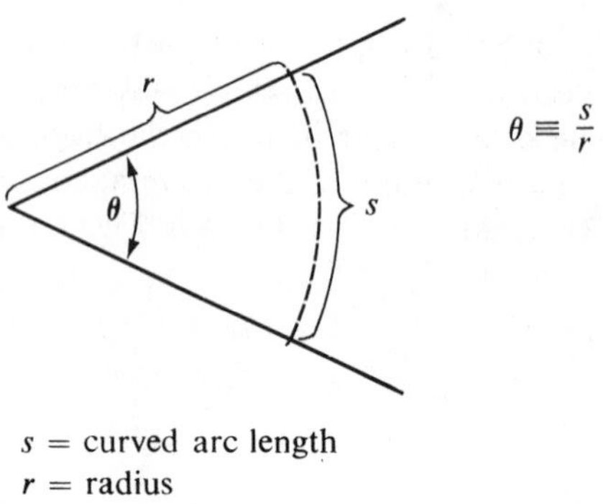

s = curved arc length
r = radius

FIGURE 2.2

The amplitude of the wave is a measure of how "large" or strong the wave is. A common way to specify the amplitude is to give the change in current or voltage from the most positive value to zero; this is I_0 in Fig. 2.1(a), and V_0 in Fig. 2.1(b). The peak-to-peak amplitude is the change in current or voltage from the most positive to the most negative value, $2I_0$ in Fig. 2.1(a) and $2V_0$ in Fig. 2.1(b). Another common measure of the amplitude is the root-mean-square ("rms") amplitude, which is defined as

$$V_{\text{rms}} \equiv \left[\frac{1}{T}\int_0^T v^2(t)\,dt\right]^{1/2}. \tag{2.1}$$

If $v(t)$ is sinusoidal $v(t) = V_0 \sin \omega t$ and

$$V_{rms} = \left[\frac{1}{T}\int_0^T V_0^2 \sin^2 \omega t \, dt\right]^{1/2}$$

$$= \frac{V_0}{\sqrt{2}} = 0.707 V_0. \tag{2.2}$$

RMS values are useful because they occur in the expression for the power in ac circuits that will be explicitly shown later in this chapter. The 110 V 60 Hz line voltage commonly available in laboratories has an rms voltage of 110 V; therefore, $V_0 = \sqrt{2}V_{rms} = 154$ V, and the peak-to-peak voltage is 308 volts! This line voltage is shown in Fig. 2.3 and can vary from about 105 V rms to 120 V

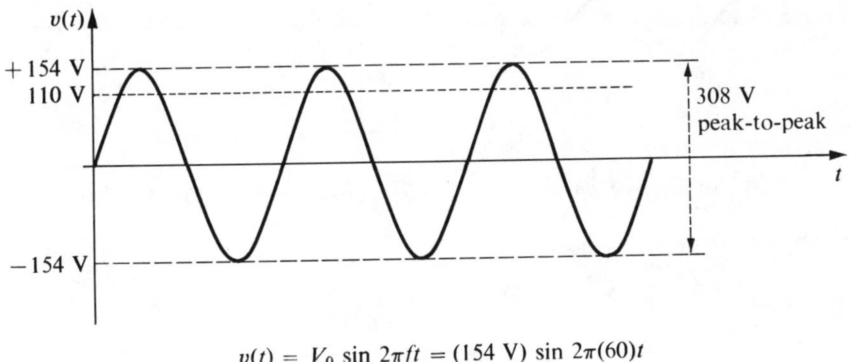

$v(t) = V_0 \sin 2\pi ft = (154 \text{ V}) \sin 2\pi(60)t$

FIGURE 2.3 110 V rms 60-"cycle" (60 Hz) line voltage.

rms depending upon the time of day and the demands made upon the electric power company.

The phase of a wave is a more subtle concept, and has meaning only when specified relative to another wave of the same frequency. The phase tells us when the wave reaches its maximum value compared to the time of the maximum for the other wave. Two waves of the same frequency are "in phase" if they reach their maximum values at exactly the same times, as shown in Fig. 2.4(a). It makes no sense to compare the phase of two waves of different frequencies. The two waves of Fig. 2.4(b) are out of phase, because their peaks do not occur at the same time. Their phase difference is seen to be $\frac{1}{4}T$ seconds or 45° because one complete period T corresponds to 360°.

Mathematically, the phase is expressed as the phase angle ϕ in the equation $v(t) = V_0 \sin(2\pi ft + \phi) = V_0 \sin(2\pi t/T + \phi)$ which is graphed in Fig. 2.5(a). The equivalence between phase in degrees (or radians) and time can be seen from the fact that a change in t by one period T or a change in ϕ by 360° or 2π radians does not change $v(t)$. Expressed in words, the phase determines the wave amplitude at the arbitrary time $t = 0$, because at $t = 0$, $v = V_0 \sin \phi$.

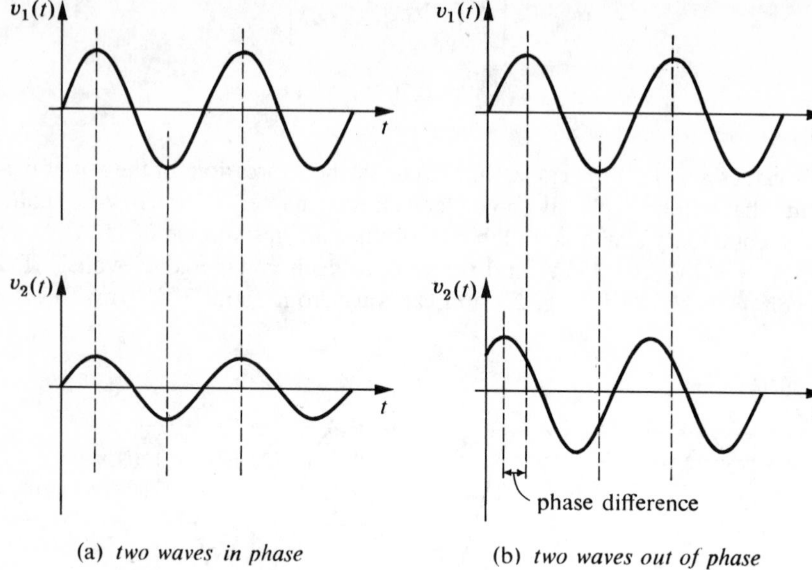

(a) *two waves in phase* (b) *two waves out of phase*

FIGURE 2.4 Phase relationships.

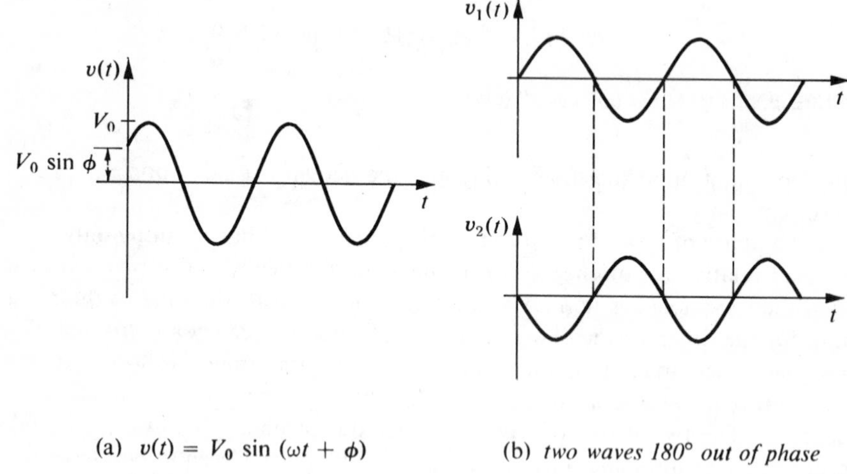

(a) $v(t) = V_0 \sin(\omega t + \phi)$ (b) *two waves 180° out of phase*

FIGURE 2.5 More phase relationships.

Phase is measured in degrees or radians and does not depend upon the amplitude or the frequency.

The voltage $v_1(t) = V_{01} \sin(\omega t + \phi)$ has a phase of ϕ radians relative to the voltage $v_2(t) = V_{02} \sin \omega t$; in other words $v_1(t)$ "leads" $v_2(t)$ by ϕ radians, or less precisely $v_1(t)$ is out of phase with respect to $v_2(t)$ by ϕ radians. If two waves (of the same frequency) are out of phase by π radians or 180°, then they appear as shown in Fig. 2.5(b). Thus two waves of equal amplitude and exactly 180° out of phase add to give exactly zero. From trigonometry, any two sine waves of the same frequency and arbitrary phases will add together to give a sine wave of the same frequency ω and a definite phase ϕ. That is, $V_{10} \sin(\omega t + \phi_1) + V_{20} \sin(\omega t + \phi_2) = V_0 \sin(\omega t + \phi)$ where V_{10} is the amplitude of the first wave, V_{20} is the amplitude of the second wave, ϕ_1 is the phase of the first wave, ϕ_2 is the phase of the second wave, V_0 is the amplitude of the resultant wave, and ϕ is the phase of the resultant wave.

It is often convenient to use cosine rather than sine waves. A cosine wave is merely a sine wave shifted in phase by 90° or $\pi/2$ radians.

$$V_0 \cos \omega t = V_0 \sin(\omega t + \pi/2)$$

The identity $\sin(a + b) = \sin a \cos b + \sin b \cos a$ has been used here.

2.2 AC POWER

Consider a sinusoidal voltage $v(t) = V_0 \sin \omega t$ applied across a resistance R. Ohm's law holds at every instant of time, and so

$$i(t) = \frac{v(t)}{R} = \frac{V_0 \sin \omega t}{R} = \left(\frac{V_0}{R}\right) \sin \omega t = I_0 \sin \omega t. \tag{2.3}$$

Thus the current through the resistance also varies sinusoidally with time at the same frequency as the current. The amplitude I_0 of the current equals V_0/R (see Fig. 2.6).

Notice that the current and the voltage are exactly in phase. The instantaneous power developed in the resistance is still given by $P = vi$ (the dc formula), but now v and i are the instantaneous values of the voltage drop across and the current through the resistance. Hence, $P(t) = I_0^2 R \sin^2 \omega t$; the *instantaneous* power $P(t)$ thus varies with time, as is shown in Fig. 2.5. Usually, however, it is much more useful to consider the *average* power, that is the power averaged over an integral number of cycles of voltages and current. The average power can be written by averaging the instantaneous power over one (or many) periods of time:

$$P_{av} = \frac{1}{T} \int_0^T P(t)\, dt \tag{2.4}$$

$$P_{av} = \frac{1}{T} \int_0^T v(t) i(t)\, dt = \frac{1}{T} \int_0^T V_0 \sin \omega t \frac{V_0}{R} \sin \omega t\, dt$$

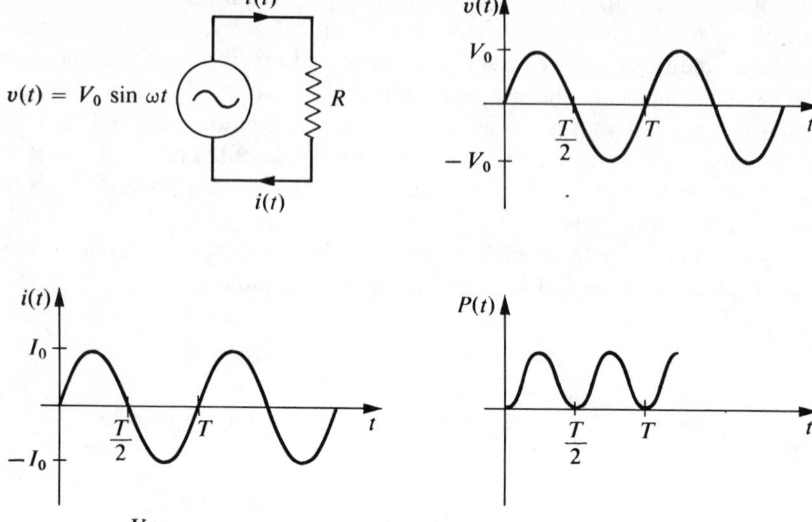

FIGURE 2.6 Alternating current and voltage through a resistance.

$$P_{av} = \frac{1}{T}\frac{V_0^2}{R}\int_0^T \sin^2 \omega t \, dt = \frac{V_0^2}{2R}. \tag{2.5}$$

But $$V_{rms} = \frac{V_0}{\sqrt{2}}. \quad \text{So: } P_{av} = \frac{V_{rms}^2}{R}. \tag{2.6}$$

And $$I_{rms} = \frac{I_0}{\sqrt{2}} = \frac{V_0}{\sqrt{2}R} = \frac{V_{rms}}{R}. \quad \text{So: } P_{av} = I_{rms}^2 R. \tag{2.7}$$

The result, then, for sinusoidal current flowing through a resistance is that the average power dissipated in the resistance equals $I_{rms}^2 R$ or V_{rms}^2/R where I_{rms} and V_{rms} are the rms values of the current and voltage, respectively as defined in Section 2.1. The occurrence of the rms values in the power formula is one reason rms values are useful.

If the current through a two terminal element is sinusoidal and the voltage across it is also sinusoidal of the same frequency but differing in phase by ϕ radians, then the instantaneous power dissipated in the element is:

$$P = v(t)i(t) = V_0 \sin(\omega t + \phi) I_0 \sin \omega t, \tag{2.8}$$

and the average power dissipated can be shown to be:

$$P_{av} = \frac{V_0 I_0}{2} \cos \phi = V_{rms} I_{rms} \cos \phi. \tag{2.9}$$

2.3 CAPACITANCE

In addition to resistance, "capacitance" is very important in ac circuits. A circuit element which has capacitance is called a capacitor or a condenser, and is represented in circuit diagrams by the symbol shown in Fig. 2.7.

V_A V_B

$$C \equiv \frac{Q}{|V_A - V_B|}$$

(a) *schematic symbol* (b) *definition*

FIGURE 2.7 Symbol for capacitance.

The definition of capacitance is the charge stored on one plate divided by the voltage difference between the two plates, or

$$C \equiv Q/V. \tag{2.10}$$

(If $+Q$ is on one plate, then $-Q$ is on the other plate because the presence of positive charge on one plate repels an equal amount of positive charge from the other plate.) The larger the capacitance, the more charge is stored on the plates for a given voltage difference; capacitance is the "amount of charge stored per volt." A capacitor has two terminals or leads, one going to each plate.

Capacitance is measured in coulombs per volt; one coulomb per volt is called a "farad". A one farad capacitor would be physically huge, so other units are used for capacitance: the microfarad (μF) which equals 10^{-6} farads, and the picofarad (pF) which equals 10^{-12} farads. A picofarad is often called a "puff" in informal conversation. The reason for the capacitance symbol in Fig. 2.7 is that a capacitor is actually made by placing two metal plates or foils parallel to and insulated from each other; the two terminals are connected to the two plates. Notice that if the insulating material between the two plates does not break down from too high a voltage being applied across the plates, then *there is no way direct current can flow through the capacitor.* In other words, a capacitor has an infinite dc resistance in the steady state. The thicker the insulating material, the higher the voltage rating of the capacitor.

However, *alternating* current can pass through a capacitor. As a positive charge surges onto the left-hand plate, the positive charge on the right-hand plate is repelled toward the right. Thus a surge of current into one terminal results in a surge of current out of the other terminal and, one half cycle later, charge surges into the right-hand terminal and out of the left-hand terminal. In other words the capacitor does pass alternating current.

There are two general types of capacitors: polarized and unpolarized. Polarized capacitors will function properly only when a definite dc voltage polarity is maintained at all times across the plates. One terminal is labeled

+ and the other −; the dc voltage on the + side must always be maintained positive with respect to the − side regardless of the ac voltage present. The most common type of polarized capacitor is the electrolytic capacitor in which the dielectric film between the two conducting plates is formed by an electrochemical reaction. The film is extremely thin; thus, large capacitances can be obtained at relatively low voltage ratings in a small volume. For example, an electrolytic capacitor as large as half a cigarette may have 100 μF capacitance with a maximum voltage rating of 25 V dc. If the dc voltage polarity across an electrolytic capacitor is reversed, the dielectric film is usually punctured, and the capacitor is permanently shorted out. Such a situation can usually be detected by measuring the dc resistance of the capacitor; a good electrolytic capacitor should yield a dc resistance of several hundred kΩ or more. In making this resistance measurement, the red or positive ohmmeter probe should be placed on the + lead of the electrolytic capacitor.

Unpolarized capacitors can function properly with either polarity dc voltage between the plates. These are of many types. The tubular capacitor shown in Fig. 2.8 consists of a thin sheet of insulating material (mylar, paper,

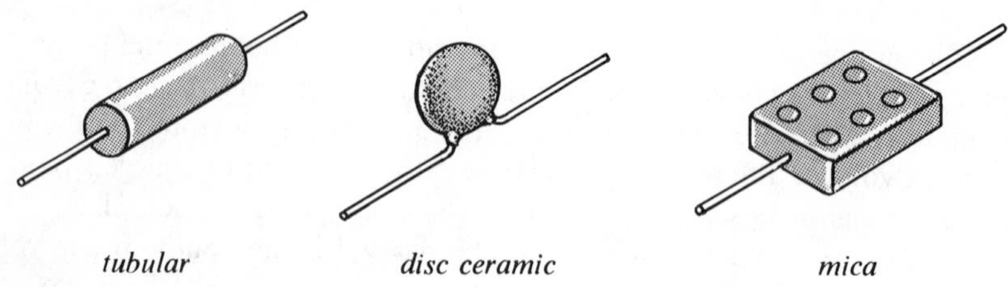

FIGURE 2.8 Various types of capacitors.

etc.) rolled up in a cylinder between two thin metal foils. One wire lead is connected to each metal foil, and the entire assembly is potted in wax or some insulating plastic. Tubular capacitors are normally used for audio frequencies, and typical capacitance values range from 0.001 μF to 1 μF.

The disc ceramic capacitor, shown in Fig. 2.8, has a thin layer of ceramic as the insulator between the electrodes and unfortunately has a rather large temperature coefficient, i.e., the capacitance changes relatively rapidly with changing temperature. Hence, they are usually used for bypass applications where only a certain minimum capacitance is required, rather than a specific value of capacitance. They are used at frequencies up to several hundred megahertz. A special type of disc ceramic, type "NPO", has a zero temperature coefficient at room temperature and is widely used. The mica capacitor, shown in Fig. 2.8, has a thin sheet of mica as the insulator, and provides the most stable, precise value of capacitance of all capacitor types. Mica capacitors are useful up to hundreds of megahertz and are also rather expensive. Another type of expensive capacitor is the tantalum capacitor which has low inductance, high capacitance per unit volume, and provides relatively high capacitance values, up to hundreds of μF. A summary of the properties of different types of capacitors is given in Appendix A.

Variable capacitors are also available; the capacitance is adjusted by turning a shaft. As the shaft turns, two sets of parallel metal plates mesh, without touching [Fig. 2.9(b)]. The larger the overlap, the higher the capacitance. The

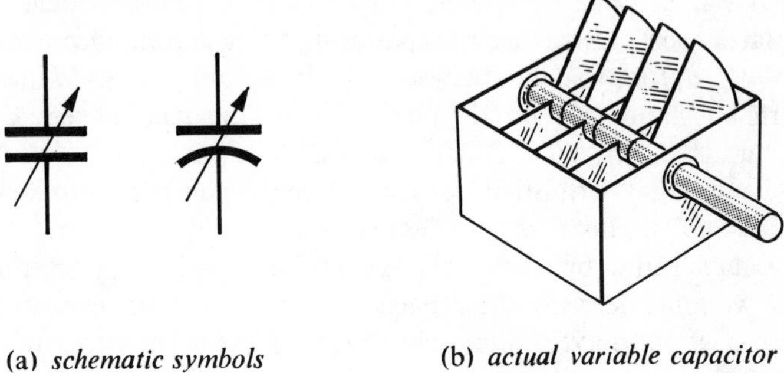

(a) *schematic symbols* (b) *actual variable capacitor*

FIGURE 2.9 Variable capacitor.

stationary set of plates is called the "stator" and the rotating set the "rotor". The rotor is connected electrically and mechanically to the shaft which is usually connected to the chassis and thus is at ground. The curved line in the schematic symbol, Fig. 2.9(a), is always the rotor and usually is grounded.

A useful way of thinking of a capacitor in terms of the water flow-current analogy is to regard the capacitor as an enlargement in the water pipe with a flexible membrane stretched across the enlargement as shown in Fig. 2.10.

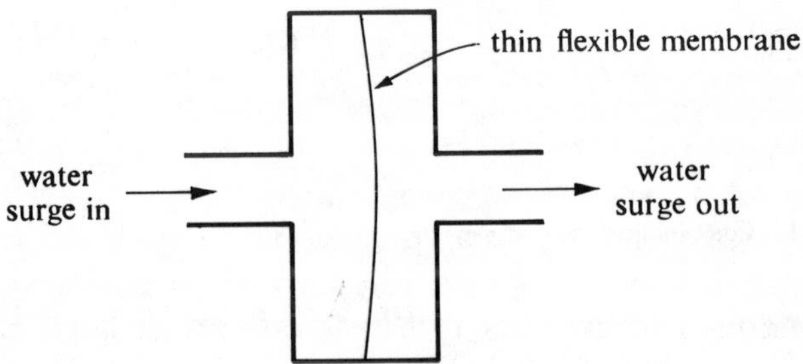

FIGURE 2.10 Water pipe analogy of capacitor.

As water surges in from the left, the membrane stretches toward the right and water surges out of the right-hand pipe. No water actually passes completely through, from left to right, but the surge of water does flow out of the right-hand pipe. In the case of the capacitor, no direct current flows through, but alternating current does because ac is really a back-and-forth surging of electrons. Note that as the water pressure between the two sides increases, the membrane stretches and water flows out of the low pressure side. This is analogous to a capacitor, which passes some current every time the voltage difference across the capacitor plates changes. A very stiff membrane corresponds to a small capacitance, and a very flexible one to a large capacitance.

Capacitance depends only on the area and separation of the plates, and on the material between the plates, not on the current or voltage. For two parallel plates each of area A and separation d, the capacitance can be shown to be $C = \epsilon_0 A/d$, for a vacuum between the plates, and $C = \epsilon A/d$ for a material of dielectric constant ϵ between the plates. If A is in square meters, d in meters, and ϵ in farads/meter, then C is in farads. The constant $\epsilon_0 = 8.85 \times 10^{-12}$ farads/meter is the "permittivity of free space" which occurs in so many formulae in mks units. Because C varies inversely with d, we see that if we obtain a higher voltage rating by using a thicker insulating material (larger d) between the plates, we must decrease the capacitance for a given volume capacitor. In other words, we can have a high voltage rating and a low capacitance, or low voltage rating and a high capacitance for a fixed physical size.

If the capacitors of capacitance C_1 and C_2 are connected in parallel as shown in Fig. 2.11(a), then the resultant can be regarded as one new capacitor. We

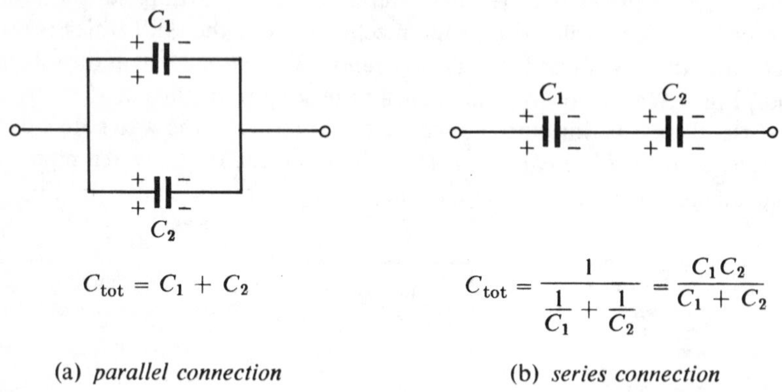

FIGURE 2.11 Combinations of two capacitors.

would expect the new capacitor to have a capacitance larger than C_1 or C_2 because the area of the left-hand plate of the new resultant capacitor is the sum of the areas of the left-hand plates of C_1 and C_2, and similarly for the right-hand plates. In fact if we do not change the plate separation of each capacitor when we connect them in parallel, since the areas add, we might then expect simply $C_{\text{total}} = C_1 + C_2$, which is the case as can be seen by the following derivation. By definition

$$C_{\text{total}} = \frac{Q_{\text{total}}}{V}. \tag{2.11}$$

where Q_{total} is the total charge on either plate and V is the voltage difference across the plates. Note that $Q_{total} = Q_1 + Q_2$ where $Q_1 =$ the charge on C_1 and $Q_2 =$ the charge on C_2. Therefore,

$$C_{total} = \frac{Q_1 + Q_2}{V}. \tag{2.12}$$

But from the definition of capacitance $Q_1 = C_1V_1$ and $Q_2 = C_2V_2$, and because C_1 and C_2 are in parallel $V_1 = V_2$. Therefore

$$C_{total} = \frac{C_1V + C_2V}{V} = C_1 + C_2. \tag{2.13}$$

For example, a 0.01 μF and a 0.05 μF capacitor in parallel act effectively like one 0.06 μF capacitor. For N capacitors in parallel it is easily shown that

$$C_{total} = C_1 + C_2 + \cdots + C_N. \tag{2.14}$$

If two capacitors of capacitance C_1 and C_2 are connected in series as shown in Fig. 2.11(b), then the resultant can be regarded as one new capacitor. If a voltage difference C is applied between terminals A and B, with A positive with respect to B, then positive charge from the right-hand plate of C_1 will be repelled over to the left-hand plate of C_2 where it must stop. Thus, $Q_1 = Q_2$, because the center wire, the right-hand plate of C_1, and the left-hand plate of C_2 must remain electrically neutral. Thus the charge on C_1 must exactly equal the charge on C_2. Let this charge be denoted by Q. By definition $C_{total} = Q/V = Q/(V_1 + V_2)$. But $V_1 = Q_1/C_1$ and $V_2 = Q_2/C_2$ and $Q_1 = Q_2 = Q$, so

$$C_{total} = \frac{Q}{\frac{Q}{C_1} + \frac{Q}{C_2}} = \frac{1}{\frac{1}{C_1} + \frac{1}{C_2}} = \frac{C_1C_2}{C_1 + C_2}. \tag{2.15}$$

Thus the total resultant capacitance of two capacitors in series is the reciprocal of the sum of the reciprocals of the capacitances. This result is the same as for two resistors in parallel. For N capacitors in series $C_{total} = 1/(1/C_1 + 1/C_2 + \cdots + 1/C_N)$. A 0.01 μF capacitor and a 0.05 μF capacitor in series act like one 0.0083 μF capacitor.

In addition to manufactured capacitors intentionally wired in a circuit, there is always some "stray" capacitance between any two elements in the circuit, for example between a wire and the metal chassis, or between the grid and plate of a vacuum tube, or between the base and collector of a transistor. Such stray capacitance is rarely shown on circuit diagrams, but nevertheless can often be extremely important, particularly at high frequencies, for example at tens or hundreds of MHz. The reason for this is that at high frequencies capacitors tend to act as "low resistances". In the limit as the frequency increases the capacitance acts like a short circuit as will be shown in the next section on capacitive reactance.

When a capacitor is charged with a voltage difference V between the plates, there is an electric field present in the region between the plates. Energy is stored in this electric field, and work must be done to create this field, i.e., work must be done to charge up the capacitor. Conversely, when the capacitor discharges, the energy stored in its electric field must go somewhere; it must be dissipated as heat or be stored in some other form such as the energy of a magnetic field. The energy stored in a capacitance C with a voltage difference V between the plates is $W = \frac{1}{2} CV^2$. The derivation is short:

$$W = \int_0^V V\, dQ = \int_0^V V\, d(CV) = C\int_0^V V\, dV = \tfrac{1}{2} CV^2. \tag{2.16}$$

Because Q, the charge on either plate, is related to V by $Q = CV$, W can be re-written as $W = \frac{1}{2} QV$ or $W = \frac{1}{2} Q^2/C$.

2.4 CAPACITIVE REACTANCE

Suppose a sinusoidal voltage $v = V_0 \cos \omega t$ is applied across a capacitor as shown in Fig. 2.12(a). Then the current through the capacitor can be obtained by taking the first derivative with respect to time of the equation $Q = Cv$, which defines capacitance.

$$i = \frac{dQ}{dt} = \frac{d}{dt}(Cv) = \frac{d}{dt}(CV_0 \cos \omega t) = -\omega CV_0 \sin \omega t = \omega CV_0 \cos (\omega t + \tfrac{1}{2}\pi) \tag{2.17}$$

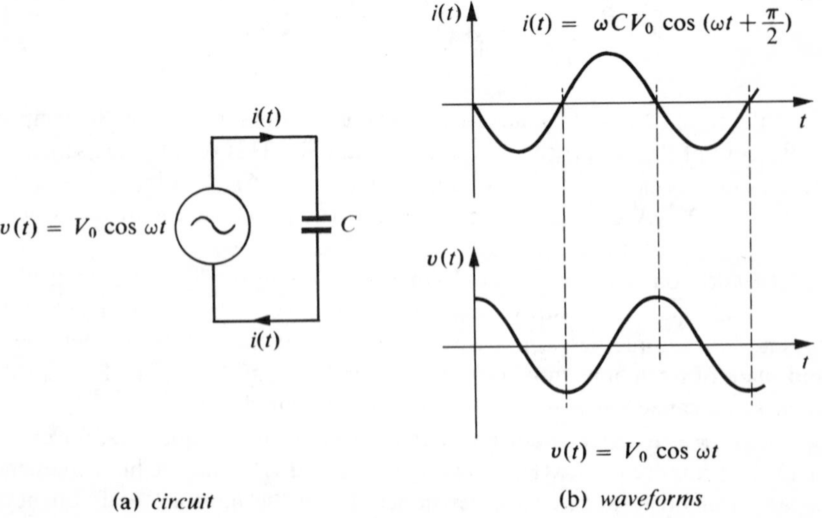

FIGURE 2.12 Current and voltage in a capacitance.

where we have used the trigonometric identity $\cos(\omega t + \pi/2) = -\sin \omega t$. Thus, we see that the current i varies sinusoidally with time at the same frequency ω as does the applied voltage, but is 90° or $\pi/2$ radians out of phase with respect to the voltage. The current through the capacitor leads the voltage across the capacitor by 90° or, equivalently, the voltage lags the current by 90° or $\pi/2$ radians, as is shown in Fig. 2.12(b). Notice also that the higher the frequency ω of the applied voltage, the larger the current for a given voltage; that is, at higher frequencies the capacitor presents *less* opposition to the current flow. All these features can be consolidated very neatly by introducing complex numbers. For a brief summary of complex numbers see Appendix E. We now regard the applied voltage $V_0 \cos \omega t$ as the real part of $V_0 e^{j\omega t}$ which we can call the complex voltage. (j^2 equals -1 and we recall $e^{j\theta} = \cos\theta + j\sin\theta$.)

$$v(t) = V_0 \cos \omega t = \textit{real part of } [V_0 e^{j\omega t}]. \tag{2.18}$$

The current can also be written as the real part of a complex number:

$$i(t) = \omega C V_0 \cos\left(\omega t + \frac{\pi}{2}\right) = Re\left[\omega C V_0 e^{j(\omega t + \pi/2)}\right], \tag{2.19}$$

where Re stands for "real part of."

Now we would like to be able to write an ac version of Ohm's law that takes into account the *phase* as well as the amplitude of the current through the capacitor. The current can be written in a form similar to Ohm's law $i = v/R$ by writing

$$i(t) = Re[\omega C V_0 e^{j\omega t} e^{j\pi/2}] = Re[j\omega C V_0 e^{j\omega t}] \tag{2.20}$$

where we have used $e^{j\pi/2} = j$. Then, $i(t)$ can be written

$$i(t) = Re\left[\frac{V_0 e^{j\omega t}}{\dfrac{1}{j\omega C}}\right]. \tag{2.21}$$

This expression says that the current through the capacitor equals the voltage divided by $1/j\omega C$ which depends only on the capacitance and the frequency of the applied voltage. In other words, the current equals the real part of the ratio of the complex voltage to the complex number $1/j\omega C$. Thus, $1/j\omega C$ plays the role of the "effective resistance" or "impedance" of the capacitor to current flow; the larger $1/j\omega C$, the smaller the current for a given voltage. The value $1/j\omega C$ is called the capacitive "reactance", usually denoted by X_C, and is measured in ohms. The capacitive reactance can be thought of as the ac frequency-dependent resistance of a capacitor. By convention we usually omit writing "real part of" and simply write:

$$i(t) = \frac{v(t)}{X_C} = \frac{V_0 e^{j\omega t}}{\dfrac{1}{j\omega C}}. \tag{2.22}$$

It is important to realize that once we define capacitance by $C = Q/V$, we are forced to the conclusion that the current and voltage are 90° out of phase and that the current through a capacitor equals the voltage across it divided by $1/j\omega C$. *The presence of the complex number j simply takes into account the phase of the current relative to the voltage.* Notice that because the current and voltage through and across a capacitor differ in phase by $\phi = 90°$, the power dissipated in the capacitor is zero from $P = I_{rms}V_{rms} \cos \phi = I_{rms}V_{rms} \cos 90° = 0$. This is in complete contrast to a resistance where the current and voltage are in phase and electrical power is dissipated as heat. Notice also that in the limiting case of zero frequency, which is direct current, the capacitive reactance goes to infinity; that is, a capacitor can pass no direct current.

It is often desired to block out the dc component of a voltage while retaining the ac component. A series capacitor called a "blocking" capacitor or "coupling" capacitor does the trick as shown in Fig. 2.13. If any dc current I_{dc}

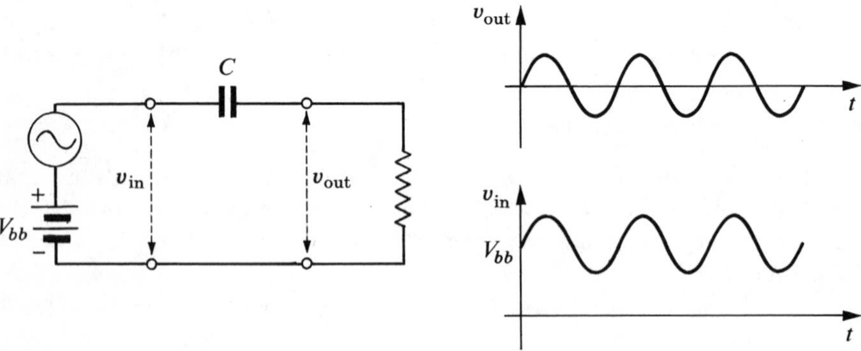

FIGURE 2.13 Coupling or blocking capacitor.

flowed through C, there would be a dc voltage drop across R equal to $I_{dc}R$. But the capacitor cannot pass any dc current, hence $I_{dc} = 0$ and $V_{dc\ out} = 0$.

2.5 INDUCTANCE

Inductance, as well as resistance and capacitance, is important in ac circuits. A circuit element that has inductance is called an inductor, a choke, or sometimes an inductance. It is represented in circuit diagrams as a coil (see Fig. 2.14).

Actual inductors are usually made by winding wire as a coil around some kind of core; the two terminals are the two ends of the coil. The definition of inductance is

$$L \equiv \frac{\phi}{I}, \tag{2.23}$$

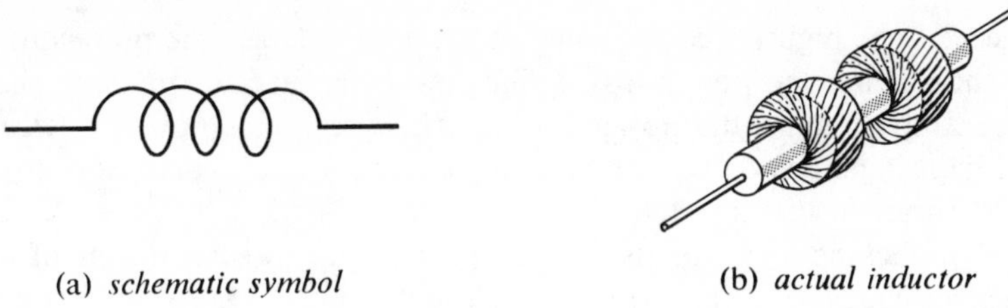

(a) *schematic symbol* (b) *actual inductor*

FIGURE 2.14 Circuit symbol for an inductance.

the ratio of magnetic flux ϕ with respect to current I. (We recall that magnetic flux equals the magnetic field induction times area, $\phi = BA$.) The current I flowing through the coil produces a magnetic flux ϕ through the cross-sectional area of the coil. The larger I is, the larger ϕ is, so L is positive. A more easily understood definition of inductance is in terms of the voltage drop v across the inductor; L is defined from the equation $v = -L\,dI/dt$. These two definitions are, of course, equivalent, as can be seen from the Faraday law expression for the voltage produced by a changing magnetic flux. Starting with Faraday's law, $v = -d\phi/dt = -d/dt(LI) = -L\,dI/dt$. If the current changes through the inductor at a rate of one ampere per second and the voltage produced across the inductor is one volt, then the inductor has an inductance of one "henry." A one henry inductor would be rather large physically; hence other units are used for inductance; the millihenry (mH) which equals 10^{-3} H, and the microhenry (μH) which equals 10^{-6} H.

The physical reason for the inductance is that the current flowing in the coil produces a magnetic flux through the coil. As the current changes ($dI/dt \neq 0$) the magnetic flux through the coil also changes $d\phi/dt \neq 0$; hence a voltage is induced by Faraday's law (induced voltage or EMF $= -d\phi/dt$) between the two ends of the coil. In other words, whenever the magnetic flux lines "cut" the wire turns of the coil, a voltage is induced in the coil. The minus sign means that the voltage induced is of a polarity to oppose the change in current which produced the voltage. It can be seen now that a direct current (I = constant and $dI/dt = 0$) will produce no voltage across the coil terminals from Faraday's law; the only voltage drop produced by direct current I through the coil will be the Ohm's law drop $V = IR$ where R is the resistance of the wire in the coil, which is usually negligibly small. For example, a low power one henry iron core inductance may have a dc resistance of about 50 ohms, hence 10 mA dc flowing through it will produce a steady voltage drop across it of only one-half volt. Inductance ideally depends only upon the number of turns and the size of the coil, and on the material inside the coil, not on the current or voltage.

Often solid or powdered iron is placed inside the coil to increase the inductance. The iron produces more magnetic flux ϕ through the coil for a given number of ampere turns. However, eddy currents always flow in the iron core which has some resistance; thus I^2R power is dissipated as heat in the core. This power comes from the ac signal in the coil, so that ac signal is attenuated or reduced. The higher the frequency of the ac, the larger the power loss. It can be shown that the power loss is linear with frequency; doubling the frequency exactly doubles the power loss to the core. This power loss can be reduced by using a powdered iron core in which very small particles of iron are cemented in an insulating glue, thus reducing the eddy currents. Powdered iron core inductances can be used at frequencies up to several tens of MHz. For higher frequencies, air core coils are usually used. One should note that iron core inductors often change their inductance if the dc current through the coil is changed because of magnetic saturation* of the core as shown in Fig. 2.15. $L = \phi/I$, and if saturation occurs, L clearly decreases.

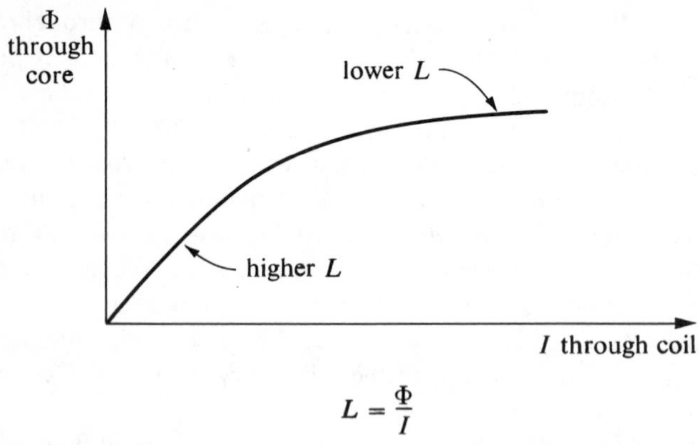

$$L = \frac{\Phi}{I}$$

FIGURE 2.15 Inductance changes with changing dc current.

If two inductors of inductance L_1 and L_2 are connected in series as shown in Fig. 2.16(a), then it can be shown that the resultant can be regarded as a single inductor with an inductance equal to $L_1 + L_2$.

This result follows from the argument

$$v = L\frac{dI}{dt} = v_1 + v_2. \tag{2.24}$$

But $v_1 = L_1(dI_1/dt)$ and $v_2 = L_2(dI_2/dt)$. And because L_1 and L_2 are in series, $I_1 = I_2 = I$ and $dI_1/dt = dI_2/dt = dI/dt$. Therefore

$$v = v_1 + v_2 = L_1\frac{dI_1}{dt} + L_2\frac{dI_2}{dt} = (L_1 + L_2)\frac{dI}{dt} = L\frac{dI}{dt}.$$

Hence,

$$L = L_1 + L_2. \tag{2.25}$$

* Saturation means a flattening or leveling off of the ϕ-versus-I curve.

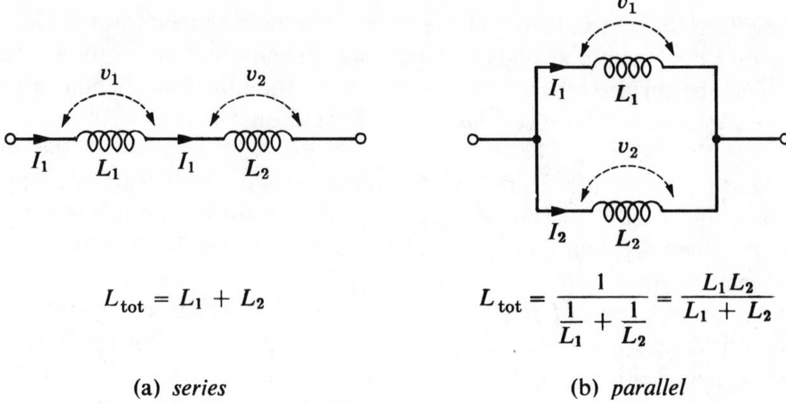

FIGURE 2.16 Series and parallel inductance.

If two inductors are connected in parallel (this is almost never done in actual practice because of magnetic coupling between the two inductors) as shown in Fig. 2.16(b), the resultant can be regarded as a single inductor of inductance $L = 1/(1/L_1 + 1/L_2)$. This result follows from $I = I_1 + I_2$. Therefore,

$$\frac{dI}{dt} = \frac{dI_1}{dt} + \frac{dI_2}{dt}.$$

By the definition of inductance $dI/dt = v/L$, $dI_1/dt = v_1/L_1$, $dI_2/dt = v_2/L_2$, and because L_1 and L_2 are in parallel $v_1 = v_2 = v$. Therefore,

$$\frac{v}{L} = \frac{v}{L_1} + \frac{v}{L_2} \quad \text{or} \quad \frac{1}{L} = \frac{1}{L_1} + \frac{1}{L_2} \quad \text{or} \quad L = \frac{1}{\frac{1}{L_1} + \frac{1}{L_2}},$$

$$L = \frac{L_1 L_2}{L_1 + L_2}. \tag{2.26}$$

It can be shown that a coil of wire of length L, radius R, with N turns, has an inductance of $L = \mu_0 (N/L)^2 LR$.

It can be shown that for a given length and cross-sectional area, a wire of rectangular cross section (a flat strip) has much less inductance than a wire of circular cross section. Hence flat metal strips are often used in very high-frequency circuits instead of ordinary round wire, particularly for ground connections. Even a perfectly straight piece of wire has an inductance because of the finite cross-sectional area of the wire. For example, a 4 inch $\cong$ 10 cm long piece of No. 18 (0.04 inch diameter) wire has an inductance of 0.1 μH. This inductance can be very important in determining circuit behavior at high frequencies in the tens or hundreds of MHz. Current flowing in one part of the wire creates a magnetic flux ϕ in other parts of the wire, and this flux changing in time affects the current flow in the wire. The inductance also will change with frequency, because at higher frequencies the current tends to be concentrated near the surface of the wire and not in the interior; this phenomenon is called the "skin effect." Electromagnetic fields do not penetrate into a good conductor. They decrease according to $e^{-x/\delta}$ where x is the depth of the conductor and δ is the "skin depth" of the conductor. It can be shown that δ is given by $\delta = (\pi f \sigma \mu)^{-1/2}$, where f is the frequency of the electromagnetic field, σ is the conductivity, and μ the permeability of the conductor. The skin depth in copper at 60 Hz is 0.85 mm, and at 1 MHz it is only 0.066 mm.

When an inductance L has a certain current I flowing through it, there is a magnetic field present in the region around the inductance. Energy is stored in this magnetic field, and work must be done to create the field, i.e., work must be done to increase the current flowing through the inductance from zero to I. Conversely, when the current through the inductance falls to zero, the magnetic field decreases to zero and the energy must go somewhere. It must be dissipated as heat or be stored in some other form such as the electric field energy in a charged capacitor. The energy stored in an inductance through which a current I is flowing is given by $W = \frac{1}{2}LI^2$. The derivation is short:

$$W = \int V\,dQ = \int \left(L\frac{dI}{dt}\right)(I\,dt) = L\int_0^I I\,dI = \tfrac{1}{2}LI^2. \tag{2.27}$$

2.6 MUTUAL INDUCTANCE

In the previous discussion of inductors in series and parallel, we have tacitly assumed that the magnetic flux lines, which cut the wires of an inductor and induce a voltage, come from the inductor itself and not from any other coil. But magnetic flux lines from *any* source will induce a voltage in a wire if they cut the wire. Thus we expect a changing current in one conductor to produce a voltage in a second conductor if the magnetic flux lines from the first conductor cut through the second.

This concept of changing magnetic flux in one conductor being produced by changing current in a different conductor leads us to the concept of *mutual* inductance between the two conductors. Consider the circuit of Fig. 2.17 where we have two coils of inductance L_1 and L_2 with N_1 and N_2 turns, respectively. Suppose a time varying current I_1 is flowing in L_1. Let ϕ_{12} be the magnetic flux through coil 2 caused by the current I_1 flowing in coil 1. The voltage v_{12} induced in L_2 will be given by Faraday's law:

$$v_{12} = -N_2\frac{d\phi_{12}}{dt}. \tag{2.28}$$

But the flux in L_2 must be proportional to I_1 from elementary electricity and magnetism; that is, $\phi_{21} = KI_1$ where K is a constant depending upon the dis-

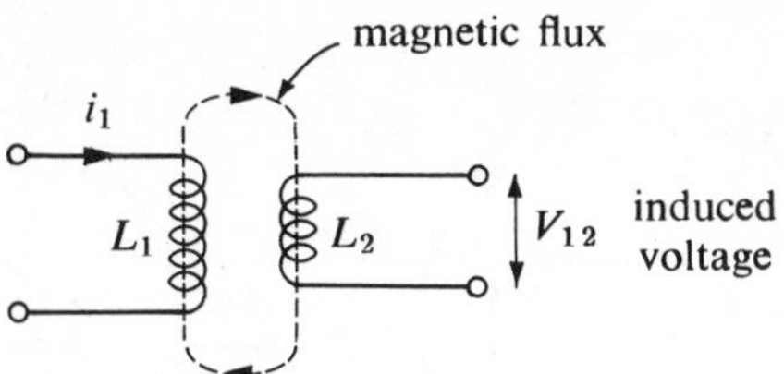

FIGURE 2.17 Mutual inductance between two coils.

tance between the two coils, the number of turns in coil 1, and their relative orientation. Therefore the voltage induced in L_2 is

$$v_{12} = -N_2K\frac{dI_1}{dt}. \tag{2.29}$$

We define $-N_2K$ as M_{12}, the "mutual inductance" between coil L_1 and L_2. In terms of M_{12}

$$v_{12} = M_{12}\frac{dI_1}{dt}. \tag{2.30}$$

M_{12} clearly has units of henrys, just like self-inductance. The mutual inductance depends upon the number of magnetic flux lines from L_1 which pass through L_2, i.e., on the geometry: the distance apart, the relative orientation of the two coils, and also on the medium between L_1 and L_2. A little thought will show that the further apart the two coils are, the smaller the mutual inductance, and if the coils are oriented at right angles the mutual inductance will be minimum. If the coil axes are parallel, the mutual inductance will be maximum. Any substance between the coils with a high magnetic permeability such as iron will increase the mutual inductance.

The self-inductance L_2 will also create an induced voltage in coil 2 due to the changing I_2. The total voltage v_2 will therefore be given by

$$v_2 = M_{12}\frac{dI_1}{dt} - L_2\frac{dI_2}{dt}. \tag{2.31}$$

A changing current in L_2 will produce magnetic flux lines which cut L_1; hence by a similar argument

$$v_1 = M_{21}\frac{dI_2}{dt} - L_1\frac{dI_1}{dt}. \tag{2.32}$$

The mutual inductance M_{21} can be shown to equal M_{12}. If two coils are wound close together, so that most of the flux lines from one pass through the other, this is called a "transformer," the circuit symbol for which is shown in Fig. 2.18. If both coils are wound on a common iron core, then almost all the magnetic

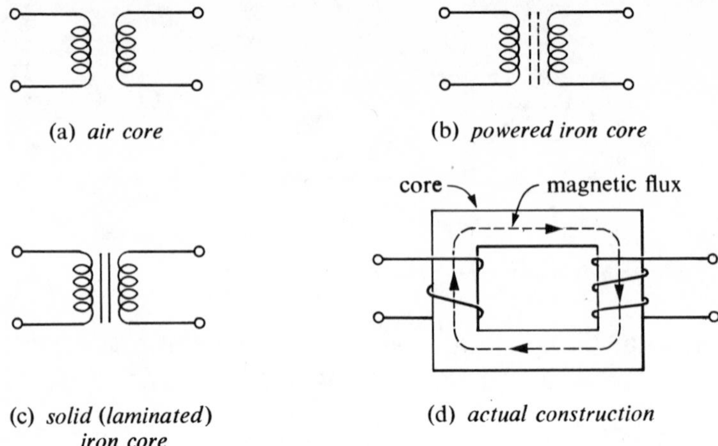

FIGURE 2.18 Transformers.

flux lines produced by a current in one coil will pass through the other coil as shown in Fig. 2.18(d). The practical details of transformers are discussed in Appendix A.

2.7 INDUCTIVE REACTANCE

Suppose a sinusoidal voltage $v = V_0 \cos \omega t$ is applied across an inductance as shown in Fig. 2.19. Then the current i through the inductance can be obtained by integrating with respect to time the $v = L\, di/dt$ equation which defines inductance:

$$v = L\frac{di}{dt} \qquad di = \frac{v}{L}\,dt \qquad i = \int di \tag{2.33}$$

$$i(t) - i(0) = \int_0^t di = \frac{1}{L}\int_0^t v\,dt = \frac{1}{L}\int_0^t V_0 \cos \omega t\,dt$$

$$i(t) - i(0) = \frac{V_0}{\omega L}\sin \omega t. \tag{2.34}$$

If we choose $t = 0$ when the current equals zero, then $i(0) = 0$ and

$$i(t) = \frac{V_0}{\omega L}\sin \omega t = \frac{V_0}{\omega L}\cos(\omega t - \pi/2) \tag{2.35}$$

where we have used the trigonometric identity $\cos(\omega t - \pi/2) = \sin \omega t$. The current i varies sinusoidally with time at the same frequency ω as does the applied voltage, and is 90° or $\pi/2$ radians out of phase with respect to the voltage. The current in the case of the inductance *lags* the voltage by $\pi/2$ radians or 90° as

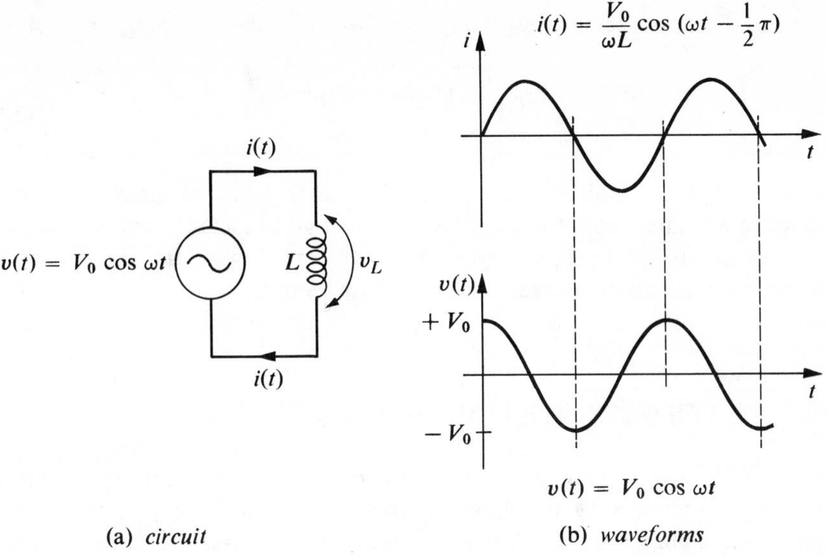

(a) *circuit* (b) *waveforms*

FIGURE 2.19 Current and voltage in an inductance.

shown in Fig. 2.19. Thus, the power dissipated in an inductance as heat is zero because of the 90° phase difference between the current and voltage. Notice that the higher the frequency of the applied voltage the smaller the current; that is, at higher frequencies the inductance presents more "opposition" to the current flow. The use of complex numbers can be used to consolidate these features. Again, we regard the applied voltage $V_0 \cos \omega t$ as the real part of $V_0 e^{j\omega t}$ and write

$$i = \frac{V_0 \cos(\omega t - \pi/2)}{\omega L} = Re\left(\frac{V_0 e^{j(\omega t - \pi/2)}}{\omega L}\right)$$

$$i = Re\left(\frac{-jV_0 e^{j\omega t}}{\omega L}\right) \quad \text{using } e^{-j\pi/2} = -j$$

$$i = Re\left(\frac{V_0 e^{j\omega t}}{j\omega L}\right) \tag{2.36}$$

where we have used $-j = 1/j$.

This expression for the current is in a form similar to Ohm's law; the current equals the real part of the complex voltage divided by the complex number $j\omega L$. Thus $j\omega L$ plays the role of the effective resistance of the inductance to current flow; the larger $j\omega L$, the smaller the current for a given voltage. The inductive reactance X_L is defined as $j\omega L$, and can be thought of as the ac frequency-dependent resistance of an inductance. The presence of the complex number j simply takes into account the *phase* of the current relative to the voltage. By convention, we usually omit writing "the real part," and simply write

$$i = \frac{V_0 e^{j\omega t}}{j\omega L} = \frac{v(t)}{X_L}. \tag{2.37}$$

Notice that in the limiting case of zero frequency, which is direct current, the inductive reactance goes to zero which means that inductance offers *no* opposition to the flow of direct current. Direct current flowing through an inductance is limited only by the resistance of the wire in the coil.

2.8 THE COMPLEX VOLTAGE PLANE

Let us now consider a few points about the representation of sinusoidal voltages by complex numbers. The voltage $V_0 \cos \omega t$ can certainly be written as the real part of the complex exponential $V_0 e^{j\omega t}$ because $V_0 e^{j\omega t} = V_0 \cos \omega t + jV_0 \sin \omega t$. We can represent $V_0 e^{j\omega t}$ by a point in the complex voltage plane, which moves counterclockwise in a circle of radius V_0 with an angular frequency ω. At any time t, the actual voltage $V_0 \cos \omega t$ is the projection of the complex voltage $V_0 e^{j\omega t}$ on the real voltage axis. Such a diagram is often referred to as a rotating vector diagram as shown in Fig. 2.20. The vector drawn from the origin out to the

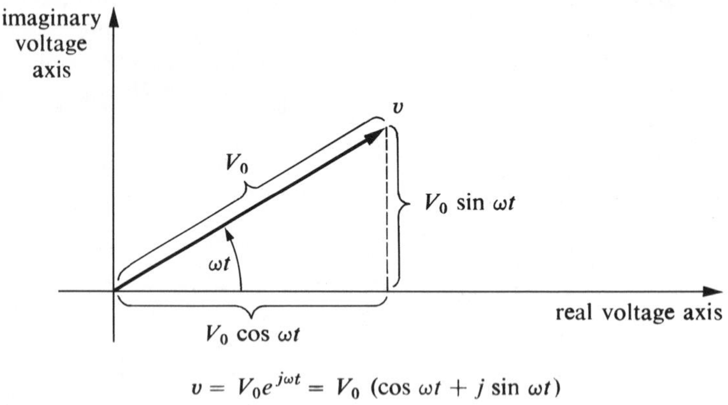

FIGURE 2.20 Basic rotating vector diagram.

point $V_0 e^{j\omega t}$ is called the voltage vector. The voltage which is found in the actual circuit is the projection of the rotating complex voltage vector on the real axis.

One of the main advantages of this kind of diagram is that the phase difference between two voltages (of the same frequency) is simply the geometrical angle between their two rotating vectors. This will be particularly useful in analyzing "phase shifter" circuits in which the output voltage has been shifted in phase with respect to the input voltage. That is, if we have two voltages $v_1 = V_{10} \cos(\omega t + \phi_1)$ and $v_2 = V_{20} \cos(\omega t + \phi_2)$ then at $t = 0$ they would be represented by the diagram in Fig. 2.21. The phase difference between v_1 and

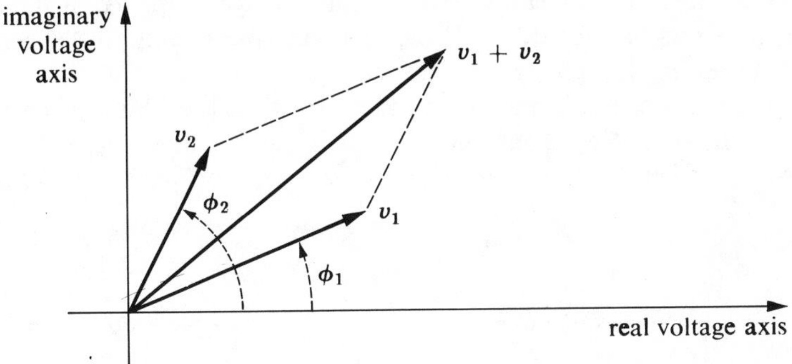

FIGURE 2.21 Two sinusoidal voltages with different phases and their sum at a certain instant of time.

v_2 is $\phi_1 - \phi_2$, the angle between v_1 and v_2; v_2 leads v_1 by $\phi_2 - \phi_1$. $v_1 + v_2$ can be calculated very easily from the rotating vector diagram by adding v_1 and v_2 using the standard parallelogram method. The phase and amplitude of $v_1 + v_2$ can be read off the diagram immediately.

2.9 RC HIGH-PASS FILTER

Consider the circuit of Fig. 2.22 in which the output is taken across the resistor. The "gain" of any circuit is defined as the output divided by the input. The "attenuation" of any circuit is defined as the reciprocal of the gain. The voltage gain of the *RC* high-pass filter equals v_2/v_1 and clearly will depend upon the frequency of the input. At very low frequencies the capacitor will present a very high reactance thus giving a small output, and at very high frequencies the

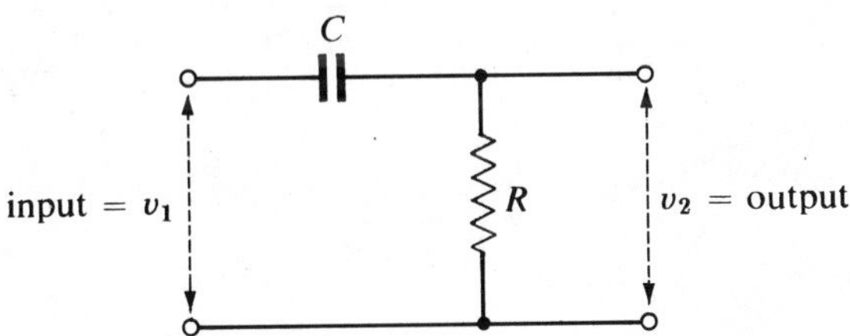

FIGURE 2.22 RC high-pass filter.

capacitor will be essentially a short circuit, thus making the output nearly equal to the input. In other words, low frequencies will be attenuated, and high frequencies will be passed without much loss in amplitude, with the gain approaching 1.0 as the frequency $\omega \to \infty$. For this reason the circuit is called a "high-pass" circuit or filter. The gain is zero for dc (zero frequency), because the capacitor passes no direct current. If a dc voltage is applied at the input, once the transients have died out all the dc input voltage will appear across C and none across R, thus giving zero output.

Let us calculate the gain assuming that a negligible amount of current is drawn from the output terminals and that the input voltage is $v_1 = V_{10} \cos \omega t = Re(V_{10}e^{j\omega t})$. Then the ac current i will be as shown in Fig. 2.23(a). For this

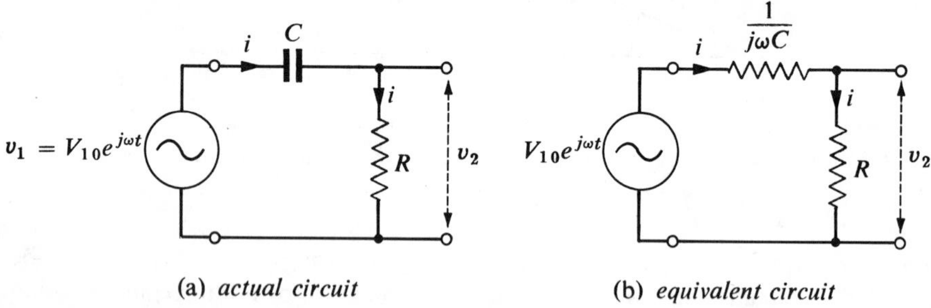

(a) *actual circuit* (b) *equivalent circuit*

FIGURE 2.23 RC high-pass filter with source.

sinusoidal input of angular frequency ω we need only replace the capacitor by its complex capacitive reactance $X_C = 1/j\omega C$ and treat the circuit as follows [see Fig. 2.23(b)]. $v_1 = i(1/j\omega C + R)$ and $v_2 = iR$. The voltage gain equals:

$$\frac{v_2}{v_1} = \frac{iR}{i(X_C + R)} = \frac{iR}{i\left(\frac{1}{j\omega C} + R\right)} = \frac{R}{\frac{1}{j\omega C} + R} = \frac{1}{1 + \frac{1}{j\omega RC}} = \frac{j\omega RC}{1 + j\omega RC}. \quad (2.38)$$

The fact that the voltage gain is complex merely means that the output differs in *phase* from the input. The magnitude or absolute value of the gain will tell us the magnitude of the output divided by the magnitude of the input:

$$\left|\frac{v_2}{v_1}\right|^2 = \left(\frac{v_2}{v_1}\right)^* \left(\frac{v_2}{v_1}\right) = \left(\frac{-j\omega RC}{1 - j\omega RC}\right)\left(\frac{j\omega RC}{1 + j\omega RC}\right) = \frac{\omega^2R^2C^2}{1 + \omega^2R^2C^2} \quad (2.39)$$

$$\left|\frac{v_2}{v_1}\right| = \frac{\omega RC}{\sqrt{1 + \omega^2R^2C^2}}. \quad (2.40)$$

* indicates complex conjugate.

Notice that this expression for the gain approaches 1.0 as the angular frequency ω of the input approaches infinity, which agrees with our intuitive feeling that the capacitor acts like a short circuit at very high frequencies.

If we plot a graph of the magnitude of the gain versus frequency on log-log paper, we obtain Fig. 2.24. The curved graph of the actual gain can be approx-

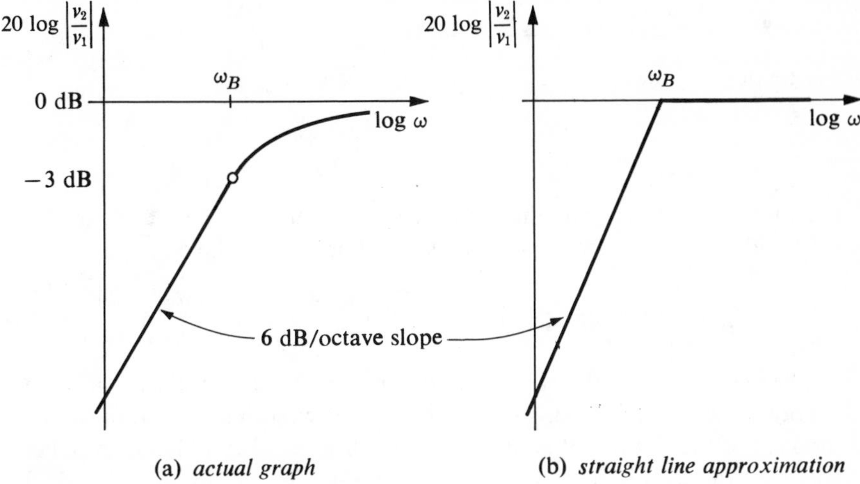

FIGURE 2.24 Gain versus frequency for RC high-pass filter.

imated quite well by two straight lines giving a "sharp break" or "knee" in the graph. The frequency at which this break occurs is given by $\omega_B = 1/RC$. This is called the "break point frequency" or usually simply the "break point." For example, a high-pass filter with $R = 10\,\mathrm{k}\Omega$ and $C = 0.1\,\mu\mathrm{F}$ will have a break point at $\omega_B = 1/RC = 1/(10^4\,\Omega)(10^{-7}\,\mathrm{F}) = 10^3$ rad/sec, or $f_B = \omega_B/2\pi = 160\,\mathrm{Hz}$. At the break point, the voltage gain $|v_2/v_1| = 0.707$, as can be seen by substituting $\omega = 1/RC$ in the expression for $|v_2/v_1|$. As a gross simplification, the *RC* high-pass filter can be thought of as passing frequencies above $\omega_B = 1/RC$ and attenuating those below ω_B.

It is common practice to express gain (or attenuation) in decibels or dB. The voltage gain v_2/v_1 in decibels is defined as

$$\frac{v_2}{v_1}_{\text{in dB}} \equiv 20 \log_{10}\left(\frac{v_2}{v_1}\right). \tag{2.41}$$

The phase difference (if any) between v_2 and v_1 does not enter in the gain expressed in decibels. Note that a negative gain in decibels merely means the output is *less* than the input; a positive gain means, of course, that the output is greater than the input.

$\left\|\frac{v_2}{v_1}\right\|$	in dB	$\left\|\frac{v_2}{v_1}\right\|$	in dB
0.01	−40	0.5	−6
0.1	−20	0.707	−3
1.0	0	1.414	+3
10.0	+20	2.0	+6
100.0	+40	4.0	+12

Often the power gain is expressed in decibels according to

$$\left(\frac{P_2}{P_1}\right)_{\text{in dB}} \equiv 10 \log_{10}\left(\frac{P_2}{P_1}\right). \tag{2.42}$$

If the impedance levels are the same at the input and output, then $P_2/P_1 = v_2^2/v_1^2$ and the voltage and power gains in dB are seen to be equivalent.

$$\left(\frac{P_2}{P_1}\right)_{\text{dB}} = 10 \log_{10}\left(\frac{P_2}{P_1}\right) = 10 \log_{10}\left(\frac{v_2}{v_1}\right)^2 = 20 \log_{10}\left(\frac{v_2}{v_1}\right). \tag{2.43}$$

Thus a power gain of 20dB means the output power is 100 times the input power.

Notice that because the log of the product of two gains is equal to the sum of the logs of the individual gains, the total dB gain of two filters, amplifiers, or whatever in series is equal to the *sum* of the individual dB gains. A 10dB amplifier driving a 30dB amplifier has a net gain of 40dB, provided the second amplifier does not "load" the first one. In other words, the second amplifier should not draw too much current from the output of the first amplifier. More on this when input and output impedance are discussed in Chapter 6.

Expressed in decibels, the voltage gain or power gain at the break point for an *RC* high-pass filter is −3dB or 3dB "down," relative to the gain of 1.0 at high frequencies. If the gain is down 3dB at the break point, the voltage gain is down by a factor of 0.707 and the power gain is down by a factor of 2. At lower frequencies, the voltage gain continues to fall off at a constant rate of 6dB per octave frequency change or equivalently 20dB per decade (×10) frequency change according to the straight line approximation to the gain curve. For example, if the gain at 1000Hz for an *RC* high-pass filter is 10dB down, then the gain at 500Hz will be 16dB down, and at 250Hz will be 22dB down, etc.

The phase of the output is different from the phase of the input, and this can be seen most easily from a rotating vector diagram. Choose the phase of the current as the reference phase, and recall that the current leads the voltage in a capacitor by 90°. Therefore, the voltage $v_R = iR$ must be 90° more counterclockwise (ahead of v_C). Thus, we have the diagram of Fig. 2.25 since $v_1 = v_R + v_C$. Thus, the output voltage across the resistor leads the input voltage v_1 by θ, which is given by $\theta = \arctan(|v_C|/|v_R|) = \arctan 1/\omega RC$. Notice that as the frequency increases the phase shift θ goes to zero, and that as θ increases the output voltage decreases. Sometimes this circuit is used to shift the phase of a sinusoidal signal; the amount of phase shift can be varied by using a variable

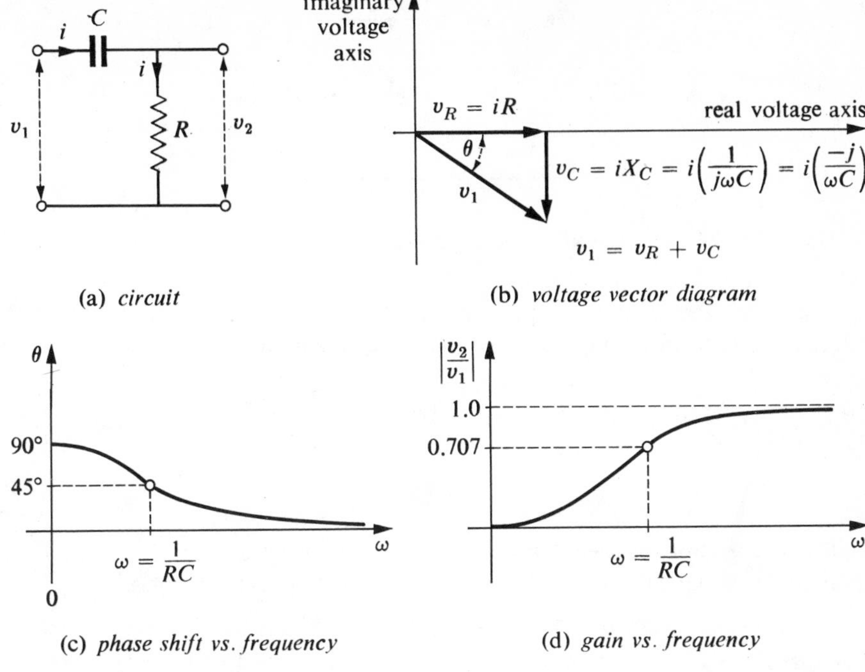

FIGURE 2.25 RC phase shifter.

resistor for R. When $R = 1/\omega C$, $\theta = 45°$ and $|v_2/v_1| = 1/\sqrt{2} = 0.707$. However, as a phase shifter this circuit has two disadvantages: the amplitude of the output voltage changes as R is changed, and also the maximum phase shift possible with this circuit is 90° in the limit as the output voltage goes to zero.

2.10 RC LOW-PASS FILTER

Consider the circuit of Fig. 2.26 in which the output is taken across the capacitor. The voltage gain v_2/v_1 clearly will go to zero at very high frequencies, since the capacitor acts like a short circuit at high frequencies. And, at zero frequency (dc) the gain will be unity if we assume no current is drawn from the output. The circuit is therefore called a low-pass filter because it passes the low frequencies and attenuates the high frequencies. For a sinusoidal input of angular frequency ω, the voltage gain will be given by (assuming no output current)

$$\frac{v_2}{v_1} = \frac{iX_C}{i(R + X_C)} = \frac{i\left(\frac{1}{j\omega C}\right)}{i\left(R + \frac{1}{j\omega C}\right)} = \frac{-j/\omega C}{R - j/\omega C} = \frac{1}{1 + j\omega RC}. \tag{2.44}$$

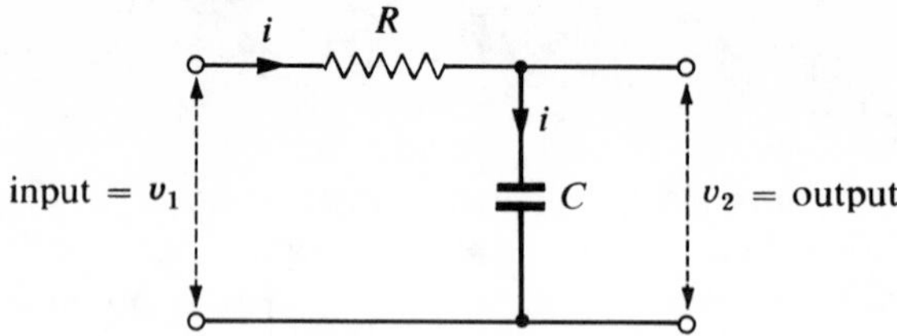

FIGURE 2.26 RC low-pass filter.

The fact that the gain is complex means that the output has been shifted in phase relative to the input. The magnitude of the gain is

$$\left|\frac{v_2}{v_1}\right| = \left[\left(\frac{v_2}{v_1}\right)^* \left(\frac{v_2}{v_1}\right)\right]^{1/2} = \left[\left(\frac{1}{1 - j\omega RC}\right)\left(\frac{1}{1 + j\omega RC}\right)\right]^{1/2} = \frac{1}{\sqrt{1 + \omega^2 R^2 C^2}} \tag{2.45}$$

which goes to unity as $\omega \to 0$, and which goes to zero as $\omega \to \infty$. If we plot the magnitude of the gain versus frequency on log-log paper we obtain Fig. 2.27.

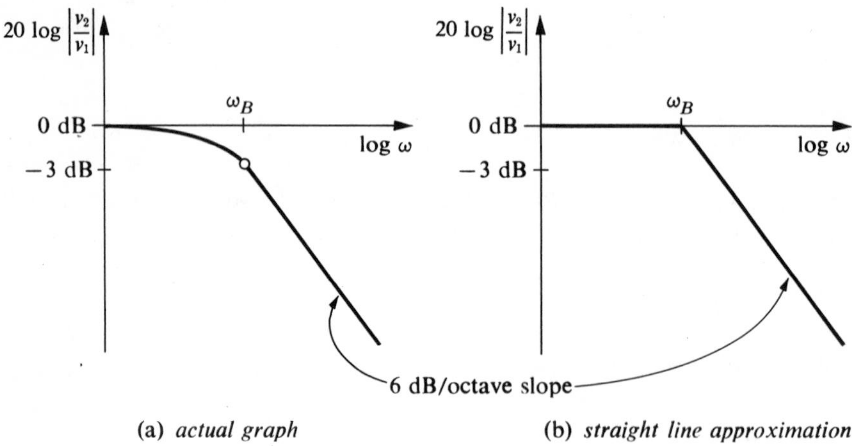

FIGURE 2.27 Gain versus frequency for RC low-pass filter.

The curved graph of the actual gain can be approximated by two straight lines giving a sharp break or knee in the curve at the frequency $\omega = 1/RC = \omega_B$, which is called the "break point." At the break point the voltage gain is down by 0.707 or 3 dB down relative to the gain of unity at zero frequency as can be seen by substituting $\omega = 1/RC$ in the expression for the gain. The slope of the straight line is 6 dB per octave. That is, if the voltage gain at 1000 Hz is down 12 dB, then the voltage gain at 2 kHz is down 18 dB.

This *RC* circuit basically accounts for the decrease in gain with increasing frequency for *all* amplifier circuits, which is why it is worth studying in detail. The capacitance *C* is often the stray capacity between the circuit wiring and the chassis, or it may be an inherent capacity built in a transistor or tube. Thus, we can see that if we are "stuck" with a certain minimum capacitance between a signal carrying wire and ground, and if we wish to maximize the high frequency response of the circuit, then we should take care that the effective series resistance *R* is as small as possible. More on this later in the section on amplifiers.

The difference in phase between the input and the output can be calculated most easily from a rotating vector diagram. Choose the phase of the current as the reference phase and recall that the current leads the voltage in a capacitor by 90°. Thus we have, assuming $i_{out} = 0$, the diagrams of Fig. 2.28. The output

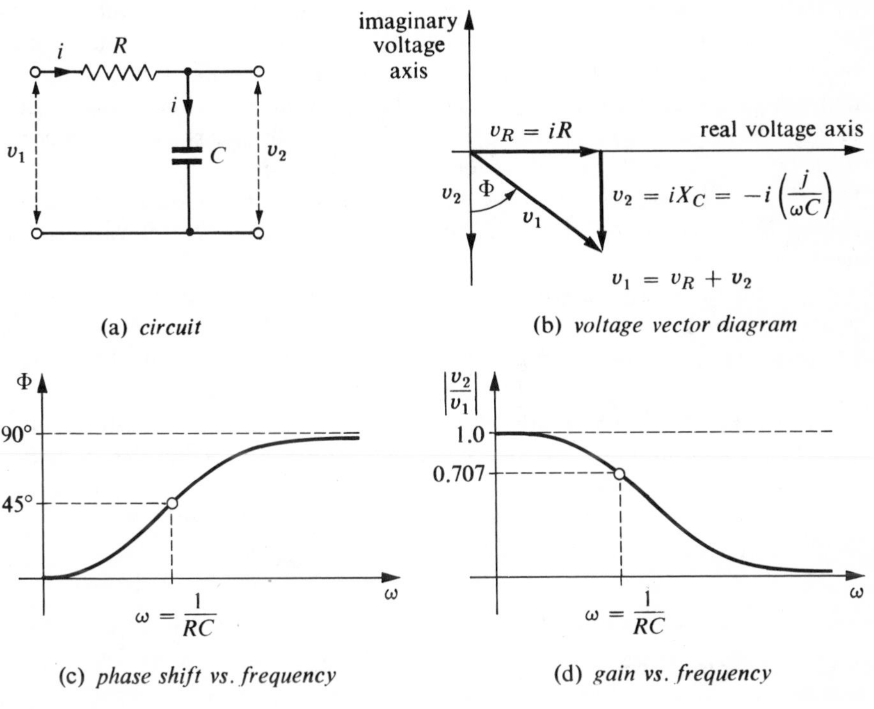

FIGURE 2.28 *RC* phase shifter.

put is seen to "lag" the input in phase by an angle ϕ given by $\phi = \arctan (v_R/v_C) = \arctan \omega RC$. Notice that as the frequency increases, the phase shift ϕ goes to 90°, and the output voltage goes to zero in amplitude. If either the resistance or the capacitance is made variable, the phase shift can be varied, but the amplitude of the output varies with the phase shift. The maximum phase shift is 90°.

Most four terminal networks composed of simple combinations of R, L, and C with a resistive source and load have gains that vary with frequency and contribute a frequency dependent phase shift; that is, the phase difference between the output and the input is a function of the frequency. If the network contributes the minimum possible phase shift for a given gain versus frequency behavior, the network is termed a "minimum phase shift network," and for such networks the gain characteristics and the phase characteristics are not independent. Knowing one characteristic enables one to calculate the other. Fortunately, most simple four terminal networks fall into this class. Examples are the simple RC and LR low- and high-pass filters.

The exact mathematic theory of minimum phase shift networks is beyond the scope of this book, and the interested reader is referred to Bode's theory in the *Radio Engineer's Handbook* (First edition) by Fred. E. Terman, McGraw-Hill (1943, p. 218). The essence of the theory is that the phase shift produced by the network at a frequency f depends on the rate of *change* of the network gain with respect to frequency evaluated at that frequency f. In other words, the network introduces the greatest phase shift in a frequency range where the network gain is rapidly changing. In the simple RC low-pass filter, for example, at high frequencies ($f \gg 1/2\pi RC$) the gain falls off linearly with frequency (6 dB per octave or 20 dB per decade) and this corresponds to a phase shift of 90°. For an LC network with the gain varying at 12 dB per octave, the phase shift is 180°. Gain and phase shift characteristics are shown for various networks in Fig. 2.29.

Considerable ingenuity must be expended usually to produce a simple network which is not a "minimum phase shift" network. One such network is the phase shifter shown in Fig. 2.30 in which the gain is absolutely constant from dc to infinite frequency, and the phase shift varies from 0° to 180°.

Another complicated phase shifter which has an output of constant amplitude and a continuously variable phase shift from 0 to 180° is shown in Fig. 2.31. It can be shown that if $1/\omega C = \frac{1}{2}\omega L$, then as R changes from 0 to infinity, the phase shift of v_{out} relative to v_{in} will vary from $-90°$ to $+90°$, and the output amplitude v_2 will remain constant.

2.11 RLC CIRCUITS

In general, practical electronic circuits consist of combinations of resistance, inductance, and capacitance as well as tubes and transistors. We will now show by a "brute force" solution of the Kirchhoff voltage equation for a simple series combination of R, L, and C, that the total impedance ("net opposition to the current flow") offered by R and L and C is given by the magnitude of the complex sum of: the resistance R, the inductive reactance $j\omega L$, and the capacitive reactance $1/j\omega C$. The "impedance" of any collection of $R\,L\,C$ is the complex algebraic sum of R, $j\omega L$, and $1/j\omega C$ treating each term as a "resistance" and

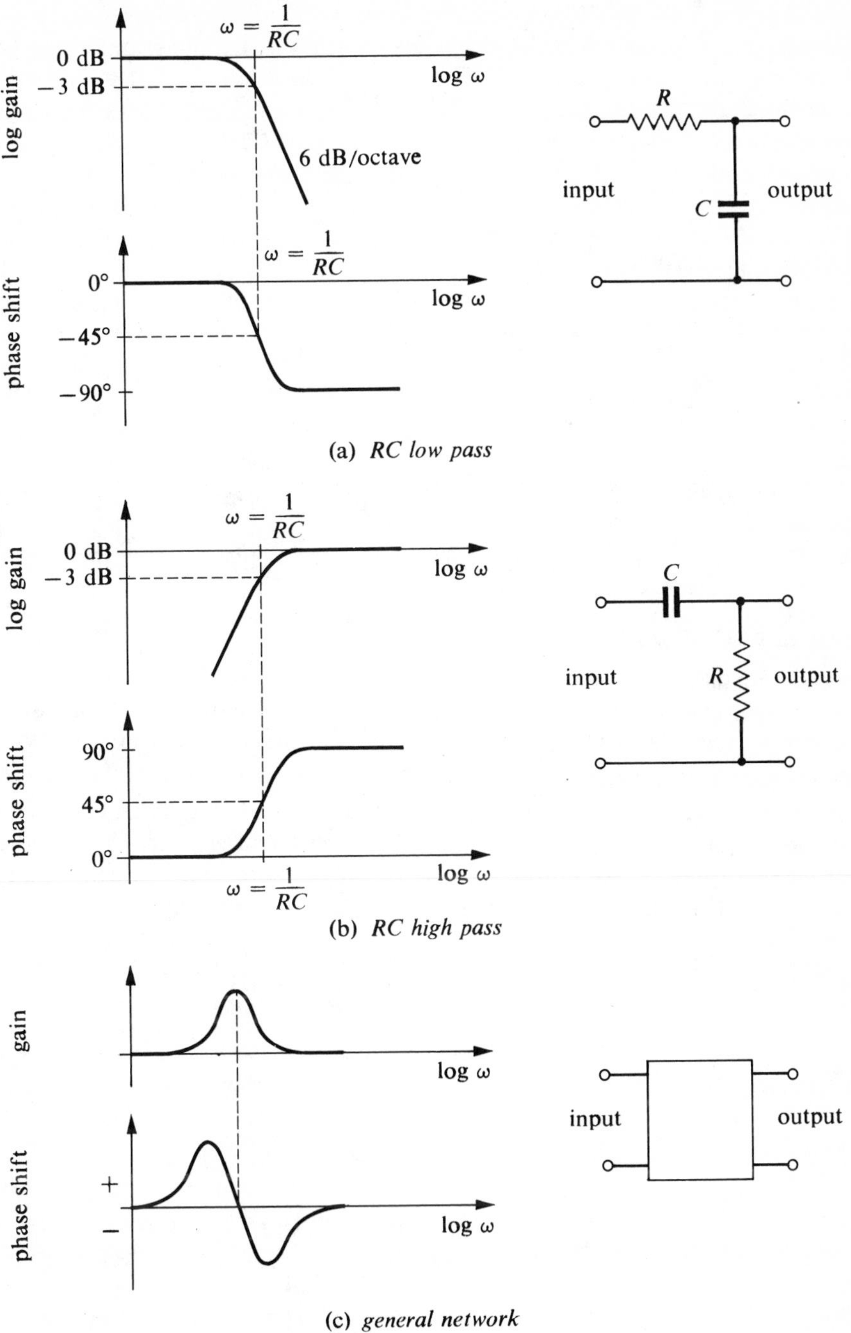

FIGURE 2.29 Gain and phase shift for three circuits.

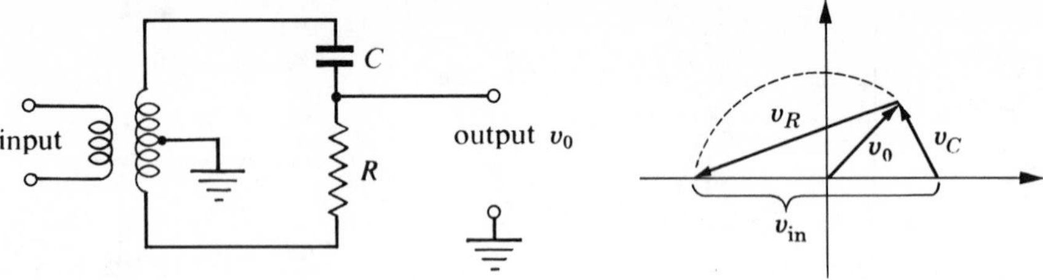

FIGURE 2.30 Phase shifter.

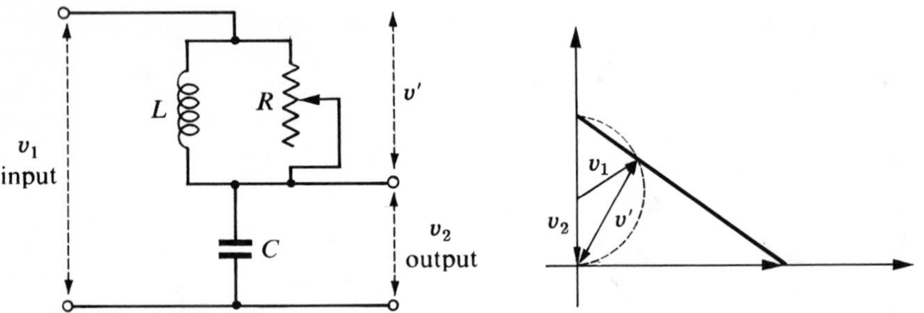

FIGURE 2.31 Phase shifter with constant output amplitude.

combining terms in series or parallel as the circuit implies. Thus, the impedance of R and L in series is $Z = R + j\omega L$, and the impedance of R and C in parallel is $R \times (1/j\omega C)/(R + 1/j\omega C)$. The magnitude i_0 of the current through R, L, and C in series is given by $i_0 = V_0/|Z|$ where $Z = R + j\omega L + 1/j\omega C$ is the total impedance, $|Z| = \sqrt{R^2 + (\omega L - 1/\omega C)^2}$, and V_0 equals the magnitude of the applied voltage. Consider the series $R\,L\,C$ circuit of Fig. 2.32. The applied voltage is $V_0 \cos \omega t$; we are given R, L, and C and are asked to calculate the current i. The Kirchhoff voltage law is:

$$V_0 \cos \omega t - v_R - v_L - v_C = 0 \qquad \text{or} \qquad V_0 \cos \omega t = iR + L\frac{di}{dt} + \frac{Q}{C} \tag{2.46}$$

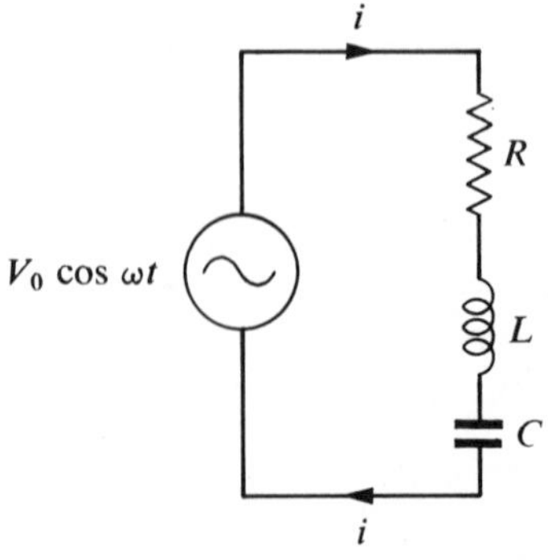

FIGURE 2.32 Series *RLC* circuit.

which when differentiated once with respect to time yields a second-order differential equation:

$$L\frac{d^2i}{dt^2} + R\frac{di}{dt} + \frac{i}{C} = -\omega V_0 \sin \omega t.$$

As we have seen, the current has the same frequency as the applied voltage in the case of the capacitor alone and the inductance alone; it is therefore a reasonable guess in the *RLC* circuit above that the current has the same frequency as the applied voltage. However, the phase of the current may be different from the phase of the voltage. To take this possibility into account we will assume a solution of the form $i = I_0 \cos (\omega t - \theta)$, and solve for I_0 and θ. I_0 and θ are both assumed to be constants. I_0 is the magnitude of the current, θ is the phase of the current relative to the voltage. If θ is positive, then the current lags the voltage. Of course, the final justification for the form of the assumed solution lies in our being able to find the solution which satisfies the differential equation and the boundary conditions. Now we substitute the assumed solution $i = I_0 \cos (\omega t - \theta)$ in the second-order differential equation in i and try to solve for I_0 and θ. We obtain one $\cos \omega t$ term and one $\sin \omega t$ term after a little algebra:

$$I_0\left[\left(\omega L - \frac{1}{\omega C}\right)\cos\theta - R\sin\theta\right]\cos\omega t$$
$$+\left[I_0\left(\omega L - \frac{1}{\omega C}\right)\sin\theta + I_0 R\cos\theta - V_0\right]\sin\omega t = 0. \quad \textbf{(2.47)}$$

Because the functions $\cos \omega t$ and $\sin \omega t$ are "orthogonal," the coefficient of each must equal zero. Another way of seeing this is to note that the above equation must hold for all values of t. In particular when $t = 0$, $\sin \omega t$ is zero, so the coefficient of $\cos \omega t$ must equal zero, and when $t = \pi/2\omega = T/4$, $\cos \omega t = \cos \frac{1}{2}\pi$ is zero, so the coefficient of $\sin \omega t$ must also equal zero. By either reasoning we obtain two equations that can be solved for I_0 and θ.

$$\left(\omega L - \frac{1}{\omega C}\right)\cos\theta - R\sin\theta = 0$$

and

$$I_0\left(\omega L - \frac{1}{\omega C}\right)\sin\theta + I_0 R\cos\theta - V_0 = 0 \quad \textbf{(2.48)}$$

Hence

$$\tan\theta = \frac{\omega L - 1/\omega C}{R} \quad \textbf{(2.49)}$$

$$I_0 = \frac{V_0}{R\cos\theta + \left(\omega L - \dfrac{1}{\omega C}\right)\sin\theta} \quad \textbf{(2.50)}$$

From equation (2.49), we find $\theta = \tan^{-1}(\omega L - 1/\omega C)/R$. That is, if $\omega L > 1/\omega C$ then the current lags behind the applied voltage by θ or equivalently, the applied voltage leads the current by θ. From equation (2.50) and the impedance triangle shown in Fig. 2.33, I_0 is found to be

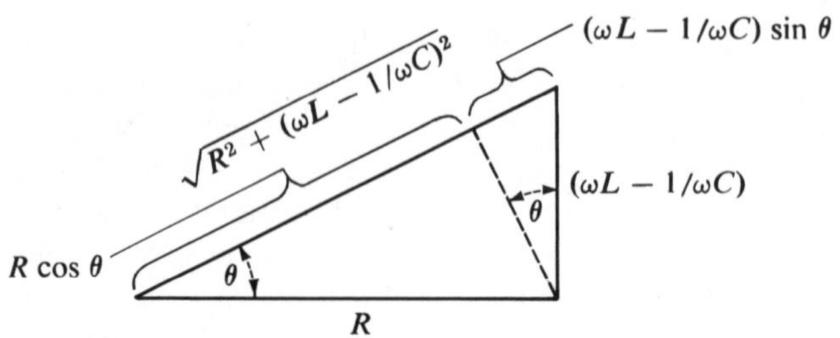

FIGURE 2.33 Impedance triangle for series *RLC* circuits.

$$I_0 = \frac{V_0}{[R^2 + (\omega L - 1/\omega C)^2]^{1/2}}, \tag{2.51}$$

which is very interesting. I_0, the magnitude of the current, equals V_0, the magnitude of the voltage, divided by $[R^2 + (\omega L - 1/\omega C)^2]^{1/2}$ *which is precisely the magnitude of the complex number* $(R + j\omega L + 1/j\omega C)$. Thus by treating the problem as a series connection of three complex reactances R, $j\omega L$, and $1/j\omega C$, we can get the magnitude of the current without even writing down the differential equation much less solving it! In other words, using the complex inductive and capacitive reactance provides us with a quick shortcut to solving the differential equation for the current amplitude.

Let us show explicitly how the use of complex impedance simplifies the job of finding the amplitude of the current. We replace the voltage $V_0 \cos \omega t$ by $V_0 e^{j\omega t}$, L by its inductive reactance $j\omega L$, and C by its capacitive reactance $1/j\omega C$ as shown in Fig. 2.34. The total impedance is thus

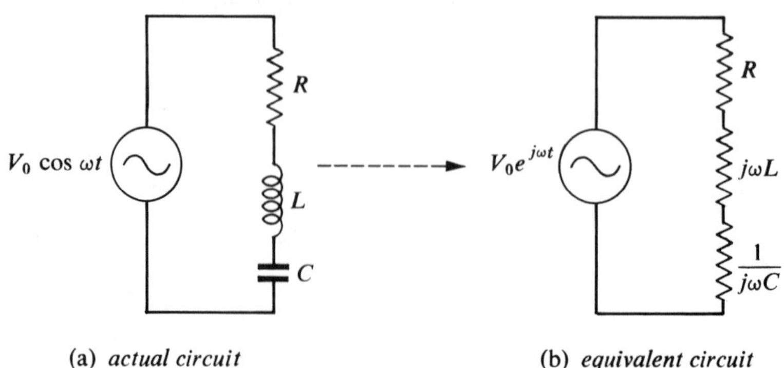

FIGURE 2.34 Series *RLC* circuit and equivalent.

$$Z = R + j\omega L + \frac{1}{j\omega C} \tag{2.52}$$

and

$$|Z| = \sqrt{R^2 + \left(\omega L - \frac{1}{\omega C}\right)^2}. \tag{2.53}$$

The current flowing in such a series circuit, Fig. 2.34(b), can be written down from Ohm's law:

$$i = \frac{V_0 e^{j\omega t}}{R + j\omega L + \dfrac{1}{j\omega C}}. \tag{2.54}$$

The magnitude of the current is obtained merely by taking the magnitude or absolute value of both sides of the above equation. If we let I_0 be the magnitude of the current, then we see that

$$I_0 = \frac{|V_0 e^{j\omega t}|}{\left|R + j\omega L + \dfrac{1}{j\omega C}\right|} = \frac{V_0}{\left[R^2 + \left(\omega L - \dfrac{1}{\omega C}\right)^2\right]^{1/2}} = \frac{V_0}{|Z|} \tag{2.55}$$

which is the same result as was obtained by the relatively laborious procedure of solving the differential equation.

To obtain the phase of the current we will draw a rotating voltage vector diagram and start by picking a reference phase. Draw $v_R = iR$ along the real voltage axis. Because R, L, and C are in series, the current through each of them is the same in magnitude and phase. Thus the voltage v_L across the inductance is given by $v_L = iX_L = ij\omega L$, which leads the current by 90°. The voltage across the capacitance is given by $v_C = iX_C = i/j\omega C$. The rotating voltage vector v_C will then be $v_C = -ji/\omega C$ which lags the current by 90°. The rotating voltage vector diagram is shown in Fig. 2.35.

Notice that the voltage across the resistance is exactly in phase with the current through the resistance. We now note that at any instant of time by Kirchhoff's voltage law the voltage v_{in} must exactly equal the instantaneous sum of v_R, v_L, and v_C. Thus we have Fig. 2.36(c). From this diagram the phase θ of the current relative to the applied voltage is given by $\theta = \arctan (\omega L - 1/\omega C)/R$. If ωL is greater than $1/\omega C$, the current lags the applied voltage by θ; if ωL is less than $1/\omega C$ the current leads the applied voltage by θ. For later times ($t > 0$), all the voltage vectors rotate counter-clockwise at the constant angular frequency ω, with θ remaining constant. And, as usual, the actual voltages measured in the circuit will be the real parts of the complex voltages we have diagrammed, that is, the projection of the voltage vectors on the horizontal real voltage axis.

To sum up the advantages of using complex numbers: they provide a shortcut to finding the currents in circuits without having to solve differential

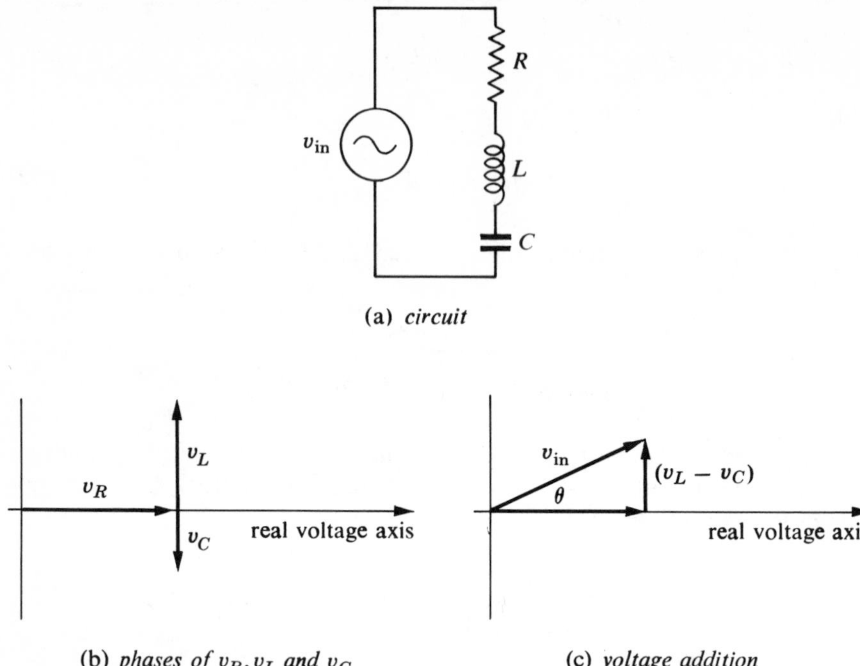

FIGURE 2.35 Phases of voltages in a series *RLC* circuit.

equations, and they automatically take care of the phase differences among the various voltages and currents.

2.12 SERIES AND PARALLEL RESONANCE

In the series *RLC* circuit $I_0 = V_0/[R^2 + (\omega L - 1/\omega C)^2]^{1/2}$; the current magnitude I_0 is maximum when $\omega L = 1/\omega C$, and the maximum value of i is given by $I_{0\,\max} = V_0/R$. A graph of I_0 as a function of angular frequency is shown in Fig. 2.36. The frequency ω_0 at which $\omega L = 1/\omega C$ is called the "resonant" frequency and is given by $\omega_0 = 1/\sqrt{LC}$ in angular frequency (radians/sec), or by $f_0 = 1/(2\pi\sqrt{LC})$ in Hertz or cycles per second. Also, v_L and v_C are equal in magnitude and 180° out of phase at resonance, and $Z = R$ at resonance, which means that the only limiting agent to the current at resonance is the resistance. Thus, if R is very small, the current at resonance will be very large. The voltage at resonance across the inductance $v_L = I_{0\,\max}\,(\omega L)$ may be extremely high (even greater than V_0), perhaps high enough to cause arcing in the inductance. Similarly, $v_C = I_{0\,\max}\,(1/\omega C)$ at resonance may be large enough to destroy the capacitor. Notice also that the phase of the current relative to the applied

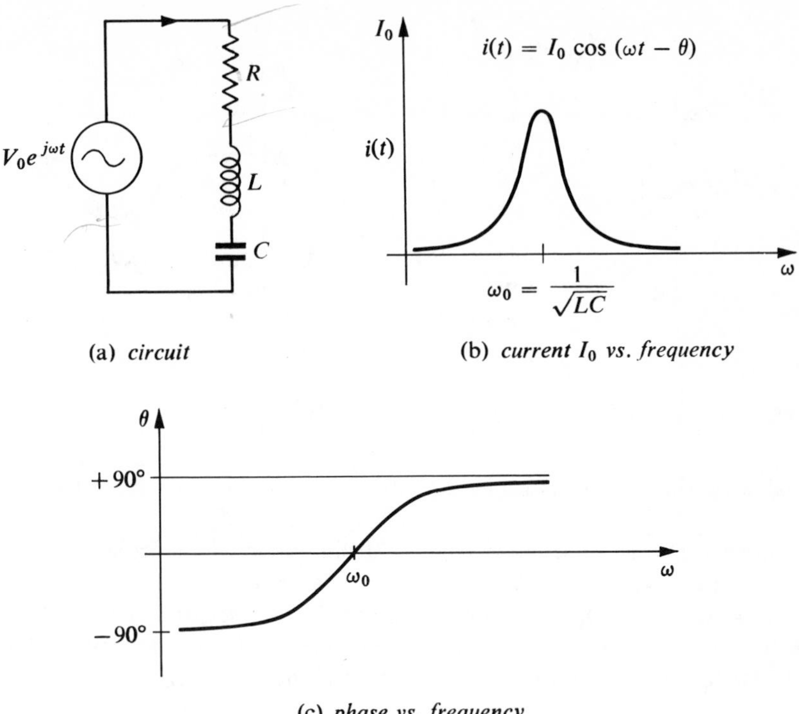

(a) *circuit*

(b) *current I_0 vs. frequency*

(c) *phase vs. frequency*

FIGURE 2.36 Series *RLC* circuit.

voltage changes from a lag to a lead as ω passes through $\omega_0 = 1/\sqrt{LC}$, because $\tan\theta = (\omega L - 1/\omega C)/R$ as shown in Fig. 2.36(c).

One application of the series *RLC* circuit is to attenuate a narrow band of frequencies around $\omega_0 = 1/\sqrt{LC}$ by the circuit shown in Fig. 2.37. The voltage gain versus frequency has a relative minimum at $\omega_0 = 1/\sqrt{LC}$ because the impedance between A and B in Fig. 2.37 is a minimum at ω_0 from equation (2.53). At ω_0 the gain equals $R/(R_1 + R) = 1/(1 + R_1/R)$ which can be much less than one if $R_1 \gg R$.

Consider the *RLC* circuit in Fig. 2.38 in which *L* and *C* are in parallel; this circuit behaves rather differently from the series *RLC* circuit. Let us calculate the current as a function of frequency. The parallel combination of *L* and *C* has an impedance of

$$Z = \frac{(j\omega L)(1/j\omega C)}{j\omega L + 1/j\omega C} = \frac{-jL/C}{\omega L - 1/\omega C} = \frac{j\omega L}{1 - \omega^2 LC} = \frac{j\omega L}{1 - \omega^2/\omega_0^2}, \qquad (2.56)$$

which approaches *infinity* as ω approaches the resonant frequency ω_0. Therefore, at resonance, the current $i = v/Z$ goes to zero and the voltage v_{ab} goes

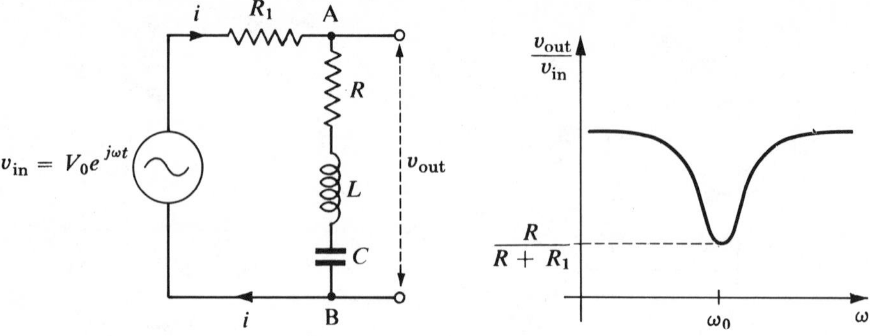

FIGURE 2.37 Series *RLC* circuit used to attenuate frequencies near ω_0.

through a maximum. Thus we have just the opposite behavior from the series circuit.

Notice that the impedance changes from inductive to capacitive as ω passes through ω_0. For $\omega < \omega_0$, $Z = j\omega L/\gamma$ and for $\omega > \omega_0$, $Z = -j\omega L/\gamma$ where $\gamma = |(1 - \omega^2/\omega_0^2)|$. Because the voltage across the parallel *LC* circuit is given by $v = iZ$, the phase of the output voltage relative to the input voltage $V_0e^{j\omega t}$ changes by 180° as ω passes through ω_0 as shown in Fig. 2.39(c).

$$i = \frac{v}{Z} = \frac{V_0e^{j\omega t}}{r + \left(\dfrac{-jL/C}{\omega L - 1/\omega C}\right)} \tag{2.57}$$

$$v_{ab} = \frac{Z_{ab}}{r + Z_{ab}} \cdot V_0e^{j\omega t} \rightarrow V_0e^{j\omega t} \text{ as } Z_{ab} \rightarrow \infty$$

The parallel *LC* circuit is sometimes called a "tank" circuit and is most often used to select one desired frequency from a signal containing many dif-

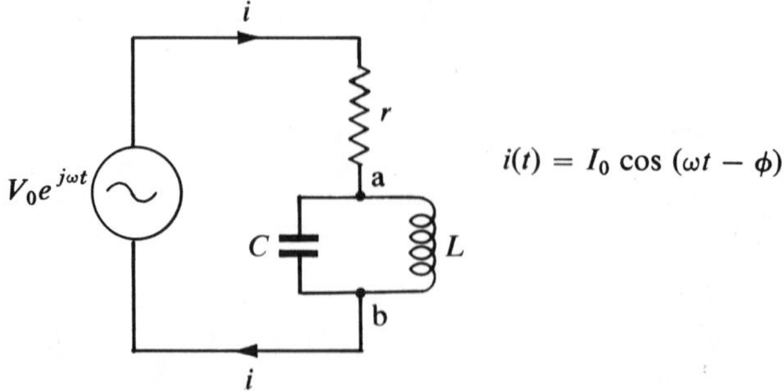

FIGURE 2.38 Parallel *LC* circuit.

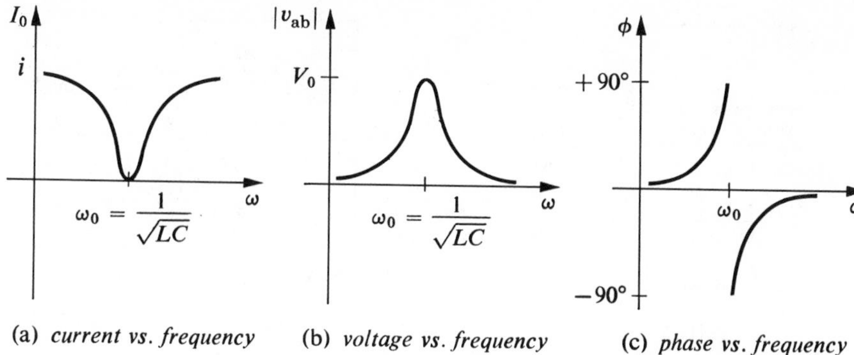

FIGURE 2.39 Current and voltage as functions of frequency in a parallel *LC* circuit.

ferent frequencies. This selection is accomplished by the following basic circuit in Fig. 2.40 in which the voltage gain ($=v_{\text{output}}/v_{\text{input}}$) is maximum ($=1$) at $\omega_0 = 1/\sqrt{LC}$. This circuit is said to be "tuned" to the frequency ω_0. Often either L or C is made variable so that different frequencies may be selected or

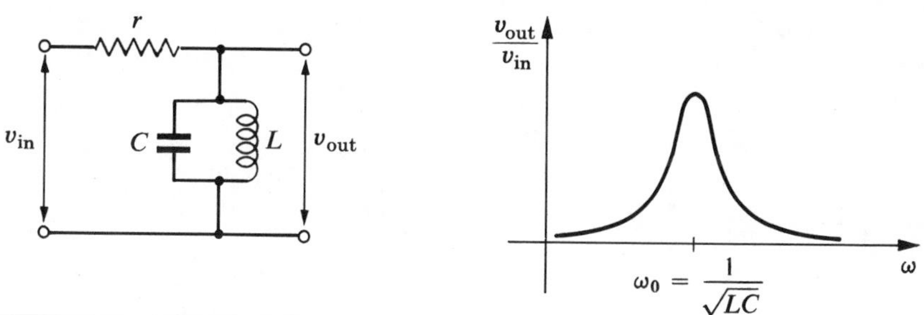

FIGURE 2.40 Parallel *LC* circuit used to select frequencies near ω_0.

tuned. This is precisely what one does when one tunes in a radio station; the tuning knob is usually a variable capacitor in a parallel *LC* circuit.

The parallel *LC* circuit will tend to oscillate as can be seen by the following physical argument. Suppose at $t = 0$ the capacitor C is fully charged and the current is zero. The energy of the circuit is then all in the electric field between the capacitor plates. The capacitor C will start to discharge through the inductance as shown in Fig. 2.41(b). The current flowing through L will create a magnetic field around L, and energy will be stored in this magnetic field. This increase in energy will be exactly balanced by the decrease in the energy stored in the electric field between the plates of C, the total energy of the *LC* circuit remaining constant. But by Lenz's law, the magnetic field lines cutting the turns in L will induce a voltage across the terminals of L of such a polarity as to *oppose* the increasing current. Eventually the current will reverse direction because of this induced voltage and will recharge C with the opposite polarity

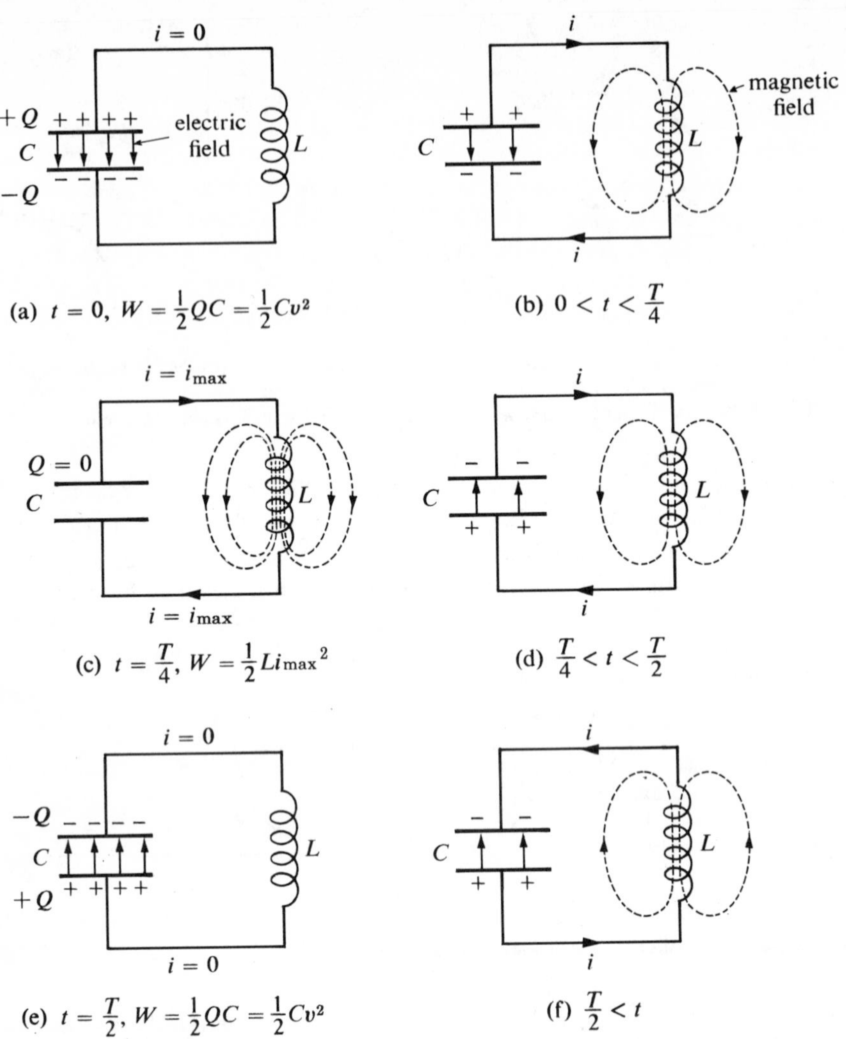

FIGURE 2.41 Oscillation of current in parallel *LC* circuit.

as in Fig. 2.41(c). The capacitor C will now discharge through L in the opposite direction as in 2.41(d), eventually returning the circuit to the state of Fig. 2.41(a) when the entire cycle starts over again. This back and forth flow of current is called oscillation, and the oscillations for an LC circuit will in principle last forever with a constant peak-to-peak amplitude. If, however, there is any *resistance* at all in the circuit, there will be a conversion of electrical energy to heat energy in the resistance at a rate i^2R. This loss of electrical energy occurs regardless of the *direction* of the current flow. Thus the electrical energy of the circuit gradually decreases, and the oscillations die out. In actual practice, of course, there is always some resistance associated with any LC circuit, mainly due to the resistance of the wire in the inductance. The only exception would be a circuit constructed entirely from superconducting material which, at this writing, must be kept at liquid helium temperatures ($\approx 4°$ K) and hence is somewhat impractical. Thus, the oscillations in a real circuit will die out eventually. If constant *amplitude* oscillations are desired, then some circuit must be devised to continuously replenish the energy lost to heat in each cycle. This is precisely what one does in building an oscillator, which will be covered in Chapter 9.

It can be shown that the current in a pure LC circuit (no resistance) is exactly sinusoidal, constant in amplitude, and of angular frequency $\omega_0 = 1/\sqrt{LC}$. To show this we simply write down the Kirchhoff voltage law for the closed loop containing L and C.

$$v_C - v_L = 0$$

$$v_C + L\frac{di}{dt} = 0$$

$$\frac{Q}{C} + L\frac{di}{dt} = 0 \qquad \textbf{(2.58)}$$

Differentiating once with respect to time gives us an equation in the current:

$$\frac{i}{C} + L\frac{d^2i}{dt^2} = 0 \qquad \frac{d^2i}{dt^2} + \frac{1}{LC}i = 0 \qquad \textbf{(2.59)}$$

where we have used $i = dQ/dt$. This is the familiar "wave equation" or "oscillator equation." The solution is

$$i = I_0 \cos(\omega_0 t + \phi) \qquad \text{or} \qquad i = I_0 \sin(\omega_0 t + \theta) \qquad \textbf{(2.60)}$$

where $\omega_0 = 1/\sqrt{LC}$. Thus, the current flowing in a pure LC circuit is sinusoidal, of constant amplitude I_0 and of angular frequency $\omega_0 = 1/\sqrt{LC}$. Precisely the same differential equation is obtained for the motion of a mass attached to a perfect, Hooke's law spring. $M\frac{d^2X}{dt^2} = -KX$ where M is the mass, X is the displacement, and K is the spring constant. Thus, $\frac{d^2X}{dt^2} + \frac{K}{M}X = 0$ which is also the familiar wave equation with solution $X = X_0 \sin(\omega_0 t + \theta) = X_0 \cos(\omega_0 t + \phi)$ where $\omega_0 = \sqrt{K/M}$. L is seen to be analogous to M, and C to $1/K$.

2.13 "Q" (QUALITY FACTOR)

The Q or quality factor of any circuit can be defined as 2π times the energy stored per cycle divided by the energy lost or dissipated (usually as heat) per cycle.

$$Q \equiv 2\pi \frac{W_S}{W_L} \tag{2.61}$$

The energy stored refers to the energy stored in the electric and magnetic fields. The energy converted from electrical energy to heat energy is "lost" in the sense that it is no longer available in the electrical current. The higher the Q, the less energy is converted into heat per cycle. An ideal circuit which dissipates no energy would have an infinite Q.

Notice that the energy has been converted from a more *ordered* form, i.e., the coherent average drift of electrons in the current flow in the wire, to the more *disordered* random thermal motion of the molecules. This conversion corresponds to an increase in entropy and is consistent with the second law of thermodynamics, which says that the reverse process will never occur with 100% efficiency, i.e., the thermal energy can never be completely converted back into the more ordered electrical energy. Thus, some (and usually all in most circuits) of the energy converted into heat is permanently lost from the electrical current.

We will now show from the above definition that the Q of a series RLC circuit can be expressed as $Q = \omega L/R$. From the definition, equation (2.61), we can write $Q = 2\pi W_S/W_H$, where W_S is the energy stored in one cycle and W_H is the energy converted to heat in one cycle. Assume $i = I_0 \cos \omega t$ is the current flowing in the circuit. The energy stored in the magnetic field of L when a current i is flowing is $W = \frac{1}{2}Li^2$. W will therefore be maximum when i is maximum. Because i and v_C are 90° out of phase, when i is maximum, $v_C = 0$. Therefore when $i = i_{\max} = I_0$, *all* the energy is stored in the magnetic field. $W = \frac{1}{2}Li_{\max}^2$, and none is stored in the electric field of the capacitor because $v_C = 0$. $W_S = \frac{1}{2}LI_0^2$ and $W_H = \int_0^T P\,dt$ where P is the instantaneous power converted into heat, $P = i^2(t)R$, and T equals the period $= 2\pi/\omega$. Thus,

$$\begin{aligned} Q &= 2\pi \frac{W_S}{W_H} = 2\pi \frac{\frac{1}{2}LI_0^2}{\int_0^T i^2(t)R\,dt} = \frac{\pi L I_0^2}{R\int_0^T I_0^2 \cos^2 \omega t\,dt} \\ &= \frac{\pi L}{R} \frac{1}{\int_0^T \cos^2 \omega t\,dt} = \frac{\pi L}{R} \cdot \frac{1}{T/2} = \frac{\omega L}{R}. \end{aligned} \tag{2.62}$$

Note at the resonance frequency $\omega_0 = 1/\sqrt{LC}$, Q can be written as $Q = (1/R)\sqrt{L/C}$. Thus the higher the ratio of L to C, the higher the Q at resonance. However, it is very difficult to make a large inductance without also increasing R. And, as expected, the larger R is the lower the Q, because a larger R will dissipate more energy as heat.

For the parallel RLC circuit of Fig. 2.42, we show that the Q is also given by $Q = \omega L/R$. Here we explicitly draw in the resistance of the wire in the inductance by including R in series with L. But we can always write $i_1(t)$ in the form $i_{1\,\max} \cos(\omega t - \theta)$ so then Q becomes:

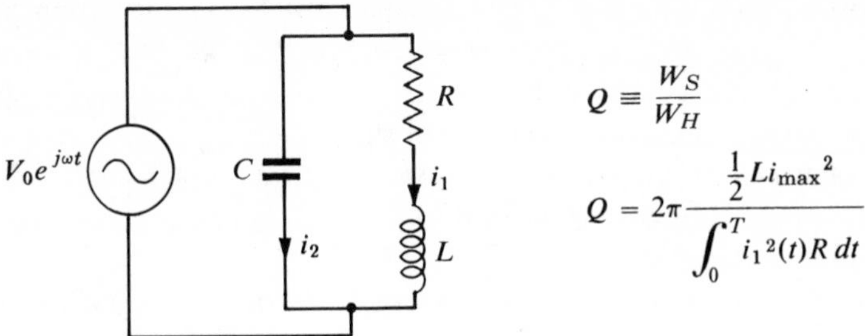

FIGURE 2.42 Q calculation for a parallel *RLC* circuit.

$$Q = \frac{\pi L i_{1\,max}^2}{\int_0^T i_{1\,max}^2 \cos^2(\omega t - \theta)\,R\,dt}$$

$$= \frac{\pi L}{R}\frac{1}{\int_0^T \cos^2(\omega t - \theta)\,dt} = \frac{\pi L}{R}\cdot\frac{1}{T/2} = \frac{\omega L}{R}, \qquad \textbf{(2.63)}$$

which is the same as for the series *RLC* circuit. Note that we could have solved explicitly for $i_1(t)$ by writing:

$$i_1(t) = \frac{V_0 e^{j\omega t}}{R + j\omega L} = \frac{V_0 e^{j(\omega t - \theta)}}{\sqrt{R^2 + \omega^2 L^2}}$$

where $\theta = \tan^{-1}\omega L/R$. Thus $i_{1\,max} = V_0/\sqrt{R^2 + \omega^2 L^2}$, but this procedure is not necessary to calculate the Q because $i_{1\,max}$ cancels out in the expression for the Q.

The resistance R in an actual *RLC* circuit is usually the resistance of the wire in the inductance L. Practically speaking, we usually can neglect any power dissipated as heat in the capacitor as we have done by not drawing any resistance associated with C. If we draw the parallel *LC* circuit as in Fig. 2.43

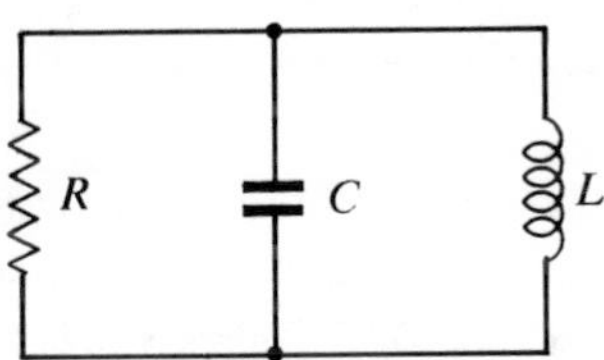

FIGURE 2.43 Parallel *LC* circuit with capacitor leakage resistance *R*.

with R representing the leakage resistance of the capacitor, then an ideal circuit with zero power dissipated as heat would have an infinite R. Thus we would expect Q to be proportional to R. A calculation from the definition of Q in fact yields the result $Q = R/\omega L$ or $Q = \omega_0 RC$ or $Q = R\sqrt{C/L}$ at resonance.

We will now show that for a series RLC circuit, the sharpness of the current versus frequency curve depends upon the Q; a high Q corresponding to a narrow, sharp curve and a low Q corresponding to a broad curve. The current as a function of frequency is given by:

$$i = \frac{V_0 e^{j\omega t}}{R + j\omega L + \dfrac{1}{j\omega C}} = I_0 e^{j(\omega t - \theta)}$$

$$I_0 = \frac{V_0}{\left[R^2 + \left(\omega L - \dfrac{1}{\omega C}\right)^2\right]^{1/2}} = \frac{V_0}{|Z|}. \tag{2.64}$$

Clearly I_0 reaches a maximum when $\omega L = 1/\omega C$, i.e., at resonance and $I_{0\,\max} = V_0/R$ as shown in Fig. 2.44.

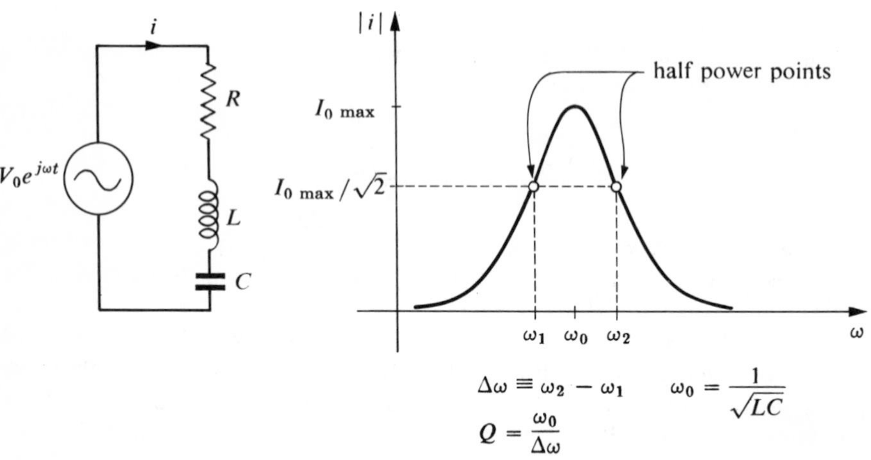

FIGURE 2.44 Frequency response of parallel LC circuit.

Let us now define the width $\Delta\omega$ of the I_0 vs ω curve as the width from ω_1 to ω_2 where $i(\omega_1) = i(\omega_2) = I_{0\,\max}/\sqrt{2}$. That is, $\Delta\omega$ is the width of the curve where the current has fallen to $1/\sqrt{2} = 0.707$ of its value at resonance. At ω_1 and ω_2 the power dissipated in R is $\frac{1}{2}$ of the power at resonance, thus the points ω_1 and ω_2 are sometimes called the "half-power" points. To find $\Delta\omega$, we note $I_{0\,\max} = V_0/R$ and i has fallen to $I_{0\,\max}/\sqrt{2}$ when $Z = \sqrt{2}\,R$. Therefore at $\omega = \omega_1$ and at $\omega - \omega_2$, $Z = \sqrt{2}R$ or $|Z(\omega_1)| = |Z(\omega_2)| = \sqrt{2}\,R$.

$$\sqrt{R^2 + (\omega_1 L - 1/\omega_1 C)^2} = \sqrt{R^2 + (\omega_2 L - 1/\omega_2 C)^2} = \sqrt{2}\,R \tag{2.65}$$

Thus, ω_1 and ω_2 are the two roots of $(\omega L - 1/\omega C)^2 = R^2$. If we assume a symmetrical curve about $\omega = \omega_0$ and that $\Delta\omega \ll \omega_0$, then substituting $\omega_2 = \omega_0 + \Delta\omega/2$ and $\omega_1 = \omega_0 - \Delta\omega/2$ we obtain $\Delta\omega = \omega_0/Q$. In other words

$$Q = \omega_0/\Delta\omega. \tag{2.66}$$

Thus a high Q circuit has a narrow width $\Delta\omega$ for a given resonant frequency ω_0, and a low Q circuit has a broad width $\Delta\omega$. A high Q circuit is said to be "sharp" or to have a high selectivity because it can be used to select a narrow band of frequencies around ω_0 and to reject other frequencies with the circuit of Fig. 2.45. The best intuitive way to think of Q is with the relationship $Q = \omega_0/\Delta\omega$.

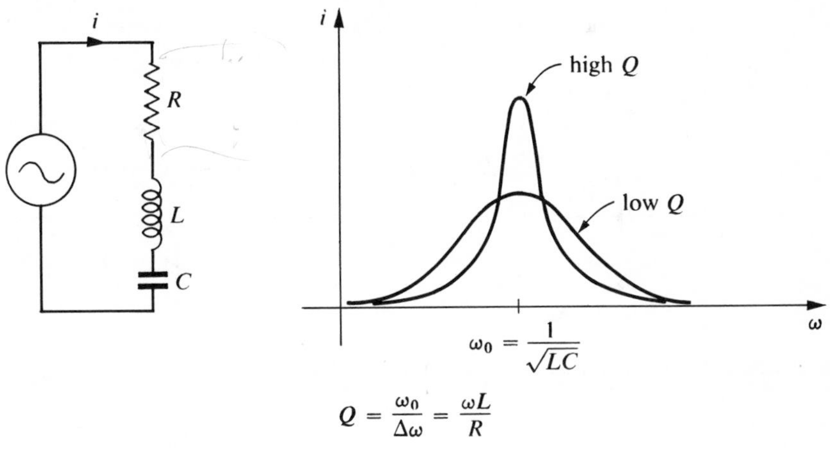

FIGURE 2.45 Frequency response of high Q and low Q series *RLC* circuit.

One can also show for the parallel *RLC* circuit if $\omega_0 L \gg R$, where R is the resistance associated with L, that $Q \cong \omega_0/\Delta\omega = \omega_0 L/R$, where ω_0 is the angular resonant frequency and $\Delta\omega$ is the full width between the half-power points. The impedance of the parallel tank at resonance is $Z_0 = Q\omega_0 L = Q/\omega_0 C$ and at the half-power points the impedance is $Z = Z_0/\sqrt{2}$. Notice that at resonance the Q effectively multiplies the impedance of the inductance or capacitance.

problems

1. Calculate the period in seconds of a 400 Hz sinusoidal voltage. If the zero-to-peak amplitude is 5 volts, calculate the maximum instantaneous rate of change of the voltage in volts per second.
2. Prove that the root-mean-square (rms) value of $V(t) = V_0 \cos \omega t$ is equal to $V_0/\sqrt{2}$. What is the root-mean-square value of the voltage of problem one?
3. Calculate the average power (in watts) dissipated as heat in a device passing

a current of $I(t) = 4 \cos \omega t$, if the voltage across the device is given by $V(t) = 10 \cos (\omega t + 30°)$.

4. Calculate the capacitance in μF between two 1 cm² conducting plates 1 mm apart in a vacuum. Repeat if the space between the plates is filled with a plastic dielectric with a dielectric constant of 8.
5. Calculate the capacitive reactance in ohms of a 0.01 μF capacitor at (a) 100 Hz, (b) 1 kHz, (c) 100 kHz, (d) 1 MHz.
6. Calculate the inductive reactance in ohms of a 2.5 mH choke at (a) 100 Hz, (b) 1 kHz, (c) 100 kHz, (d) 1 MHz.
7. Express L dimensionally in terms of ohms and seconds. Repeat for C.
8. Calculate the energy in joules stored in a 2000 μF capacitor charged to 100 volts. Physically, how is the energy stored?
9. A one henry inductance carries a current of 500 mA. The wire breaks and in 10^{-3} seconds the current drops to zero. What would happen?
10. Calculate the impedance Z_{ab} in the form $a + jb$ for:

11. Calculate the impedance Z_{ab} in the form $a + jb$ for:

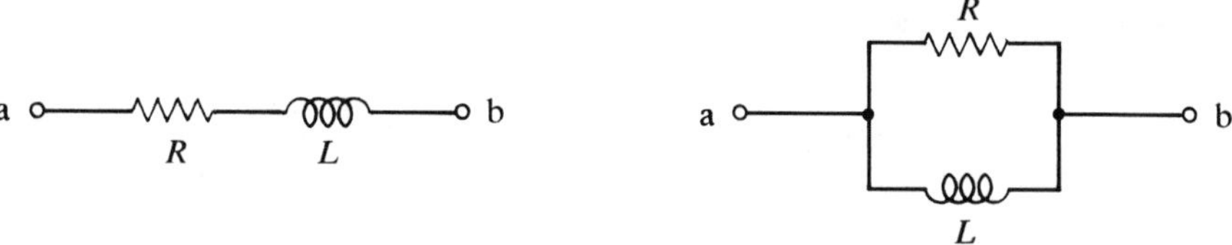

12. Calculate the impedance Z_{ab} of the following circuit.

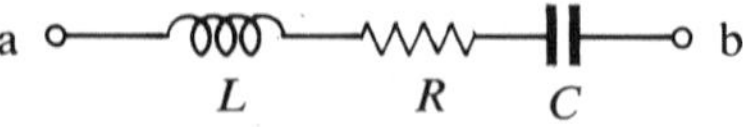

13. Design a high-pass RC filter with a break point at 100 kHz. Use a 1 kΩ resistance. Explain in words why the high-pass filter attenuates the low frequencies.
14. Design a low-pass RC filter that will attenuate a 60 Hz sinusoidal voltage by 12 dB relative to the dc gain. Use a 100 ohm resistance. Explain in words why the low-pass RC filter attenuates the high frequencies.
15. Prove for a low-pass RC filter that (a) at the frequency $\omega = 1/RC$, the voltage

gain equals 0.707 = $1/\sqrt{2}$; (b) the rise time of the output pulse equals $2.2RC$ for a zero rise time input pulse.

16. Calculate the slope of the gain versus frequency curve for a low-pass RC filter in dB per octave and also in dB per decade for high frequencies, i.e., for frequencies $\omega \gg 1/RC$. (One decade frequency change is a factor of ten, e.g., from 10 Hz to 100 Hz, or from 50 kHz to 500 kHz.)

17. Carefully sketch the voltage vector diagram for a high-pass RC filter, and calculate the phase of the output voltage relative to the phase of the input.

18. Derive an expression for the voltage gain and phase shift of the following LR circuit.

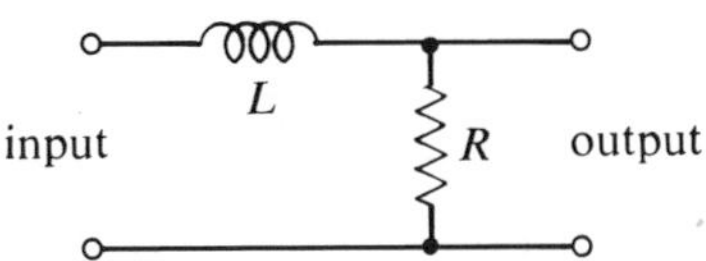

19. Design a parallel LC resonant circuit or "tank" to resonate at 1 MHz. Assume the inductance $L = 100\,\mu$H and has a dc resistance of 10 ohms. What is the Q of this circuit at resonance?

20. Carefully sketch the rotating voltage vector diagram for a series RLC circuit at resonance. If the circuit has a Q of 100, calculate the voltage ratings L and C must have.

21. Sketch a graph of the magnitude of the impedance versus frequency for a series and a parallel RLC circuit. State the change in phase of the impedance as the frequency passes through resonance.

22. Sketch the approximate gain versus frequency curve for the following circuit. You may treat the circuit as being composed of two independent RC filters.

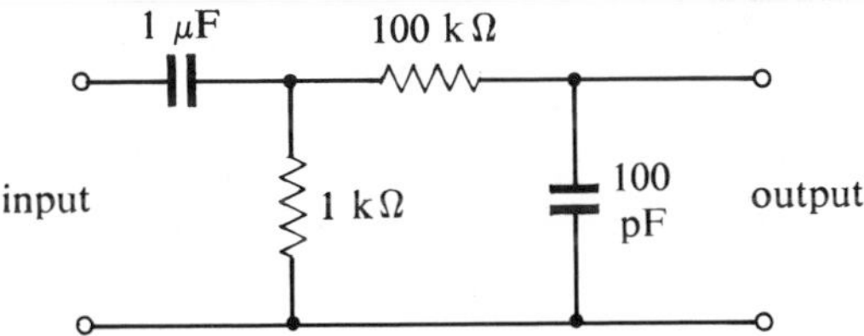

23. Prove for a high Q parallel RLC circuit, that the $Q = \omega_0/\Delta\omega$ where ω_0 is the (angular) resonant frequency and $\Delta\omega$ is the width at the half-power points.

24. Prove for a high Q parallel RLC circuit, that at resonance the impedance equals the Q times the inductive reactance at resonance. Calculate the impedance at resonance for $L = 100\,\mu$H, $C = 0.001\,\mu$F, $R = 5\,\Omega$.

3

FOURIER ANALYSIS AND PULSES

3.1 INTRODUCTION

In many situations we encounter voltage changes which are not sinusoidal in shape. Any relatively rapid change in voltage is usually referred to as a "pulse." For example, every time an ionizing particle such as a proton or electron passes through a Geiger Mueller tube, a small voltage pulse several microseconds long will be produced at the output of the tube. This pulse must then be amplified and recorded in some way in order to count the protons or electrons. And, every time an ionizing particle passes through a scintillation crystal such as NaI (sodium iodide), a light flash of very short duration is produced in the crystal or scintillator. This light flash is converted to a voltage pulse lasting about 10^{-6} sec by a photo-multiplier tube, and this voltage pulse must then be amplified and counted. Voltage pulses may also be used to indicate the beginning and the end of a time interval, and hence must be amplified and recorded. The electronic signals in modern computers also consist of voltage pulses.

3.2 DESCRIPTION OF A PULSE

An ideal single pulse is rectangular in shape as is shown in Fig. 3.1(a). This pulse can be described by giving its amplitude V_0, and its width or duration τ. An actual pulse encountered in the laboratory will always appear somewhat rounded if observed with a sufficiently fast oscilloscope as shown in Fig. 3.1(b). The steepness of the sides of this pulse is described by giving the "rise time", which is defined as the time for the pulse to rise from 10% to 90% of its maximum value. The "fall time" of the pulse is defined as the time for the pulse to fall from 90% to 10% of its maximum value. An ideal pulse has zero rise and fall time.

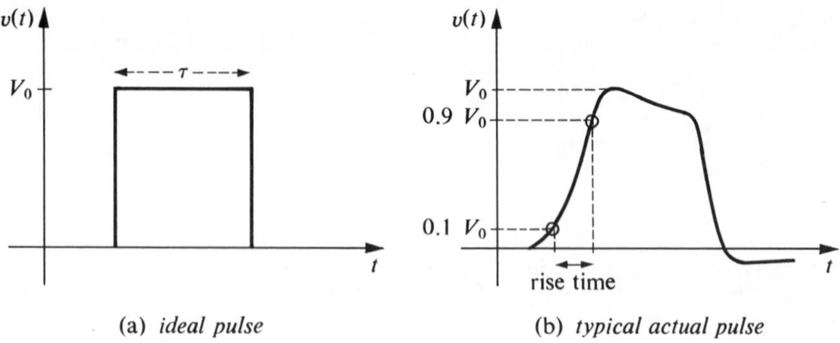

FIGURE 3.1 Pulses.

If a series of pulses occurs regularly in time as is shown in Fig. 3.2, we can describe the pulses completely by giving: the number of pulses per second which is called the "frequency" or the repetition rate, the pulse width τ, the pulse amplitude V_0, and the phase. In Fig. 3.2, the frequency or repetition rate is 200 kHz, the pulse width is $1\,\mu$ sec, and the amplitude is 10 V. Notice that if the zero voltage level were at the top of the $1\,\mu$ sec pulses, we would say that we

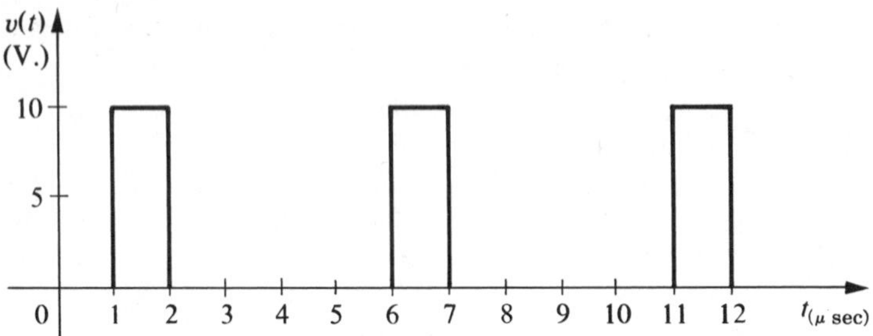

FIGURE 3.2 Train of rectangular pulses.

had 10 volt negative pulses of pulse width $4\,\mu$ sec with the same frequency, 200 kHz.

3.3 FOURIER ANALYSIS

In this section we will outline the general idea of Fourier analysis and apply it to the problem of analyzing a single rectangular pulse. Fourier analysis basically says that any function can be written as a sum of sine and cosine functions of different frequencies and amplitudes. In this book we shall be interested in analyzing voltage wave forms, so from now on we shall speak about the problem of analyzing a voltage $v(t)$. The Fourier analysis theorem says that if the voltage $v(t)$ is defined from $t = -T/2$ to $t = T/2$, then $v(t)$ can be written as the following sum of sines and cosines:

$$v(t) = \frac{a_0}{2} + \sum_{n=1}^{\infty} a_n \cos n\omega t + \sum_{m=1}^{\infty} b_m \sin m\omega t$$

where $$\omega = 2\pi/T. \tag{3.1}$$

Or:

$$v(t) = \frac{a_0}{2} + a_1 \cos \frac{2\pi t}{T} + a_2 \cos \frac{4\pi t}{T} + \cdots$$
$$+ b_1 \sin \frac{2\pi t}{T} + b_2 \sin \frac{4\pi t}{T} + \cdots .$$

In equation (3.1) a_0, a_n ($n = 1, 2, 3, \ldots$), and b_m ($m = 1, 2, 3, \ldots$) are constants which can be calculated from $v(t)$ with the following equations:

$$a_n = \frac{2}{T} \int_{-T/2}^{T/2} v(t) \cos n\omega t \, dt \qquad b_m = \frac{2}{T} \int_{-T/2}^{T/2} v(t) \sin m\omega t \, dt. \tag{3.2}$$

If we start with the voltage $v(t)$ and proceed to express it as the sum of the sines and cosines, then we say this process is one of "analysis"; if we start with the sines and cosines and add them together to form the voltage $v(t)$ we say this process is one of "synthesis."

Usually we know the response of any circuit to an input of sines and/or cosines. Thus we can determine the response of a circuit to a complicated input voltage by analyzing the input wave form into sines and cosines and then calculating the response of the circuit to the sine and cosine components. Strictly speaking, this procedure will work only for a linear circuit, but many circuits encountered are linear and any circuit at all will act in a linear manner if the input voltage is sufficiently small.

Let us now consider the meaning of the a and b Fourier coefficients. The first term $a_0/2$ represents simply the average dc level of the voltage from $t = 0$ to $t = T$, since $a_0/2 = 1/T \int_{-T/2}^{T/2} v(t)\, dt$. Thus if the area of the $v(t)$ curve above the $v = 0$ horizontal axis equals the area below the axis, then $a_0/2 = 0$. This will be the case, for example, whenever $v(t)$ has been passed through a series capacitor, because a capacitor can pass no direct current. a_1 is the coefficient of the term $\cos 2\pi t/T = \cos 2\pi f t$; thus a_1 is the amplitude of the Fourier cosine component of frequency f. Similarly, a_2 is the amplitude of the cosine component of frequency $2f$, and so on. The point to remember here is that the shape of the voltage $v(t)$ determines just what the coefficients a_n and b_m are. The voltage can be described equally well by giving its shape versus time $v(t)$, or by giving all the a_n and b_m Fourier coefficients (see Fig. 3.3).

If we assume the Fourier series expansion exists and converges to the voltage waveform $v(t)$, then it is relatively easy to show mathematically that the Fourier coefficients a_n and b_m must be given by equation (3.2). Starting with the Fourier expansion,

$$v(t) = \frac{a_0}{2} + \sum_{n=1}^{\infty} a_n \cos \frac{2\pi nt}{T} + \sum_{m=1}^{\infty} b_m \sin \frac{2\pi mt}{T}, \tag{3.3}$$

we multiply by $\cos (2\pi kt)/T$ where k is some positive integer and integrate with respect to time from $t = -T/2$ to $t = T/2$.

$$\int_{-T/2}^{T/2} v(t) \cos \frac{2\pi kt}{T}\, dt = \int_{-T/2}^{T/2} \left(\frac{a_0}{2} + \sum_{n=1}^{\infty} a_n \cos \frac{2\pi nt}{T} + \sum_{m=1}^{\infty} b_m \sin \frac{2\pi mt}{T} \right) \cos \frac{2\pi kt}{T}\, dt. \tag{3.4}$$

There are three types of integrals:

$$\int_{-T/2}^{T/2} \frac{a_0}{2} \cos \frac{2\pi kt}{T}\, dt = \frac{a_0}{2} \int_{-T/2}^{T/2} \cos \frac{2\pi kt}{T}\, dt = 0 \tag{3.5}$$

$$\int_{-T/2}^{T/2} \left(\sum_{m=1}^{\infty} b_m \sin \frac{2\pi mt}{T} \right) \cos \frac{2\pi kt}{T}\, dt = \sum_{m=1}^{\infty} b_m \int_{-T/2}^{T/2} \sin \frac{2\pi m}{T} t \cos \frac{2\pi k}{T} t\, dt = 0 \tag{3.6}$$

$$\int_{-T/2}^{T/2} \sum_{n=1}^{\infty} a_n \cos \frac{2\pi nt}{T} \cos \frac{2\pi kt}{T}\, dt = \sum_{n=1} a_n \int_{-T/2}^{T/2} \cos \frac{2\pi nt}{T} \cos \frac{2\pi kt}{T}\, dt \tag{3.7}$$

All the integrals involving the product of a sine and a cosine are zero. The only integral which is not zero is the one for which $n = k$ in the product of two cosine terms. This integral works out to be:

$$\int_{-T/2}^{T/2} \cos \frac{2\pi kt}{T} \cos \frac{2\pi kt}{T}\, dt = \tfrac{1}{2}T. \tag{3.8}$$

Thus we have shown that

$$\int_{-T/2}^{T/2} v(t) \cos \frac{2\pi kt}{T}\, dt = a_k \frac{T}{2}$$

or

$$a_k = \frac{2}{T} \int_{-T/2}^{T/2} v(t) \cos \frac{2\pi kt}{T}\, dt. \tag{3.9}$$

which is the expression for the a coefficient of equation (3.2) except that the subscript k has been used in place of n. A similar argument will yield the result that

$$b_k = \frac{2}{T}\int_{-T/2}^{T/2} v(t) \sin \frac{2\pi kt}{T}\, dt. \tag{3.10}$$

A few mathematical points might be in order here: there are certain restrictions on the function $v(t)$ in order to ensure that the expansion exists. Namely, $v(t)$ can contain only a finite number of discontinuities and must remain bounded, but any voltage waveform encountered in electronics will satisfy these conditions. Also, if we plot the Fourier expansion versus time for times outside the $-T/2$ to $T/2$ interval, we will merely get a periodic extension of $v(t)$ with period T. Thus, a periodic voltage, such as a square wave or a sawtooth wave can be Fourier analyzed for t from 0 to ∞. At a point of discontinuity in $v(t)$, the Fourier expansion "splits the difference"; for example, if $v(t)$ jumps discontinuously at the instant of time t_1 from 2 V to 4 V, then the Fourier expansion will give 3 volts as the value of $v(t)$ at t_1.

Let us consider a specific example; let us analyze the square wave shown in Fig. 3.3 with pulse width τ and frequency $1/T$. The voltage waveform to be

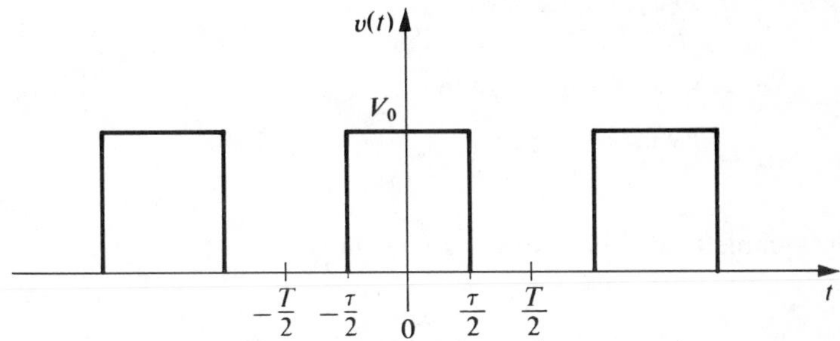

FIGURE 3.3 Square wave with pulse width τ and period T.

analyzed equals V_0 from $-\tau/2$ to $+\tau/2$ and equals zero from $\tau/2$ to $T/2$ and $-\tau/2$ to $-T/2$. Note $v(t)$ is even and so all the b_m sine coefficients equal zero.*

---

* 1. A function is even if $v(-t) = v(t)$.

2. A function is odd if $v(-t) = -v(t)$.

3. $\int_{-T}^{T} v(t)f(t)\,dt = 0$ if the integrand is odd, i.e., if $v(t)$ is even and $f(t)$ is odd, or if $v(t)$ is odd and $f(t)$ is even.

4. $\int_{-T}^{T} v(t)f(t)\,dt = 2\int_{0}^{T} v(t)f(t)\,dt$ if the integrand is even, i.e., if both $v(t)$ and $f(t)$ are even or both are odd.

5. If $v(t)$ is even, all the b_m coefficients $= 0$; if $v(t)$ is odd all the a_n coefficients $= 0$. If $v(t)$ is neither even nor odd, then in general we will have both a_n and b_m coefficients in the Fourier expansion of $v(t)$.

The a_n cosine coefficients are given by:

$$a_n = \frac{2}{T}\int_{-T/2}^{T/2} v(t)\cos n\omega t\, dt = \frac{4}{T}\int_0^{T/2} v(t)\cos n\omega t\, dt,$$

since $v(t)\cos n\omega t$ is *even*.

$$a_n = \frac{4}{T}\int_0^{\tau/2} V_0 \cos n\omega t\, dt = \frac{4V_0}{T}\left(\frac{\sin n\omega t}{n\omega}\bigg|_0^{\tau/2}\right) = \frac{4V_0}{n\omega T}\sin\frac{n\omega\tau}{2}$$

$$a_n = \frac{2V_0}{\pi n}\sin\frac{n\pi\tau}{T} \qquad \text{using } \omega \equiv \frac{2\pi}{T}$$

$$a_0 = \frac{2}{T}\int_{-T/2}^{T/2} v(t)\, dt = \frac{4}{T}\int_0^{\tau/2} V_0\, dt = 2V_0\frac{\tau}{T}. \tag{3.11}$$

Thus, the Fourier coefficients depend upon the ratio of the pulse width to the period, τ/T. Notice that a_n is nonzero for *all* values of n, $n = 0, 1, 2, 3, \ldots$. This means that the square wave contains very high frequency Fourier components.

If $\tau/T = \frac{1}{2}$ which corresponds to a symmetrical square wave, then we have:

$$a_n = \frac{2V_0}{n}\sin n\frac{\pi}{2} = \frac{2V_0}{\pi n}(-1)^{n+1} \qquad \text{for } n = 1, 3, 5, \ldots$$

$$a_n = 0 \qquad \text{for } n = 2, 4, 6, \ldots$$

$$\frac{a_0}{2} = \frac{V_0}{2}. \tag{3.12}$$

Thus the symmetrical square wave expansion is

$$v(t) = V_0/2 + \sum_n^{\infty}\frac{2V_0}{\pi n}(-1)^{n+1}\cos\frac{2\pi nt}{T}. \tag{3.13}$$

In words, the symmetrical square wave equals an average dc level of $\frac{1}{2}$ the pulse amplitude plus a series of cosine waves having frequencies of odd integral multiples of $2\pi/T$, the fundamental frequency. Notice that the amplitude a_n of the odd harmonics decreases with increasing frequency.

If τ/T is much less than unity, then we can make the following approximations:

$$a_0 = 2V_0\frac{\tau}{T}$$

$$a_n = \frac{2V_0}{\pi n}\sin n\pi\frac{\tau}{T} \cong \frac{2V_0}{\pi n}n\pi\frac{\tau}{T} \qquad \text{using } \sin\theta \cong \theta \text{ for small } \theta. \tag{3.14}$$

$$a_n \cong 2V_0\frac{\tau}{T},$$

which tells us that the amplitude a_n of the cosine Fourier components remains constant regardless of how high the frequency is. In other words, very narrow pulses ($\tau/T \ll 1$) contain large amplitude, high frequency Fourier components. Therefore, to amplify a narrow pulse requires an amplifier capable of amplifying high frequencies.

Fourier analysis can be extended to cover the problem of analyzing a voltage waveform which extends from $t = -\infty$ to $t = +\infty$. The argument will not be given here, but the result is expressed most concisely in complex notation as follows:

$$v(t) = \frac{1}{\sqrt{2\pi}} \int_{-\infty}^{\infty} g(\omega) e^{-j\omega t}\, d\omega \tag{3.15}$$

$$g(\omega) = \frac{1}{\sqrt{2\pi}} \int_{-\infty}^{\infty} v(t) e^{+j\omega t}\, dt \tag{3.16}$$

The $g(\omega)$ is the Fourier coefficient; it is the amplitude of the Fourier component of angular frequency ω. The integral over ω is analogous to the summation over the different frequencies $(n)(2\pi t)/T$. $v(t)$ and $g(\omega)$ are often called a Fourier transform pair.

The above is referred to as the Fourier integral theorem, and the interested reader is referred to Margenau and Murphy, *The Mathematics of Physics and Chemistry*. Notice that the basic idea is still the same; an arbitrary voltage $v(t)$ is expressed as a sum (an integral is just a sum of an infinite number of infinitesimally small terms) of waves of different frequencies. In this case, the waves are written in the form of complex exponentials, but their real and imaginary parts are the familiar trigonometric cosines and sines.

$$e^{j\omega t} = \cos \omega t + j \sin \omega t \tag{3.17}$$

Let us now analyze a single rectangular pulse of amplitude V_0, width τ, and zero rise and fall time as shown in Fig. 3.4(a). The voltage pulse being

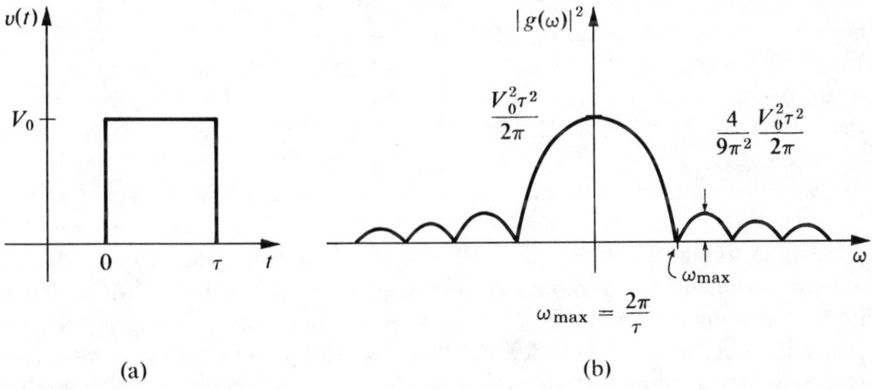

FIGURE 3.4 Single rectangular voltage pulse and Fourier components.

analyzed is zero for t less than 0 and for t greater than τ, and it equals V_0 for $0 \leqq t \leqq \tau$. Therefore, the Fourier coefficient $g(\omega)$ is given by:

$$g(\omega) = \frac{1}{\sqrt{2\pi}} \int_{-\infty}^{\infty} v(t)e^{j\omega t}\, dt = \frac{1}{\sqrt{2\pi}} \int_0^{\tau} V_0 e^{j\omega t}\, dt = \frac{V_0}{\sqrt{2\pi}} \frac{e^{j\omega t}}{j\omega}\Bigg|_0^{\tau}$$

$$g(\omega) = \frac{V_0}{j\omega\sqrt{2\pi}}(e^{j\omega\tau} - 1). \tag{3.18}$$

The coefficient $g(\omega)$ is the amplitude in volts per radian/sec bandwidth of the Fourier component of angular frequency ω. It is more convenient to consider the *power* of the Fourier component which is proportional to the amplitude squared, so we will calculate $|g(\omega)|^2$.

$$|g(\omega)|^2 = g^*(\omega)g(\omega) = \frac{V_0}{-j\omega\sqrt{2\pi}}(e^{-j\omega\tau} - 1)\frac{V_0}{j\omega\sqrt{2\pi}}(e^{j\omega\tau} - 1)$$

$$|g(\omega)|^2 = \frac{V_0^2}{2\pi\omega^2}[2 - (e^{j\omega\tau} + e^{-j\omega\tau})] = \frac{V_0^2}{2\pi\omega^2}[2 - 2\cos\omega\tau]$$

$$|g(\omega)|^2 = \frac{2V_0^2}{\pi\omega^2}\sin^2\frac{\omega\tau}{2}$$

using $1 - \cos\omega\tau = 2\sin^2\omega\tau/2$ or, rewriting,

$$|g(\omega)|^2 = \frac{V_0^2\tau^2}{2\pi}\left[\frac{\sin\dfrac{\omega\tau}{2}}{\dfrac{\omega\tau}{2}}\right]^2. \tag{3.19}$$

Thus, we see that the power in the various Fourier components varies with frequency as $(\sin^2\omega\tau/2)/\omega^2$, which is plotted in Fig. 3.4(b). Notice that $g(\omega)^2$ falls to zero when $\omega = 2\pi/\tau \equiv \omega_{\max}$. Subsequent zeros occur when $\omega = 4\pi/\tau$, $6\pi/\tau$ . . . . The area under the $g(\omega)^2$ curve is proportional to the power in the Fourier components. It can be seen by an inspection of Fig. 3.5(b) that most of the power occurs for frequencies from zero to $f_{\max} = 1/\tau$, the frequency of the first zero. Thus we would expect a circuit capable of passing frequencies from zero to $f_{\max} = 1/\tau$ to pass the voltage pulse without serious distortion of the pulse shape, since most of the energy of the pulse lies in this range of frequencies. This result is extremely important practically speaking, because it is impossible to build a circuit that will pass all frequencies from zero to infinity without attenuation. The gain of any real circuit will decrease if the signal frequency is increased enough. The range of frequencies that a circuit will pass with a certain gain is called the "bandwidth." The preceding result tells us that a bandwidth of approximately $1/\tau$ (or more) is necessary to pass a pulse of pulse width τ. For example, a bandwidth of one MHz is necessary to pass a one microsecond pulse. The narrower the pulse width (the "faster" the pulse), the larger the bandwidth required. In actual electronic circuit design, many ingenious techniques are used to increase the bandwidth of circuits in order to pass fast pulses with minimum distortion.

In order to pass a perfectly rectangular pulse (zero rise and fall time) with no distortion, a circuit must have an infinite bandwidth, from dc to infinity. No real circuit can have an infinite bandwidth, and so we are led to the question: What will the output pulse look like for a circuit of finite bandwidth? One's physical intuition (properly trained) can answer this question to a surprisingly high degree of accuracy. It is intuitively reasonable (and correct) that any rapidly changing part of a voltage waveform, such as the steep leading or trailing edge of a pulse, will contain *high* frequency Fourier components, while a slowly changing portion of the waveform such as the flat top of a pulse will contain *low* frequency Fourier components. Thus, if a zero rise time pulse is fed into a circuit of finite bandwidth we would expect the leading and trailing edges of the pulse to be rounded off, because the high frequency components in these edges are not passed by the circuit. This is illustrated in Fig. 3.5. The smaller

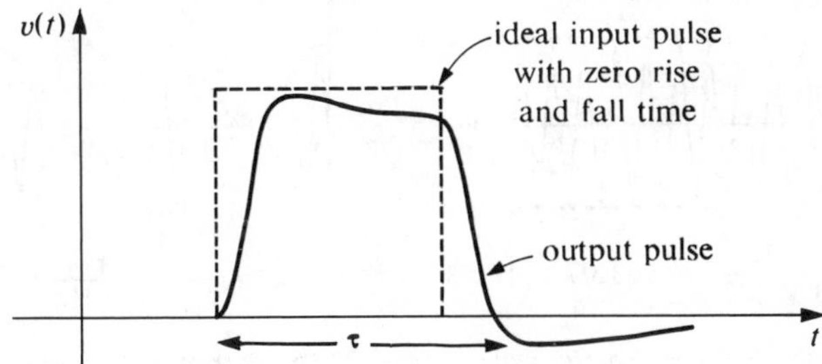

FIGURE 3.5 Rounding of output pulse due to finite bandwidth.

the bandwidth the more rounding of the pulse will occur; in fact if the bandwidth is much less than from 0 to $f = 1/\tau$, then the amplitude of the output pulse will be appreciably less than that of the input pulse.

It is a general feature of Fourier analysis that any rapid change in the voltage waveform being analyzed contains high frequency Fourier components. For example, note the leading and trailing edges of pulses, the sharp "corners" of the triangular or sawtooth waves, and the discontinuity of the slope of a rectified sine wave as indicated in Fig. 3.6.

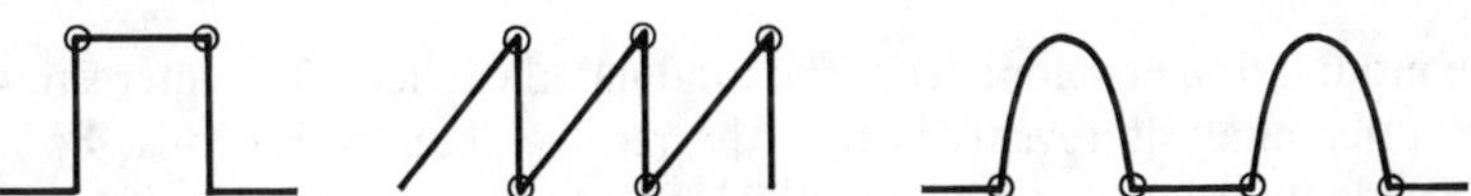

FIGURE 3.6 Examples of rapid voltage changes (circled) which contain high frequency Fourier components.

Thus, a circuit incapable of passing the high frequency Fourier components will round off any sharp corners or rapid changes in the input waveform; the output will be distorted and may even be much less in amplitude than the input.

An interesting example of Fourier analysis occurs in ordinary AM radio broadcasting. The "AM" stands for "amplitude modulation" and refers to the fact that the amplitude of the high frequency electromagnetic wave broadcast by the radio station is changed or modulated according to the type of audible sound transmitted. For example, an AM station at 930 on your dial broadcasts an electromagnetic wave at a frequency $f_c = 930\,\text{kHz}$. This wave is called the "carrier" wave, and if it were of constant amplitude, no audible sound would be heard through the receiver. If a sound wave of frequency $f_m = 400\,\text{Hz}$ is to be transmitted via this high frequency carrier, then the 400 Hz sound wave is first converted into a 400 Hz voltage by a microphone, and the 400 Hz voltage is used to change the amplitude of the carrier, i.e., to "modulate" the carrier as shown in Fig. 3.7. The receiver now must somehow extract the

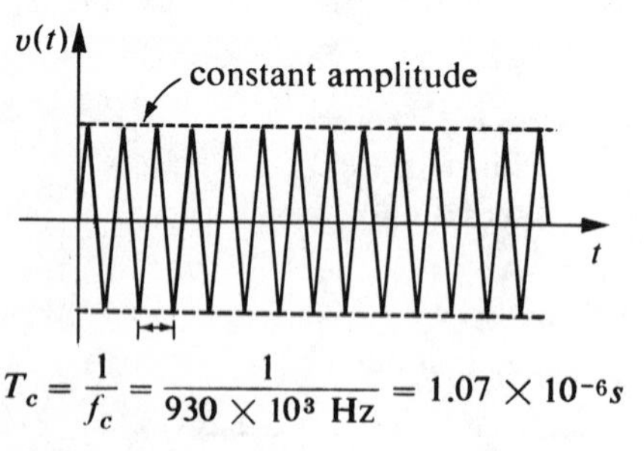

$$T_c = \frac{1}{f_c} = \frac{1}{930 \times 10^3\ \text{Hz}} = 1.07 \times 10^{-6} s$$

(a) *unmodulated 930 kHz carrier wave*

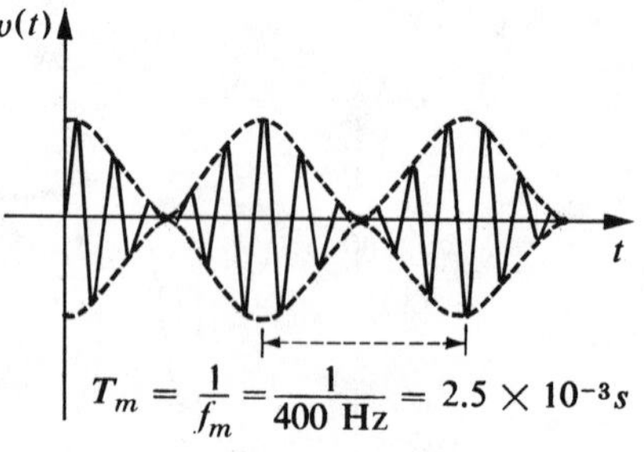

$$T_m = \frac{1}{f_m} = \frac{1}{400\ \text{Hz}} = 2.5 \times 10^{-3} s$$

(b) *carrier with 100% 400 Hz amplitude modulation*

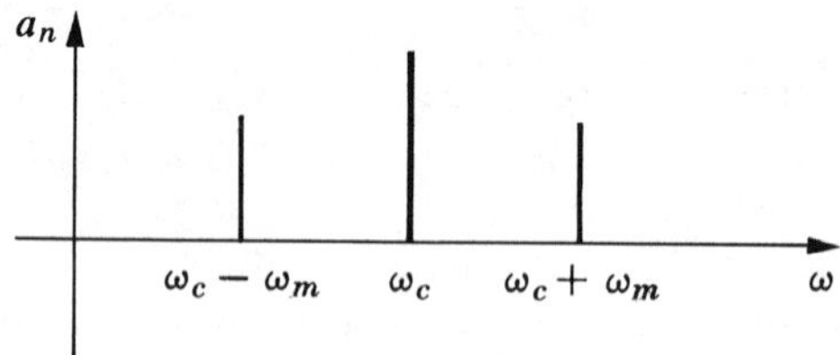

(c) *Fourier components of 100% modulated carrier of* (b)

FIGURE 3.7 Amplitude modulation.

400 Hz modulation envelope from the modulated carrier. The interesting thing that Fourier analysis reveals is that the receiver bandwidth must be at least 400 Hz wide in order to receive the 400 Hz audio signal. That is, if the receiver bandwidth were too narrow, say from 929.9 to 930.1 kHz, a bandwidth of 0.2 kHz or 200 Hz, no 400 Hz signal would be present in the receiver.

Let us Fourier analyze the amplitude modulated carrier.

$$v(t) = V_0(1 + M\cos\omega_m t)\cos\omega_c t \tag{3.20}$$

where $\omega_m = 2\pi f_m$, $f_m = 400$ Hz, $\omega_c = 2\pi f_c$, $f_c = 930k$Hz and M is the degree of modulation $0 \leqq M \leqq 1$. M is 1.0 in figure 3.7(6). The modulated carrier wave $v(t)$ is an even function, so only cosine terms will be present in the Fourier expansion. Also, we can see that $a_0 = 0$, because the average dc level is zero. Thus

$$v(t) = \sum_{n=1}^{\infty} a_n \cos n\omega t \tag{3.21}$$

where

$$a_n = \frac{2}{T}\int_{-T/2}^{T/2} v(t)\cos n\omega t \, dt. \tag{3.22}$$

Choose $T = 2\pi/\omega_m \equiv T_m$; this is the shortest time interval over which $v(t)$ is periodic. Substituting (3.20) in (3.22) yields with $\omega = \omega_m$

$$a_n = \frac{4}{T_m}\int_0^{T_m/2} V_0(1 + M\cos\omega_m t)\cos\omega_c t \cos n\omega_m t dt.$$

A trigonometric identity is used to change the integrand product of three cosine factors into a product of two cosines:

$$\cos\alpha \cos\beta = \frac{1}{2}[\cos(\alpha + \beta) + \cos(\alpha - \beta)] \tag{3.23}$$

Let $\alpha = \omega_c t$, let $\beta = \omega_m t$. Thus a_n becomes

$$a_n = \frac{4V_0}{T_m}\int_0^{T_m} \cos\omega_c t \cos n\omega_m t dt$$

$$+ \frac{4V_0 M}{T_m}\int_0^{T_m/2} \cos(\omega_c + \omega_m) t \cos n\omega_m t dt$$

$$+ \frac{4V_0 M}{T_m}\int_0^{T_m/2} \cos(\omega_c - \omega_m) t \cos n\omega_m t dt \tag{3.24}$$

Each integral equals zero unless the arguments of the two cosine factors are equal. Thus, there are three non zero Fourier components, one for each integral of (3.24). The frequencies $n\omega_m$ of these three components are $n\omega_m = \omega_c$ (the carrier frequency), $n\omega_m = \omega_c - \omega_m$ (ω_m lower than the carrier frequency), and $n\omega_m = \omega_c + \omega_m$ (ω_m higher than the carrier frequency).

The Fourier coefficients work out to be

$$a_n = V_0 \text{ for } n\omega_m = \omega_c \tag{3.25}$$

and

$$a_n = MV_0/2 \text{ for } n\omega_m = \omega_c \pm \omega_m \tag{3.26}$$

Thus the Fourier spectrum of the amplitude modulated carrier consists of a central component at the carrier frequency ω_c and two other equal components called "sidebands" spaced equally on either side of the carrier, one ω_m above and the other ω_m below the carrier frequency. For $M = 1$ or 100% modulation as shown in figure 3.7(b), the sidebands are each one half the amplitude of the carrier, or are 6dB "down" with respect to the carrier. For the receiver to hear the 400Hz modulation signal, it must receive from ω_c to $\omega_c + \omega_m$ or from ω_c to $\omega_c - \omega_m$, i.e. the receiver bandwidth must be at least ω_m wide.

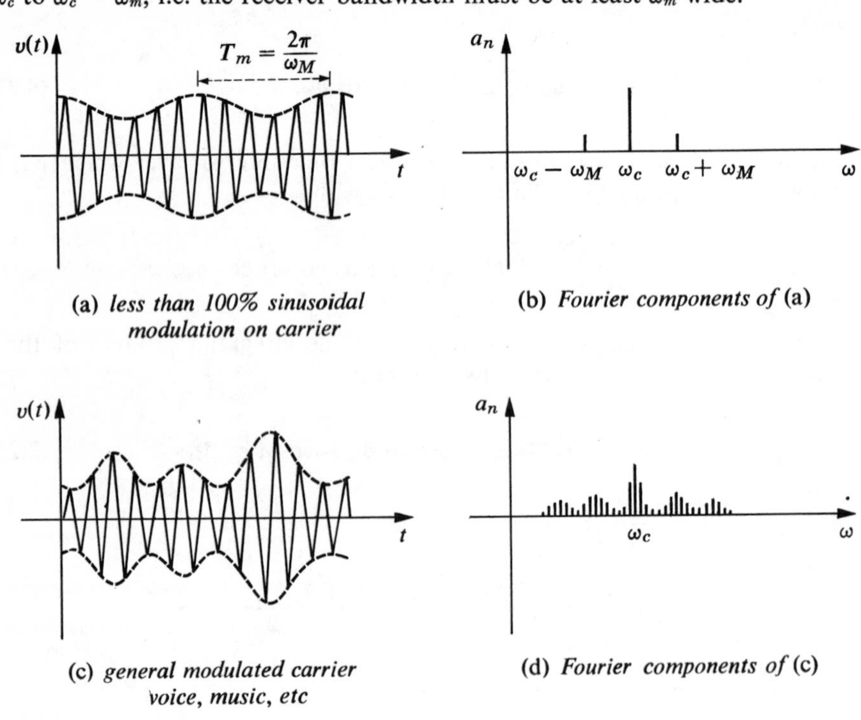

(a) *less than 100% sinusoidal modulation on carrier*

(b) *Fourier components of* (a)

(c) *general modulated carrier voice, music, etc*

(d) *Fourier components of* (c)

FIGURE 3.8 Amplitude modulation waveforms.

If the carrier waveform is not modulated to zero, i.e. if $M < 1$, as shown in Fig. 3.8(a), then the sidebands are smaller as shown in Fig. 3.8(b).

In actual practice, the modulation signal is not usually a pure sine wave but a complicated audio waveform as shown in Fig. 3.8(c) with frequency components ranging from 20 Hz to about 20 kHz, the range of response of the human ear. A Fourier analysis of a carrier whose envelope is amplitude modulated by such a complicated audio waveform will show that the sidebands (Fourier components) lie in the range from $\omega_c - \omega_{max}$ to $\omega_c + \omega_{max}$ as shown in

Fig. 3.8(d). ω_{max} is the maximum frequency component of the audio modulation signal. To broadcast high fidelity sound, ω_{max} should equal the maximum frequency audible to the human ear $\omega_{max} = 2\pi \times 20\,\text{kHz}$. Thus the receiver must have a bandwidth from $\omega_c - \omega_{max}$ to $\omega_c + \omega_{max}$, 40 kHz, for high fidelity. In order to pack more AM stations into the AM band, the maximum modulation frequency ω_{max} is limited by law to 10 kHz, so an AM receiver bandwidth must be a maximum of 20 kHz. The reason AM radio music is not high fidelity is simply that only audio frequencies up to 10 kHz can be broadcast. The crisp, high pitched sounds from 10 kHz to 20 kHz are simply not present in AM broadcasts.

Another type of amplitude modulation widely used in long range noncommercial radio communication is "single sideband," abbreviated "SSB." In this technique, only one set of sidebands is actually transmitted, not the carrier or the other set of sidebands. Both sets of sidebands contain the same information (the modulating audio signal) so only one set need be broadcast. The carrier itself, being a wave of constant amplitude, really contains no information so it is wasteful to broadcast it, especially because the carrier power is many times the power in one set of sidebands. At the receiver, the carrier is generated at a low power level and used to demodulate the sidebands which have been received. That is, the carrier at ω_c is beat against the received sidebands at $\omega_c + \omega_m$ and the difference beat frequency created is the desired audio signal at ω_m.

It should be pointed out that a waveform which is amplified (or attenuated) by a *nonlinear* circuit will have an entirely different Fourier spectrum at the output compared to its spectrum at the input. A linear circuit is one in which the output is simply a multiple of the input. A nonlinear circuit essentially changes the shape of the waveform and thus the Fourier spectrum. The output versus input graph for a linear circuit is a straight line, for a nonlinear circuit a curved line as shown in Fig. 3.9. Such graphs need not go through the origin but usually do; that is, there is usually zero output for zero input.

To see how extra frequency components are generated, we write $v_{out} = Av_{in}$, and think of the gain A as being a function of the input voltage.

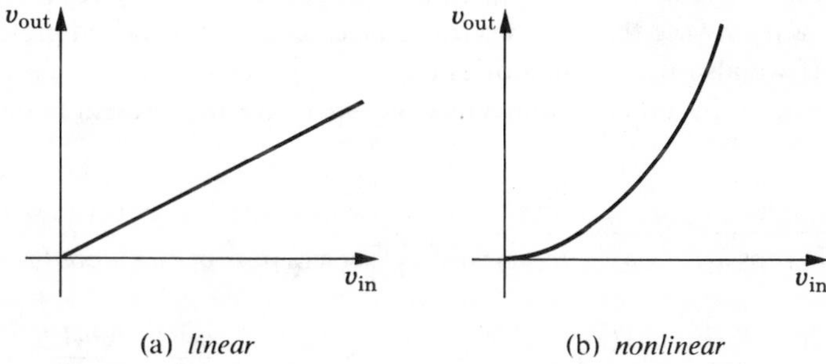

FIGURE 3.9 Input–output graphs.

Then we expand the gain A in a power series in v_{in} using a standard Taylor series expansion.

$$A = A(0) + \left.\frac{dA}{dv_{\text{in}}}\right|_0 v_{\text{in}} + \frac{1}{2!}\left.\frac{d^2A}{dv_{\text{in}}^2}\right|_0 v_{\text{in}}^2 + \cdots. \tag{3.27}$$

Thus,

$$v_{\text{out}} = \left[A(0) + \left(\left.\frac{dA}{dv_{\text{in}}}\right|_0\right) v_{\text{in}} + \frac{1}{2!}\left(\left.\frac{d^2A}{dv_{\text{in}}^2}\right|_0\right) v_{\text{in}}^2 + \cdots\right] v_{\text{in}}, \tag{3.28}$$

which can be rewritten as

$$v_{\text{out}} = a_1 v_{\text{in}} + a_2 v_{\text{in}}^2 + a_3 v_{\text{in}}^3 + \cdots = \sum_{n=1}^{\infty} a_n v_{\text{in}}^n, \tag{3.29}$$

where $a_1, a_2, a_3, \ldots$ are constants:

$$a_1 = A(0), \quad a_2 = \left.\frac{dA}{dv_{\text{in}}}\right|_0, \quad \text{etc.}$$

Thus the constants a_n are seen to depend upon the exact shape of the output versus input graph. Any particular device or amplifier will have a certain set of constants a_n.

If the input contains the sum of two different frequencies $v_{\text{in}} = V_{10} \cos \omega_1 t + V_{20} \cos \omega_2 t$ then the $a_2 v_{\text{in}}^2$ term will generate *new* frequencies of $\omega_1 + \omega_2$ and $\omega_1 - \omega_2$. Let us neglect a_3 and higher a_k's, which is a good approximation for small inputs. Then,

$$v_{\text{out}} \cong a_1 v_{\text{in}} + a_2 v_{\text{in}}^2. \tag{3.30}$$

$$v_{\text{out}} \cong a_1(V_{10} \cos \omega_1 t + V_{20} \cos \omega_2 t) + a_2(V_{10} \cos \omega_1 t + V_{20} \cos \omega_2 t)^2$$

$$v_{\text{out}} \cong a_1 V_{10} \cos \omega_1 t + a_1 V_{20} \cos \omega_2 t + a_2(V_{10}^2 \cos^2 \omega_1 t + 2V_{10}V_{20} \cos \omega_1 t \cos \omega_2 t + V_{20}^2 \cos^2 \omega_2 t). \tag{3.31}$$

Using the trigonometric identity $\cos \omega_1 t \cos \omega_2 t = \frac{1}{2}[\cos(\omega_1 + \omega_2)t + \cos(\omega_1 - \omega_2)t]$, we see that the two frequencies $\omega_1 \pm \omega_2$ are indeed present in the output. If we also use the identities $\cos^2 \omega t = \frac{1}{2}(1 + \cos 2\omega t)$, we see that the new frequencies, $2\omega_1$ and $2\omega_2$, and a dc component are also present in the output:

$$v_{\text{out}} = \overbrace{a_1 V_{10} \cos \omega_1 t}^{\omega_1} + \overbrace{a_1 V_{20} \cos \omega_2 t}^{\omega_2} + \overbrace{\frac{a_2 V_{10}^2}{2} \cos 2\omega_1 t}^{2\omega_1} + \overbrace{a_2 V_{10} V_{20} \cos(\omega_1 + \omega_2)t}^{(\omega_1 + \omega_2)}$$
$$+ \underbrace{a_2 V_{10} V_{20} \cos(\omega_1 - \omega_2)t}_{(\omega_1 - \omega_2)} + \underbrace{\frac{a_2 V_{20}^2}{2} \cos 2\omega_2 t}_{2\omega_2} + \underbrace{\frac{a_2 V_{10}^2}{2} + \frac{a_2 V_{20}^2}{2}}_{\text{dc component}} \tag{3.32}$$

A circuit specifically built to generate the sum and difference frequencies $\omega_1 \pm \omega_2$ for two inputs is usually called a "mixer." The two inputs at ω_1 and ω_2 are said to be "mixed" together. In superheterodyne radio receivers, the signal picked up from the antenna at the carrier frequency ω_C plus any sidebands due to the audio modulation of the carrier is mixed or "beat against" a frequency ω_{LO} generated by the "local oscillator" in the receiver (see Fig. 3.10). The dif-

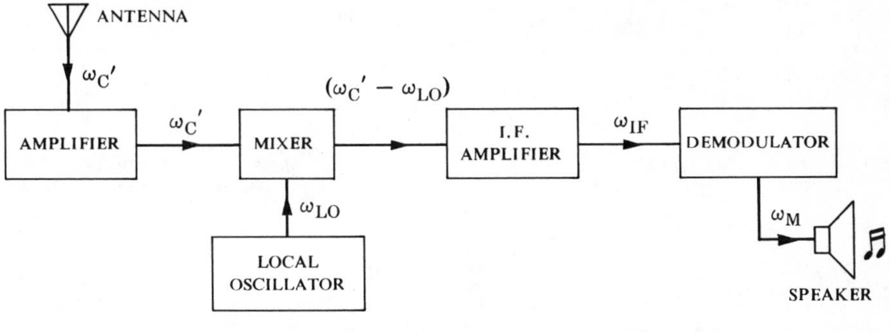

ω_C' = *carrier frequency plus modulation sidebands*
ω_{LO} = *local oscillator frequency*
ω_{IF} = *intermediate frequency* = $\omega_C' - \omega_{LO}$

FIGURE 3.10 Superheterodyne radio receiver block diagram.

ference frequency or "beat" frequency $\omega_{LO} - \omega_c$ contains the modulation sidebands and is amplified and demodulated to recover the audio modulation. Usually the local oscillator is tuned so that the difference frequency (called the intermediate frequency or "i.f.") $\omega_{LO} - \omega_c$ is constant. Usually $(\omega_{LO} - \omega_c)/2\pi = 455\,\text{kHz}$ for AM radios. Thus, if $\omega_c/2\pi = 930\,\text{kHz}$, $\omega_{LO}/2\pi$ must equal 1385 kHz. Details of the detection process are beyond the scope of this book; the interested reader is referred to any book on commercial radio.

The sum beat frequency $\omega_C' + \omega_{LO}$ is usually well outside the bandpass of the mixer and i.f. amplifier; hence it is strongly attenuated. In the microwave region of thousands of MHz, a circuit designed to generate the difference frequency $\omega_1 - \omega_2$ is often called a "down converter," because the output frequency $\omega_1 - \omega_2$ has been shifted down or reduced compared to the input frequency ω_1.

3.4 INTEGRATING CIRCUIT (LOW-PASS FILTER)

The *RC* filter with the output taken across the capacitor is a low-pass filter as we have already seen in section 2.10. Let us now consider the effect of such a filter on a rectangular pulse of width τ. The *RC* low-pass filter, its gain–frequency curve, and the Fourier power spectrum of the pulse are shown in Fig. 3.11.

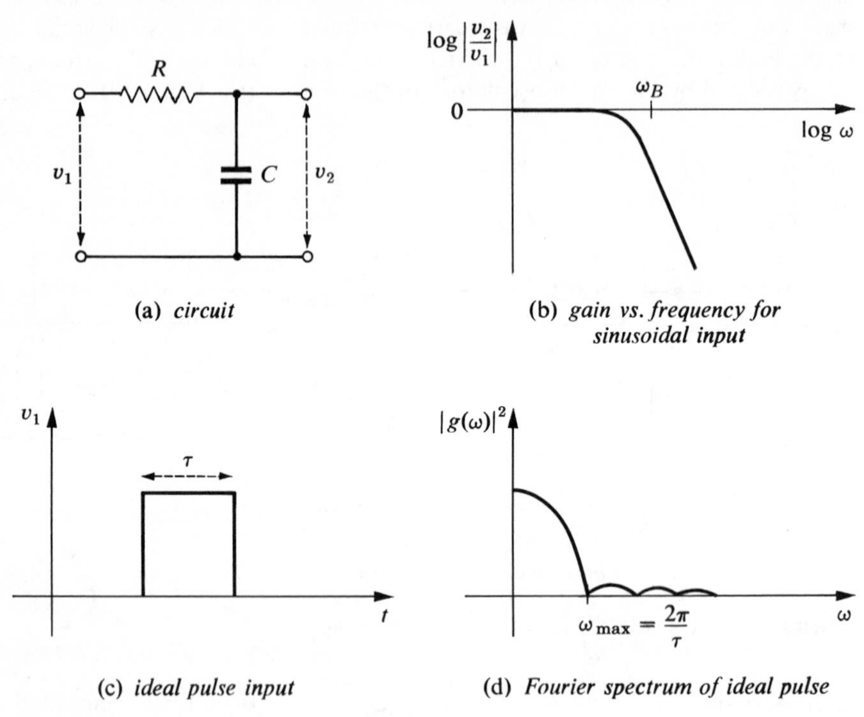

FIGURE 3.11 RC low-pass filter or integrating circuit.

Without doing any calculations at all, we can predict the qualitative effect of the *RC* low-pass filter on a rectangular pulse of width τ by comparing the break point frequency of the filter ω_B with the ω_{max} frequency of the pulse. This is true because most of the energy of the pulse lies in the Fourier frequency components below ω_{max}, and the filter will pass frequencies below its break point frequency ω_B. Thus, if ω_{max} of the pulse is *less* than the break point frequency of the filter, then the pulse will be passed by the filter with little rounding off or distortion. If ω_{max} of the pulse is *greater* than the break point frequency of the filter, then the pulse will be appreciably rounded off and attenuated. Because $f_{max} = 1/\tau$ and $\omega_B = 1/RC$, $\omega_{max} \ll \omega_B$ implies $\tau \gg RC$, and the pulse will suffer little rounding off. $\omega_{max} \gg \omega_B$ implies $\tau \ll RC$ and the pulse will be seriously rounded off. These results are shown in Fig. 3.12. This is what we would expect intuitively; a large capacitor is slow to charge up and discharge and thus a large capacitor (large *RC*) would result in a rounded output pulse.

The *RC* low-pass filter circuit is often known as an "integrating circuit" because if $RC \gg \tau$ the output voltage is essentially the integral (with respect to

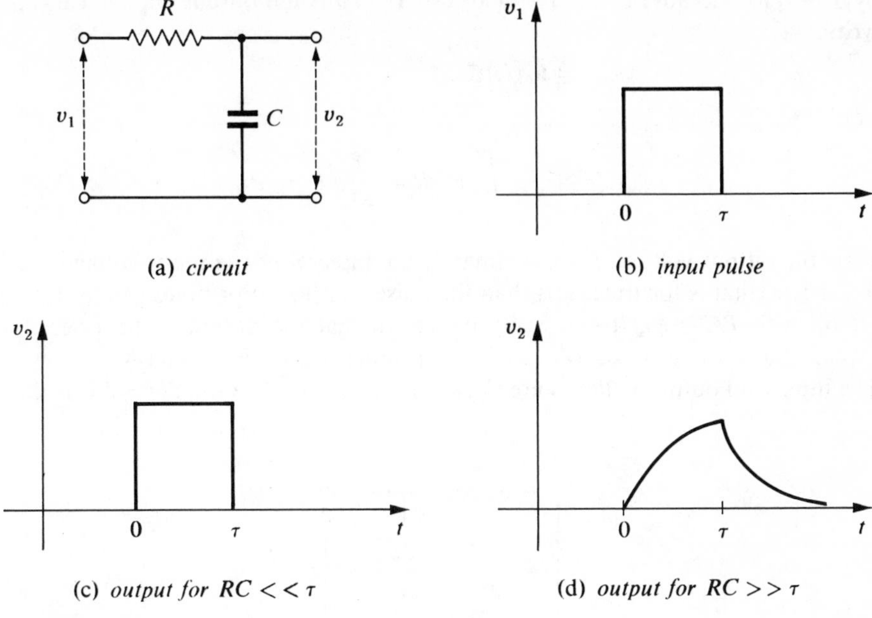

(a) *circuit*

(b) *input pulse*

(c) *output for* $RC << \tau$

(d) *output for* $RC >> \tau$

FIGURE 3.12 Pulse distortion in an RC low-pass filter.

time) of the input voltage pulse. This result can be shown by writing down the Kirchhoff voltage equation for the circuit for $0 \leqq t \leqq \tau$:

$$v_{\text{in}} = V_0 = v_R + v_C = iR + Q/C. \tag{3.33}$$

Then we substitute $i = dQ/dt$ and solve the resulting differential equation for $Q(t)$ because we wish to obtain an expression for v_{out} which equals the voltage across the capacitor Q/C.

$$v_{\text{in}} = V_0 = \frac{dQ}{dt}R + \frac{1}{C}Q \quad \text{or} \quad \frac{dQ}{dt} + \frac{1}{RC}Q = \frac{V_0}{R} \tag{3.34}$$

whose solution is:

$$Q = CV_0(1 - e^{-t/RC}). \tag{3.35}$$

The exact solution for the output voltage on the capacitor is thus $v_{\text{out}} = Q/C$ or

$$v_{\text{out}} = V_0(1 - e^{-t/RC}), \tag{3.36}$$

which satisfies the boundary condition $Q = 0$ when $t = 0$. Now if $RC \gg \tau$, then $t/RC \ll 1$ and in the above expression for v_{out} we may approximate $e^{-t/RC}$

by $1 - t/RC$ because $e^{-x} \cong 1 - x$ if $x \ll 1$. Thus the output voltage can be written as:

$$v_{out} \cong V_0 t/RC$$

or

$$v_{out} \cong \frac{1}{RC}\int_0^t V_0\,dt = \frac{1}{RC}\int_0^t v_{in}\,dt. \tag{3.37}$$

Thus the output voltage is approximately the integral of the input voltage, for $0 \leqq t \leqq \tau$, that is for times less than the pulse width τ. For times greater than τ (still with $RC \ll \tau$), it can easily be shown that the output voltage decays exponentially to zero according to the formula: $v_{out} = V_0\tau/(RC)\, e^{[-(t-\tau)/RC]}$. The input and output voltages are shown in Fig. 3.13. The smaller τ/RC is the

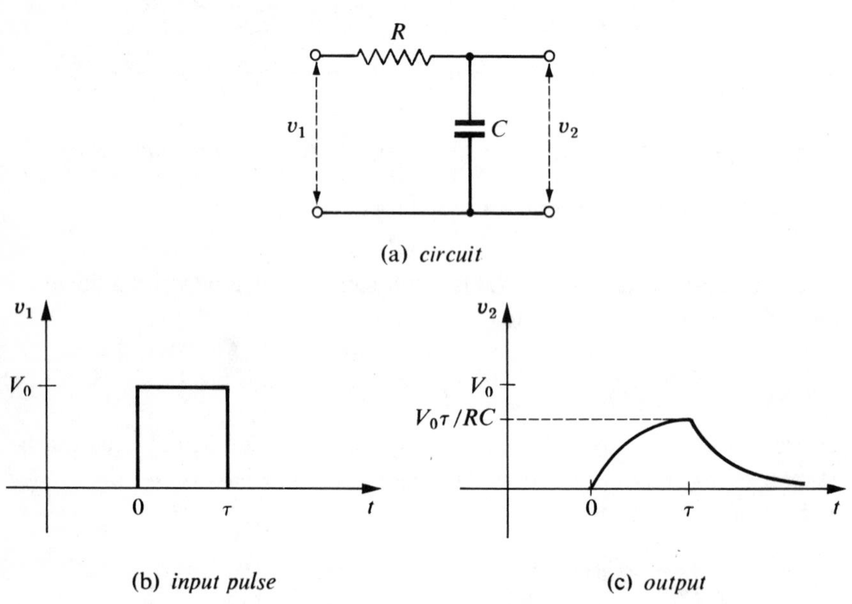

FIGURE 3.13 Pulse response of integrating circuit (*RC* low-pass filter).

more nearly linear the output voltage is from $t = 0$ to $t = \tau$, but the smaller the output amplitude. With $\tau \ll RC$ an integrating circuit is often used to generate an approximately linear output voltage or "sawtooth" from an input square pulse.

If an ideal rectangular voltage pulse with zero rise time is fed into an *RC* low-pass filter circuit with $RC \ll \tau$ as shown in Fig. 3.12, then the output pulse can be shown to have a rise time of 2.2*RC* from $v_{out} = V_0(1 - e^{-t/RC})$ and the definition of rise time. Thus, in agreement with our previous results, the time constant *RC* should be as small as possible for the sharpest possible output pulse. It can be shown that for any circuit with a bandwidth B Hz the output rise time in seconds will be (for a zero rise time input) given by R.T. $= 0.35/B$. This result is intuitively reasonable because the larger the bandwidth, the smaller the rise time; for an ideal amplifier with an infinite bandwidth the output rise time would be zero. The relationship between rise time and bandwidth can be derived approximately by the following argument. A sinusoidal wave goes from zero to peak amplitude in one quarter of a period. Thus for a circuit with bandwidth B, capable of passing frequencies up to B Hz, a rectangular voltage pulse could rise from zero to its final amplitude in a time no shorter than one quarter the period of the *maximum* frequency B, i.e., $1/4B$. Thus we would expect an output rise time of:

$$\text{R.T.} = \frac{1}{4B} = \frac{0.25}{B} \tag{3.38}$$

A more accurate calculation gives $R.T. = 0.35/B$ (see Fig. 3.14).

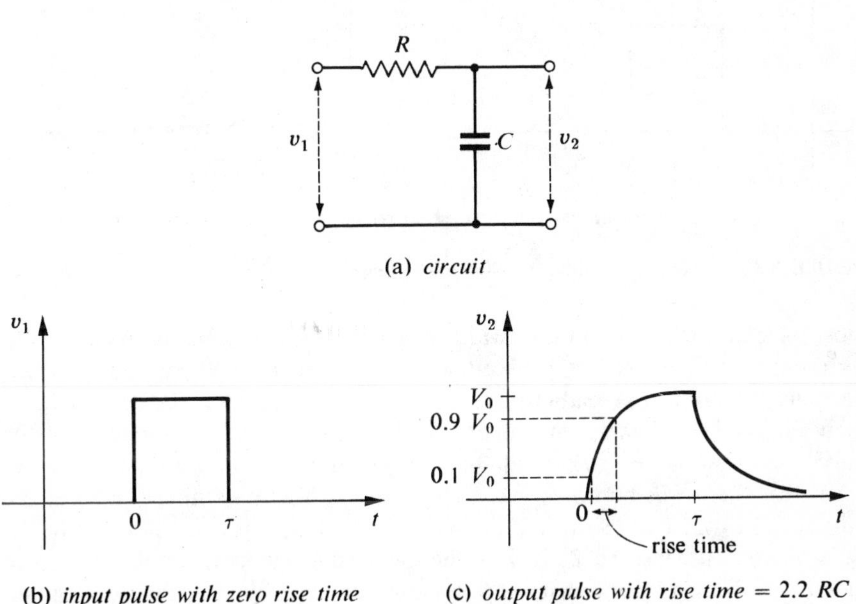

FIGURE 3.14 Effect of RC low-pass filter on rise time if $RC \ll \tau$.

3.5 DIFFERENTIATING CIRCUIT

The *RC* high-pass filter with the output taken across the resistance and a pulse input behaves quite differently from the low-pass filter. Let us first try to predict qualitatively the output waveform from a consideration of the Fourier components of the input pulse. The gain versus frequency curve of the high-pass RC filter is shown in Fig. 3.15; the break point is at $\omega_B = 1/RC$. Because

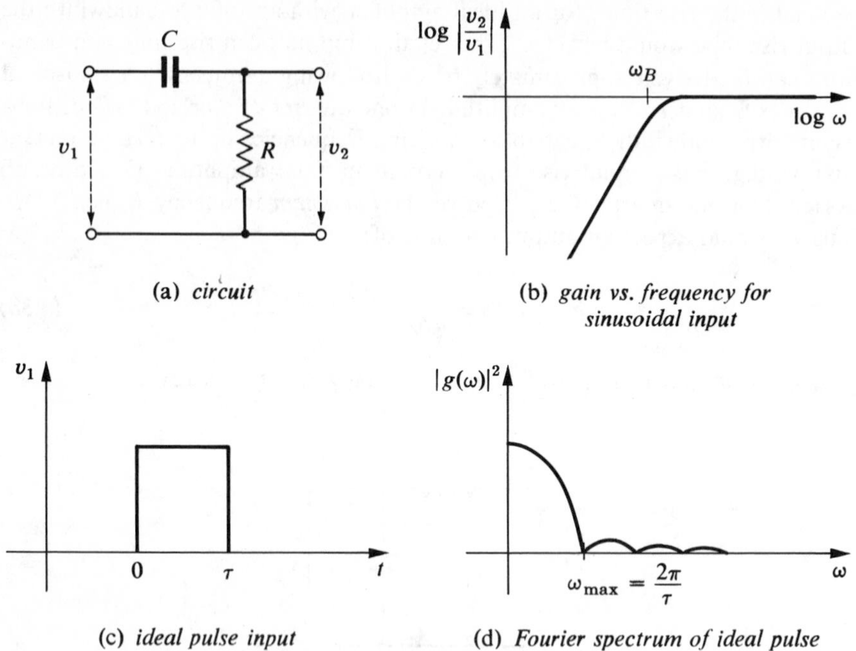

(a) *circuit*

(b) *gain vs. frequency for sinusoidal input*

(c) *ideal pulse input*

(d) *Fourier spectrum of ideal pulse*

FIGURE 3.15 High-pass RC filter or differentiating circuit.

most of the energy of a pulse of width τ occurs in frequency components from zero (dc) to $f_{max} = 1/\tau$, we can see that in order for the high-pass filter to pass the pulse without appreciable distortion, the break point must be much less than $f_{max} = 1/\tau$. Even then, the extremely low frequency components of the pulse (less than $\omega_B = 1/RC$) will be attenuated. And because of the series capacitor, there will be no dc passed by the RC high-pass filter. Hence, the output pulse will be lacking the dc and the low frequency components below $\omega_B = 1/RC$. Because of the lack of the low frequency components, the top of the pulse will "sag" somewhat—it takes low frequency components to maintain the output voltage near V_0 for any length of time. Because of the zero gain of the filter at dc, the net charge passed by the filter must be zero, which means that the total area under the output voltage versus time curve must be exactly zero.

$$v_{out} = iR$$

$$\int_0^\infty v_{out}\,dt = \int_0^\infty iR\,dt = R\int_0^\infty i\,dt = RQ \tag{3.39}$$

But $Q = \int i\,dt =$ *the total net charge passed through the capacitor C*, which must be exactly zero. Therefore, $\int_0^\infty v_{out}\,dt = 0$. Thus, for $\omega_B \ll \omega_{max}$ or

$1/RC \ll 1/\tau$ or $RC \gg \tau$, the input and output pulses will be as shown in Fig. 3.16. Because $\int_0^\infty v_{\text{out}}\, dt = 0$, the area of the pulse above the $v = 0$ axis must

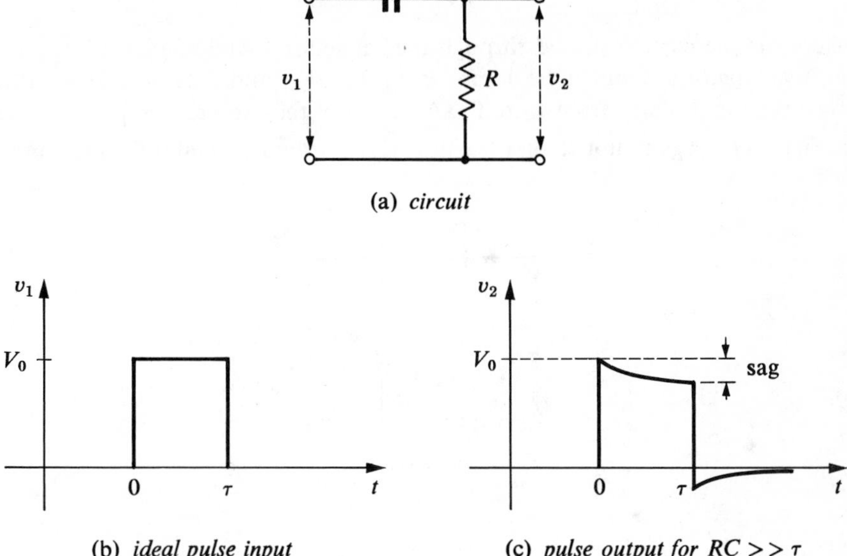

FIGURE 3.16 Input and output pulses for differentiating circuit when $RC \gg \tau$.

exactly equal the area below the axis. Thus there is *always* a long negative tail accompanying every positive pulse passed through an *RC* high-pass filter. The presence of this long negative tail may give rise to severe problems if a large number of positive pulses are fed into the input. This situation is commonly encountered in nuclear pulse amplifiers when high nuclear radiation levels produce many pulses per second, i.e., a high counting rate. Each positive pulse then will appear at the output of the high-pass *RC* filter superimposed on top of the negative tails of all the preceding pulses. This may result in the base line of the pulses falling well below zero volts; the pulses may go from -2 to $+1$ V instead of from 0 to $+3$V. The addition of a diode across the resistance is a simple remedy and is discussed in Chapter 4. Techniques to counteract this are beyond the scope of this book, but are generally referred to as "baseline restoration" or "dc restoration."

If, on the other hand, the break point frequency $\omega_B = 1/RC$ of the filter is close to or higher than $f_{\max} = 1/\tau$, then only a small fraction of the pulse energy will be passed by the filter, namely the energy in the highest frequency components of the pulse. We would thus expect to get a narrow spike or pulse out, a positive spike out when the input pulse jumps from 0 to V_0, and a negative spike out when the input pulse jumps from V_0 back to 0 because the high frequency components occur when the input voltage changes quickly. This result can also be arrived at intuitively by saying that the capacitor will pass only rapid changes in voltage because

$$Q = Cv_C \qquad i = C\,dv_C/dt \qquad v_{\text{out}} = iR = RC\,dv_C/dt \tag{3.40}$$

where i is the current passed through the capacitor. And, dv_2/dt is large and positive when the input pulse jumps from 0 to V_0, and is large and negative when the input jumps from V_0 to 0. A sketch of the expected output is shown in Fig. 3.17. Again, notice that because $\int_0^\infty v_{\text{out}}\,dt$ must equal 0 the area under

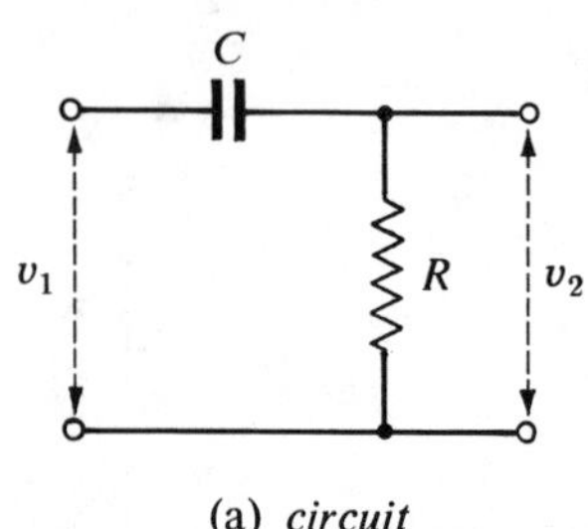

(a) *circuit*

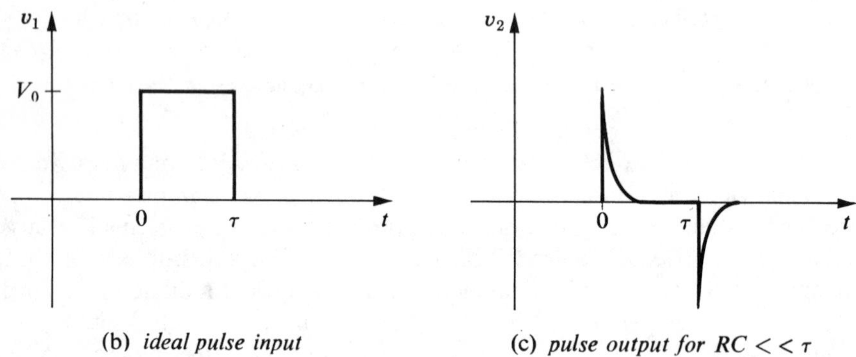

(b) *ideal pulse input* (c) *pulse output for RC << τ*

FIGURE 3.17 Input and output pulses for differentiating circuit when $RC \ll \tau$.

the positive spike must exactly equal the area under the negative spike.

To determine the output more quantitatively, we note that *if* $RC \ll \tau$ then $v_R \ll v_C$. This follows from (neglecting any current drawn from the output):

$$\frac{|v_R|}{|v_C|} = \frac{|iR|}{|iX_C|} = \frac{R}{1/\omega C} = \omega RC. \tag{3.41}$$

For an input pulse of width τ, the maximum Fourier component angular frequency $\omega \cong 1/\tau$. Therefore, the *maximum* value for v_R/v_C is:

$$\left(\frac{|v_R|}{|v_C|}\right)_{\max} = \omega_{\max} RC = \frac{RC}{\tau} \ll 1 \tag{3.42}$$

Thus, $v_R/v_C = RC/\tau \ll 1$ for $\omega_{\max}$. If ω is less than $\omega_{\max}$, the ratio v_R/v_C is even smaller. Therefore the Kirchhoff voltage law can be written with the approximation $v_R \ll v_C$.

$$v_{\text{in}} = v_C + v_R \cong v_C = \frac{Q}{C} \tag{3.43}$$

This equation can be solved easily for i by differentiating with respect to time.

$$\frac{dv_{\text{in}}}{dt} \cong \frac{i}{C}$$

$$i \cong C\, dv_{\text{in}}/dt \tag{3.44}$$

The output voltage is simply $v_{\text{out}} = iR$

$$v_{\text{out}} = iR \cong RC\, dv_{\text{in}}/dt \tag{3.45}$$

The output is therefore approximately equal to the time constant RC multiplied by the time derivative of the input. Hence, the circuit is called a "differentiating" circuit.

The result that the output is the derivative of the input can be shown quantitatively by considering the circuit response to an input pulse with a finite slope as shown in Fig. 3.18.

The Kirchhoff voltage equation is as usual (again neglecting any current drawn from the output):

$$v_{\text{in}} = v_C + v_R = \frac{Q}{C} + iR. \tag{3.46}$$

Differentiating yields:

$$\frac{dv_{\text{in}}}{dt} = \frac{i}{C} + R\, di/dt.$$

Thus, for $0 \leq t \leq t_1$, $v_{\text{in}} = KT$, $dv_{\text{in}}/dt = K$. The differential equation is then:

$$K = \frac{i}{C} + R\frac{di}{dt} \quad \text{or} \quad \frac{di}{dt} + \frac{1}{RC}i = K/R,$$

which has the solution,

$$i = KC(1 - e^{-t/RC}).$$

The output voltage across the resistor is then given by:

$$v_{\text{out}} = iR = KRC(1 - e^{-t/RC}), \tag{3.47}$$

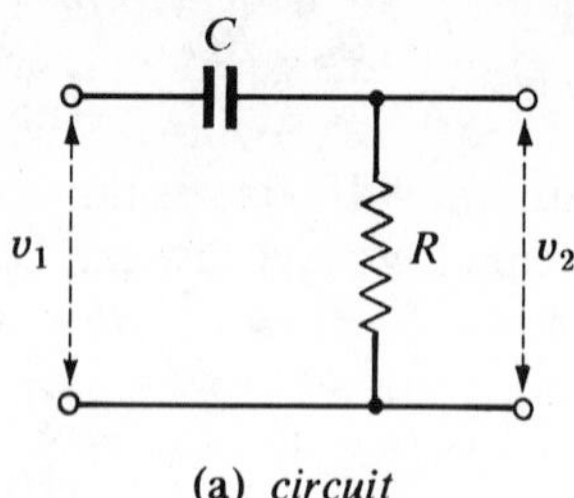

(a) *circuit*

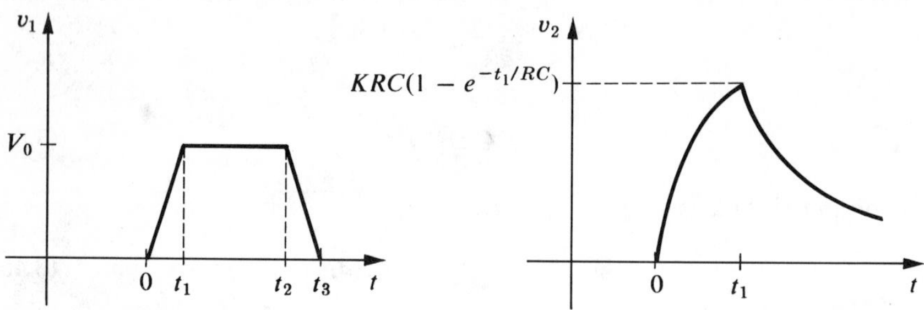

(b) *input pulse with finite slopes* (c) *output leading edge (magnified)*

FIGURE 3.18 Differentiating circuit waveforms.

which is shown greatly magnified in Fig. 3.19(c). The output voltage thus rises exponentially toward $RCK = RC\,dv_{in}/dt$ during the time interval $0 < t < t_1$. If RC is much less than t_1, then the output will rise to a voltage very close to RCK in a time less than t_1. The rise time of the input pulse is equal to 0.8 t_1, so if RC is much less than the pulse rise time, the output will rise to RCK in the time interval from 0 to t_1. Finally, because $dv_{in}/dt = K$ for $0 < t < t_1$, v_{out} rises to $RC\,dv_{in}/dt$; hence, the name differentiating circuit.

For the time interval $t_1 < t < t_2$ the input voltage is equal to V_0, and the differential equation is

$$v_{in} = V_0 = \frac{Q}{C} + iR \quad \text{or} \quad \frac{di}{dt} + \frac{1}{RC}i = 0,$$

the solution to which is

$$i = i_1 e^{-(t-t_1)/RC} \quad \text{or} \quad v_{out} = iR = Ri_1 e^{-(t-t_1)/RC}, \tag{3.48}$$

where i_1 is the current at the time $t = t_1$. The output voltage thus falls to zero with a time constant RC. In the above expression for the output voltage t is in the interval $t_1 < t < t_2$, or $t - t_1 < \tau$. Thus, if τ/RC is large (in other words if $RC \ll \tau$), the output falls rapidly to zero. For the negative going portion of the pulse, the same sort of thing happens, only the sign of the output voltage is changed (see Fig. 3.19).

To summarize our results: If RC is much less than the rise time, then the output voltage across the resistor will rise rapidly to $RC\,dv_{\text{in}}/dt$ and will then fall rapidly to zero as $e^{-t/RC}$, thus producing a sharp positive spike whose amplitude is RC times the derivative of the input. When the input pulse drops from V_0 back to zero, the output voltage can similarly be shown to drop rapidly to $RC\,dv_{\text{in}}/dt$ (which is negative because $dv_{\text{in}}/dt < 0$) and then to rise rapidly back to zero, thus producing a sharp negative output spike which is shown in Fig. 3.18.

The use of a differentiating circuit to produce sharp positive and negative pulses from rectangular pulses is extremely common. The time constant RC of the circuit is usually chosen to be at least ten times less than the pulse width. The smaller RC is the sharper the output spike, but the smaller its amplitude. For example, a pulse with a one microsecond pulse width would be differentiated into sharp positive and negative spikes by a differentiating circuit with $RC = 0.1$ microsecond, e.g., with $R = 10\,\text{k}\Omega$ and $C = 10\,\text{pF}$. It is best to choose C no smaller than 10 pF. This is because the typical stray capacitance in a circuit may be of the order of 5 pF, and the desired capacitance should always be greater than the (uncertain) stray capacity of the circuit, particularly if many circuits are to be made.

Sometimes in order to achieve extremely sharp voltage spikes, a rectangular pulse is differentiated twice, the output of the first differentiating circuit being fed into a second differentiating circuit as is shown in Fig. 3.19.

The output from R_2C_2 will be sharper than from just one differentiating

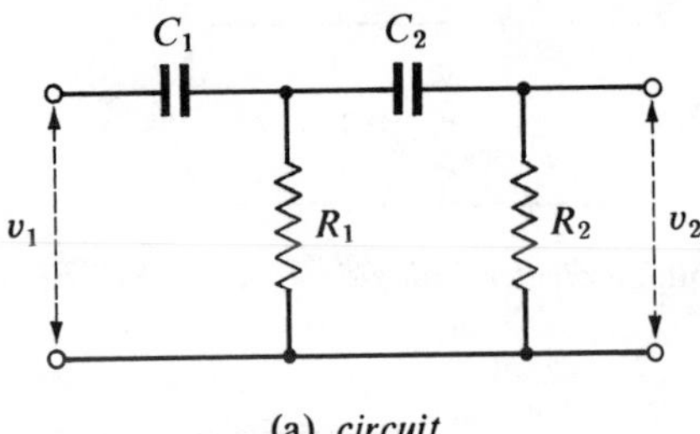

(a) *circuit*

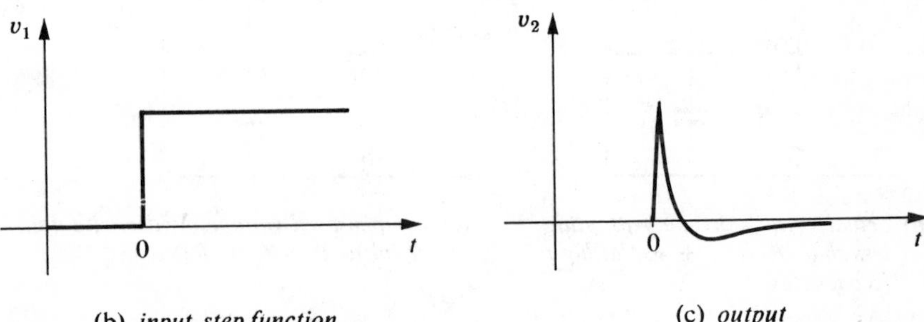

(b) *input step function*

(c) *output*

FIGURE 3.19 Double differentiation.

circuit because the derivative has been taken twice. Also, the output from R_2C_2 will go negative because of the negative slope of the output from R_1C_1. The net result is an output pulse which is narrower or "sharper," smaller in amplitude, and which has a long negative tail. This technique is commonly used in pulse amplifiers in nuclear physics.

3.6 COMPENSATED VOLTAGE DIVIDER

We recall from Chapter 1 that the simple two resistor network of Fig. 3.20(a) acts as a voltage divider with a division ratio of $v_{out}/v_{in} = R_1/(R_1 + R_2)$ if negligible current is drawn from the output. However, if very narrow or "fast" pulses are fed into such a divider, they are severely rounded off as well as attenuated; that is, the output pulse rise time is substantially larger than the input pulse rise time which is generally an undesirable situation. The reason for this rounding off of the input pulses is that there is always some stray or unavoidable circuit capacitance C_1 across R_1; this capacitance is shown as a dotted capacitor in Fig. 3.20(b). The very high frequency Fourier components of the input pulse are simply shorted out by C_1, thus rounding off the output.

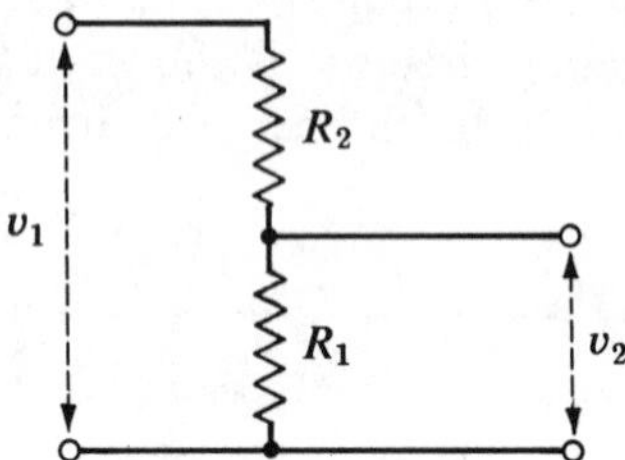

(a) *ideal voltage divider with gain equal to* $R_1/(R_1 + R_2)$

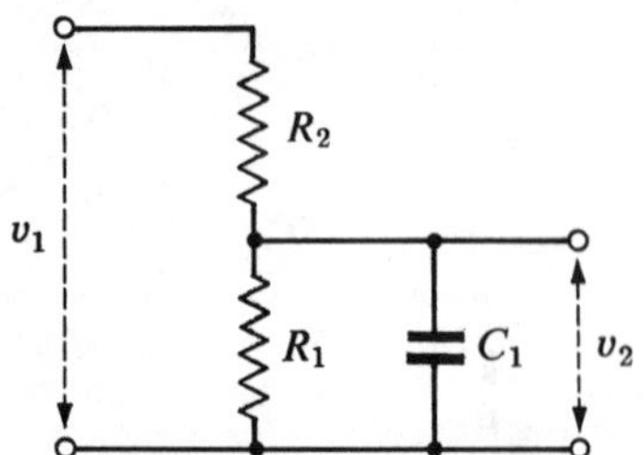

(b) *actual voltage divider with gain less than* $R_1/(R_1 + R_2)$ *at high frequencies*

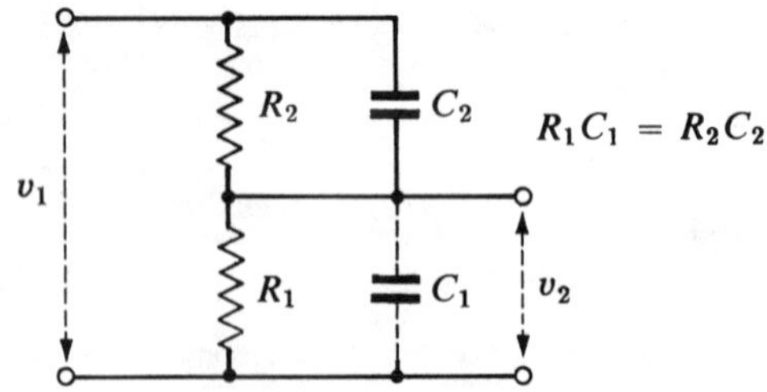

(c) *compensated voltage divider with gain equal to* $R_1/(R_1 + R_2)$

FIGURE 3.20 Voltage divider.

A calculation of the division ratio including C_1 (again assuming negligible output current) yields the result:

$$\frac{v_{\text{out}}}{v_{\text{in}}} = \frac{(R_1 \| C_1)}{R_2 + (R_1 \| C_1)} = \frac{\dfrac{R_1 X_{C1}}{R_1 + X_{C1}}}{R_2 + \dfrac{R_1 X_{C1}}{R_1 + X_{C1}}}$$

$$= \frac{R_1}{R_1 + R_2\left(1 + \dfrac{R_1}{X_{C1}}\right)} = \frac{R_1}{R_1 + R_2(1 + j\omega R_1 C_1)} \qquad \textbf{(3.49)}$$

The division ratio is seen to equal $R_1/(R_1 + R_2)$ only when the frequency is low enough such that $R_1 \ll X_{C1} = 1/\omega C_1$, or $\omega R_1 C_1 \ll 1$. And as the frequency ω increases, the division ratio becomes much smaller, i.e., there is more attenuation. The frequency at which the division ratio begins to differ appreciably from the dc value of $R_1/(R_1 + R_2)$ is approximately that frequency at which $R_1 \cong X_{C1} = 1/\omega C_1$ or $\omega = 1/R_1 C_1$. The larger R_1 or C_1, the lower the frequency at which this takes place. For example, a typical value of C_1 is 10pF, so if $R_1 = 100\,\text{k}\Omega$, then frequency components of the order of $\omega = 1/(10^5)(10^{-11}) = 10^6$ radians/sec or $f = 160\,\text{kHz}$ or higher will be seriously attenuated. In terms of pulse width τ, the principal Fourier components are from 0 to $1/\tau$, so pulses narrower than $\tau = 1/f = 1/160\,\text{kHz} = 6.3\,\mu\text{sec}$ would be seriously rounded off.

The way to avoid this rounding off is to make the division ratio a constant value independent of frequency. The capacitance C_1 cannot be eliminated, so we must add a frequency dependent impedance to the R_2 part of the divider. Because C_1 tends to short out the high frequency components, we wish to add a component to R_2 to make the impedance of the top half of the divider decrease as frequency increases. A little thought will show that adding another capacitance C_2 in parallel with R_2 will accomplish this as shown in Fig. 3.20(c). The capacitance C_2 tends to short out R_2 partially as the frequency increases.

The division ratio will be independent of frequency if the ratio of R_1 to R_2 is exactly the same as the ratio of the capacitive reactance of C_1 to C_2; that is, if:

$$\frac{R_1}{R_2} = \frac{X_{C1}}{X_{C2}} = \frac{1/\omega C_1}{1/\omega C_2} = \frac{C_2}{C_1},$$

which implies that

$$R_1 C_1 = R_2 C_2. \qquad \textbf{(3.50)}$$

If we desire a 10 : 1 division ratio, then a suitable choice could be $R_1 = 1\,\text{M}\Omega$, $R_2 = 9\,\text{M}\Omega$, $C_1 = 72\,\text{pF}$, and $C_2 = 8\,\text{pF}$. Then C_1 is the total capacitance across the output, e.g., if the stray capacitance is 10pF, an additional 62pF would be added in parallel to give a total of 72pF for C_1. Usually C_2 is made adjustable; it is commonly a small ceramic trimmer capacitor which is tuned to $R_1 C_1 / R_2$ by a screwdriver adjustment. If C_2 is too large (compared to R_1C_1/R_2), the high frequency Fourier components will be attenuated less than the low frequency ones and the output will appear slightly differentiated as shown in Fig. 3.21(a). If C_2 is too small (compared to $(R_1/R_2)\,C_1$), the high frequency Fourier components will be attenuated more than the low frequencies and the output will appear rounded as in Fig. 3.21(b).

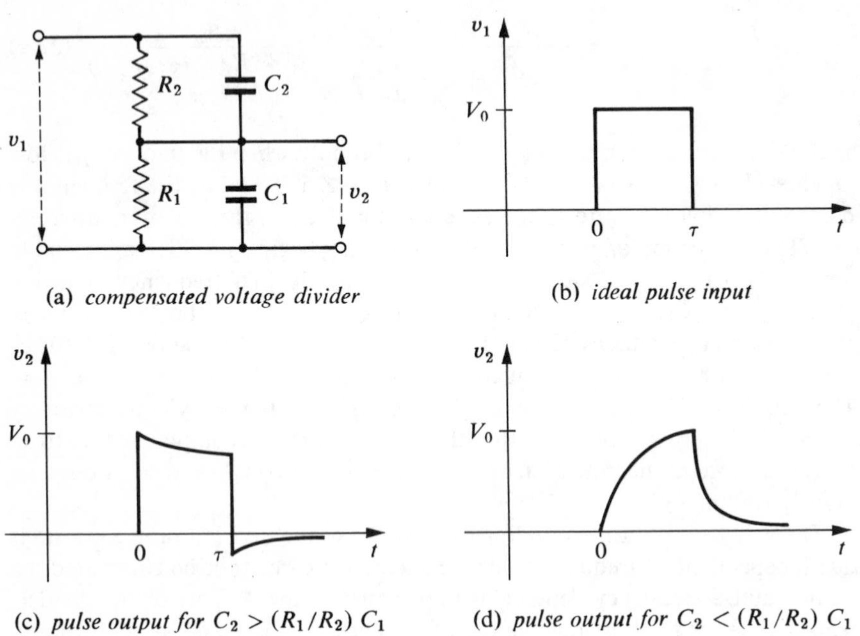

FIGURE 3.21 Pulse output from compensated voltage divider.

The principal advantage of such a compensated divider is that it presents a very high impedance to the input. It draws little current into the input and therefore loads the circuit being measured very little. It can be shown that the compensated divider is equivalent to a single *RC* parallel combination as shown in Fig. 3.22, with:

$$R_{\text{eff}} = R_1 + R_2 \qquad \text{and} \qquad C_{\text{eff}} = \frac{R_1}{R_1 + R_2}\,C_1 \tag{3.51}$$

Thus the 10 : 1 compensated divider with $R_2 = 9\,\text{M}\Omega$, $R_1 = 1\,\text{M}\Omega$, $C_1 = 72\,\text{pF}$, $C_2 = 8\,\text{pF}$ is electrically equivalent to a $10\,\text{M}\Omega$ resistor in parallel with only a 7.2pF capacitor. Not only is the dc input impedance of the compensated divider high (10MΩ), but the effective capacitance presented to the input pulse is only 7.2pF. Thus, the effective shunt capacitance of the divider as seen by the input signal is less than the original stray capacitance C_1 (10pF) present. The

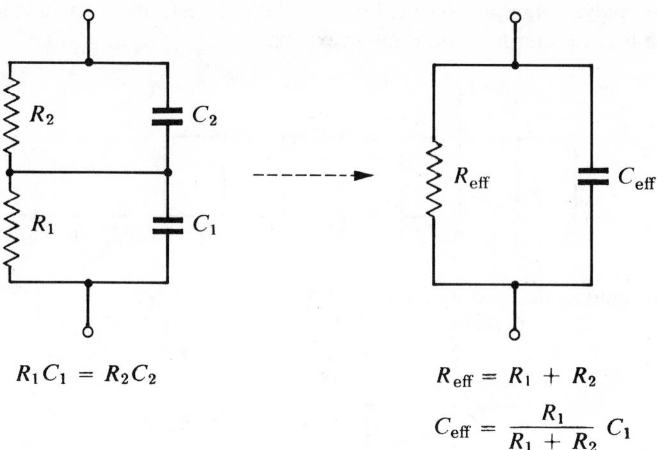

FIGURE 3.22 Compensated voltage divider equivalent.

price one pays for this is the reduced gain, but this can be overcome by feeding the output into a suitable high frequency amplifier.

Almost all measurement of fast pulses with oscilloscopes are made with such a compensated voltage divider scope probe. The input to the divider is the pulse taken from the circuit under investigation, and the output from the divider is fed directly into the oscilloscope. The oscilloscope input capacitance is usually of the order of 47 pF, and this appears across R_1. Thus, to make C_1 total 72 pF, only 25 pF need be added across R_1. The effective shunt capacitance of the divider is still only 7.2 pF which is less by a factor of about seven than the input capacitance of the oscilloscope. The only thing one must remember in using such a probe is that all signals are attenuated by a factor of ten; that is, a 2.0 V pulse in the circuit under investigation will appear as a 0.2 V pulse on the oscilloscope.

problems

1. Given the pulses shown below, determine the (a) pulse width, (b) pulse frequency or repetition rate, (c) period, (d) amplitude.

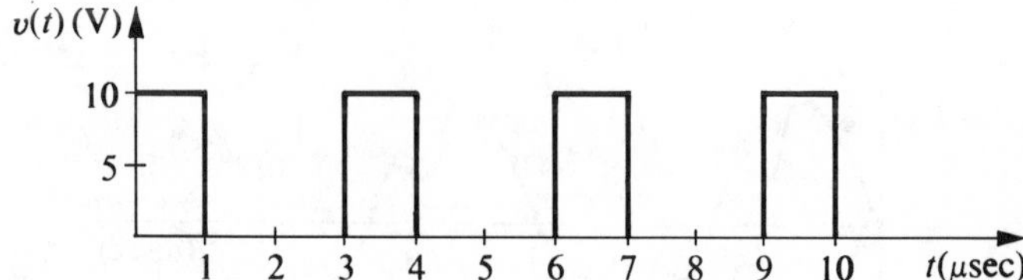

2. Explain briefly, yet clearly, in words the meaning of the Fourier analysis theorem.

3. Fourier analyze the square wave shown. Without performing any integrals, calculate the first (dc) term in the Fourier expansion.

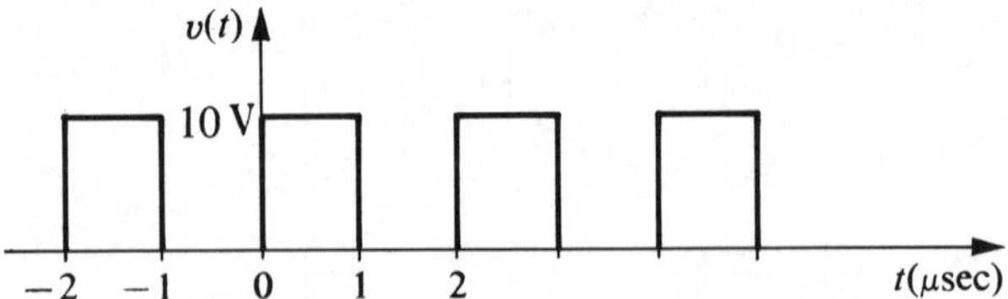

4. Fourier analyze the sawtooth wave shown.

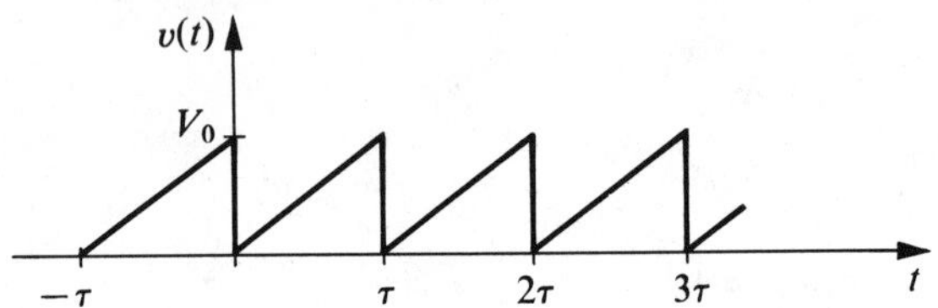

5. What approximate bandwidth would be required to amplify randomly occurring pulses of 50 nsec width?

6. Calculate how long after the switch S is closed it will take the condenser to charge up to 3 V to 3.78 V. The condenser is initially uncharged.

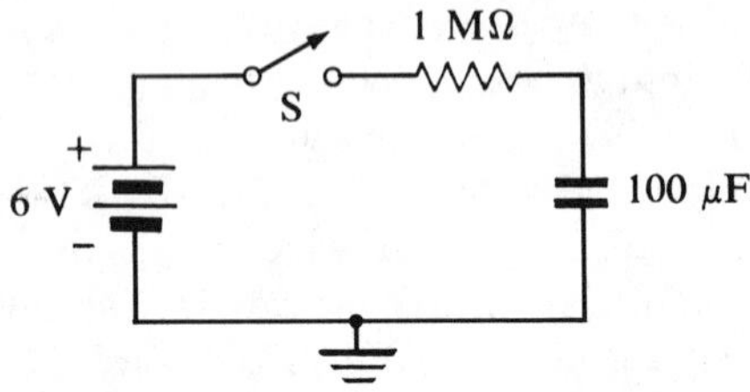

7. Prove that the rise time for the RC low-pass filter is $2.2RC$ for a perfect step function input.

8. The rise time is independent of the input pulse amplitude—true or false? The output voltage will reach a specific voltage (say +3 V) in a shorter time if the amplitude of the input step function is increased—true or false?

9. If the audio signal V_A to be carried by an AM radio station is shown below, carefully (a) sketch the modulated carrier wave. (b) Sketch the expected sidebands. (HINT: What harmonics would give this wave shape?)

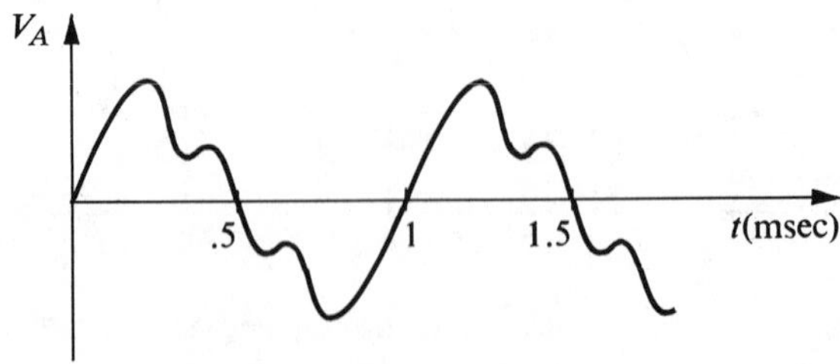

10. Compensate the voltage divider shown.

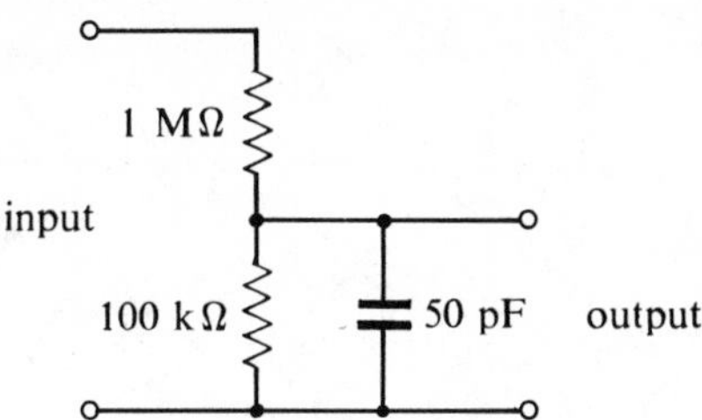

11. Sketch the approximate output from a differentiating circuit with $R = 10\text{k}\Omega$, $C = 0.01\,\mu\text{F}$ for the following inputs:

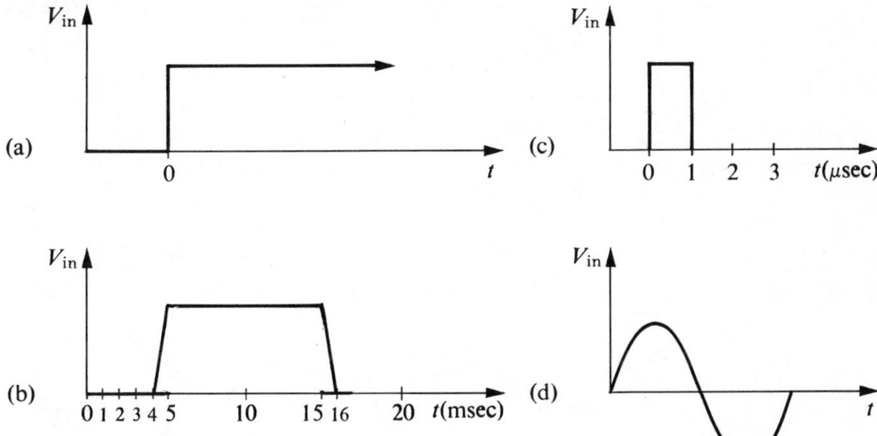

12. Sketch the approximate output from an integrating circuit with $R = 10\text{k}\Omega$, $C = 0.001\,\mu\text{F}$ for the inputs of Problem 11.

13. Derive the expression $v_{out} \cong RC\, dv_{in}/dt$ for a differentiating circuit. Be careful to state your assumptions.

14. Sketch the approximate scope trace you would see for the following input if the scope is on ac coupling.

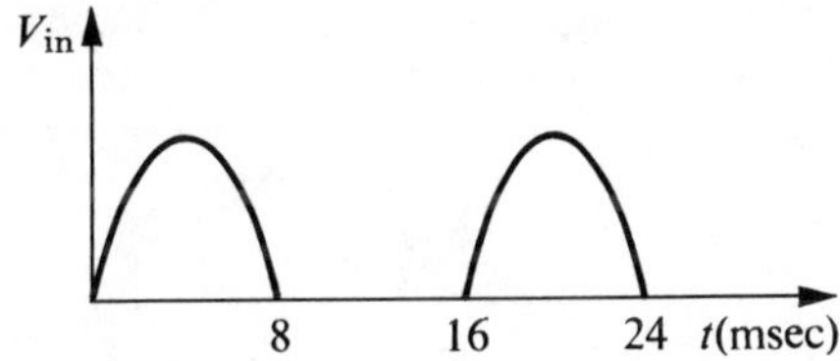

4

SEMICONDUCTOR PHYSICS

4.1 INTRODUCTION

In this chapter we will consider some of the basic features of the physics of semiconductors which will enable us to understand physically why a transistor can be made to amplify signals, why transistor parameters depend so strongly upon temperature, and the magnitudes of the various resistances, voltages, and currents in transistor circuits. Most semiconductor devices are made from either germanium or silicon crystals; hence particular emphasis will be placed upon these elements.

4.2 ENERGY LEVELS

In classical physics, energy can be thought of as the ability to do work. It has units of joules (in mks units) or ergs (in cgs units) or electron-volts. One electron volt (eV) is equal to 1.6×10^{-12} erg or 1.6×10^{-19} joule and is defined as the energy an electron gains when it falls through a potential drop (or voltage drop) of one volt. The electron volt is commonly used to measure energy in nuclear, atomic, and semiconductor physics. A useful number to remember is $1/40\,\text{eV} = 0.025\,\text{eV}$ which is the approximate average kinetic energy an atom or free electron has in thermal equilibrium at room temperature.

Classically, there is no restriction on the amount of energy an object may have. However, in quantum physics where we are dealing with very small objects such as nuclei, atoms, electrons, or molecules the energy an object may have is often restricted by the basic quantum-mechanical laws which apply in such situations. If we calculate the energy possible for a single isolated atom according to the quantum theory, we find that there are a number of sharp or discrete energy values or "levels" possible. These energy levels depend mostly upon the orbital electronic properties, and not upon the nuclear properties because the nuclear states are usually so stable. The higher levels are usually more closely spaced than the lower levels, but the main point to remember is that the energy levels are discrete. It is impossible for the atom to have any energy between these discrete energy levels, which are usually spaced of the order of electron volts apart. The energy level diagram for a hydrogen atom, for example, is given in Fig. 4.1. The horizontal axis is essentially meaningless.

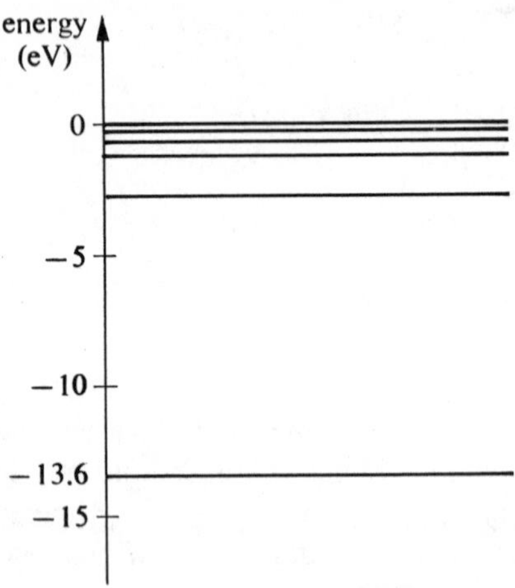

FIGURE 4.1 Energy levels for a single hydrogen atom.

Larger atoms containing more electrons in general have more complicated energy level diagrams, but the levels are still discrete with forbidden gaps between them and look basically like those in Fig. 4.1.

4.3 CRYSTALS

A crystal is a solid in which the atoms are arranged in a regular periodic way. Most elements, if extremely pure, will form crystals when cooled carefully from the liquid. The exact geometry of the atoms depends upon the chemical bonds formed among the outer or "valence" electrons of the atoms. The electrons in the inner shells ordinarily take no part in bonds between atoms, because they are so tightly bound to their own nucleus. Thus, the crystal structure depends upon the atomic number and the valence of the atom involved. Some compounds, particularly the simpler ones, also form crystals easily; for example, NaCl, KCl, GaAs, etc. If the regular, periodic arrangement of atoms is unbroken throughout the entire piece of material, then we have a perfect "single" crystal. A solid is called "polycrystalline" if it consists of a large number of smaller single crystals oriented randomly with respect to one another. In general, a substance takes on a polycrystalline form upon solidifying from the liquid, and special experimental techniques and extremely pure samples must be used to make a single crystal of an appreciable size. In fact, much of the recent progress in transistor technology has come from improvements in the purification techniques for germanium and silicon.

The regular geometrical arrangement of atoms is called a crystal "lattice." One example of a crystal is a lattice in which all the atoms occur at the corners of cubes of identical size, as shown in Fig. 4.2. Such an arrangement is called

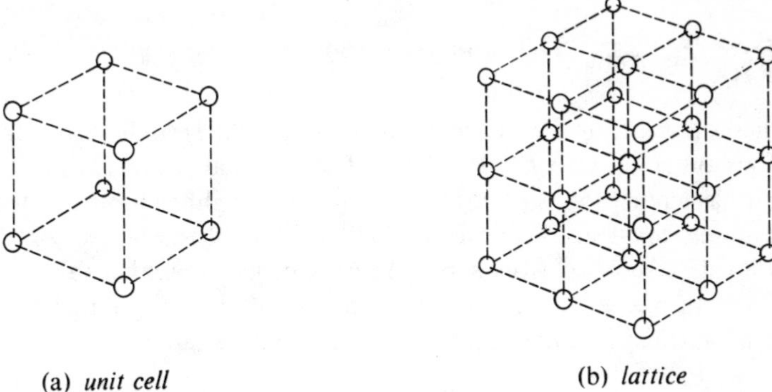

(a) *unit cell* (b) *lattice*

FIGURE 4.2 Cubic crystal lattice.

a "cubic" lattice. The cube is called a unit cell of this crystal, because the entire crystal lattice can be built up conceptually by a regular "stacking-up" of cubes.

There are many types of imperfections or defects present in any real single crystal: there can be mechanical breaks or discontinuities in the crystal structure extending over many unit cells, atoms can be missing, an extra atom can be present—in an unusual position in a unit cell, or an atom of a different element (with the "wrong" number of valence electrons) may be present in a regular lattice position. The type and number of such imperfections play an extremely important role in determining the mechanical and electrical properties of the crystal. Most transistors and diodes today are made from extremely small single crystals of either germanium or silicon, although some special purpose semiconductor devices are made from single crystals of compounds such as GaAs. The unit cell of germanium and silicon is a face-centered cubic cell, as shown in Fig. 4.3, in which the atoms occur at the 8 corners of a cube and in the centers of the 6 faces of the cube. Thus there are 14 atoms per unit cell. The length of one side of the cube is called the "lattice constant" and is 5.66 Å in germanium and 5.43 Å in silicon (1 Å $= 10^{-8}$ cm $= 10^{-10}$ m).

If several face-centered cubic cells are drawn together, as they occur in either a germanium or a silicon crystal, a very complicated interlocking structure results in which each atom has four nearest neighbors. Both germanium

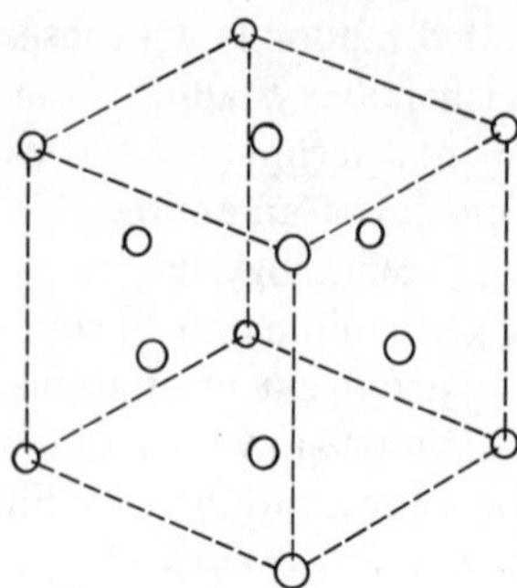

FIGURE 4.3 Face-centered cubic cell (Ge and Si).

and silicon are Group IV elements in the chemical periodic table and hence have four outer or valence electrons. The four valence electrons enter into covalent electronic bonds with the four nearest neighbor atoms in the lattice. The resulting covalent chemical bonds literally hold the lattice together, and energy must be supplied from some source in order to break a valence electron loose from such a bond. The inner electrons are much more tightly bound to the atom's nucleus and are never free to move in the lattice.

4.4 ENERGY LEVELS IN A CRYSTAL LATTICE

If we calculate, from the quantum theory, the possible energy levels for an electron of an atom in a crystal containing a large number of atoms (e.g., 10^{20} atoms for a tiny crystal comprising a transistor), we find that the allowed energy levels are very closely spaced together and that there are large forbidden energy gaps between groups of closely spaced energy levels. A group of many closely spaced energy levels is called an energy "band." The most important two bands for the purpose of understanding transistors are the "valence" band and the "conduction" band shown in Fig. 4.4. The gap between the valence and the conduction bands is 0.72 eV in germanium and 1.09 eV in silicon. Note that these energy gaps are much greater than kT which is equal to 0.025 eV at room temperature. The electron energy levels in either band are so closely spaced in energy that the band may be regarded as a continuum for most purposes.

The electrons in the conduction band are bound only very weakly to individual atoms in the lattice; hence they are essentially free to move throughout the lattice and take part in conduction of electrical current. Electrons in the valence band are those electrons which form the covalent bonds between atoms of the lattice. They are "valence electrons" in chemical terminology, hence the name "valence" band. Electrons in atomic shells inside the valence elec-

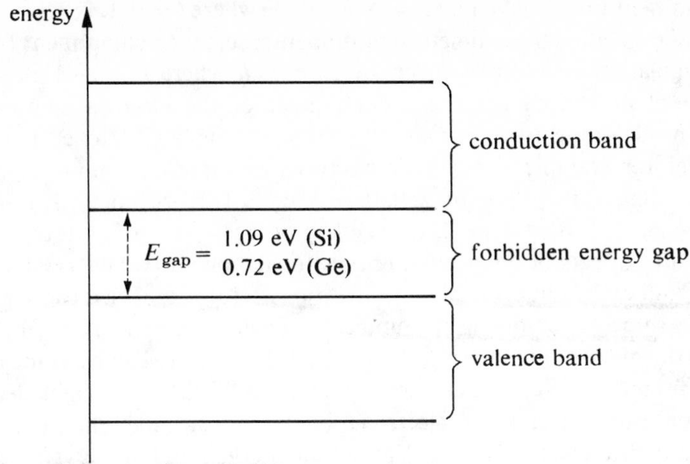

FIGURE 4.4 Energy bands in a semiconductor crystal.

trons lie in energy bands of the order of electron volts below the valence band. These electrons play no part in conduction at temperatures under several thousand degrees, and so from now on we will consider only the valence and conduction band electrons.

The physical reason for the energy gap is simply that it takes a certain amount of energy or work E_{gap} to pull an electron loose from a bond between lattice atoms and make it free to move through the crystal. Thus, the energy gap is physically the amount of energy which must be given an electron in a bond to free it from that bond. We can now see that at absolute zero, when all the electrons must fall into the bonds between the lattice atoms, all the electrons must lie in the valence band and *none* in the conduction band.

The problem of conduction in a crystal is more complicated than we have just stated, because of two quantum-mechanical laws which we have not yet considered: the Pauli exclusion principle and the Fermi-Dirac statistics.

4.5 PAULI EXCLUSION PRINCIPLE

An energy level is characterized by a set of quantum numbers which come from a quantum mechanical calculation of the physical system's allowed energies. In the simplest case of a hydrogen atom, each electron has four quantum numbers: n, l, m_l, and m_s. It can be shown from quantum mechanics, that the energy of a hydrogen atom depends only upon n, the principle quantum number according to the formula $E = -\mu e^4/2\hbar^2 n^2$. e equals the electronic charge, μ is the reduced mass of the electron and nucleus, $\hbar = h/2\pi$, where h is Planck's constant. l is called the azimuthal quantum number; the electron's orbital angular momentum L is given by $L = \sqrt{l(l+1)}\hbar$ where $l = 0, 1, 2, 3, \ldots, n - 1$. The symbol m is called the magnetic quantum number; the z component L_z of the orbital angular momentum is given by $L_z = m\hbar$ where $m = -1, \ldots, +1$. The value of m_s determines the z component of the electron spin angular momentum according to $S_z = m_s\hbar$, and $m_s = -\frac{1}{2}$ or $+\frac{1}{2}$. Hence the $n = 2$ energy level, for example, can contain electrons with quantum numbers n, l, m_l, and m_s equaling $2, 0, 0, +\frac{1}{2}$ or $2, 0, 0, -\frac{1}{2}$ or $2, 1, 1, +\frac{1}{2}$ or $2, 1, 1, -\frac{1}{2}$ or $2, 1, 0, +\frac{1}{2}$ or $2, 1, 0, -\frac{1}{2}$ or $2, 1, -1, +\frac{1}{2}$ or $2, 1, -1, -\frac{1}{2}$, a total of eight possible states or sets of quantum numbers for the one $n = 2$ energy level.

The Pauli exclusion principle says that no two (or more) electrons can have exactly the same set of quantum numbers. Thus, in an energy level characterized by the quantum numbers l, l, m_l, m_s, if $n = 2$ there can be at most eight electrons, no more. We say that the $n = 2$ level is "filled" with eight electrons. A hydrogen atom never has 8 electrons, but a hydrogenlike energy level can exist in many heavy atoms. In the case of an energy band in a crystal, each of the closely spaced energy levels has its own set of quantum numbers. Electrons in the crystal will naturally tend to fill up the lowest energy levels first, but always subject to the restriction that no two electrons can occupy the same state, i.e., no two can have the same set of quantum numbers.

The most important implication of the exclusion principle is that a completely full energy band of electrons cannot conduct. This can be seen by realizing that conduction implies an electron is raised from a lower to a slightly higher energy state. However in a completely full band the higher energy state is already occupied by an electron, and the exclusion principle prevents a second electron from jumping up to this higher state. Thus in order for conduction to occur, an energy band must be only partially full; there must be empty higher energy states for the electrons to jump up to. Another implication of the exclusion principle is that electrons for conduction come mainly from near the top of the partially filled band because these electrons need only a small increase in energy to jump up to an unfilled level in the band.

4.6 FERMI-DIRAC STATISTICS

In order to understand the electrical behavior of a semiconductor, we clearly must know how many electrons are in the conduction band and how many are in the valence band, because only those electrons in the conduction band can move and create a current flow in response to an applied voltage. The basic problem is then to calculate the distribution of the electrons in the crystal among the various allowable energy levels. We have already stated that a quantum-mechanical calculation of the energy levels in a crystal lattice yields the result that the allowable energy levels lie in two bands, the conduction and the valence band, separated by a forbidden energy region or gap. The next thing to calculate is the probability $F(E)$ that a given energy level E somewhere in a band is actually occupied by an electron. If we let $N(E)\,dE$ equal the number of electrons per unit volume between E and $E + dE$, then we write $N(E)\,dE$ as the product of the number $\rho(E)\,dE$ of allowable or available energy states per unit volume in the range E to $E + dE$ [$\rho(E)$ is often called the "density of states"], times the probability $F(E)$ that the energy level E is actually occupied.

$$N(E)\,dE = \rho(E)\,dE\,F(E) \tag{4.1}$$

The $\rho(E)$ and therefore $N(E)\,dE$ will be zero in the energy gap and positive in the conduction and valence bands. The exact form of $\rho(E)\,dE$ comes from the quantum-mechanical solution to the problem of calculating the allowable energy levels for electrons in a periodic crystal lattice. The result is:

$$\rho(E) = \frac{2^{7/2}m^{3/2}\pi}{h^3}E^{1/2} \tag{4.2}$$

for electrons not too near the top of the band. $E = 0$ represents the energy of the bottom of the band;* m is the mass of the electron; h equals Planck's constant.

The Fermi function $F(E)$ tells us the probability that the energy level E is actually occupied by an electron. It seems intuitively reasonable that $F(E)$ should depend upon temperature, because the higher the temperature the more thermal energy is available to be distributed among the electrons and the nuclei of the lattice atoms. And, the more energetic the electrons the greater the probability that a higher energy state is occupied. At absolute zero when there is zero** thermal energy available, all vibrational motion of the atoms in the lattice should cease; and all the electrons should be lying in the lowest possible energy levels, subject only to the restriction of the Pauli exclusion principle. Thus we would expect the Fermi function $F(E)$ to be temperature dependent; and, at absolute zero, to be a constant value from $E = 0$ up to some maximum energy which is the energy of the highest filled state.

$F(E)$ is calculated by an involved statistical argument. The problem is to calculate the probability that an energy level E is occupied subject to four conditions: (1) that the total number of electrons is constant (conservation of charge), (2) that the total energy of all the electrons remains constant (conservation of energy), (3) that no two electrons can be distinguished from one another, and (4) that no two (or more) electrons can lie in the same quantum state. This total energy depends on the temperature, which is assumed to be constant, i.e., the particles are assumed to be in thermal equilibrium. Once the probability is known the maximum probability is calculated, because we know nature in

* Near the top of the band electrons undergo Bragg reflections from atoms in the lattice, and the net result is that fewer energy levels are allowable than are predicted from equation (4.2).

** Actually at absolute zero, there is a nonzero, fixed minimum value of energy available to the electrons and atoms; this is called the "zero-point" energy and comes from a careful consideration of the uncertainty principle of quantum mechanics.

equilibrium always assumes the most probable configuration. There are clearly many ways to distribute a fixed total amount of energy among N electrons; this calculation gives us the most *likely* or most probable distribution of electrons among the various energy levels.

The net result of the calculation is that the Fermi function is given by:

$$F(E) = \frac{1}{e^{\frac{E-E_F}{kT}} + 1} \tag{4.3}$$

where E_F is a constant energy called the Fermi energy, k is Boltzmann's constant equal to 1.38×10^{-16} ergs/deg $= 1.38 \times 10^{-23}$ joule/deg, and T is the temperature in degrees Kelvin. In order to see the physical meaning of $F(E)$ let us graph it for various temperatures; see Fig. 4.5. At absolute zero, the

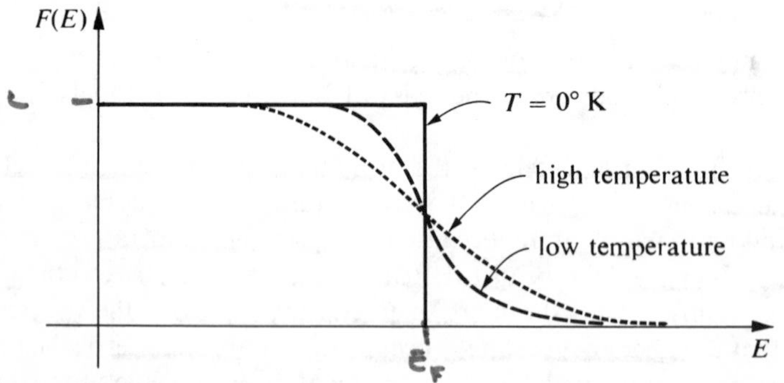

FIGURE 4.5 The Fermi function versus E.

Fermi function $F(E)$ is 1.0 from $E = 0$ up to $E = E_F$, the Fermi energy, and for energies above E_F, $F(E) = 0$. This says that at absolute zero all the allowable energy levels from 0 up to E_F are filled and that no levels above E_F are filled, which is just what we expect physically. The Fermi energy E_F represents the highest energy level filled at absolute zero.

The physical meaning may become clearer if we refer to Fig. 4.6 which is a simple energy-level picture of a hypothetical solid with evenly spaced, allowed energy levels. The density of states $n(E)$ gives the number of these energy levels per unit energy. At absolute zero, the electrons have cascaded down to the lowest energy states possible subject to the restriction of the Pauli exclusion principle; namely, that no more than one electron can occupy any one energy level with a specific set of quantum numbers. In Fig. 4.6 we have assumed that the spin quantum number does not affect the energy; i.e., the energy does not depend upon the quantum number m_s. Thus two electrons, one with spin "up" ($m_s = +\frac{1}{2}$) and one with spin "down" ($m_s = -\frac{1}{2}$), can occupy the same energy level. The Fermi energy will be at the topmost occupied energy level at 0°K. Notice that the only way to change the Fermi energy would be to add more

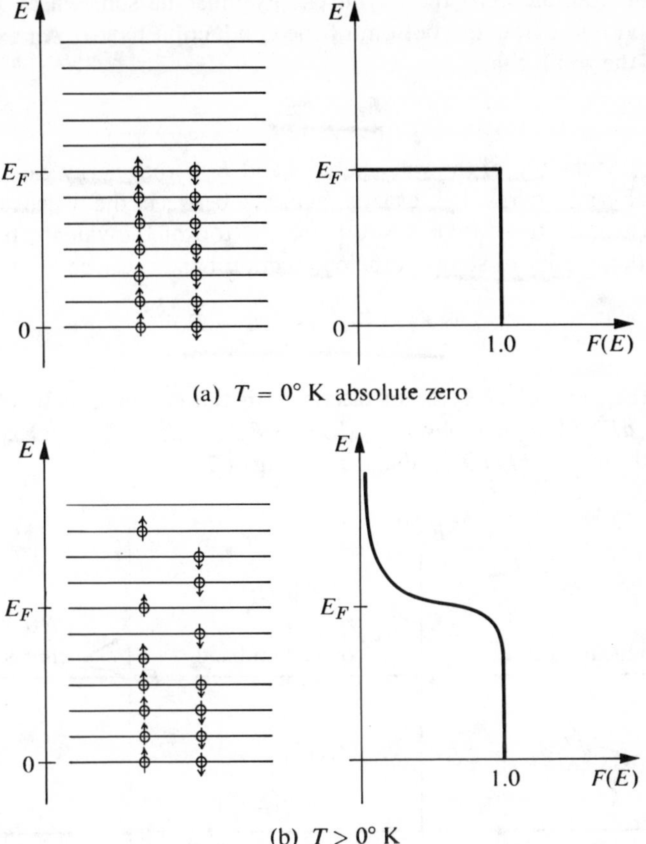

FIGURE 4.6 Simple energy-level picture and Fermi function.

electrons or to change the spacing of the energy levels. Figure 4.6(b) shows the distribution of the electrons at a temperature above absolute zero. The "density" of the electrons per unit energy increases for energies above the Fermi level as the temperature is raised.

4.7 ELECTRON ENERGY DISTRIBUTION

Before we can multiply the density of states $\rho(E)\,dE$ times the Fermi function to obtain the number of electrons actually between E and $E + dE$, we must first know the Fermi energy E_F. If we note that at absolute zero all the electrons in the lattice will be bound in the covalent bonds between the atoms, we see that at absolute zero the conduction band must be empty and the valence band full. Therefore, since the Fermi level is the maximum energy an electron can have at absolute zero the Fermi energy must be somewhere above the valence band and below the bottom of the conduction band. An exact treatment gives the result that

$$E_F = \tfrac{1}{2} E_g \tag{4.4}$$

where $E = 0$ is the top of the valence band and E_g is the energy gap. In other words the Fermi energy lies exactly halfway between the valence and the conduction band. If we take $E = 0$ to be the top of the valence band, then from (4.2) the density of states in the conduction band must be

$$\rho(E) = \frac{2^{7/2} m^{3/2} \pi}{h^3} (E - E_g)^{1/2}. \tag{4.5}$$

Thus the number $N(E)\,dE$ of electrons in the conduction band between E and $E = dE$ will be given by $N(E)\,dE = \rho(E)F(E)\,dE$. $F(E)$, $\rho(E)$, and $N(E)$ are graphed on an energy-level diagram in Fig. 4.7.

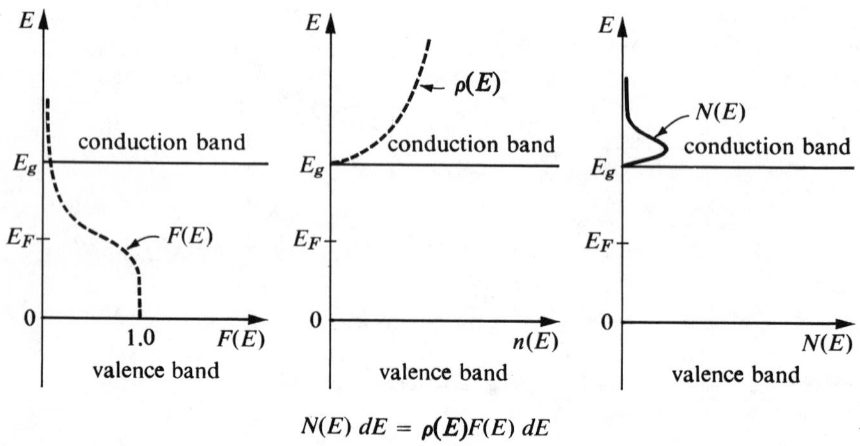

FIGURE 4.7 Density of electrons in conduction band.

Substituting for $\rho(E)$ and $F(E)$ we obtain:

$$N(E)\,dE = \frac{2^{7/2} m^{3/2} \pi}{h^3} (E - E_g)^{1/2} \frac{1}{e^{\frac{E-E_F}{kT}} + 1}\,dE. \tag{4.6}$$

For pure silicon or germanium with E in the conduction band, $E - E_F \geqq E - E_{g/2} \geqq E_{g/2}$ which is much larger than kT. Hence,

$$e^{\frac{E - E_F}{kT}} + 1 \cong e^{\frac{E - E_F}{kT}}.$$

So we may write

$$N(E)\,dE \cong \frac{2^{7/2}m^{3/2}\pi}{h^3}(E - E_g)^{1/2}\,e^{-\frac{E-E_F}{kT}}\,dE. \tag{4.7}$$

The total number N_{cb} of electrons in the conduction band can be obtained by integrating $N(E)\,dE$ from the bottom of the conduction band to the top which can be taken to be $E = \infty$ for all practical purposes. The result is:

$$N_{cb} = \int_{E_g}^{\infty} N(E)\,dE = \frac{2^{5/2}(m\pi kT)^{3/2}}{h^3}\,e^{-E_g/2kT} \tag{4.8}$$

where we have set $E_F = \frac{1}{2}E_g$. It is useful to rewrite N_{cb} as:

$$N_{cb} = AT^{3/2}e^{-E_g/2kT} \tag{4.9}$$

where $A \equiv \dfrac{2^{5/2}(m\pi k)^{3/2}}{h^3} = 4.6 \times 10^{15}\,\dfrac{\text{electrons}}{\text{cm}^3}\,(\text{deg})^{-3/2}$. The most important thing to remember is that the total number of electrons in the conduction band depends upon the temperature according to $T^{3/2}$ times $e^{-E_g/2kT}$. The following table gives some numerical values for various temperatures. Note that the exponential factor $e^{-E_g/kT}$ is much smaller for silicon because of the larger value of E_g. Therefore, the number of electrons thermally excited to the conduction band is much less for silicon than for germanium.

TABLE 4-1. N_{cb} EQUALS THE NUMBER OF THERMALLY EXCITED ELECTRONS/CM³ IN THE CONDUCTION BAND FOR VARIOUS TEMPERATURES IN SILICON AND GERMANIUM

	T	$T^{3/2}$	$e^{-E_g/2kT}$	$AT^{3/2}e^{-E_g/2kT} = N_{cb}$
	20°C = 293°K	5000	6.21×10^{-7}	1.51×10^{13}
Ge	30°C = 303°K	5280	10.3×10^{-7}	2.65×10^{13}
($E_g = 0.72$ eV)	40°C = 313°K	5560	16.2×10^{-7}	4.39×10^{13}
	50°C = 323°K	5800	28.8×10^{-7}	8.15×10^{13}
	20°C = 293°K	5000	3.52×10^{-10}	0.858×10^{10}
Si	30°C = 303°K	5280	7.09×10^{-10}	1.82×10^{10}
($E_g = 1.1$ eV)	40°C = 313°K	5560	13.56×10^{-10}	3.67×10^{10}
	50°C = 323°K	5800	26.1×10^{-10}	7.36×10^{10}

Also notice that the exponential factor $\exp(-E_g/2kT)$ is a much stronger function of temperature than $T^{3/2}$. Therefore, the gap energy E_g mainly determines the number of electrons in the conduction band for pure germanium and silicon.

From the graphs in Fig. 4.8 we see that, starting at 20°C, the electron density in the conduction band N_{cb} doubles for a 13°C temperature rise for ger-

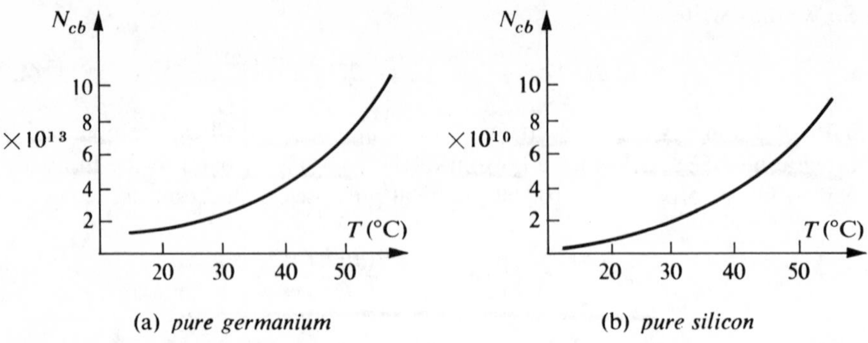

FIGURE 4.8 Conduction band electron density N_{cb} as a function of temperature.

manium; and for silicon, N_{cb} doubles for only an 8°C rise. However, there are far fewer electrons in absolute number in silicon for the same temperature.

4.8 CONDUCTION IN SEMICONDUCTORS

The Fermi level in pure semiconductors lies halfway between the top of the valence band and the bottom of the conduction band. Hence at absolute zero all the electrons are in the valence band, which is then full and thus can carry no current. At higher temperatures some electrons acquire enough thermal energy to break loose from a valence bond in the lattice, thus jumping from the valence band up to the conduction band. The conductivity of a semiconductor then increases with increasing temperature.

In a metal, the Fermi level lies *in* the conduction band, which is only partially full. Hence the metal is a good conductor at any temperature because of the many empty energy levels in the conduction band above the filled levels. There is no energy gap that the electrons must jump in order to get into the conduction band.

In a crystalline insulator, the energy-level picture is qualitatively the same as for a semiconductor, but the energy gap is much larger. For example, in diamond the energy gap is 5 eV which is so large that an enormous temperature (approximately 60,000°C!) is required to thermally excite an electron from the valence band all the way up to the conduction band where it may then contribute to conduction. (See Fig. 4.9.)

If we are to use semiconductors as practical components in electrical circuits, then they must be able to carry a reasonable amount of current. From the preceding section the total number, N_{cb}, of electrons per unit volume in the conduction band of a pure semiconductor was given by

$$N_{cb} = AT^{3/2}e^{-E_g/2kT}$$

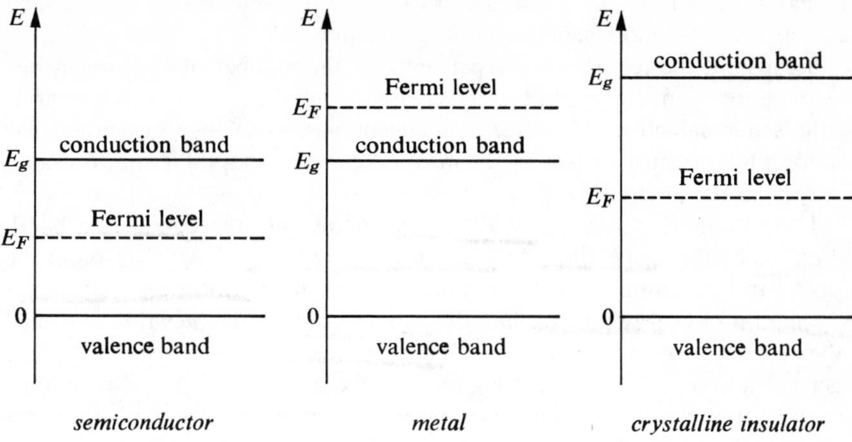

FIGURE 4.9 Location of Fermi level in various materials.

Substituting in numerical values gives us:

1. $N_{cb} = 1.5 \times 10^{13}$ electrons/cm^3 for Ge at room temperature.
2. $N_{cb} = 8.6 \times 10^{9}$ electrons/cm^3 for Si at room temperature.

Are there enough electrons in either pure germanium or pure silicon to carry a practical amount of current? The answer is no, for the following reasons. The maximum current I which can be passed is the number of electrons flowing per second multiplied by the charge per electron. The number flowing per second is the number per unit volume times the area times the speed. Thus, $I = N_{cb} Aev$ where e is the electronic charge. It can be shown that the average speed v of electrons flowing in germanium is of the order of 4×10^4 cm/sec for a 10 V/cm electric field (voltage gradient) applied.* If we assume a generous cross-sectional area of 1 mm × 1 mm, then the current I is:

$$I = N_{cb}Aev = (1.5 \times 10^{13}\text{cm}^{-3})(0.1 \text{ cm})^2(1.6 \times 10^{-19}\text{C})(4 \times 10^4\text{cm sec}^{-1})$$
$$I = 9.6 \times 10^{-4}\text{C/sec}$$
$$I = 960\mu\text{A} = 0.96\text{mA}.$$

A value of 0.96mA is too small a current for many practical applications. In addition, the actual current will be less than that calculated because not every electron in the conduction band will contribute to conduction. The current passed by a silicon cube is even less because the number of electrons in the

* The speed is calculated from the electron "mobility" μ which is defined as the electron drift speed per unit electric field.

$$\mu \equiv \frac{v \text{ cm/sec}}{E \text{ V/cm}}$$

	Ge	Si	
$\mu =$	3600	1200	for electrons
$\mu =$	1700	250	for holes

conduction band is smaller, and also because for a given applied electric field the electrons move more slowly than in germanium.

Because of the temperature dependence of the number of electrons available for conduction, one might try to increase the number of electrons by warming the semiconductor. However, a quick numerical calculation shows that even for a temperature of 100°C, the maximum current passed is still too small for practical use in circuits.

There is another way of exciting electrons up into the conduction band, and that is by the absorption of electromagnetic radiation. An electron in the valence band can jump up to the conduction band if it absorbs an electromagnetic photon of energy greater than the gap, $h\nu \geqq E_g$, where ν is the frequency of the photon. This process is called "photoconduction." However, this is not in general a practical way of getting appreciable numbers of electrons up into the conduction band, although some devices ("photocells") are made to detect light using this process.

The practical way of increasing the number of electrons in the conduction band so that the semiconductor can carry a reasonable amount of current is to effectively add more electrons by introducing "impurity" or "donor" atoms to the lattice. The process of adding impurity atoms is called "doping" the crystal. Consider a germanium lattice at absolute zero. All the valence electrons are locked in covalent bonds between the atoms, so none is available for conduction. Each atom has four valence electrons which bond to the four nearest atoms in the lattice. If, now, we introduce a donor atom with *five* valence electrons in a regular lattice site in place of a germanium atom, then there clearly will be one electron left over, i.e., not bound in a covalent bond with the four nearest germanium neighbors. This one electron is only very *loosely* bound to its atom, and therefore can easily be pulled loose. Once freed, it can migrate through the crystal under the influence of an applied electric field and contribute to conduction. In other words, only a small amount of energy is necessary to free this fifth electron from its donor atom and raise it up into the conduction band.

On the energy-level diagram for germanium, the fifth electron when loosely attached to its donor atom then lies in an energy level in the gap only slightly below the bottom of the conduction band (see Fig. 4.10). The concentration of these donor atoms is such that they are spaced of the order of 30 Å apart in the lattice. Thus, each donor atom has four germanium atoms as its nearest neighbors in the lattice. Notice that the donor atom, once it loses its fifth electron, is a positively charged ion and is still locked in the crystal lattice by the four remaining valence electrons—being bonded to the adjacent four germanium atoms. Thus, it is only the electrons, not the positive ions, which contribute to conduction. The atom which gives up its electron is called a "donor" atom, because it *donates* an electron to the conduction band. The donor energy level in germanium lies only about 0.01 eV below the conduction band, so at room temperature (300°K) where the average thermal agitation energy $kT = 0.025$ eV, almost all the donor atoms are ionized and contribute an electron to the con-

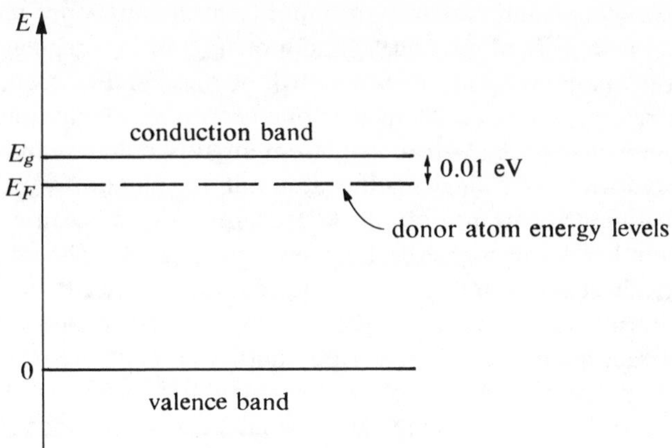

FIGURE 4.10 n-type semiconductor energy-level diagram.

duction band. A typical concentration of donor impurity atoms is $10^{16}/\text{cm}^3$, so there will be of the order of 10^{16} electrons/cm^3 in the conduction band from the ionized donor atoms. Thus a current of the order of amperes can be carried by such a semiconductor. For every electron in the conduction band which has been thermally excited up from the valence band, there will be of the order of 10^3 electrons in the conduction band from the donor atoms.

In the conduction band of germanium at 300°K, there will thus be roughly 1.5×10^{13} electrons/cm^3 from the valence band and about 10^{16} electrons/cm^3 from the ionized donor atoms. There will be 10^{16} positively charged donor atoms/cm^3 fixed in the lattice. There will also be 1.5×10^{13} singly positive ionized germanium atoms/cm^3 in the valence band, because breaking one valence bond and lifting one electron up to the conduction band must leave behind exactly one singly ionized germanium atom. This positively ionized germanium atom is fixed in the lattice, but it can attract an electron from an adjacent germanium atom. Thus, the positive charge appears to move in the opposite direction from the motion of the electron it attracted. These positive vacancies or charges are called "holes" and move in the valence band quite freely, although not quite so rapidly as do electrons in the conduction band. A hole may be thought of as the absence of an electron, or something like a bubble in a fluid.

If an external electric field is applied to such a doped semiconductor, most of the current flow will be due to the electrons in the conduction band, and only roughly *one part in* 10^3 will be due to the positive holes in the valence band. Because the current is primarily carried by the negatively charged electrons, the semiconductor is called an "n-type" semiconductor. The electrons are called the "majority" carriers, and the positive holes are called the "minority" carriers. Any impurity atom having five valence electrons, such as arsenic, antimony, or phosphorus, will produce an n-type semiconductor when added to either pure germanium or pure silicon.

If, on the other hand, we add an impurity atom with only *three* valence electrons such as boron, aluminum, gallium, or indium in a regular lattice site in place of a germanium atom, then there will be one too few valence electrons to make the required four bonds to the four nearest germanium atoms. The impurity atom will pull an electron over from an adjacent germanium atom to fill the empty bond, thus leaving the adjacent germanium atom positively charged. The impurity atoms are called "acceptor" atoms because they accept electrons. In other words a positive hole has been created. The impurity atom is now negatively charged, but it remains locked in the lattice by its four bonds to adjacent germanium atoms. The positively charged adjacent germanium atom may attract a valence electron from another germanium atom, and then the positive charge is transferred to the new atom, and so on. In this way, the positive charge or hole will move through the lattice as the valence electrons are attracted from one atom to another. If the concentration of the impurity atoms is of the order of $10^{16}\,\text{cm}^{-3}$, then the hole concentration will greatly exceed the concentration of conduction band electrons which have been thermally excited across the energy gap. To give some numbers for germanium, there will be 10^{16} holes/cm^3 in the valence band from the impurity atoms, 1.5×10^{13} electrons/cm^3 in the conduction band, and 1.5×10^{13} holes/cm^3 in the valence band from thermally broken Ge–Ge bonds. Thus conduction will take place primarily due to motion of the positively charged holes in the valence band. Hence, this type of doped semiconductor is called a "p-type" semiconductor. The holes are called the *majority* carriers, and the electrons are called the *minority* carriers.

On the energy-level diagram of germanium (see Fig. 4.11), the acceptor atoms have energy levels about 0.01 eV above the top of the valence band. When a valence electron from a neighboring germanium lattice atom jumps over to the acceptor atom, this corresponds to an electron jumping up from the valence band to the acceptor energy level. This process is very probable at

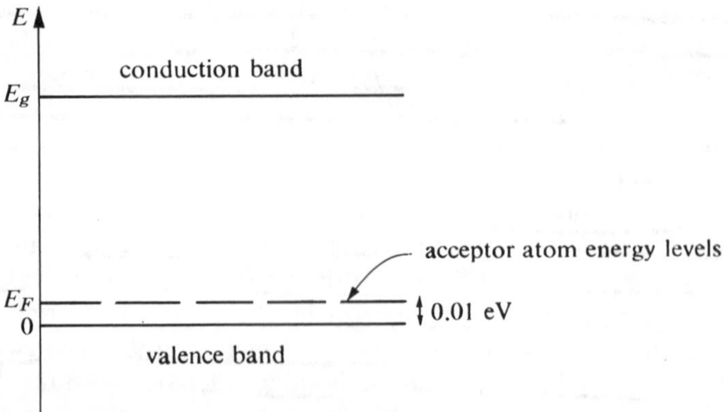

FIGURE 4.11 p-type semiconductor energy-level diagram.

room temperature, because the average thermal agitation energy $kT = 0.025$eV is greater than the energy the electron needs to make the jump.

The presence of the impurity atoms, either donor or acceptor, radically changes the value of the Fermi energy. In n-type material, the Fermi level is just above the donor atom energy levels and below the bottom of the conduction band. This position follows from the fact that the Fermi energy physically means the maximum occupied energy level at absolute zero. At absolute zero the valence band and all the donor levels must be filled, because as the crystal is cooled to absolute zero all the electrons in the conduction band must fall down to the lowest possible energy levels, thereby filling the donor levels and the valence band. In p-type material the Fermi level can be shown to be below the acceptor energy levels and above the valence band. A more refined treatment shows that the Fermi level is slightly temperature dependent. An increase in temperature will slightly lower the Fermi level in n-type material and raise the Fermi level in p-type material.

It is worth pointing out explicitly that if the electric field E applied to a semiconductor points from left to right as shown in Fig. 4.12, then electrons

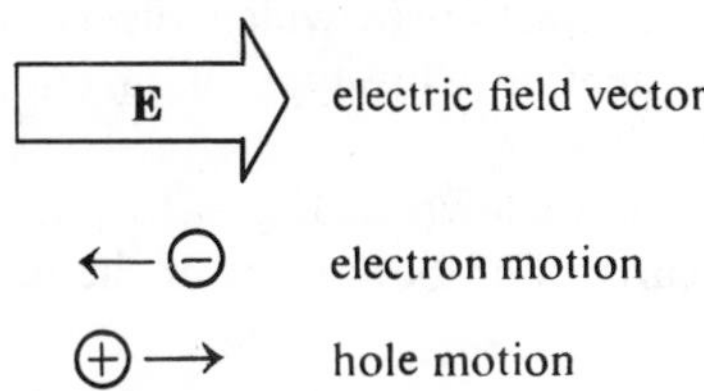

FIGURE 4.12 Electron and hole motion.

will move from right to left against the electric field and holes will move from left to right. A positive hole moving from left to right actually involves electrons moving from right to left, but from now on we will speak of the holes as if they were positively charged particles capable of moving in the direction of the applied electric field. It can be shown that one can treat the motion of the holes exactly by treating them as positively charged particles with an effective mass slightly larger than the electron mass. The larger hole mass simply means that for a given applied electric field, a hole will accelerate less rapidly than will an electron.

To sum up: In an n-type semiconductor, the impurity or donor atoms have one more valence electron than the atoms comprising the crystal lattice. These donor atoms readily donate their extra valence electron to the conduction band, thus producing mobile electrons in the conduction band to carry current and positively charged donor atoms or ions locked in the lattice. In a p-type semiconductor, the impurity or acceptor atoms have one less valence electron than the atoms comprising the lattice. These acceptor atoms readily attract electrons up out of the valence band, thus producing mobile positive holes in the valence band to carry current and negatively charged acceptor atoms or ions locked in the lattice. In both n- and p-type semiconductors, at room temperature, there are always present a few ($10^{13}\,cm^{-3}$ in Ge and $10^{10}\,cm^{-3}$ in Si) holes in the valence band and electrons in the conduction band caused by thermal breaking of lattice bonds.

4.9 p-n JUNCTIONS

If a p-type semiconductor is joined to an n-type semiconductor to form a "good" junction, that is, a junction at which there are few breaks or imperfections in the lattice structure, then this junction will act as a rectifier—it will conduct current readily in one direction and only very poorly in the other direction. Such a junction cannot be formed by merely pressing together a p-type and an n-type semiconductor; this procedure would produce a poor junction with gaps at the junction larger than the lattice spacing. A good junction is usually made by changing over the type of impurity from p- to n-type while the crystal is being grown. We will consider a junction in which the type of impurity changes abruptly as we cross the junction; this type of junction is called an abrupt junction.

An abrupt p-n junction is shown in Fig. 4.13. Let us first consider the behavior of the majority carriers on both sides of the junction. Majority car-

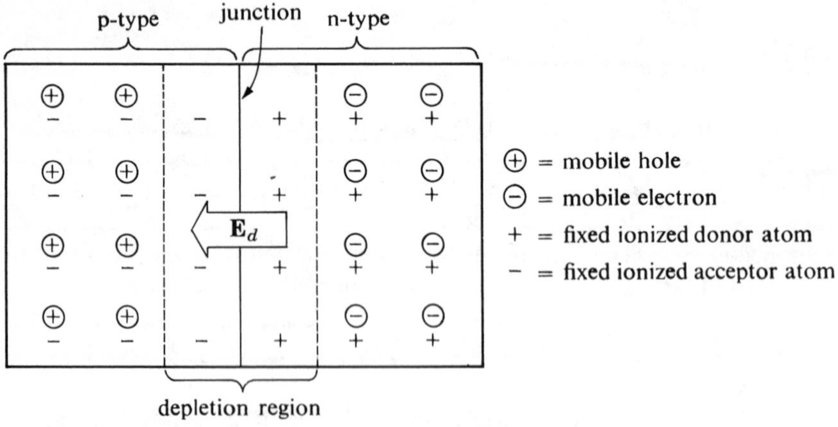

FIGURE 4.13 p-n Junction (only majority carriers shown).

riers are mobile and are continually diffusing around in the lattice due to thermal motion. If holes in the p-type material diffuse away from the junction back into the p-type material, then a concentration gradient will build up, thus tending to diffuse the holes back toward the junction. Diffusion always results in a flow from regions of higher concentration towards regions of lower concentration. Similarly, if electrons in the n-type material diffuse away from the junction the concentration gradient thus established will tend to diffuse the electrons back towards the junction. Near the junction some holes will diffuse across the junction from the p-type into the n-type material where they will recombine with the mobile electrons of the n-type material. Similarly some electrons will diffuse across the junction from the n-type into the p-type material and will recombine with the mobile holes of the p-type material. Recombination of a hole and an electron merely means that the electron drops from the conduction band down into the hole in the valence band. Notice that after recombination neither the electron nor the hole can conduct, because the electron is no longer in the conduction band, and the valence band hole is filled.

The result of this recombination is the formation of a thin region at the junction where there are essentially no mobile charge carriers. This is called the "depletion region" because the mobile charge carriers have been depleted here. The only charges remaining in the depletion region are the ionized acceptor and donor impurity atoms which are fixed in the crystal lattice, and of course the electrons in the filled valence band. These fixed charges produce an electric field $\mathbf{E}_d$ in the depletion region pointing from the n-type toward the p-type material. This field tends to sweep any electrons near the junction back toward the n-type and any holes back toward the p-type material. Thus, the depletion region has no free charge carriers, and the electric field produced tends to keep the depletion region free of any mobile charge carriers which are either created in it by thermal excitation or which may diffuse into it. In effect, the depletion region is a thin slab of insulator sandwiched between the p- and n-type material. A typical thickness d for the depletion region is of the order of one micron which is 10^{-4}cm or 10,000 Å. If the doping of the p- and n-type material is equal, that is, if the number of acceptor atoms per cm^3 [p] equals the number of donor atoms per cm^3[n], then the depletion region extends equally into the p- and n-type material on either side of the p-n interface. However, if p $\neq$ n, then the depletion region extends unequally into the p- and n-type material. It can be shown that $d_n/d_p = [p]/[n]$, where d_n is the depletion region thickness in the n-type material and d_p is the depletion region thickness in the p-type material. The total depletion region thickness $d = d_n + d_p$. This result is reasonable, because the electrons and holes *diffuse* across the p-n interface and then recombine. If there are many more donor atoms per cm^3 in the n-type material than acceptor atoms per cm^3 in the p-type material ($[n] \gg [p]$), then the electrons diffusing into the p-type material will have to diffuse a long way before they all recombine with the holes. Thus we expect a large d_p. Conversely, the few holes from the p-type material which diffuse over into the n-type material very quickly recombine with the plentiful electrons there; thus a small d_n.

The effective voltage V_c developed between the p- and n-type materials is called the "contact potential" and, if we assume parallel plate geometry, is given by $V_c = E_d d$ where d is the thickness of the depletion region. V_c is about 0.2 V for a germanium junction and 0.5 V for a silicon junction. The electric field in the depletion region points from the n- toward the p-type material. This field thus tends to keep the majority carriers *out* of the depletion region, but notice that any *minority* carriers thermally generated in the depletion region will be swept across the depletion region by this same electric field.

When equilibrium is established for an isolated p-n junction, there is no net current flowing across the junction. For every majority hole from the p-type material diffusing across against the electric field to the n material, there will be a minority hole from the n-type material accelerated back by the electric field to the p-type material. A similar two way flow will occur for electrons.

If a battery is connected to a p-n junction with a polarity to make [see Fig. 4.14(a)] the p-type material negative with respect to the n-type material, only a very small current will flow and the junction is said to be *reverse* biased.

It is only in the depletion region that the p-n junction has a high resistance due to the absence of mobile charge carriers; hence the battery will apply an electric field $\mathbf{E}_B$ across the depletion region only. In the reverse bias configuration this electric field from the battery $\mathbf{E}_B$ is in the same direction as $\mathbf{E}_d$ and will tend to pull the mobile charge carriers away from the junction, thus increasing the thickness of the charge depletion region which acts like an insulating slab, and little current will flow through the junction. Actually the preceding argument applies only to the majority charge carriers. The minority carriers (electrons in the p-type and holes in the n-type) will be attracted *toward* the junction by the electric field introduced by the battery. Thus at the junction the minority carriers will recombine and the junction will pass current. More minority carriers are continually being created by thermal excitation on both sides of the junction, so this current passed due to the minority carriers will continue to be passed. This is called the reverse current. In a germanium junction at room temperature the reverse current is on the order of several microamperes (10^{-6} A) while in a silicon junction it is on the order of ten nanoamperes (10^{-8} A). As the number of minority carriers created depends upon thermal excitation across the energy gap, we expect the reverse current of a p-n junction to be temperature dependent. At room temperature a 10°C increase in temperature will approximately double the reverse current in a germanium junction; a 6°C increase will approximately double the reverse current in a silicon junction. At 70°C a typical reverse current for a germanium junction is 100 μA, for a silicon junction only 1 μA. Thus if low reverse currents are important, particularly at high temperatures, one *must* use silicon.

Because of the fixed charges present in the depletion layer due to the ionized donor and acceptor atoms, the depletion layer acts like a charged capacitor. Even though the depletion region as a whole is electrically neutral, there is positive charge on the n-type side of the p-n interface due to ionized donor atoms and negative charge on the p-type side due to the ionized acceptor atoms. Thus we have two layers of charge in a parallel plate configuration which comprises a capacitor. The situation is somewhat different from a parallel plate

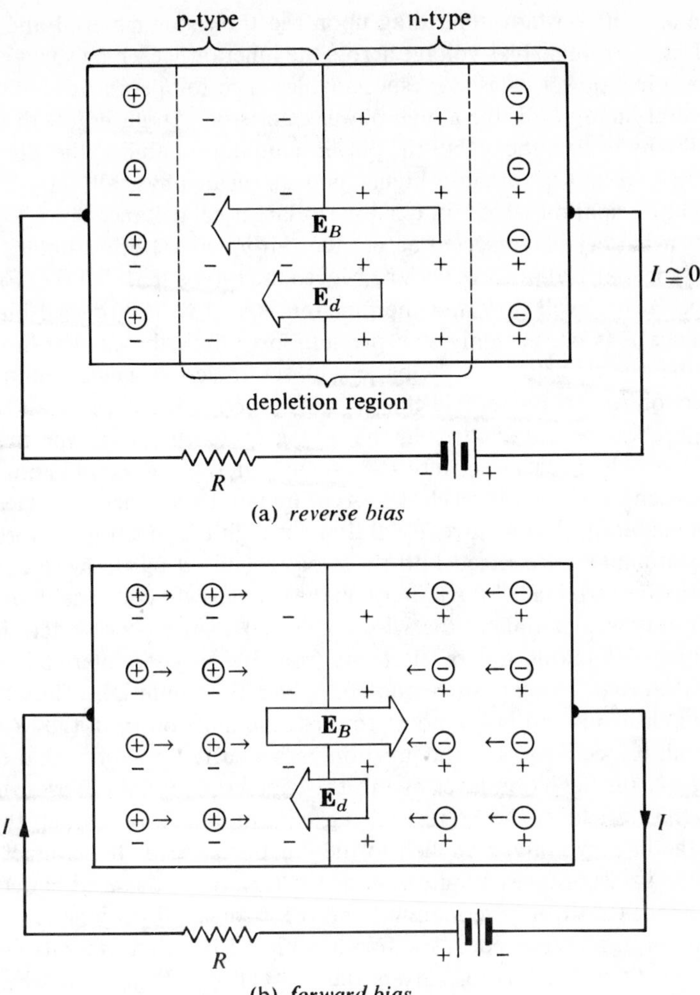

FIGURE 4.14 Biased p-n junction.

capacitor where the insulating region between the plates contains no charge. In the reverse biased junction, the charges are distributed throughout the volume of the depletion region. As the reverse bias increases in magnitude, the depletion region grows thicker (the capacitor "plates" become further apart) and the capacitance decreases. An exact calculation shows that the junction capacitance for an abrupt junction is given by:

$$C_J = \frac{\mathcal{K}}{\sqrt{V}} \tag{4.10}$$

where $\mathcal{K}$ is a constant depending upon the transistor material and geometry, and V is the reverse bias voltage across the junction. C_J may vary from several pF for a low power transistor especially designed to operate at high frequencies to several hundred pF for a high power transistor. In general with transistors, as with any device, the higher the power handling capability, the worse the high frequency response, because higher powers mean physically larger structures for heat dissipation which, in turn, mean larger capacitances.

If a battery is connected across the p-n junction with a polarity to make the p-type material positive with respect to the n-type material [see Fig. 4.14(b)] a large current will flow and the junction is said to be forward biased. The resistance R is present only as a precaution to limit the current flow. In this case, the electric field from the battery $\mathbf{E}_B$ set up across the depletion layer is in a direction to pull the mobile charge carriers *across* the depletion layer; that is, electrons will be attracted from the n-type over into the p-type material and holes from the p-type over into the n-type material. Recombination will take place when the electrons reach the p-type material and when the holes reach the n-type material, thus a current will flow through the junction. More electrons will continually be injected into the n-type material by the wire connected to the negative terminal of the battery, and electrons will continually be taken out of the p-type material by the wire connected to the positive terminal of the battery. The taking out of electrons from the p-type material is electrically equivalent to injecting positive holes into the p-type material. Thus, a continual flow of electrons and holes moves towards the junction from the n- and p-type material, respectively, and the junction passes current. Notice that the electric field $\mathbf{E}_d$ in the depletion layer due to the contact potential opposes the forward bias electric field of the battery, so no current will flow through the junction until the battery voltage applied to the junction exceeds the contact potential. In other words, it takes a finite voltage to "turn-on" a forward biased p-n junction, about 0.2V for a germanium and 0.5V for a silicon junction. Once the turn-on voltage is exceeded, the forward biased junction presents essentially a short circuit to the flow of current, i.e., $I \cong (V_{bb} - V_{\text{turn on}})/R$ where R is the external resistance in series with the forward biased junction.

The turn-on voltage V_{to} is slightly temperature dependent because of the weak temperature dependence of the Fermi energy. In a pure or "intrinsic" semi-conductor with no doping, the Fermi level is in the energy gap, halfway between the valence and the conduction band. In a doped n-type semiconductor, the Fermi level is between the donor levels and the bottom of the conduction band (see Fig. 4.10).

Thus, an increase in temperature lowers the Fermi energy for n-type material because at higher temperatures a larger fraction of the mobile charge carriers are from electrons thermally excited across the gap from the valence to the conduction band. A similar argument shows that the Fermi level is raised for an increase in temperature in p-type material. The net result is that if the temperature increases, a smaller forward bias voltage is required to maintain a constant current flowing through a forward biased p-n junction. If the temperature rises, and the forward bias voltage is maintained constant by the circuit, then the current flowing through the junction will increase. In other words a temperature rise will lower the effective "turn-on" voltage. Empirically, for both germanium and silicon p-n junctions near room temperature, the turn-on voltage decreases by approximately 2.5 mV for every degree centigrade temperature rise.

In the reverse biased configuration, the electric field within the semiconductor exists only in the depletion region. Thus, the motion of the mobile charge carriers in the rest of the semiconductor is mainly governed by diffusion—there simply is little or no electric field present to "hurry" the charges along. Charge will flow through the semiconductor because of the concentration gradients set up by the depletion layer absorbing charge carriers and the wire contacts injecting more carriers. The charge carriers will then diffuse from regions of greater concentration toward regions of lesser concentration. Thus, there is a certain definite lag in the propagation of a current through a semiconductor, but this lag or "transit time" does not become important until frequencies of the order of tens or hundreds of megahertz or higher are considered. And, the diode manufacturers make the physical size of the semiconductor material used as small as possible in order to minimize the time necessary for a charge carrier to diffuse from the connecting wire to the depletion layer. In some diodes the conductivity of the semiconductor material is deliberately made less so as to have an electric field exist inside the material; this field moves the charge carriers faster than by diffusion alone.

In summary, we have seen that a p-n semiconductor junction will conduct current very well in one direction (the "forward" direction) and very poorly in the other direction (the "backward" direction). This "one-way" type of conduction is called rectification, and the device is called a diode.

An ideal or perfect diode would present zero resistance in the forward direction and infinite resistance in the backward direction. If we plot a graph of current I passed by an ideal diode vs V the voltage difference across it, we would get a 90° break in the curve as is shown in Fig. 4.15.

The current–voltage curve for a real semiconductor diode differs from the ideal diode curve in several ways: the reverse current is not exactly zero because the thermally generated minority carriers will pass current even when the diode is reverse biased, the forward current is finite because the semiconductor material comprising the diode has some resistance, and a minimum voltage (the "turn-on" voltage) must exist across the diode before it will pass appreciable current. The reverse current increases with increasing temperature, because there are more minority carriers at higher temperatures. Typical curves for germanium and silicon diodes are shown in Fig. 4.15. It can be shown that the current–voltage equation for a p-n junction is

$$I = I_0(e^{eV_B/kT} - 1) \tag{4.11}$$

where V_B is the applied bias voltage, e is the electronic charge, k is Boltzmann's

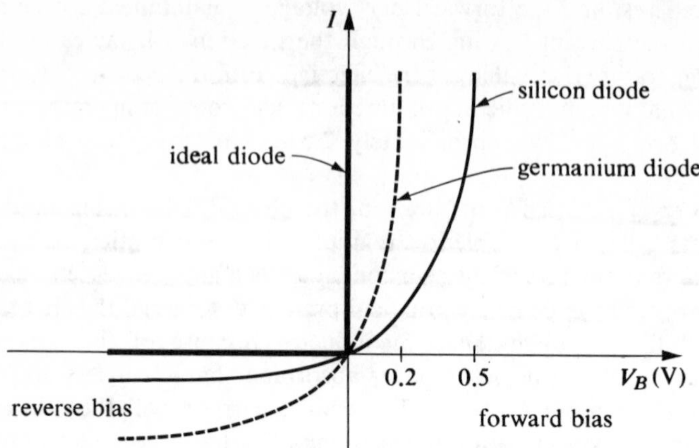

FIGURE 4.15 Current–voltage curves for diodes.

constant, and T is the absolute temperature. V_B is positive for forward bias and negative for reverse bias. I_0 is the reverse current for large reverse bias. Notice that the silicon diode passes much less reverse current than does the germanium diode and that the turn-on voltage differs appreciably between the two types. Therefore, in applications requiring an extremely small reverse current silicon is used, and if one has only a small voltage to turn on the diode germanium is used. Silicon diodes can be used at temperatures up to about 200°C, whereas germanium diodes can be used only up to about 85°C.

The explanation for the turn-on voltage and the equation for the current–voltage curve for a p-n junction can be obtained from a more detailed energy-level picture of both sides of the junction. We recall that the Fermi energy for n-type material lies just above the donor levels and in p-type material lies just below the acceptor levels. When any two materials—metal, semiconductor, or insulator—are brought together, their Fermi levels must equalize. If the Fermi levels are not initially equal, then electrons will flow from the material having the higher Fermi level into the other material, until the Fermi levels are equalized as shown in Fig. 4.16.

From reference to Fig. 4.16, a number of conclusions can be drawn. The conduction band of the p-type material is higher in energy than that of the n-type material by an amount ΔE equal to the difference in the Fermi energies, which is approximately equal to the gap energy E_g. The electric field $\mathbf{E}_d$ in the depletion region (pointing from n- to p-type) simply arises from the slope of the bottom of the conduction band as we go across the junction, $E_d \cong \Delta E/ed$. Thus, an electron at the bottom of the conduction band in the n-type material must somehow get ΔE energy in order to climb up the potential hill and arrive in the conduction band of the p-type material. The "turn-on" voltage V_{to} of the junction is approximately given by $eV_{to} = \Delta E$, because in order for any electrons to flow from the n to the p side they must attain a minimum of ΔE energy,

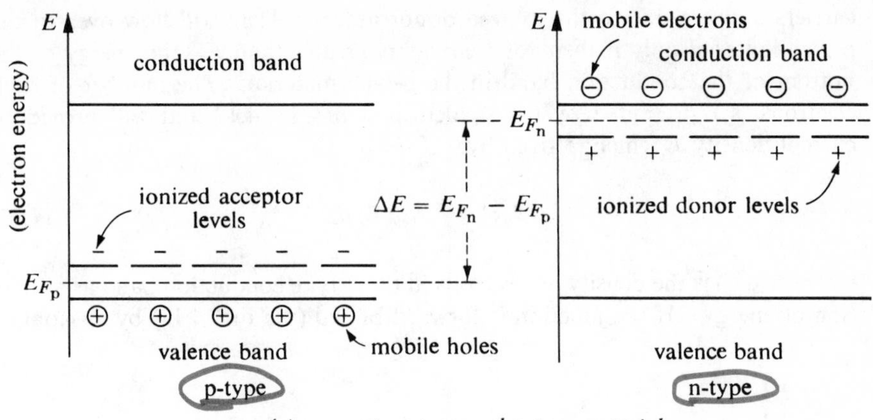

(a) *separate* p-*type and* n-*type materials*

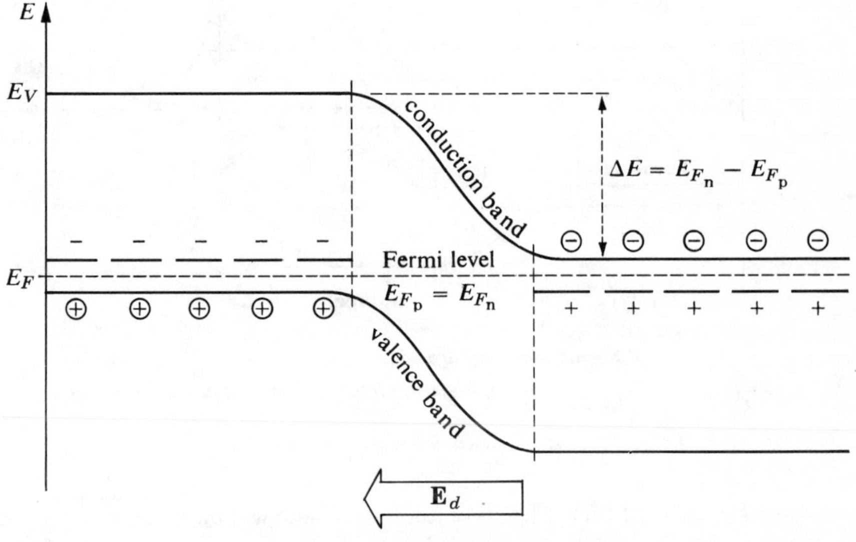

(b) *joined* p-*type and* n-*type materials*

FIGURE 4.16 Energy levels in an unbiased p-n junction.

which means a forward bias of at least $V_B = \Delta E/e$ volts must be applied. The reason for the larger turn-on voltage of silicon as compared to germanium p-n junctions is simply that the energy gap (and therefore ΔE) is larger for silicon than for germanium.

There are four types of current which can flow across the junction, two from majority carriers and two from minority carriers. Let J represent current densities due to majority carriers and j represent current densities due to minority carriers.

Any electrons in the conduction band of the n-type material are majority carriers and are due to the ionized donor atoms. They will flow over to the p-type material only if their total energy is greater than E_v, the energy of the bottom of the conduction band in the p-type material. The number of such electrons is represented by the shaded area in Fig. 4.17 and will provide a current density J_e which is given by:

$$J_e \propto \int_{E_v}^{\infty} N_{cb}(E)\, dE \tag{4.12}$$

where $N_{cb}(E)$ is the density of electrons in the n-type conduction band as a function of energy. If the junction is forward biased (see Fig. 4.17) by an applied

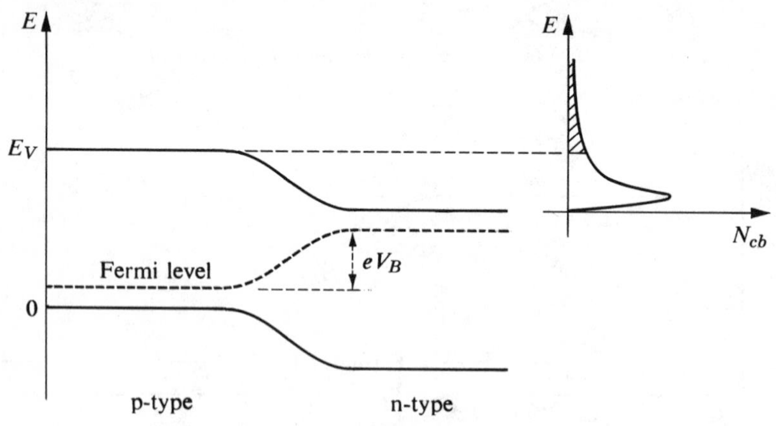

V_B = *applied forward bias voltage*
N_{cb} = *number of electrons/cm³ in conduction band*

FIGURE 4.17 Forward biased p-n junction energy-level diagram.

voltage V_B, then all energy levels on the n-type side will be raised by an amount eV_B assuming the p side is grounded. (Grounding the p side merely fixes all the energy levels in the p-type material.) Therefore,

$$J_e \propto \int_{(E_v - eV_B)}^{\infty} N_{cb}(E)\, dE \propto e^{-(E_v - eV_B)/kT} \propto e^{eV_B/kT}. \tag{4.13}$$

Let J_{e_0} be the current density when no bias is applied ($V_B = 0$). Then J_e can be written simply as

$$J_e = J_{e_0} e^{eV_B/kT}. \tag{4.14}$$

Any electrons in the p-type material thermally excited from the valence to the conduction band will be minority carriers and will tend to "fall down" the potential hill, thus flowing from the p- to the n-type material. The resulting current density j_e will be proportional only to the number of minority carriers and hence to the temperature so long as there is a downhill potential slope over to the n-type material. The net current flow from these two *electron* currents will then be

$$J_e - j_e$$

where positive current means effective positive charge flowing from the p- to the n-type side of the junction.

Any holes in the valence band of the p-type material are majority carriers and are due to the ionized acceptor atoms. They will flow over to the n-type material only if their energy is less than E_v' of Fig. 4.17. An argument similar to that given above for the majority carrier electrons in the n-type shows that the resulting current density J_h from these majority carrier holes is given by

$$J_h = J_{h_0} e^{eV_B/kT}. \tag{4.15}$$

Any holes in the valence band of the n-type material are minority carriers and are produced by thermal excitation. They give rise to a voltage independent current density j_h because the holes spontaneously flow up the potential hill from n- to p-type side of the junction. The net current flow from these two *hole* currents is then $J_h - j_h$ with the same sign convention.

The net current density flowing across the junction is the algebraic sum of these four current densities:

$$J_{\text{net}} = J_h - j_h + J_e - j_e. \tag{4.16}$$

If we now realize that the net current J must equal zero when no bias is applied ($V_B = 0$), then we must have

$$J_{h_0} - j_h + J_{e_0} - j_e = 0. \tag{4.17}$$

It can be shown that the hole and electron currents must separately equal zero in equilibrium, so

$$J_{h_0} = j_h \quad \text{and} \quad J_{e_0} = j_e. \tag{4.18}$$

Therefore,

$$J_{\text{net}} = J_{h_0} e^{eV_B/kT} - J_{h_0} + J_{e_0} e^{eV_B/kT} - J_{e_0}. \tag{4.19}$$

Rewriting yields

$$J = (J_{h_0} + J_{e_0})(e^{eV_B/kT} - 1), \tag{4.20}$$

or letting

$$J_0 = J_{h_0} + J_{e_0}$$

we have

$$J = J_0(e^{eV_B/kT} - 1). \tag{4.21}$$

Equation (4.21) is called the "diode equation" or the "rectifier equation" and tells us that as the forward bias $V_B > 0$ is increased the total current passed through the junction increases rapidly with V_B. And, for large reverse bias ($V_B < 0$) the current will decrease to a value of $-J_0$. For large forward bias the increase of J will be somewhat slower than implied by equation (4.21) because of the inherent resistance of the semiconductor material which we have neglected in this treatment.

Diodes can also be made with a sharp junction between a metal and a semiconductor, the metal acting like a p-type material. Such diodes are called "point contact" diodes. They are superior to junction diodes at high frequencies, but can carry far less current than either Ge or Si junction diodes.

It should be pointed out that all metal-semiconductor contacts do not act as rectifying junctions. Only extremely sharp p-n junctions will rectify. The gradual junctions found between the metal leads and the semiconductor material of diodes and transistors do not rectify; they pass current equally well in both directions. These contacts are usually made by depositing a thin metal coating on the semiconductor and soldering the lead wire to the metal coating.

Another type of diode, called the "zener" or "avalanche" or "reference" diode can be made to break down at a specified reverse voltage V_z. The breakdown process occurs, briefly, because an electron or hole obtains enough energy (from the reverse bias electric field) between collisions to break a covalent bond in the crystal lattice and thereby create two new charge carriers which in turn are accelerated and create more charge carriers. This process occurs very sharply at a certain voltage V_z. If one tries to increase the voltage drop across the diode above V_z, then enough more charge carriers are produced in the diode to increase the voltage drop across whatever resistance is in series with the diode. The net result is that the zener diode draws just enough current to keep the voltage drop across it constant at essentially V_z volts. The zener diode circuit symbol and current–voltage curve are shown in Fig. 4.18. Its principal use is in regulating voltages, for if its reverse current changes from I_A to I_B, the voltage across the diode will remain essentially constant at the zener voltage V_z. In a typical zener diode, a one milliampere change in current will result in only a one millivolt change in the voltage across the diode, corresponding to a dynamic resistance of only about one ohm. Special zener diodes can be made with a very low temperature coefficient to provide an extremely stable reference voltage.

If the amount of impurities is increased to the order of $10^{19}\,\text{cm}^{-3}$, then the impurity energy levels merge together to form a small band and the diode can conduct due to a quantum-mechanical "tunneling" process for small forward biases on the order of several tenths of a volt. The tunneling falls to zero for larger forward biases, yielding the current–voltage curve shown in Fig. 4.19. This device is, appropriately, called a "tunnel" diode and is useful not as a rectifier, but because it exhibits a negative dynamic resistance from point A to

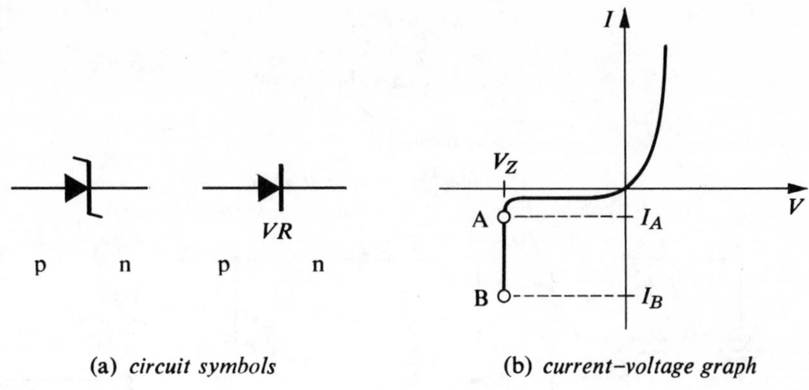

(a) *circuit symbols* (b) *current–voltage graph*

FIGURE 4.18 The zener or voltage regulator diode.

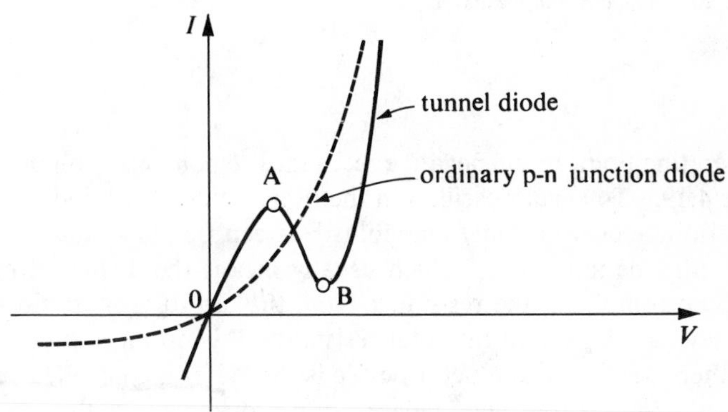

FIGURE 4.19 Tunnel diode current–voltage curve.

B on the current–voltage curve. The peak current I_P is typically from 1 to 10 mA; the valley current I_V is 1 mA or less, and the voltage at the peak V_P is only about 50 mV. Oscillators and amplifiers can be made utilizing this negative resistance phenomenon. Because the tunneling occurs with majority rather than minority carriers, it can be shown that the tunnel device will operate at extremely high frequencies—up to 10^{10} Hz (10 GHz), which is well up in the so-called microwave or radar region. The tunnel diode is also useful because from points 0 to A its current increases more rapidly with voltage than for ordinary p-n diodes.

A simple tunnel diode oscillator circuit is shown in Fig. 4.20 along with the tunnel diode schematic symbol. The oscillation frequency is determined by L and C, $\omega_0 = (1/LC)^{1/2}$, and the setting of the voltage divider R places the tunnel

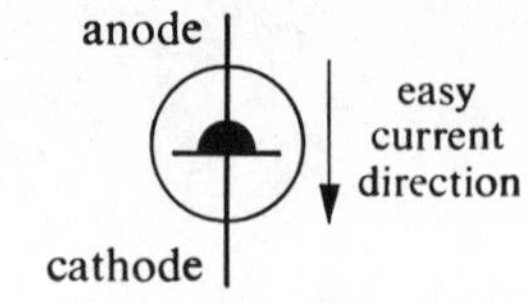

tunnel diode circuit symbol

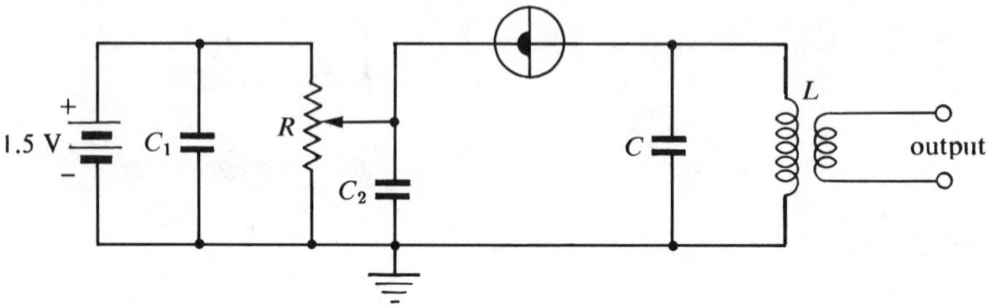

FIGURE 4.20 Tunnel diode oscillator.

diode operating point in the negative resistance region between points A and B of Figure 4.19. To ensure oscillation the tunnel diode load line must intersect the tunnel diode curve at only *one* point between points A and B. Thus, the load line must be fairly steep which usually means the divider R must be set "near the bottom," i.e., the resistance from the tap to ground must be of the order of several ohms, and the total resistance R is no more than 100Ω.

Another useful semiconductor device is the "silicon controlled rectifier" or SCR. The SCR is a four layer device p-n-p-n or n-p-n-p with three terminals called the anode, the cathode, and the gate as shown in Fig. 4.21. The SCR will not conduct in the forward direction until the anode-cathode voltage V_{ac} exceeds a certain value which depends upon the (small) gate current I_G. Once the SCR conducts or "fires", V_{ac} drops to a very small value, and the SCR current I_A is independent of the gate current—the SCR acts just like a forward biased diode. The SCR current I_A drops to zero only when the polarity of V_{ac} is reversed. The diode can be made to conduct again only when the anode is made positive with respect to the cathode, the exact value which will fire the diode depends upon the gate current. SCR's are commonly used to control high power ac circuits such as electric motors. One SCR can control currents of tens of amperes. The SCR is a solid-state equivalent of a thyratron vacuum tube.

The basic operation of an SCR in controlling an ac current is to fire the SCR at various phases of the ac voltage across the SCR as shown in Fig. 4.22. If a trigger circuit supplies a pulse of current to the SCR gate at t_1, the SCR will fire and conduct so long as its anode voltage remains positive with respect to its

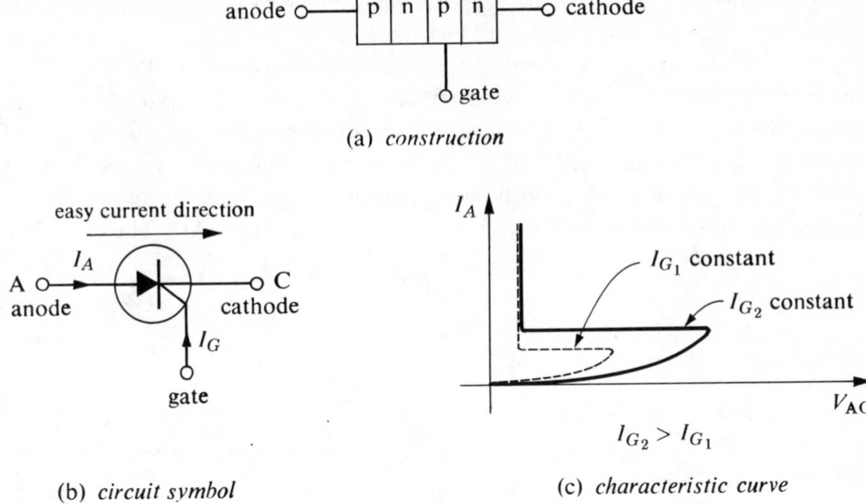

FIGURE 4.21 Silicon controlled rectifier (SCR).

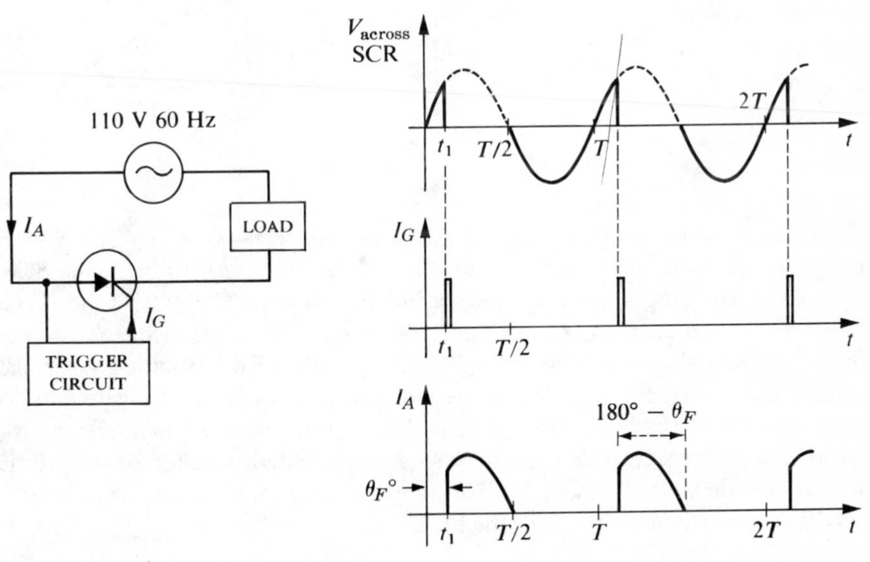

FIGURE 4.22 SCR operation.

cathode. The phase angle corresponding to t_1 is often called the "firing angle," θ_F; the number of electrical degrees during which the SCR conducts equals $180° - \theta_F$. The trigger pulse must last approximately 30μsec or more. The amplitude of the trigger pulse will determine at what voltage the SCR will fire. The current passed by the SCR will then last from t_1 to $T/2$ on each cycle. By varying the time t_1 at which the trigger pulses occur, the average current passed by the SCR can be adjusted. SCR trigger circuits commonly use small neon lamps which will not conduct until the voltage across the lamp terminals reaches approximately 80 volts. One such circuit is shown in Fig. 4.23. The capacitor

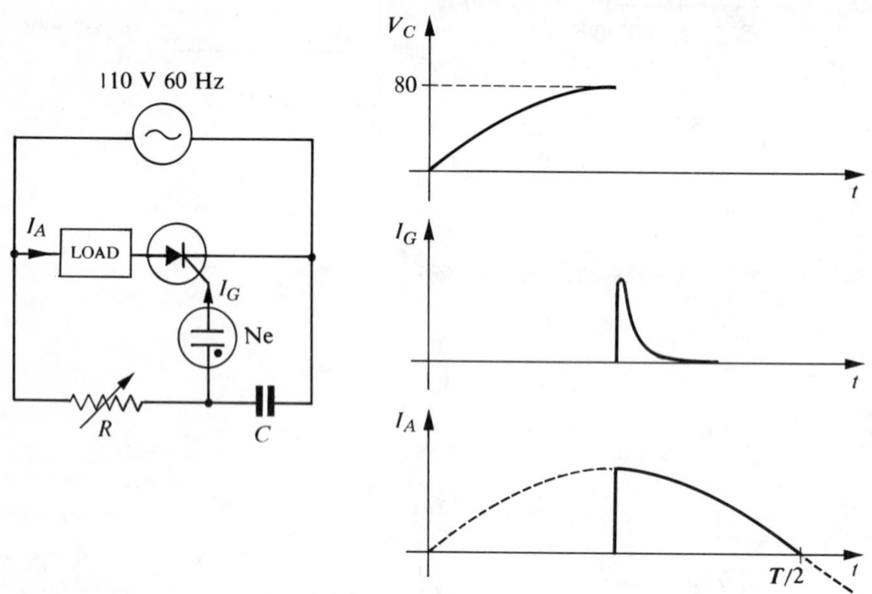

FIGURE 4.23 SCR controlled load.

C is charged up positively through R by the 110V 60Hz. When V_C reaches the firing voltage of the neon lamp, it fires and C is discharged through the SCR gate, thus providing a pulse of gate current to turn the SCR on. C should be large enough so that the I_G pulse is large enough to turn on the SCR even for a very low anode-cathode voltage, which would be the situation when the firing angle is close to 180°. When the 110V ac reverses polarity, a gate pulse is again applied to the SCR, but the SCR does not fire because its anode is now negative with respect to its cathode. As R is increased, it takes a longer time for C to charge up to the neon bulb firing voltage, and θ_F is larger, resulting in a smaller average current through the load.

4.10 DIODE APPLICATIONS

One of the most common uses of the diode is in power supplies where the standard 110V 60 Hz ac line voltage is converted into a dc voltage. A simple diode-resistance circuit shown in Fig. 4.24(a) will convert the input ac voltage $v_{in} = V_{AB}$

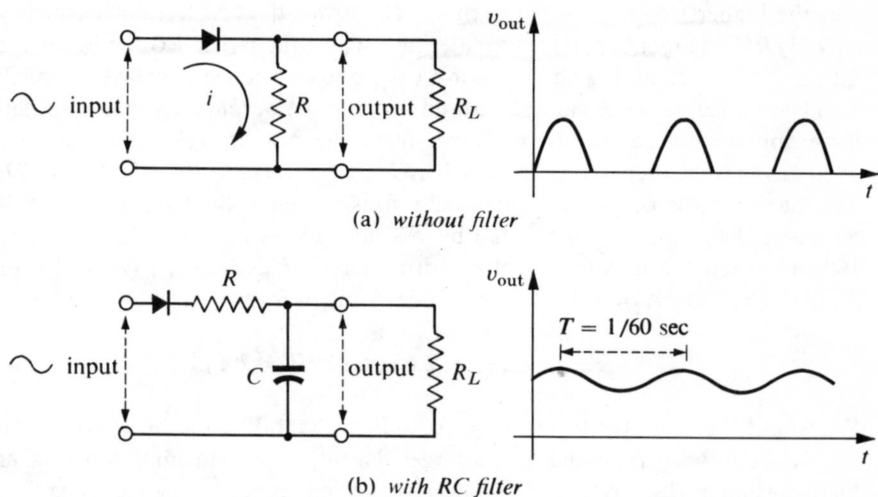

FIGURE 4.24 Half-wave diode rectification.

$= V_0 \sin \omega t$ into a pulsating dc voltage. The explanation is simply that the diode conducts only on the positive half cycles and not on the negative half cycles of the input. We assume in this section that the diode has an infinite resistance in the reverse direction and has zero turn-on voltage. Thus the current i passed by the diode is unidirectional; i.e., it is pulsating direct current always flowing in the direction shown in Fig. 4.24(a). The positive output voltage is the ir voltage drop across r. Notice that when the 60 Hz input voltage makes point A negative with respect to point B the diode is reverse biased, and essentially all the 60 Hz voltage appears across the diode because $i = 0$. For the circuit to function properly, the diode must be capable of withstanding this reverse voltage without breaking down. The appropriate diode rating is the "peak reverse voltage" PRV or "peak inverse voltage" PIV. Modern silicon rectifier diodes come with PIV ratings up to 500 V or more. Diodes are also rated according to how much current they may safely pass when forward biased. The maximum current rating usually refers to the average dc current which the diode may safely pass. The peak or surge current rating is generally much higher and refers to the maximum peak current the diode may safely pass in a very short interval of time. Typical modern silicon rectifier diodes cost less than $1 each, can carry average currents of one ampere and peak currents of tens of amperes, and have peak inverse ratings of 500V.

If a steady dc output voltage is desired, an RC low-pass filter is added as shown in Fig. 4.24(b). The resistance R also limits the current surge through the diode when the circuit is turned on, when C is completely uncharged. The capacitance C is charged up and serves to smooth out the amplitude variations in the output. Another view of the filter is that it passes dc and the lower frequency Fourier components of the pulsating voltage of Fig. 4.24(a) and attenuates the higher frequency components. The lower the break point frequency $\omega_B = 1/RC$, the more effective the filtering. Thus, the larger RC the better the filtering. The remaining ac variation in the output voltage is called "ripple." Too large a resistance R cannot be used because the output dc voltage will fall if an appreciable dc current is drawn from the output. For example, if $f = \omega/2\pi = 60$ Hz, then we desire $\omega_B = 1/RC \ll 2\pi \times 60$ or $1/RC \ll 120\pi = 377$. The maximum dc output current usually fixes an upper limit for R. Thus, if we desire the output voltage to fall by less than 1V as I_{out} increases from 0 to 100 mA, then the maximum voltage drop across R is $(I_{out\,max})\,(R) = 1$V or $R_{max} = 1\text{V}/100\,\text{mA} = 10\Omega$. Thus C is determined from

$$C \gg \frac{1}{377R} = \frac{1}{3770} = 2.65 \times 10^{-4}\text{F} = 265\,\mu\text{F}.$$

We would therefore try to use $C = 1000\,\mu$F or $2000\,\mu$F if the budget permits.

It should also be noted that a larger dc output current implies a smaller load resistance R_L. When the diode is not conducting, the capacitance C is discharging through R_L. Thus, the smaller R_L, the more rapid the discharge of C and the more the output voltage falls before the next positive half cycle when D conducts and charges up C again. In other words, the larger the dc output current, the larger the ripple.

Practical power supply circuits are considerably more complicated than the half-wave circuit of Fig. 4.24. Usually, both halves of the input ac voltage are utilized, either in a full-wave circuit shown in Fig. 4.25 or in the full-wave bridge circuit of Fig. 4.26. Usually a transformer is used to change the 110V 60Hz line voltage to the approximate desired output voltage. V_{AB} is the secondary voltage of the transformer. Notice that the half-wave circuit ripple is 60Hz, while the full-wave ripple is 120Hz. Also, the half-wave circuit requires a center tap "CT" on the transformer, while the full-wave bridge circuit does not. For the same transformer secondary voltage, the half-wave and the full-wave bridge circuits give twice the output voltage that the full-wave circuit does. More current can be drawn from the half-wave circuit than the full-wave circuit without overheating the transformer because current flows through the half-wave circuit transformer only during 50% of the time; i.e., the "duty cycle" is 50%. A complete schematic for a practical regulated power supply is given in Appendix I.

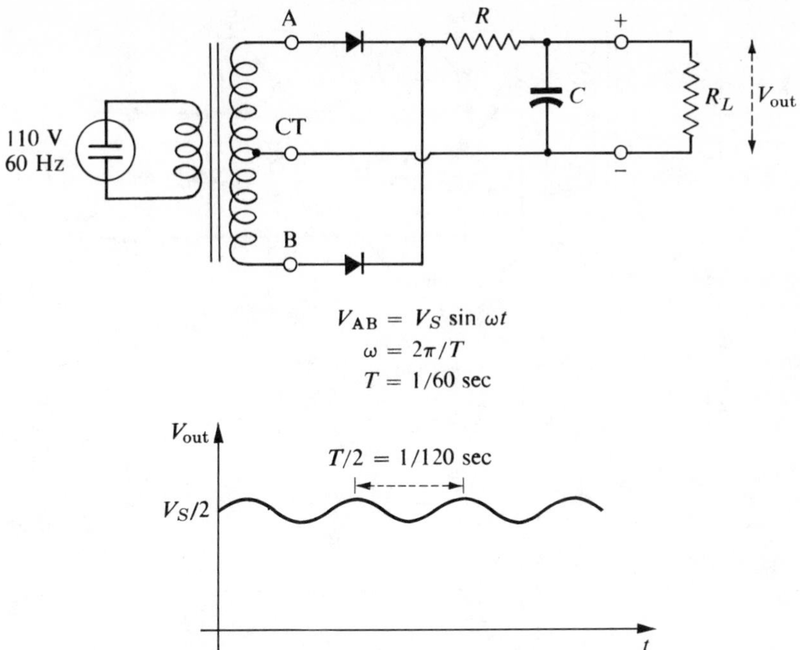

FIGURE 4.25 Full-wave rectifier.

Another common use of diodes is in clipping circuits where the amplitude of a signal must be limited or clipped. In Fig. 4.27(a) the diode conducts only on the negative half-cycle of the input; thus for the negative portion of the input, the output is limited to the turn-on voltage. The positive portion of the input is passed through to the output provided only that $R \ll R_r$ where R_r is the reverse resistance of the diode. If a battery V_{bb} is added as shown in Fig. 4.27(b), then the diode will not start to conduct until the input is more negative than $-(V_{bb} + V_{to})$. Thus the negative excursion of the output is clipped off at $-(V_{bb} + V_{to})$ volts. The circuit in Fig. 4.27(c) will pass voltages of either polarity only if the amplitude is less than approximately the diode turn-on voltage, i.e., this circuit rejects large amplitude inputs and passes small amplitude inputs.

High voltage transformers are expensive and impractical at voltages much above several thousand volts. A useful diode power supply circuit with no high voltage transformer is a voltage doubler shown in Fig. 4.28. The dc output voltage will be equal to twice the zero-to-peak input voltage between terminals A and B which is usually the voltage from the secondary of a transformer. This can be seen by the following argument. If A is *negative* with respect to B, then diode D_2 conducts and diode D_1 does not. Current i flows through D_2 to charge up C_2 as shown in Fig. 4.27(b). The voltage across C_2 will be V_s, the

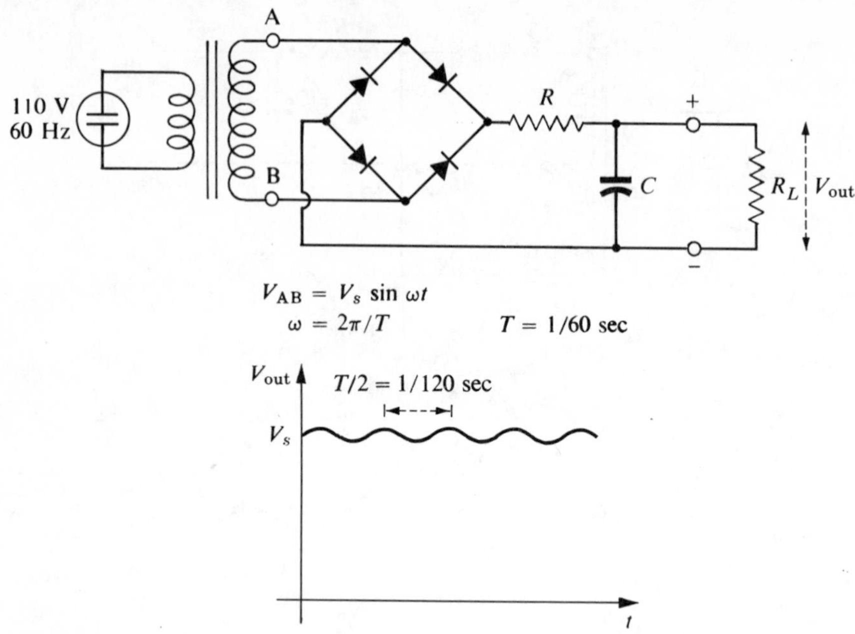

FIGURE 4.26 Full-wave bridge rectifier.

peak secondary transformer voltage. On the next half-cycle when A is *positive* with respect to B, D_1 conducts and D_2 does not. Thus current i' flows as shown in Fig. 4.27(c). C_1 is thus charged up to a voltage $2V_s$ because the peak transformer secondary voltage $V_{AB} = V_s$ and the voltage on $C_2 = V_s$ *add.* C_2 can be an electrolytic capacitor, because the voltage drop across it is always of the same polarity. The value of V_s depends upon the turns ratio of the transformer; it can be up to several kV.

Higher output voltages can be achieved with similar circuits. A voltage tripler is possible, and a voltage quadrupler is shown in Fig. 4.29. In general, these circuits can be extended to yield higher output voltages limited only by the breakdown of the capacitors and the diodes (when reverse biased). It is interesting to note that the first successful "atom smashing" experiment was performed in 1932 by Cockroft and Walton in the Cavendish Laboratory, England, using a voltage multiplier containing vacuum tube diodes and capacitors as the source of high dc voltage. Protons were accelerated through a voltage drop of 250,000 V dc and struck a lithium target producing disintegration of the lithium nuclei.

A simple diode clipping circuit provides for dc baseline restoration as mentioned in Chapter 2. Recall that when a positive pulse is passed through an *RC* coupling network (an *RC* high-pass filter), the output pulse area above and below the zero voltage axis must be equal. For negligible distortion of the pulse

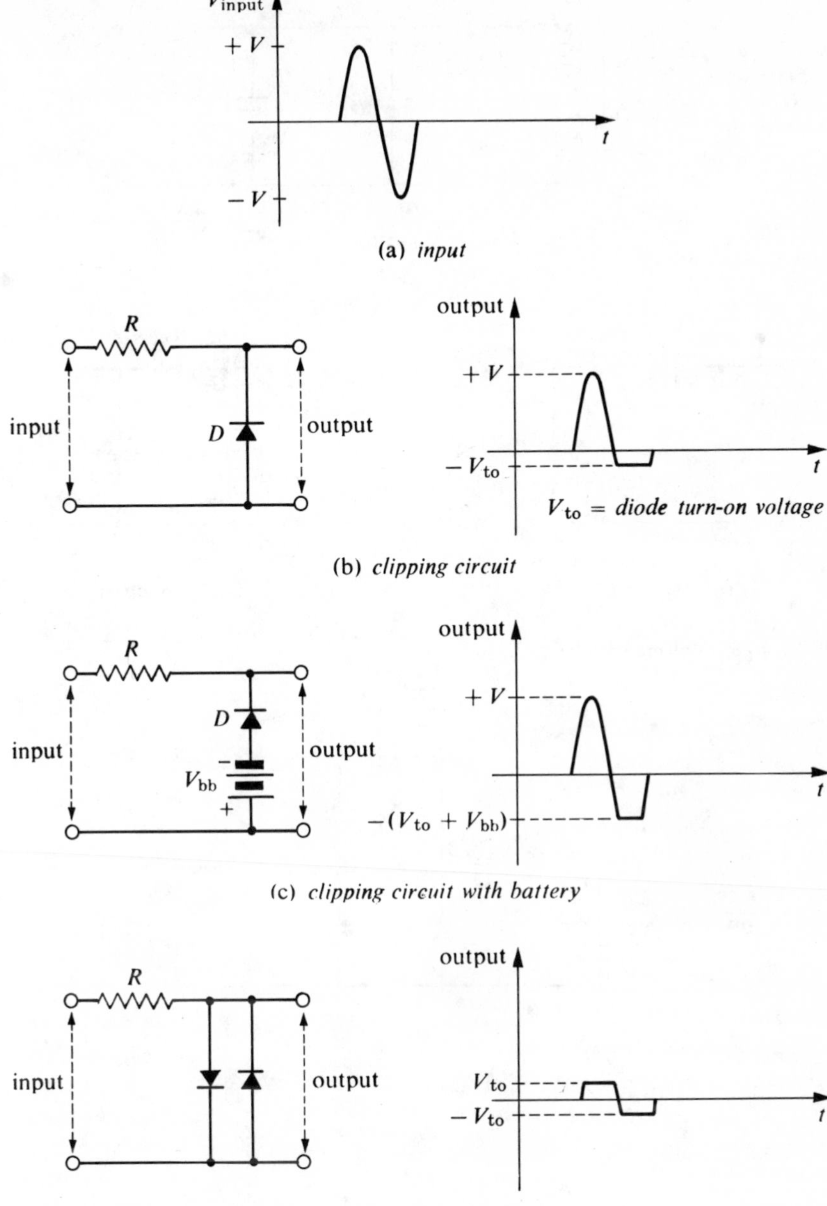

FIGURE 4.27 Diode clipping circuits.

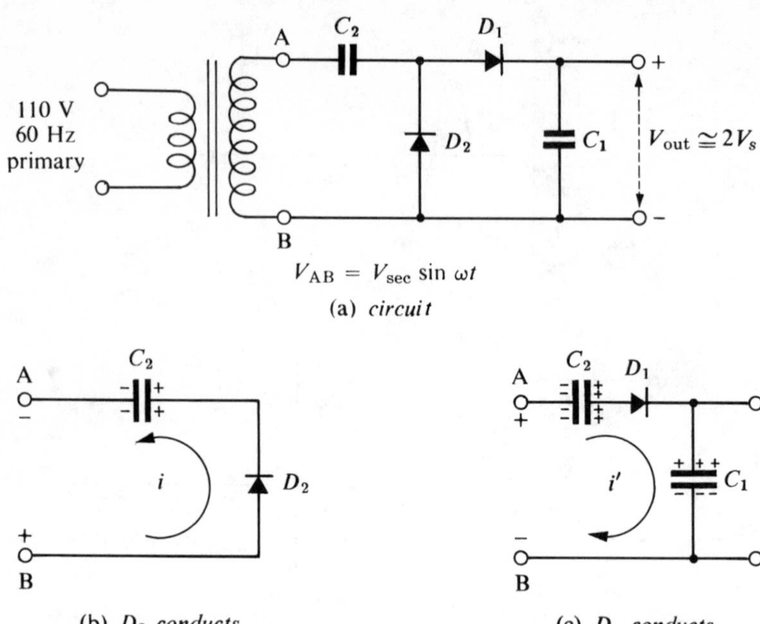

FIGURE 4.28 Voltage doubler.

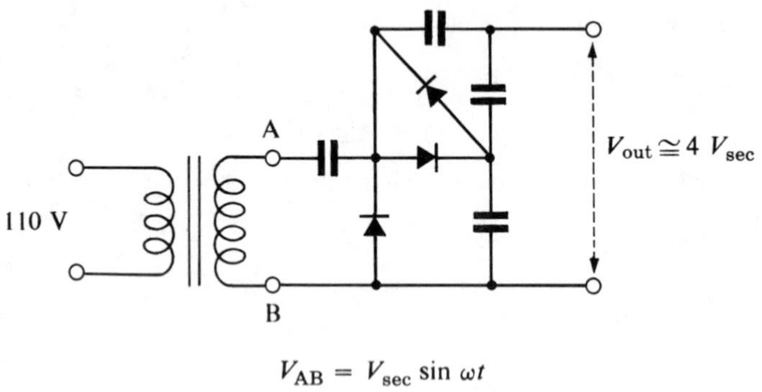

FIGURE 4.29 Voltage quadrupler.

shape, the time constant of the circuit should be much larger than the pulse width T. The output pulse, shown in Fig. 4.30(a), then has a long negative tail which decays with the characteristic time constant RC sec. If many pulses come along at the input in a time of the order of RC sec or less, then the cumulative effect of all the negative pulse tails is to appreciably depress the dc voltage level

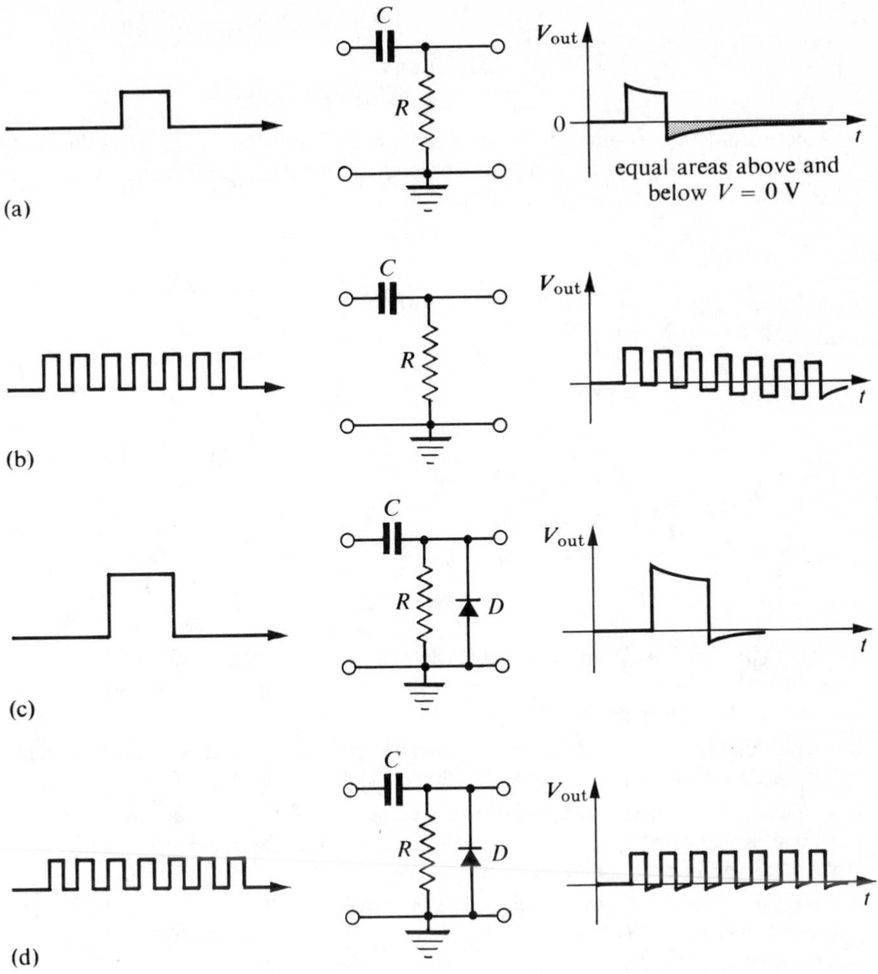

FIGURE 4.30 Elementary diode baseline restoration circuit.

or baseline of the output as shown in Fig. 4.30(b). The addition of a diode across the output as shown in Fig. 4.30(c) and (d) will essentially cure the problem by providing a low impedance path to ground for negative outputs. There are many other possible diode clipping or limiting circuits. Several diode circuits useful in computers are discussed in Chapter 12.

problems

1. A new atom "Confusium" is discovered with the following energy-level diagram. How much work must be done (how much energy must be expended) in order to ionize a Confusium atom in the ground state?

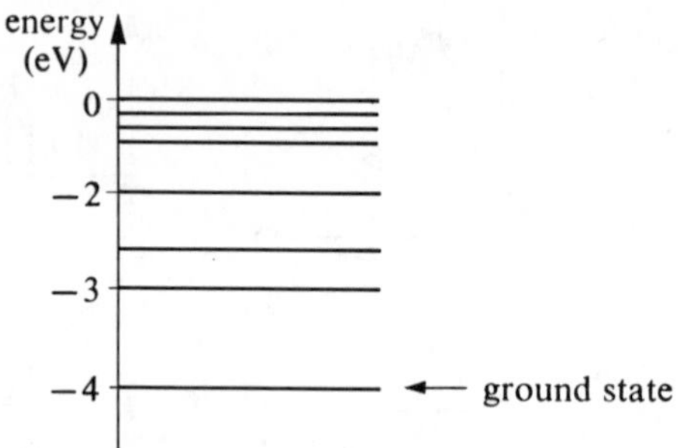

2. Approximately how large would the energy gap in a pure semiconductor have to be for it to be a nonconductor at room temperature? What is the physical significance of the energy gap?
3. In the energy-level diagram of "Confusium" in question, at what temperature (approximately) would you expect to have the energy level at −3 eV populated?
4. Calculate an approximate speed for (a) an electron in thermal equilibrium with a silicon lattice at room temperature, (b) an electron in thermal equilibrium with a germanium lattice at room temperature.
5. Calculate an approximate speed for the random thermal motion of a silicon atom (atomic weight = 28) in thermal equilibrium at room temperature. Repeat for germanium (atomic weight = 73) at room temperature.
6. Calculate the total number of mobile charge carriers in a pure semiconductor if 10^{10} valence bonds/cm^3 are broken on the average due to thermal agitation. Does one distinguish between "majority" and "minority" charge carriers in a pure semiconductor?
7. At roughly what temperature would you expect large numbers of electrons to be excited from the valence band to the conduction band in pure silicon? In pure germanium?
8. A silicon crystal is doped with pentavalent arsenic. Is the resulting semiconductor p- or n-type? If doped with trivalent phosphorus?
9. Calculate the average distance between the arsenic doping atoms in a silicon crystal if there are 10^{16} As atoms per cm^3.
10. Briefly explain physically why the donor atom energy levels in an n-type semiconductor lie so close to the conduction band.

11. Briefly give a physical interpretation of the Fermi energy. Why can't the Fermi energy lie in the conduction band for a pure semiconductor? What does the Fermi factor mean physically?
12. Briefly explain why the Fermi energy cannot lie below the donor energy levels in an n-type semiconductor.
13. Calculate an approximate temperature at which most of the donor atoms would be ionized in a doped n-type semiconductor with the following energy-level diagram.

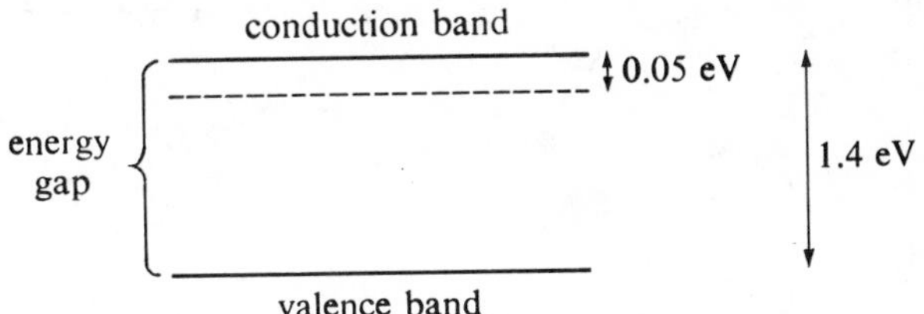

14. At roughly what temperature would you expect a conventional silicon transistor to cease operating as it is cooled down? Explain briefly.
15. Explain why a positively ionized donor atom does not contribute to conduction in an n-type semiconductor.
16. In an abrupt p-n junction with *no* applied bias, why will there be a small depletion region formed at the junction?
17. Why does an n-type doped semiconductor have many more mobile negative charge carriers than positive? Give typical approximate numbers for silicon and germanium.
18. Carefully state the Pauli exclusion principle. Why does this principle imply that a completely filled band cannot conduct?
19. Describe in words and sketch a diagram of the motion of *minority* charge carriers in a reverse-biased abrupt p-n junction. Include the depletion region.
20. Distinguish among a metal, a semiconductor, and an insulator on the basis of the location of the Fermi level.
21. Set up (do not evaluate) an expression for the number of electrons per cm^3 in the conduction band of an n-type semiconductor with energies more than 0.1 eV above the bottom of the conduction band. Define all symbols carefully.
22. Sketch the energy-level diagram (to scale) for an abrupt p-n silicon junction with a 0.3 V forward bias applied. Include the positions of the Fermi levels and the various bands.
23. Sketch a full-wave diode rectifier circuit to produce a -12 V dc output voltage relative to ground. Include approximate numerical values for the transformer, and the other components used.
24. Repeat Problem 23, only use a full-wave bridge rectifier circuit.
25. Design a diode clipping circuit to clip off negative going pulses so that the output is never more negative than -3 V.
26. Sketch the output waveform to scale for the following.

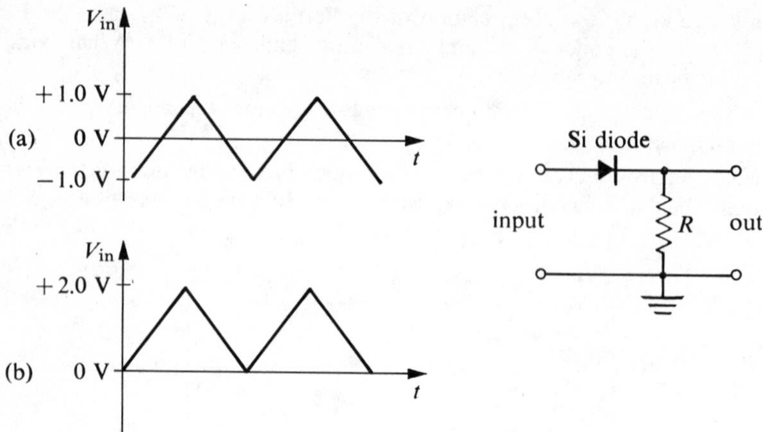

27. Sketch the output waveform to scale for the following.

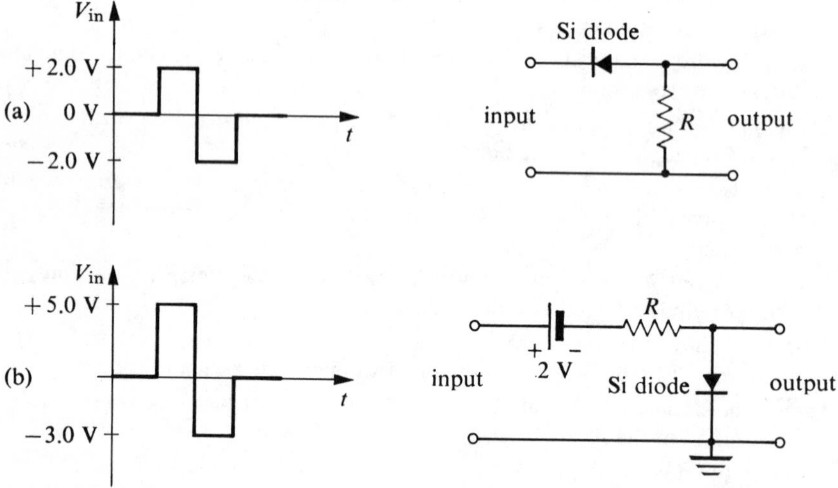

28. Explain why the ripple amplitude in a power supply increases as the load current drawn from it is increased.

5

TRANSISTOR ACTION AND SIMPLIFIED AMPLIFIER DESIGN

5.1 INTRODUCTION

In this chapter we will consider the physical construction of a transistor, explain transistor properties on the basis of how forward and reverse biased p-n junctions work, and design a common emitter transistor amplifier circuit.

5.2 TRANSISTOR CONSTRUCTION

A transistor consists of three layers of doped semiconductor material in the order p-n-p, which is called a "pnp" transistor or in the order n-p-n which is called "npn" transistor. Figure 5.1 shows the construction and the circuit symbols used for pnp and npn transistors. The center layer is always called the "base." The outside layers are called the "emitter" and the "collector" and look identical in Fig. 5.1 but in practice are not. Briefly, the collector has different doping and is thermally connected to the outside world for better heat dissipation while the emitter is not.

The boundary between the different types of semiconductors must be abrupt; that is, the type of impurity atom must suddenly change from donor to acceptor as we go from n- to p-type and vice versa from p- to n-type. Also, the crystal lattice structure must be undistorted or unbroken as we go from p- to n-type material or vice versa. One cannot make a transistor by gluing or clamping together separate pieces of doped semiconductor, because there would be too many imperfections in the crystal structure at the interface. The entire transistor must be grown from one single crystal with the type of impurity changed abruptly to create the p-n junctions.

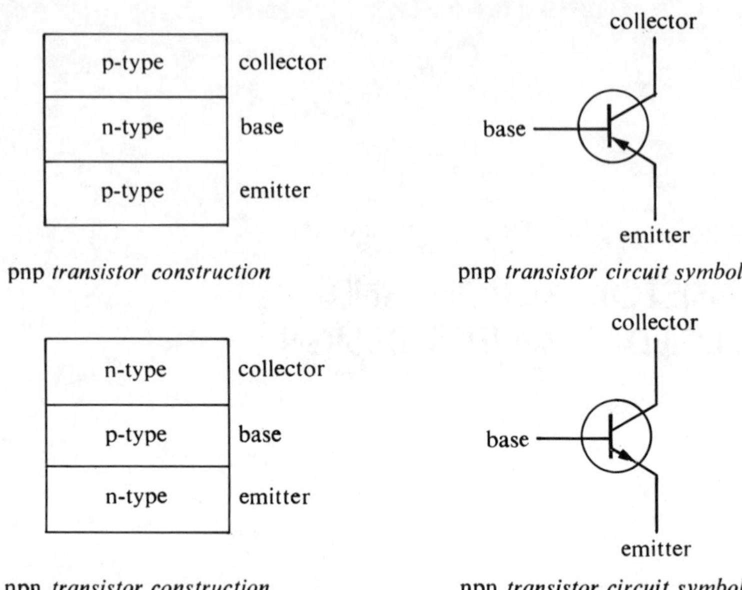

FIGURE 5.1 Transistor construction and circuit symbols.

An ohmmeter can be used to distinguish between a pnp and an npn transistor by measuring the resistance between the base and emitter for different polarities of the ohmmeter probes. In most ohmmeters, the red probe is positive in voltage with respect to the black probe. Hence for a pnp transistor the red lead on the emitter lead and the black lead on the base lead should forward bias the base-emitter junction and therefore should yield a low resistance, somewhere between 1 and 10 ohms. If the leads are reversed the measured resistance should be roughly a thousand times higher; on the R x 1 scale the resistance should appear infinite. For an npn transistor the black lead on the emitter lead and the red lead on the base lead should yield a low resistance, and a high resistance with the leads reversed. Similar arguments apply to the base-collector junction, etc.

A transistor which has been "burned out" or damaged in a circuit by excessively high voltage can also be identified with an ohmmeter by a similar procedure. One should measure *six* resistances to be absolutely sure: the base-emitter, base-collector, and collector-emitter resistances, each with two polarities of the ohmmeter leads on the R x 1 scale. A good transistor will have the following readings. R_{BE} means the base positive with respect to the emitter, R_{EB} the emitter positive with respect to the base, etc. Both R_{CE} and R_{EC} are high for any transistor because in either case there is one reverse biased junction.

Because the operation of the transistor depends upon the nonlinear characteristics of the p-n junctions, we repeat the p-n current–voltage curve here for

TABLE 5-1. R x 1 OHMMETER SCALE READINGS FOR A GOOD TRANSISTOR

	R_{BE}	R_{EB}	R_{BC}	R_{CB}	R_{CE}	R_{EC}
pnp	high	low	high	low	high	high
npn	low	high	low	high	high	high

convenience in Fig. 5.2.

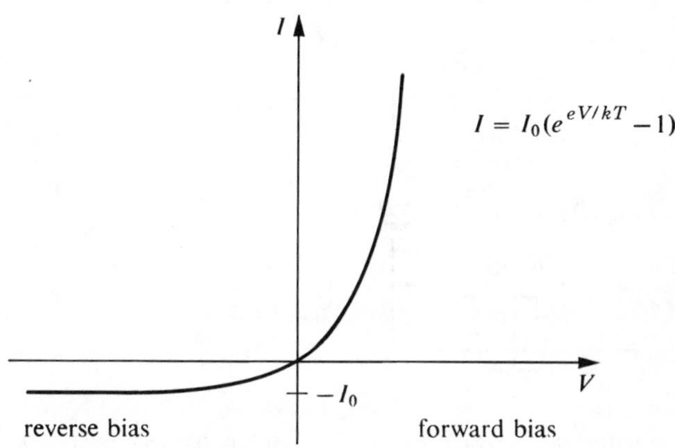

FIGURE 5.2 p-n junction current–voltage graph.

5.3 BIASING AND CURRENT FLOW INSIDE A TRANSISTOR

Reading from bottom to top in Fig. 5.1, the three layers of semiconductor material are called the "emitter," the "base," and the "collector," for both the pnp transistor and the npn transistor. We will now confine our attention to a pnp transistor. The two p-type regions are not exactly identical as we will see later, hence they go by different names.

Let us first consider the problem of biasing the junctions. In this section "bias" or "bias voltage" will mean a dc voltage which is applied to the transistor at all times and which we will see is necessary for the transistor to function properly. The base-emitter junction is biased in a *forward direction*; that is, the p-type emitter is positive in voltage with respect to the n-type base. Thus, a large current flows from the emitter into the base. This could be accomplished by connecting up a battery of voltage V_{bb1} and a resistor R_1 as shown in Fig. 5.3(a). The base-collector junction is biased in a *reverse direction*; that is, the

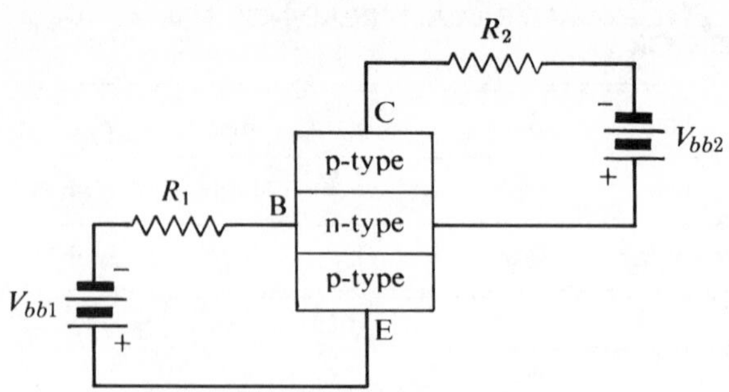

(a) pnp *transistor bias circuit*

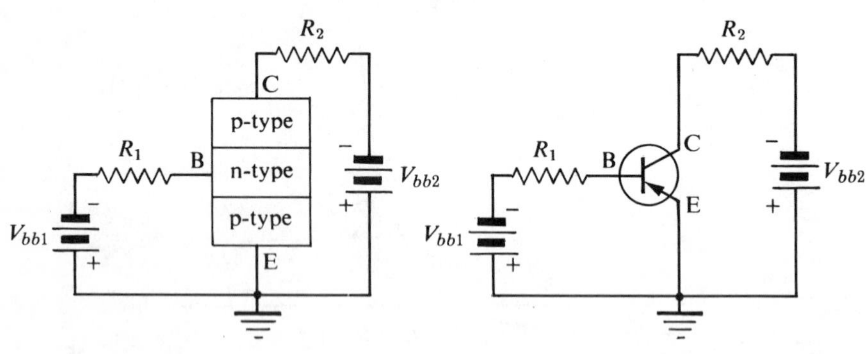

(b) *practical* pnp *bias circuit*

(c) pnp *bias circuit with standard transistor symbol*

FIGURE 5.3 Transistor biasing.

n-type base is positive with respect to the p-type collector. A battery of voltage V_{bb2} (typically from 3 to 30 V) and a resistor R_2, also shown in Fig. 5.3(a), would be suitable for this bias. A more common and convenient arrangement is to connect the battery V_{bb2} and resistor R_2 between the emitter and the collector as shown in Fig. 5.3(b). This arrangement is more convenient because the two batteries have one point in common which can be grounded. The two circuits provide essentially the same base-collector reverse-bias voltage because the base-emitter junction is forward-biased, so the voltage difference between the base and emitter is only 0.2 to 0.5 V. In other words the base-emitter junction has a very low resistance compared to the high resistance of the reverse-biased base-collector junction. The contacts between the external wire leads and the transistor are not rectifying; they conduct current equally well in both directions.

Because of the reverse bias between base and collector, there is a depletion region formed between the base and the collector with an electric field $\mathbf{E}_d$ pointing from the base *towards* the collector as shown in Fig. 5.4. The magnitude of the electric field is simply $\mathbf{E}_d \cong V_{CE}/d$ where d is the thickness of the depletion region if we assume that the base-collector junction resistance is large compared to the base-emitter junction resistance. The point here is that most of the collector-

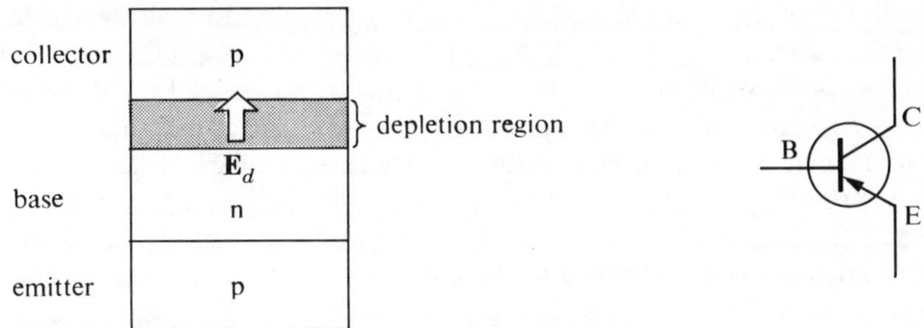

FIGURE 5.4 Depletion regions and field at base-collector junction of pnp transistor.

emitter voltage V_{CE} appears across the reverse-biased base-collector junction. Notice that the electric field tends to sweep majority carriers in the base and collector material away from rather than across the base-collector junction. Outside the depletion region, there is no appreciable electric field in the transistor material, and the motion of charge carriers, both holes and electrons, is governed solely by diffusion—which is simply the statistical tendency for particles with random thermal motion to migrate from regions of higher concentration towards regions of lower concentration.

Because of the conservation of current we must have $I_E = I_B + I_C$; that is, the current I_E flowing in the emitter must exactly equal the current I_B leaving via the base, plus the current I_E leaving via the collector. Again remember that we are talking about positive currents consisting of the flow of positively charged holes. We would expect the effective resistance of the base-emitter junction to be very small because of its *forward* bias; and similarly, we expect the resistance of the base-collector junction to be relatively high because of its *reverse* bias. One might be tempted at this point to conclude that the base current I_B would be large compared to the collector current, but this conclusion is absolutely wrong. In fact exactly the opposite situation occurs: $I_B \ll I_C$. To explain this we must consider the current flow inside the transistor more carefully.

Let us consider in detail the motion of positive holes injected from the emitter into the base. The base is essentially a field-free region so the holes move around by diffusion. These holes in the base can do one of three things: (1) They can recombine with the mobile electrons in the base (the majority carriers of the n-type base material). (2) They can diffuse through the base and recombine with electrons injected into the base region from the wire going to R_1. (3) They can diffuse across the base region into the depletion layer of the reverse-biased base-collector junction where the electric field $\mathbf{E}_d$ sweeps them over into the collector. Once in the collector the holes diffuse around and eventually recombine with the electrons injected into the collector from the wire connected to R_2. In other words a positive current I_C flows out of the collector.

Positive holes are continuously created in the emitter because the wire going to the positive terminal of the battery V_{bb2} continuously draws electrons out of the emitter, thus creating holes in the emitter. The recombination of the holes and electrons in the base region forms the base current I_B, and the holes which are swept into the collector region form the collector current, I_C.

The single most important feature of a transistor which makes it a useful device is the thin design of the base region so that process (3) is much more probable than either (1) or (2). That is, very few of the holes injected from the emitter into the base recombine in the base; most of them are swept into the collector region by the electric field $\mathbf{E}_d$ existing at the base-collector junction. This is done by making the thickness of the base region small compared to the "mean free path" or the "diffusion length" of the holes in the base. In such a case, the holes in the base simply do not have much time to recombine with electrons before they are swept into the collector by the electric field $\mathbf{E}_d$ at the base-collector junction.

In an actual transistor, 98% or 99% of the holes injected into the base will be swept into the collector. Thus, the base current will be only 2% or 1% of the emitter current. If we let α be the ratio I_C/I_E ($\alpha = 0.98$ or 0.99), then the conservation of current equation $I_E = I_B + I_C$ becomes

$$\frac{I_C}{\alpha} = I_B + I_C$$

or

$$\frac{I_C}{I_B} = \frac{\alpha}{1 - \alpha}. \tag{5.1}$$

Also,

$$I_B = (1 - \alpha)I_E \tag{5.2}$$

If $\alpha = 0.99$, then $I_C/I_B = 0.99/(1 - 0.99) = 99$, $I_B = 0.01\ I_E$, and $I_C = 0.99\ I_E$. If $\alpha = 0.98$, then $I_C/I_B = 0.98/(1 - 0.98) = 49$, $I_B = 0.02\ I_E$, and $I_C = 0.98\ I_E$.

Notice that the ratio I_C/I_B is a very strong function of α. A 1% change in α will double the I_C/I_B ratio. It is common notation to let $\beta \equiv \alpha/(1 - \alpha)$. Thus, $I_C/I_B = \beta$. The β is usually on the order of 20 to 200 for most transistors. Notice also that the collector current is much larger than the base current and that the collector and emitter currents are essentially equal. For example, a low power transistor may have an emitter current $I_E = 5.0\,\text{mA}$, a base current of $50\,\mu\text{A}$, and a collector current of $4.95\,\text{mA}$ if $\alpha = 0.99$. Also notice that if α remains constant, then we will obtain amplification if we can use I_B as an input and I_C as an output because $I_C/I_B = \beta \gg 1$. More on this later.

Alpha, and therefore beta, vary with emitter current. At low emitter currents, an appreciable fraction of the holes injected from the (p-type) emitter into the (n-type) base may recombine. At higher emitter currents a smaller fraction of the emitter current is lost to recombination because the number of electrons in the base is now much less than the number of holes injected from the emitter. Thus alpha and beta increase with increasing emitter current. Beta may double as the emitter current is increased from 0.5 mA to several mA for a typical transistor. Alpha and beta decrease slightly (by about 10% to 20%) as the emitter current is increased from several mA to around 100 mA due to increased base conductivity resulting from the larger number of charge carriers in the base.

As we have just mentioned, the main feature of a transistor is the thinness of the base region, which enables most of the holes injected from the emitter into the base to reach the collector and form the collector current. Another physical feature of transistor construction is that the doping of the emitter is made larger than that of the base. This is done so that the concentration of holes injected from emitter to base is larger than the concentration of electrons in the base, thus producing minimum recombination of holes and electrons in the base and thus higher values of α and β for the transistor.

A third feature of transistor construction is that the collector is physically larger than the emitter. There is more power dissipated as heat in the collector than in the emitter because the base-collector voltage V_{BC} is much larger (typically several volts) than the base-emitter voltage V_{BE} (typically 0.2 V for germanium and 0.5 V for silicon). This is simply because the base-collector junction is *reverse* biased and the emitter-base junction is *forward* biased. The power dissipated as heat in the base-collector region is $I_C V_{BC}$, while the power dissipated in the emitter-base region is only $I_E V_{BE}$. For example, a germanium transistor with $\alpha = 0.99$, $I_E = 5$ mA, and $V_{BC} = 10$ V has a collector current of $I_C = \alpha I_E = 4.95$ mA, and will have $I_E V_{BE} = (5\text{ mA})(0.2\text{ V}) = 1.0$ mW and $I_C V_{BE} = (4.95\text{ mA})(10\text{ V}) = 49.5$ mW. A typical low power transistor can safely dissipate approximately several hundred milliwatts. Because of these differences between the emitter and the collector, one cannot blithely interchange the emitter and collector leads even though both emitter and collector are p-type material. The dopings of the collector and emitter differ and the breakdown voltage of the emitter-base junction is much less (typically 5 V) than for the base-collector junction (typically 30 V). The collector is often connected to the metal case of the transistor for better heat dissipation. The circuit symbol distinguishes between the emitter and collector; the arrow of the symbol in Fig. 5.1 always denotes the emitter in either pnp or npn transistors.

5.4 AMPLIFICATION

Amplification occurs in a device when the output is greater than the input. The amplification or gain A usually is defined as the output divided by the input, i.e., the voltage gain is defined as $A_v \equiv V_{out}/V_{in}$ or the current gain as $A_i \equiv I_{out}/I_{in}$. If we regard the base current as the input and the collector current as the output, the transistor will amplify so long as most of the holes injected into the base from the emitter are collected in the collector region. This can be seen by the following argument and reference to Fig. 5.5. A change in the

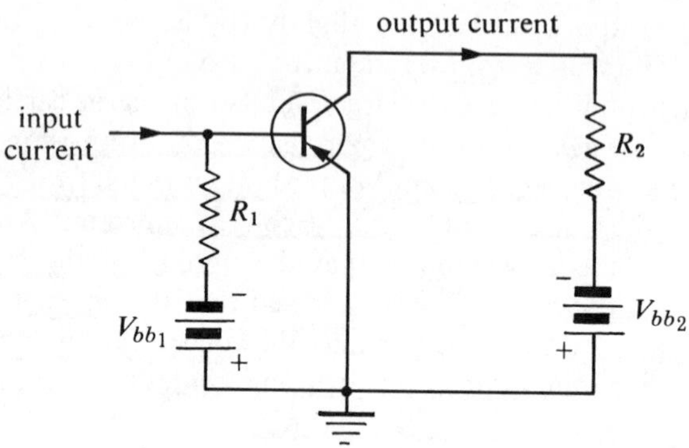

FIGURE 5.5 Transistor amplification.

input current changes the base current and therefore changes the base-emitter voltage V_{BE}. The emitter-base junction is forward biased, so a very *small* change in V_{BE} will result in a very *large* change in the current flowing from the emitter into the base. This follows from the steepness of the current-voltage curve (Fig. 5.2) for a forward-biased junction. A fraction α (on the order of 0.99) of this large *increase* in emitter-base current will then flow in the collector lead. Thus, a small change ΔI_B in input current I_B produces a large change ΔI_C in collector current I_C. And every hole passing from the base to the collector is accelerated by the electric field $\mathbf{E}_d$ in the depletion region of the base-collector junction. In other words, every hole going from the base to the collector falls through a voltage drop V_{BC} with $V_{BC} = \mathbf{E}_d d$ where d is the width of the depletion region. The power output is the power dissipated in the external resistance R_2: $P_{out} \cong I_C^2 R_2$. The input power is much smaller, as it is given approximately by $P_{in} \cong I_{in} V_{BE}$. Thus the power gain A_p would be equal to $I_C^2 R_2 / I_{in} V_{BE}$. For a typical low power transistor with $\alpha = 0.99$, a $10\,\mu$A change in input current will produce a 1 mA change in collector current; R_2 is several kilohms, and $V_{BE} = 0.2$ V for a germanium transistor. Hence $A_p \cong (1\text{ mA})^2(2\text{ k}\Omega)/(10\,\mu\text{A})(0.2\text{ V}) = 1000$. If the output voltage is measured between the collector and ground V_{CE}, it will change according to $\Delta V_{CE} = \Delta I_C R_2$. The change in the input voltage V_{BE} will be the very small voltage change across the forward-biased base-emitter junction, on the order of 10 millivolts at most. Hence the voltage gain will be approximately:

$$A_v \cong \frac{\Delta V_{CE}}{\Delta V_{BE}} = \frac{\Delta I_C R_2}{\Delta V_{CE}} \cong \frac{(1\,\text{mA})R_2}{10\,\text{mV}} = 200$$

if $R_2 = 2\,\text{k}\Omega$.

The net result of the transistor construction and biasing is that a very small change in the base-emitter voltage produces a large increase in the current flowing from the emitter to the base, because the base-emitter junction is *forward biased*. In other words we are working on the very steep forward-bias portion of the p-n junction current-voltage curve of Fig. 5.2. The bias is necessary to ensure we are on the steep portion of the curve rather than the flatter portion near the origin. In other words, the bias is necessary to turn the base-emitter junction on. The large fraction of holes collected by the depletion electric field $\mathbf{E}_d$ at the base-collector junction ensures that most of the large increase in the emitter-base current shows up as collector current.

5.5 BIASING AND GRAPHICAL TREATMENT

A transistor has three terminals, so if we have two input and two output terminals one of the three terminals must be common between the input and the output. These three possible configurations are called the "common emitter," the "common collector," and the "common base" configurations. In this section we will restrict ourselves to the common emitter configuration shown in Fig. 5.6,

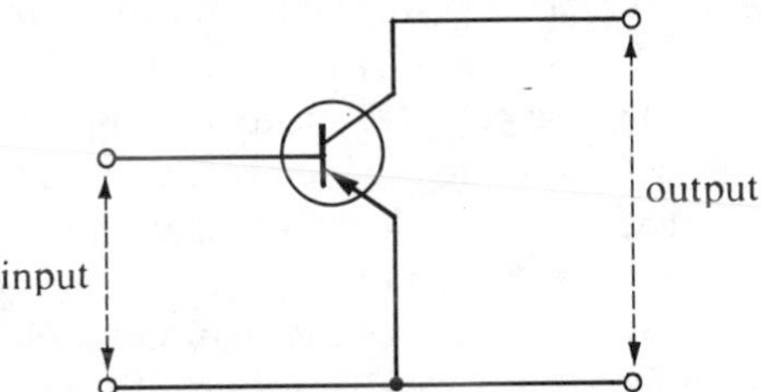

FIGURE 5.6 Common emitter configuration.

because it is the most useful configuration for most amplifiers. The other two configurations will be treated in Chapter 6. The base-emitter junction is forward-biased, so it is easiest to control this junction by controlling and measuring the base current I_B rather than the voltage V_{BE}, because V_{BE} is essentially constant at 0.2 volts for a germanium transistor (0.5 volts for silicon)—whereas I_B may vary by a factor of ten or more from $5\,\mu\text{A}$ to $50\,\mu\text{A}$. Because it is reverse-biased, it is easiest to control the collector-base junction by controlling the collector-base, voltage V_{BC}, which is approximately equal to the collector-emitter voltage V_{CE}. If we measure the current-voltage curves of these two junctions for a typical low power germanium transistor, we obtain curves shown in Fig. 5.7.

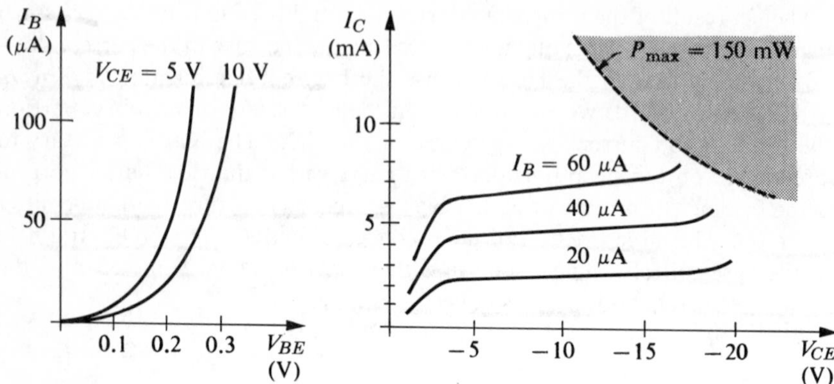

FIGURE 5.7 Input and output graphs for a 2N1309, low cost (@ $.90) germanium low power pnp transistor in the common emitter configuration (V_{CE} negative merely means the collector is negative with respect to the emitter).

The input graph of V_{BE} versus I_B is merely the graph of the forward-biased p-n junction similar to Fig. 5.2. It can be seen that the base-emitter voltage remains substantially constant at approximately 0.2–0.25 V if the base current exceeds about 20 μA. For a silicon transistor, the input graph would be similar except that the base-emitter voltage would be approximately constant at 0.5 V.

The output graph of I_C versus V_{CE} is the current-voltage graph for a re-versed-biased p-n junction; the collector current is approximately independent of V_{CE}. However, the collector current increases drastically if the base current increases, because an increase in base current means the forward-biased base-emitter junction has been turned on more.

The equation $P_{max} = I_C V_{CE} = 150\,\text{mW}$ has been plotted on the output graph to indicate the maximum allowable values for the collector current and collector-emitter voltage. Every transistor is rated for the maximum power it may dissipate without overheating; for the 2N1309, $P_{max} = 150\,\text{mW}$. The product of the I_C and V_{CE} must *never* exceed 150 mW even for a very short time; this means that the shaded region on the output graph corresponding to $I_C V_{CE}$ greater than 150 mW is a forbidden region for transistor operation. If the power dissipated in the transistor exceeds the maximum power rating, the transistor will quietly and quickly burn out.

At this point it might be well to ask exactly what would happen if the transistor were overheated. The answer is that as the temperature increases, the impurity atoms tend to diffuse through the semiconductor from points of high concentration toward points of low concentration, i.e., acceptor atoms from the p-type emitter and collector would diffuse into the n-type base, and donor atoms would diffuse from the base into the emitter and collector. This diffusion destroys the sharpness of the p-n junctions, and produces gradual junctions which have current-voltage graphs totally different from those of Fig. 5.2. If the temperature is lowered, the impurity atoms will *not* diffuse back into their original positions. Such a process goes from a less ordered (more probable) to a more ordered (less probable) state and would violate the second law of thermodynamics. Thus, a transistor (or any semiconductor device containing p-n junctions) can be irreversibly damaged by overheating. It is also possible for "thermal runaway" to occur; in which a rise in temperature produces an increase in collector current which increases the power dissipated in the transistor, which in turn increases the temperature still more, and so on. This may burn out the transistor in a very short time, of the order of milliseconds. A detailed treatment of this calamity and its prevention will be given later in this chapter.

Let us now consider the problem of biasing a common emitter amplifier using a 2N1309 transistor whose input and output curves are shown in Fig. 5.7. In order to make the transistor function as an amplifier we must: (1) bias the base-emitter junction in the *forward* direction, (2) bias the base-collector junction in the *reverse* direction, and (3) arrange the circuit so that neither temperature changes nor the input signal disturbs these bias conditions. Also, if we are going to make many copies of the same circuit, we want the circuit behavior to be essentially independent of transistor parameter changes we might encounter when we replace transistors. In general there is a wide sample-to-sample variation of transistor parameters. The parameter $\beta = \alpha/(1 - \alpha)$ may vary by $\pm 50\%$ or more for inexpensive transistors. Specially selected and more expensive transistors are available with smaller tolerances on β and other parameters. The β for the transistor whose curves are shown in Fig. 5.7 is approximately equal to 100, from $\beta = I_C/I_B$.

The base-collector junction can be biased in the reverse direction by hooking up a resistance R_C and a battery of voltage V_{bb} as shown in Fig. 5.8(a). The Kirchhoff voltage law for V_{bb}, R_C, and V_{CE} is:

$$-V_{CE} - I_C R_C + V_{bb} = 0, \tag{5.3}$$

which can be solved for I_C,

$$I_C = \frac{V_{bb}}{R_C} - \left(\frac{1}{R_C}\right) V_{CE}. \tag{5.4}$$

Equation 5.4 can be plotted on the output graph for the transistor as is shown in Fig. 5.8(b). The plot is a straight line with slope $-1/R_C$, a vertical collector current intercept of V_{bb}/R_C, and a horizontal voltage intercept of V_{bb}. This straight line is called the "load line," and its significance is that regardless of the behavior of the transistor the collector current I_C and the collector emitter voltage V_{CE} must *always* lie on the load line. In other words the linear relation of I_C and V_{CE} is forced upon us by Kirchhoff's voltage law and Ohm's law. In a

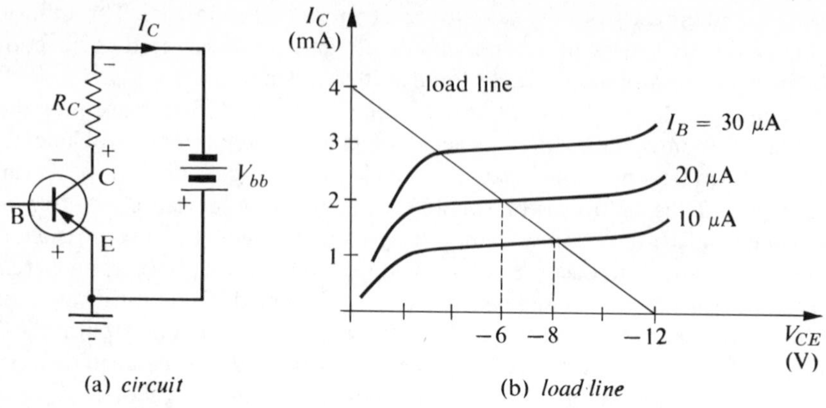

FIGURE 5.8 Common emitter circuit and load line.

practical circuit, the load line should be kept below the $P_{\max}$ graph to avoid burning out the transistor. Notice also that the intercepts have a simple physical meaning: the vertical intercept $I_C = V_{bb}/R_C$ corresponds to all the battery voltage appearing across the resistor R_C and none across the transistor. This means the transistor has a resistance of zero ohms; it is fully conducting or "turned on." The horizontal intercept $V_{CE} = V_{bb}$ means that all the battery voltage appears across the transistor and none across R_C; thus, no current I_C can flow through R_C. The transistor has an effective resistance of infinity, in other words it is completely "turned off" or nonconducting. For a point on the load line with certain values of I_C and V_{CE}, the base current is determined by the intersection of the constant base current line with the load line. As a final point, notice that this is a dc load line relating dc voltages and currents. The ac load line which is appropriate for the ac collector current and ac collector-emitter voltage may be completely different because of various capacitances and inductances. More on this later.

Once the load line is drawn, the current amplification can be read off the graph. If the base current changes from $10\,\mu A$ to $20\,\mu A$, then the collector current must change from 1 mA to 2 mA, giving a current gain of 100. If the base current varies sinusoidally from $10\,\mu A$ to $20\,\mu A$, the collector current will vary approximately sinusoidally from 1 mA to 2 mA. The voltage gain will also be large because only a very small change in the base-emitter voltage is necessary to change I_B from $10\,\mu A$ to $20\,\mu A$ while V_{CE} changes from -8 V to -6 V. If a large change in I_C is desired, the load line should be steep, meaning R_C should be small; if a large change in V_{CE} is desired the load line should be flat, meaning R_C should be large. In other words for a large current gain choose R_C small, for a large voltage gain choose R_C large.

It can also be seen that the magnitude of the change in collector current or collector-emitter voltage for a given change in base current will depend upon what the initial base current is with no input. The "operating" point or the "quiescent" point on the load line is the point which represents I_C, V_{CE}, and I_B for no input signal.

The problem of fixing the operating point is basically the problem of setting the base-emitter forward bias so that the dc base current has the desired value with no signal input. If the input signal that is to be amplified has roughly equal positive and negative swings, then the operating point should be chosen approximately in the center of the load line. To be specific, suppose $V_{bb} = 12\text{V}$ and $R_C = 3\,\text{k}\Omega$. Then the load line has a vertical intercept of $12\text{V}/3\,\text{k}\Omega = 4\,\text{mA}$, and a horizontal intercept of -12V. If we choose the operating point to be $I_C = 2\,\text{mA}$, then $V_{CE} = 6\text{V}$, which implies that the base current must be $I_B = I_C/\beta = 20\,\mu\text{A}$.

Thus, our biasing problem is to supply a steady current of $20\,\mu\text{A}$ to the base even when there is no input signal. The value of base current either can be read off the graph of I_C versus V_{CE} or can be calculated from $I_B = I_C/\beta$. This biasing could be done by connecting up a 1.5V battery and a $65\,\text{k}\Omega$ resistor as shown in Fig. 5.9(a). The current I_B flowing in the base will be given by I_B

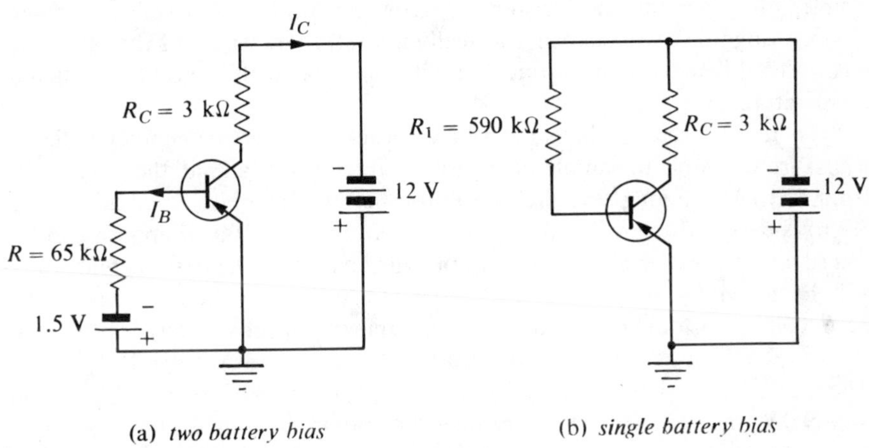

(a) *two battery bias* (b) *single battery bias*

FIGURE 5.9 Common emitter biasing circuits.

$= (1.5\text{V} - 0.2\text{V})/R$. Thus $R = 1.3\text{V}/I_B = 1.3\text{V}/20\,\mu\text{A} = 65\,k\Omega$. We have set the base-emitter junction voltage to 0.2V, because the junction is turned on and we have assumed a germanium transistor.

However, this biasing circuit of the battery and the $65\,\text{k}\Omega$ resistor has three disadvantages. It requires a separate battery in addition to the 12V battery biasing the collector-base junction. A second, separate battery is expensive, heavy, and bulky. Secondly, this circuit is unstable with respect to changes in temperature. And finally, if a different transistor of the same type (2N1309) but with a different β is put in the same circuit, the operating point may change appreciably; that is, the operating point is a sensitive function of the transistor β. One possible circuit using only one battery is shown in Fig. 5.9(b).

We recall $\beta = 100$ from the graph of Fig. 5.7; if the graph is not available one looks up the value of β for the particular transistor in a transistor manual. The value of R_1 can be calculated as follows. For the transistor to be "turned on," i.e., to conduct an appreciable (>1 mA) amount of collector current, the base-emitter junction must be forward-biased. If the transistor is germanium the base must be approximately 0.2 volts negative with respect to the emitter. In Fig. 5.9(b), the emitter is grounded so its voltage is exactly zero. Therefore the base should be at -0.2V. At the desired operating point the transistor will draw 2 mA of current with no signal input. Thus $I_C = 2\,\text{mA}$, and $I_B = I_C/\beta = 2\,\text{mA}/100 = 20\,\mu\text{A}$. R_1 is now calculated from Ohm's law. We know the current through R_1 and the voltage at each end:

$$R_1 = \frac{V_B - V_{bb}}{I_B} = \frac{-0.2\text{V} - (-12\text{V})}{20\,\mu\text{A}} = \frac{11.8\text{V}}{20\,\mu\text{A}} = 590\,\text{k}\Omega.$$

The biasing arrangement for the base shown in Fig. 5.9(b) is certainly simple, it involves only one resistor R_1. However, it still has two disadvantages: its operating point will change drastically if another transistor of the same type but with a different β is substituted, and it is unstable with respect to changes in temperature.

There is a considerable spread or variation in transistor characteristics as we go from sample to sample of the same type. For example, the β may vary from 70 to 150 among several 2N1309 transistors. The biasing circuit of Fig. 5.9(b) is designed to work with a transistor with a β of 100. Suppose another transistor with a β of 50 were used in the same circuit. The base current would still be given by $I_B = (V_B - V_{bb})/R_1 = 11.8\text{V}/590\,\text{k}\Omega = 20\,\mu\text{A}$, because the base voltage of the new transistor will remain approximately constant at $V_B = -0.2$V. But now $I_C = \beta I_B = 50 \times 20\,\mu\text{A} = 1.0\,\text{mA}$ instead of 2.0 mA. Therefore, $V_{CE} = -V_{bb} + I_C R_C = -12 + (1.0\,\text{mA})(3\,\text{k}\Omega) = -12 + 3.0\text{V} = -9.0\text{V}$, instead of -6V. The operating point has shifted from $I_C = 2\,\text{mA}$, $V_{CE} = 6\text{V}$ to $I_C = 1.0\,\text{mA}$, $V_{CE} = 9.0\text{V}$, just by replacing a $\beta = 100$ (2N1309) with a $\beta = 50$ (2N1309).

A two-resistor voltage divider as shown in Fig. 5.10(a) can also be used to bias the transistor with one battery, and it has the additional advantage that the bias voltage tends to change less due to temperature induced changes in R_1 and R_2. That is, the bias in Fig. 5.10(a) depends upon the ratio of R_1 to R_2, which is less sensitive to temperature changes than the single-resistor bias circuit of Fig. 5.9. Also R_1 and R_2 in Fig. 5.10 are usually much less than R_1 in Fig. 5.9(b), thus reducing the change in the operating point caused by temperature induced changes in the base current. This is covered in the next several paragraphs. The voltage at the base V_B is given by $V_B = I_D R_1$ where I_D is the current flowing through R_1. And $V_{bb} = (I_D + I_B)R_2 + I_D R_1$ or $I_D = (V_{bb} - I_B R_2)/(R_1 + R_2)$. Thus, $V_B = I_D R_1$ becomes

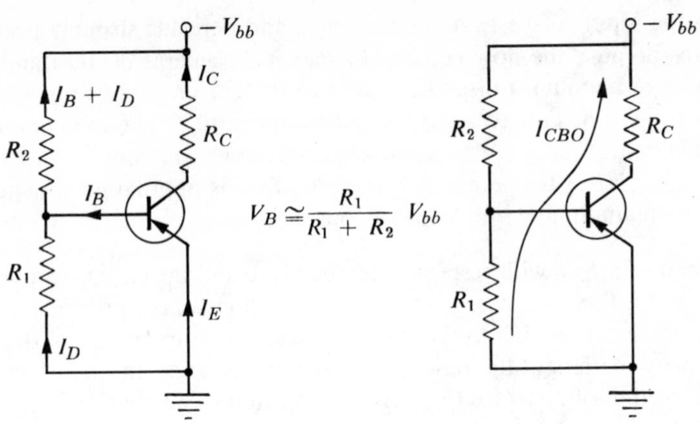

(a) $R_1 R_2$ *voltage divider bias* (b) I_{CBO} *temperature dependent current*

FIGURE 5.10 Two-resistor bias circuit.

$$V_B = \frac{V_{bb} - I_B R_2}{R_1 + R_2} R_1. \tag{5.5}$$

Usually R_1 and R_2 are made sufficiently small so that $I_B \ll I_D$ or $I_B R_2 \ll V_{bb}$. Then $I_D \cong V_{bb}/(R_1 + R_2)$, and

$$V_B = \frac{R_1}{R_1 + R_2} V_{bb}. \tag{5.6}$$

However, the bias circuit of Fig. 5.10 is still unstable with respect to temperature fluctuations. The temperature instability arises from the thermal generation of minority carriers in the semiconductor material of the transistor. Any increases in temperature, from whatever source, will increase the number of covalent bonds broken in the semiconductor lattice and thereby will increase the number of minority carriers in all three regions of the transistor. The relative increase of majority carriers will be small because almost all the donor or acceptor atoms are already ionized at room temperature. Let us focus our attention on the thermally generated minority carriers in the base region and the collector region. A higher temperature means more mobile holes in the base and electrons in the collector for a pnp transistor.

The voltage drop between the base and collector is a reverse voltage only for the majority carriers, but is a *forward* voltage for minority carriers. Therefore, the thermally generated holes in the base and electrons in the collector will flow across the base-emitter junction and constitute a current from base to collector. This reverse current is usually denoted by I_{CBO} or I_{CO}, where the subscripts stand for collector to base with the emitter open. This current flows even when the input to the transistor is open, and depends strongly upon the temperature, because the flow consists of *thermally* generated holes and electrons. For a germanium transistor, I_{CBO} at room temperature will increase by a factor of two (from $1\,\mu$A to $2\,\mu$A) for only about a 10°C rise in temperature, and for a silicon transistor will increase by a factor of two (from 5 nA to 10 nA) for only a 6°C rise in temperature. At 70°C, I_{CBO} is approximately equal to $100\,\mu$A for germanium and 50 nA for silicon.

The change in I_{CBO} with respect to temperature and the change in the turn-on voltage for the forward-biased emitter-base junction are the principal sources of temperature instability in transistors. As the temperature increases, the number of minority carriers in the base and collector increases, and I_{CBO} increases. From the direction of I_{CBO} (see Fig. 5.10(b)), it can be seen that an increase in I_{CBO} will make the base voltage more negative, thus increasing the forward bias voltage between base and emitter. In other words, an increasing temperature turns the transistor on more. This increases the emitter-base current and also the collector current. The increase in collector current increases the power dissipated in the transistor and thereby raises the transistor temperature. This temperature rise generates more minority carriers, and the cycle starts over again. If not checked, the transistor can eventually destroy itself from overheating.

The turn-on voltage or V_{BE} for a transistor tends to change with temperature, the transistor turning on more as the temperature increases because of the thermal generation of minority carriers in the base, and the resulting change is the Fermi level as discussed in Chapter 4. Equivalently, the base-emitter voltage required for a given collector current decreases with increasing temperature. For both germanium and silicon transistors, near room temperature, the base-emitter voltage decreases at approximately -2.5 mV/°C, although one would expect slightly less temperature dependence for silicon because of the larger energy gap.

The only link in the chain just mentioned that can be broken is the change in the base-emitter voltage. There is absolutely no way to prevent an increase in temperature from increasing the number of minority carriers. However, if we can devise a bias circuit for the base that will hold the base-emitter voltage constant at the desired voltage, regardless of the increase in I_{CBO}, then the collector current will remain constant except for the small increase due to the increasing I_{CBO}. There is no way to prevent V_B from becoming more negative due to the increase in I_{CBO}, but if we also can make the emitter voltage go more negative at the same time, then the forward bias V_{BE} will not increase nearly so much. A little thought will show that a resistor R_E placed in series with the emitter as shown in Fig. 5.11 will do the trick. For then, as the temperature increases, I_{CBO} increases, V_B goes more negative, and I_C and I_E increase. But now $V_E = -I_E R_E$ also goes more negative, so $V_B - V_E = V_{BE}$ remains much more nearly constant.

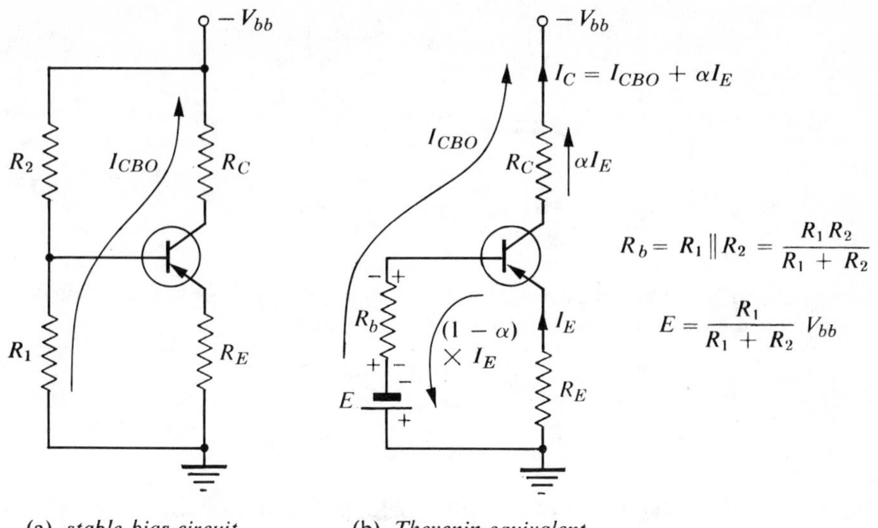

(a) *stable bias circuit* (b) *Thevenin equivalent*

FIGURE 5.11 Stable common emitter bias circuit.

5.6 STABILITY FACTOR

In order to calculate exactly how large a resistor one must place in the emitter lead to stabilize the common emitter amplifier against changes in temperature, we must carefully write down the Kirchhoff equations for the circuit and solve for I_C in terms of I_{CBO} and V_{BE}. We will then calculate the change in I_C due to a change in I_{CBO} and in V_{BE}.

First we replace the R_1R_2 voltage divider by its Thevenin equivalent as shown in Fig. 5.11(b). The R_b represents the parallel resistance of R_1 and R_2; in most circuits $R_1 \ll R_2$ so $R_b \cong R_1$. Now we can write the Kirchhoff voltage equation for the loop containing the base-emitter junction of the transistor:

$$-E + (1 - \alpha)I_E R_b - I_{CBO} R_b + V_{BE} + I_E R_E = 0. \tag{5.7}$$

Using $I_E = (I_C - I_{CBO})/\alpha = (\beta + 1)/\beta(I_C - I_{CBO})$ to eliminate I_E and $\alpha = \beta/(\beta + 1)$ to eliminate α, we obtain:

$$I_C = \frac{R_b + R_E}{R_E + \dfrac{R_b}{\beta + 1}} I_{CBO} + \frac{E - V_{BE}}{\dfrac{\beta + 1}{\beta} R_E + \dfrac{R_b}{\beta}}. \tag{5.8}$$

We now can see how changes in I_{CBO} and in V_{BE} will affect the collector current I_C. A small change ΔI_{CBO} and a small change ΔV_{BE} will produce, to first order, a change ΔI_C in the collector current:

$$\Delta I_C = \left(\frac{\partial I_C}{\partial I_{CBO}}\right)_{V_{BE}} \Delta I_{CBO} + \left(\frac{\partial I_C}{\partial V_{BE}}\right)_{I_{CBO}} \Delta V_{BE}. \tag{5.9}$$

Using (5.8) we obtain:

$$\Delta I_C = \frac{R_b + R_E}{R_E + \dfrac{R_b}{\beta + 1}} \Delta I_{CBO} - \frac{\beta}{(\beta + 1)R_E + R_b} \Delta V_{BE}. \tag{5.10}$$

Equation (5.10) can be simplified by noting that β typically is of the order of 100. Thus $\beta/(\beta + 1) \cong 1$ and we get:

$$\Delta I_C \cong \frac{R_b + R_E}{R_E + \dfrac{R_b}{\beta + 1}} \Delta I_{CBO} - \frac{1}{R_E + \dfrac{R_b}{\beta}} \Delta V_{BE}. \tag{5.11}$$

It is customary to define a stability factor S as the rate at which I_C changes with respect to changes in I_{CBO} alone, i.e., with V_{BE} being assumed constant.

$$S \equiv \left(\frac{\partial I_C}{\partial I_{CBO}}\right)_{V_{BE}} = \frac{R_b + R_E}{R_E + \dfrac{R_b}{\beta + 1}} \tag{5.12}$$

In terms of S, ΔI_C becomes:

$$\Delta I_C \cong S\,\Delta I_{CBO} - \frac{S}{R_b + R_E} \Delta V_{BE}. \tag{5.13}$$

At this point, we should explicitly point out that for a temperature increase ΔI_{CBO} is positive, but ΔV_{BE} is negative. Thus, both terms in (5.12) tend to *increase* I_C for a temperature rise. The value of ΔI_{CBO} is about 1 to 2μA for a 10°C temperature rise in germanium, and about 10 to 20 nA for a 10°C temperature rise in silicon. For both germanium and silicon devices, $\Delta V_{BE} \cong -2.5$ mV/°C as discussed in Chapter 4. We now see that a small stability factor S is desirable to minimize the change in collector current caused by temperature changes. If $R_b/(\beta + 1) \ll R_E$, the stability factor is simply:

$$S \cong 1 + \frac{R_b}{R_E}. \tag{5.14}$$

If we want ΔI_C due to ΔI_{CBO} to be only 1% of the total emitter current, which is usually from 1 to 2 mA, for a 10°C temperature rise, then we can calculate the required stability factor S for a germanium transistor from equation 5.12:

$$\Delta I_C = \frac{I_C}{100} = S\,\Delta I_{CBO}\big|_{10^\circ} \tag{5.15}$$

$$S = \frac{0.01 I_C}{\Delta I_{CBO}\big|_{10^\circ}} \cong \frac{0.01\text{ mA}}{1\,\mu\text{A}} = 10.$$

For a silicon transistor $\Delta I_{CBO} \cong 10\,\text{nA}$, and $S = 1000$ which is very easy to achieve. If we want ΔI_C due to ΔV_{BE} to be only 1% of I_C, we have for a 10°C temperature rise:

$$\Delta I_C = \frac{I_C}{100} = \frac{S}{R_b + R_E} \Delta V_{BE}$$

$$S = \frac{0.01 I_C (R_b + R_E)}{\Delta V_{BE}} = \frac{(0.01\,\text{mA})(R_b + R_E)}{(2.5\,\text{mV/deg})(10\,\text{deg})}$$

$$S = \frac{(0.01\,\text{mA})(R_b + R_E)}{25\,\text{mV}} = 4 \times 10^{-4}(R_b + R_E). \tag{5.16}$$

Equation (5.16) applies to both silicon and germanium devices.

In a typical circuit, because of the power supply voltages and the current carrying capabilities of the transistor, R_E is usually from 100Ω to 1 kΩ. To make I_D, the current through the $R_1 R_2$ bias network, large compared to I_B, $R_b = R_1 \| R_2 \cong R_2$ is usually from 5 kΩ to 50 kΩ. Thus, we can obtain a stability factor of $S = 10$ by choosing $R_b \cong 10\,\text{k}\Omega$, $R_E \cong 1\,\text{k}\Omega$. This satisfies (5.15) and makes $S \cong 4$ from (5.15). Thus, with $S \cong 10$ a 10°C rise in a germanium transistor will produce an increase of collector current given by:

$$\Delta I_C = S\,\Delta I_{CBO} + \frac{S}{R_b + R_E}|\Delta V_{BE}| = 10(1\,\mu\text{A}) + \frac{10}{11\,\text{k}\Omega} 25\,\text{mV}$$

$$= 10\,\mu\text{A} + 25\,\mu\text{A} = 35\,\mu\text{A},$$

which will be small compared to the normal dc collector current of several mA. For a silicon transistor with $S \cong 10$ and a 10°C rise, we will have

$$\Delta I_C = S\,\Delta I_{CBO} + \frac{S}{R_b + R_E}|\Delta V_{BE}| = 10(10\,\text{nA}) + \frac{10}{11\,\text{k}\Omega} 25\,\text{mV}$$

$$= 0.1\,\mu\text{A} + 25\,\mu\text{A} \cong 25\,\mu\text{A}.$$

Thus for both germanium and silicon transistors at room temperature where $\Delta I_{CBO} \cong 1/\mu\text{A}$ and 10 nA, respectively, the major source of collector current increase due to temperature changes is due to the ΔV_{BE} term, i.e., it is due to the change in the turn-on voltage of the base-emitter junction at the rate of 2.5 mV/°C. Only for higher temperatures does the ΔI_{CBO} term become dominant.

The net result of this analysis is that we should design the bias network that the stability factor $S = 1 + R_b/R_E$ is not larger than approximately 10 and also keep $R_b + R_E$ about 10 kΩ or more in order to minimize the temperature induced changes in collector current.

If $R_E = 0$, however, the situation is disastrously different. Going back to equation (5.11) and setting $R_E = 0$, we obtain:

$$\Delta I_C \cong (\beta + 1)\,\Delta I_{CBO} + \frac{\beta}{R_b}|\Delta V_{BE}|. \tag{5.17}$$

In the previous example with $\beta = 100$, the 10°C temperature change in a germanium transistor would now result in a change in collector current of approximately

$$\Delta I_C = (101)(1\,\mu\text{A}) + \frac{101}{10\,\text{k}\Omega}\,25\,\text{mV} = 101\,\mu\text{A} + 250\,\mu\text{A} = 351\,\mu\text{A} = 0.351\,\text{mA},$$

which is an appreciable fraction of the usual collector current (~1 mA) flowing. The resulting increase in collector current will increase the power dissipated in the base-collector junction and this could easily raise the junction temperature another 10°C, thus doubling I_{CBO} and changing V_{BE} by 25 mV, and possibly leading to thermal runaway and a burned out transistor. The junction itself, as distinct from the transistor case, is extremely small and thus has a very low mass. Hence a small increase in power dissipated in the junction may raise its temperature appreciably. The much larger transistor case always is considerably cooler than the junction.

The moral is clear—to ensure a reasonably constant operating point (I_C, V_{CE}) for a germanium common emitter amplifier, we must include a resistor R_E in the emitter lead, and we would furthermore try to keep the ratio $R_b/R_E < 10$ and $R_b + R_E > 10\,\text{k}\Omega$, where R_b is the parallel resistance $R_1R_2/(R_1 + R_2)$ of the two base biasing resistors R_1 and R_2. It should be pointed out further that for silicon transistors, with their much smaller value of I_{CBO} (~10 mA versus ~1 μA for Ge), the matter is not quite so serious.

The temperature induced change in collector current caused by the change in the base-emitter or turn-on voltage can be minimized by the introduction of a diode as shown in Fig. 5.12 for a pnp transistor. The diode D should have

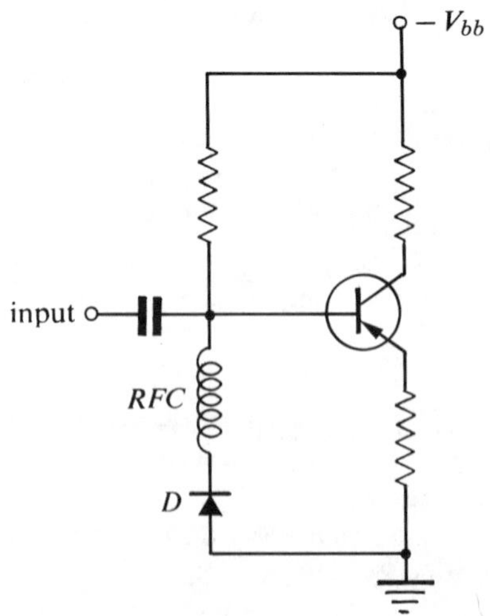

FIGURE 5.12 Diode temperature stabilization.

very nearly the same temperature coefficient as the transistor for best stabilization. Thus, a silicon diode should be used for a silicon transistor, etc. The principle of operation is simply that as the temperature changes the base-emitter voltage of the transistor will change, and the voltage across the diode will also change, but in the same direction. Thus the base voltage of the transistor will change in such a way as to minimize the change in the base-emitter or turn-on voltage. The radio frequency choke *RFC* in the circuit is merely to provide a high ac impedance from the base to ground for the signal input; hence this circuit is principally used at "high" frequencies where the inductive reactance of the *RFC* is several kΩ or more.

5.7 COMMON EMITTER AMPLIFIER DESIGN

Let us now quickly go through a "seat of the pants" design of a common emitter amplifier stage and illustrate the assumptions and approximations that go into an actual design problem. A more detailed common emitter design problem will be treated in a later chapter with attention paid to the input and output impedances and the exact gain desired. We first choose a transistor. For frequencies up to several hundred kHz and an environment under 70°C we may choose an inexpensive germanium transistor such as the 2N1308 or 2N1309. For operation at higher temperatures a silicon transistor would be necessary, such as the 2N3565. For higher frequency operation we would need a more expensive high frequency transistor and/or negative feedback that will be discussed in a later chapter.

Let us now focus our attention on the 2N1309, a pnp germanium transistor costing approximately $1.00 in small quantities. We will choose the voltage divider bias circuit (Fig. 5.13) for reasons of stability discussed in the previous section.

The 2N1309 characteristics available in an inexpensive transistor manual are:

1. Maximum power at 25°C = 125 mW.
2. Maximum collector-base voltage = 30 V.
3. Maximum collector current = 300 mA.
4. Maximum junction temperature = 100°C.
5. Minimum β = 150.

We choose a power supply dc voltage less than the 30 V maximum V_{CB}; let us choose $V_{bb} = -12$ V. Let us now choose a load line on the I_C versus V_{CE} graph. The Kirchhoff voltage law implies $-I_E R_E - V_{CE} - I_C R_C + V_{bb} = 0$. But $I_C = \alpha I_E$ and $\alpha \cong 1$, so we set $I_E \cong I_C$. Hence $I_C = (V_{bb} - V_{CE})/(R_E + R_C)$, which is the load line equation. This equation is graphed in Fig. 5.14, and the two intercepts are at $(I_C, V_{CE}) = (0, V_{bb}) = (0, -12\text{V})$ and $(I_C, V_{CE}) = [V_{bb}/(R_E + R_C), 0]$. We have already chosen $V_{bb} = -12$ V, so a choice of

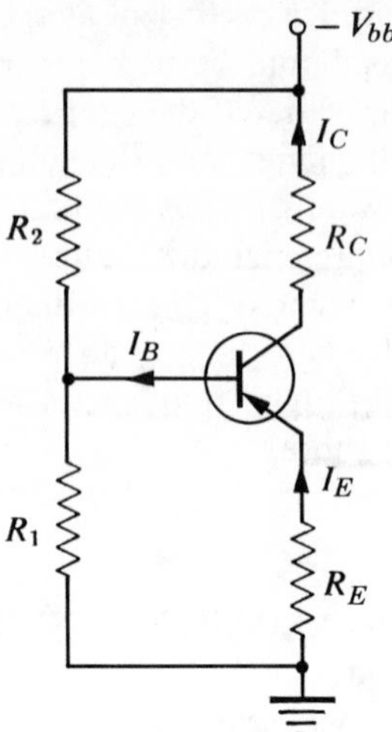

FIGURE 5.13 Common emitter amplifier.

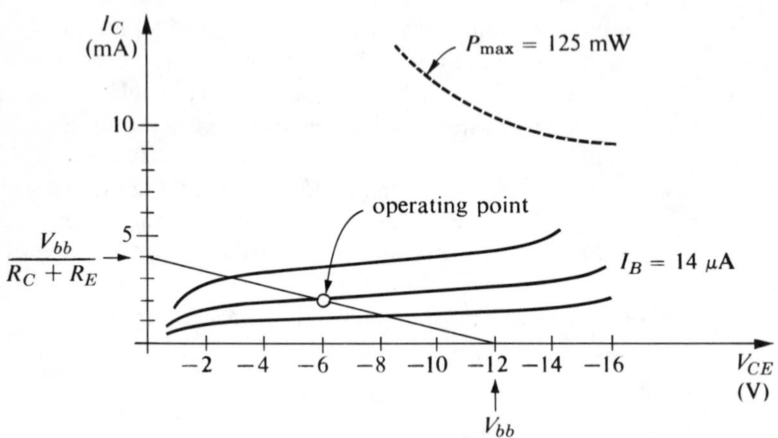

FIGURE 5.14 Load line.

$(R_E + R_C)$ will fix the other intercept and the load line. We must keep the load line below the $P_{max} = 125$ mW curve to avoid burning out the transistor. As we argued before, a flat load line (large $R_E + R_C$) will give a higher voltage gain, and a steep load line (small $R_E + R_C$) a higher current gain. Let us choose the vertical intercept at 4 mA. Then,

$$4\,\text{mA} = \frac{V_{bb}}{R_E + R_C} = \frac{12\,\text{V}}{R_E + R_C}.$$

Thus $R_E + R_C = 3\,\text{k}\Omega$. We still have not yet uniquely determined the four resistors, though.

The operating point must now be chosen. This is the point on the load line representing the transistor's dc collector current and dc collector-emitter voltage when there is no input signal applied. In general transistors function best if I_C is greater than about 1 mA and $|V_{CE}|$ is greater than about 1 V. If these conditions are not met, the gain of the transistor is usually decreased. So let us pick the operating point $(I_C, V_{CE}) = (2\,\text{mA}, -6\,\text{V})$. We now must pick R_E or R_C and then R_1 and R_2 will be determined by the bias requirements. $R_E + R_C = 3\,\text{k}\Omega$, so clearly neither R_E nor R_C can exceed $3\,\text{k}\Omega$. It can be shown that R_C should be as large as possible for a large gain. So choose $R_C = 2.2\,\text{k}\Omega$ and $R_E = 0.8\,\text{k}\Omega$. At the operating point $I_E = 2\,\text{mA}$, so the voltage at the emitter must be $V_E = -I_E R_E = -(2\,\text{mA})(0.8\,\text{k}\Omega) = -1.6\,\text{V}$. The base-emitter junction must be turned on so the base must be 0.2 V more negative than the emitter for a germanium transistor (0.5 V more for silicon). Thus, $V_B = -0.2\,\text{V} - 1.6\,\text{V} = -1.8\,\text{V}$.

We now must calculate R_1 and R_2 to fix the base voltage at $-1.8\,\text{V}$, but we must also make sure that the resulting circuit is stable against temperature fluctuations; that is, we must have $R_b = (R_1 \parallel R_2) \leqq 10 R_E$. See Fig. 5.15. This

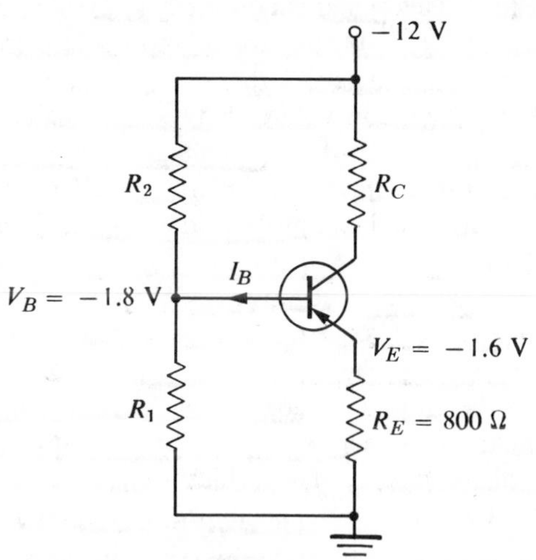

FIGURE 5.15 Circuit for calculation of R_1 and R_2.

requirement places an upper limit on $R_1 \parallel R_2$ and therefore on R_1. However, it is generally desirable to have a lower limit of roughly $2\,\text{k}\Omega$ on R_1 to avoid decreasing the gain. Too small a value of R_1 will tend to draw too much current from the signal source which is often another transistor stage; in other words, a low value for R_1 will then load down the signal source. More on this in Chapter 6.

Let us choose $R_b = (R_1 \| R_2) = 9R_E$ to make the stability factor $S = 1 + R_b/R_E = 10$. If we now assume the current drawn through the R_1, R_2 divider chain I_D is large compared to the base current I_B, we can write $R_1/(R_1 + R_2) = 1.8\text{V}/12\text{V}$. This assumption is easy to satisfy because $I_B = I_C/\beta = 2\,\text{mA}/150 = 14\,\mu\text{A}$ at the operating point. Then R_1 and R_2 are uniquely determined from the two equations:

$$R_b = (R_1 \| R_2) = 9R_E = (9)(800\,\Omega) = 7.2\,\text{k}\Omega$$

and

$$\frac{R_1}{R_1 + R_2} = \frac{1.8\,\text{V}}{12\,\text{V}}.$$

Solving for R_1 and R_2 yields $R_1 = 8.5\,\text{k}\Omega$ and $R_2 = 48\,\text{k}\Omega$. As a check that $I_D \gg I_B$, $I_D = V_B/R_1 = 1.8\,\text{V}/8.5\,\text{k}\Omega = 212\,\mu\text{A}$, which is $\gg I_B = 14\,\mu\text{A}$. Also, $R_1 = 8.5\,\text{k}\Omega$ is larger than the minimum desired value of roughly $2\,\text{k}\Omega$, and $R_b + R_E = 7.2\,\text{k}\Omega + 0.8\,\text{k}\Omega = 8.0\,\text{k}\Omega$, which is close to the desired $R_b + R_E \cong 10\,\text{k}\Omega$.

We will now show that the presence of the emitter resistor, R_E, tends to decrease the gain of the amplifier. We assume that the transistor has the proper dc bias voltages applied. Suppose an input signal of $-2\,\text{mV}$ is applied to the base of the transistor. This signal makes the base 2 mV more negative than the emitter, thus turning the (pnp) transistor on more and increasing I_E. But, the voltage V_E at the emitter equals $-I_E R_E$, so as I_E increases V_E becomes more negative. Hence the change in voltage between the base and the emitter due to the signal is less than 2 mV. The net voltage change V_{BE} present across the base-emitter junction is the input the transistor "sees" and thus determines how much the transistor turns on, i.e., how large the output is. Because V_E also goes negative, as V_B does, V_{BE} is less than the signal input. Hence, the gain is decreased. This is an example of negative feedback which will be discussed in a later chapter. To prevent this, we must prevent the voltage of the emitter from changing as the base voltage changes due to the signal input. The addition of a capacitor C_E from the emitter to ground will accomplish this by making the emitter an ac ground, i.e., by smoothing out any variations in the emitter voltage due to the signal at the base. The dc bias is not affected because C_E will pass no direct current. If we make the capacitance large enough so that its reactance is small compared to R_E, then the emitter will be an ac signal ground, i.e., $X_C = 1/\omega C_E \ll R_E$. This inequality is hardest to satisfy at the lowest frequency of operation. For an audio amplifier, for example, the lowest frequency the human ear can hear is about 20 Hz, so we want $X_C \ll R_E$ at 20 Hz. Therefore,

$$C_E \gg \frac{1}{\omega_{\min} R_E} = \frac{1}{2\pi f_{\min} R_E} = \frac{1}{(2\pi)(20\,\text{Hz})(800\,\Omega)} = 9.9\,\mu\text{F}.$$

A $100\,\mu\text{F}$ electrolytic capacitor of a low voltage rating (5 V to 10 V) will serve nicely, because the dc voltage drop across R_E is only $I_E R_E = (2\,\text{mA})(800\,\Omega)$

= 1.6V. The minus terminal of C_E should be at the emitter because the emitter dc voltage is always negative with respect to ground for our pnp transistor. Notice that if the minimum frequency to be amplified is larger a smaller capacitance will suffice.

The only problem remaining is to feed the signal to be amplified into the base terminal and take the amplified output out from the collector terminal. To avoid changing the dc bias voltages shown in Fig. 5.15, we must not allow any dc current to flow in the input or out the output. The simplest way to do this is to place capacitors, called dc "blocking" or "coupling" capacitors, in series with the input and output. The resulting circuit is shown in Fig. 5.16.

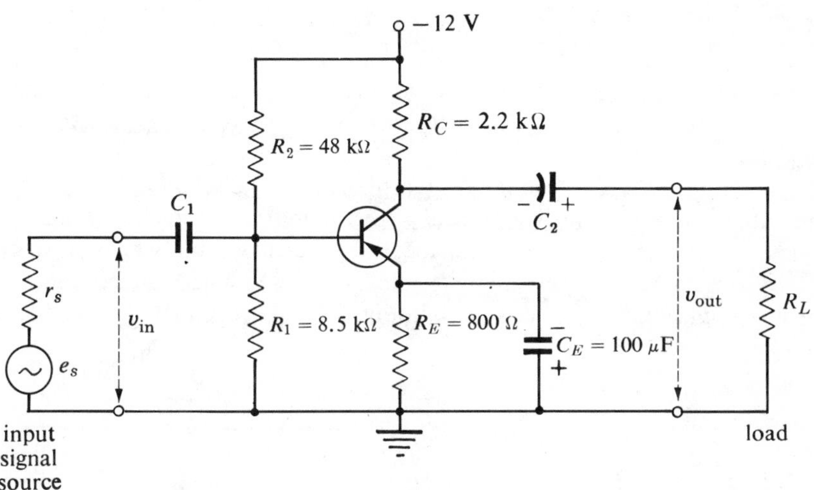

FIGURE 5.16 Complete common emitter amplifier.

However, C_1 and C_2 for high-pass RC filters (see section 2.8), we recall that the break point for a high-pass RC filter is at $\omega_B = 2\pi f_B = 1/RC$, as shown in Fig. 5.17. At frequencies higher than f_B, the filter gain is essentially unity; that is, a negligible signal voltage drop occurs across the capacitor. Thus, we choose the capacitance such that the break point is appreciably less than the lowest frequency the amplifier will be required to amplify, 20 Hz for our audio amplifier. The lower the frequency to be amplified, the larger C_1 and C_2 must be.

The resistance R in the output high-pass filter is merely R_L, the load resistance which the amplifier is driving. This load may be the input of another amplifier or it may be a loudspeaker, or an actual resistance. To be specific, suppose $R_L = 10\,\text{k}\Omega$. Then we must choose $1/R_LC_2 = \omega_B \ll 2\pi \times 20\,\text{Hz}$ or

$$C_2 \gg \frac{1}{2\pi 20 R_L} = \frac{1}{(2\pi)(20\,\text{Hz})(10^4\Omega)} = 1.6 \times 10^{-6}\text{F}$$

or $C_2 \gg 1.6\mu$F. A 10μF, 25V capacitor would be adequate. Unless there is

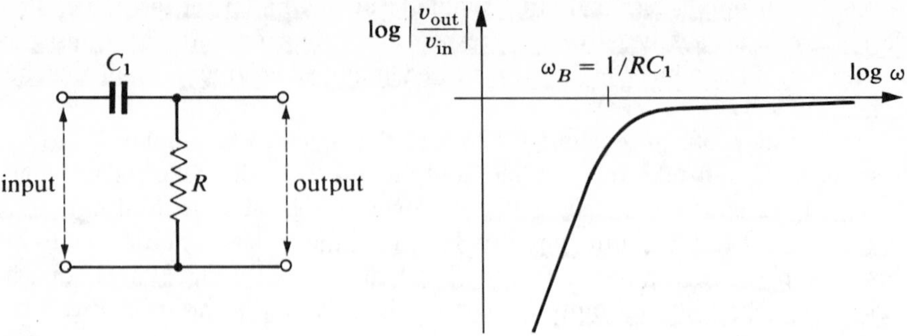

FIGURE 5.17 High-pass *RC* filter.

some voltage at the top of R_L more negative than $V_C = -6\text{V}$, the polarity of C_2 should be as shown in Fig. 5.16 with the negative side connected to the collector.

The resistance R in the input high-pass filter containing C_1 is not just R_1, but rather the total effective ac resistance (or impedance really) which the ac signal sees. That is, we must draw the ac equivalent circuit, which is obtained by realizing that the power supply terminal at -12V dc is an ac ground. Thus the ac equivalent circuit is as shown in Fig. 5.18. Notice that R_E is not present

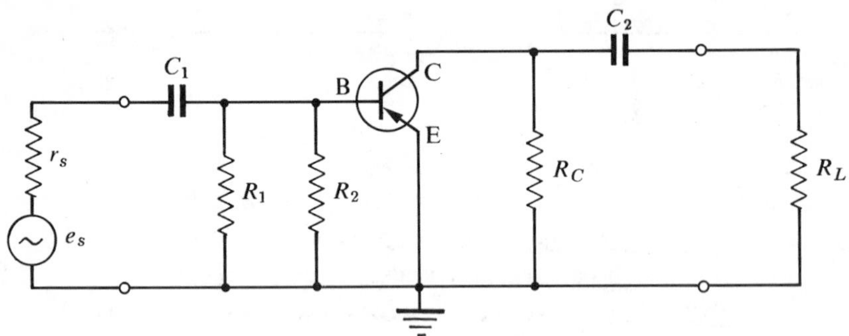

FIGURE 5.18 ac equivalent circuit of the common emitter amplifier of Fig. 5.16.

in the ac equivalent circuit because we have $X_{CE} \ll R_E$; the capacitance C_E is an ac short circuit. The input high-pass filter therefore consists of C_1 and the parallel combination of R_1, R_2 and the effective input resistance R_{BE} between the transistor base and emitter terminals, which is usually anywhere from several hundred ohms to ten thousand ohms. If we assume $R_{BE} = 1\,\text{k}\Omega$, then the effective resistance in the filter is equal to $R_1 \| R_2 \| R_{BE} = 800\Omega$ as shown in Fig. 5.19. Thus the input high-pass filter consists of C_1 and an 800Ω resistor. Therefore to pass frequencies down to 20 Hz we should have:

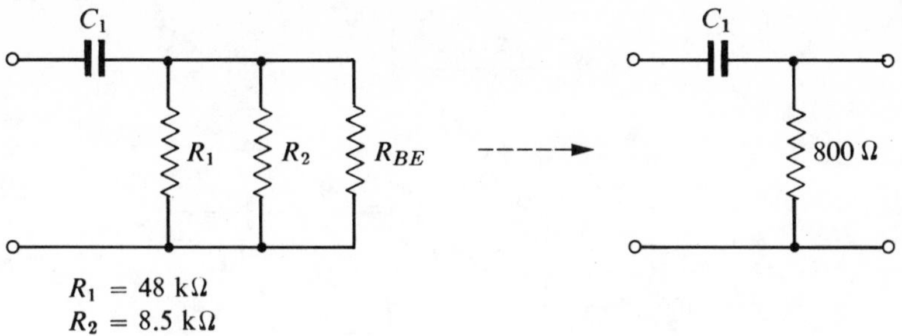

FIGURE 5.19 Input high-pass RC filter circuit.

$$\omega_B = 2\pi f_B = \frac{1}{R'C_1} \ll 2\pi(20\,\text{Hz}) \quad \text{where } R' = R_1 \| R_2 \| R_{BE} = 800\Omega,$$

which implies

$$C_1 \gg \frac{1}{R'2\pi(20\,\text{Hz})} = \frac{1}{(2\pi)(20\,\text{Hz})(800\Omega)} \cong 10\mu\text{F}.$$

Hence a 50 or 100 microfarad capacitor is needed for C_1. Fortunately, the dc voltage C_1 must withstand is not too high, so we may use a low voltage electrolytic capacitor for C_1, which will be inexpensive. If the dc voltage of the input is more positive than -1.8 V, then we hook up the electrolytic capacitor C_1 as shown in Fig. 5.20(a). If the dc voltage of the input is more negative than -1.8 V (as it would be for the input taken from the collector of a similar pnp amplifier stage), then the polarity of C_1 would be reversed as is shown in Fig. 5.20(b). If we expect a normal dc voltage of approximately 5 V to exist across C_1, then we would choose a 10 V or a 25 V rating for C_1. The final amplifier circuit is shown in Fig. 5.20 with the dc voltages given for various points in the circuit.

problems

1. Describe the construction of a pnp and an npn transistor. Why will the collector-to-emitter resistance of a good transistor always read high regardless of the polarity of the ohmmeter probes?

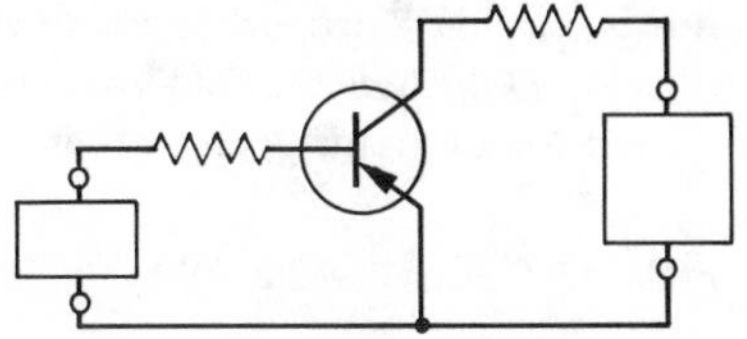

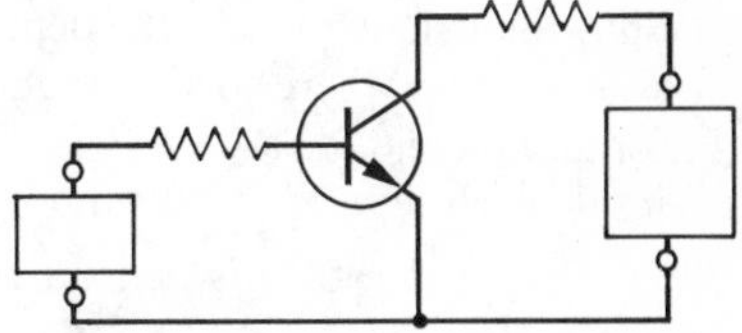

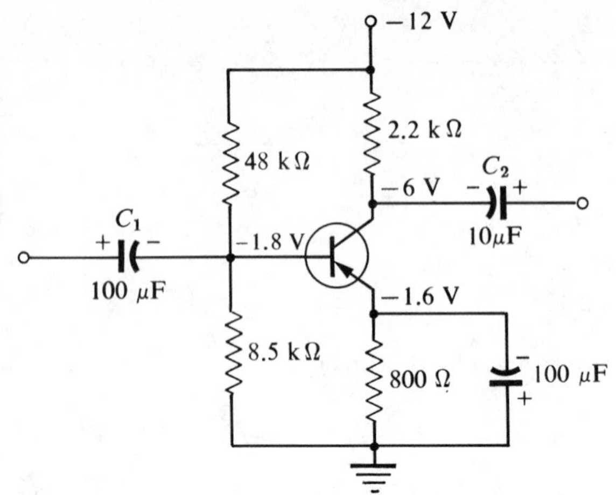

(a) *input more positive than* $-1.8\text{ V} = V_B$

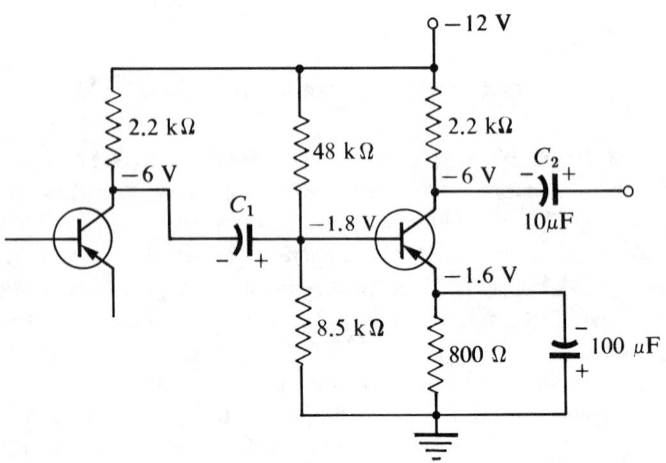

(b) *input more negative than* −1.8 V

FIGURE 5.20 Polarity of C_1 input capacitor.

2. Label the polarities of the voltage supplies.
3. The base-emitter junction of a transistor, either pnp or npn, is always ________________ biased. The base-collector junction of a transistor, either pnp or npn, is always ________________ biased. (Fill in the blanks.)
4. Show in a sketch where the depletion region exists inside a properly biased transistor. Show where an appreciable electric field exists. Include the field's direction for an npn and for a pnp transistor. Show the motion of majority charge carriers inside each type of transistor.
5. Prove that β, which is defined as I_C/I_B, equals $\alpha/(1 - \alpha)$. Also solve for α in terms of β.

6. Explain why the base of a transistor is purposely made thin. Would increasing the doping concentration of the base tend to increase or decrease the α? Explain.
7. Sketch a graph of the collector current versus base-emitter voltage for a silicon transistor. Repeat for the base current versus base-emitter voltage. Include approximate numerical values for the voltages and currents. Assume the transistor $\beta = 50$.
8. Explain briefly why a transistor amplifies when connected in the common emitter configuration.
9. Consider a transistor with a maximum power dissipation of 200 mW and a 20 V power supply. On a graph of I_C versus V_{CE}, sketch the maximum power curve and shade in the forbidden region of operation. Also draw the dc load line for a 2 kΩ collector resistor. Is this a safe load line? Repeat for a 400 Ω collector resistor. Is this load line safe?
10. Calculate I_C and V_{CE}. The transistor is silicon and has a β of 100.

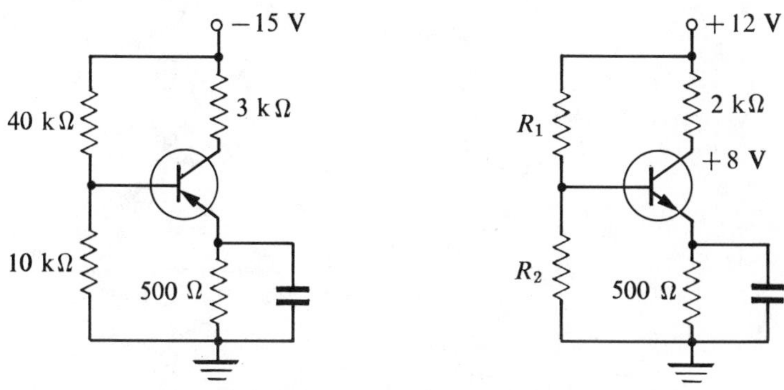

11. Calculate R_1 and R_2 if the maximum current drawn from the power supply is 10 mA. The transistor is silicon and has a β of 75.
12. Design a bias circuit using two batteries and two resistors for a silicon transistor with a $\beta = 100$. The desired operating point is $I_C = 2$ mA, $V_{CE} = 5$ V. Calculate numerical values for the battery voltage and resistances. ($R_E = 0$.)
13. Design a bias circuit using one battery and two resistors for a silicon common emitter transistor amplifier with $\beta = 75$. The desired operating point is $I_C = 3$ mA, $V_{CE} = 5$ V. Calculate numerical values for the battery voltage and resistances. ($R_E = 1$ kΩ.)
14. Design a bias circuit using one battery and three resistors for a silicon transistor with $\beta = 100$. The desired operating point is $I_C = 2$ mA, $V_{CE} = 7$ V. Calculate numerical values for the battery voltage and the resistances. How would you stabilize this circuit against temperature induced changes in the operating point?
15. Calculate the input and output coupling capacitors for the single-stage common emitter amplifier of Fig. 5.20 if the desired bandwidth is from 100 Hz to 4 kHz. Assume a 10 kΩ load connected to the output.
16. Show by a graphical argument exactly how much raising the supply voltage will increase the voltage gain for a transistor. Assume that the transistor has a $\beta = 100$ and that the operating point remains fixed at $I_C = 2$ mA, $V_{CE} = 5$ V for the two different supply voltages which you may take as 10 V and 20 V.

6

h PARAMETER EQUIVALENT CIRCUIT AND AMPLIFIERS

6.1 INTRODUCTION

In this chapter we will use a more detailed approach in designing amplifiers. Specifically, we will first introduce equivalent circuits to represent the transistor and then deduce the various circuit gains and impedances by a mathematical treatment of the equivalent circuits and the associated components. A certain loss of physical reality results, but one does get a deeper insight into such concepts as input and output impedance and high frequency operations, as well as a more precise knowledge of transistor amplifier operation. To minimize the mathematical complexity of the calculations, a simplified equivalent circuit will be shown for most cases yielding results precise to roughly $\pm 20\%$, which is adequate in many cases. The complicated expressions for gains and impedances resulting from the complete equivalent circuit will also be given.

6.2 TRANSISTOR EQUIVALENT CIRCUITS

A transistor equivalent circuit is basically a circuit consisting of ideal voltage and/or current "sources" or "generators" and passive components (R, L, and C) and acting electrically exactly like the transistor. In other words the transistor can be replaced by an appropriate collection of generators and passive components of the equivalent circuit. The advantage of the equivalent circuit is that one can predict the transistor's exact behavior (gain, etc.) by writing down the Kirchhoff voltage and current laws for the various loops and junctions and then by using only algebra and Ohm's law solve for the desired quantities. A simple example will show how the calculations are performed. Suppose a certain transistor's equivalent circuit is simply an ideal voltage gen-

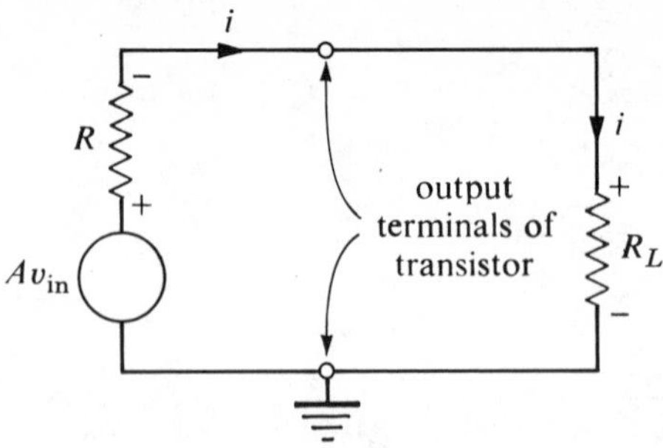

FIGURE 6.1 Simple equivalent circuit.

erator of magnitude Av_{in} in series with a resistance R as shown in Fig. 6.1. Algebraically, A is a positive number, v_{in} is the amplitude of the input voltage to the transistor. The R_L is the load resistance connected to the output terminals. The voltage generator produces a voltage A times as large as the input, so one might intuitively think of A as the voltage gain at this stage of the calculation. However, the voltage gain is:

$$A_v = \frac{v_{out}}{v_{in}} = \frac{iR_L}{v_{in}}, \tag{6.1}$$

and the current i can be obtained from the Kirchhoff voltage equation for the loop containing the generator, R and R_L.

$$Av_{in} - iR - iR_L = 0 \qquad \text{or} \qquad i = \frac{Av_{in}}{R + R_L} \tag{6.2}$$

Thus the voltage gain becomes:

$$A_v = \frac{v_{out}}{v_{in}} = \frac{iR_L}{v_{in}} = \frac{Av_{in}R_L}{(R + R_L)v_{in}} \tag{6.3}$$

or

$$A_v = A\left(\frac{R_L}{R + R_L}\right). \tag{6.4}$$

We see that the voltage gain depends upon A, R, and R_L, and only as R_L becomes very large compared to R do we get $A_v \cong A$.

Let us now develop a perfectly general equivalent circuit which will apply to *any* four-terminal device, including a transistor, as shown in Fig. 6.2. I_1 and I_2 are the currents flowing into the input and output, respectively; and V_1 and V_2 are the voltage differences across the input and the output terminals, respectively. We have four variables I_1, V_1, I_2 and V_2, which represent the total, instantaneous values of the currents and voltages. There are two Kirchhoff

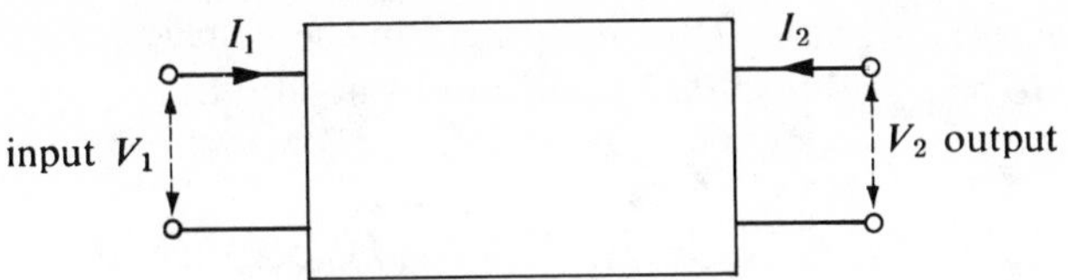

FIGURE 6.2 General four-terminal black box.

voltage equations we can write—one for the input and one for the output loop. In general they can be expressed in the form:

$$f(I_1, V_1, I_2, V_2) = 0 \tag{6.5}$$

and

$$g(I_1, V_1, I_2, V_2) = 0 \tag{6.6}$$

where f and g are mathematical functions whose exact form depends upon the internal structure of the black box. We now choose (arbitrarily) to solve for the input voltage V_1 and the output current I_2 in terms of the other two variables I_1 and V_2. In mathematical language we are treating I_1 and V_2 as the independent variables and I_2 and V_1 as the dependent variables. This can always be done by solving (6.5) for V_1 and substituting the resulting expression for V_1 into (6.6), thus obtaining an equation not involving V_1.

$$g(I_1, I_2, V_2) = 0$$

This equation can then be solved for I_2 in terms of I_1 and V_2:

$$I_2 = I_2(I_1, V_2). \tag{6.7}$$

Similarly (6.6) can be solved for I_2 and substituted in (6.5), which can then be solved for V_1:

$$V_1 = V_1(I_1, V_2). \tag{6.8}$$

In general we are interested in the response of the transistor to ac signals, so we will take the differential of (6.7) and (6.8) to obtain an expression for the change in I_2, dI_2, and the change in V_1, dV_1.

$$dI_2 = \left(\frac{\partial I_2}{\partial I_1}\right)_{V_2} dI_1 + \left(\frac{\partial I_2}{\partial V_2}\right)_{I_1} dV_2 \tag{6.9}$$

$$dV_1 = \left(\frac{\partial V_1}{\partial I_1}\right)_{V_2} dI_1 + \left(\frac{\partial V_1}{\partial V_2}\right)_{I_1} dV_2 \tag{6.10}$$

Let us change to a notation useful for considering ac signals or any change in the currents and voltages. Let $i_2 = dI_2$, $i_1 = dI_1$, $v_1 = dV_1$, and $v_2 = dV_2$.

That is, lowercase v's and i's refer to *changes* in the voltages and currents or, equivalently, to ac signal *amplitudes*. With this notation:

$$i_2 = \left(\frac{\partial I_2}{\partial I_1}\right)_{V_2} i_1 + \left(\frac{\partial I_2}{\partial V_2}\right)_{I_1} v_2 \tag{6.11}$$

$$v_1 = \left(\frac{\partial V_1}{\partial I_1}\right)_{V_2} i_1 + \left(\frac{\partial V_1}{\partial V_2}\right)_{I_1} v_2 . \tag{6.12}$$

We now define the h parameters for our four-terminal black box in terms of the partial derivatives:

$h_{21} \equiv \left(\frac{\partial I_2}{\partial I_1}\right)_{V_2}$ *a pure number*

$h_{22} \equiv \left(\frac{\partial I_2}{\partial V_2}\right)_{I_1}$ *a conductance in mhos or siemens* (1 mho = 1 ohm^{-1} = 1 siemen)

$h_{11} \equiv \left(\frac{\partial V_1}{\partial I_1}\right)_{V_2}$ *a resistance in ohms*

$h_{12} \equiv \left(\frac{\partial V_1}{\partial V_2}\right)_{I_1}$ *a pure number*

The h parameters have a variety of dimensions; hence, the name "hybrid" parameters. With this notation we have:

$$i_2 = h_{21} i_1 + h_{22} v_2 \tag{6.13}$$

$$v_1 = h_{11} i_1 + h_{12} v_2 . \tag{6.14}$$

Equations (6.13) and (6.14) relating the dependent variables i_2 and v_1 to the independent variables i_1 and v_2 via the h parameters determine the equivalent circuit. The term $h_{21}i_1$ means there is a current generator of magnitude h_{21} times i_1; the term $h_{22}v_2$ means the voltage v_2 appears across a conductance h_{22} (or equivalently across a resistance of $1/h_{22}$ ohms). The term $h_{11}i_1$ means the current i_1 flows through an effective resistance of h_{11} ohms; the term $h_{12}v_2$ means there is a voltage generator of magnitude $h_{12}v_2$. Therefore, we can draw the equivalent circuit of Fig. 6.3. Equation (6.13) is seen to be merely the Kirchhoff current equation for the output, and equation (6.14) is merely the Kirchhoff voltage equation for the input. The important point here is that the perfectly general mathematical treatment which led to equations (6.13) and (6.14) implies the equivalent circuit of Fig. 6.3.

Some physical feeling for the h parameters can be obtained from the equivalent circuit of Fig. 6.3. The h_{11} is a resistance in the input circuit; it is usually called the "input resistance." The term $h_{12}v_2$ is the amplitude of a voltage generator in the input; it represents how much of the output voltage v_2 is transferred or fed back to the input, and h_{12} is called the "reverse voltage transfer ratio." The word "reverse" is used to denote the transfer from the output back to the input. The h_{21} represents how much of the input current i_1 is transferred to the output; h_{21} is called the "forward current transfer ratio." The higher the

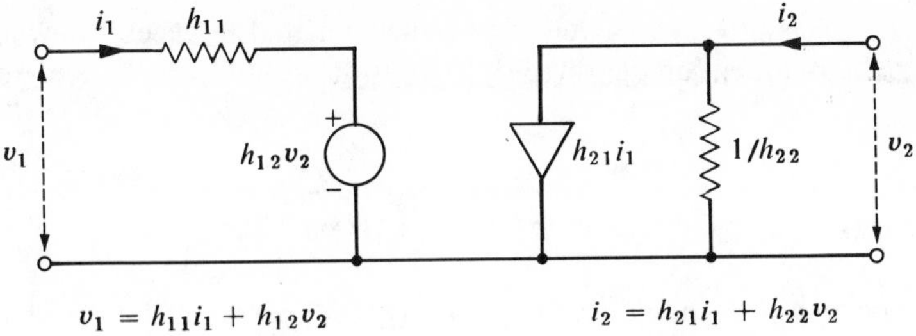

FIGURE 6.3 *h* parameter equivalent circuit.

value of h_{21}, the larger the change in output current for a given input current change. We call h_{22} the "output admittance," because it is an admittance or conductance directly across the output terminals.

The preceding development is entirely mathematical and is exact; that is, no approximations have been made except that v and i must refer to small signals because equations (6.11) and (6.12) hold exactly only for infinitesimal i's and v's. However, we have not yet shown that the equivalent circuit of Fig. 6.3 is, in fact, a representation of a *real* transistor. To do so, we must look at the *experimental* input and output curves for a transistor and see if by choosing numerical values for the h parameters the curves and thus the transistor can be accurately represented by the equivalent circuit. For a valid equivalent circuit, we would like a constant set of h parameters to represent the experimental curves over a *wide* range of currents and voltages. The equivalent circuit is simply not useful if the h parameters are strong functions of current and voltage.

A transistor has three terminals, while our black box from which the equivalent circuit was developed has four. Hence, to apply the equivalent circuit to a transistor one transistor terminal must be common between the input and the output. This can be either the emitter, the collector, or the base, called respectively the "common emitter" (*CE*), the "common collector" (*CC*), or the "common base" (*CB*) configuration.

The h parameters will be different for the three configurations, and letter subscripts are usually used to distinguish one from another according to the following table. The second letter of the subscript denotes the common ter-

TABLE 6-1. *h* PARAMETER NOTATION FOR THE THREE TRANSISTOR CONFIGURATIONS: *CE* = COMMON EMITTER, *CC* = COMMON COLLECTOR, *CB* = COMMON BASE

h Parameter	*CE*	*CC*	*CB*
h_{11}	h_{ie}	h_{ic}	h_{ib}
h_{12}	h_{re}	h_{rc}	h_{rb}
h_{21}	h_{fe}	h_{fc}	h_{fb}
h_{22}	h_{oe}	h_{oc}	h_{ob}

minal, "e" for emitter, etc. One must remember that in general the values of the parameters differ for the three configurations, e.g., $h_{fe} \neq h_{fb}$ and so forth.

6.3 COMMON EMITTER CONFIGURATION

Let us first consider the common emitter ("CE") configuration in which $I_1 \rightarrow -I_B$, $V_1 \rightarrow -V_{BE}$, $I_2 \rightarrow -I_C$, and $V_2 \rightarrow -V_{CE}$ as shown in Fig. 6.4. The

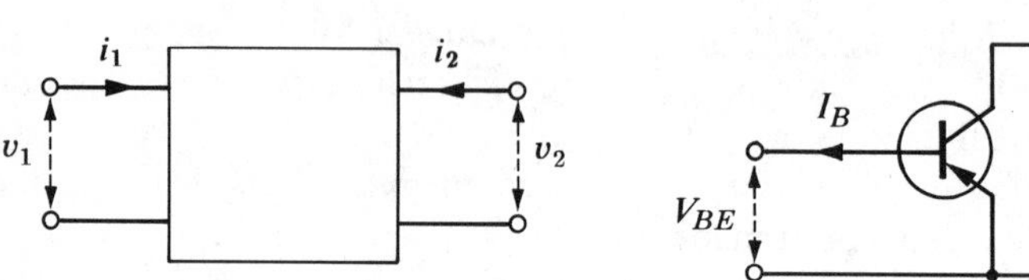

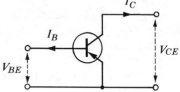

FIGURE 6.4 Common emitter configuration.

equivalent circuit equations (6.11) and (6.12) have the input current i_1 and the output voltage v_2 as the dependent variables, so we must look at the CE curves of $I_C = I_C(I_B, V_{CE})$ and $V_{BE} = V_{BE}(I_B, V_{CE})$. These experimental curves for a low power germanium pnp transistor are shown in Fig. 6.5. The four h parameters can now be interpreted geometrically by reference to Fig. 6.5.

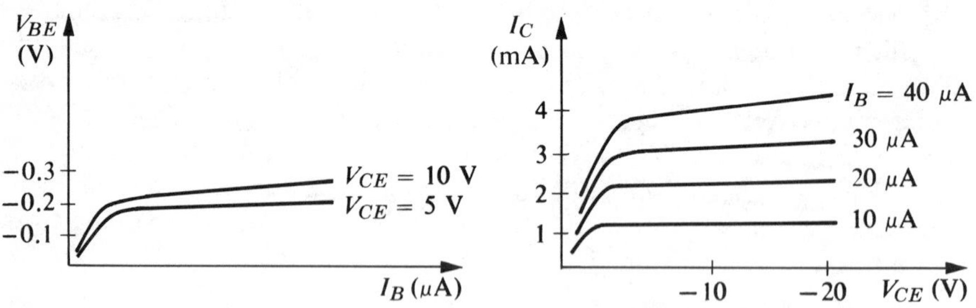

FIGURE 6.5 Experimental input and output graphs for low power germanium transistor in the common emitter configuration.

$$h_{11} \equiv \left(\frac{\partial V_1}{\partial I_1}\right)_{V_2} = \left(\frac{\partial V_{BE}}{\partial I_B}\right)_{V_{CE}}$$

is seen to be the slope of the input graph along a line of constant V_{CE} and is approximately constant as long as the base current exceeds about 10μA. But this region $I_B > 10\mu$A is precisely the region in which the transistor is turned on, i.e., the region of normal operation. The resistance h_{11} is of the order of 1000–3000 Ω, and is only a weak function of V_{CE}.

From Chapter 4 we recall that the current-voltage equation for the base-emitter junction is

$$I_E = I_0(e^{eV_{BE}/kT} - 1) \tag{6.15}$$

where e is the electronic charge, T is the absolute temperature of the junction, k is Boltzmann's constant, and I_0 is the (constant) reverse current which the junction will pass. If we regard $\beta = I_C/I_B \cong I_E/I_B$ as constant, we can write:

$$I_B = \frac{I_0}{\beta}(e^{eV_{BE}/kT} - 1). \tag{6.16}$$

Then the h_{11} parameter can be expressed as:

$$h_{11} = \left(\frac{\partial V_{BE}}{\partial I_B}\right) = \frac{1}{\left(\dfrac{\partial I_B}{\partial V_{BE}}\right)} = \frac{1}{\dfrac{\partial}{\partial V_{BE}}\left[\dfrac{I_0}{\beta}(e^{eV_{BE}/kT} - 1)\right]}$$

or

$$h_{11} = \frac{\beta kT}{eI_0} e^{-eV_{BE}/kT}. \tag{6.17}$$

If $e^{eV_{BE}/kT} \gg 1$, which is true if $V_{BE} > 0.06\,\text{V}$, or equivalently if $I_B > 0.1\,\mu\text{A}$ then (6.16) becomes:

$$I_B \cong \frac{I_0}{\beta} e^{eV_{BE}/kT}. \tag{6.18}$$

Equations (6.18) and (6.17) imply:

$$h_{11} \cong \frac{kT}{eI_B} = \frac{\beta kT}{eI_C} = 2.6\,\text{k}\Omega, \tag{6.19}$$

for $\beta = 100$, $I_C = 1\,\text{mA}$, $T = 300°\text{K}$. The point here is that if the base-emitter junction is forward-biased more than $V_{BE} = 0.06\,\text{V}$, h_{11} will depend inversely on the dc *collector current* I_C. If $h_{11} = 2.6\,\text{k}\Omega$ for $I_C = 1\,\text{mA}$, then changing the dc operating point to $I_C = 0.1\,\text{mA} = 100\,\mu\text{A}$ will increase h_{11} to $26\,\text{k}\Omega$. Conversely, changing to $I_C = 10\,\text{mA}$ will lower h_{11} to $260\,\Omega$. These results can be expressed very conveniently in the formula:

$$h_{11}(\text{in k}\Omega) \cong \frac{2.6}{I_C(\text{in mA})} \tag{6.20}$$

provided $I_C > 10\,\mu\text{A}$.

$$h_{12} \equiv \left(\frac{\partial V_1}{\partial V_2}\right)_{I_1} = \left(\frac{\partial V_{BE}}{\partial V_{CE}}\right)_{I_B}$$

is seen to be the rate of change of the base-emitter voltage V_{BE} with respect to V_{CE} for a constant base current. The term h_{12} is of the order of 10^{-3} to 10^{-4} in most modern transistors. Note that h_{12} is roughly constant for $I_B > 10\,\mu A$.

From the output graph,

$$h_{21} \equiv \left(\frac{\partial I_2}{\partial I_1}\right)_{V_2} = \left(\frac{\partial I_C}{\partial I_B}\right)_{V_{CE}}$$

is the rate of change of collector current with respect to base current and is approximately 100 for the graph in Fig. 6.5. The value of h_{21} is roughly constant for $|V_{CE}| > 1\,V$, although increasing slightly as I_C increases. Notice also that h_{21} decreases drastically for $|V_{CE}|$ much below one volt. It will shortly be shown that the gain of a *CE* amplifier depends upon the value of h_{21}, so this is the fundamental reason why we should choose an operating point with V_{CE} greater than 1 V as was stated in Chapter 5.

$$h_{22} \equiv \left(\frac{\partial I_2}{\partial V_2}\right)_{I_1} = \left(\frac{\partial I_C}{\partial V_{CE}}\right)_{I_B}$$

is seen to be the slope of the output graph along a line of constant base current. For $V_{CE} > 1\,V$, h_{22} is seen to be essentially constant and equal to roughly 2×10^{-5} mhos or siemens, although h_{22} does increase as I_C increases.

In summary, the four h parameters are seen to have essentially constant values for V_{CE} greater than one volt and for I_B greater than about $10\,\mu A$ (or equivalently, I_C greater than about 1.0 mA), that is, with the transistor base-emitter junction turned on. Hence we should be able to predict accurately the transistor circuit behavior by a straightforward mathematical analysis of the equivalent circuit of Fig. 6.3, if we use the constant values for the h parameters obtained from the graph regions where $|V_{CE}| > 1\,V$ and $I_B > 10\,\mu A$.

Let us now calculate the voltage gain, the current gain, and the power gain for the common emitter amplifier shown in Fig. 6.6(a) with an ideal voltage source e_s and a resistance r_s supplying the input and a load resistor R_L across the output.

We first replace the actual circuit of Fig. 6.6(a) with the ac equivalent circuit, shown in Fig. 6.6(b). We assume that the reactance of C_1, C_2, and C_E may be neglected; that is, they act as short circuits as far as the ac signal is concerned. We also use the fact that the power supply or battery is an ac short circuit. The transistor of Fig. 6.6(b) is now replaced by its h parameter equivalent circuit as shown in Fig. 6.7. The circuit of Fig. 6.7 can now in principle be completely analyzed by a straightforward application of Kirchhoff's laws and Ohm's law.

Let us simplify the circuit by combining parallel resistances. Let $R_{CL} = R_C \| R_L$, and let $R_b = R_1 \| R_2$; R_b and r_s, the source resistance, can be combined by replacing the e_s voltage source with its equivalent Norton current source as shown in Fig. 6.8(a). Now the Norton resistance r_s is in parallel with

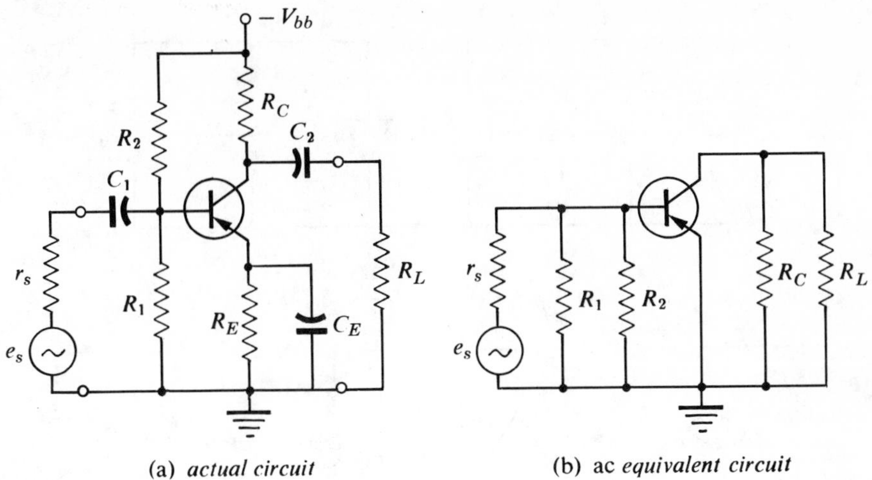

FIGURE 6.6 Common emitter amplifier.

R_b, so they can be combined as one resistance $R_s = r_s \| R_b$. Or, Thevenin's theorem can be used to obtain the e_sR_s source directly from Fig. 6.8.

For convenience, we convert the i_s current source back to a voltage source as shown in Fig. 6.9.

The Kirchhoff voltage law for the input is:

$$e_s = i_1(R_s + h_{11}) + h_{12}v_2 \tag{6.21}$$

or in terms of v_1,

$$v_1 = h_{11}i_1 + h_{12}v_2 . \tag{6.22}$$

The Kirchhoff current law for the output is:

$$i_2 = h_{21}i_1 + h_{22}v_2 . \tag{6.23}$$

We also have

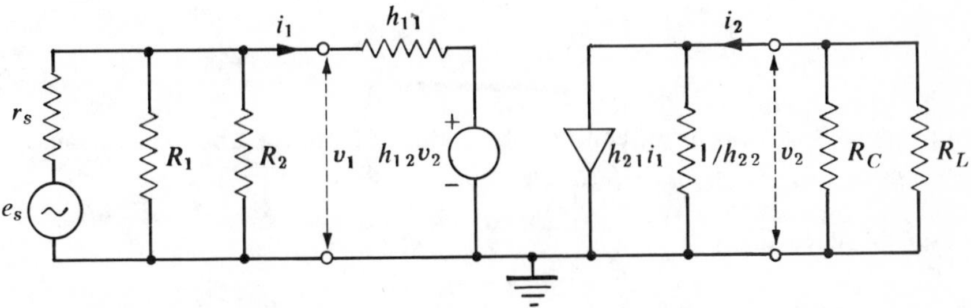

FIGURE 6.7 *h* parameter ac equivalent circuit for common emitter amplifier.

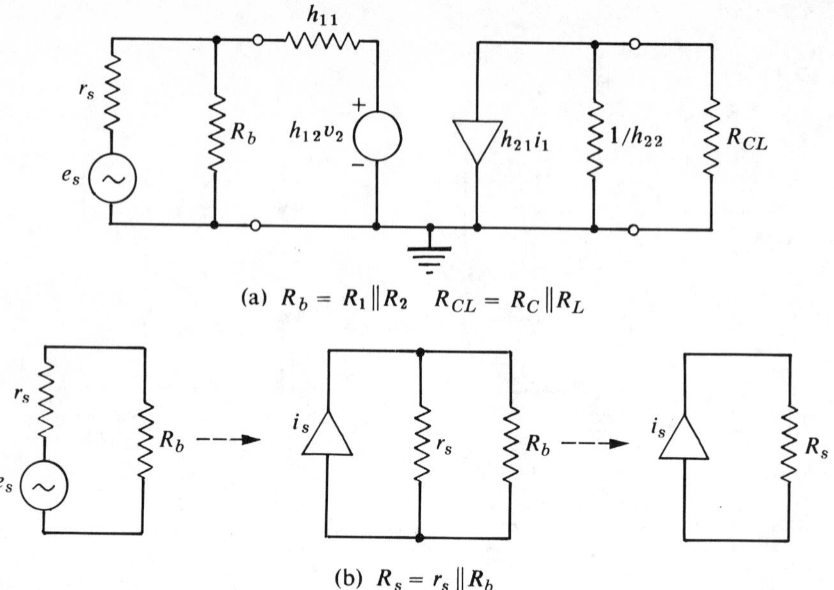

FIGURE 6.8 Simplified common emitter amplifier h parameter ac equivalent circuit.

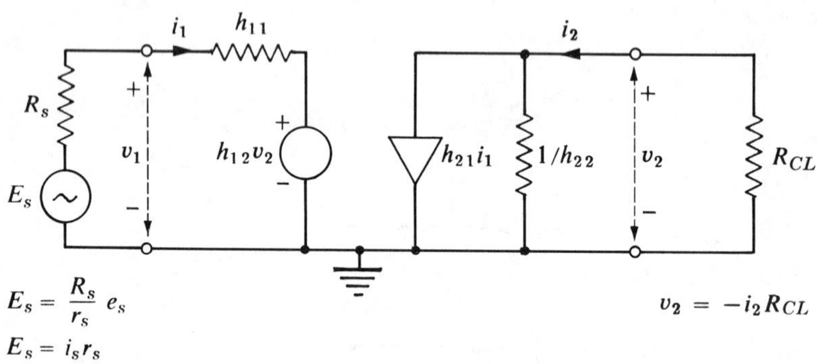

FIGURE 6.9 Final common emitter amplifier circuit ready for Kirchhoff law analysis.

$$v_2 = -i_2 R_{CL}. \tag{6.24}$$

All properties of the circuit can now be deduced from the above equations.

6.3.1 Voltage Gain

The voltage gain $A_v = v_{out}/v_{in} = v_2/v_1$ is obtained by eliminating i_2 from (6.23) using (6.24), and then eliminating i_1 using (6.22), viz., (6.23) becomes $i_2 = -v_2/R_{CL} = h_{21}i_1 + v_2 h_{22}$ and $i_1 = (v_1 - h_{12}v_2)/h_{11}$ from (6.22). The minus sign in equation (6.24) perhaps deserves a brief comment. Positive i_2 by convention flows upward through R_{CL} and into the output (collector) terminal of the transistor, thus making the top of R_{CL} negative. However, positive v_2 means by convention that the top of R_{CL} is positive with respect to the bottom; hence a positive i_2 means a negative v_2 and vice versa. Therefore,

$$\frac{-v_2}{R_{CL}} = \frac{h_{21}}{h_{11}}(v_1 - h_{12}v_2) + v_2h_{22}.$$

Solving for A_v gives:

$$A_v = \frac{v_{\text{out}}}{v_{\text{in}}} = \frac{v_2}{v_1} = \frac{h_{21}}{h_{21}h_{12} - h_{11}h_{22} - \dfrac{h_{11}}{R_{CL}}}. \tag{6.25}$$

If we substitute in typical numerical values for a low power transistor of $h_{11} = 2\,\text{k}\Omega$, $h_{21} = 100$, $h_{12} = 10^{-4}$, $h_{22} = 2 \times 10^{-5}$ mhos, we see that the h_{11}/R_{CL} term in the denominator is the largest if $R_{CL} \cong 2\,\text{k}\Omega$ which is commonly the case. Thus we may write:

$$\boxed{A_v \cong \frac{-h_{21}}{h_{11}} R_{CL}.} \tag{6.26}$$

Substituting in numerical values yields $A_v \cong -h_{21} = -100$. The minus sign in the voltage gain expression is no cause for alarm; it merely indicates that the output is 180° out of phase with respect to the input. A positive input produces a negative output, and vice versa.

6.3.2 Current Gain

The current gain can be calculated from (6.23) by substituting $-i_2R_{CL}$ for v_2:

$$i_2 = h_{21}i_1 + (-i_2R_{CL})h_{22}.$$

Thus,

$$\boxed{A_i = \frac{i_2}{i_1} = \frac{h_{21}}{1 + h_{22}R_{CL}}.} \tag{6.27}$$

Again using the typical numerical values $h_{22} = 2 \times 10^{-5}$ mhos, $R_{CL} = 2\,\text{k}\Omega$, we see that

$$A_i \cong h_{21} = 100. \tag{6.28}$$

Thus the $h_{21} = h_{fe}$ parameter is often colloquially referred to as the "gain" or the "current gain" of the transistor. It is clear now that the larger the h_{fe} or β of the transistor, the higher the gain of the amplifier. In general, high power transistors capable of carrying amperes of current tend to have lower values of β, from 20 to 40. Specially selected low power transistors are available with β of the order of 200–500, but a typical value of β is approximately 100.

6.3.3 Input Impedance

The input impedance of the transistor is the effective impedance the input signal "sees" when it is applied across the two input terminals of the transistor. Thus the input impedance is defined as the input signal voltage v_1 appearing across the two input terminals divided by the input current i_1.

$$Z_{\text{in}} \equiv \frac{v_1}{i_1} \tag{6.29}$$

If v_1 and i_1 are in phase, then Z_{in} will be a real number and will be a resistance; we may then speak of the input "resistance." If v_1 and i_1 are out of phase, Z_{in} will be complex, and we must speak of an input "impedance." In general Z_{in} will be complex only when there are inductive and/or capacitive components in the equivalent circuit. This is generally true only for frequencies above 100 kHz or so. From now on we will restrict ourselves to the equivalent circuit of Fig. 6.9, which contains only resistances and voltage generators in phase with one another; hence, Z_{in} will be a resistance. This analysis will apply quite well below 100 kHz.

From the equivalent circuit of Fig. 6.9, we have:

$$v_1 = h_{11}i_1 + h_{12}v_2 \tag{6.22}$$

$$i_2 = h_{21}i_1 + h_{22}v_2 \tag{6.23}$$

$$v_2 = -i_2R_{CL}. \tag{6.24}$$

We can calculate $Z_{\text{in}} = v_1/i_1$ by eliminating v_2 from (6.22). Equations (6.23) and (6.24) can be combined to eliminate i_2 as follows: substituting (6.24) in (6.23) to eliminate i_2 yields

$$\frac{-v_2}{R_{CL}} = h_{21}i_1 + h_{22}v_2.$$

Solving for v_2:

$$v_2 = -\frac{h_{21}i_1}{h_{22} + 1/R_{CL}}. \tag{6.30}$$

Substituting (6.30) in (6.22) gives:

$$v_1 = h_{11}i_1 - \frac{h_{21}i_1}{h_{22} + 1/R_{CL}}h_{12},$$

which can be solved for $Z_{in} = v_1/i_1$:

$$Z_{in} = h_{11} - \frac{h_{12}h_{21}}{h_{22} + 1/R_{CL}}. \tag{6.31}$$

As we might have expected, h_{11} is contained in Z_{in} because h_{11} is a resistance in series with the input of the transistor equivalent circuit. The second term in Z_{in} decreases the effective input impedance because of the $h_{12}v_2$ voltage generator in the input. As v_1 increases positively i_1 increases, but v_2 becomes larger and *more negative* because of the 180° phase inversion between input and output (a positive input voltage produces a negative output). Thus more voltage drop appears across h_{11} for a given v_1, and i_1 *increases*, thus lowering Z_{in}.

If we substitute in typical numerical values for a low power germanium transistor, $h_{11} = 2\,\text{k}\Omega$, $h_{12} = 10^{-4}$, $h_{21} = 100$, $h_{22} = 2 \times 10^{-5}$ mhos, we obtain:

$$Z_{in} = 2\,\text{k}\Omega - \frac{(10^{-4})(100)}{2 \times 10^{-5} + 1/R_{CL}}.$$

A common value of $R_{CL} = R_C \parallel R_L$ is roughly 2kΩ, so $Z_{in} = 2\text{k}\Omega - 20\Omega \cong 2\text{k}\Omega$. Thus, the input impedance can be taken to be h_{11} for all practical purposes. This result could have been obtained immediately by neglecting the voltage feedback, i.e., setting $h_{12} = 0$ which would immediately give for the input voltage equation:

$$v_i = h_{11}i_1,$$

which directly gives

$$Z_{in} = \frac{v_1}{i_1} = h_{11} \cong 2\,\text{k}\Omega. \tag{6.32}$$

Once again we should remind ourselves that h_{11} varies inversely as the collector current: for $I_C = 100\,\mu\text{A}$, $h_{11} \cong 26\,\text{k}\Omega$, and for $I_C = 1\,\text{mA}$, $h_{11} \cong 2.6\,\text{k}\Omega$, etc.

At this point we should note that Z_{in} is the impedance seen looking into the *transistor*. But the source drives the transistor *and* the bias network of R_1 and R_2. Hence the impedance which the source e_s sees is Z_{in} *in parallel with* $(R_1 \parallel R_2)$ as shown in Fig. 6.10. Thus if $Z_{in} = 1\,\text{k}\Omega$, $R_1 = 8.5\,\text{k}\Omega$, and $R_2 = 47\,\text{k}\Omega$,

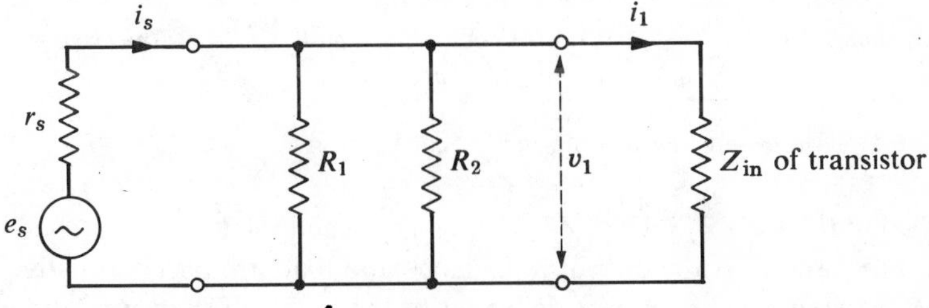

FIGURE 6.10 Input impedance circuit.

then $R_1 \| R_2 = 7.2\,\text{k}\Omega$ and the impedance seen by the source is $(R_1 \| R_2) \| Z_{\text{in}} = 880\,\Omega$. If R_1 were $1\,\text{k}\Omega$, then $R_1 \| R_2 \cong 1\,\text{k}\Omega$, and the source would see this $1\,\text{k}\Omega$ in parallel with the $2\,\text{k}\Omega$ transistor input impedance or a net circuit input impedance of only about $670\,\Omega$.

Notice also that the current i_s from the source will divide between $R_b = (R_1 \| R_2)$ and Z_{in}; $(i_s - i_1)$ will flow through R_b and i_1 through the transistor according to:

$$\frac{i_1}{i_s - i_1} = \frac{R_b}{Z_{\text{in}}} = \frac{R_1 \| R_2}{Z_{\text{in}}}.$$

We clearly want i_1 to be as large as possible for a given i_s, so it behooves us to make $R_1 \| R_2 \gg Z_{\text{in}}$ for the transistor if possible. Then, of course, Z_{in} for the circuit $\cong Z_{\text{in}}$ for the transistor. If $R_1 \| R_2$ is small compared to Z_{in} of the transistor, then most of the input signal current i_s will flow through $R_1 \| R_2$, thus lowering the input current i_1 to the transistor and consequently lowering the transistor output, i.e., lowering the transistor gain.

6.3.4 Output Impedance

The output impedance of any circuit is the impedance seen looking back into the two output terminals and is important for two reasons: First, it determines the high frequency response of the circuit, particularly when the output is driving a strongly capacitive load such as a long cable; and secondly, it determines how much power is transferred to the load connected across the output.

At this point we repeat the power matching theorem. We represent the circuit output by a Thevenin voltage generator e with Z_{out} the output impedance, as shown in Fig. 6.11(a). Then for a *fixed* Z_{out}, maximum power is

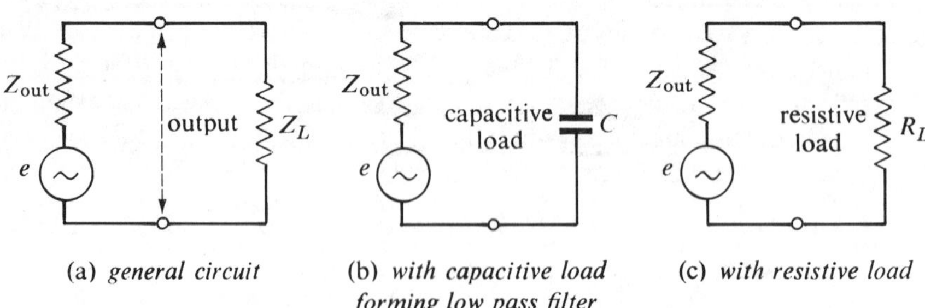

(a) *general circuit* (b) *with capacitive load forming low pass filter* (c) *with resistive load*

FIGURE 6.11 Thevenin equivalent of amplifier output.

dissipated in the load Z_L, when $Z_L = Z_{\text{out}}^*$, i.e., when the load is matched to the output impedance. If the output impedance and load are purely resistive, then $R_L = R_{\text{out}}$ maximizes the power delivered to the load. If, however, the *load* is fixed, maximum power is delivered to the load when $Z_{\text{out}} = 0$.

The high frequency response of a circuit is important when either high frequency sinusoidal signals or any signals containing appreciable high frequency Fourier components are being handled. An example of the latter case is when "fast" or narrow pulses must be amplified; a pulse of width τ seconds requires an amplifier with a bandwidth of at least $1/\tau$ Hz to avoid rounding off and attenuating the pulse. For example, the typical plastic scintillator used in nuclear physics to detect charged particle radiation, such as protons or alpha particles, emits a pulse of light lasting only about 4×10^{-9} sec or 4 nanoseconds. A good photomultiplier tube will convert this light pulse into a voltage pulse of about 4 nanosecond duration. Hence, to amplify such a pulse with minimum distortion, we need an amplifier with a bandwidth on the order of $1/\tau = (4 \times 10^{-9})^{-1} = 250\,\text{MHz}$.

With a capacitive load, the output impedance affects the high frequency response because it is essentially the resistance in a low-pass filter as shown in Fig. 6.11. If Z_{out} is real, i.e., a resistance, then $Z_{\text{out}}(= R_{\text{out}})$ and a capacitive load form a low-pass RC filter with a break point of $\omega_B = 1/R_{\text{out}}C$. Thus the lower R_{out} for a fixed C, the higher the break-point frequency and the larger the bandwidth. In actual practice, considerable effort is exerted to make C as low as possible, but often one still has to design a circuit with a low Z_{out} in order to achieve sufficient bandwidth. One example of a capacitive load commonly encountered is a length of coaxial cable. Particular care must be taken to drive the cable with a circuit of low output impedance in order to get a fast pulse of reasonable amplitude and minimum distortion out of the other end of the cable. For example, if a 1 μsec wide pulse is to be transmitted down 10 feet of RG58/U coaxial cable, the bandwidth needed is on the order of bandwidth $\cong 1/\tau = 1/10^{-6} = 10^6\,\text{Hz} = 1\,\text{MHz}$. The cable has a capacitance of 28 pF/foot, so 10 feet represents 280 pF capacitance. If we set the desired bandwidth of 10^6 Hz equal to the break point frequency, we can calculate the maximum output impedance allowable.

$$R_{\text{out max}} = \frac{1}{\omega_B C_{\text{cable}}} = \frac{1}{(2\pi \times 10^6\,\text{Hz})(280 \times 10^{-12}\text{F})} = 565\,\Omega \qquad \textbf{(6.33)}$$

In actual practice one would try to design for as small an output impedance as possible.

Actually the problem of cable capacitance is not quite the same as the problem of a pure shunt capacitance equal to the length of the cable times the cable capacitance per unit length. In addition to the cable capacitance between the inner conductor and the outer shield there is also a certain small amount of inductance per unit length. Thus the equivalent circuit for a cable consists of inductances in series and capacitances in parallel, something like the circuit of Fig. 6.12. The net result of the series inductance distributed along the cable is to make a voltage pulse propagate down the cable *without* having to charge up the entire cable capacitance. If the cable is terminated with the proper or characteristic resistance (50 Ω for RG58/U cable), then no electrical energy

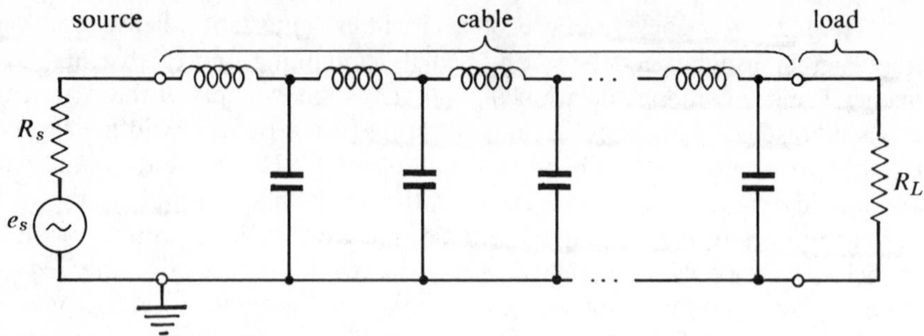

FIGURE 6.12 Cable equivalent circuit.

will be reflected back from the termination, and optimum transmission of the pulse will occur. The cable should also be driven by a source whose output impedance matches the characteristic cable impedance for minimum reflections of the pulse. Thus, a desirable source for a 50Ω cable is a circuit with a 50Ω output impedance. This topic is covered more fully in Chapter 10.

The relative size of the load and the output impedance determines the voltage across the load, the current through the load, and the power transferred to the load. By reference to Fig. 6.11(c), we can see for a fixed output equivalent circuit (fixed e and Z_{out}) that the output voltage across the load will be maximum when $R_L \gg Z_{out}$ and the output current through the load will be maximum when $R_L \ll Z_{out}$. In other words, for a high voltage gain $R_L \gg Z_{out}$, and for a high current gain $R_L \ll Z_{out}$. Practically speaking, it does no good whatsoever to design a circuit with a large voltage gain if the intended load resistance R_L is small compared to the output impedance of the circuit. Similarly, it does no good to design for a high current gain if R_L is large compared to the output impedance.

In practice we are usually presented with a fixed load R_L, and the problem is to design a circuit to drive the load. Thus we usually must design for as low an output impedance as possible, because then the voltage across the load, the current through the load, and the power transferred to the load are all maximized. In other words an ideal voltage amplifier would have zero output impedance.

If the output voltage source can be represented by a Thevenin voltage generator with an output impedance Z_{out} as in Fig. 6.13(a), then the output impedance can be measured in several ways. First, it can easily be shown that $Z_{out} = V_{open\ circuit}/i_{short\ circuit}$ because $V_{open} = e$ and $i_{short\ circuit} = e/Z_{out}$. Experimentally it is easy to measure the open circuit voltage, but extremely difficult to measure the short circuit current, so this method is not practical. Second, if the output terminal voltage v_2 is plotted against the output current i_2, then the slope of the graph equals $-Z_{out}$. Thus, any two measurements of v_2 and i_2 will yield the slope and therefore Z_{out}. The first method for obtaining Z_{out} is really a special case of this technique. The open circuit voltage is v_2

when $i_2 = 0$, and the short circuit current is i_2 when $v_2 = 0$, i.e., these two points are the end points of the graph in Fig. 6.13(b). The open circuit voltage measurement must be made with a high impedance voltmeter or an oscilloscope. The important consideration is that the voltmeter or oscilloscope impedance must be large compared with the circuit impedance between the two terminals.

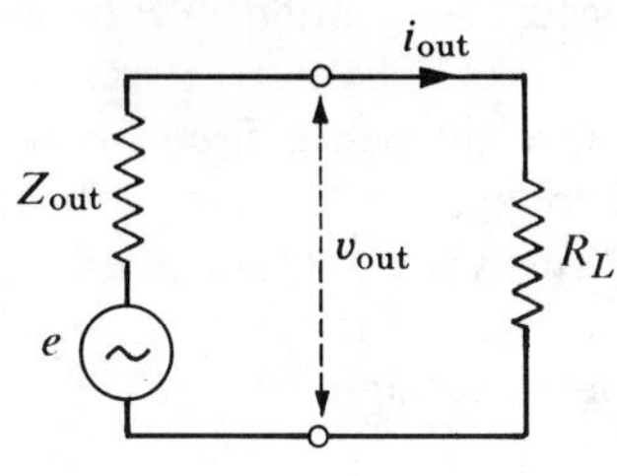

(a) *Thevenin output equivalent circuit*

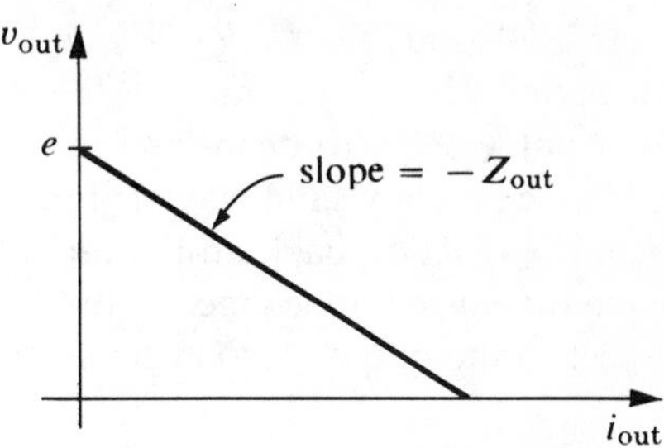

(b) *output voltage vs. output current*

FIGURE 6.13 Output impedance.

Another special case of measuring the slope is to vary R_L until the output voltage v_2 equals one-half of the open circuit voltage. Then $R_L = Z_{out}$. This method is usually the most convenient experimentally.

The output impedance of a transistor is more difficult to measure and calculate than the input impedance, because the output voltage generator or current source is not constant in magnitude unlike the Thevenin circuit of Fig. 6.13. The output side of the h parameter equivalent circuit is not a Thevenin equivalent, strictly speaking, because the output current generator $h_{21}i_1$ depends upon the input current i_1 which depends upon v_2 through the reverse voltage transfer ratio generator $h_{12}v_2$. (v_2 and $h_{21}i_1$ are connected in (6.35), the output equation.)

Let us now calculate the output impedance in terms of the h parameters. We repeat the necessary equations for convenience:

$$e_s - i_1R_s = v_1 = h_{11}i_1 + h_{12}v_2 \tag{6.34}$$

$$i_2 = h_{21}i_1 + h_{22}v_2. \tag{6.35}$$

If the input signal is set equal to zero to avoid making the output depend upon the applied input signal ($e_s = 0$), then the output impedance v_2/i_2 can be obtained by eliminating i_1 from:

$$-i_1R_s = h_{11}i_1 + h_{12}v_2 \tag{6.36}$$

$$i_2 = h_{21}i_1 + h_{22}v_2. \tag{6.37}$$

The result is:

$$Z_{out} = \frac{v_2}{i_2} = \frac{h_{11} + R_s}{(h_{11} + R_s)h_{22} - h_{12}h_{21}},$$

or

$$\boxed{Z_{out} = \frac{h_{11} + R_s}{\Delta h + h_{22}R_s}} \tag{6.38}$$

where Δh stands for $(h_{11}h_{22} - h_{12}h_{21})$, which occurs very often in h-parameter calculations.

Another way to define the output impedance of such a circuit is to consider an extremely small change in the output voltage and current. This minimizes the reaction back on the input which changes the output. We can then define the output impedance as the partial derivative of the output voltage with respect to the output current for a constant source voltage.

$$Z_{out} = \left(\frac{\partial v_2}{\partial i_2}\right)_{e_s}$$

Because of the coupling between the input and output, we expect the output impedance to depend upon resistances in the input side. The problem is now algebraic—to obtain an equation involving only v_2 and i_2. We first eliminate v_1 in (6.34) by writing:

$$e_s - i_1R_s = h_{11}i_1 + h_{12}v_2. \tag{6.39}$$

Now i_1 can be eliminated by solving (6.39) for i_1, and substituting in (6.35):

$$i_1 = \frac{e_s - h_{12}v_2}{h_{11} + R_s}.$$

Therefore,

$$i_2 = h_{21}\left(\frac{e_s - h_{12}v_2}{h_{11} + R_s}\right) + h_{22}v_2.$$

Solving for v_2 yields:

$$v_2 = \frac{i_2 - \dfrac{h_{21}e_s}{h_{11} + R_s}}{h_{22} - \dfrac{h_{12}h_{21}}{h_{11} + R_s}}.$$

Therefore Z_{out} is given by:

$$Z_{out} = \frac{\partial v_2}{\partial i_2} = \frac{1}{h_{22} - \dfrac{h_{12}h_{21}}{h_{11} + R_s}} = \frac{h_{11} + R_s}{(h_{11} + R_s)h_{22} - h_{12}h_{21}}, \tag{6.40}$$

which is identical to (6.38). Notice that if the reverse voltage generator is not present ($h_{12} = 0$), then $Z_{out} = 1/h_{22}$ which is reasonable. Notice also that the input resistance h_{11} and the effective source resistance R_s appear, which is what we expected because of the coupling between the input and the output due to h_{12}.

Typical numerical values for a low power transistor give:

$$Z_{out} = \frac{2\,\text{k}\Omega + R_s}{(2\,\text{k}\Omega + R_s)(2 \times 10^{-5}) - 10^{-4}(10^2)}$$

If the signal source has a low impedance, say $R_s = 50\,\Omega$, then

$$Z_{out} \cong 67\text{k}\Omega.$$

If $R_s = 10\,\Omega$, then $Z_{out} \cong 52\,\text{k}\Omega$. It can be seen that the output impedance approaches $1/h_{22} = 50\,\text{k}\Omega$ as R_s increases. However, in most applications, the source impedance R_s is less than $10\,\text{k}\Omega$. The main point here is that the output impedance of a transistor in the common emitter configuration is relatively high, of the order of $50\,\text{k}\Omega$ or more. Therefore, it is extremely inefficient to use a common emitter amplifier to drive a low impedance load, because most of the signal power will be dissipated within the amplifier due to the impedance mismatch. Nevertheless, for loads of the order of $10\,\text{k}\Omega$, the common emitter amplifier provides high gain, and it is, in fact, the most commonly used amplifier configuration.

A final point should be made here. The collector resistance R_C is in parallel with the output impedance of the transistor, so if $Z_{out} \cong 50\,\text{k}\Omega$ for the *transistor*, then the net output impedance of the *amplifier* is $Z_{out} \| R_C$. If $R_C = 2.2\,\text{k}\Omega$, then $Z_{out\ amp} = 50\,\text{k}\Omega \| 2.2\,\text{k}\Omega = 2.1\,\text{k}\Omega$. If R_C is usually from 2 to $20\,\text{k}\Omega$ then the *amplifier* output impedance is almost always approximately equal to the collector resistance R_C, thus making the common emitter amplifier unsuitable for driving low impedance loads. In order to drive a low impedance load with a common emitter amplifier, some sort of impedance matching device must be used to step down the relatively high output impedance of the amplifier to the low impedance of the load. This device can be a step down transformer (see Fig. 2.7) or a common collector amplifier which will be described in the next section.

A circuit diagram for a practical common emitter amplifier using an inexpensive 2N1309 germanium transistor is shown in Fig. 6.14. The first design objective is to achieve a fairly high voltage gain. A voltage gain of 260 can be achieved by using a $15\,\text{k}\Omega$ collector resistor, but this raises the output impedance to $10.8\,\text{k}\Omega$. A compromise must be chosen between high voltage gain and low output impedance; one cannot have both. If a lower output impedance is more important, then one might choose $R_C = 2.2\,\text{k}\Omega$, which gives an output impedance of $2.2\,\text{k}\Omega$ with a voltage gain of 50. Notice that the measured output impedance of the circuit is less than the collector resistance, because the *circuit* output impedance is basically the collector resistance in parallel with the *transistor* output impedance which for this circuit is approximately $45\,\text{k}\Omega$.

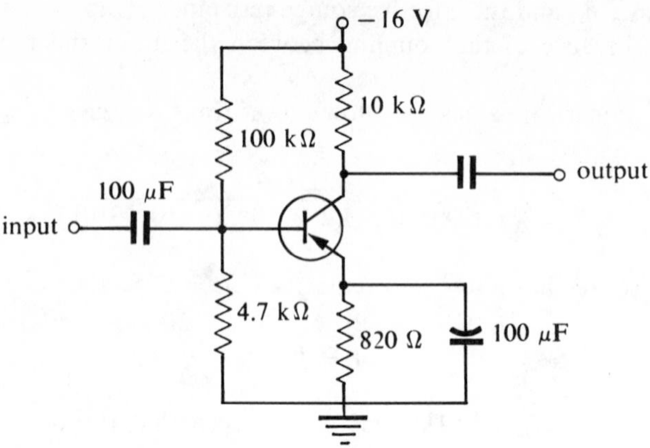

FIGURE 6.14 Common emitter amplifier.

6.4 COMMON COLLECTOR CONFIGURATION

The *h* parameter equivalent circuit can also be applied to the common collector configuration shown schematically in Fig. 6.15. Notice that the collector terminal of the transistor is common to both the input and the output in Fig. 6.15(a); hence the name "common collector." In Fig. 6.15(b), the collector is also common to the input and output for ac signals, because the power supply terminal labeled $-V_{bb}$ is an ac ground.

Let us now quickly go through the dc bias design. We use a divider net-

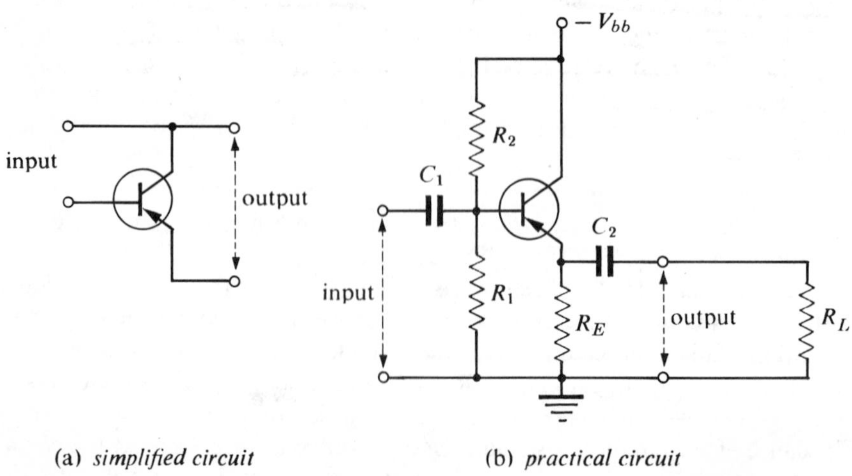

FIGURE 6.15 Common collector amplifier.

work of two resistors R_1 and R_2 to fix the operating point as shown in Fig. 6.15(b). A load line similar to that in the common emitter amplifier design implies that $R_E = 3\,\text{k}\Omega$. The same operating point ($V_{CE} = -6\,\text{V}$, $I_C = 2\,\text{mA}$) implies $V_E = I_E R_E = I_C R_E = (2\,\text{mA})(3\,\text{k}\Omega) = -6\,\text{V}$. Therefore, $V_B = -6.2\,\text{V}$ to turn on a pnp germanium transistor. If we choose $R_1 = 50\,\text{k}\Omega$, then R_2 must be determined by $R_2/R_1 = 5.8\,\text{V}/6.2\,\text{V}$, or $R_2 = 46\,\text{k}\Omega$. The divider current is approximately $I_D = V_{bb}/(R_1 + R_2) = 12\,\text{V}/96\,\text{k}\Omega \cong 0.120\,\text{mA} = 120\,\mu\text{A}$, which will not overly tax the power supply; and we also see that $I_D \gg I_B = I_C/\beta = 2\,\text{mA}/100 = 20\,\mu\text{A}$.

The capacitors C_1 and C_2 are chosen large enough not to attenuate the lowest signal frequency $\omega_{\min}$ encountered. That is, $C_1 > 1/\omega_{\min}(R_1 \| R_2 \| R_{\text{in}})$ and $C_2 > 1/\omega_{\min} R_L$. If the signal to be amplified consists of a relatively long pulse, then C_1 and C_2 may have to be fairly large to avoid an unacceptable amount of "sag" in the top of the output pulse. For a pulse width of $\tau = 1\,\text{msec} = 1000\,\mu\text{sec} = 10^{-3}\,\text{sec}$, the amplifier should be capable of handling a frequency down to around $f_{\min} = 1/10\,\tau = 100\,\text{Hz}$ to avoid sag in the top of the pulse.

If we assume the minimum frequency of operation is 20 Hz, which would be the case for an *audio* amplifier, then setting the break point at 10 Hz implies that $C_1 = 16\,\mu\text{F}$ and $C_2 = 320\,\mu\text{F}$ if $R_L = 50\,\Omega$; or $C_2 = 1.6\,\mu\text{F}$ if $R_L = 10\,\text{k}\Omega$.

We can calculate the voltage and current gain and the input and output impedances by using the common emitter h parameter equivalent circuit for the transistor and solving the Kirchhoff voltage and current equations. The ac equivalent circuit is shown in Fig. 6.16. Point "E" is the emitter terminal of the transistor, "B" the base terminal, and "C" the collector terminal. The power supply terminal has been grounded to obtain Fig. 6.16. The equations we need to solve are:

$$i_E + i_{\text{out}} = i_1 + i_2 \qquad \textit{(from current conservation at junction E)} \tag{6.41}$$

$$v_{\text{out}} = i_E R_E \tag{6.42}$$

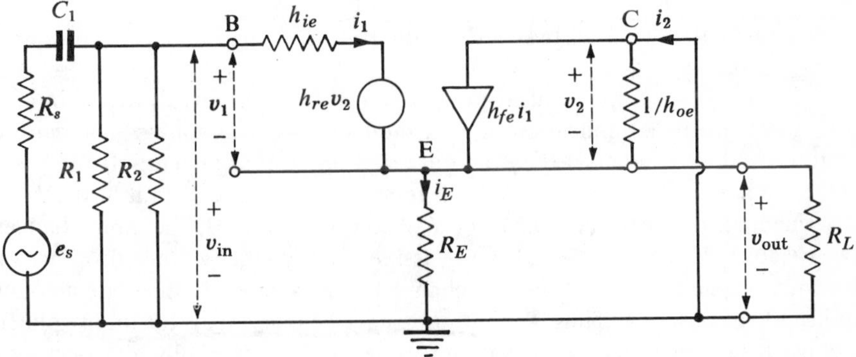

FIGURE 6.16 Common collector amplifier, ac h parameter equivalent circuit.

$$v_{\text{out}} = i_{\text{out}} R_L \tag{6.43}$$

$$v_{\text{in}} = v_1 + v_{\text{out}} \tag{6.44}$$

$$v_{\text{out}} = -v_2 , \tag{6.45}$$

and the two h parameter equations,

$$i_2 = h_{fe} i_1 + h_{oe} v_2 \tag{6.46}$$

$$v_1 = h_{ie} i_1 + h_{re} v_2 . \tag{6.47}$$

There are seven equations in eight unknowns (i_1, i_2, i_E, i_{out}, v_1, v_2, v_{in}, v_{out}) which can be solved for the ratio of any two of the unknowns: the voltage gain $A_v = v_{\text{out}}/v_{\text{in}}$, the current gain $A_i = i_{\text{out}}/i_1$, the output impedance $Z_{\text{out}} = v_{\text{out}}/i_{\text{out}}$ with $e_s = 0$, or the input impedance $Z_{\text{in}} = v_{\text{in}}/i_1$. However, the resulting algebra is tedious, and the results are too complicated to be really useful.

The results are:

$$A_i = \frac{i_{\text{out}}}{i_1} = \frac{-(1 + h_{fe})}{1 + \dfrac{R_L}{R_E} + h_{oe} R_1} \cong -h_{fe} \tag{6.48}$$

$$A_v = \frac{v_{\text{out}}}{v_{\text{in}}} = \frac{(1 + h_{fe}) R_E}{h_{ie} + R_E[(1 + h_{fe})(1 - h_{re}) + h_{oe} h_{ie}]} \cong 1 \tag{6.49}$$

$$Z_{\text{out}} \cong \frac{1}{\dfrac{1}{R_E} + h_{oe} + \dfrac{1 + h_{fe}}{R_{bs} + h_{ie}}} \cong \frac{1}{\dfrac{1}{R_E} + \dfrac{1}{\left(\dfrac{R_{bs} + h_{ic}}{h_{fe}}\right)}}$$

$$\text{where } R_{bs} = R_1 \| R_2 \| R_s \tag{6.50}$$

$$Z_{\text{in}} = \frac{1}{\dfrac{1}{R_b} + \dfrac{1}{h_{ie} + (1 + h_{fe}) R_E}} \cong \frac{1}{\dfrac{1}{R_b} + \dfrac{1}{h_{ie} + h_{fe} R_E}} . \tag{6.51}$$

Essentially the same results can be obtained from the circuit shown in Fig. 6.17 and the simple application of Kirchhoff's laws. The voltage gain can be seen to be almost exactly 1.0 by the following physical argument. If the base-emitter junction is turned on, V_{BE} will remain essentially constant at 0.2 V for a germanium transistor and 0.5 V for a silicon transistor. If the input changes by ΔV volts, V_B changes by ΔV volts assuming negligible voltage drop across the input capacitor C_1, and V_E will change by approximately the same amount to keep V_{BE} constant. Thus, the output voltage change equals the input voltage change assuming negligible voltage drop across C_2. Actually, V_E does not change quite as much as V_B, so typical voltage gains are 0.98 or 0.99. Notice

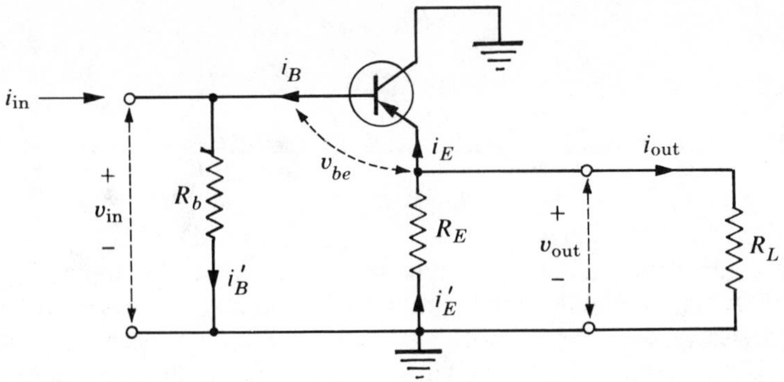

FIGURE 6.17 Common collector amplifier ac equivalent circuit.

also that as the base voltage goes negative so does the emitter voltage. In other words the emitter voltage or output voltage is in phase with the input; the output "follows" the input. Hence this circuit is often referred to as an "emitter follower." The appropriate Kirchhoff equations for Fig. 6.17 are:

$$i_{B'} = i_{in} + i_B \tag{6.52}$$

$$i_{E'} = i_E + i_{out} \tag{6.53}$$

$$v_{in} = v_{BE} + v_{out} \tag{6.54}$$

and Ohm's law implies:

$$v_{in} = i_{B'}R_b \tag{6.55}$$

$$v_{out} = -i_{E'}R_E = -i_E R_{EL} = i_{out}R_L\,. \tag{6.56}$$

We also need

$$i_E/\beta = i_B\,. \tag{6.57}$$

The input impedance is obtained from $Z_{in} = v_{in}/i_{in} = i_{B'}R_b/i_{in}$. Using (6.52) and (6.57)

$$Z_{in} = \frac{\left(\dfrac{i_E}{\beta} + i_{in}\right)R_b}{i_{in}} = \left(\frac{i_E}{\beta i_{in}} + 1\right)R_b\,. \tag{6.58}$$

Eliminating i_E using (6.56) gives:

$$Z_{in} = \left(\frac{-v_{out}}{i_{in}\beta R_{EL}} + 1\right)R_b\,. \tag{6.59}$$

Assuming the voltage gain is unity, $v_{out} = v_{in}$ and

$$Z_{\text{in}} = \left(\frac{-v_{\text{in}}}{i_{\text{in}}\beta R_{EL}} + 1\right) R_b,$$

or

$$Z_{\text{in}} = \frac{v_{\text{in}}}{i_{\text{in}}} = \frac{1}{\dfrac{1}{R_b} + \dfrac{1}{\beta R_{EL}}} = \frac{\beta R_{EL} R_b}{\beta R_{EL} + R_b}. \tag{6.60}$$

The input impedance of the *transistor* is seen to be βR_{EL}. The input impedance of the *circuit* is the parallel combination of R_b and βR_{EL}. In other words the emitter resistor R_E in parallel with the load resistor R_L appears multiplied by $\beta = h_{fe} \gg 1$ in the input parallel with R_b. Thus the input impedance can be very high, on the order of 50–100 kΩ if $R_b \gtrsim 50\,\text{k}\Omega$. The only approximation made in this derivation is $v_{\text{in}} \cong v_{\text{out}}$ which is usually true to within one or two percent. Notice that the input resistance depends upon the emitter resistance R_E *and* on the load resistance R_L. If $R_b = 50\,\text{k}\Omega$ and $R_{EL} \cong 1\,\text{k}\Omega$, then $Z_{\text{in}} \cong 33\,\text{k}\Omega$ for $\beta = 100$.

The current gain $A_i = i_{\text{out}}/i_{\text{in}}$ for the *circuit* can be obtained from (6.52), (6.55), (6.56), (6.57), and $v_{\text{out}} \cong v_{\text{in}}$. We have $v_{\text{out}} = -i_E R_{EL} \cong v_{\text{in}} = i_{B'} R_b = (i_{\text{in}} + i_B) R_b$ or

$$-i_E R_{EL} \cong \left(i_{\text{in}} + \frac{i_E}{\beta}\right) R_b.$$

Eliminating i_E yields:

$$i_{\text{out}} \frac{R_L}{R_{EL}} \left(\frac{R_b}{\beta} + R_{EL}\right) = -i_{\text{in}} R_b.$$

Thus the current gain for the *circuit* is:

$$A_i = \frac{i_{\text{out}}}{i_{\text{in}}} = \frac{-\beta}{\left(1 + \dfrac{R_L}{R_E}\right)\left(1 + \dfrac{\beta R_{EL}}{R_b}\right)}. \tag{6.61}$$

If $R_b \gg \beta R_{EL}$, then

$$A_i \cong \frac{-\beta}{\left(1 + \dfrac{R_L}{R_E}\right)}, \tag{6.62}$$

which approximately agrees with the exact expression (6.48) from the equivalent circuit.

The output impedance can be obtained from the following argument: If we assume the base-emitter voltage remains constant, then v_{BE} (the ac change in the base-emitter voltage) equals zero. In other words $v_{\text{in}} = v_{\text{out}}$, which is equivalent to setting the voltage gain equal to unity.

We also have:

$$i_B R_{bs} = v_{\text{out}} \qquad \text{where } R_{bs} = R_b \| r_s$$

$$i_B = \frac{i_C}{\beta} \cong \frac{i_E}{\beta} \qquad \text{so } \frac{i_E}{\beta} R_{bs} \cong v_{\text{out}}.$$

Using $i_E = i_E' + i_{\text{out}}$, we have $[(i_E' + i_{\text{out}})/\beta]R_{bs} = v_{\text{out}}$. But $i_E' = -v_{\text{out}}/R_E$. Hence, $\{[(-v_{\text{out}}/R_E) + i_{\text{out}}]/\beta\} R_{bs} = v_{\text{out}}$. Thus,

$$Z_{\text{out}} = \frac{v_{\text{out}}}{i_{\text{out}}} = \frac{R_{bs}/\beta}{1 + \dfrac{R_{bs}}{\beta R_E}} = \frac{1}{\dfrac{\beta}{R_{bs}} + \dfrac{1}{R_E}} = \frac{1}{\dfrac{1}{R_{bs}/\beta} + \dfrac{1}{R_E}}. \tag{6.63}$$

The output impedance is seen to be simply the parallel combination of R_E, the emitter resistor, with R_{bs}/β. The resistor R_{bs} divided by β appears in parallel with R_E. Thus the output impedance can be made very low, of the order of R_{bs}/β which is 0.5 Ω for $R_{bs} = 50\,\Omega$ and $\beta = 100$. This treatment neglects the $h_{11} = h_{ie}$ input resistance of the transistor, which is really quite important. The exact expression is

$$Z_{\text{out}} = \frac{1}{h_{oe} + \dfrac{1}{R_E} + \dfrac{1}{\left(\dfrac{R_{bs} + h_{ie}}{\beta}\right)}}, \tag{6.64}$$

which gives $Z_{\text{out}} \cong h_{ie}/\beta = 20\Omega$ for $R_E = 1\,\text{k}\Omega$, $h_{oe} = 2 \times 10^{-5}$ mhos, $R_{bs} \cong r_s = 50\Omega$, and $h_{ie} = 2\,\text{k}\Omega$, which agrees better with the experimental values for the output impedance.

A useful equivalent circuit of an emitter follower is shown in Fig. 6.18 in which we have used the approximate expressions: voltage gain $A_v = 1$, input impedance $Z_{\text{in}} = R_b \| (\beta R_{EL} + h_{ie})$ and $Z_{\text{out}} \cong R_E \| [(R_{bs} + h_{ie})/\beta]$.

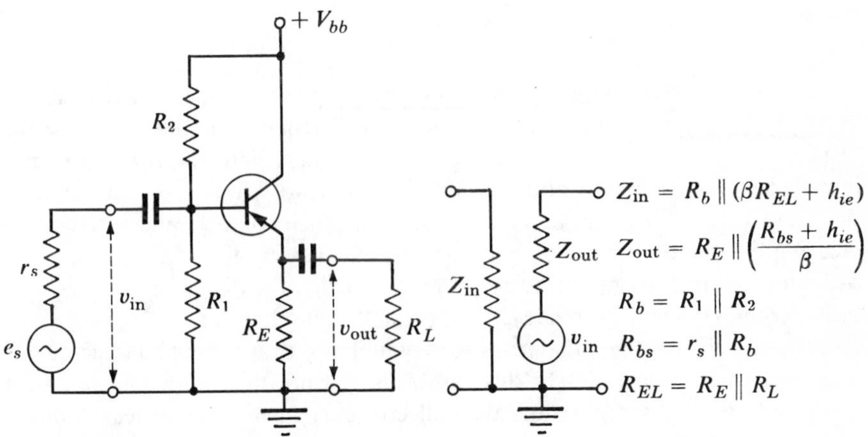

FIGURE 6.18 Emitter follower and equivalent circuit.

Most modern high fidelity power amplifiers use an emitter follower power output stage to drive the low impedance loudspeaker load. In the past, the technique to match the high amplifier output impedance to the low impedance loudspeaker was to use an expensive, high power, high fidelity impedance matching transformer.

A practical emitter follower amplifier using an inexpensive 2N1309 germanium transistor is shown in Fig. 6.19.

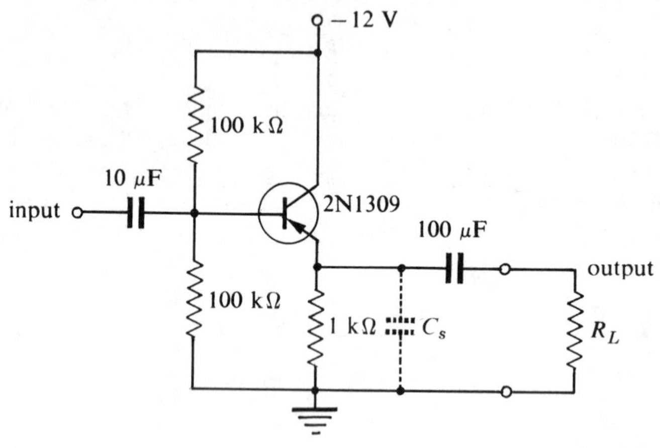

C_s = stray capacitance

Measured Circuit Performance at 1 kHz ($R_L >> R_E$)

voltage gain = 0.99
circuit output impedance = 10 Ω
circuit input impedance = 50 kΩ
3 dB frequency > 10 MHz

FIGURE 6.19 Common collector or emitter follower amplifier.

Notice that the voltage gain is essentially unity and that the output impedance is rather low, 10 ohms. The *circuit* input impedance of 50 kΩ is the parallel resistance of the two 100 kΩ bias resistors in parallel with the *transistor* input impedance βR_{EL}, which is of the order of hundreds of kilohms depending upon the exact value of R_L and β. The input impedance could be increased by increasing the value of the two bias resistors, but only up to a certain point. In order for the operating point to be set by R_1 and R_2, the divider current I_D must be larger than the base current I_B. For example, one couldn't use $R_1 = R_2 = 10$ megohms, because then the transistor would not draw enough collector current to operate properly. (In this case $I_B \gg I_D$ and the base would not be at -6 V dc.) For most transistors the collector current must be at least 1 mA in order to achieve a reasonably high value for β. Notice also that the bandwidth is of the order of 10 MHz for this circuit.

It should be mentioned that if a rectangular pulse is fed into an emitter follower, the output rise time will differ significantly from the output fall time. For the pnp circuit of Fig. 6.19, the *positive* going edge of the input pulse will turn the transistor *off* and the output voltage at the emitter can rise only by the stray capacitance C_s between the output terminal and ground charging through the emitter resistor R_E. Thus, the output time constant for the positive going edge will be $R_E C_s$. The value of C_s is typically 5 pF–10 pF, so for $R_E = 1\text{ k}\Omega$, $R_E C_s = (10^3\,\Omega)(10^{-11}\text{F}) = 10^{-8}$ sec. However, for the *negative* going edge of the input pulse, the pnp transistor will be turned on, and the output voltage will fall by the stray capacitance C_s discharging through the transistor that, being turned on, presents a very small output resistance, of the order of one hundred ohms or less usually. Thus the time constant for the negative going edge of the output pulse will be an order of magnitude less than for the positive going edge. For an npn transistor, the reverse situation holds. The time constant for the positive going edge will be less than for the negative going edge of the output.

An interesting and useful variation of a common collector amplifier or emitter follower is the "double emitter follower" or "Darlington" amplifier shown in Fig. 6.20. The Darlington circuit consists of two transistors connected together such that the emitter current of the first transistor provides the entire base current of the second transistor. The advantage of this arrangement is that the two transistors may be regarded as one "super transistor" with a β essentially equal to the *product* of the β's of the two individual transistors. This can be seen from the following approximate argument:

$$I_{E_1} = I_{B_2} \quad \therefore \quad I_{B_1} \cong \frac{I_{E_1}}{\beta_1} = \frac{I_{B_2}}{\beta_1} = \frac{I_{E_2}/\beta_2}{\beta_1} = \frac{I_{E_2}}{\beta_1\beta_2}. \tag{6.65}$$

Thus the base current of the input transistor T_1 is the emitter current of the output transistor T_2 divided by $\beta_1\beta_2$, the product of the β's of the two transistors. The same equations hold approximately for the ac signal currents: $i_{B_1} \cong i_{E_2}/\beta_1\beta_2$. An overall β of the order of $100 \times 100 = 10{,}000$ may be achieved, thus producing a circuit with an extremely high input impedance and an extremely low output impedance when connected as an emitter follower.

In actual practice, the principal difficulty is to make the emitter current of T_1 large enough to turn T_1 on and the collector current of T_2 small enough to keep T_2 from overheating. Thus we would probably never use the same transistor type for T_1 and T_2, because assuming $\beta = 100$, if we turn on T_1 by making $I_{E_1} = 1$ mA, then $I_{B_2} = I_{E_1} = 1$ mA, and $I_{C_2} = \beta_2 I_{B_2} = 100$ mA, which would probably destroy T_2. Or, if we choose $I_{C_2} = 10$ mA, to avoid overheating T_2, then $I_{B_2} = I_{C_2}/\beta_2 = 10\text{ mA}/100 = 100\,\mu\text{A} = I_{E_1}$, which would probably not be enough to turn T_1 on. And if T_1 is not turned on, its gain is extremely low. Hence, T_1 should be a transistor especially designed to operate at low collector currents of the order of $100\,\mu$A, or else T_2 must be capable of handling large collector currents of the order of 100 mA.

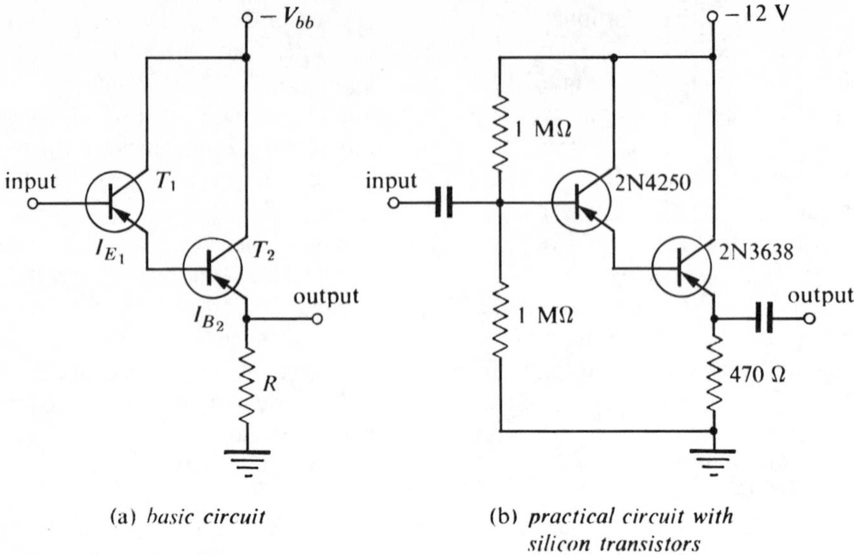

FIGURE 6.20 Darlington amplifier or double emitter follower.

The circuit shown in Fig. 6.20(b) contains two inexpensive silicon transistors, the 2N4250 for T_1 and the 2N3638 for T_2. The 2N4250 is especially designed to have a high β and low noise when drawing only 100μA of collector current. Because of its low collector current (and therefore low base current) one may choose large values for R_1 and R_2 thus producing a high circuit input impedance of 450 kΩ. If the 2N4250 collector current is increased by changing the operating point, the current gain and bandwidth will increase at the expense of slightly greater noise. The 2N3638 was chosen because it was an available, inexpensive, high speed silicon pnp transistor. A 2N3638A would have been a better choice, because it has a higher β.

6.5 COMMON BASE CONFIGURATION

The common base amplifier configuration is shown in simplified form in Fig. 6.21(a). Let us first consider the problem of biasing. The base-emitter junction must be biased in the forward direction, i.e., the base must be negative with respect to the emitter for a pnp transistor. And the base-collector junction must be biased in the reverse direction, with the collector negative with respect to the base for a pnp transistor. The circuit of Fig. 6.21(b) accomplishes this with two resistors R_1 and R_2 and two batteries V_{bb_1} and V_{bb_2}. Notice that the base terminal of the transistor is still common to the input and the output for ac signals, because the two batteries are ac short circuits. At this point we might

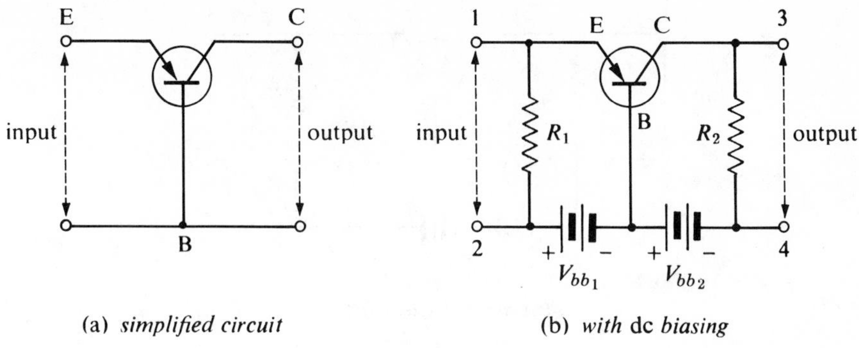

FIGURE 6.21 Simplified common base amplifier circuit.

think that we must ground point B implying two separate batteries, which is a major disadvantage. However, *any* one point in a circuit may be grounded. Thus if we ground point 2 instead of point B, we may use a voltage divider consisting of R_3 and R_4 to obtain the proper bias voltages for the base and collector as shown in Fig. 6.22(a). If the divider current I_D is much larger than the dc base current I_B, then the voltage divider resistors R_3 and R_4 are chosen according to $R_3/R_4 = V_{bb_1}/V_{bb_2}$. To make the base common to the input and output as far as the ac signal is concerned, we make the base an ac ground by adding a capacitor C_3 across R_3 such that the capacitor is an ac short circuit at the lowest (angular) signal frequency:

$$\frac{1}{\omega_{\min} C_3} \ll R_3 .$$

Terminal 4 is an ac ground because V_{bb} is an ac short, so the output may be taken between terminal 3 and ground instead of between terminals 3 and 4. The resulting circuit is shown in Fig. 6.22(b) with the addition of an input and an output coupling capacitor C_1 and C_2 to prevent the source or load from affecting the dc bias voltages. R_1C_1 is a high-pass filter, so we must choose the filter break point $\omega_B = 1/R_1C_1$ much less than the lowest (angular) signal frequency $\omega_{\min}$. Similarly C_2 and the load form a high-pass filter and the break point $\omega_B = 1/R_LC_2$ should be chosen much less than the lowest (angular) signal frequency $\omega_{\min}$.

Let us assume $V_{bb} = 18\text{V}$, that the desired load line implies $R_1 + R_2 = 10\,\text{k}\Omega$, and that the desired operating point on the load line is $V_{CE} = 8\text{V}$, $I_C = 1\,\text{mA}$. If we choose $R_2 = 10\,\text{k}\Omega$ and $R_1 = 1\,\text{k}\Omega$, then $V_E = -I_ER_1 \cong -I_CR_1 = -(1\,\text{mA})(1\,\text{k}\Omega) = -1\text{V}$. Therefore, $V_B = -1.2\text{V}$ to turn a germanium transistor on, or $V_B = -1.5\text{V}$ for a silicon transistor. Assuming $I_D \gg I_B = I_C/\beta \cong 1\,\text{mA}/100 = 10\,\mu\text{A}$, then R_3 and R_4 are determined by $I_DR_3 = V_B$ and $I_D(R_3 + R_4) = V_{bb}$. Choose $I_D = 0.1\,\text{mA}$. For a silicon transistor:

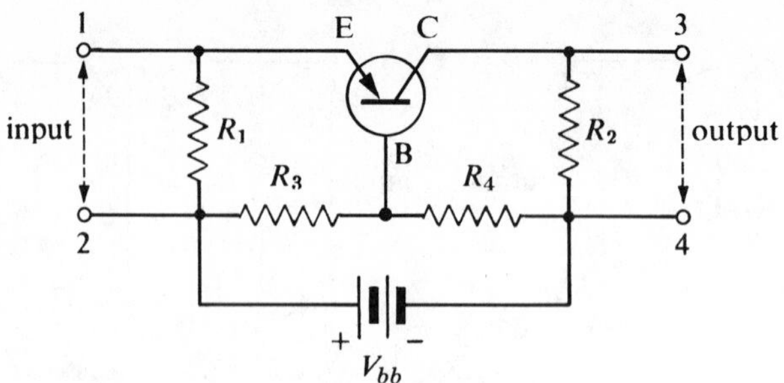

(a) *single battery bias*

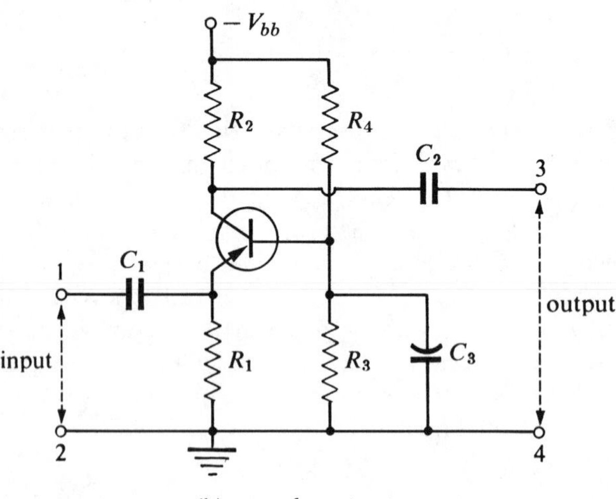

(b) *complete circuit*

FIGURE 6.22 Common base amplifier circuit.

$$R_3 = \frac{V_B}{I_D} = \frac{1.5\,\text{V}}{0.1\,\text{mA}} = 15\,\text{k}\Omega$$

$$R_3 + R_4 = \frac{V_{bb}}{I_D} = \frac{18\,\text{V}}{0.1\,\text{mA}} = 180\,\text{k}\Omega\,.$$

Therefore,

$$R_4 = 180 - 15 = 165\,\text{k}\Omega\,.$$

To avoid low frequency signal attenuation at frequencies down to 100 Hz, C_1 must be greater than 1.6μF from:

$$\omega_B = \frac{1}{R_1 C_1} \ll 2\pi(100\,\text{Hz}) \qquad \text{or} \qquad C_1 \gg \frac{1}{(2\pi)(100\,\text{Hz})(10^3\,\Omega)} = 1.6\mu\text{F},$$

and C_2 must be larger than a certain value depending upon R_L:

$$\omega_B = \frac{1}{R_L C_2} \ll 2\pi(100\text{ Hz}) \qquad \text{or} \qquad C_2 \gg \frac{1}{2\pi(100\text{ Hz})R_L}.$$

An exact analysis of the common base amplifier voltage gain, current gain, power gain, input impedance, and output impedance in terms of the common emitter h parameters is very complicated. However, the results can be obtained from straightforward algebra and transformations between the common base and the common emitter parameters.

The general h parameter equivalent circuit is repeated for convenience in Fig. 6.23. For this general circuit, one can calculate:

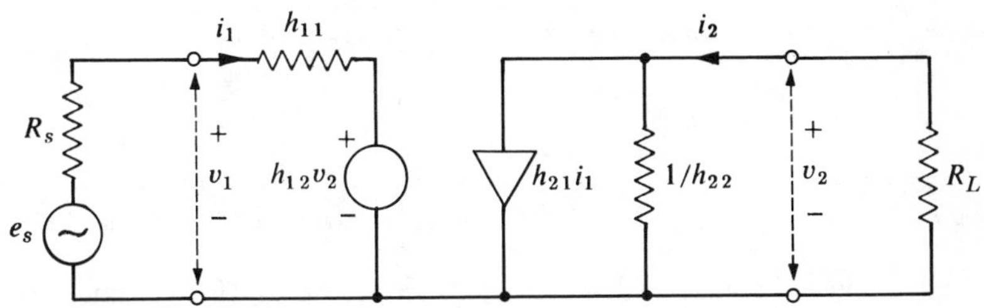

FIGURE 6.23 General h parameter transistor equivalent circuit.

Voltage Gain $$A_v = \frac{-h_{21}}{\Delta h R_L + h_{11}} \tag{6.66}$$

Current Gain $$A_i = \frac{h_{21}}{1 + h_{22}R_L} \tag{6.67}$$

Input Impedance $$Z_{\text{in}} = \frac{\Delta h R_L + h_{11}}{1 + h_{22}R_L} \tag{6.68}$$

Output Impedance $$Z_{\text{out}} = \frac{h_{11} + R_s}{\Delta h + h_{22}R_s} \tag{6.69}$$

where $\Delta h = h_{11}h_{22} - h_{12}h_{21}$.

Equations (6.66)–(6.69) can be used to find the gains and impedances for the *CE*, *CC*, or *CB* configuration by substituting in the *CE*, *CC*, or the *CB* parameters, respectively. For example, if we know the common emitter parameters then substitution of h_{ie} for h_{11}, h_{re} for h_{12}, h_{fe} for h_{21}, and h_{oe} for h_{22} will yield the gains and impedances for the common emitter configuration. However, to use the expressions (6.66)–(6.69) for the common base circuit, we must know the common base h parameters h_{ib}, h_{rb}, h_{fb}, and h_{ob}. Manufacturers usually list only the common emitter h parameters in their transistor manuals or data sheets, so in general we don't know the *CC* or *CB* h parameters.

Thus to calculate the gains and impedances for the common base amplifier we must first find the *CB* h parameters. This can be done by a straightforward algebraic process. The results are shown in Table 6-2 for the common base

TABLE 6-2.* COMMON BASE AND COMMON COLLECTOR h PARAMETERS IN TERMS OF THE COMMON EMITTER h PARAMETERS

$$h_{ib} = \frac{h_{ie}}{1 + h_{fe}} \qquad h_{ic} = h_{ie} \tag{6.70}$$

$$h_{rb} = \frac{h_{ie}h_{oe}}{1 + h_{fe}} - h_{re} \qquad h_{rc} = 1 - h_{re} \tag{6.71}$$

$$h_{fb} = \frac{-h_{fe}}{1 + h_{fe}} \qquad h_{fc} = 1 + h_{fe} \tag{6.72}$$

$$h_{ob} = \frac{h_{oe}}{1 + h_{fe}} \qquad h_{oc} = h_{oe} \tag{6.73}$$

* From *Basic Theory and Applications of Transistors,* U.S. Army Technical Manual TM 11–690.

and the common collector h parameters in terms of the common emitter parameters. To emphasize the differences among the *CE, CC,* and *CB* h parameters we list here the approximate numerical values for the h parameters in the three configurations for a low power transistor.

TABLE 6-3. NUMERICAL VALUES OF h PARAMETERS

	CE	*CC*	*CB*
h_{11}	$h_{ie} = 500\Omega$–$3\text{k}\Omega$	$h_{ic} = 500\Omega$–$2\text{k}\Omega$	$h_{ib} = 5\Omega$–20Ω
h_{12}	$h_{re} \cong 10^{-3}$–10^{-4}	$h_{rc} \cong 1$	$h_{rb} \cong 2 \times 10^{-4}$
h_{21}	$h_{fe} \cong 100$	$h_{fc} \cong -100$	$h_{fb} \cong -0.99$
h_{22}	$h_{oe} \cong 4 \times 10^{-5}$ mhos	$h_{oc} \cong 2 \times 10^{-5}$ mhos	$h_{ob} \cong 2 \times 10^{-7}$ mhos
Δh	$\Delta_e h \cong 0.02$	$\Delta_c h \cong 100$	$\Delta_b h \cong 4 \times 10^{-4}$

The calculations for the common base gains and impedances give the following:

Voltage Gain

$$A_v = \frac{-h_{fb}}{\Delta_b h R_L + h_{ib}} = \frac{h_{fe}R_L}{h_{ie} + \Delta_e h R_L} \cong \frac{h_{fe}}{h_{ie}} R_L \tag{6.74}$$

Current Gain

$$A_i = \frac{h_{fb}}{1 + h_{ob}R_L} = \frac{h_{fe}}{1 + h_{fe} + h_{oe}R_L} \cong 1 \tag{6.75}$$

Input Impedance

$$Z_{\text{in}} = \frac{\Delta_b h R_L + h_{ib}}{1 + h_{ob}R_L} = \frac{h_{ie} + \Delta_e h R_L}{1 + h_{fe} + h_{oe}R_L} \cong \frac{h_{ie}}{h_{fe}} \tag{6.76}$$

Output Impedance

$$Z_{\text{out}} = \frac{h_{ib} + R_s}{\Delta_b h + h_{ob} R_s} = \frac{h_{ie} + R_s(1 + h_{fe})}{\Delta_e h + h_{oe} R_s} \cong \frac{R_s h_{fe}}{\Delta_e h}. \tag{6.77}$$

If we substitute in numerical values we find for the common base configuration:

	$R_s = 100\,\Omega$, $R_L = 10\,\text{k}\Omega$	$R_s = 1\,\text{k}\Omega$, $R_L = 10\,\text{k}\Omega$	$R_s = 100\,\Omega$, $R_L = 1\,\text{k}\Omega$	$R_s = 1\,\text{k}\Omega$, $R_L = 1\,\text{k}\Omega$
Voltage Gain	500	500	50	50
Current Gain	1	1	1	1
Input Impedance	$23\,\Omega$	$23\,\Omega$	$20\,\Omega$	$20\,\Omega$
Output Impedance	$5\,\text{M}\Omega$	$2\,\text{M}\Omega$	$5\,\text{M}\Omega$	$2\,\text{M}\Omega$

Thus we see that the common base amplifier has a high voltage gain with no phase inversion, a low (unity) current gain, a low input impedance, and a high output impedance. In a sense, it is almost the exact opposite of the common collector (emitter follower) circuit which has a low (unity) voltage gain, a high current gain, and a high input and low output impedance. The common base amplifier is widely used at high frequencies (above 10 MHz) because of the excellent isolation between the input and the output provided by the grounded base. The isolation means that the amplifier is stable; it does not tend to break into unwanted oscillation.

The same procedure of transforming from the *CC h* parameters to the *CE h* parameters could have been used to calculate the exact gain and impedance expressions for the common collector amplifier in terms of the common emitter *h* parameters. The gain and impedance expressions for all three configurations are repeated here in Table 6-4 for convenience in terms of the available *CE h* parameters.

A practical common base amplifier circuit using an inexpensive 2N1309 pnp germanium transistor is shown in Fig. 6.24. As in the case of the common emitter amplifier circuit of Fig. 6.14, the output impedance is slightly less than the collector resistance, but in the common base circuit the input impedance is substantially lower—43 Ω compared to 2.2 kΩ for the common emitter amplifier. The 3 dB frequency, at which the gain drops to $1/\sqrt{2}$ of its mid frequency value, is seen to be considerably greater for the common base circuit than for the common emitter circuit.

In summary, it is interesting to compare the measured characteristics of the three practical amplifiers given in this chapter. It should be added that the three amplifiers were constructed on a breadboard circuit with no attempt made to minimize lead lengths or to otherwise enhance the high frequency performance. Variations of up to ±50% in circuit performance may be expected if

TABLE 6-4. EXACT GAIN AND IMPEDANCE EXPRESSIONS FOR THE THREE TRANSISTOR CONFIGURATIONS IN TERMS OF THE COMMON EMITTER h PARAMETERS

	CE	CC	CB
Voltage Gain	$\dfrac{-h_{fe}R_L}{\Delta_e h R_L + h_{ie}}$	$\sim\dfrac{(1+h_{fe})R_L}{h_{fe}R_L + h_{ie}}$	$\dfrac{h_{fe}R_L}{\Delta_e h R_L + h_{ie}}$
Current Gain	$\dfrac{h_{fe}}{1 + h_{oe}R_L}$	$-\dfrac{1 + h_{fe}}{1 + h_{oe}R_L}$	$\dfrac{h_{fe}}{1 + h_{oe}R_L + h_{fe}}$
Input Impedance	$\dfrac{\Delta_e h R_L + h_{ie}}{1 + h_{oe}R_L}$	$\sim\dfrac{h_{fe}R_L + h_{ie}}{1 + h_{oe}R_L}$	$\dfrac{\Delta_e h R_L + h_{ie}}{1 + h_{oe}R_L + h_{fe}}$
Output Impedance	$\dfrac{h_{ie} + R_s}{\Delta_e h + h_{oe}R_s}$	$\sim\dfrac{h_{ie} + R_s}{h_{fe} + h_{oe}R_s}$	$\dfrac{h_{ie} + R_s(1 + h_{fe})}{\Delta_e h + h_{oe}R_s}$

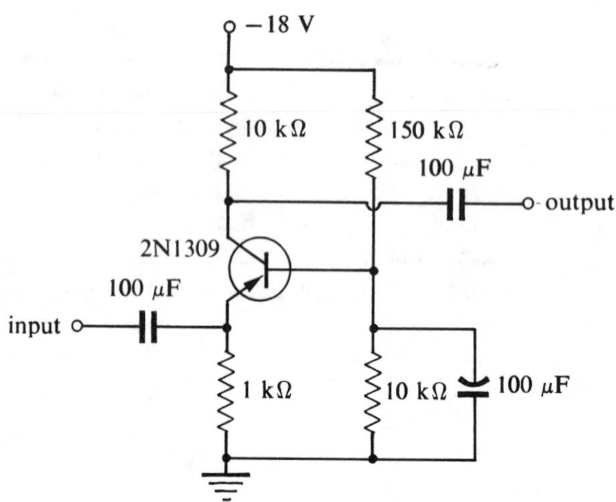

FIGURE 6.24 Common base amplifier.

Circuit	1 kHz *Voltage Gain*	1 kHz *Circuit Input Impedance*	1 kHz *Circuit Output Impedance*	3 dB *Frequency*
(Fig. 6.14) *Common Emitter Amplifier*	200	2.2 kΩ	8.2 kΩ	175 kHz
(Fig. 6.19) *Common Collector Amplifier*	0.99	50 kΩ	10 Ω	>10 MHz
(Fig. 6.24) *Common Base Amplifier*	200	43 Ω	9 kΩ	400 kHz

these circuits are constructed with different transistors of the same type and 20% tolerance resistors.

problems

1. In the simple equivalent circuit shown, calculate the current gain.

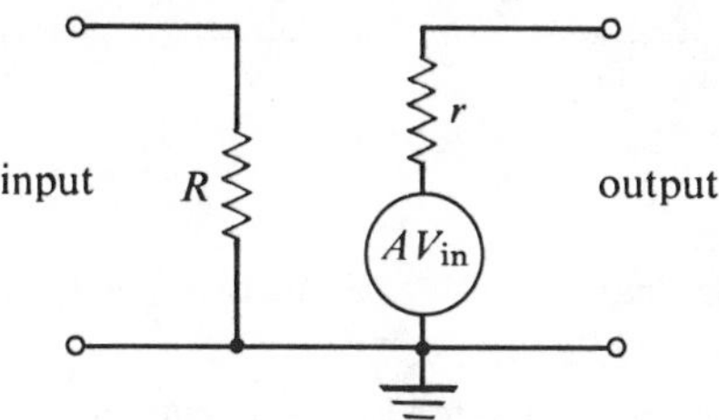

2. Explain why the h-parameter equivalent circuit can be applied to any four-terminal device so long as small signals are concerned.
3. State the units of the four h parameters.
4. Derive the h-parameter equivalent circuit of the following resistive "black box."

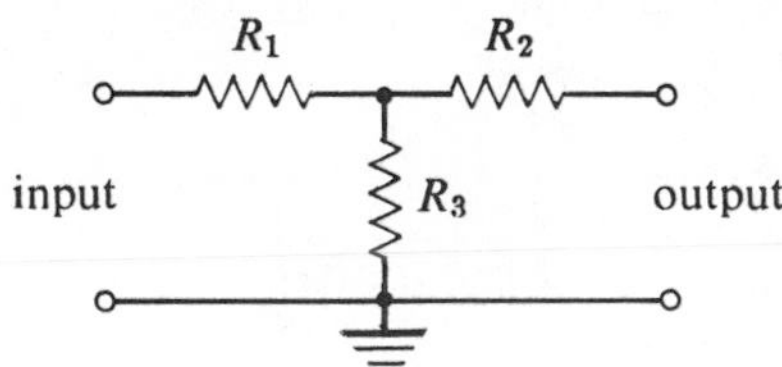

5. Derive the h-parameter equivalent circuit of the following.

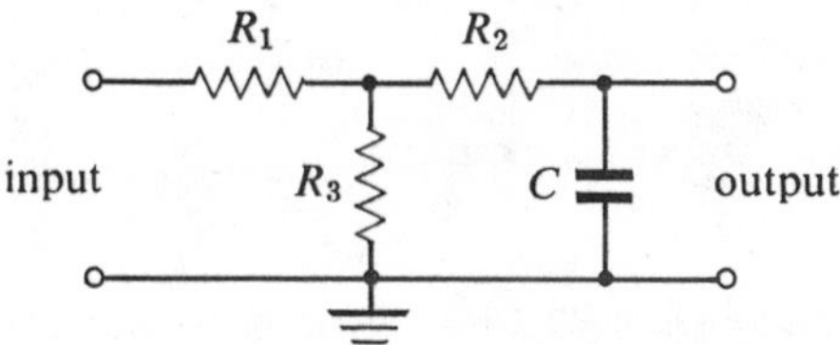

6. Give typical values for h_{ie}, h_{re}, h_{fe}, and h_{oe}. What would be the values for an ideal transistor?
7. (a) Calculate the four h parameters and β for the transistor whose input and output characteristic curves are shown below if the dc operating point is $I_C = 2\,\text{mA}$, $V_{CE} = 10\,\text{V}$. (b) As the collector current is increased, how do h_{fe} and h_{oe} change?

(c) Compare h_{fe} with $\beta = \alpha/(1 - \alpha)$. (d) Explain why V_{CE} at the operating point should be greater than (1 mA, 1 V).

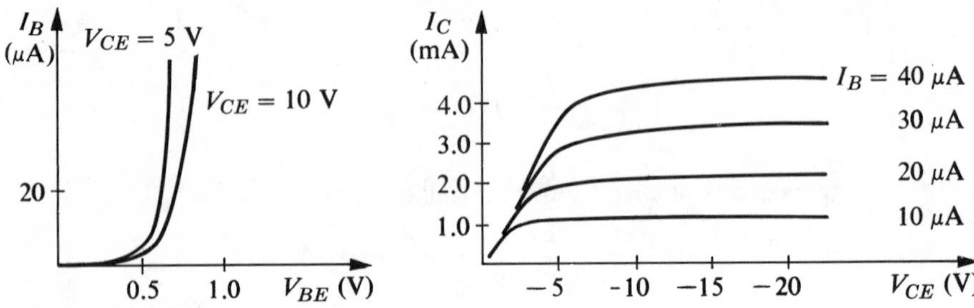

8. Show that the input impedance for a transistor connected in the common emitter configuration is equal to h_{ie} if $h_{re} = 0$.

9. Calculate an approximate value for h_{ie} for a transistor at room temperature with $h_{fe} = 100$ and an operating point $V_{CE} = 10\text{V}$, $I_C = 1\text{ mA}$. Repeat if the operating point is changed to $V_{CE} = 5\text{V}$, $I_C = 100\,\mu\text{A}$.

10. From the definition of h_{ie} and the current–voltage graph for the emitter–base junction, show (by a graphical argument) that the value of h_{ie} decreases for increasing emitter current. Sketch a rough graph of h_{ie} versus I_E.

11. Calculate the approximate 1 kHz ac input impedance to the *circuit* (not the transistor alone) between terminals A and B. $h_{ie} = 2\text{k}\Omega$, $h_{re} = 10^{-4}$, $h_{fe} = 100$, $h_{oe} = 10^{-5}$ mhos.

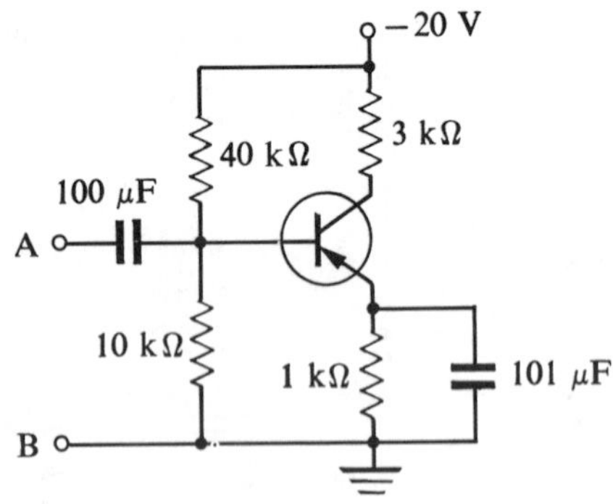

12. Calculate the voltage gain and current gain for the amplifier of Problem 11.

13. Calculate the output impedance for the amplifier of Problem 11.

14. Explain why R_1 should not be much less than h_{ie} for a common emitter amplifier. (R_1 is the resistance from base to ground.)

15. What is the effective output impedance of an amplifier driven by a constant input voltage whose output versus load resistance is given in the following table?

R_L	V_{out}
1 MΩ	500 mV
100 kΩ	500 mV
10 kΩ	450 mV
1 kΩ	240 mV
100 Ω	50 mV

16. What can you say about the ac output impedance of the following common emitter amplifier stage?

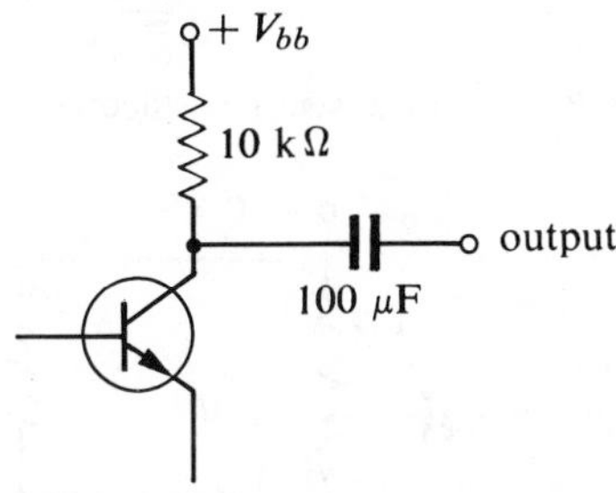

17. Explain briefly, *without* using h parameters, why one would expect the voltage gain of an emitter follower or common collector amplifier to be less than unity.

18. Calculate I_C and V_{CE} for the emitter follower shown. You may take the transistor $\beta = 100$.

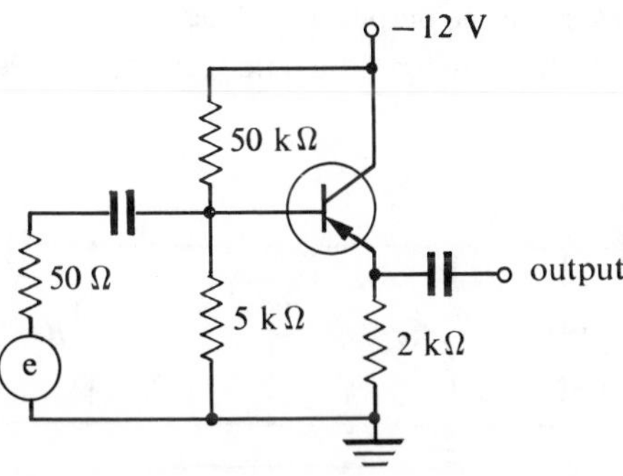

19. Calculate the voltage gain and input and output impedance of the emitter follower of Problem 18. Repeat if the emitter resistor is 20 kΩ (assume $\beta = 100$).

20. Calculate R_1, R_2, and R_E for an emitter follower if the desired operating point is $V_{CE} = 10\text{V}$, $I_C = 3\text{mA}$, and the desired circuit input impedance is 30 kΩ. Use $V_{bb} = 18\text{V}$.

21. Calculate the collector currents of T_1 and T_2 in the following Darlington amplifier. $\beta_1 = 100$, $\beta_2 = 50$. T_1 and T_2 are silicon.

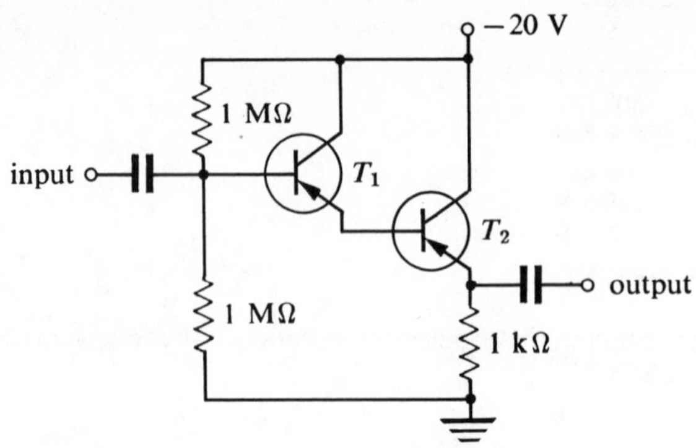

22. Calculate C_3, R_3, and R_4. The transistor is silicon.

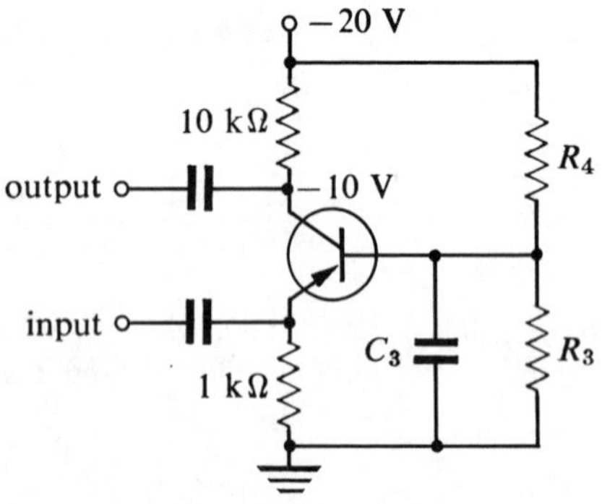

What would you estimate the output impedance to be?

23. What is wrong with the common base amplifier shown below? Why? Correct the mistakes.

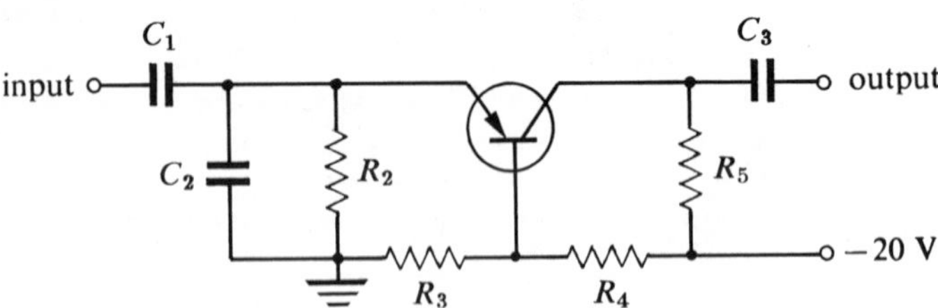

7

THE FIELD EFFECT TRANSISTOR (FET)

7.1 INTRODUCTION

The field effect transistor or FET is a special type of transistor which offers a number of advantages in some applications over the ordinary bipolar pnp or npn transistors described in earlier chapters. Its advantages include considerably higher input impedance, lower noise, and a higher resistance to nuclear radiation damage. The high input impedance is important in applications where the signal source driving the FET has a high output impedance or where very small currents must be amplified. If the FET input impedance is much higher than the source output impedance, then little signal is lost due to impedance mismatch, i.e., the source is not appreciably "loaded." In general, however, the gain available from an FET is less than from a carefully selected bipolar transistor. Because an FET carries current only by the flow of majority carriers, it is sometimes termed a "unipolar" transistor, although the term FET is more common. An ordinary transistor is often termed a "bipolar" transistor to distinguish it from an FET, because it carries current by the flow of both majority and minority carriers. In a pnp transistor, for example, the holes are majority carriers in the emitter and the collector, but are minority carriers in the base. In this chapter we discuss the construction of various types of FET's, the y parameter equivalent circuit, and give examples of several practical FET circuits.

7.2 FET CONSTRUCTION

A field effect transistor is basically a small slab of doped semiconductor, either p- or n-type, with the opposite type impurity diffused in either side of the slab as shown in Fig. 7.1. The main slab is called the "channel" and the two small

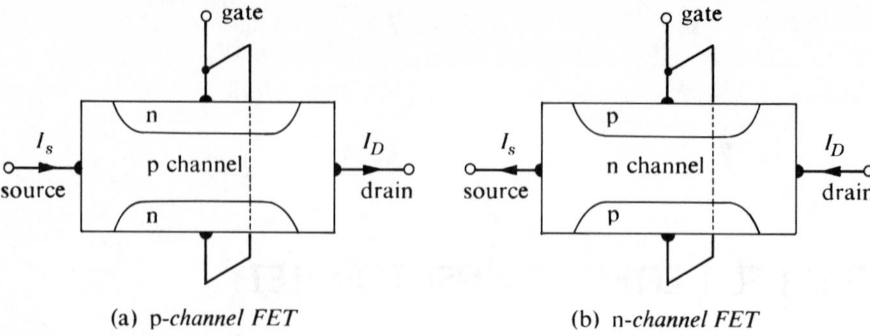

(a) p-*channel FET* (b) n-*channel FET*

FIGURE 7.1 FET construction.

regions of opposite type semiconductor material on the sides of the channel are connected together and called the "gate."

In Fig. 7.1(a), the channel is made of p-type material and the gate of n-type material; this device is termed a "p channel FET" or a "p channel junction FET" to be more precise. An n channel junction FET is shown in Fig. 7.1(b). From now on we will confine our attention to a p channel FET.

Ohmic contacts are made on each end of the channel, and one is called the "source" and the other the "drain." That is, the source and drain terminals are merely the two ends of the channel. If a voltage difference is applied between the source and the drain, with the drain negative with respect to the source, then the majority carrier holes in the p-type channel will flow from the source towards the drain. A battery V_{DD} and a resistor R_D will accomplish this as shown in Fig. 7.2. If the gate terminal is unconnected to anything, then the current I_S

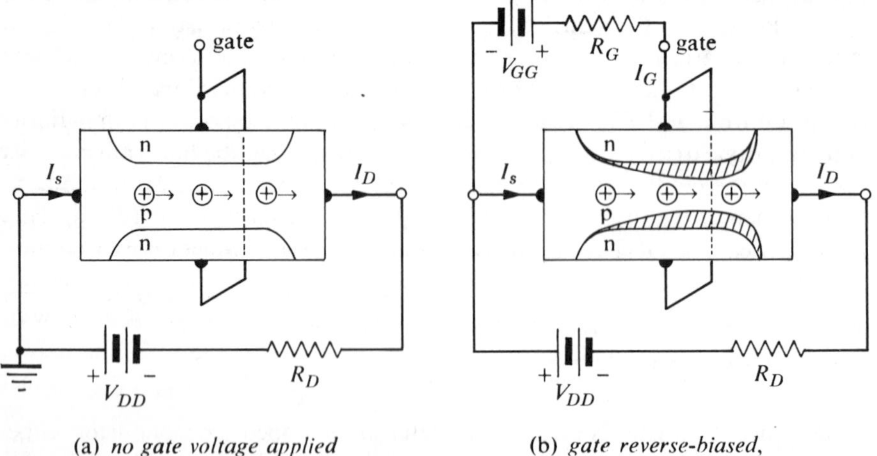

(a) *no gate voltage applied* (b) *gate reverse-biased, depletion region shaded*

FIGURE 7.2 Current flow in a p channel FET.

flowing into the source must exactly equal the current I_D leaving the drain. In this situation with the gate floating, the current is limited only by the resistivity of the channel and, of course, the external resistance R_D. Recall that the resistance of a slab of material of cross-sectional area A, length L, and resistivity ρ is given by

$$r = \rho \frac{L}{A}. \tag{7.1}$$

We can see that the drain current I_D will be given by

$$I_D = \frac{V_{DD}}{r + R_D} = \frac{V_{DD}}{\dfrac{\rho L}{A} + R_D}. \tag{7.2}$$

In other words, the FET is acting just like a resistor, with its resistance being the resistance of the channel material with a constant length L and cross-sectional area A.

However, if the gate material is reverse-biased with respect to the channel, as shown in Fig. 7.2(b), then the situation changes drastically. As we recall from Chapter 4, a reverse-biased p-n junction will have a "depletion region" formed at the junction. This depletion region contains no mobile charge carriers, but only fixed ionized impurity atoms, and is shown as a shaded region in Fig. 7.2(b). The larger the reverse bias voltage applied to the gate, the thicker the depletion region, and the smaller the cross-sectional area of the channel through which the hole current I_D can flow. Notice that the hole current in the channel continues to flow only through p-type material, unlike the flow of holes in a pnp transistor where the mobile holes find themselves minority carriers in the field-free n-type base material. One should also notice that because of the ohmic voltage drop along the channel material as we go from the source toward the drain, the gate-channel reverse bias voltage is larger near the drain end than near the source end. Hence, the depletion layer is thicker nearer the drain than near the source.

If the voltage V_{DS} between the drain and the source is kept fixed, then as the gate reverse bias is increased, i.e., as the gate is made more positive, the drain current will decrease as the depletion regions enlarge and narrow the cross-sectional area of the channel available for drain current flow. Eventually, the two depletion regions will meet and effectively block the channel, thereby reducing the drain current to zero. In a very real physical sense the gate terminal, when reverse-biased, acts like a "gate" in restricting and cutting off the flow of current through the channel. When the two depletion regions meet and reduce the drain current to zero, this is referred to as "pinch off," because the channel has literally been pinched off by the two depletion regions. Typical values of gate-source voltage required to achieve pinch off are several volts. The exact value of the pinch-off voltage depends upon the dielectric constant ϵ of the channel material, the number of impurity atoms p per unit volume in the channel, and the physical thickness H of the channel measured perpendicular to the channel current, i.e., from gate to gate. An exact analysis yields the result:

$$V_{\text{pinch off}} = \frac{epH^2}{8\epsilon} \tag{7.3}$$

As expected, a thicker channel implies a larger pinch-off voltage.

The pinch-off voltage can be derived from an elementary Gauss's law argument. Figure 7.3 shows a cross section of a p channel FET with the depletion

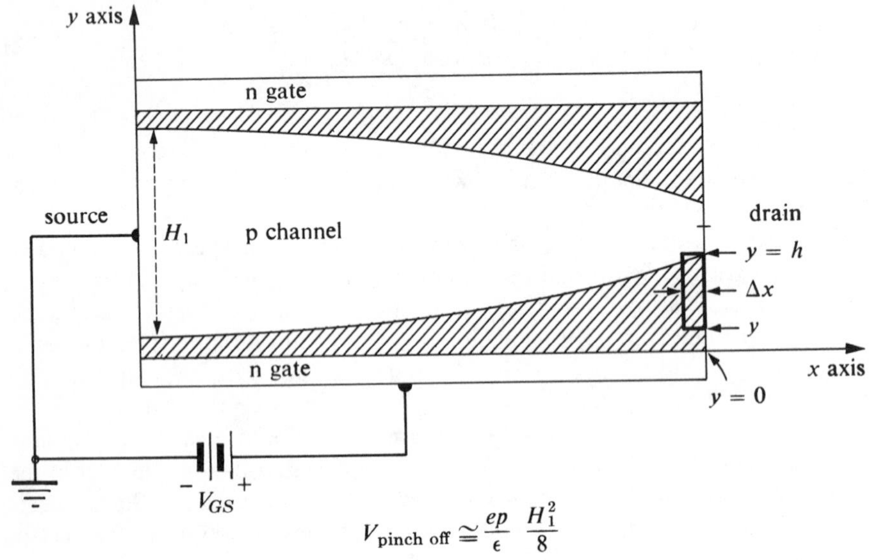

FIGURE 7.3 FET diagram for pinch-off voltage derivation.

region shaded. Voltage V_{GS} is the reverse bias applied between the n-type gate and the p-type channel. We have assumed here that the n-type impurity concentration in the gate greatly exceeds the p-type impurity concentration in the channel, so that the depletion region formed extends mainly into the channel and not into the gate. Let h be the maximum height of the depletion region at the drain end of the channel. Consider a small rectangular volume element at the drain with height $(h - y)$ in the y direction, Δx in the x direction, and Δz in the z direction. The total charge in this volume is given by:

$$Q = ep\,\Delta x(h - y)\,\Delta z, \tag{7.4}$$

where e is the electronic charge and p is the number of p-type impurities per unit volume in the channel material (p is assumed to be constant). The electric field in the depletion region will be in the y direction and will be zero at the top of this volume, i.e., at $y = h$. Thus Gauss's law says:

$$\int \mathbf{E} \cdot \mathbf{n}\, dA = \frac{Q}{\epsilon},$$

where ϵ is the dielectric constant of the channel material and Q is the total net charge inside the rectangular volume.

$$E\, \Delta x\, \Delta z = \frac{ep\, \Delta x(h - y)\, \Delta z}{\epsilon}$$

or

$$E = \frac{ep(h - y)}{\epsilon}. \tag{7.5}$$

Setting E equal to $-dV/dy$ where V is the potential or voltage in the depletion region, and integrating with respect to y from $y = 0$ to $y = h$, we have:

$$-\int_0^h \frac{dV}{dy}\, dy = \int_0^h \frac{ep(h - y)\, dy}{\epsilon}$$

$$V(0) - V(h) = \frac{ep}{\epsilon}(hy - y^2/2)\Big|_0^h \tag{7.6}$$

or

$$V_{GS} - V(h) = \frac{ep}{\epsilon} \cdot \frac{h^2}{2},$$

where we have set $V(0) = V_{GS}$, since the voltage at the gate edge of the channel is essentially equal to the gate-source bias V_{GS}. Pinch-off occurs when the depletion regions from each gate side of the channel meet, or in other words, when $h = \frac{1}{2}H$ where H is the total width of the channel material. The value of V_{GS} that produces this value of h is termed the pinch-off voltage.

$$V_{\text{pinch off}} - V(h) = \frac{ep}{\epsilon} \cdot \frac{H^2}{8}$$

The final step is to set $V(h) = 0$ when pinch off occurs, because there is then no *IR* voltage drop along the channel, and $V(h)$ equals the voltage at the source end of the channel which is grounded. Thus the pinch-off voltage is given by:

$$V_{\text{pinch off}} \cong \frac{ep}{\epsilon} \cdot \frac{H^2}{8}. \tag{7.7}$$

If the voltage on the gate is kept constant, and the drain-source voltage V_{DS} is increased, then the drain current increases. But, as V_{DS} is increased, the depletion region near the *drain* end of the gate will increase in thickness even though the gate remains at a constant voltage, because of the increasing negative voltage at the drain end of the channel caused by the *IR* drop along the channel.

In other words, as V_{DS} is increased the cross-sectional area of the channel is decreased, increasing the resistance of the channel. Thus the drain current will *not* increase linearly with V_{DS} as for an ordinary resistor obeying Ohm's law, but rather will increase at a slower and slower rate as V_{DS} is increased. The resulting graph of I_D versus V_{DS} for a fixed gate-source reverse bias voltage is shown in Fig. 7.4. As V_{DS} is increased the depletion regions nearly meet, and the cross-sectional area of the channel available to pass drain current is very small. The resistance of the channel is then very large (of the order of 10^4 to $10^5 \Omega$), and further increases of V_{DS} result in only a very small increase in drain current. This region of operation is on the flat portion of the I_D versus V_{DS} graph in Fig. 7.4, and is sometimes referred to as the "constant current," "satu-

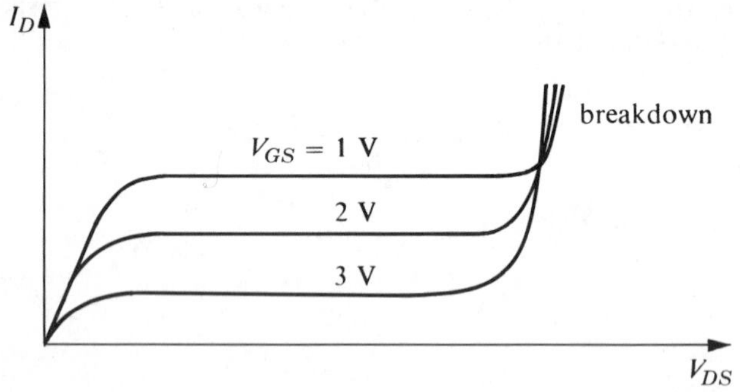

FIGURE 7.4 FET drain current versus drain-source voltage for constant gate-source voltage.

ration" region, or the "pinch-off" region. If the drain-source voltage is increased far enough, the depletion region blocking the channel will break down and the drain current will rise sharply, possibly destroying the FET. Breakdown usually occurs for the drain-source voltages of the order of 20 to 50 volts. In practical FET circuits the drain-source voltage is kept well below the breakdown voltage.

For low values of the drain-source voltage V_{DS}, the drain current I_D increases approximately linearly with V_{DS}. In this region the FET acts like an ohmic resistance that can be controlled by varying the gate-source voltage V_{GS}, a higher value of V_{GS} resulting in a higher value of the FET channel resistance. When operated like this, the FET is termed a "voltage controlled resistance."

As with any reverse-biased, p-n junction, there is a certain reverse current I_G which flows in the gate terminal as shown in Fig. 7.2(b). We immediately see that conservation of current requires $I_S + I_G = I_D$. The reverse current I_G arises from the thermal generation of minority carriers in the depletion region between the gate and the channel, and for a silicon junction is on the order of 10 nA, doubling roughly for every 5 °C rise in temperature. Most FET's are silicon, so we immediately see why they have such large input impedances if the gate is the input. The input impedance is simply the impedance of the reverse-biased, gate-channel silicon junction which may be 100 megohms or more.

The generally accepted circuit symbol for a junction p channel FET is shown in Fig. 7.5(a) and for a junction n channel FET in Fig. 7.5(b). The direction of the arrow in the gate lead points away from the channel for a p channel device and toward the channel for an n channel device; this is consistent with the diode symbol. That is, the arrow indicates the direction of "easy" current flow when the junction is forward-biased. It should be emphasized again that the FET gate-channel junction is always *reverse* biased so that current should *never* flow in the direction of the arrow in the FET symbol.

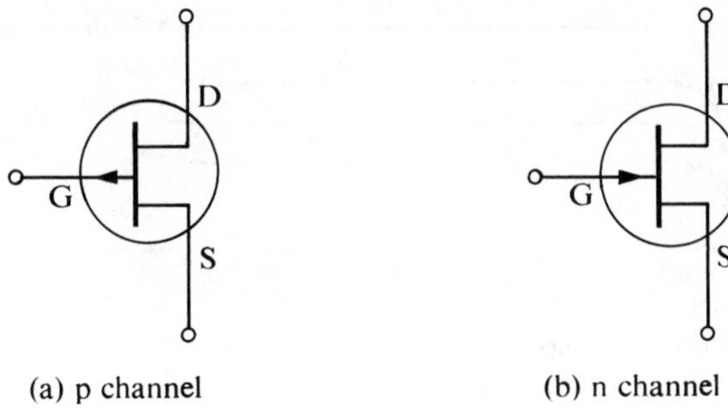

FIGURE 7.5 FET circuit symbols.

7.3 FET CHARACTERISTIC CURVES

We have just seen that if the gate is made more positive with respect to the channel for a p channel device, then the channel is constricted and the drain current I_D is reduced. If the gate-source voltage V_{GS} is kept constant and the drain-source voltage V_{DS} is increased, the drain current at first increases linearly with V_{DS} and then curves off to approach a nearly constant value as V_{DS} is increased more. Such behavior of an FET can most conveniently be presented on a graph of drain current versus drain-source voltage for various constant values of the gate-source voltage. Such a set of characteristic curves for a typical p channel FET is shown in Fig. 7.6; these curves are sometimes referred to as the "output" curves because the drain current is often regarded as the output of the FET. The negative value of V_{DS} for p channel FET's merely means that the drain is negative with respect to the source. The fact that V_{GS} is positive means that the gate is positive with respect to the source which is a reverse bias for a p channel FET. The negative value for I_D is due to the convention that a positive current flows *into* the drain terminal. This will be the convention used in developing an FET equivalent circuit.

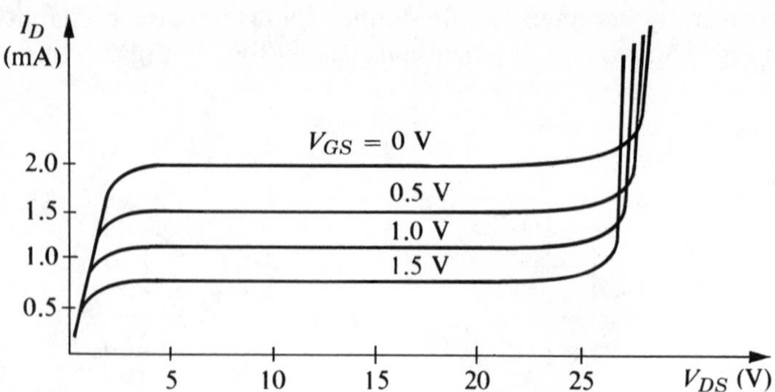

FIGURE 7.6 FET output characteristic curves.

If the drain current is plotted versus the gate-source voltage for the flat region of the I_D versus V_{DS} curves, we get a curve like the one of Fig. 7.7.

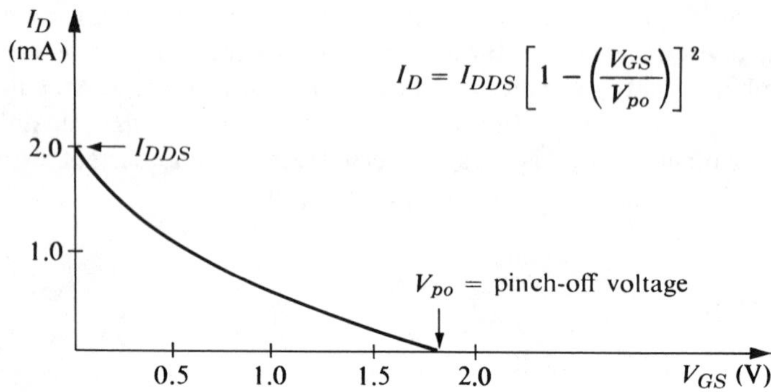

FIGURE 7.7 FET drain current versus gate-source voltage.

A careful analysis shows that:

$$I_D = I_{DSS}[1 - V_{GS}/V_{po}]^2, \qquad \textbf{(7.8)}$$

where V_{po} is the pinch-off voltage and I_{DSS} is the maximum drain current when $V_{GS} = 0$. The triple subscript notation is explained in section 7.6. For the graph of Fig. 7.7, the pinch-off voltage V_p is seen to be approximately 1.8 volts, and the rate of change of drain current with respect to gate/source voltage, dI_D/dV_{GS}, is seen to be larger for low values of V_{GS}, and smaller for larger V_{GS} when the FET is nearer pinch off. In fact dI_D/dV_{GS} approaches zero as V_{GS} approaches the pinch-off voltage V_{po}.

It is desirable to have a large numerical value for dI_D/dV_{GS}, because this means a very small signal voltage change applied to the gate will produce a large change in the drain current. This derivative dI_D/dV_{GS} has the dimensions of conductance and is measured in siemens or mhos or micromhos (1 mho $= 1\,\text{ohm}^{-1}$). The dI_D/dV_{GS} is called the "transconductance" for an FET and is usually denoted by the symbols g_m or y_{21} or y_{fs} and typical numerical values for FET's range from $1000\,\mu$mhos to $5000\,\mu$mhos. Strictly speaking the value of g_m depends upon the value of the drain-source voltage V_{DS}. Thus the precise definition of the transconductance is:

$$g_m \equiv \left(\frac{\partial I_D}{\partial V_{GS}}\right)_{V_{DS}} \tag{7.9}$$

So long as we restrict ourselves to the flat portion of the I_D versus V_{DS} curves of Fig. 7.4, g_m is only a weak function of V_{DS}. However if V_{DS} is small enough so that we are below the knee of the I_D versus V_{DS} curve, that is, for V_{DS} below approximately 2 V, then g_m decreases sharply with decreasing V_{DS}. The moral of this behavior is: To obtain a reasonably high value for the transconductance g_m of an FET, one should bias the FET so that V_{DS} exceeds two or three volts.

7.4 FET y PARAMETER EQUIVALENT CIRCUIT

We will now treat the FET as a four-terminal "black box" and derive an equivalent circuit for use in more precise FET circuit design. Consider the four-terminal black box or network of Fig. 7.8.

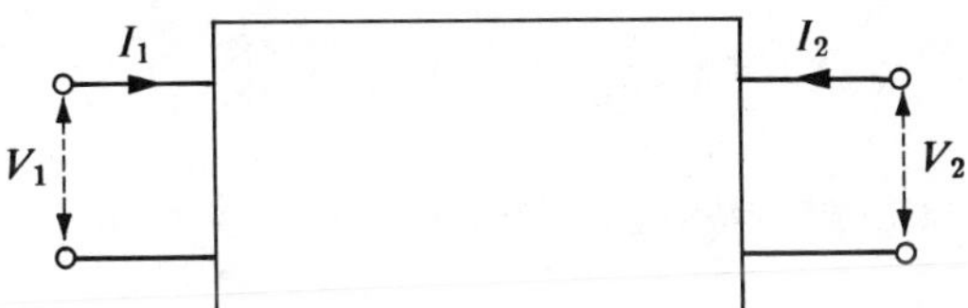

FIGURE 7.8 FET four-terminal network.

By the same argument as was used in section 6.2 in developing the h parameter bipolar transistor equivalent circuit, we can choose any two of the four variables I_1, I_2, V_1, V_2 as our independent variables (which we think of intuitively as "causes") and the other two as dependent variables (which we think of as "effects"). For an FET, we choose V_1 and V_2 as the independent variables and I_1 and I_2 as the dependent variables. We then can write the appropriate Kirchhoff equations and express I_1 and I_2 each in terms of V_1 and V_2:

$$I_1 = I_1(V_1, V_2) \tag{7.10}$$

$$I_2 = I_2(V_1, V_2) \tag{7.11}$$

Different FET's will have different functions $I_1(V_1, V_2)$ and $I_2(V_1, V_2)$.

We now consider the effect on I_1 and I_2 of small changes ΔV_1 and ΔV_2:

$$\Delta I_1 = \left(\frac{\partial I_1}{\partial V_1}\right)_{V_2} \Delta V_1 + \left(\frac{\partial I_1}{\partial V_2}\right)_{V_1} \Delta V_2 \tag{7.12}$$

$$\Delta I_2 = \left(\frac{\partial I_2}{\partial V_1}\right)_{V_2} \Delta V_1 + \left(\frac{\partial I_2}{\partial V_2}\right)_{V_1} \Delta V_2 . \tag{7.13}$$

The notation is now changed: $i_1 = \Delta I_1$, $i_2 = \Delta I_2$, $v_1 = \Delta V_1$, $v_2 = \Delta V_2$. The lower case i's and v's now represent *changes* in current and voltage respectively; one can think of them as signal amplitudes, that is, peak-to-peak current or voltage swings. In terms of the new notation:

$$i_1 = \left(\frac{\partial I_1}{\partial V_1}\right)_{V_2} v_1 + \left(\frac{\partial I_1}{\partial V_2}\right)_{V_1} v_2 \tag{7.14}$$

$$i_2 = \left(\frac{\partial I_2}{\partial V_1}\right)_{V_2} v_1 + \left(\frac{\partial I_2}{\partial V_2}\right)_{V_1} v_2 . \tag{7.15}$$

We now define four "y parameters" as the four partial derivatives occurring in (7.14) and (7.15):

$$\begin{aligned} y_{11} &\equiv \left(\frac{\partial I_1}{\partial V_1}\right)_{V_2} & y_{12} &\equiv \left(\frac{\partial I_1}{\partial V_2}\right)_{V_1} \\ y_{21} &\equiv \left(\frac{\partial I_2}{\partial V_1}\right)_{V_2} & y_{22} &\equiv \left(\frac{\partial I_2}{\partial V_2}\right)_{V_1} \end{aligned} \tag{7.16}$$

In terms of the four y parameters, (7.14) and (7.15) become:

$$i_1 = y_{11}v_1 + y_{12}v_2 \tag{7.17}$$

$$i_2 = y_{21}v_1 + y_{22}v_2 . \tag{7.18}$$

Notice that all the y parameters have dimensions of conductance or admittance; they are all measured in siemens or mhos or micromhos.

Equations (7.17) and (7.18) imply the equivalent circuit. The y_{11} is a conductance across which the input voltage appears, or in other words, v_1 appears across an impedance $1/y_{11}$. $y_{12}v_2$ is an ideal current generator in the input. The equation $i_1 = y_{11}v_1 + y_{12}v_2$ is seen to be simply the Kirchhoff current law applied to junction "A" of Fig. 7.9. The equation $i_2 = y_{21}v_1 + y_{22}v_2$ is the Kirchhoff current law applied to junction "B" of Fig. 7.8, because $y_{22}v_2$ is the current flowing through an impedance $1/y_{22}$ across which a voltage v_2 appears, and $y_{21}v_1$ can be regarded as an ideal current generator.

This equivalent circuit can apply to *any* four-terminal network, not just to

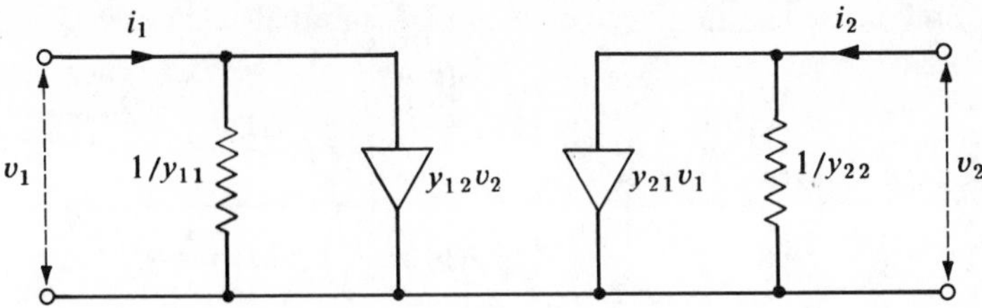

FIGURE 7.9 y parameter FET equivalent circuit.

an FET. Of course, as with any equivalent circuit, the problem is to find or choose the circuit that accurately describes the actual four-terminal network behavior over a wide range of currents and voltages for one *fixed* set of the four y parameters.

Because an FET, like an ordinary bipolar transistor, has only three terminals, one terminal must be common between the input and the output of Fig. 7.8. The three possibilities are: the source terminal common, the drain terminal common, or the gate terminal common. This gives rise, respectively, to the "common source," the "common drain," or the "common gate" configuration. These three configurations are shown in bare outline without any dc biasing in Fig. 7.10.

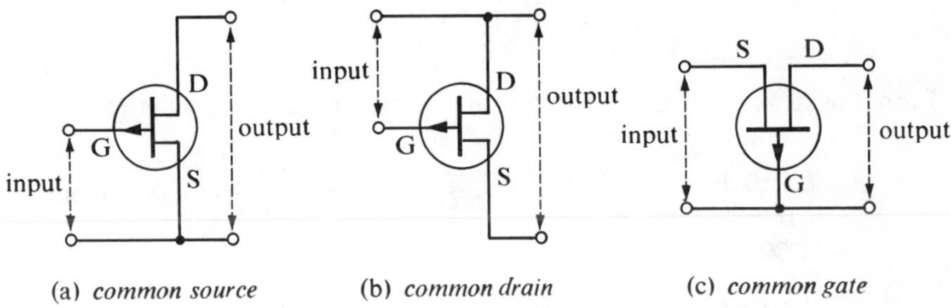

FIGURE 7.10 FET configurations (p channel).

As we have already seen, the gate of an FET corresponds to the base of a bipolar transistor, and the source and drain to the emitter and collector, respectively. Hence, the FET common source is analogous to the common emitter configuration, the FET common drain to the common collector or emitter follower configuration, and the FET common gate to the common base configuration.

We will now focus our attention on the common source configuration which is shown with appropriate dc biasing in Fig. 7.11. This circuit is the one which was used to explain the general behavior of the FET in section 7.2. The output curves of Fig. 7.11(b) are precisely those curves needed to evaluate y_{21} and y_{22}.

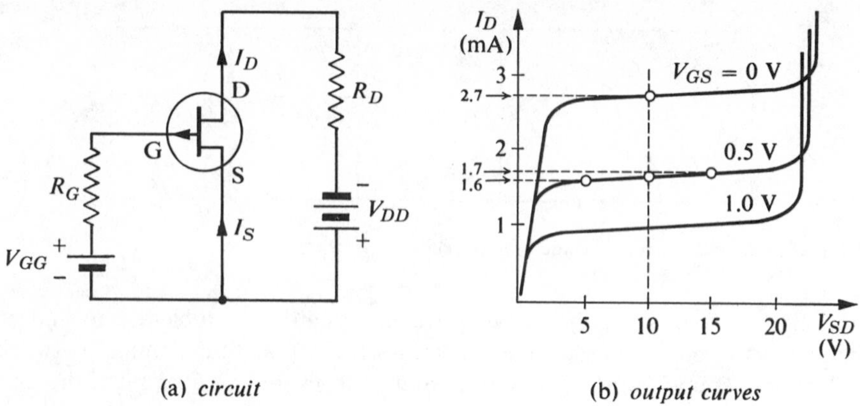

(a) *circuit* (b) *output curves*

FIGURE 7.11 FET common source configuration including biasing for p channel FET.

In the common source configuration: $I_2 \to -I_D$, $V_2 \to -V_{DS}$, $V_1 \to -V_{GS}$, $I_1 \to -I_G$. Thus, the y parameters from (7.18) are given by:

$$y_{21} \equiv \left(\frac{\partial I_2}{\partial V_1}\right)_{V_2} = \left(\frac{\partial I_D}{\partial V_{GS}}\right)_{V_{DS}}$$
$$y_{22} \equiv \left(\frac{\partial I_2}{\partial V_2}\right)_{V_1} = \left(\frac{\partial I_D}{\partial V_{DS}}\right)_{V_{GS}} \tag{7.19}$$

From the curves of Fig. 7.11(b), we can evaluate y_{21} and y_{22} numerically:

$$y_{21} \cong \left(\frac{\Delta I_D}{\Delta V_{GS}}\right)_{V_{DS}=-10\mathrm{V}} = \frac{2.7\,\mathrm{mA} - 1.6\,\mathrm{mA}}{0.5\,\mathrm{V}} = \frac{22\,\mathrm{mA}}{\mathrm{V}} = 0.0022\,\mathrm{mhos}$$

$$y_{21} = 2200\,\mu\mathrm{mhos}$$

$$y_{22} \cong \left(\frac{\Delta I_D}{\Delta V_{DS}}\right)_{V_{GS}=+0.5\mathrm{V}} = \frac{1.7\,\mathrm{mA} - 1.6\,\mathrm{mA}}{15\,\mathrm{V} - 5\,\mathrm{V}} = 0.01\,\mathrm{mA/V}$$

$$y_{22} = 10^{-5}\mathrm{mhos} \qquad \text{or} \qquad \frac{1}{y_{22}} = 10^5\Omega = 100\,\mathrm{k}\Omega.$$

The gain of an FET amplifier can be shown to be proportional to y_{21}, so the higher the value of y_{21}, the better the FET. A modern high gain FET can have y_{21} as large as 10,000μmhos, although inexpensive FET's typically have y_{21} in the neighborhood of 1000 to 2000μmhos.

Notice that y_{21}, which is called the "forward transconductance" and which is also denoted by the symbols g_m, g_{fs}, or y_{fs}, decreases as the drain current decreases. A more careful analysis of the constriction of the channel by a changing gate voltage shows that the transconductance varies approximately inversely as the square root of the drain current. A decrease in drain current from 1.0 mA to 0.5 mA will approximately decrease the transconductance by a factor of $1/\sqrt{2} = 0.707$, for example, from 2000 μmhos to 1400 μmhos. The y_{21} may be 2000 μmhos for a 3 mA drain current, but may typically fall to 1000–1500 μmhos for a drain current of around 1 mA. Thus, the drain current at the operating point can affect the gain of an FET by as much as a factor of two because of the variation of y_{21} with I_D. The y_{22} also depends upon the drain current; it decreases as I_D decreases. Typical values of y_{22} range from 10^{-5} to 10^{-6} mhos, corresponding to $1/y_{22} = 100\,\text{k}\Omega$ to $1\,\text{M}\Omega$.

The other two y parameters y_{11} and y_{12} could be obtained from the input characteristic curves of the FET, which are curves of the gate current I_G versus gate-source voltage V_{GS} for various values of drain-source voltage. These curves are almost never given in handbooks for FET's, but they would look roughly like the curve of Fig. 7.12. Notice the extremely small gate current

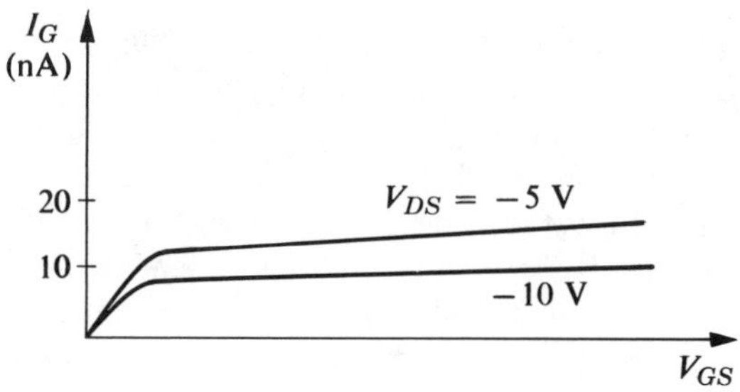

FIGURE 7.12 Input curve for FET.

which flows, $I_G \sim 10\,\text{nA}$. This is, of course, because the gate-channel junction is a reverse-biased silicon junction.

$$y_{11} = \left(\frac{\partial I_G}{\partial V_{GS}}\right)_{V_{DS}}$$

is the slope of the graph of Fig. 7.12 and is seen to be extremely small. Typical values for y_{11} are 10^{-8} to 10^{-9} mhos, which correspond to an effective resistance for the gate-channel junction of $1/y_{11} = 100$ to $1000\,\text{M}\Omega$.

$$y_{12} = \left(\frac{\partial I_G}{\partial V_{DS}}\right)_{V_{GS}}$$

is a measure of how much the gate current changes as the drain/source voltage is changed. The value of y_{12} is extremely small, of the order of 10^{-10} mhos or less, and is usually set equal to zero in most calculations. The simplified y parameter FET equivalent circuit used is shown in Fig. 7.13 with typical values for the y parameters.

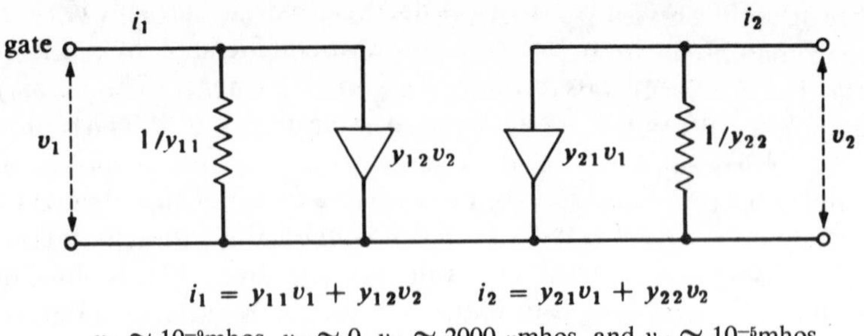

$$i_1 = y_{11}v_1 + y_{12}v_2 \qquad i_2 = y_{21}v_1 + y_{22}v_2$$

$y_{11} \cong 10^{-9}$mhos, $y_{12} \cong 0$, $y_{21} \cong 2000\ \mu$mhos, and $y_{20} \cong 10^{-5}$mhos

FIGURE 7.13 FET y parameter equivalent circuit.

7.5 FET TEMPERATURE EFFECTS

We have already seen in Chapter 4 that temperature changes will affect the operation of ordinary bipolar transistors, principally by turning the forward-biased, base-emitter junction more on or off. A temperature increase tends to turn the base-emitter junction on more and thus increase the collector current, whereas a temperature decrease turns the transistor off more. The net effect of a temperature change is to change the base voltage by about 2.5 mV/°C. In an FET the gate-channel junction, which is analogous to the base-emitter junction of a bipolar transistor, is reverse-biased, however; and so a temperature increase merely increases the number of thermally generated minority carriers in the depletion region and thereby increases the gate current. Typical gate currents at room temperature (25°C) range from 1 to 10 nanoamperes, and a temperature rise to 100°C may increase the gate current to 50 nanoamperes. Typically the gate is connected to ground through a resistance R_G as shown in Fig. 7.14; hence, the gate voltage will change by $\Delta V_G = \Delta I_G R_G$ where ΔI_G is the change in gate current due to a temperature change. This change in gate voltage increases the drain

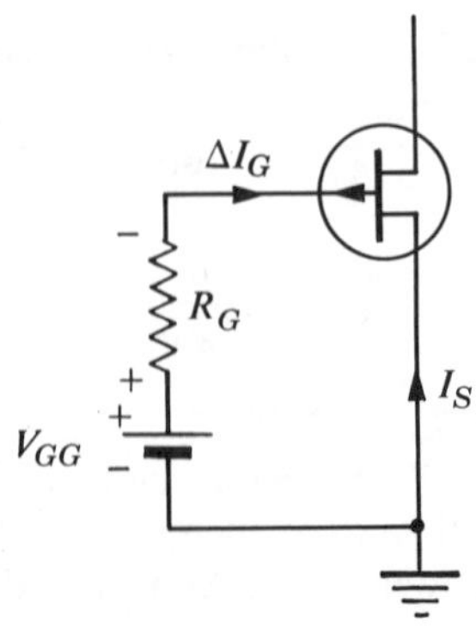

FIGURE 7.14 FET gate current (p channel).

current because it decreases the reverse gate-channel bias. Usually R_G is from 1 to 10 megohms, so at most $\Delta V_G = \sim(50\,\text{nA})(10\,\text{M}\Omega) = 0.5$ volts for a 75°C temperature rise. A smaller temperature rise and/or a smaller value for R_G will, of course, produce a smaller change in gate voltage, e.g., a 10°C temperature rise with $R_G = 1$ megohm will produce a change in gate voltage of only 0.007 volts. If the temperature-induced gate voltage change is small compared to the gate/source dc reverse-bias voltage at the operating point, then the operating point will not change appreciably. Typically the gate-source bias is of the order of 0.5 to 1.0 volt, so major changes in the operating point will occur only when large gate resistors (10 megohms) are used and extreme temperature changes (>20°C) are encountered.

There are two other mechanisms which will change the operating point due to temperature changes. One is the width of the depletion layer between the gate and the source; this will change with temperature, a temperature increase resulting in a shrinking of the depletion layer and thus an increase in the drain current. The net effect is to change the drain current by an amount equivalent to about a 2.5 mV/°C change in gate voltage. The larger the transconductance y_{fs}, the larger the change in drain current, as

$$\Delta I_D = \left(\frac{\partial I_D}{\partial V_{GS}}\right) \Delta V_{GS} = y_{fs}\,\Delta V_{GS}. \tag{7.20}$$

$$\Delta V_{GS} = \left(\frac{\partial V_{GS}}{\partial T}\right) \Delta T = (2.5\,\text{mV}/^\circ\text{C})\Delta T. \tag{7.21}$$

The other mechanism is the change of channel majority carrier mobility with temperature. As the temperature rises, the mobility decreases and the drain current decreases. The drain current changes by less than 1%/°C temperature change due to this mobility factor. It is possible with careful design to choose the operating point so that these two factors almost cancel one another out, thus leaving the drain current independent of temperature, but the details are beyond the scope of this book.

As we will see in the next section on biasing, the presence of a resistor R_S in series with the source will stabilize the circuit greatly against temperature-induced changes in drain current.

7.6 BIASING AND PRACTICAL CIRCUITS

Before we take a detailed look at the problem of biasing an FET and the design of actual amplifier circuits, we first must understand the various parameters used to describe an FET. We will consider the following parameters:

Equivalent circuit y parameters in the common source configuration (notice that the second subscript *s* stands for the common terminal, the source):

$y_{11} = y_{is}$ input admittance

$y_{12} = y_{rs}$ reverse transconductance (*usually taken as zero*)

$y_{21} = y_{fs}$ (*or* g_{fs} *or* g_m) forward transconductance

$y_{22} = y_{os}$ output admittance.

Various breakdown voltages:

BV_{GDS} gate to drain, source connected to drain

BV_{GSS} gate to source, drain connected to source

BV_{DGO} drain to gate, source open

BV_{DGS} drain to gate, source connected to gate

BV_{DSX} drain to source, gate biased beyond the cutoff.

Currents:

I_{DSS} drain current, gate connected to source

I_{GSS} gate current, drain connected to source

I_{DGO} drain to gate current, source open.

Pinch-off voltage:

V_p sometimes also called $V_{GS}(off)$ gate-source cutoff voltage.

Capacitance:

C_{iss} short circuit input capacitance

C_{rss} reverse transfer capacitance.

The notation is more complicated than for bipolar transistors. In particular, when three subscripts are present, an "*S*" as the third subscript means "shorted" or "connected" (*not* source) and refers to the condition of the terminal not represented by either of the first two subscripts relative to the second subscript which is the common point. For example, BV_{GSS} means the breakdown voltage between the gate and the source (the first two subscripts) with the drain shorted or connected to the source (the second subscript). I_{GSS} means the gate leakage or reverse current, i.e., the gate-to-source current (the first two subscripts) with the drain connected to the source (the second subscript).

There are several two subscript maximum ratings often used; the condition of the third terminal is unspecified. We also include several one subscript ratings.

Maximum ratings:

V_{DS} drain-source voltage

V_{DG} drain-gate voltage

$V_{GS(r)}$ reverse gate-source voltage

I_G gate current

P_D total device dissipation (*usually in* 25°C *air*), derate approximately 3mW/°C above 25°C.

To simplify the preceding mélange of symbols, one needs usually to look at only several ratings in the initial stages of choosing a particular FET. The most important ratings are the transconductance y_{fs} or g_m, in micromhos, the device power dissipation, usually in mW, the maximum drain-source and drain-gate voltage, and I_{DSS}. One can never exceed these maximum voltage ratings, even for a very short time, usually without destroying the FET. The I_{DSS} represents the maximum drain current for the device. If one tries to make the drain current exceed I_{DSS}, one would have to *forward* bias the gate, which would drastically lower the input impedance.

Let us now consider an inexpensive silicon n channel junction FET, the 2N5459 which costs less than $1 each and is packaged in a plastic case. Its parameters are:

Maximum ratings:

$$V_{DS} = 25\,\text{V dc}$$
$$V_{DG} = 25\,\text{V dc}$$
$$V_{GS(r)} = 25\,\text{V}$$
$$I_G = 10\,\text{nA dc}$$
$$P_D = 310\,\text{mW at } 25\,°\text{C}.$$

Other ratings:

$$V_{GS(\text{off})} = 2.0\,\text{V dc minimum (for } I_D = 10\,\text{nA)}$$
$$8.0\,\text{V dc maximum}$$

y_{fs} = 4500 μmhos typical (2000–6000)
C_{iss} = 4.5 pF typical (7.0 pF max) input capacitance
C_{rss} = 1.5 pF typical (3.0 pF max) reverse transfer capacitance.

The FET analog of the common emitter bipolar transistor amplifier is the common source amplifier. Like the common emitter amplifier, it has a relatively high power, voltage, and current gain with a moderately high output impedance. We recall that the common emitter bipolar transistor amplifier has a low input impedance due to its forward-biased, base-emitter junction. But, with an FET common source amplifier, a much higher input impedance can be achieved, because the gate-source junction in parallel with the input is always *reverse*-biased. A two-battery n channel FET common source amplifier is shown in Fig. 7.15. The resistance R_G can be as high as several megohms, and the input impedance of the amplifier *circuit* is R_G in parallel with the gate-source impedance, which is usually of the order of 10 megohms or more. Thus, the input impedance will usually be essentially equal to R_G if R_G is appreciably less than the gate-source impedance.

The biasing arrangement of Fig. 7.15 has, however, two serious disadvantages. First, a rise in temperature, by increasing the number of minority charge carriers in the FET semiconductor material, will result in an increase in the reverse current flowing through the gate-source junction. As explained in section 7.5, the gate current may increase by as much as a factor of 50 for a temperature rise from 25 °C to 100 °C, reducing the reverse gate-channel bias by as much as 0.5 volt. Usually this large a change in the gate voltage will produce a drastic increase in the drain current with an accompanying increase in the transconduct-

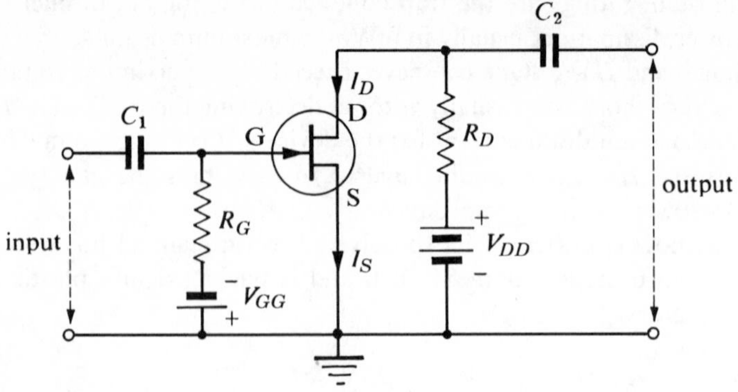

FIGURE 7.15 FET common source amplifier (n channel FET).

ance and a change in the operating point. The second disadvantage is that two, rather than one, bias batteries are required.

The remedy for both these disadvantages is to obtain the bias from the voltage drop across a resistance in series with the source as is shown in Fig. 7.16

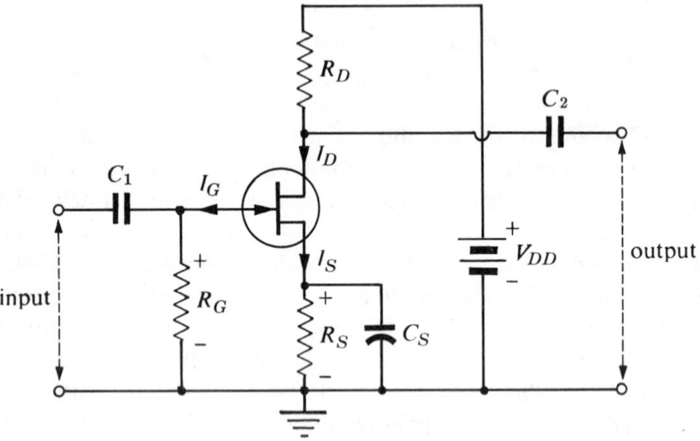

FIGURE 7.16 Common source n channel FET amplifier with source bias.

for an n channel FET. The dc voltage drop $V_S = I_S R_S$ is of such a polarity as to reverse bias the gate-source junction. The Kirchhoff voltage equation for the loop containing R_G, the gate-source junction and R_S is:

$$+I_G R_G - V_{GS} - I_S R_S = 0$$

or

$$V_{GS} = -I_S R_S + I_G R_G \cong -I_S R_S .$$

The gate current I_G, we recall, is extremely small; thus if R_G is not too large $I_G R_G \ll I_S R_S$, and the gate-source bias is simply the $I_S R_S$ voltage drop across the source resistance R_S with the gate negative with respect to the source, i.e., a reverse-bias. In value, I_G is of the order of 10^{-8} to 10^{-9} amperes, and $I_S R_S$ is usually on the order of one volt, so as long as $R_G \leqq 10$ megohms, $I_G R_G \ll I_S R_S$. The source resistance also decreases the temperature instability, because as the temperature rises and I_G increases the gate voltage goes positive, thus tending to increase the source and drain current. However, as the source current increases, the voltage at the source terminal *also* goes positive, because $V_S = +I_S R_S$. Thus the bias voltage *between* the gate and the source tends to remain constant, and the source and drain current remain relatively constant also. This is the same stabilization effect that the presence of an emitter resistor has in the case of a bipolar transistor common emitter amplifier. The source resistance is bypassed at signal frequencies to allow the gate-source voltage to respond to a varying signal voltage applied to the gate through the coupling capacitor C_1. Thus, we desire $X_{CS} = 1/\omega C_S \ll R_S$ at the lowest signal frequency.

We also desire $X_{C_1} = 1/\omega C_1 \ll Z_{in}$ at the lowest signal frequency, because C_1 and Z_{in} form a high-pass filter. Notice because the input impedance of the FET can be as large as 10 megohms that very low signal frequencies can be passed with small coupling capacitors. For example, with $R_{in} = 10$ megohms, and $C = 0.01\,\mu$F, the break point frequency is only $f = 1/2\pi \cdot 1/R_{in}C_1 \cong 1.6$ Hz. By comparison, for a bipolar transistor common emitter amplifier, $R_{in} \cong 1\,\text{k}\Omega$ and one would need a coupling capacitor of $C_1 = 100\,\mu$F to achieve a 3 dB point of 1.6 Hz. Thus, if extremely low frequencies must be amplified, large coupling capacitors can be avoided by using FET amplifiers with their high input impedances. We will now consider in detail the three possible FET amplifier configurations: common source, common drain, and common gate.

7.7 THE COMMON SOURCE FET AMPLIFIER

The common source FET amplifier is analogous to the common emitter bipolar transistor amplifier. Hence we expect a moderate voltage and current gain, and a relatively high output impedance. We expect, however, an input impedance much higher than for a bipolar transistor circuit, because the FET input (gate-channel) junction is reverse-biased rather than forward-biased as for a bipolar transistor.

A practical common source FET amplifier is shown in Fig. 7.17 for an inexpensive (less than $1) n channel FET. Perhaps the first difference that is apparent between the FET common source amplifier characteristics and those of the bipolar transistor common emitter amplifier is that the FET voltage gain is substantially less and the input impedance is much higher. A better understanding of these and other characteristics can be obtained by looking at the

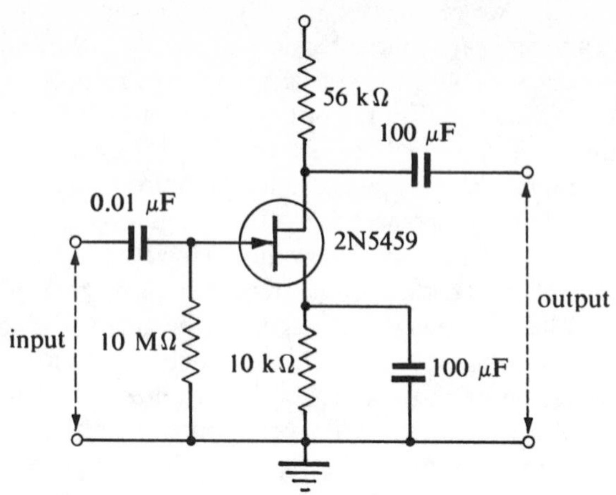

FIGURE 7.17 An n channel FET common source amplifier.

gain and impedance expressions for the FET calculated from the y parameter equivalent circuit. The FET y parameter common source equivalent circuit that will be used is repeated in Fig. 7.18 for convenience along with the two equations that were used to develop the y parameters.

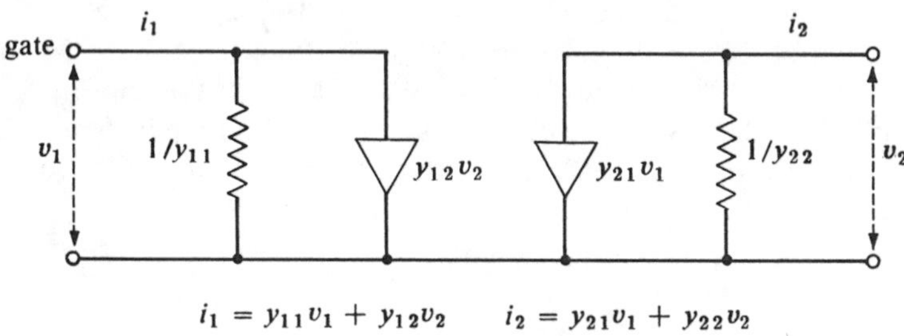

FIGURE 7.18 Low frequency common source FET y parameter equivalent circuit.

$$i_1 = y_{11}v_1 + y_{12}v_2 \cong y_{11}v_1 \tag{7.22}$$

$$i_2 = y_{21}v_1 + y_{22}v_2 \tag{7.23}$$

The common source FET amplifier with signal source or generator and load connected is shown in Fig. 7.19(a), and its equivalent circuit in Fig. 7.19(b) with the simplifying (and very good) assumption that $y_{12} = 0$. R_{DL} is the parallel combination of R_D and R_L, and we have also used $v_{\text{in}} = v_1$ and $v_{\text{out}} = v_2$. We also have:

$$v_0 = -i_2R_{DL}. \tag{7.24}$$

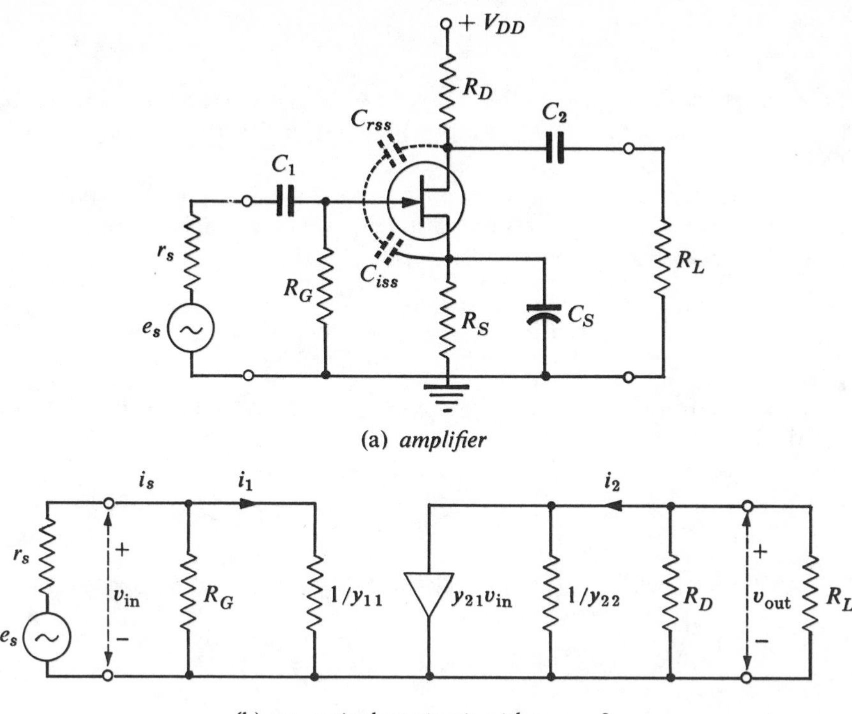

FIGURE 7.19 Common source FET amplifier circuit and equivalent circuit.

In drawing the equivalent circuit, we have assumed the capacitive reactances of C_1, C_2, and C_S are negligible (i.e., the frequency is high enough) and that any stray or shunt capacities in (C_{rss}, C_{iss}) in the circuit or the FET are too small to be significant (i.e., the frequency is low enough). In other words the equivalent circuit is for "intermediate" frequencies that are usually from several hundred Hz to 100 kHz, depending upon the exact size of the capacitances involved. The reader should be warned, however, that at frequencies as low as 10 kHz a small capacitance (C_{rss}) from the drain to the gate (1 pF to 5 pF) may have a serious effect on lowering the input impedance due to the "Miller effect" which is discussed in Chapter 10.

The voltage gain can be calculated from equations (7.22) and (7.23). Elimination of i_2 quickly yields the voltage gain:

$$A_v = \frac{v_{out}}{v_{in}} = \frac{-y_{21}}{y_{22} + \dfrac{1}{R_{DL}}} = \frac{-y_{21}R_{DL}}{1 + y_{22}R_{DL}} \cong -y_{21}R_{DL}.$$

For $y_{21} = 2000\,\mu$mhos, $R_{DL} = 50\,\text{k}\Omega$, and $y_{22} = 10^{-5}$ mhos:

$$A_v = \frac{-(2 \times 10^{-3})(5 \times 10^4)}{1 + (10^{-5})(5 \times 10^4)} = -67.$$

The minus sign simply means that there is a 180° phase reversal between the input and the output; a positive input produces a negative output and vice versa.

The current gain can be calculated from equations (7.22), (7.23), and (7.24). Elimination of v_{in} and v_{out} yields the ratio i_2/i_1, which is really the current gain for the FET rather than the amplifier circuit.

$$A_{i\text{FET}} = \frac{i_2}{i_1} = \frac{y_{21}/y_{11}}{1 + y_{22}R_{DL}} \tag{7.25}$$

The current gain for the amplifier circuit is i_L/i_S, the ratio of the load current i_L to the total input current from the source.

$$\frac{i_L}{i_S} = \frac{i_2}{i_1}\frac{i_L}{i_2}\frac{i_1}{i_S}$$

A quick Ohm's law calculation shows:

$$A_{i\text{circuit}} = \frac{i_L}{i_S} = \frac{y_{21}/y_{11}}{1 + y_{22}R_{DL}} \cdot \frac{R_D}{R_D + R_L} \cdot \frac{R_G}{\dfrac{1}{y_{11}} + R_G}. \tag{7.26}$$

The factor of $R_D/(R_D + R_L)$ reflects the fact that the FET drain current i_2 is split between the drain resistance R_D and the load resistance R_L, which are in parallel with respect to the ac signal current. The larger the ratio R_D/R_L, the larger the fraction of drain current which flows through the load R_L. The factor of

$$\frac{R_G}{1/y_{11} + R_G} = \frac{1}{1 + 1/y_{11}R_G}$$

reflects the splitting of the current input to the amplifier i_S between $1/y_{11}$ and R_G. The larger $R_G/(1/y_{11})$ is, the larger the current gain. Substituting in $y_{21} = 2000\,\mu$mhos, $y_{22} = 10^{-5}$ mhos, $R_{DL} = 50\,\text{k}\Omega$, $R_L = R_D = 100\,\text{k}\Omega$, $R_G = 10$ MΩ, $y_{11} = 10^{-8}$ gives $A_i = 6000$.

The power gain is the product of the voltage gain and the current gain:

$$A_p = A_vA_i = \frac{y_{21}R_{DL}}{1 + y_{22}R_{DL}} \cdot \frac{y_{21}/y_{11}}{1 + y_{22}R_{DL}} \cdot \frac{R_D}{R_D + R_L} \cdot \frac{R_G}{R_G + \dfrac{1}{y_{11}}}.$$

Substitution of numerical values gives $A_p \cong 400{,}000$. Notice that the voltage gain varies as y_{21}, the current gain as y_{21}/y_{11}, and the power gain as y_{21}^2/y_{11}. For all these reasons, the higher the forward transconductance $y_{21} = y_{fs} = g_m$ the higher the gain.

The input impedance of the FET is seen to be $1/y_{11}$ by inspection from the equation $v_{in} = y_{11}i_1$. The input impedance of the *amplifier circuit* is the parallel combination of the gate-to-ground resistance R_G and $1/y_{11}$. Thus:

$$R_{in} = R_G \Big\| \frac{1}{y_{11}} = \frac{R_G}{1 + y_{11}R_G}. \tag{7.27}$$

Usually R_G is much smaller than $1/y_{11}$, so $R_{in} \cong R_G$. Typical values for R_G are from 1 to 10 megohms. The larger the FET input impedance $1/y_{11}$, the higher the amplifier input impedance can be made by increasing R_G. It should be pointed out that $1/y_{11}$ represents the impedance of the reverse biased gate-channel junction ($10^8\ \Omega$) in parallel with the capacitance C_{iss}. For $C_{iss} = 4$ pF, its capacitive reactance is only approximately 40 kΩ at 1 MHz and less at higher frequencies of course.

The output impedance of the amplifier which the load sees is the parallel combination of the FET output impedance and R_D the drain resistance. The output impedance of the FET can be seen to be $1/y_{22}$ from equation (7.23) and the definition v_2/i_2 with the condition that $e_s = 0$. For, with $e_s = 0$, there are no voltage sources in the input (gate) side of the circuit, and thus $v_1 = 0$. Hence equation (7.23) directly yields $v_2/i_2 = 1/y_{22}$. The output impedance for the entire amplifier circuit is then:

$$R_{out} = \left(\frac{1}{y_{22}}\right) \| R_D = \frac{R_D}{1 + y_{22}R_D}. \tag{7.28}$$

Usually $1/y_{22}$ is much larger than R_D, so that the output impedance equals R_D.

7.8 THE COMMON DRAIN FET AMPLIFIER

The common drain FET amplifier or "source follower" is analogous to the common collector or "emitter follower" amplifier made with bipolar transistors. Hence, we expect a low output impedance, a broad bandwidth, a voltage gain close to unity, and a high current gain.

A practical common drain amplifier is shown in Fig. 7.20 along with measured circuit parameters. A higher voltage gain of 0.9 can be obtained at the expense of a higher output impedance of 1 kΩ by increasing the source resistance R_S from 1 kΩ to 10 kΩ. A y parameter analysis of the circuit gives the following expression for the output impedance:

$$Z_{out} \cong \frac{1}{1/R_S + y_{21}}, \tag{7.29}$$

which is seen to be the parallel combination of the source resistance R_S, and a resistance equal to the reciprocal of the forward transconductance

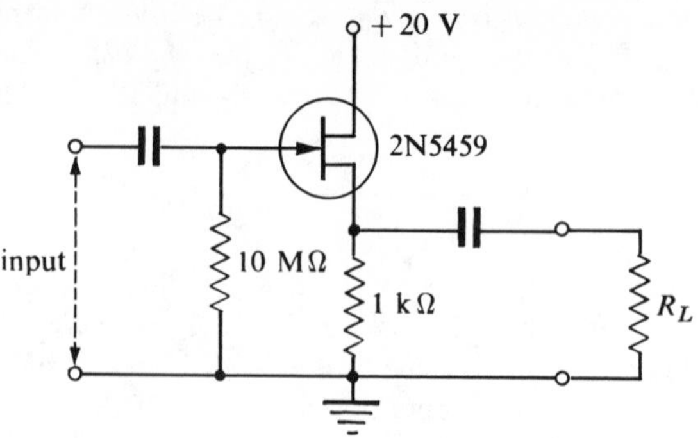

Measured Circuit Performance

1 kHz input impedance	1 megohm
1 kHz output impedance	300 Ω
1 kHz voltage gain	0.67
upper 3 dB frequency	9.6 MHz

FIGURE 7.20 Common drain FET amplifier (source follower).

$1/y_{fs}$ ($= 1/y_{21} = 1/g_m$). Thus we see that for a "high gain" FET with a large forward transconductance y_{fs}, the output impedance is approximately equal to $1/y_{fs}$, which result will be familiar to those already conversant with vacuum tube cathode followers. Substitution of $y_{fs} = 2000\,\mu$mhos, $R_S = 1\,\text{k}\Omega$ gives $Z_{\text{out}} = 330\Omega$, which is quite close to the experimental value. It should be pointed out that it is difficult to achieve an output impedance much below 200Ω with a FET source follower, because the maximum FET transconductance is of the order of $5000\,\mu$mhos. With a bipolar transistor emitter follower, however, output impedances of the order of 50Ω are easily obtainable, albeit with a lower input impedance than that of an FET source follower.

The input impedance expression from a y parameter analysis is:

$$Z_{\text{in}} \cong R_G \,\|\, (y_{21}/y_{11})R_{SL}, \tag{7.30}$$

which is the parallel combination of the gate-to-ground resistance R_G, and $(y_{21}/y_{11})R_{SL}$ where R_{SL} is the parallel combination of the source resistance R_S and the load resistance R_L. For a typical FET $(y_{21}/y_{11})R_{SL}$ is of the order of 100 megohms, so the input impedance is essentially equal to R_G, which can be made as high as 10 megohms without any difficulty. The experimental value for the input impedance was found to be equal to R_G.

The voltage gain expression from a y parameter analysis is:

$$A_v \cong \frac{y_{21}}{y_{21} + \dfrac{1}{R_{SL}}}, \tag{7.31}$$

where R_{SL} again is the parallel combination of the source resistance R_S and the load resistance R_L. Notice that to achieve a larger voltage gain for a given FET with a constant transconductance y_{21}, one must increase R_{SL}, and the voltage gain approaches 1 in the limit as R_{SL} becomes much larger than $1/y_{21}$.

Substitution of $y_{21} = 2000\,\mu$mhos and $R_{SL} = 1\,\text{k}\Omega$ gives $A_v = 0.67$ in excellent agreement with the measured voltage gain. It should be pointed out that the agreement with experiment is perhaps coincidental, as the spread in y_{21} for a 2N5459 FET is from 2000 to 6000$\,\mu$mhos according to the manufacturers' data sheets. Also, for a given FET, we should remember that the transconductance varies as the square root of the drain current.

The current gain from a y parameter analysis is:

$$A_i = \frac{y_{21} + y_{11}}{y_{11}(1 + y_{22}R_{SL})} \cdot \frac{R_S}{R_S + R_L}, \tag{7.32}$$

or with $y_{21} = 2000\,\mu$mhos, $y_{11} = 10^{-8}$ mhos, $y_{22} = 10^{-5}$ mhos, $R_{SL} = 10^3\Omega$,

$$A_i \cong \frac{y_{21}}{y_{11}} \frac{R_S}{R_S + R_L}. \tag{7.33}$$

The factor y_{21}/y_{11} represents the total ac source current i_S divided by the input gate current i_1, and the $R_S/(R_S + R_L)$ factor represents how the source current is split between the source resistance R_S and the load resistance R_L. y_{21}/y_{11} is of the order of 10^5 for a typical FET, so large current gains are indeed obtained. Intuitively, the magnitude of the current gain comes from the physical fact that the gate-channel input junction is reverse biased and therefore has a high impedance and draws only an extremely small gate or input current. This is represented by the y_{11} term in the denominator of the current gain expression. The power gain is the product of the voltage gain and the current gain and is of the order of 10^5.

An ingenious and common way to increase the voltage gain of a source follower is to replace the source resistance R_S, which is usually from 1$\,$kΩ to 10$\,$kΩ, by a transistor constant current source as shown in Fig. 7.21. The transistor acts as a constant current source, because its base voltage and current are fixed by the resistance network of R_1 and R_2. In other words, the transistor acts as a low impedance dc source resistance, but as an extremely high ac resistance to the signal, thus producing a voltage gain of unity. Some care must be taken in biasing the transistor. If the dc collector voltage is too low, the transistor will not be on the constant current portion of its characteristic curve, while if the collector voltage is too high the FET may be pinched off. In the latter case, the FET gate voltage may have to be raised by adding a resistor from the gate to the +15$\,$V supply in order to keep V_{GS} of the order of 2 or 3 volts.

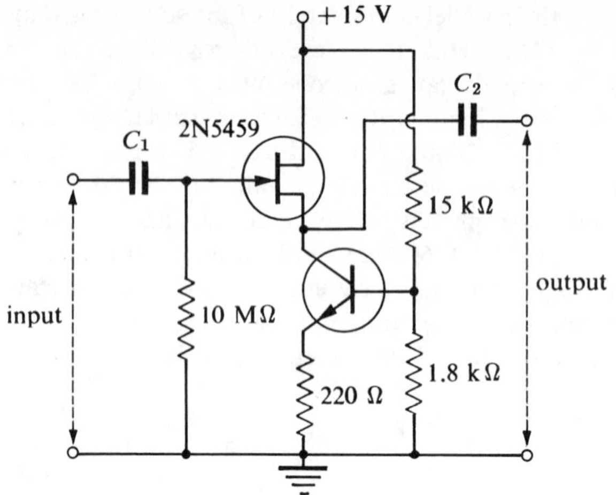

Measured Circuit Performance

1 kHz voltage gain	1.00
1 kHz output impedance	380 Ω
upper 3 dB frequency	7 MHz

FIGURE 7.21 FET source follower with transistor.

7.9 THE COMMON GATE FET AMPLIFIER

The common gate FET amplifier is similar to the common base bipolar transistor amplifier, so we expect similar characteristics: a high voltage gain, a low input impedance, and a high output impedance. The circuit of such an amplifier is shown in Fig. 7.22. Notice that there is no voltage divider chain necessary for the gate as there was for the base of the bipolar transistor common base amplifier of Fig. 6.22, because the gate of an FET is reverse biased. Thus we can use the dc voltage drop across the source resistance to bias the FET. The high (56 kΩ) output impedance makes for very poor high frequency gain; with only a 50 pF capacitance between the output terminal and ground, the upper 3 dB frequency is only about 300 kHz.

The common gate FET amplifier provides excellent isolation between the input at the source and the output at the drain as does the common base bipolar transistor amplifier. Hence, the common gate amplifier is mainly used at frequencies of 100 MHz or higher where unwanted feedback from output to input often causes oscillations. The FET is superior to the bipolar transistor in that it handles overloads better.

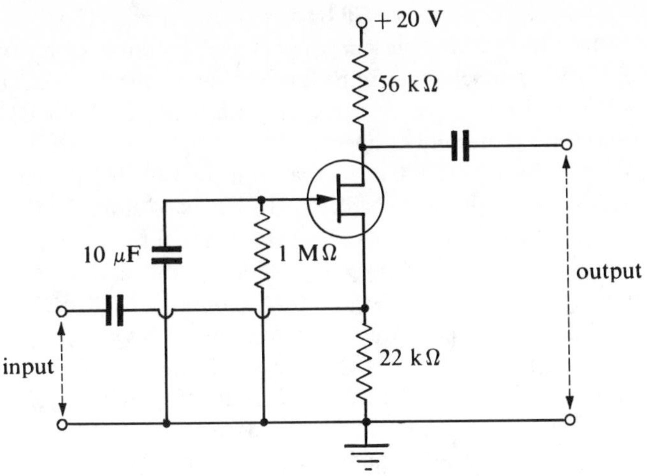

Measured 1 kHz *Circuit Performance*

voltage gain	36
input impedance	1.5 kΩ
output impedance	56 kΩ
upper 3 dB frequency	300 kHz

FIGURE 7.22 Common gate FET amplifier.

7.10 THE METAL OXIDE SEMICONDUCTOR FIELD EFFECT TRANSISTOR (MOSFET)

The ordinary junction FET which we have considered up to now contained only p- and n-type semiconductor material. For an n channel FET, for example, the n-type channel is bounded on each side by p-type regions that are connected together to form the gate. The interface between the gate and the channel is a p-n junction; hence the name "junction" FET. The gate is reverse biased with respect to the channel, thereby producing a high gate-channel impedance on the order of 10^8 to 10^{10} ohms. This large value of impedance arises from the small reverse leakage current in silicon on the order of nanoamperes, which flows from the channel to the gate. We recall from Chapter 4 that this leakage current arises from the thermal generation of minority charge carriers in the depletion region formed at the junction between the gate and the channel.

The metal oxide semiconductor field effect transistor (MOSFET), sometimes called the insulated gate field effect transistor (IGFET), is similar to the junction FET, but differs in two important ways: First, the MOSFET has smaller internal parasitic capacitances and hence will operate at higher frequencies than the ordinary junction FET. Secondly, the gate of a MOSFET is metal and is separated from the channel by a thin layer of extremely good insulating material, usually silicon dioxide SiO_2 for a silicon FET. The SiO_2 is a far better insulator than the reverse-biased gate-channel junction of an ordinary FET, so the MOSFET gate-channel impedance is far larger. Impedances on the order of 10^{12} to 10^{15} ohms can be achieved with modern MOSFET's. Because the gate terminal is usually connected to the input, we usually say that the input impedance of a MOSFET is on the order of 10^{12} to 10^{15} ohms. The gate-to-channel capacitance is very small, on the order of several pF, so large voltages can be created between the gate and the channel by the accumulation of only small amounts of charge ($V = Q/C$). Hence, great care must be taken to avoid building up electrostatic charge on the gate of a MOSFET to avoid high voltage breakdown and permanent damage to the SiO_2 layer. It is, in fact, standard practice for MOSFET's to come from the manufacturer with a thin shorting wire wound around the leads. This shorting wire is left in place until the MOSFET is installed in the circuit. Soldering irons used on MOSFET circuits should also be externally grounded, with a ground wire attached to the tip itself to keep 60 Hz leakage voltages off the tip.

There are two types of MOSFET, the depletion type and the enhancement type. In the depletion type, shown in Fig. 7.23(b), the channel is open when the gate-source voltage is zero, whereas in the enhancement type shown in Fig. 7.24 the channel is normally closed. Current will flow through the channel

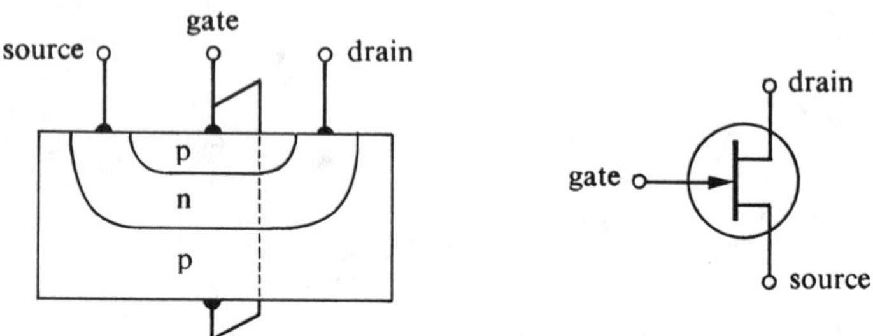

(a) *junction* FET n *channel construction and schematic symbol*

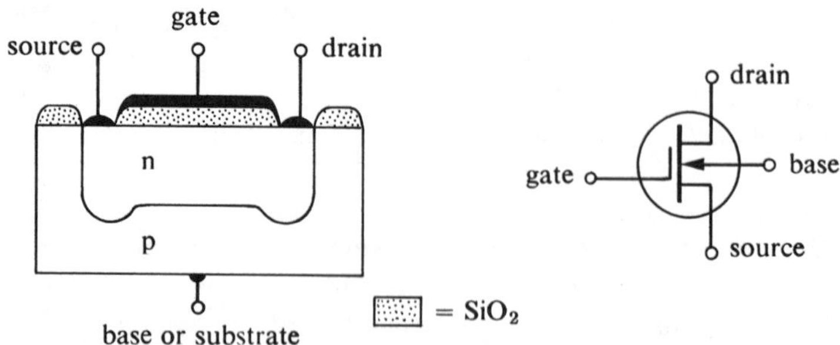

(b) MOSFET n *channel depletion type construction and schematic symbol*

FIGURE 7.23 Junction FET and n channel depletion type MOSFET construction.

of a depletion type MOSFET with no dc bias voltage applied between the gate and the channel. For the n channel enhancement MOSFET shown in Fig. 7.24(a), the gate must be dc biased positive with respect to the channel in order

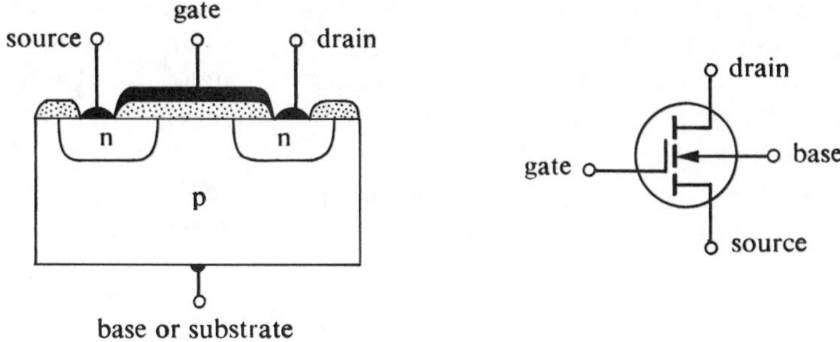

(a) MOSFET n *channel enhancement type construction and schematic symbol*

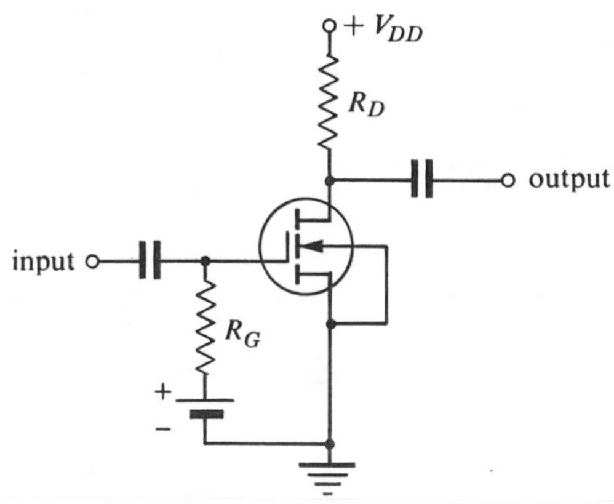

(b) *amplifier with biasing*

FIGURE 7.24 n channel enhancement type MOSFET.

to open the channel; making the gate positive repels mobile positive charge carriers away from the gate and draws mobile electrons from the n-type regions near the drain and source into the region near the gate, thus creating an n-type channel joining the drain and source. The necessary biasing arrangement is shown in Fig. 7.24(c).

The p-type material at the bottom of either type of n channel MOSFET is called the "base," "substrate," or sometimes the "bulk" gate. Notice that in the MOSFET symbol, the gate line does not touch the vertical channel line between the source and drain. This symbolizes the fact that the gate is insulated from the channel by a SiO_2 layer. The name metal oxide semiconductor (MOS) refers, respectively, to the materials of the gate, the SiO_2 insulating layer, and the channel. The conductivity of the ordinary junction FET is modulated by changes in the gate-channel reverse bias voltage; such changes produce in the channel a depletion region of varying thickness that tends to pinch off the channel. The conductivity of the MOSFET channel is also modulated by changing the gate-channel voltage; the electric field of the charge applied to the gate electrode partially penetrates the channel and affects the channel conductivity. However, unlike the junction FET, the voltage applied to the gate of a depletion type MOSFET can be of *either* polarity; there is no junction which must be kept dc *reverse-biased.* So long as the SiO_2 insulating layer remains intact, the gate terminal is effectively insulated from the channel. Typical drain current versus source-drain voltage curves for a variety of gate-source voltages are shown in Fig. 7.25 for both depletion and enhancement type MOSFET.

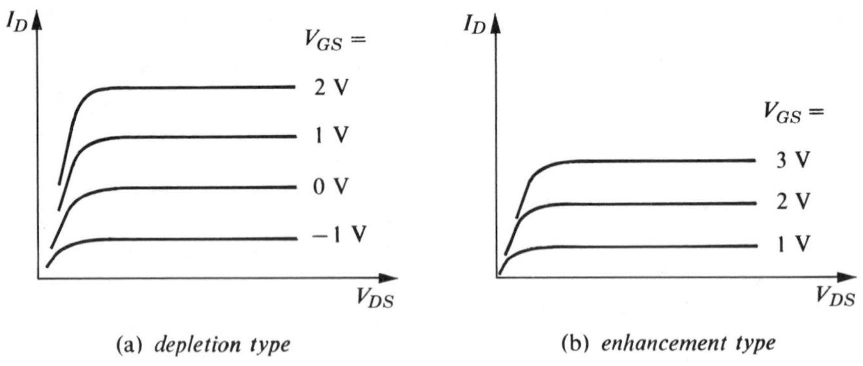

FIGURE 7.25 Curves for n channel MOSFET.

In the case of the n channel depletion type, notice the gate-source voltage V_{GS} can be of either polarity for the depletion type; if the gate is made more negative with respect to the channel, the channel width is decreased and the drain current decreases. If the gate is made more positive the channel enlarges and the drain current increases.

One advantage of the fact that either polarity gate-source voltage will affect the channel conductivity and hence the drain current for a depletion type MOSFET is that there is no necessity for a dc bias; the gate-source voltage can vary symmetrically about 0 volts. A simple common source amplifier using a depletion type MOSFET with zero bias is shown in Fig. 7.26(b). Notice that in the MOSFET circuit, the base or substrate is connected directly to the source. In the two other basic amplifier configurations the base is not connected to the source, but rather, is grounded; the appropriate circuits are shown in Fig. 7.27. The main point in biasing the base is that it must be reverse-biased with respect to the channel material, because there exists a p-n junction at the interface between the base and the channel. A further treatment of FET circuits will be found in Chapter 9 after the subject of feedback is covered.

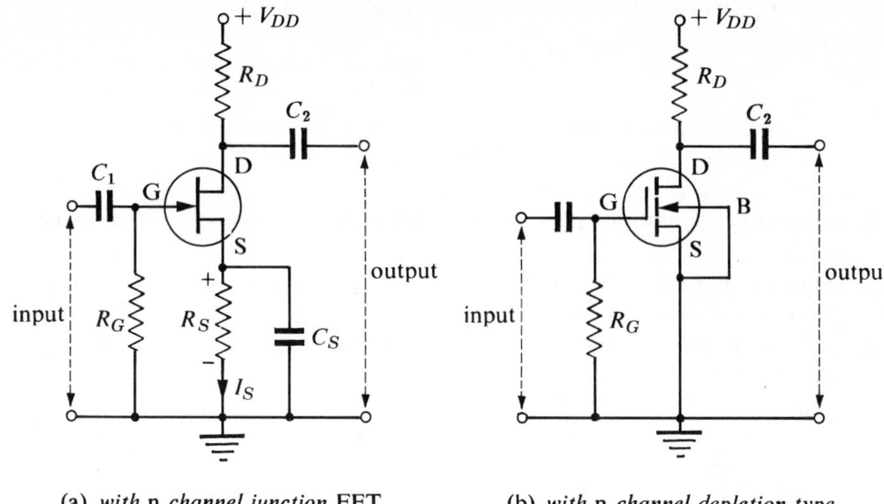

(a) *with* n *channel junction* FET dc *gate bias* = $I_S R_S$

(b) *with* n *channel depletion type* MOSFET *no* dc *bias required*

FIGURE 7.26 Common source amplifier (n channel).

To sum up, the MOSFET provides extremely high input impedances, up to $10^{15}\Omega$, and finds its principal use as the first stage for an amplifier which is driven by a high impedance signal source or where very small currents must be amplified. Because the voltage gain available from FET's is considerably less than that from bipolar transistors, combinations of FET and bipolar tran-

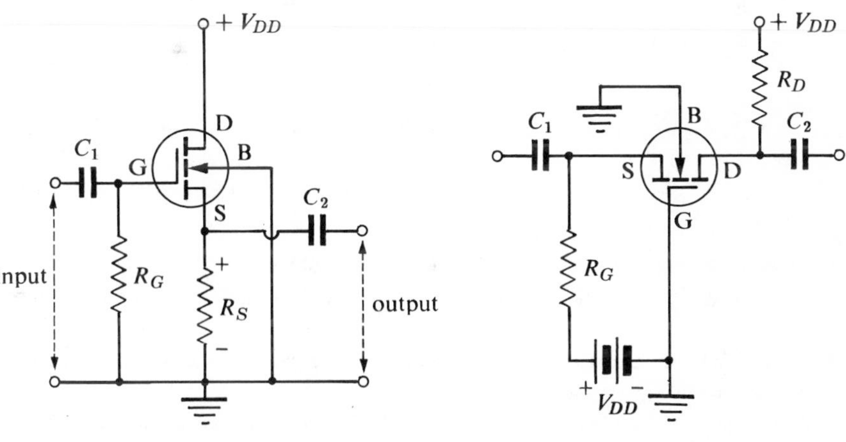

(a) *common source configuration*

(b) *common gate configuration*

FIGURE 7.27 MOSFET amplifier circuits.

sistors are often used, the FET to obtain a high input impedance and the bipolar transistors to obtain voltage gain.

problems

1. Identify the majority carriers in the channel for n channel and for p channel FET's.
2. Explain why the depletion region between the gate and the channel is thicker at the drain end than at the source end.
3. Fill in the polarities of the power supplies and the voltage drops across the source and drain resistors.

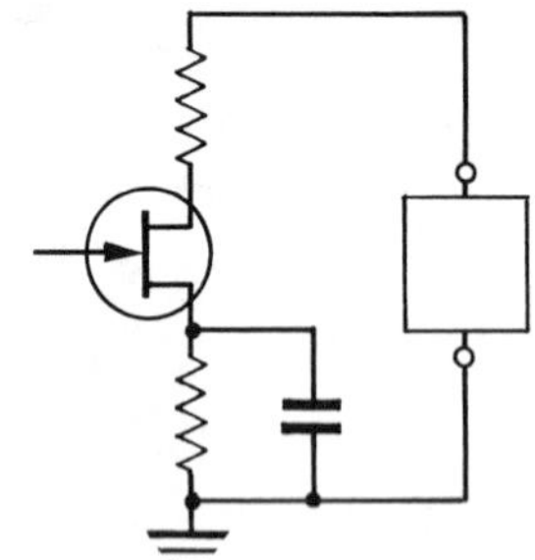

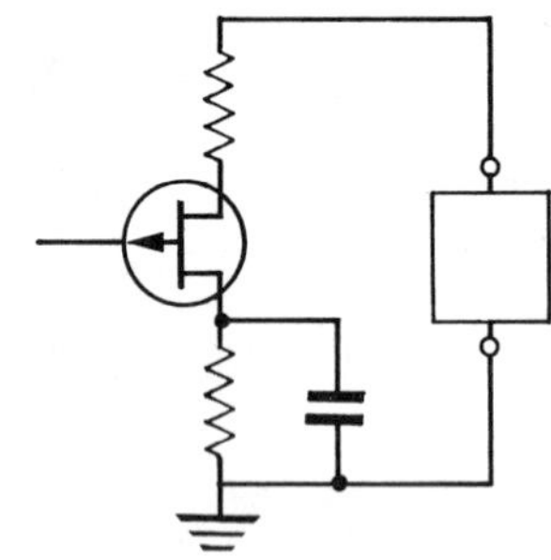

4. (a) Sketch the load line. (b) Determine V_S, V_D, and I_D at the operating point.

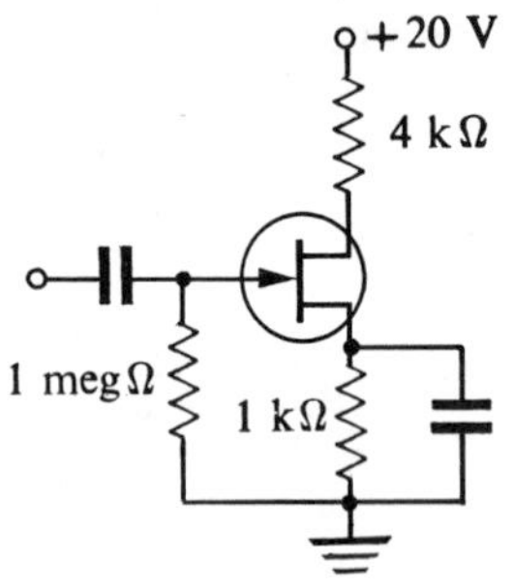

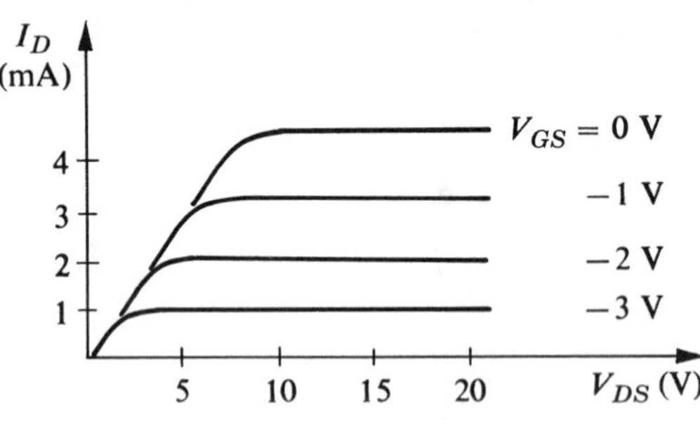

5. Explain the physical meaning of the pinch-off voltage.
6. Explain why you should not design an FET circuit to have a drain current larger than I_{DSS}.
7. Explain why the transconductance of an FET decreases as the pinch-off condition is approached.
8. Calculate $y_{21} = g_m = y_{fs}$ and y_{22} for the FET whose output curves are given in Problem 4.
9. State the expected resistance measurements taken with an ohmmeter on the $R \times 1$ scale for (a) gate–source, (b) source–gate, (c) source–drain, (d) drain–source.
10. For the FET amplifier shown below, (a) explain why the dc gate voltage is very

close to zero volts; (b) calculate the operating point (I_D, V_{DS}). Use the curve of Problem 4.

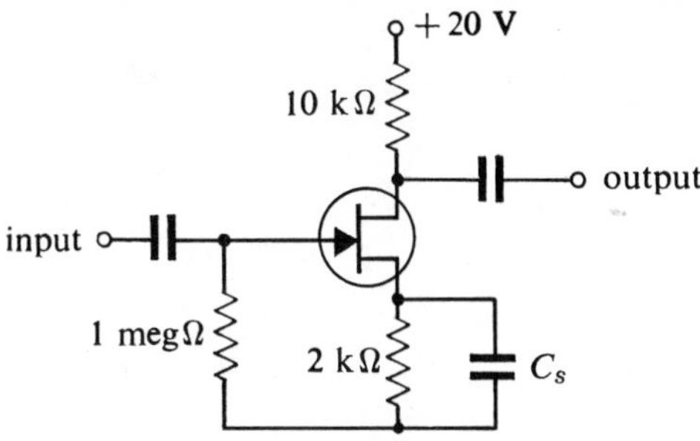

11. For an FET whose curves are given in Problem 4, (a) calculate R_D and R_S if the desired operating point is $I_D = 2.0\,\text{mA}$, $V_{DS} = 10\,\text{V}$; (b) calculate C_s if the amplifier must amplify signals from 1 kHz to 30 kHz.
12. Derive the current gain equation for an FET common source amplifier *circuit* with a load R_L connected to the drain.
13. Explain why the output impedance of a common source FET amplifier is almost exactly equal to the drain resistance R_D. Is it slightly higher or lower than R_D?
14. Design a common drain or source follower FET amplifier using an FET with $y_{21} = y_{fs} = 4000\mu\text{mhos}$ when the drain current is 3 mA. The FET curve is that of Problem 4, and the desired output impedance is 250 ohms or less.
15. Explain why the input impedance of a MOSFET can be even higher than that of a junction FET.
16. Explain (a) how the base or substrate for a MOSFET is different from the base of a bipolar junction transistor, (b) how the MOSFET base must be dc biased.

8

VACUUM TUBES

8.1 INTRODUCTION

The science of electronics now deals almost exclusively with transistors and other solid state devices. However, vacuum tubes were the principal building blocks of electronic circuits until approximately 1955, and for this reason alone it is appropriate to discuss vacuum tube electronics. Briefly, a vacuum tube consists of several metal electrodes of various shapes all packaged inside a glass or metal envelope which is highly evacuated. Vacuum tubes are often called "valves" in British scientific literature. A red hot metallic electrode (the filament or cathode) emits electrons which are attracted to a positively charged electrode called the plate or anode. The electrons pass through the spaces in a metallic grid electrode on their way to the plate, and the voltage on the grid controls how many electrons reach the plate.

At this writing, vacuum tubes are still widely used in oscilloscopes, television sets, high power high frequency radio transmitters, and in some special low noise amplifiers. However, every year sees a larger number of applications being transistorized. It is probably safe to say that this trend will continue in the future, as there is presently a great deal of technological development being put into solid state electronics and rather little being put into vacuum tube electronics.

As a general rule, vacuum tubes are inferior to modern solid state devices in many ways. Vacuum tubes are much larger. They require considerably more electric power to operate, both power to heat the red hot filament and power to supply appropriate voltages to the other electrodes. Vacuum tubes are also considerably more fragile than transistors, and they are more susceptible to microphonic noise, that is, noise arising from the mechanical vibration of the electrodes in the tube. However, they can handle high voltages and high powers at high frequencies somewhat more easily than solid state devices. They are also capable of withstanding temporary overloads in voltage or current which would permanently destroy a solid state device and then returning to normal operation.

8.2 THERMIONIC EMISSION

In a transistor, electric current is carried by either positive or negative charge carriers—by holes or electrons which arise from ionized acceptor or donor impurity atoms in the semiconductor lattice. These charge carriers then move through the lattice under the influence of electric fields that are produced by the various voltages applied to the terminals of the transistor. In a vacuum tube, however, electric current is carried only by electrons moving through the evacuated region inside the tube envelope.

In a tube, the source of the electrons is the red hot filament which quite literally boils electrons off its surface. The electrons are then accelerated over to the other electrodes by electric fields produced by voltages applied to the electrodes. The filament is heated red hot by passing a large current on the order of several hundred mA through it; this current heats the filament up to a temperature of 700–1000°C. At this temperature the average thermal energy of the electrons in the filament is larger than the electron binding energy which holds them in the filament. Hence, the electrons literally boil off the filament surface out into the vacuum. The binding energy of an electron in the filament depends on the filament material. For example, the binding energy per electron in pure tungsten is equal to 4.5 electron volts (1 electron volt $= 1.6 \times 10^{-19}$ joules). No positive ions are boiled off the filament due to the thermal energy, because the positive ions are bound much more tightly to the filament material than are the conduction electrons. It is only the outer or conduction electrons in the conduction band of the filament that are boiled off; the electrons in the valence and other inner bands remain in the filament material.

The higher the work function of the filament material, the hotter the filament must be before electrons are boiled off. A pure tungsten filament must be heated to over 1000°C for appreciable thermionic emission to take place. However, most commercial filaments are made in a layered structure with a central metal core on which is deposited a layer of barium and strontium oxide. This type of filament has a considerably lower electron binding energy or work function than pure tungsten. Such an oxide coated filament will glow a dull red and emit copious quantities of electrons at a temperature of only 700°C. Thus, oxide coated filaments require less electrical power to achieve thermionic emission than do pure tungsten filaments.

In most low power vacuum tubes, the filament is indirectly heated as shown in Fig. 8.1(a). The filament current which supplies the heat energy does not actually flow through the barium and strontium oxide layer, but rather through a separate wire which is inside a small cylinder coated with the barium and strontium oxide. The oxide coated cylinder is heated by radiation from the hot

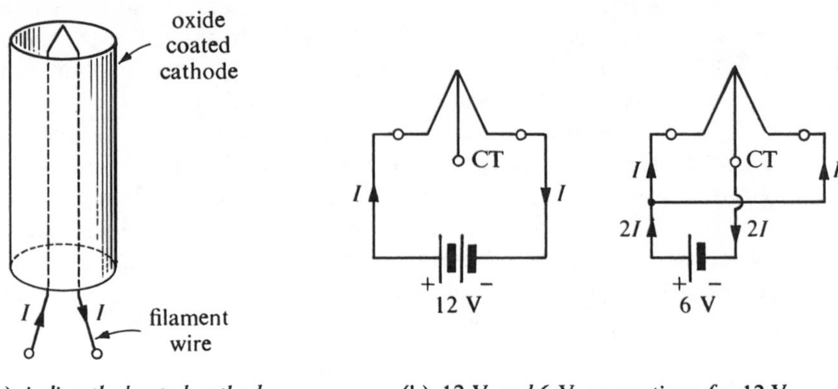

(a) *indirectly heated cathode*

(b) 12 V *and* 6 V *connections for* 12 V *center tapped filaments*

FIGURE 8.1 Vacuum tube filaments.

inner filament wire. This heating process is very efficient because of the very close spacing between the inner filament wire and the oxide coated cylinder.

Different vacuum tubes are designated by type numbers just as in the case of transistors. However in the case of vacuum tube numbers, the first number invariably designates the voltage which must be applied to the filament in order to heat it to the temperature at which sufficient electrons are emitted. For example, a 6AK5 tube requires 6 volts to heat its filament, and a 12AX7 tube requires 12 volts. Most 12 volt filaments have a center tap ("CT") so that they can be heated with a 6 volt supply by the appropriate connection shown in Fig. 8.1(b). The typical filament power for low power vacuum tubes such as used in receivers and low level amplifiers is of the order of one or two watts. A typical low power 6 volt filament draws anywhere from 0.15 to 0.3 amperes corresponding to a filament power of 0.9 to 1.8 watts. A low power 12 volt tube will draw perhaps 0.15 amperes corresponding to 1.8 watts filament power. Notice that this filament power is an order of magnitude larger than the maximum power rating for most low power transistors. In most tubes the filament is heated with 60 Hz ac current obtained from a separate 6 volt or 12 volt secondary winding on the power supply transformer. There is always a slight amount of 60 Hz noise introduced on to the signal from an ac filament, so in very high quality low noise amplifiers the filaments are usually heated with dc current, thus necessitating an extra dc supply of 6 or 12 volts.

The tube envelope, as already has been mentioned, must be highly evacuated; the remaining gas pressure in the tube envelope must be at a pressure of 10^{-6} Torr or less. The reason for this low gas pressure is so that the electrons, once emitted from the filament surface, will move freely to the other tube electrodes without suffering many collisions with gas molecules. If there is an excess of residual gas in the tube, due to a leaky tube envelope for example, then the electrons emitted from the filament will strike the gas molecules and produce ion pairs. An ion pair is a negatively charged electron and a positively charged heavy ion which are separated from one another by the impact of the moving electron. The electrons thus produced will flow in the same direction as the electrons emitted from the filament, but the heavy positive ion will be attracted back toward the filament by the same electric field which accelerates the negatively charged electrons away from the filament. The heavy positive ions striking the filament will actually damage the surface of the filament, and if enough of this ion bombardment takes place the thermionic emission can be appreciably reduced, thus essentially turning the tube off. A tube with too high a pressure is called "gassy" and can sometimes be identified by the presence of a violet glow from the gas in the tube.

8.3 THE VACUUM TUBE DIODE

The vacuum tube diode consists basically of two electrodes inside the evacuated envelope, the "cathode" which is the name given to the filament wire plus oxide surface which emits the electrons and the "anode" or plate. The construction is basically cylindrical with the small cylindrical cathode located inside the larger cylindrical anode, as shown in Fig. 8.2(a). The schematic symbol for

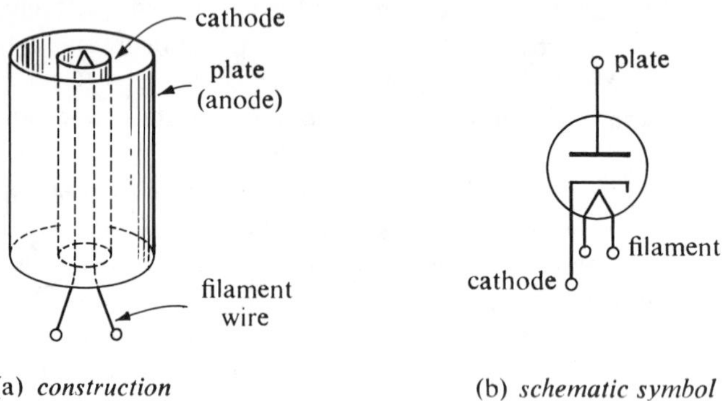

(a) *construction* (b) *schematic symbol*

FIGURE 8.2 The vacuum tube diode.

the diode is shown in Fig. 8.2(b), and the circle represents the metal or glass envelope with the indirectly heated cathode shown by the "Γ" and inverted "v" in Fig. 8.2(b). The fact that the hot filament wire does not actually touch the oxide surface that emits the electrons is symbolized by the space between the filament wire and the "Γ" symbol for the oxide surface. The plate is made positive in voltage with respect to the cathode by anywhere from 10 to 100 volts; this voltage produces an electric field pointed from the plate toward the cathode that accelerates the emitted electrons away from the cathode toward the plate. If the plate is made negative in voltage with respect to the cathode, then the diode will pass *no* current. Any electrons boiled off the cathode surface will be repelled back near the cathode.

For most vacuum tube diodes, the plate current increases roughly linearly as the voltage difference between the anode and the cathode. For example, if the plate-cathode voltage V_{PK} is 10 volts, the plate current will be approximately 10–20 mA while if V_{PK} is 50 volts, the plate current will be 50–100 mA. A vacuum tube diode will also pass some current even when the plate-cathode voltage is zero, because some of the electrons boiled off the cathode will have sufficient energy to reach the plate.

These figures should be compared with those for a semiconductor diode. For a Ge diode the turn-on voltage between the p and the n side of the junction is about 0.25 volt regardless of whether the diode conducts 1 mA or 100 mA, while for a silicon diode the turn-on voltage is only 0.6 volt for diode currents up to several hundred mA. In other words the turn-on voltage for the vacuum tube diode is considerably larger than for a semiconductor diode.

If a vacuum tube diode is designed to withstand very high voltages, perhaps 500 volts or more, then the spacing between the cathode and the anode must be made larger to prevent an electrical breakdown, and this weakens the electric field between the anode and the cathode. Thus, for an anode-cathode voltage of about 500 volts, the plate current may be only of the order of 25 mA.

The principal use of vacuum tube diodes is in rectification, that is, changing ac to dc current, which must be done in all power supplies designed to run off the ac line voltage. Typical half-wave and full-wave power supply circuits using vacuum tube diodes are shown in Fig. 8.3.

The power supply circuits are essentially identical to those using semiconductor diodes. The basic principle in both types of circuit is that the diodes conduct current in only one direction, only when the plate is positive in voltage with respect to the cathode for vacuum tube diodes, and only when the p-type material is positive with respect to the n-type material for semiconductor diodes. Notice in the vacuum tube diode circuit that a separate secondary transformer winding is required to heat the filaments. Most rectifier vacuum tube diodes have 5 volt filaments. Notice also that the filament winding on the transformer must be insulated to withstand the secondary high voltage of the transformer. Another difference is that *LC* filters are generally used to reduce the ac ripple in vacuum tube power supplies. An *LC* filter is more effective in ripple reduction than an *RC* filter and wastes less power as heat. However, it will generate high voltage transients if the current is interrupted suddenly. Such transients are easily handled by vacuum tubes, but will quickly and permanently destroy semiconductor diodes or transistors. Hence *LC* filters are found mostly in vacuum tube circuits. In general, semiconductor diodes are much better than vacuum tube diodes and have replaced vacuum diodes in almost all power supply circuits at this writing. Semiconductor diodes are cheaper, waste less

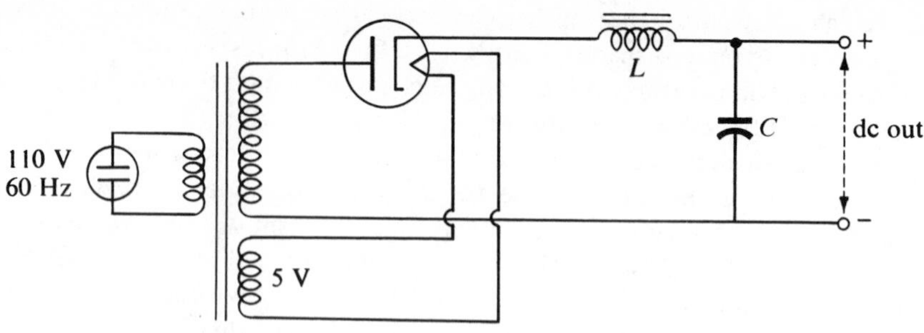

(a) *Half wave rectifier with LC filter*

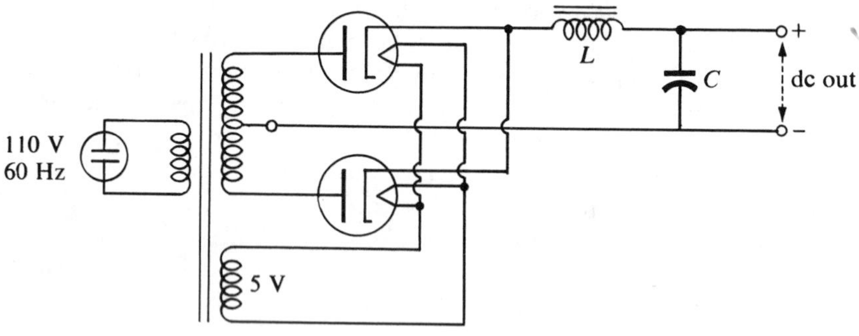

(b) *Full wave rectifier with LC filter*

FIGURE 8.3 Power supply circuits using vacuum tube diodes.

power because of their very small voltage drop, do not waste any filament power because they have no filament, and are mechanically much more rugged.

8.4 THE VACUUM TUBE TRIODE

The vacuum tube triode consists of three electrodes inside the highly evacuated envelope: the cathode, the plate, and a third element called the grid. The grid is a mesh of fine metallic wire wound between the cathode and the plate. The construction and schematic symbol for a triode are shown in Fig. 8.4. The grid is electrically insulated from both the cathode and the plate and is usually supported by either mica or glass supports. The dotted line for the grid symbolizes the fact that electrons can pass through the grid without striking the grid wire. The grid supports can vibrate if the tube is subjected to mechanical shock, and this produces a type of noise called microphonic noise which is entirely absent in transistors because of the solid construction of transistors. The grid is basically the control electrode in a triode, analogous to the base of a bipolar transistor or the gate of a field effect transistor. The grid is usually dc biased so that it is negative with respect to the cathode; hence, electrons emitted from the cathode do not strike the grid wires but rather pass through the gaps between the grid wires and continue on to the plate which is positively charged with respect to the cathode.

If the grid voltage is made more negative, the plate current is decreased,

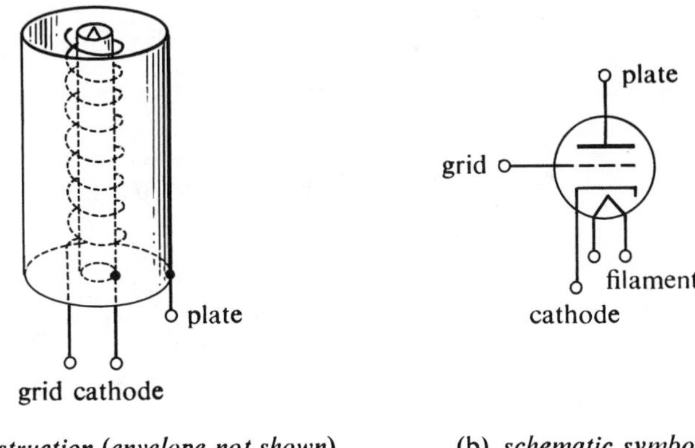

(a) *construction (envelope not shown)* (b) *schematic symbol*

FIGURE 8.4 Vacuum tube triode.

and if the grid voltage is made less negative the plate current increases. The grid wire is physically close to the cathode, and thus a very small change in grid voltage will result in a large change in the current arriving at the plate. Thus the grid voltage in a very real sense controls the amount of plate current and is sometimes called a "control" grid.

Because the grid is always kept at a negative dc voltage with respect to the cathode, it draws little or no current. Hence the grid presents a very high impedance, of the order of 1000 megohms or more, to the incoming signal voltage applied to the grid. In this sense, the grid is quite analogous to the reverse-biased gate of a field effect transistor. Because of the high grid-cathode impedance, the input power to the grid is extremely small, and thus the triode can act as an amplifier. The large fluctuations in plate current yield a very high output power at the plate.

The exact effect of changing grid-cathode voltage V_{GK} on the plate current I_P of a triode is usually determined by referring to a set of characteristic curves for the particular triode in question. Such curves are shown in Fig. 8.5 for a commonly used high gain triode, the 12AX7.

The plate current I_P is plotted against the plate-cathode voltage V_{PK} for various constant values of the grid-to-cathode voltage V_{GK}. Notice that the more negative the grid is with respect to the cathode, the less the plate current.

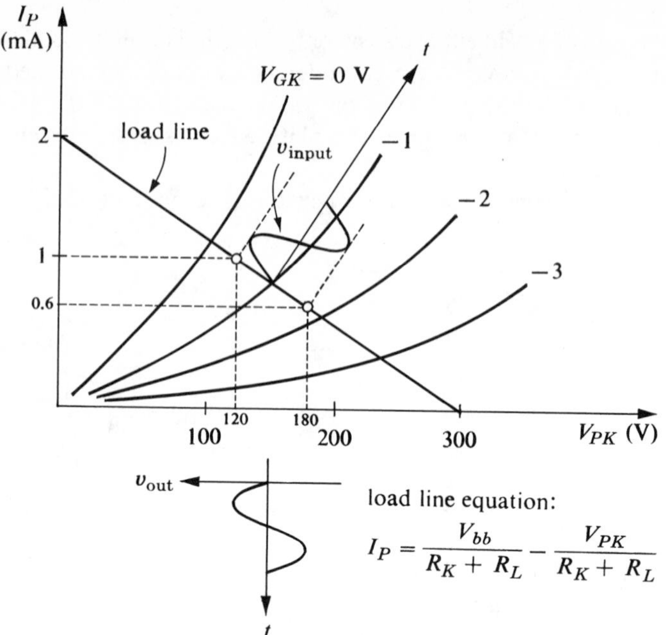

FIGURE 8.5 Characteristic curves for 12AX7 vacuum tube triode.

If the grid-cathode voltage is made more negative than about 3.5 volts, then no plate current flows, and the tube is said to be "cut off."

The effect of a sinusoidal change in the grid-to-cathode voltage V_{GK} is shown in Fig. 8.5. As V_{GK} swings from -0.5V to -1.5V along the load line, the plate current I_P swings from 1 mA to 0.6 mA, and the plate voltage swings from 120V to 180V. The voltage gain for the tube whose characteristics are shown in Fig. 8.5 is the change in plate voltage divided by the change in grid-to-cathode voltage or:

$$|A_v| = \frac{180\,\text{V} - 120\,\text{V}}{1.5\,\text{V} - 0.5\,\text{V}} = \frac{60\,\text{V}}{1.0\,\text{V}} = 60. \qquad \textbf{(8.1)}$$

A single-stage *RC* coupled voltage amplifier circuit using a triode is shown in Fig. 8.6. This voltage amplifier circuit is very similar to the common emitter amplifier circuit for a transistor, and the first thing we must determine is the appropriate biasing. We recall that the grid must be always at a negative dc bias voltage with respect to the cathode. One obvious way of accomplishing this is shown in Fig. 8.6(a) with a separate bias battery V_{cc} and a resistance R_G connected between ground and the grid. Remembering that the grid current is equal to 0 for the grid negative with respect to the cathode, we see that there is no grid current through or voltage drop across the resistor R_G; and therefore, the voltage between the grid and the cathode is equal to the bias battery voltage V_{cc}. The plate is made positive with respect to the cathode by the battery

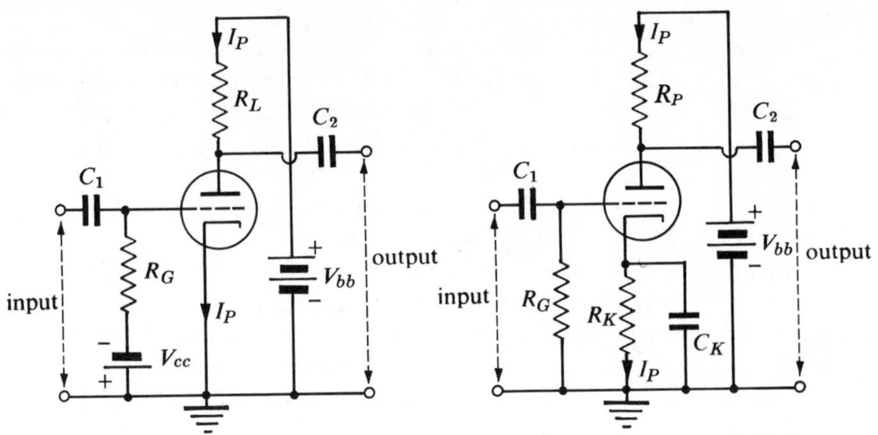

(a) *with separate bias battery* V_{cc} (b) *with self bias or cathode bias*

FIGURE 8.6 Basic triode voltage amplifier.

V_{bb}, the positive terminal of which is connected to the plate through a resistance R_L.

A more commonly used biasing circuit is shown in Fig. 8.6(b); this biasing arrangement is called "cathode bias" or "self bias." The grid is connected directly to ground through the resistance R_G, and an additional resistor R_K is placed in series with the cathode. The grid current is 0, and therefore the current flowing in the cathode equals the current flowing in the plate circuit. Therefore the voltage drop across R_K in the cathode circuit equals $I_P R_K$. Notice that the polarity of this voltage drop is such that the cathode is positive with respect to ground. Remembering, however, that the grid is at dc ground potential, we see that the grid is therefore negative with respect to the cathode, which is the desired biasing condition. The grid-cathode voltage difference simply equals the voltage drop across R_K. The presence of the coupling capacitors C_1 and C_2 is necessary to prevent any external circuitry connected to the input or the output from changing the dc bias voltages.

Let us now consider the characteristic curves of Fig. 8.5 and the operation of the self-biased triode voltage amplifier of Fig. 8.6(b). Let us first write the Kirchhoff voltage equation for the amplifier starting at ground and going clockwise around the loop containing R_K, the triode vacuum tube, R_L, and the battery V_{bb}.

$$I_K R_K + V_{PK} + I_P R_P - V_{bb} = 0, \tag{8.2}$$

where I_K is the cathode current.

Setting $I_K = I_P$ and solving for the plate current I_P yields:

$$I_P = \frac{V_{bb}}{R_K + R_P} - \frac{1}{R_K + R_P} V_{PK}. \tag{8.3}$$

Thus we see that the plate current I_P plotted versus the plate-cathode voltage V_{PK} is a straight line with a negative slope $1/(R_K + R_P)$ and a vertical intercept $V_{bb}/(R_K + R_P)$. This equation is called the load line equation and can be drawn directly on the characteristic curves of Fig. 8.5. It is important to remember that once the battery voltage V_{bb} is fixed and the sum $(R_K + R_P)$ is fixed, then the load line is determined. Regardless of what input signal is applied to the grid through C_1, the plate current I_P and the plate-cathode voltage V_{PK} must define a point which lies precisely on the load line.

Let us now calculate the values for the various resistances and capacitors for the triode voltage amplifier circuit of Fig. 8.6(b). Looking at the characteristic curves for the 12AX7 triode, we draw a convenient load line from a vertical intercept of 2 mA down to a horizontal intercept of 300 volts. From the vertical intercept of the load line, we immediately calculate that $(R_K + R_P) = V_{bb}/I_{P\,max} = 300\,\text{V}/2\,\text{mA} = 150\,\text{k}\Omega$.

We now must choose an operating point for the tube; that is, we must choose the dc plate current and plate-to-cathode dc voltage that will exist in the absence of any incoming signal. We choose an operating point (which must be on the load line) at $I_P = 1$ mA and $V_{PK} = 150$ V. This operating point plus the load line we have drawn implies that the grid-to-cathode voltage must equal approximately -1.1 V, because the $V_{GK} = -1.1$ V curve of constant grid-cathode voltage intersects the load line at the operating point. We now can calculate the resistance in the cathode R_K from the equation $V_{GK} = I_P R_K$. Or $R_K = V_{GK}/I_P = 1.1$ volts/1 mA $= 1.1\,\text{k}\Omega$. A $1\,\text{k}\Omega$ resistor would be used in practice. The resistance R_P in the plate circuit is therefore equal to $150\,\text{k}\Omega - 1.1\,\text{k}\Omega = 148.9\,\text{k}\Omega$. A $150\,\text{k}\Omega$ resistor would be used in practice.

From inspection of the characteristic curves and the operating point, we see that the grid voltage cannot swing more than about 1 volt positive without driving the grid positive with respect to the cathode. This would result in the grid drawing current as some of the electrons emitted from the cathode would then be attracted to the grid rather than repelled from it. This would drastically distort the output voltage waveform. When the grid goes positive with respect to the cathode the tube is essentially turned on completely and acts almost like a short circuit, thus dropping the voltage at the plate to a very low voltage. If too much current is drawn the grid may in fact be overheated and permanently damaged, because it is made of very fine wire. We also see from inspection of the characteristic curves and our operating point, that the grid voltage cannot swing more negative than about -4 V without cutting the tube off completely, that is, without decreasing the plate current to zero. If this happens then the tube is cut off and the voltage at the plate equals the power supply voltage V_{bb}. The output voltage waveform at the plate will then be badly distorted.

We must add a capacitor C_K in parallel with R_K in order to keep the cathode voltage constant as the current passed by the triode fluctuates in response to the input signal. This is very similar to bypassing the emitter resistor in a common emitter transistor amplifier. If the grid becomes more positive due to a positive going input signal, the tube will draw more current, thus making the cathode more positive. This decreases the effective input signal to the tube, which is the *difference* in voltage between the grid and the cathode. If the grid goes negative in response to an input signal, the current through the triode will decrease, and the cathode voltage will become less positive or more negative, again decreasing the grid-cathode voltage. In other words the cathode voltage will *follow* the grid voltage just as the voltage on the emitter of a transistor tends to follow the base voltage. To prevent this following action from decreasing the amplitude of the grid-to-cathode input voltage, we merely make the cathode an ac ground by bypassing it to ground through a large capacitor C_K. The criterion for this capacitor is simply that $1/\omega C_K$ should be much less than R_K at the lowest signal frequency.

The two coupling capacitors C_1 and C_2 are chosen large enough to pass the lowest signal frequency of interest just as in the case of an RC coupled transistor amplifier. This is true because C_1 and C_2 each form the capacitor of a high-pass RC filter: C_1 with R_G, and C_2 with whatever load resistance is attached to the output to the triode amplifier.

The ac equivalent circuit of a triode comes from a "black box" parameter analysis similar to that used for transistors and is shown in Fig. 8.7. If V_{GK} and

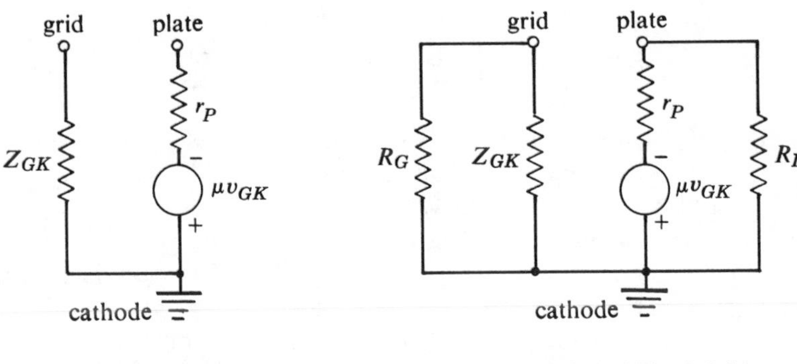

(a) *for triode alone* (b) *for amplifier of* Fig. 8.6 (b)

FIGURE 8.7 Triode equivalent circuit.

I_P are chosen as the independent variables, then the plate voltage can be written as $V_P = V_P(V_{GK}, I_P)$. Taking differentials yields:

$$dV_P = \left(\frac{\partial V_P}{\partial V_{GK}}\right)_{I_P} dV_{GK} + \left(\frac{\partial V_P}{\partial I_P}\right)_{V_{GK}} dI_P \tag{8.4}$$

or

$$v_P = -\mu v_{GK} + r_P i_P, \tag{8.5}$$

where the lowercase letters represent differential or ac signal amplitudes, and μ the "amplification factor" is defined as $(-\partial V_P/\partial V_{GK})_{I_P}$, and r_P the "plate resistance" is defined as $(\partial V_P/\partial I_P)_{V_{GK}}$. The equivalent circuit for the output follows from (8.5), and the input equivalent circuit can be shown to be simply the grid-to-cathode impedance. From the definitions it follows that $\mu = g_m r_P$ where $g_m = (\partial I_P/\partial V_{GK})$ is called the transconductance. We see that the equivalent circuit for the triode is merely a voltage generator $= \mu v_{GK}$ in series with a resistance r_P. The minus sign in the voltage generator merely represents the 180° phase inversion between the plate and the grid; that is, a positive input at the grid produces a negative output at the plate and vice versa. μ is the amplification factor of the tube, and can be looked up in tube handbooks for various triodes. For the 12AX7, $\mu = 100$; for other triodes, μ may vary from 20 to 100. r_P is called the plate resistance, and for the 12AX7 is anywhere from 60 kΩ to 80 kΩ depending upon the plate current. Thus we see that the triode tube inherently has a very high output impedance when connected in the common cathode configuration of Fig. 8.6. This is similar to the common emitter amplifier transistor configuration, which also has a high output impedance.

We can quickly obtain an expression for the voltage gain of the triode amplifier of Fig. 8.6(b) by using the ac equivalent circuit for the amplifier as shown in Fig. 8.7(b). The voltage gain equals:

$$A_v = \frac{v_{\text{out}}}{v_{\text{in}}} = \frac{i_P R_P}{v_{\text{in}}} = \frac{\dfrac{-\mu v_{GK} R_P}{R_P + r_P}}{v_{GK}} = \frac{-\mu R_P}{r_P + R_P}.$$

Substituting in $\mu = 100$, $r_P = 80\,\text{k}\Omega$, $R_P = 150\,\text{k}\Omega$ yields $A_v = 65$. Notice that voltage gain from the amplifier circuit is always less than μ, the amplification factor for the triode tube used. Only as the plate load resistance R_P becomes large in comparison to r_P, the plate resistance of the tube, does the voltage gain approach μ; and this situation is extremely rare. The input impedance of the amplifier circuit is seen to be R_G in parallel with Z_{GK} the tube grid-cathode impedance. Usually $R_G \ll Z_{GK}$, so the amplifier input impedance equals the external grid-to-ground resistance R_G, which is typically on the order of 1 MΩ. The output impedance of the amplifier is the parallel combination of r_P and R_P.

8.5 THE VACUUM TUBE PENTODE

The vacuum tube pentode basically consists of five electrodes inside the evacuated tube envelope. Starting from the cathode and going outward, we have the cathode, the first grid (usually called the control grid), a second grid called a screen grid, a third grid called a suppressor grid, and finally the anode or plate. Some tubes are made without a suppressor grid; that is, these contain just the cathode, the control grid, the screen grid, and the plate. Such tubes are called tetrodes and will not be considered here in any detail. We have already seen that the purpose of the first or control grid is literally to control the flow of electrons from the cathode to the plate. If the grid is made more negative with respect to the cathode, then the electron current is decreased and vice versa. The purpose of the second or screen grid is to decrease the space charge effect produced by a high density of electrons in the region between the cathode and the plate. A large density of electrons will effectively prevent more electrons from being emitted from the cathode and attracted to the plate. The screen grid is usually operated at a constant dc voltage which is of the order of 100 V positive with respect to the cathode, thus decreasing the repulsive effect of the electron cloud. However, the screen grid is less positive than the plate; typically, the plate is 150 or 200 volts positive with respect to the cathode. The screen grid also enables the pentode to operate at much higher frequencies than the triode. In a triode the control grid-to-plate capacitance is effectively amplified by the voltage gain of the tube and effectively limits the triode to frequencies of 100 kHz or less. This phenomenon is called the Miller effect and is discussed in greater detail in Chapter 10. In the pentode the screen grid, which is maintained at a constant dc potential, effectively screens or shields the control grid-to-plate capacitance. Pentodes can operate at frequencies up to tens and hundreds of megahertz with careful circuit design.

The third or suppressor grid is usually connected to the cathode and therefore is at a voltage negative with respect to the plate. If large numbers of electrons of several hundred electron volts energy strike the plate, then some secondary electrons will be emitted from the plate; the purpose of the suppressor grid is to repel these electrons back to the plate. The schematic symbol for a vacuum tube pentode is shown in Fig. 8.8.

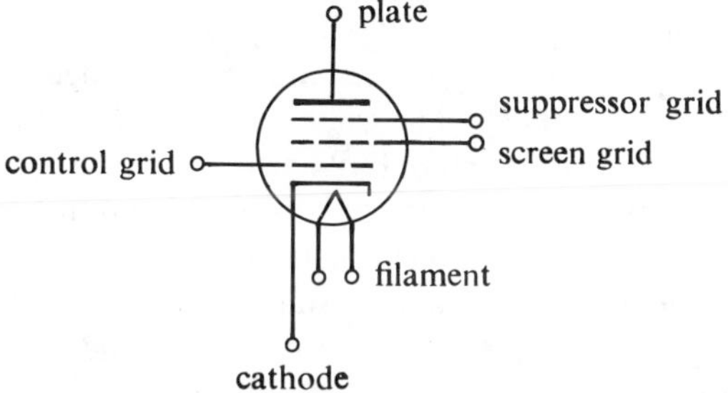

FIGURE 8.8 Vacuum tube pentode schematic symbol

The exact behavior of a vacuum tube pentode is difficult to predict, and in practice one refers to the manufacturer's characteristic curves which relate the plate current to the plate-cathode voltage for various values of control grid voltage. A typical pentode characteristic curve is shown in Fig. 8.9 with the plate current I_P plotted versus the plate-cathode voltage V_{PK} for various values of the control grid-cathode voltage. These characteristic curves are given in vacuum tube handbooks, and the fixed dc voltage on the screen grid and the suppressor grid are specified for the curve. The plate current in a pentode may be from 1 to 10 mA. The screen grid is positive with respect to the

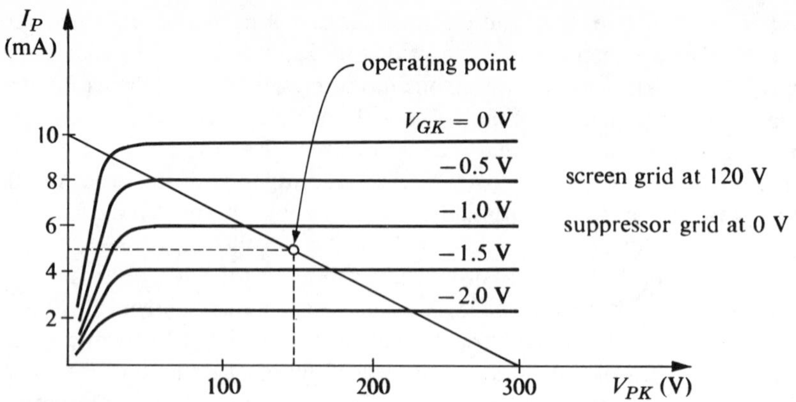

FIGURE 8.9 6AU6 pentode characteristic curve.

cathode and therefore draws current, usually on the order of 2 or 3 mA. Hence the conservation of current requires that $I_K = I_P + I_{G2}$; that is, the current flowing in the cathode equals the plate current plus the screen grid current. The suppressor grid does not draw any current because it is at the same potential as the cathode; that is, it does not attract electrons. A single-stage voltage amplifier using a vacuum tube pentode is shown in Fig. 8.10.

Let us now consider the biasing of the pentode. The control grid is returned directly to ground through a resistor R_G of about 1 megohm. Because the control grid draws negligible current, there is negligible voltage drop across R_G, and thus the grid is at dc ground. The grid is made negative with respect to the cathode by the self-bias or the cathode bias arrangement with a resistor R_K in series with the cathode. The voltage drop across the cathode resistor R_K is equal to $I_K R_K$ and is of such polarity as to make the cathode positive with respect to the control grid or, in other words, the control grid negative with respect to the cathode. The cathode resistor is bypassed by a capacitor C_K in order to fix the cathode potential at a constant value, so that the full input signal voltage applied to the grid appears between the grid and the cathode just as in the triode amplifier and the common emitter transistor amplifier.

The screen grid is fixed at its appropriate voltage V_{G2} which we look up in the vacuum tube manual, by a resistor R_{G2} going from the screen grid to the V_{bb} supply. For a given pentode, the screen current is fixed at a particular value depending on the construction of the tube, and this screen current value is listed in the tube handbook. Thus, the value of the resistor R_{G2} is determined by Ohm's law: $R_{G2} = (V_{bb} - V_{G2})/I_{G2}$. For example, if $V_{bb} = 300$ volts, and the tube handbook specifies $I_{G2} = 2\,\text{mA}$ and $V_{G2} = 120$ volts, then $R_{G2} = (300 - 120)/2 = 90\,\text{k}\Omega$. It is desirable to run the screen grid at a constant voltage regardless of the signal fluctuations; therefore, the screen grid is bypassed to ground with a capacitor C_{G2} such that the capacitive reactance of C_{G2} is very small at the lowest signal frequency.

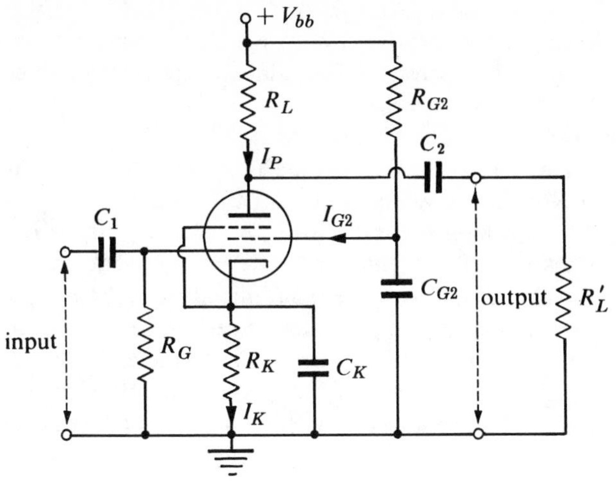

(a) *pentode amplifier circuit*

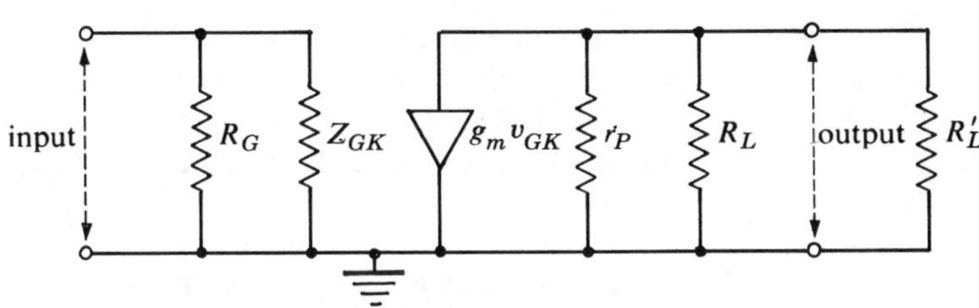

(b) *pentode amplifier equivalent circuit*

FIGURE 8.10 Vacuum tube pentode amplifier.

The Kirchhoff voltage equation for the loop containing the cathode resistor R_K, the plate-to-cathode voltage V_{PK}, the plate resistor R_P, and the power V_{bb} implies a load line equation just as in the case of the vacuum tube triode.

$$(I_P + I_{G2})R_K + V_{PK} + I_P R_P - V_{bb} = 0.$$

If $I_P \gg I_{G2}$, then $I_P = V_{bb}/(R_K + R_P) - V_{PK}/(R_K + R_P)$. We usually look at the characteristic curves for the pentode we are using and draw a convenient load line. On the curves of Fig. 8.9, we draw the load line from 10 mA down to 300 volts; we have chosen the power supply V_{bb} to equal 300 volts. The sum of the cathode resistance R_K and the plate resistance R_P is determined by the vertical 10 mA intercept: $(R_L + R_K) = V_{bb}/I_{P\max} = 300 \text{ volts}/10 \text{ mA} = 30 \text{ k}\Omega$. The cathode resistor R_K is chosen by the operating point. If we choose (5 mA, 150 volts) as the operating point, then we see from the characteristic curves that the grid-to-cathode voltage must be -1.2 volts. Therefore, the voltage drop across the cathode resistor must be -1.2 volts, and R_K is given by

$R_K = V_{GK}/I_K = V_{GK}/(I_P + I_{G2}) = 1.2\text{V}/(5\,\text{mA} + 2\,\text{mA}) = 170\,\Omega$. Notice that the current flowing in the cathode resistor equals the plate current of 5 mA *plus* the 2 mA screen grid current. The plate resistance R_P therefore equals $30\,\text{k}\Omega - 170\,\Omega = 29.8\,\text{k}\Omega$. In actual practice a $27\,\text{k}\Omega$ or a $33\,\text{k}\Omega$ resistor would probably be used.

The equivalent circuit for the pentode amplifier is shown in Fig. 8.10(b) and is seen to be a Norton equivalent circuit for the output with a constant current generator equal to $g_m V_{GK}$ in parallel with the plate resistance r_P. g_m is called the transconductance of the pentode, and for a typical pentode will range from 2000 to 6000 micromhos. The input impedance for the equivalent circuit is approximately the external grid-to-ground resistance, R_G, because so long as the control grid is biased negative with respect to the cathode the control grid-cathode tube impedance Z_{GK} is extremely high, of the order of 1000 megohms. Thus the grid-to-ground-to-impedance Z_{GKA} of the amplifier is given by:

$$Z_{GKA} = \frac{R_G Z_{GK}}{R_G + Z_{GK}} \cong R_G .$$

The voltage gain can be calculated quickly from inspection of the equivalent circuit. The current flowing in the output is equal to $g_m V_{GK}$, and this current is divided among the plate resistance of the pentode r_P, the resistance R_P in series with the plate, and the load resistance R_L which the amplifier drives. Therefore, the voltage gain is given by:

$$A_v = \frac{v_{\text{out}}}{v_{\text{in}}} = \frac{-g_m V_{GK}(r_P \,\|\, R_P \,\|\, R_L)}{V_{GK}} = -g_m \frac{r_P R_{PL}}{r_P + R_{PL}},$$

if we let $R_{PL} = R_P \,\|\, R_L$. (Notice that the plate resistance for a pentode is much higher than for a triode.) For a 6AU6 pentode, $g_m = 4800\,\mu$mhos and $r_P \cong 1$ megohm. $R_P = 30\,\text{k}\Omega$ in our amplifier circuit, and if $R_L \gg 30\,\text{k}\Omega$ then $R_{PL} \cong 30$ kΩ. Thus the numerical voltage gain is $A_v \cong (4800 \times 10^{-6})\,(30 \times 10^3) = 144$.

The output impedance of the amplifier will be the parallel combination of r_p and R_P. Usually $r_p \gg R_P$, so the amplifier output impedance is essentially equal to the resistance R_P in the plate lead, $29.8\,\text{k}\Omega$ in this case.

8.6 REPRESENTATIVE VACUUM TUBE CIRCUITS

There are three possible configurations for vacuum tube amplifiers just as for transistors; namely, the common cathode configuration which we have already considered, the common plate configuration (usually called the cathode follower), and the common grid configuration. The common plate configuration is analogous to the common collector transistor amplifier configuration and has the same characteristics as the emitter follower; namely, high input impedance, low output impedance, and a voltage gain of the order of unity. The common grid configuration is used mainly in high frequency amplifiers and will not be discussed here.

A practical cathode follower circuit is shown in Fig. 8.11, which utilizes

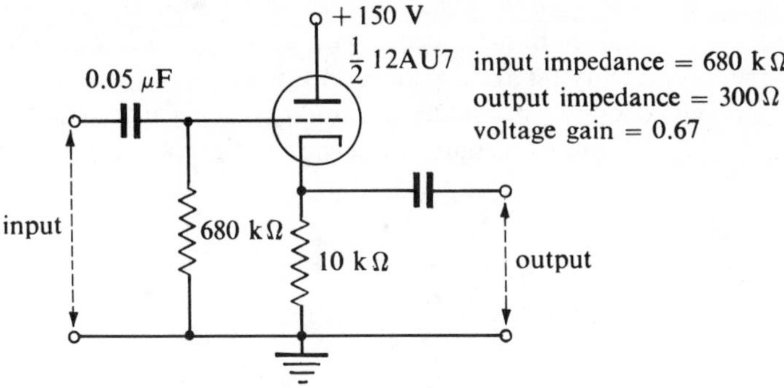

FIGURE 8.11 Triode vacuum tube cathode follower.

one of the two triodes in a 12AU7 vacuum tube. Notice that the input impedance in the circuit will be essentially equal to the grid-to-ground resistor of 680 kΩ and that the voltage gain is roughly unity, 0.67. A careful analysis of a cathode follower amplifier shows that the output impedance is approximately given by $1/g_m$ where g_m is the transconductance of the tube.

A two-stage triode voltage amplifier suitable for audio frequencies is shown in Fig. 8.12.

A 5751 low noise, highly reliable tube is used; it is a special version of the less expensive 12AX7. A single 5751 or 12AX7 tube contains two separate

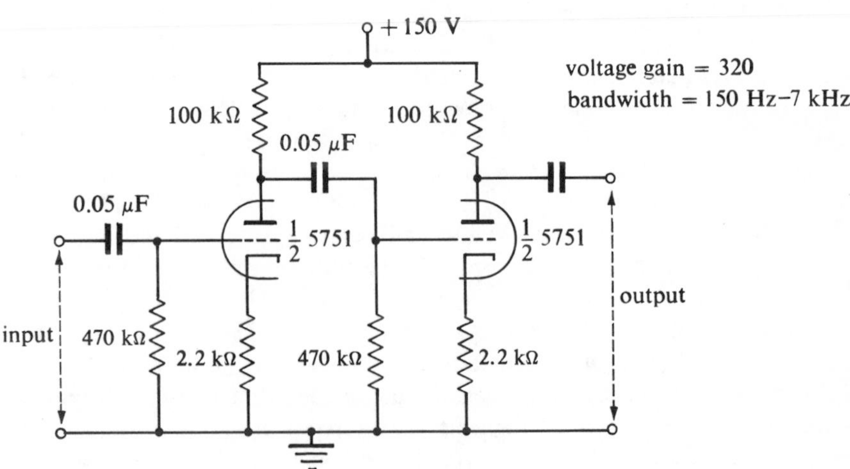

FIGURE 8.12 Two-stage triode amplifier.

triodes inside the same evacuated tube envelope; such tubes are often referred to as "twin" triodes. The cathode resistors have been left un-bypassed in this particular circuit to lower the gain a bit and to provide some stabilization which will be discussed at greater length in Chapter 9. The voltage gain of this amplifier is about 300 over a limited range of audio frequencies 150 Hz to 7 kHz. Such a bandwidth will reproduce human speech with adequate fidelity but would not be considered high fidelity for the reproduction of music.

A pentode video amplifier designed to amplify fast negative pulses is shown in Fig. 8.13. The voltage gain is only 4, but the rise time is 0.035 μsec assuming

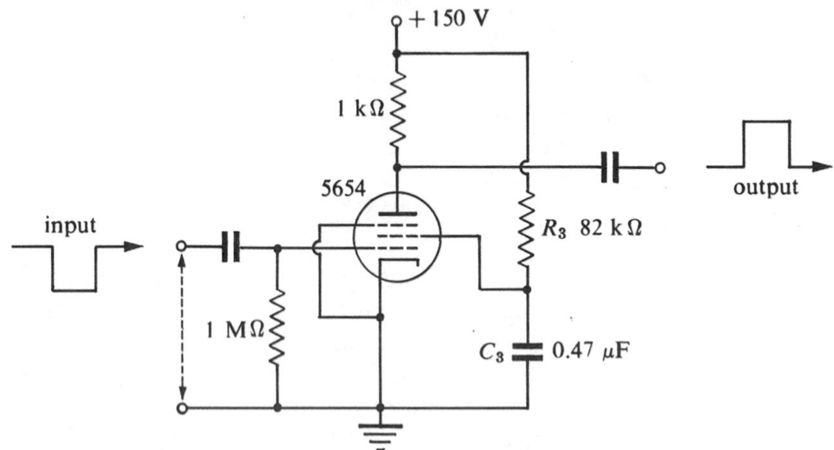

FIGURE 8.13 Pentode video amplifier.

a total output to ground capacitance of 16 pF including output and wiring capacitance. The voltage gain could be raised by increasing the plate resistance above 1 kΩ, but at the expense of smaller bandwidth and consequently longer rise time. With the 1 kΩ plate resistance the bandwidth is approximately 10 MHz. The 5654 pentode is a "reliable" or preferred military equivalent of the commercial 6AK5. The resistance R_3 provides the proper screen grid voltage and C_3 bypasses the screen to make it an ac ground. No cathode–bias resistor is necessary in this amplifier as it is designed for only negative input pulses.

A vacuum tube Colpitts oscillator designed to operate at 3.5 MHz is shown in Fig. 8.14. The circuit can be tuned by adjustment of the 150 pF variable capacitor across the 4.3 μH choke in the grid circuit. The grid-to-cathode negative bias is obtained in this circuit by means of the 47 kΩ, 100 pF *RC* parallel combination. As the tube oscillates the grid draws a slight amount of electron current once each cycle, and this negative charge builds up on the 100 pF capacitor gradually bleeding off through the 47 kΩ resistor and the 4.3 μH inductance to ground. The net effect of the 47 kΩ, 100 pF combination is to keep the dc grid voltage slightly negative with respect to the cathode. Such an arrangement is called "grid-leak" bias.

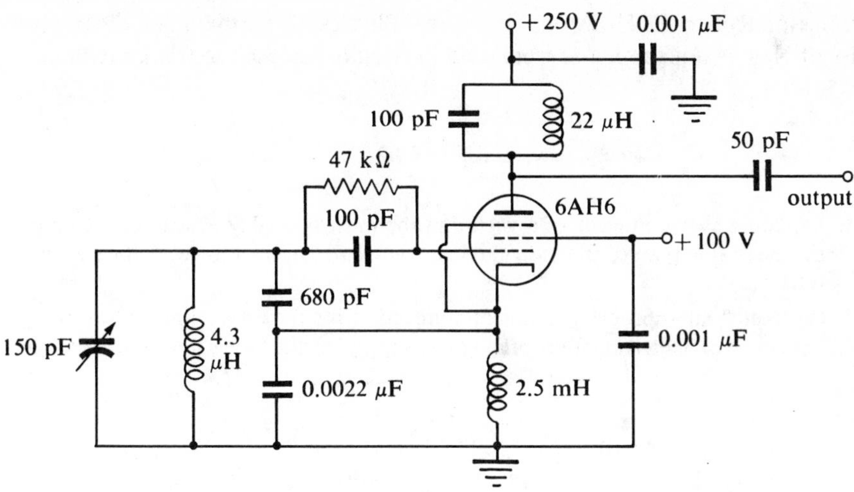

FIGURE 8.14 Practical 3.5 MHz pentode vacuum tube Colpitts oscillator.

Another practical vacuum tube circuit is the cathode follower shown in Fig. 8.15. This circuit will operate at frequencies up to above 10 MHz and was designed to drive a variety of loads, anything from an amplifier to several thousand feet of RG58/U coaxial cable, by adjustment of L and R_1. The 1 kΩ and 0.01 μF RC filter in the plate lead serves to make the plate a true ac ground even at high frequencies. The circuits of Figs. 8.14 and 8.15 were kindly supplied by

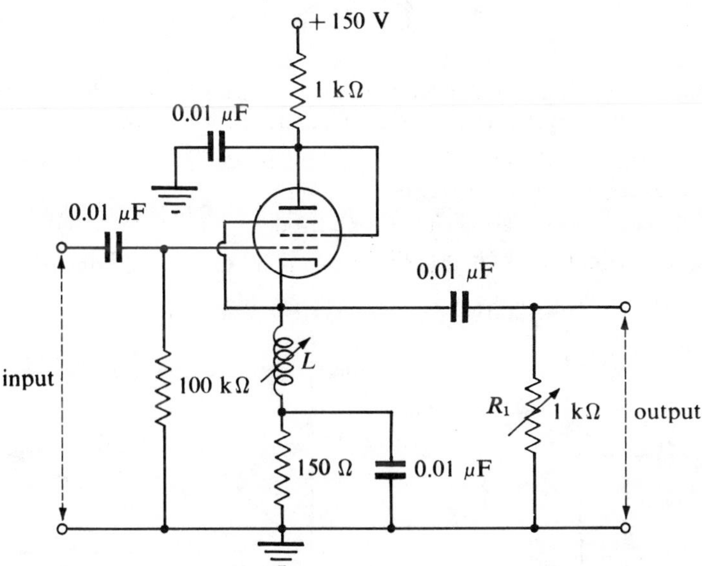

FIGURE 8.15 Practical pentode vacuum tube cathode follower.

Professor Robert E. Houston, Jr., of the Physics Department at the University of New Hampshire, and were used in circuits designed to track satellites.

problems

1. Explain why the interior of a vacuum tube must be highly evacuated. Compare the mean free path of the individual gas molecules in the tube with the size of the tube.
2. (a) Sketch an appropriate filament connection for the 6AK5 tube shown using a battery. (b) Sketch an appropriate plate supply for the tube using another battery.

3. Sketch a circuit to run the filament of a 12AX7 tube off a 6V supply.

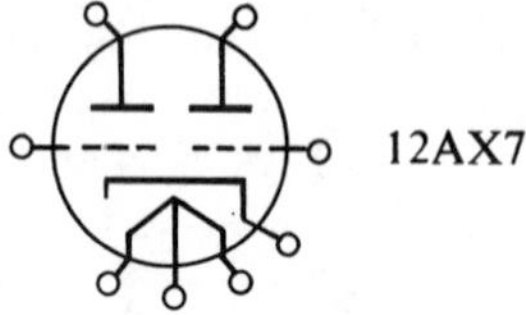

4. Compare a vacuum tube diode with a semiconductor junction diode. Include the schematic diagram symbols and the directions of current flow.
5. (a) Explain why the grid of a vacuum tube triode draws little current when the tube is biased in the usual way.
   (b) Show a typical biasing circuit for a triode.
6. Sketch the load line and find the operating point for the following triode.

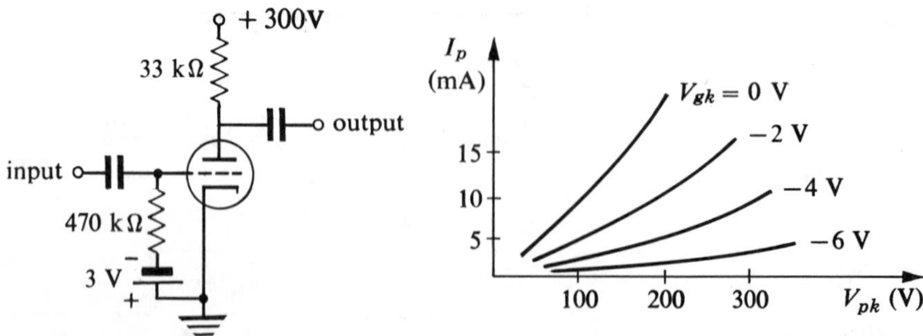

7. (a) If a triode draws 5 mA plate current and is self or cathode biased, calculate the cathode resistor if the grid-to-cathode voltage is 4 V.
   (b) Calculate the value of the cathode bypass capacitance if the triode must amplify audio frequency signals from 20 Hz to 20 kHz.
8. Explain why the cathode resistance is bypassed.
9. (a) Sketch the shape of the output waveform for a pure sinusoidal input of 4 V pk–pk amplitude if the operating point of the amplifier is at "A".

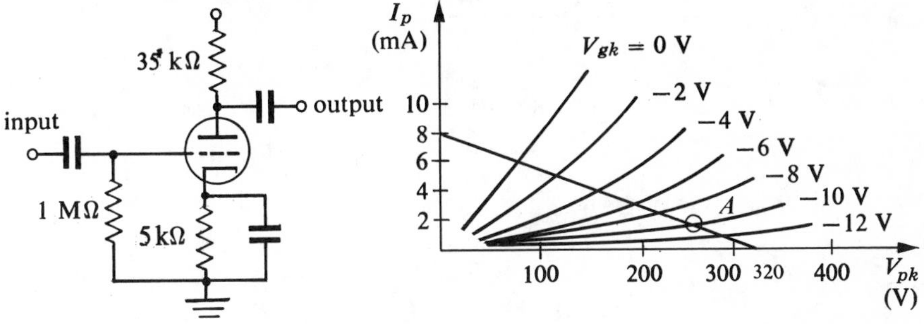

   (b) Where would be a more desirable operating point for less distortion of the output?
   (c) What changes would you make in the circuit to achieve this new operating point?
10. Using the equivalent circuit for the triode in Fig. 8.7, calculate (a) the *circuit* output impedance; (b) the *circuit* input impedance.
11. Explain the purpose of the suppressor grid in a pentode.
12. Explain the purpose of the screen–grid in a pentode, and why its presence reduces the plate-to-cathode capacitance.
13. (a) Calculate R_K, R_{SG}, C_1, C_2, C_K for the pentode amplifier shown below if the operating point is 3 mA, 200 V, the screen voltage is 100 V, and the screen current is 2 mA.

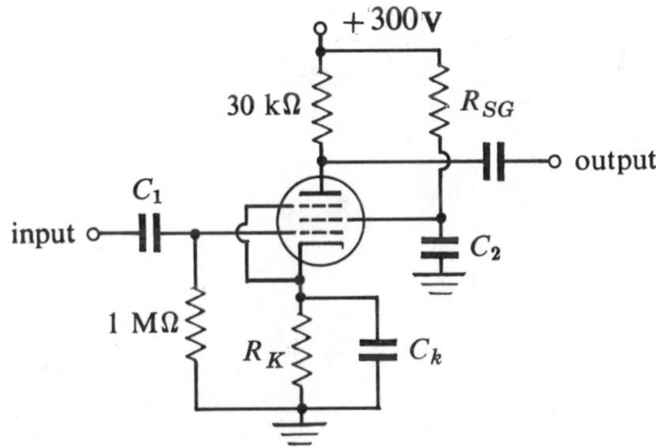

   (b) Calculate the voltage gain. (The pentode curves are given in Fig. 8.9, and the amplifier is to amplify a 100 kHz signal.)

14. Using the pentode equivalent circuit of Fig. 8.10, calculate (a) the *circuit* output impedance; (b) the *circuit* input impedance.

15. For the cathode follower circuit shown below, calculate the value of R_K and R'_K if the desired operating point is (120V, 0.6mA). The triode curves are shown in Fig. 8.5.

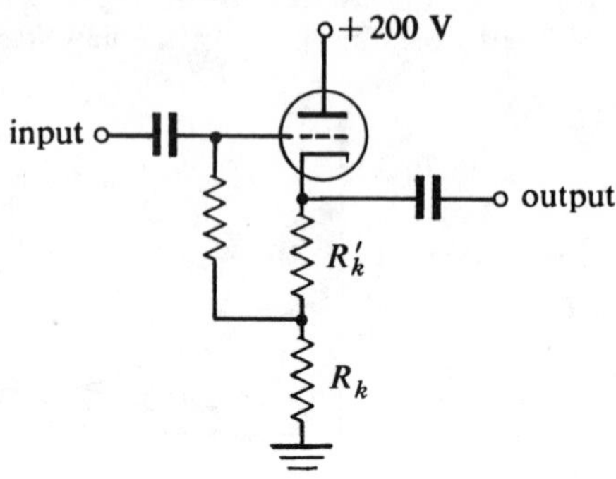

9

FEEDBACK AND PRACTICAL AMPLIFIER CIRCUITS

9.1 INTRODUCTION

In almost all practical circuits "feedback" is employed. Feedback may be defined as the taking of a portion of the output of a circuit and coupling or feeding it back into the input. If the portion of the output that is fed back is *in* phase with respect to the input, then the feedback is termed "positive feedback." If the output fed back is *out* of phase with respect to the input, then the feedback is termed "negative feedback." Positive feedback is usually used in oscillators and to increase the gain of a circuit as in super-regenerative receivers. With positive feedback, a circuit can be made to generate an output with no external input; a random noise voltage or voltage transient (when the circuit is turned on) at the input creates an output, part of which is fed back to reinforce the input, and the cycle repeats itself until some nonlinearity of the circuit results in a constant or repetitive output.

Negative feedback is used in almost all amplifiers. Its only disadvantage is that the gain of the amplifier is reduced, because the fed-back voltage subtracts from the input. The gain can usually be increased to the desired value by merely adding more stages of amplification, however. The advantages of negative feedback are numerous: increased circuit stability against almost *any* type of disturbance, (e.g., power supply or temperature fluctuation and component aging) increased input impedance, and decreased output impedance. It also provides an increased frequency bandwidth for a constant gain, and decreased distortion introduced by the circuit. Negative feedback is sometimes referred to as "bootstrapping," because the voltage fed back, usually to the emitter of the first stage, goes in the same direction as the input voltage at the base. For example, if the signal input voltage at the base becomes more positive, negative feedback will result in a positive going feedback voltage being applied to the emitter, thus decreasing the base-emitter voltage difference, which is the effective input to the amplifier. The first stage amplifier "reaches down and pulls up" its emitter voltage, like a man pulling up on his bootstraps.

9.2 NEGATIVE FEEDBACK

Let us consider negative feedback in which a fraction $\beta < 1$ of the output voltage is subtracted from the input voltage, so that the effective input to the amplifier is $v' = v - \beta v_{\text{out}}$. This type of feedback is called negative voltage feedback and is shown schematically in Fig. 9.1.

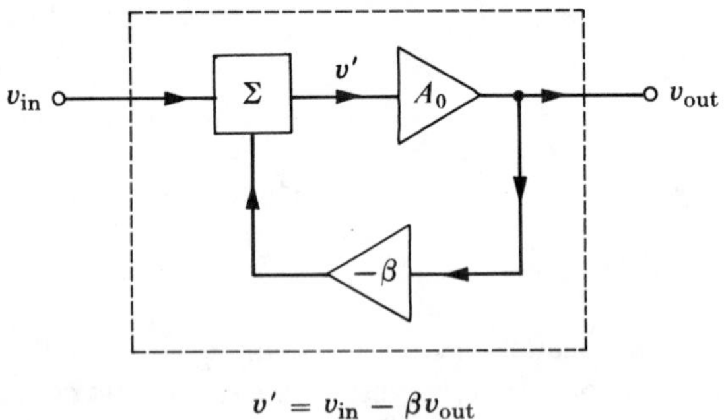

FIGURE 9.1 Generalized negative voltage feedback amplifier.

The triangle A_0 represents an amplifier with a gain A_0, that is, $A_0 = v_{\text{out}}/v'$; where v' is the input to A_0. The triangle $-\beta$ represents a circuit that is an amplifier with a gain of $-\beta$; that is, its output is $-\beta$ times its input. β is sometimes called the "feedback factor." The minus sign means the output is out of phase with respect to its input. The square $\sum$ represents a summing or addition circuit whose output $v' = v_{\text{in}} - \beta v_{\text{out}}$. The entire amplifier is the system in the dotted rectangle, and its gain with the negative voltage feedback is by definition:

$$A_f \equiv \frac{v_{\text{out}}}{v_{\text{in}}}, \tag{9.1}$$

and can be calculated from the two equations:

$$A_0 = \frac{v_{\text{out}}}{v'} \tag{9.2}$$

$$v' = v_{\text{in}} - \beta v_{\text{out}}. \tag{9.3}$$

Eliminating v' between (9.2) and (9.3) yields the following expression for the amplifier gain:

$$A_f = \frac{v_{out}}{v_{in}} = \frac{A_0}{1 + A_0\beta}. \tag{9.4}$$

As we expect, the overall gain A_f of the amplifier depends upon A_0 and β. We also see from (9.4) that the gain A_f with negative feedback is *less* than the gain A_0 without negative feedback. Usually, the fraction β of the output voltage which is fed back is much less than unity, and the gain without negative feedback A_0 is made so large that $A_0\beta \gg 1$. For example, $\beta = 0.1$ and $A_0 = 1000$ would make $A_0\beta = 100$. With the approximation $1 + A_0\beta \cong A_0\beta$, (9.4) becomes simply:

$$A_f = \frac{A_0}{1 + A_0\beta} \cong \frac{1}{\beta}. \tag{9.5}$$

We thus have an astounding result; the gain of the entire amplifier depends only upon β, the fraction of the output voltage fed back, and not at all upon the gain A_0 of the amplifier without feedback. For example, if β were 0.01 corresponding to only 1% of the output voltage being fed back to the input and if $A_0 = 10{,}000$, then $A_0\beta = (10{,}000)\,(0.01) = 100$, which is much greater than unity. Therefore the overall gain $A_f = 1/0.01 = 100$.

The beauty of this situation is that β can be made extremely constant. For example, β can be determined by the ratio of two extremely stable resistors, R_1 and R_2, with the circuit of Fig. 9.2. Thus the amplifier gain is given by

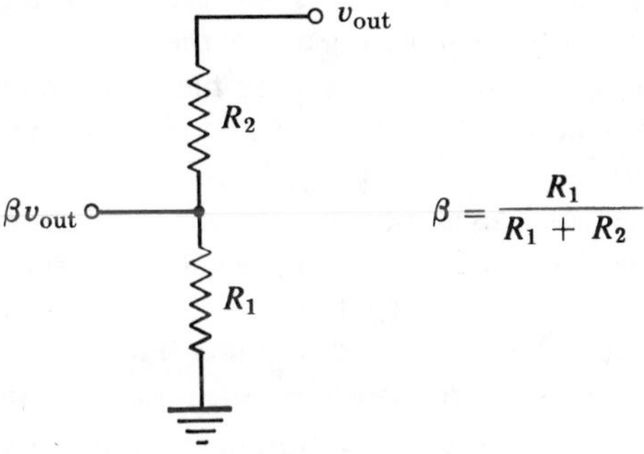

FIGURE 9.2 Resistor divider to obtain feedback voltage.

$A_f = 1/\beta = (R_1 + R_2)/R_1$. If a gain of $A_f = 100$ is desired, typical values might be $R_1 = 100\Omega$, $R_2 = 9{,}900\Omega \cong 10\,\text{k}\Omega$. Even if the gain A_0 varies widely, from 5,000 to 20,000 say, the gain of the amplifier, A_f, will remain essentially fixed at $1/\beta$. Substitution of $A_0 = 5{,}000$, 10,000, and 20,000 into the exact gain expression (9.4) shows how constant the gain A_f is. Thus, if A_0 changes by a factor of two, A_f changes by less than 1%. Notice that this argument holds regardless of *what* factor causes the change in A_0, temperature changes, power supply voltage fluctuations, aging of transistors, etc. The gain with feedback A_f re-

	A_0	$A_f = A_0/(1 + A_0\beta)$
$\beta = 0.01$ *fixed*	5,000	98.3
	10,000	99.0
	20,000	99.6

mains constant so long as $A_0\beta$ remains very large compared to one. A gain change of a factor of two in A_0 is, of course, rather unusual. A change in A_0 of, say, 10% would be more likely. If A_0 changes by 10%, from 10,000 to 11,000, A_f changes only from 99.0 to 99.1, a change of only 0.1%. A quick calculation shows that an absolute change in A_0 results in a change in A_f smaller by approximately a factor of $(1/A_0\beta)^2$; for:

$$\frac{\partial A_f}{\partial A_0} = \frac{\partial}{\partial A_0}\left(\frac{A_0}{1 + A_0\beta}\right) = \frac{(1 + A_0\beta) - A_0\beta}{(1 + A_0\beta)^2} = \frac{1}{(1 + A_0\beta)^2}$$

$$\frac{\partial A_f}{\partial A_0} \cong \frac{1}{(A_0\beta)^2}. \tag{9.6}$$

In the case just discussed, $A_0\beta = (10{,}000)(0.01) = 100$, so $\partial A_f/\partial A_0 \cong 1/(A_0\beta)^2 = 1/(100)^2 = 10^{-4}$, which is indeed small. It is easy to show a relative or percentage change in A_0 results in a relative change in A_f smaller by a factor of $1/A_0\beta$. Notice also that so long as $A_0\beta \gg 1$, the gain of the amplifier, $A_f = 1/\beta$, is completely independent of the h parameters of the individual transistors. This is an exceedingly comforting thought, because the usual transistor manuals do not contain all four h parameters; they usually contain only $h_{21} = h_{fe}$ along with the maximum power, currents and voltages.

The input impedance of an amplifier without negative feedback is *increased* by the addition of negative feedback. This effect seems intuitively reasonable if we realize that the part of the output voltage fed back to the input opposes the input voltage. Thus, the net effective input voltage applied to the amplifier is reduced. Hence we expect a smaller current to flow into the input terminals of the amplifier, which means that the amplifier input effectively presents a higher impedance to the source. We will now show that the input impedance with feedback $Z_{\text{in}f}$ is larger than the input impedance without feedback $Z_{\text{in}o}$ by a factor of $(1 + A_0\beta)$, that is:

$$Z_{\text{in}f} = (1 + A_0\beta)Z_{\text{in}o}.$$

Consider an amplifier with gain A_0 without feedback, an input impedance $Z_{\text{in}o}$, and an output impedance Z_{out}, as shown in Fig. 9.3(a). The output of the amplifier is represented by an ideal Thevenin voltage generator of magnitude $A_0 v_{\text{in}}$ in series with the output impedance Z_{out}. The input terminals have an impedance of $Z_{\text{in}o}$, and we have assumed there is negligible coupling of the out-

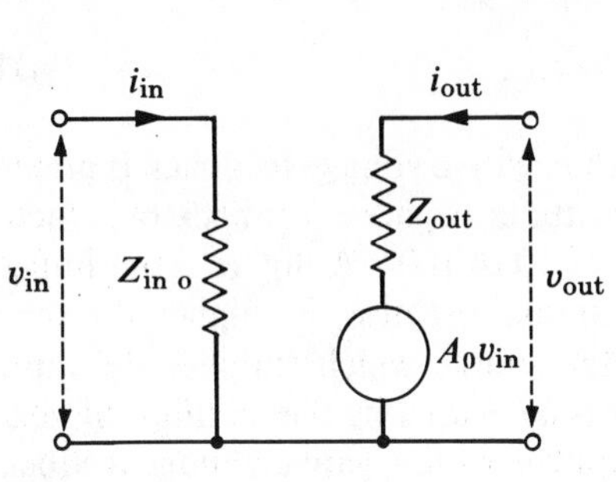

(a) *without feedback*

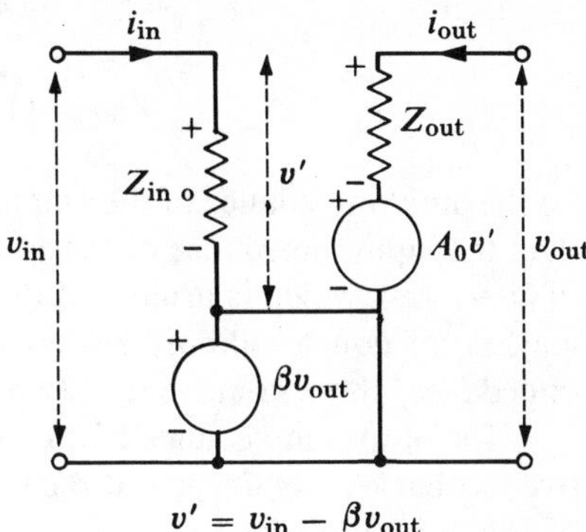

(b) *with negative voltage feedback*

FIGURE 9.3 Feedback amplifier.

put back to the input in the amplifier of Fig. 9.3(a), by assuming there is no voltage generator in series with the input terminals. In short, the amplifier of Fig. 9.3(a) is ideal.

If we now add negative voltage feedback by adding a voltage generator βv_{out} in series with the input as shown in Fig. 9.3(b), we can write the following equations for the amplifier with negative feedback:

$$v_{in} = v' + \beta v_{out} \tag{9.7}$$

$$v' = i_{in} Z_{ino} . \tag{9.8}$$

Because of the negative voltage feedback the overall amplifier gain is given by:

$$A_f = \frac{v_{out}}{v_{in}} = \frac{A_0}{1 + A_0 \beta} \cdot \tag{9.9}$$

We can find the input impedance of the amplifier with feedback from the definition $Z_{inf} = v_{in}/i_{in}$ by eliminating v' and v_{out} from equations (9.7), (9.8) and (9.9). Substituting (9.8) in (9.7) to eliminate v' gives:

$$v_{in} = i_{in} Z_{ino} + \beta v_{out} . \tag{9.10}$$

And solving (9.9) for v_{out} and substituting in (9.10) to eliminate v_{out} gives:

$$v_{in} = i_{in} Z_{ino} + \beta \frac{A_0}{1 + A_0 \beta} v_{in} .$$

Thus the input impedance is given by:

$$Z_{\text{in}f} = \frac{v_{\text{in}}}{i_{\text{in}}} = \frac{Z_{\text{in}o}}{1 - \dfrac{A_0\beta}{1 + A_0\beta}}$$

$$Z_{\text{in}f} = (1 + A_0\beta)Z_{\text{in}o}\,. \tag{9.11}$$

So the input impedance of the amplifier with negative voltage feedback is greater than the input impedance of the amplifier without negative feedback by a factor of $(1 + A_0\beta)$, which is usually of the order of 10 to 100. A higher input impedance is, of course, almost always an advantage because the higher the input impedance, the less current drawn from the source which supplies the input.

The output impedance of an amplifier is *decreased* by the addition of negative feedback. Let us look at the ideal amplifier with negative voltage feedback of Fig. 9.3(b); we have:

$$v' = v_{\text{in}} - \beta v_{\text{out}}$$

$$v_{\text{out}} = A_0 v' + i_{\text{out}} Z_{\text{out}}\,. \tag{9.12}$$

Recalling from section 6.3.4 that the output impedance is defined as the ratio of the ac output voltage to the ac output current with zero input, we have (setting $v_{\text{in}} = 0$):

$$v' = -\beta v_{\text{out}}$$

$$v_{\text{out}} = A_0 v' + i_{\text{out}} Z_{\text{out}}\,.$$

Or, eliminating v':

$$v_{\text{out}} = A_0(-\beta v_{\text{out}}) + i_{\text{out}} Z_{\text{out}}\,,$$

or

$$Z_{\text{out}f} = \frac{v_{\text{out}}}{i_{\text{out}}} = \frac{Z_{\text{out}}}{1 + A_0\beta}\,. \tag{9.13}$$

Thus, the output impedance with feedback $Z_{\text{out}f}$ is less than the output impedance without feedback by a factor $1/(1 + A_0\beta)$, which is usually from 10 to 100 in most circuits. A low output impedance is generally a significant advantage in a circuit: the voltage drop is smaller across $Z_{\text{out}f}$ internal to the circuit and a faster rise time is possible when charging an external capacitive load, i.e., the high frequency response of the circuit is improved.

Negative feedback also decreases the distortion in an amplifier. Distortion here means any change in the shape of the input signal introduced by the amplifier. A high fidelity amplifier is one in which the output is a faithful replica, only larger, of the input. A high fidelity amplifier has low distortion, and in fact all high fidelity amplifiers have large amounts of negative feedback.

We can regard any distortion of the input signal by the amplifier as an "extra" voltage generated within the amplifier. It is a general rule as we have seen that any change inside the feedback loop (inside the amplifier), for example a change in A_0, results in a very small change in the gain with feedback A_f.

So we should not be too surprised to find that any distortion voltage generated *within* the amplifier is effectively decreased by the presence of negative feedback.

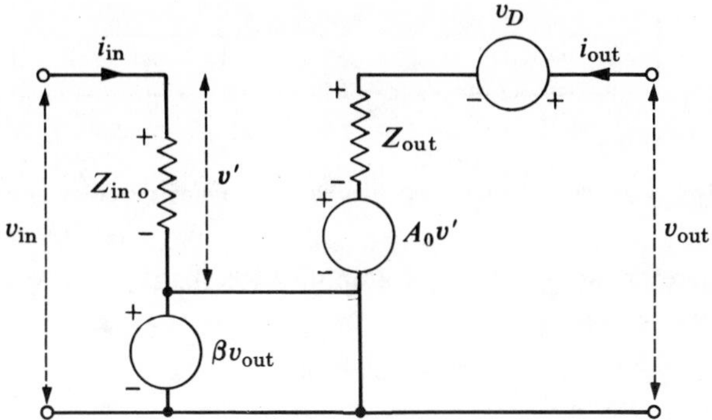

FIGURE 9.4 Negative-feedback amplifier containing distortion voltage v_D.

Suppose we represent the distortion voltage simply by an ideal voltage generator v_D in series with the output as in Fig. 9.4. The equations are:

$$\begin{aligned} v_{out} &= A_0 v' + i_{out} Z_{out} + v_D \\ v_{in} &= v' + \beta v_{out}. \end{aligned} \tag{9.14}$$

Eliminating v' gives us:

$$v_{out} = A_0(v_{in} - \beta v_{out}) + i_{out} Z_{out} + v_D.$$

Neglecting the $i_{out} Z_{out}$ term, i.e., assuming negligible loading of the output:

$$v_{out} \cong A_0(v_{in} - \beta v_{out}) + v_D$$

$$v_{out}(1 + A_0\beta) = A_0 v_{in} + v_D$$

or

$$v_{out} = \frac{A_0}{1 + A_0\beta} v_{in} + \frac{1}{1 + A_0\beta} v_D. \tag{9.15}$$

We see that the gain for the input signal v_{in} is larger by a factor of A_0 than the gain for the distortion voltage v_D.

In an actual amplifier, of course, the distortion is not generated in series with the output, but rather may be generated at several points closer to the input. To be specific, consider a two-stage amplifier, shown in Fig. 9.5. The v_{D1} represents a distortion voltage introduced at the output of the first stage whose gain is A_{01}, and v_{D2} represents a distortion voltage introduced at the output of the second stage whose gain is A_{02}. It seems clear that the distortion v_{D1} will affect the output more than v_{D2}, because v_{D1} occurs further toward the

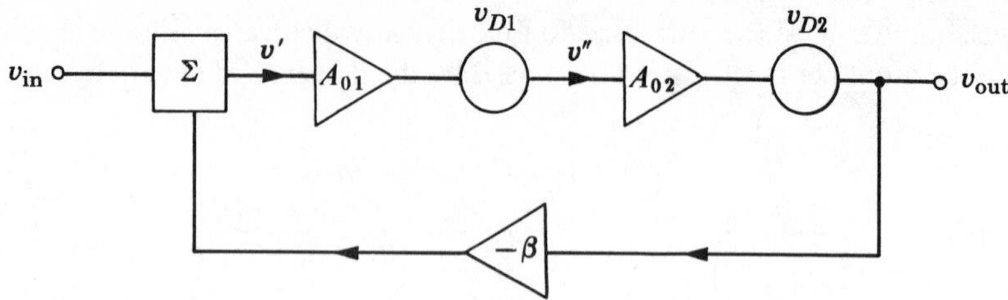

FIGURE 9.5 Two-stage, negative-feedback amplifier containing distortion at two points.

input and is therefore subject to more amplification than v_{D2}. For this two-stage amplifier, we can write:

$$v' = v_{in} - \beta v_{out} \tag{9.16}$$

$$v'' = A_{01}v' + v_{D1} \tag{9.17}$$

$$v_{out} = A_{02}v'' + v_{D2}. \tag{9.18}$$

Eliminating v' and v'', we obtain:

$$v_{out} = A_{02}A_{01}(v_{in} - \beta v_{out}) + A_{02}v_{D1} + v_{D2}$$
$$v_{out}(1 + A_{02}A_{01}\beta) = A_{02}A_{01}v_{in} + A_{02}v_{D1} + v_{D2},$$

or

$$v_{out} = \frac{A_{02}A_{01}}{1 + A_{02}A_{01}\beta}v_{in} + \frac{A_{02}}{1 + A_{02}A_{01}\beta}v_{D1} + \frac{1}{1 + A_{02}A_{01}\beta}v_{D2}. \tag{9.19}$$

The overall gain of the amplifier with negative feedback is $A_{02}A_{01}/(1 + A_{02}A_{01}\beta)$ as we would expect, but the gain applied to the distortion voltage v_{D1} is only $A_{02}/(1 + A_{02}A_{01}\beta)$ or a factor of A_{01} less. And the gain for the distortion v_{D2} is only $1/(1 + A_{02}A_{01}\beta)$ or a factor of $A_{02}A_{01}$ less. As we expected, the distortion voltage v_{D1} generated near the front end (input) of the amplifier is amplified more. In general, any unwanted voltage generated near the front end of an amplifier presents a more serious problem than those unwanted voltages introduced nearer the output, regardless of whether these voltages are a distortion of the input, random noise voltages generated by transistors, resistors, or faulty contacts, or voltages picked up from external sources. The conclusion to be drawn is that most efforts to reduce noise, pickup, and distortion should go into the *first* stage of the amplifier.

9.3 EXAMPLES OF NEGATIVE-FEEDBACK AMPLIFIER CIRCUITS

Let us now illustrate the use of negative voltage feedback by giving several actual amplifier circuits. First, consider the circuit of Fig. 9.6. The trans-

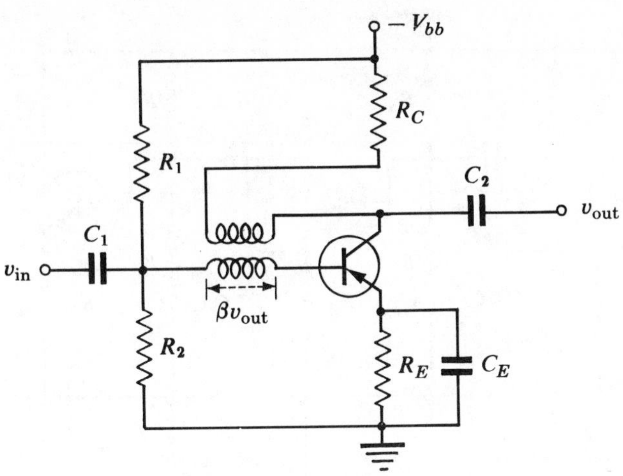

FIGURE 9.6 Transformer coupled negative-feedback amplifier.

former takes a fraction β of the output voltage and couples it in series with the input so that $v_{in} = \beta v_0 + v'$, or $v' = v_{in} - \beta v_0$, which is the necessary condition for negative feedback. Here β depends upon the turns ratio and the coupling of the transformer. However, there is one huge disadvantage of this transformer-coupled feedback; it uses a transformer. Transformers are in general relatively narrow band devices and mechanically bulky at low frequencies where large numbers of turns and iron cores are required. They are also mechanically fragile at higher frequencies where a small number of turns will suffice. Also at higher frequencies the unavoidable stray capacitances of the various wires will introduce undesirable LC resonant circuits. Notice also that if the leads to either the primary or the secondary are interchanged, the phase of the feedback voltage is changed by 180°, and we go from negative to positive feedback, which in general will produce an oscillator rather than an amplifier. This topic is discussed further in section 9.5.

The common collector or emitter follower amplifier has a great deal of negative feedback; the entire output voltage across the emitter resistor is fed back and subtracted from the input, i.e., $\beta = 1$. Thus we get a voltage gain of $1/\beta \cong 1$, and higher input and lower output impedances.

Another example of a negative-feedback amplifier is a conventional common emitter amplifier with the emitter resistor un-bypassed; the voltage fed back equals $i_E R_E$.

A more practical amplifier circuit with negative voltage feedback is shown in Fig. 9.7. It consists of two common emitter amplifiers in series to produce gain plus a final common collector stage to give a low output impedance. In this circuit, the feedback network consists of the two resistors R_1 and R_2, and $\beta \cong R_1/(R_1 + R_2)$. Hence if the condition $A_0\beta \gg 1$ is satisfied, the overall gain of the amplifier will be:

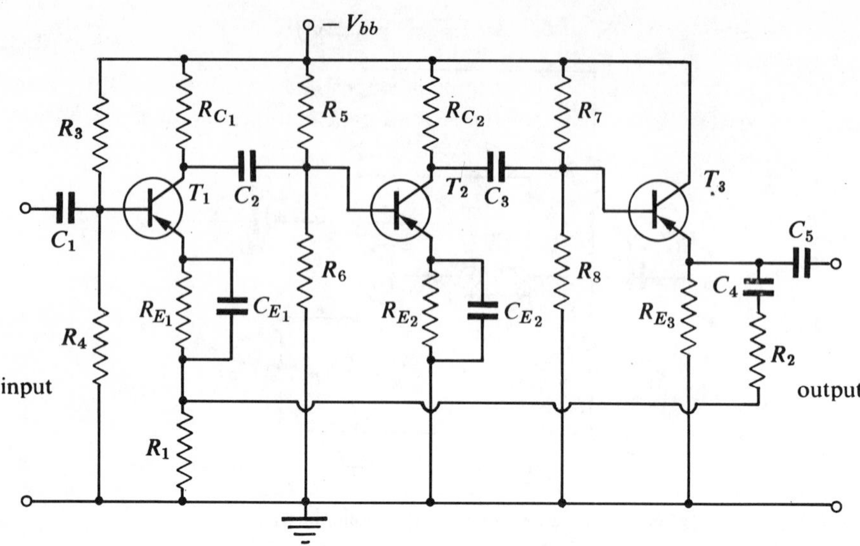

FIGURE 9.7 Multi-stage, negative-feedback amplifier circuit.

$$\frac{v_{\text{out}}}{v_{\text{in}}} = A_f = \frac{1}{\beta} = \frac{R_1 + R_2}{R_1}. \tag{9.20}$$

$R_1 + R_2$ should be large enough (i.e., much larger than the output impedance of T_3) so as not to load the output of T_3 too much, for example, $R_1 + R_2 > 1\,\text{k}\Omega$. One can reasonably expect a voltage gain of about 50 with no feedback from each of the common emitter stages and of 1 from the common collector stage, so the gain A_0 without feedback will be approximately:

$$A_0 \cong A_1 A_2 A_3 = 50 \cdot 50 \cdot 1 = 2500.$$

Therefore, $A_0\beta = 2500\beta$ and the desired condition $1 + A_0\beta \gg 1$ can be satisfied for β down to around $1/50$. Thus a gain of up to $A_f = 50$ can be expected from this amplifier with negative feedback. For $A_f = 1/\beta = (R_1 + R_2)/R_1 = 50$, we might choose $R_1 = 50\Omega$ and $R_2 = 2.5\,\text{k}\Omega$ for example. If a voltage gain of only 20 is desired, then we could choose $R_1 = 50\,\Omega$ and $R_2 = 1\,\text{k}\Omega$ to make $A_f = 1/\beta = (R_1 + R_2)/R_1 = 20$.

At this point the reader who has actually constructed a single-stage common emitter amplifier may recall that one can get a voltage gain of up to 250 with careful design rather than only 50 as just mentioned. Thus, one might expect a gain A_0 without feedback for two such common emitter stages to be $A_0 = (250)(250) = 62{,}500$, which would enable us to make β much smaller and the overall gain with feedback much larger. For example, if $A_0 = 62{,}500$, β could easily be equal to $1/1000$, which would make $1 + A_0\beta \gg 1$ and $A_f = 1/\beta = 1000$.

The reason one can not get $A_0 = 62{,}500$ from the circuit of Fig. 9.7 is that one necessarily throws away a great deal of voltage gain due to the impedance mismatch between the high output impedance of the first common emitter stage and the low input impedance of the second common emitter stage. Consider two ideal amplifiers connected in series as shown in Fig. 9.8. The input voltage

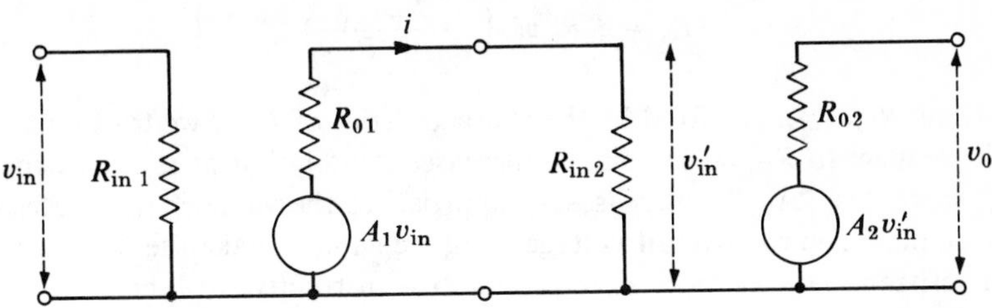

FIGURE 9.8 Two ideal voltage amplifiers connected in series.

v'_{in} to the second stage is always less than $A_1 v_{in}$ because it is split between $R_{out\,1}$, the output resistance of the first stage, and $R_{in\,2}$, the input resistance of the second stage. We have:

$$A_1 v_{in} = i(R_{01} + R_{in\,2}) \qquad \textbf{(9.21)}$$
$$v'_{in} = iR_{in\,2}.$$

Thus,

$$v'_{in} = A_1 v_{in}\left(\frac{R_{in\,2}}{R_{01} + R_{in\,2}}\right).$$

Only the fraction $R_{in\,2}/(R_{01} + R_{in\,2})$ of the maximum output $A_1 v_{in}$ of the first stage is available as the input to the second stage, and this fraction approaches unity only if $R_{in\,2} \gg R_{01}$. For a common emitter amplifier, R_{01} is the parallel combination of the collector resistor R_{C_1} (see Fig. 9.7) with the output impedance of the transistor, which is typically of the order of 50 kΩ. Because the collector resistor R_{C_1} is usually considerably less than 50 kΩ, R_{01} is essentially equal to R_{C_1}, the collector resistor, which is usually from 5 kΩ to 10 kΩ. The input impedance of the common emitter second stage, $R_{in\,2}$, is the parallel combination of the bias divider resistors R_5 and R_6 with the input impedance of the transistor T_2, which is approximately equal to $h_{ie} \cong 1\,\text{k}\Omega$ if $I_C \cong 1\,\text{mA}$. Usually $R_1 \| R_2 \gg 1\,\text{k}\Omega$, so $R_{in\,2}$ is usually around $h_{ie} \cong 1\,\text{k}\Omega$. Thus the fraction of the output of stage #1 passed on to stage #2 is:

$$\frac{R_{in\,2}}{R_{01} + R_{in\,2}} \cong \frac{1\,\text{k}\Omega}{5\,\text{k}\Omega + 1\,\text{k}\Omega} = \frac{1}{6} \qquad \text{if } R_{C_1} = 5\,\text{k}\Omega.$$

This fraction can be increased by either lowering R_{01} or raising $R_{in\,2}$. However, R_{01} can be lowered only by decreasing the collector resistance R_{C_1}, which will also lower the gain of the first stage. And $R_{\text{in}\,2} \cong h_{ie}$ can not be changed without reducing the collector current of transistor T_2, which usually reduces the gain of T_2. Thus, in practice we are faced with a fixed $R_{\text{in}\,2}$; the only variable is R_{01}. Assuming the voltage gain of the first stage is proportional to R_{C_1}, $A_1 = KR_{C_1}$, then the voltage input v'_{in} to the second stage is:

$$v'_{\text{in}} = KR_{C_1}v_{\text{in}}\left(\frac{R_{\text{in}\,2}}{R_{C_1} + R_{\text{in}\,2}}\right). \tag{9.22}$$

Regarding $R_{\text{in}\,2} \cong h_{ie}$ as fixed by the characteristics of T_2, if we try to maximize v'_{in} with respect to R_{C_1} we see that v'_{in} increases steadily but at a decreasing rate as R_{C_1} increases. As R_{C_1} increases, v'_{in} approaches the value $v'_{\text{in max}} = Kv_{\text{in}}R_{\text{in}\,2}$. Thus to maximize the overall voltage gain, we should make the first stage collector resistance R_{C_1} as *large* as possible. It is instructive to tabulate the value of $R_{C_1}R_{\text{in}\,2}/(R_{C_1} + R_{\text{in}\,2})$ as a function of increasing R_{C_1}/R_{in}. The moral we

$R_{C_1}/R_{\text{in}\,2}$	$R_{C_1}R_{\text{in}\,2}/(R_{C_1} + R_{\text{in}\,2})$
0.1	$R_{\text{in}}/11$
0.5	$R_{\text{in}}/3$
1.0	$R_{\text{in}}/2$
2.0	$\frac{2}{3}R_{\text{in}}$
5.0	$\frac{5}{6}R_{\text{in}}$
10.0	$\frac{10}{11}R_{\text{in}}$

draw is that the first stage collector R_{C_1} should be at least 5 or 10 times larger than the input impedance of the second stage for maximum voltage gain without feedback. Increasing R_{C_1} beyond $10R_{\text{in}}$ will increase the voltage gain by only about 10%. Thus, for $R_{\text{in}} \cong h_{ie} \cong 1\,\text{k}\Omega$, the first stage collector resistance should be $5\,\text{k}\Omega$ or $10\,\text{k}\Omega$. However, R_{C_1} cannot be increased without limit, because for a fixed supply voltage, increasing R_{C_1} means reducing the collector current I_{C_1}. And if I_{C_1} falls below approximately 1 mA, the gain of T_1 will decrease significantly.

An actual amplifier circuit using two *RC* coupled common emitter amplifiers is shown in Fig. 9.9. This amplifier was designed to operate at a fixed frequency of 100 kHz so no attempt was made to extend the low frequency response by using large coupling capacitors. The amplifier was intended to operate under laboratory conditions where the temperature remains relatively constant, so a large amount of negative feedback was not considered necessary. At 100 kHz the gain without feedback is $A_0 = 1300$; and with $\beta = R_1/(R_1 + R_2) = 220\,\Omega/27{,}220\,\Omega = .00813$ we have $A_0\beta = 10.6$. Thus the approximation $A_0\beta \gg 1$ is good to about 10%. The expected gain is

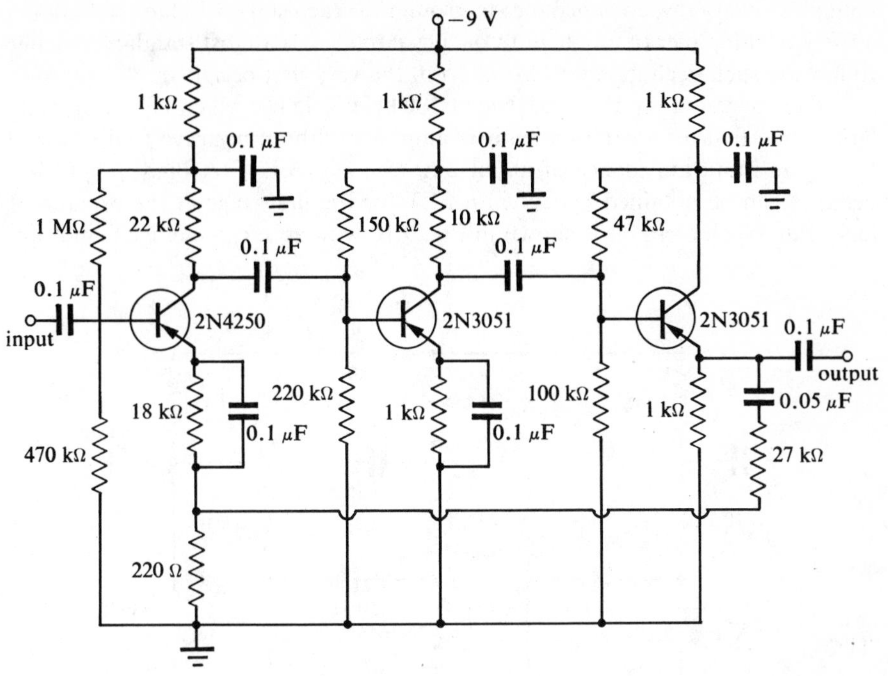

Measured Circuit Performance

voltage gain	100
input impedance	25 kΩ
output impedance	10 Ω
upper 3 dB frequency	1 MHz

FIGURE 9.9 Multi-stage, negative-feedback amplifier.

$A_f = A_0/(1 + A_0\beta) = 1300/[1 + 1300(.00813)] = 112$, which is in good agreement with the experimentally measured value of 100. If stability were more important, then it would be appropriate to use $\beta = 0.1$, thus making $A_0\beta = 130$; because one gets a perfectly stable gain $A_f = 1/\beta$ only in the limit as $A_0\beta \gg 1$. The lower gain $A_f = 1/\beta = 10$ would mean that two such stages would then be necessary to achieve a gain of 100. The first stage transistor is a 2N4250 which is specially designed to have a high gain while operating at low collector currents of about 100 μA. Such a low collector current generally means less transistor noise, but with most transistors, the gain (h_{fe}) falls off rapidly with collector currents much below 1 mA. The subject of noise will be treated at greater length in Chapter 11.

Notice the presence of the 1 kΩ − 0.1 μF *RC* decoupling network in the collector circuit of each transistor. Such networks were found necessary to eliminate a high frequency oscillation which was caused by unwanted positive feedback through the power supply leads. They essentially provide a pure dc voltage at a very low ac impedance to ground for each stage. In fact, in building a feedback amplifier or any high frequency circuit, it is almost standard practice to include such decoupling networks from the very first design.

The measured input impedance at 100 kHz is 25 kΩ, which is considerably larger than that for a common emitter amplifier without negative feedback and having an input impedance of about 2 kΩ (see Fig. 6.13). A higher input impedance can be obtained by using an FET for the first stage at the expense of less gain. Such a circuit is shown in Fig. 9.10 with an n channel FET and npn

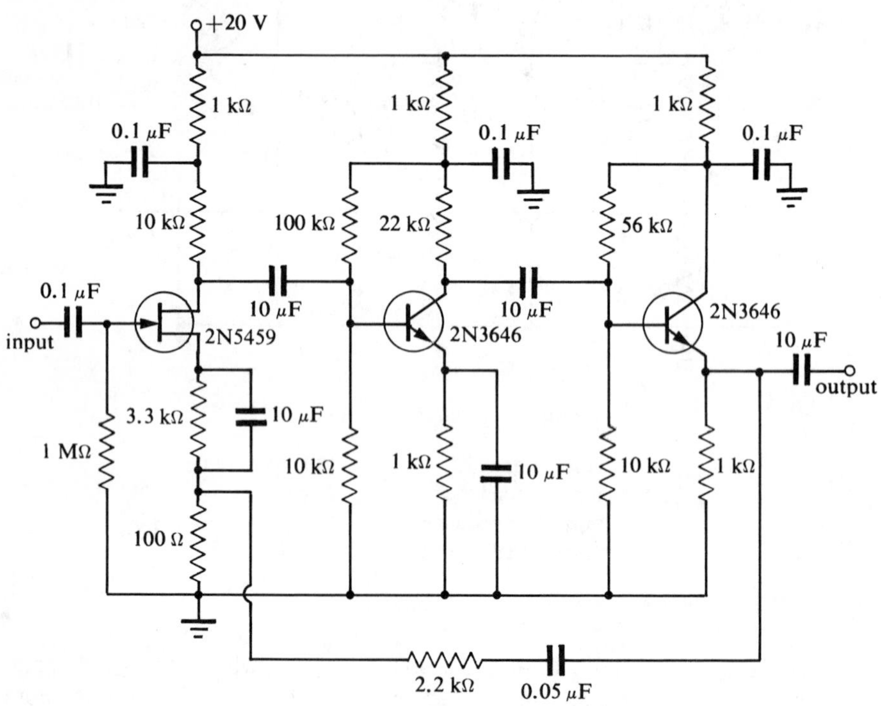

Measured Circuit Performance

voltage gain	21
input impedance	1 MΩ
output impedance	7 Ω
upper 3 dB frequency	4.8 MHz

FIGURE 9.10 Negative-feedback amplifier with FET input stage.

transistors. The negative feedback comes from the 2.2 kΩ resistor from the output to the un-bypassed 100Ω resistor in the source of the FET. Because less gain A_0 is available without feedback, due to the FET gain being less than that for the 2N4250, it is harder to satisfy the condition $A_0\beta \gg 1$ with a small value of β. Thus, the overall gain $A_f = 1/\beta$ must be smaller than in the circuit of Fig. 9.11 with the 2N4250 bipolar junction transistor input stage. But, the input impedance is much higher because of the FET.

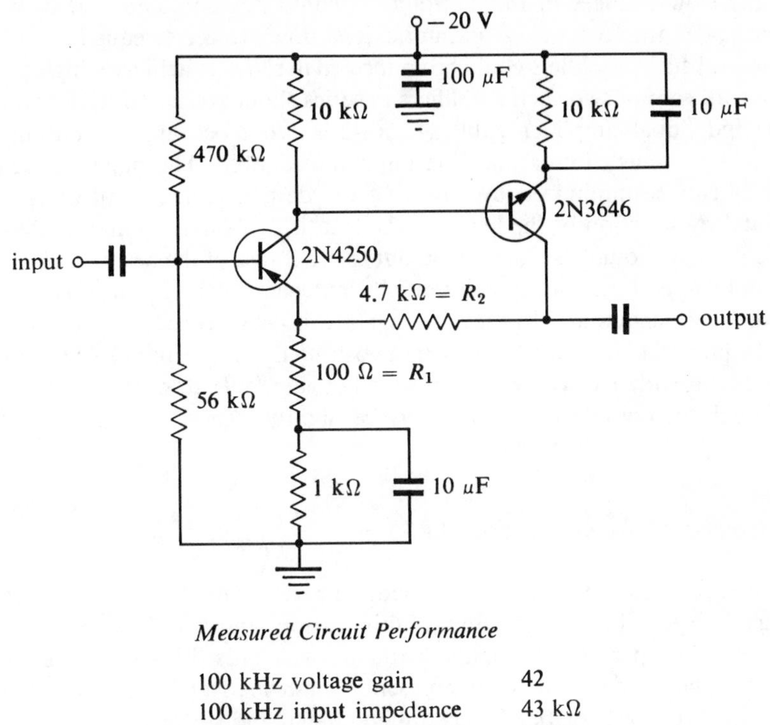

Measured Circuit Performance

100 kHz voltage gain	42
100 kHz input impedance	43 kΩ
100 kHz output impedance	200 Ω
upper 3 dB frequency	>10 MHz

FIGURE 9.11 Negative-feedback amplifier.

It should also be noted that the bandwidth is much larger for the lower gain amplifier of Fig. 9.10; in fact, the bandwidth increased by almost exactly the factor by which the gain decreased. This illustrates a general rule that the product of gain times bandwidth for a given amplifier configuration tends to remain constant. One can have a low gain and a high bandwidth or vice versa, but not both unless one radically redesigns the amplifier or cascades stages, etc. The gain–bandwidth product is treated in detail in Chapter 10.

Another amplifier with negative feedback and direct coupling between a pnp and a npn transistor is shown in Fig. 9.11.

The theoretical gain with feedback is $1/\beta = (R_1 + R_2)/R_1 = 4.8\,\text{k}\Omega/100\,\Omega = 48$, which is rather close to the experimental gain 42. The gain without feedback is 800 at 100 kHz, so $A_0\beta = (800)(1/42) = 19$. This is greater than unity, thus producing a high degree of stability. It also should be pointed out that because of the dc coupling between the two stages, any extremely slow change in the dc voltages will produce a feedback signal which will tend to minimize these dc changes. In other words, the negative feedback extends right down to dc (zero frequency). If there were series coupling capacitors between any two stages, then any slow changes in the dc voltages would not be compensated by the feedback; i.e., the feedback would not extend down to zero frequency.

Several such amplifiers could be connected in series to achieve a higher gain. The overall gain of two such amplifiers in series should be equal to the product of the individual amplifier gains, $42 \times 42 = 1764$, because the output impedance is so much lower than the input impedance. The noise in such an amplifier can be minimized by adjusting the dc collector current of the first stage and by selection of the lowest noise transistor from a number of 2N4250 transistors. It should also be pointed out that because of the negative feedback, the amplifier performance is substantially independent of the transistor type, provided the maximum electrical ratings are not exceeded. For example, a 2N3638 pnp silicon transistor can be substituted for the pnp 2N4250 in the above circuit with essentially no change in the gain, output, or input impedance—although the noise level will probably be slightly higher.

9.4 OPERATIONAL AMPLIFIERS

Negative-feedback amplifiers can be made to perform mathematical operations such as differentiation, integration, multiplication, and addition. Such amplifiers are called operational amplifiers and are widely used in analog computers where a signal voltage is made to represent a mathematical function which must be operated on mathematically, i.e., differentiated and so forth.

Operational amplifiers use a type of negative feedback somewhat different from that we have just treated in section 9.3. An operational amplifier has the feedback voltage added in *parallel* with the input signal rather than in series. The general diagram of an operational amplifier is shown in Fig. 9.12. The amplifier with gain $-A_0$ is assumed to be dc coupled, and its output is 180° out of phase with respect to its input as indicated by the minus sign in the gain. A_0 is usually referred to as the "open loop" gain, i.e., the gain with the negative-feedback loop open. The feedback voltage is coupled from the output to the input through the feedback resistor R_f. We will now calculate the overall voltage gain for the operational amplifier. The Kirchhoff current law applied to point S (the "summing point") implies:

$$i_{\text{in}} = i_f + i'. \tag{9.23}$$

Using Ohm's law, we can rewrite (9.23) in terms of voltages and resistances:

$$\frac{v_{\text{in}} - v'}{R} = \frac{v' - v_{\text{out}}}{R_f} + \frac{v'}{Z_{\text{in}}}. \tag{9.24}$$

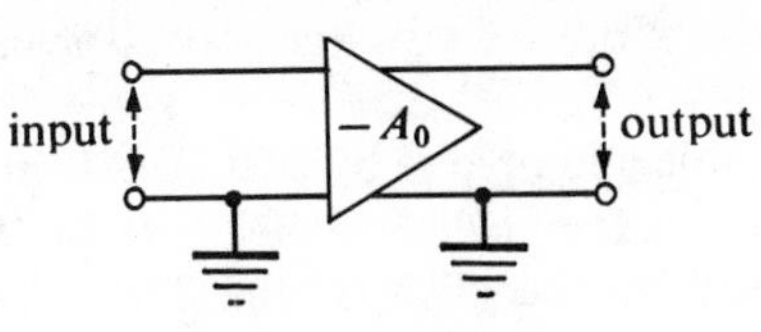

(a) *amplifier with no feedback*

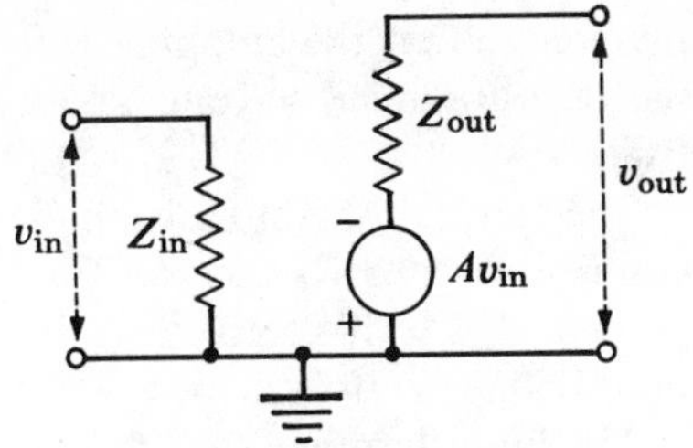

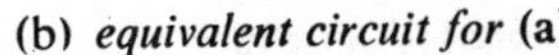

(b) *equivalent circuit for* (a)

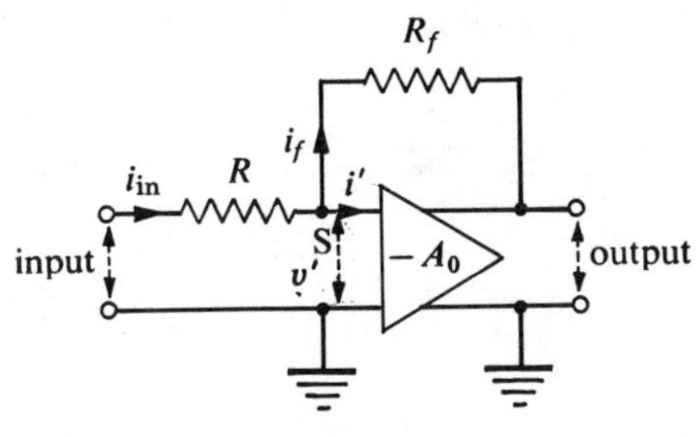

(c) *amplifier with external feedback network of R and R_f*

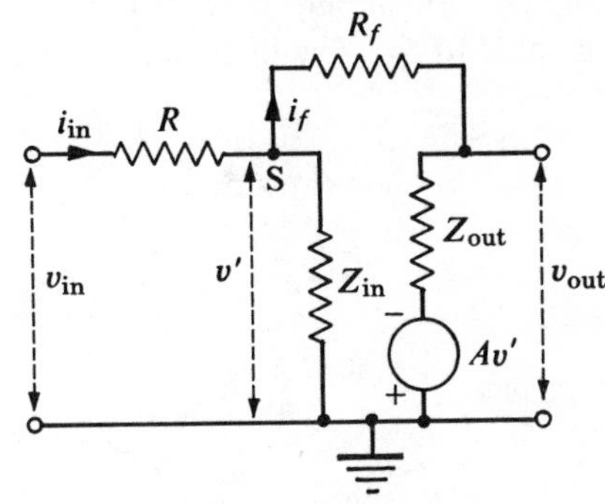

(d) *equivalent circuit for* (c)

FIGURE 9.12 General operational amplifier.

If we assume the input impedance of the amplifier Z_{in} is very large, we may neglect the $i' = v'/Z_{in}$ term. Thus,

$$\frac{v_{in} - v'}{R} \cong \frac{v' - v_{out}}{R_f}. \tag{9.25}$$

The gain of the amplifier implies:

$$v_{out} = -A_0 v'. \tag{9.26}$$

Eliminating v' between equations (9.25) and (9.26) and solving for the amplifier voltage gain yields:

$$\text{Voltage gain} = \frac{v_{out}}{v_{in}} \cong -\frac{A_0 R_f}{A_0 R + R_f}. \tag{9.27}$$

If we assume the amplifier gain is very high, so that $A_0 R \gg R_f$, then the voltage gain reduces to:

$$A_v = \frac{v_{out}}{v_{in}} \cong -\frac{R_f}{R}. \tag{9.28}$$

Thus, we see that the amplifier gain is determined only by the ratio of the two resistances R_f and R so long as our two assumptions $i' \ll i_f$ and $A_0R \gg R_f$ are satisfied.

The gain can therefore be made very stable and precise by using stable precision resistors for R_f and R. For example, if $A_0 = 10^4$, $R = 10\,\text{k}\Omega$, $R_f = 100\,\text{k}\Omega$; then $A_0R = 10^4 \times 10\,\text{k}\Omega = 10^5\,\text{k}\Omega$, which is much larger than R_f. The overall voltage gain is then $A_v = R_f/R = -100\,\text{k}\Omega/10\,\text{k}\Omega = -10$.

The input terminal "S" to the amplifier A in Fig. 9.12(c) is often referred to as a "virtual" ground. This means that this point is effectively a ground and immediately implies that the input impedance to the entire operational amplifier is equal to R. The impedance of point S to ground is:

$$Z_S = \frac{v'}{i_{\text{in}}} = \frac{v'}{i_f} \cong \frac{v'}{\dfrac{v' - v_0}{R_f}} = \frac{v'}{\dfrac{v' + A_0 v'}{R_f}}$$

or

$$Z_S = \frac{R_f}{1 + A_0}. \tag{9.29}$$

This impedance is usually rather small. For example, using the numerical values $R_f = 100\,\text{k}\Omega$, $A = 10^4$, then $Z_S = 10\,\Omega$. Physically this results from the negative feedback almost cancelling out the signal voltage at S, thus making S a "virtual" ground. Notice that this result holds even for dc because the feedback through R_f extends down to dc. Point S is not exactly at ground potential, and the actual potential at S is termed the "offset voltage," and is of the order of millivolts. The offset voltage can be reduced to less than one millivolt by a special feedback network or a trim potentiometer.

An operational amplifier can be used to perform the operations of addition, subtraction, multiplication, differentiation, and integration. Multiplication can be performed by merely using the circuit of Fig. 9.12, the multiplicative factor being the voltage gain $A_v = -R_f/R$.

Addition of two or more input signals can be performed with the circuit of Fig. 9.13. Each of the three inputs sees an input impedance of R (not $R/3$) because the summing point S is essentially a ground. Because S is a virtual ground, there is no interaction among the various inputs. A different gain can be utilized by changing the resistors in series with the inputs. If R_1 is in series with v_{in_1}, R_2 in series with v_{in_2}, R_3 in series with v_{in_3}, then:

$$v_{\text{out}} = -\left[\frac{R_f}{R_1} v_{\text{in}_1} + \frac{R_f}{R_2} v_{\text{in}_2} + \frac{R_f}{R_S} v_{\text{in}_3}\right]. \tag{9.30}$$

The derivation is simple:

$$i_1 + i_2 + i_3 = i_{\text{in}} = i' + i_f.$$

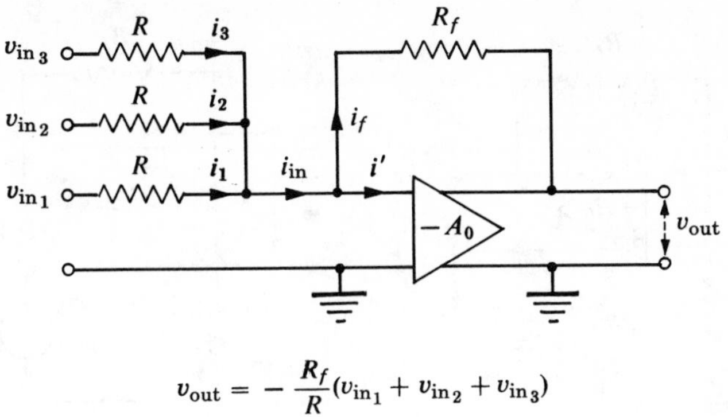

$$v_{out} = -\frac{R_f}{R}(v_{in_1} + v_{in_2} + v_{in_3})$$

FIGURE 9.13 Operational amplifier used to add.

Again we neglect i' compared to i_f if the input impedance is very high. Thus,

$$i_1 + i_2 + i_3 \cong i_f$$

or

$$\frac{v_{in_1}}{R_1} + \frac{v_{in_2}}{R_2} + \frac{v_{in_3}}{R_3} \cong \frac{v' - v_{out}}{R_f}.$$

And we say $v' \ll v_0$ because S is a virtual ground. Then,

$$\frac{v_{in_1}}{R_1} + \frac{v_{in_2}}{R_2} + \frac{v_{in_3}}{R_3} \cong \frac{-v_{out}}{R_f}.$$

Or,

$$v_{out} = -\frac{R_f}{R_1}v_{in_1} - \frac{R_f}{R_2}v_{in_2} - \frac{R_f}{R_3}v_{in_3}.$$

This circuit can clearly be extended to four or more different inputs.

If neither input terminal is grounded as is shown in Fig. 9.14, more flexibility is obtained. Such an input is called a "differential" input, as compared to a simple ended input at terminal 1 when terminal 2 is grounded. The amplifier output is given by $v_{out} = -A_0(v_1 - v_2)$. Input terminal 1 is called the inverting input; input terminal 2 is called the noninverting input. In particular, a noninverting amplifier can be made with the addition of two feedback resistors R_f and R_1 as shown in Fig. 9.14. The input $v_{in} = v_2$ is applied to terminal 2, and the overall amplifier gain is:

$$A_v = +\frac{R_1 + R_f}{R_1}. \tag{9.31}$$

The derivation uses the familiar assumption that negligible current i' flows into terminal 1 or 2 because of the high input impedance. Thus,

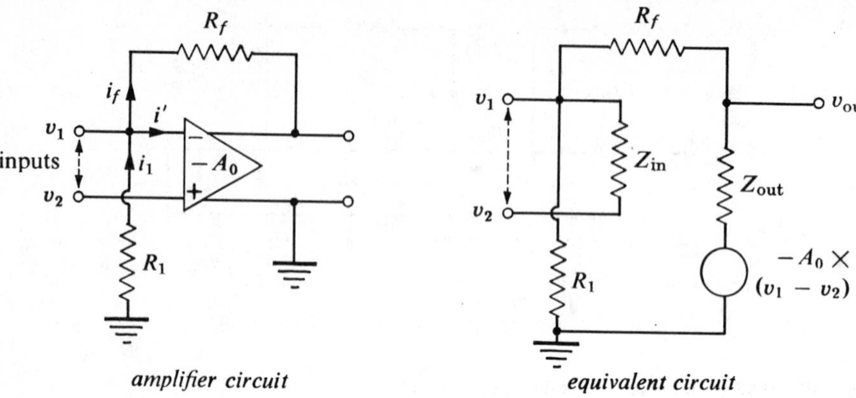

FIGURE 9.14 Non-inverting operational amplifier.

$$i_1 = i_f + i' \cong i_f$$

$$\frac{-v_1}{R_1} \cong \frac{v_i - v_{out}}{R_f}. \tag{9.32}$$

We also have, assuming negligible voltage drop across Z_{out}:

$$v_{out} = -A_0(v_1 - v_2). \tag{9.33}$$

Eliminating v_1 between (9.32) and (9.33) and solving for the overall amplifier voltage gain yields:

$$A_v = \frac{v_{out}}{v_2} = \frac{A_0(R_1 + R_f)}{(A_0 + 1)R_1 + R_f} \cong +\frac{R_1 + R_f}{R_1}.$$

Notice the plus sign in the gain; the output v_{out} is *in phase* with respect to the input v_2. The voltage gain can be made equal to unity if $R_1 \gg R_f$, or if $R_f \to 0$ and $R_1 \to \infty$. The latter case is more practical, and the resulting circuit has a gain of $A_0/(1 + A_0) \cong 1$ and is often referred to as a "voltage follower." See Fig. 9.15.

An operational amplifier can be made to differentiate by using a capacitor and resistor feedback network as shown in Fig. 9.16. Using the fact that pin S is a virtual ground, we have:

$$v_{in} = v_C + v' \cong v_C. \tag{9.34}$$

Setting $i' \cong 0$ gives us:

$$i_{in} = i_f + i' \cong i_f. \tag{9.35}$$

Using $v_C = Q/C$ where Q is the charge on the capacitor, C, in equation (9.34) gives us:

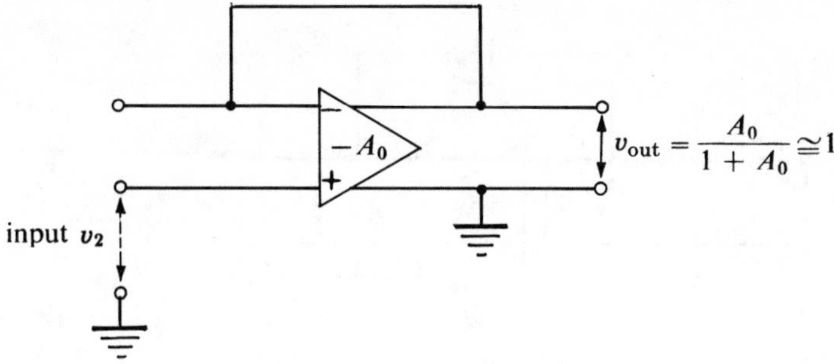

FIGURE 9.15 Operational amplifier voltage follower.

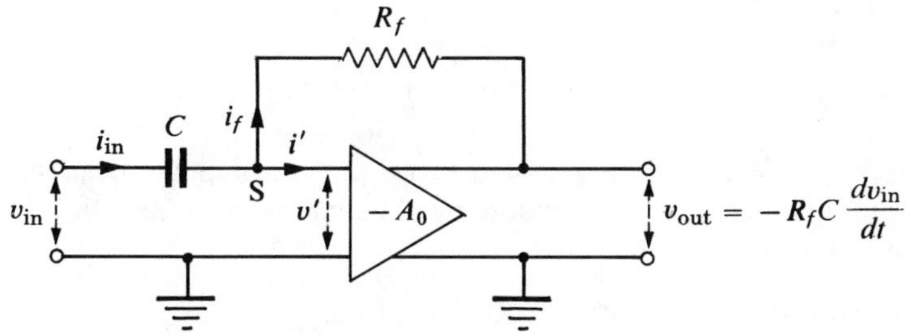

FIGURE 9.16 Operational amplifier differentiator.

$$v_{in} = \frac{Q}{C}. \tag{9.36}$$

Using $i_f = (v' - v_{out})/R_f \cong -v_{out}/R_f$ and $i_{in} = dQ/dt$ in equation (9.35) gives us:

$$i_{in} = \frac{dQ}{dt} = \frac{-v_{out}}{R_f}. \tag{9.37}$$

Eliminating Q between (9.36) and (9.37) quickly yields the result that the output voltage is proportional to the time derivative of the input voltage.

$$v_{out} = -R_f C \frac{dv_{in}}{dt}. \tag{9.38}$$

An operational amplifier can be made to integrate the input signal by the circuit of Fig. 9.17. Using the same approximations $i' \cong 0$, $v' \cong 0$, we can write:

$$v_{in} \cong i_{in} R, \qquad i_{in} \cong i_f = -\frac{dQ}{dt}, \qquad \text{and} \qquad Q \cong C v_{out}, \tag{9.39}$$

where Q is the charge on the capacitor C. Eliminating Q and i_{in} yields the

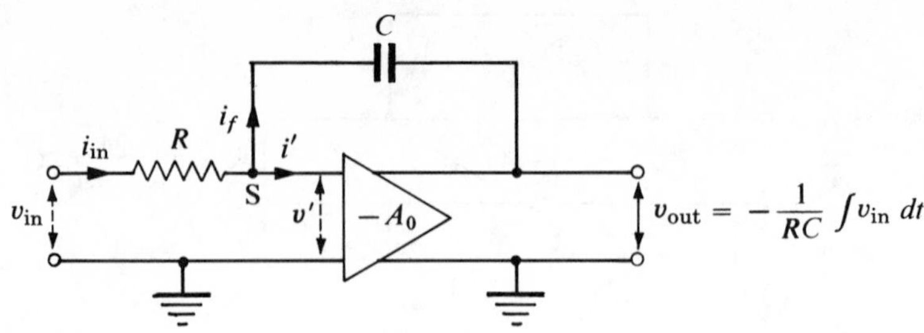

FIGURE 9.17 Operational amplifier integrator.

result that the output voltage is proportional to the time integral of the input voltage.

$$v_{out} = -\frac{1}{RC}\int v_{in}\,dt. \tag{9.40}$$

A recently developed, low-cost, general-purpose bipolar transistor operational amplifier (model 118) made by Analog Devices has the following specifications:

Open Loop Gain $A_0 = 2.5 \times 10^5$.
Gain decreases @ 20 dB/decade.
Unity Gain at 1.5 MHz.
Maximum Input Offset Voltage at 25°C, ± 5 mV (*adjustable to zero with trim pot*).
Maximum Input Noise from 10 Hz *to* 10 kHz, 2 μV rms.
Input Impedance 1 MΩ *for differential input,*
1000 MΩ *for common mode.*
Power Supply Required: ±15 V @ 4 mA.
Cost (1–9 units) $11 *each.*

Operational amplifiers are often made with FET's as the input stage to achieve a higher input impedance. One such integrated circuit amplifier is the model AD-540K made by Analog Devices. Its specifications are:

Open Loop Gain $A_0 = 5 \times 10^4$.
Unity Gain at 1 MHz.
Maximum Input Offset Voltage at 25°C, 20 mV.
Maximum Slewing Rate 6 V/μsec.
Input Bias Current 25 pA *max.*
Input Impedance 10^{10} Ω *in parallel with* 2.0 pF *differential mode.*
10^{11} Ω *in parallel with* 2.0 pF *common mode.*
Power Supply Required: ±15 V @ 7 mA.
Cost (1–9 units) $8.95 *each.*

An operational amplifier can be used to measure magnetic flux densities or magnetic fields by using it as an integrator in conjunction with a search coil.

The theory is briefly as follows [see J. E. Gordon and M. Javier Marin, *Amer. J. Phys. 38*, 94 (1970)]. From Faraday's law, a coil of N turns in a changing magnetic flux ϕ has a voltage induced in it given by $v(t) = -N\,d\phi/dt$. The ϕ represents the magnetic flux passing through the coil. If A is the area of the coil, the magnetic flux density B is, by definition, $B = \phi/A$. If the coil is suddenly removed from a field B_1 to a region where the field is B_2, then the integrated voltage induced in the coil is proportional to the change in flux ϕ, or to the change in flux density B.

$$v(t)\,dt = -N\,d\phi$$

$$\int_{t_1}^{t_2} v(t)\,dt = -\int_{\phi_1}^{\phi_2} N\,d\phi = -N(\phi_2 - \phi_1). \tag{9.41}$$

Using $\phi_1 = B_1A$ and $\phi_2 = B_2A$, we have:

$$\int_{t_1}^{t_2} v(t)\,dt = -NA(B_2 - B_1). \tag{9.42}$$

Notice that the speed with which the coil is moved from a region of field B_2 to a region of field B_1 has no effect on the *integral* of the voltage; the integral depends only upon the net change in the field, $B_2 - B_1$. If the coil is quickly moved from B_2 to B_1, then a relatively high voltage will be induced in the coil for a short period of time $(t_2 - t_1)$; whereas if the coil is slowly moved, a smaller voltage will be induced over a longer period of time. In either case, the time integral of the voltage will be the same: $NA(B_2 - B_1)$. It is easy to measure the area and the number of turns of the coil; so if the field B_2 is known, then the field B_1 can be determined from a measurement of the time integral of the voltage induced in the coil.

A "flip coil" is a special case of a search coil. With a flip coil, the coil is first carefully aligned perpendicular to the magnetic field B_1 and then flipped through 180° while remaining in the same spot in the field. The flux change through the coil is then exactly twice the initial flux $\int N\,d\phi = \int Nd(BA) = NA \int dB = 2NAB_1$. The integral of the induced voltage in the flip coil is then:

$$\int_{t_1}^{t_2} v(t)\,dt = -2NAB_1. \tag{9.43}$$

As we have just seen, an operational amplifier with an RC feedback network produces an output which is the time integral of its input voltage. Hence, if as in Fig. 9.18 we hook up the search coil to the input of such an integrating operational amplifier, we will be able to measure the change in the magnetic field in the search coil. We recall that the amplifier output is:

$$v_{\text{out}} = -\frac{1}{RC}\int v_{\text{in}}\,dt.$$

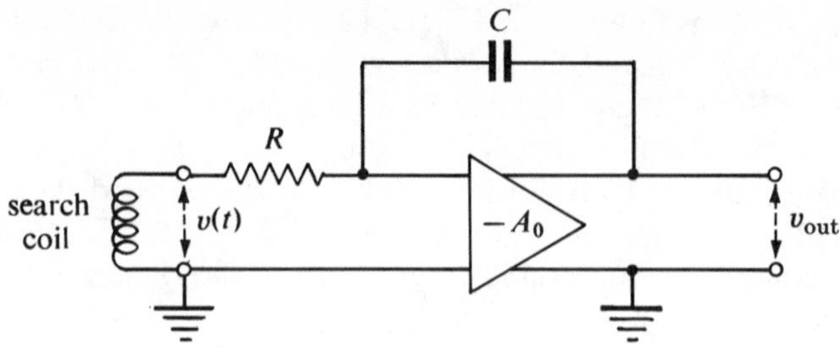

FIGURE 9.18 Integrator used to measure magnetic field.

Hence,

$$v_{out} = -\frac{1}{RC}[-NA(B_2 - B_1)] = \frac{NA}{RC}(B_2 - B_1). \tag{9.44}$$

The measurement is taken by inserting the coil in the field to be measured, shorting the capacitor C, waiting for any output transient voltages to die away, unshorting the capacitor, and then removing the coil from the field B_1 to the field B_2. Then the change in field is related to the dc output voltage V_0 by:

$$(B_2 - B_1) = \frac{RC}{NA} v_{out}. \tag{9.45}$$

Usually, a high gain stable dc amplifier is used to measure the dc output voltage v_0.

For a 300 turn coil of outer diameter 1.65 cm, inner diameter 1.05 cm, $R = 1\,\text{M}\Omega$, $C = 0.02\,\mu\text{F}$, and $B_1 = 2500$ gauss; $B_2 = 0.3$ gauss (earth's field), and $v_{out} = 0.520\,\text{V}$, which is relatively easy to measure.

Errors of the order of $\pm 5\%$ can be obtained with modern inexpensive operational amplifiers. As might be expected, the higher the open loop gain of the amplifier, the less the error. An exact analysis shows that the larger RC is, the smaller the error. However, the larger RC, the smaller the dc output voltage to be measured, so one can not increase RC without limit.

The general use of operational amplifiers in analog computers will be illustrated by showing how they can be used to solve the following differential equation:

$$\frac{d^2x}{dt^2} + a\frac{dx}{dt} + bx = f(t). \tag{9.46}$$

This is probably the most common second-order differential equation in all of science. Examples can easily be drawn from electrical and mechanical areas. The most familiar electrical example is the driven LRC circuit shown in Fig. 9.19(a). Another common mechanical example is shown in Fig. 9.19(b): a mass is attached to an ideal Hooke's law spring, subjected to a viscous damping force proportional to the velocity $F = b\,dx/dt$, and driven by an external force $f(t)$.

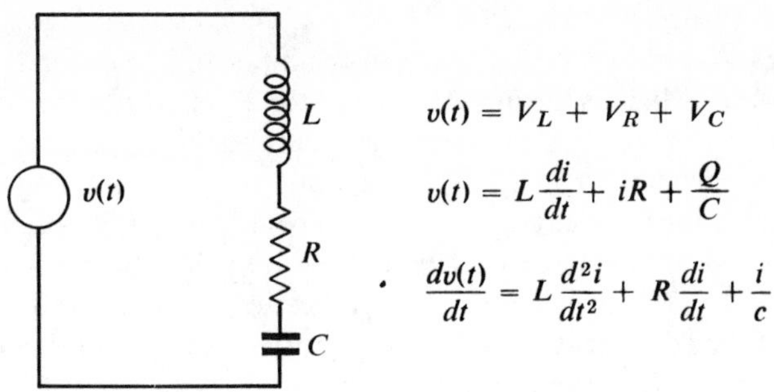

L
v(t)
R
C
$v(t) = V_L + V_R + V_C$
$v(t) = L\frac{di}{dt} + iR + \frac{Q}{C}$
$\frac{dv(t)}{dt} = L\frac{d^2i}{dt^2} + R\frac{di}{dt} + \frac{i}{c}$

(a) *LRC circuit and differential equation*

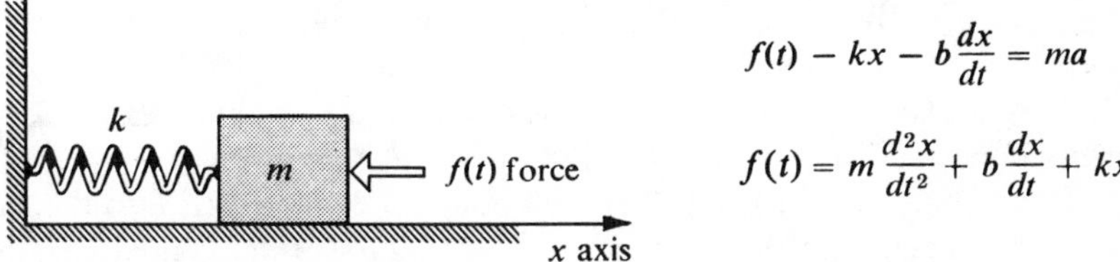

(b) *damped harmonic oscillator and differential equation*

FIGURE 9.19 Examples of second-order differential equations.

We assume that we know the driving "force," $f(t)$, the constants a and b, and the initial values ("boundary conditions") of x and dx/dt. We then are trying to find the function $x(t)$ which satisfies equation (9.46). It is easier to build integrators rather than differentiators, so the circuit starts with d^2x/dt^2 and integrates it twice rather than starting with x and differentiating it twice. Figure 9.20 shows the analog computer circuit to solve equation (9.46). Three types of operational amplifiers are used in the computer: an integrator, a summer or

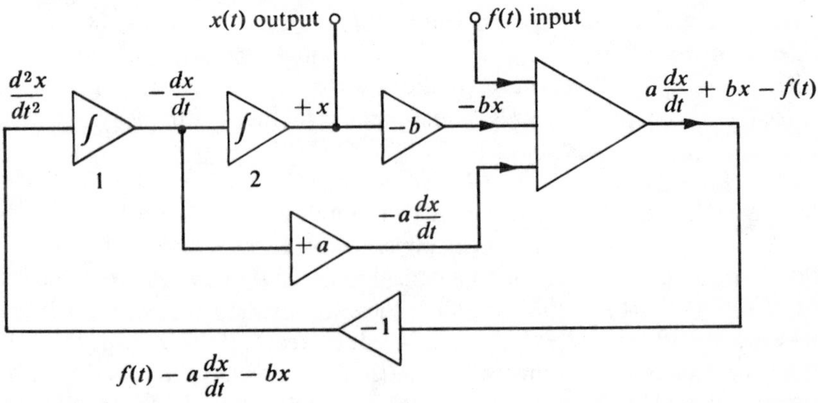

FIGURE 9.20 Analog computer circuit to solve $d^2x/dt^2 + a\,dx/dt + bx = f(t)$.

adder, and a multiplier. The circuit works as follows. Voltages are proportional to $x(t)$. Assume $d^2x(t)/dt^2$ exists at the input to integrator 1. Its output will be $-dx(t)/dt$, which is the input to integrator 2; and the output of integrator 2 will be $+x(t)$. (Remember an integrating operational amplifier also multiplies the input by -1.) The voltage $x(t)$ is then multiplied by $-b$ and fed into the summer. The voltage $-dx(t)/dt$ is taken from the output of integrator 1, multiplied by $+a$ to give $+a\,dx(t)/dt$, and fed into the summer. The summer has three inputs: $f(t)$, which must be synthesized, $-bx(t)$, and $-a\,dx(t)/dt$. The output of the summer is then $-f(t) + ax(t) + bx(t)$. This is multiplied by minus one, and fed into the input of integrator 1, since it must equal d^2x/dt^2 from the differential equation which can be written $d^2x(t)/dt^2 = f(t) - a\,dx(t)/dt - bx(t)$.

To run the problem we must supply three input voltages or functions: V_0, the value of dx/dt at $t = 0$; x_0, the value of x at $t = 0$; and $f(t)$. V_0 and x_0 are applied to integrators 1 and 2, respectively, with the integrator feedback networks disconnected. This corresponds to the initial conditions at $t = 0$. Then at $t = 0$, the integrator feedback loops are connected, $f(t)$ is fed into the summer, and the voltages are allowed to vary according to the differential equation. The solution $x(t)$ can be read directly at the output of integrator 2 as the voltage output there as a function of time. The function $-dx(t)/dt$ can also be read off the output of integrator 1 if desired, and $d^2x(t)/dt^2$ off the input to integrator 1. The parameters a, b, $f(t)$, V_0, and x_0 can easily be changed to create a different equation of the same form. Often, in practice, a and b are adjusted to achieve a desired solution $x(t)$.

9.5 POSITIVE FEEDBACK

For the negative-feedback amplifier of Fig. 9.21(a), the feedback voltage βv_{out} *subtracts from* the input signal v_{in}. The net input to the amplifier of gain A_0 is $v' = v_{in} - \beta v_{out}$, and the overall gain is: $A = v_{out}/v_{in} = A_0/(1 + A_0\beta)$. As β is increased from 0, the overall gain drops from A_0 and levels out at $A = 1/\beta$ in the limit as $A_0\beta \gg 1$. However, in the positive-feedback circuit of Fig. 9.21(b), the feedback voltage βv_{out} *adds to* the input signal v_{in}. The net input to the amplifier of gain A_0 is $v' = v_{in} + \beta v_{out}$, and the overall gain is $A = v_{out}/v_{in} = A_0/(1 - A_0\beta)$. Notice that as β increases from 0 in the case of positive feedback, the gain starts at A_0 and blows up to infinity when $A_0\beta = 1$.

What does an infinite gain mean physically? There are no infinities in practical electronics. What the mathematical result $A \to \infty$ as $A_0\beta \to 1$ means is that *if* the gain A_0 remains constant, then the overall gain goes to infinity. However, as the input $v' = v_{in} + \beta v_{out}$ keeps increasing, the gain A_0 falls off in all practical circuits; in other words, A_0 is the gain only for small inputs v'. Stated another way, as the net input v' increases in magnitude, the inherent nonlinearities in the amplifier become significant, and the output tends to approach

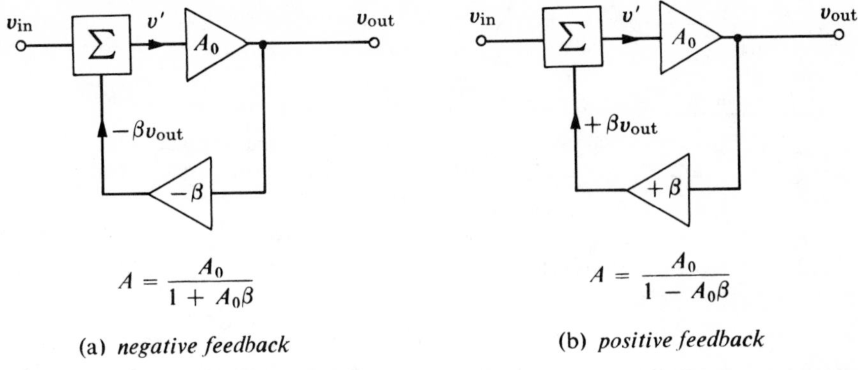

FIGURE 9.21 Feedback amplifiers.

a constant magnitude. (These nonlinearities are usually negligible in an amplifier designed for small signal inputs.) The final result is that if $A_0\beta \geqq 1$, the output of the circuit has a constant magnitude and is often quite independent of the input v_{in}. If the input v_{in} is removed, the circuit still can generate an output, i.e., we have an oscillator. Another way of looking at an infinite gain is to say that a finite output is generated with zero input.

Positive feedback is used in two ways. First, and most commonly, to make an oscillator or signal generator, i.e., to produce oscillating waveforms, sine or square waves, etc., with no input. For this purpose $A_0\beta \geqq 1$. The output waveform depends upon the type of feedback network used as will be explained in detail in a later paragraph. The second purpose is to increase the gain of an amplifier by having $A_0\beta < 1$. However, if either β or A_0 increases to the point where $A_0\beta = 1$, the amplifier will break into oscillation.

A sinusoidal oscillator of constant amplitude can be made by taking a physical system which tends naturally to oscillate when perturbed, and adding just enough energy to the system to compensate for the energy lost (by friction, radiation, etc.) in each cycle of oscillation. There are two requirements: First, the amount of energy added per cycle must exactly equal the amount lost per cycle in order to produce a constant magnitude oscillation. Secondly, this energy must be added in phase with the oscillation, i.e., at just the right time to increase rather than decrease the amplitude of the oscillation. In other words, the feedback must be positive.

Let us now consider a mechanical oscillator consisting of a simple pendulum mounted on a frictionless bearing and swinging back and forth so that the bob of mass m passes through a viscous oil bath as shown in Fig. 9.22. During each cycle as the bob passes back and forth through the oil bath, mechanical energy of oscillation is lost (as heat) due to the viscous retarding force which the oil exerts on the bob. If this energy loss is not too large, the frequency of the pendulum is given by $f = (1/2\pi)\sqrt{g/\ell}$, and the amplitude of the oscilla-

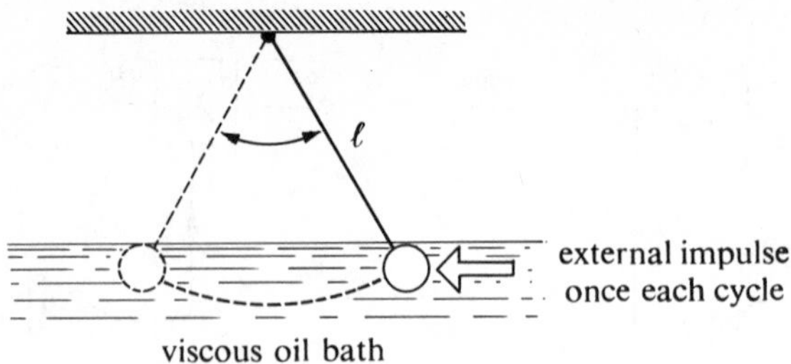

FIGURE 9.22 Mechanical oscillator.

tions will slowly fall to zero.* To keep the amplitude constant, exactly the same amount of energy must be added each cycle by some external agency as is lost to heat in the oil bath each cycle. This can be accomplished by applying an impulse every T seconds; that is, the impulse must be periodic and must have exactly the same frequency as the pendulum oscillation. In order for the impulse to *add* energy to the pendulum, it must be applied to the left when the bob is at the extreme right or when it is moving from right to left. If the impulse is applied too soon in the cycle, i.e., when the bob is moving toward the right, energy will be taken *from* the pendulum and the amplitude will decrease. In other words, the point in the oscillation cycle when the impulse is applied, or the phase, is important.

If the applied impulse supplies more energy to the pendulum each cycle than is lost to heat in the oil bath, then the amplitude of oscillation will increase. The velocity of the bob through the oil will also increase, and the energy loss as heat in the oil bath will increase until a new equilibrium situation is reached when the energy input from the external impulse each cycle exactly equals the energy loss in the oil bath each cycle. This is analogous to the electrical situation when $A_0\beta > 1$; the amplitude of the oscillations grows until nonlinearities in the amplifier decrease the gain A_0 so that $A_0\beta$ falls to unity. However, it should be emphasized that an oscillator "pushes itself" by means of the positive feedback, coupling some of the amplified output back into the input with the proper phase.

Another example of oscillation can be found in the wild snowshoe hare and Canadian lynx population in the North American Arctic. The hare population naturally oscillates with a period of about 10 years. The amplitude of the population fluctuation is roughly a factor of four from maximum to minimum years. As the hares become more numerous due to natural reproduction, they

* An exact analysis will show that the amplitude of the oscillations is given by:

$$x(t) = x_0 \sin\left\{\omega_0 t\left[1 - \left(\frac{1}{2\omega_0\tau}\right)^2\right]^{1/2}\right\} e^{-t/2\tau},$$

where τ is called the relaxation time and is defined by $\tau \equiv m/\gamma$ where γ is the viscous retarding force constant: $-\gamma\, dx/dt = F_{\text{viscous}}$. ω_0 equals the natural oscillation frequency with no friction ($\gamma = 0$).

overgraze their feed and the lynx begin to increase in number. The Canadian lynx is the principal predator for the hare. The larger number of lynx plus the reduced food available per hare reduces the hare population; and with fewer hare the lynx decrease in number, and the hare browse becomes more abundant. These factors in turn lead to an increase in the hare population, and the cycle begins again. As might be expected, the lynx population also oscillates at the same frequency, and slightly out of phase with respect to the hare population—a peak in hare population precedes a peak in the lynx population. Five- and ten-year oscillations are actually common in many forms of wildlife. These oscillations tend to be most violent in northern latitudes, because the arctic ecosystems tend to be simpler—there are fewer predators which prey on only one species. The harsh arctic environment also tends to reduce greatly the diversity of species able to survive.

If the feedback signal is precisely in phase with the input of an amplifier, oscillation will result if $A_0\beta \geqq 1$. Often the phase shift depends upon the frequency, due to *RC* or *LC* networks in the circuit. For example, in an *RC* coupled amplifier, an *RC* high-pass filter is used at the input and output; and at very low frequencies, when $\omega < 1/RC$, there is an appreciable phase shift as well as attenuation. The phase difference between the output across *R* and the input is given by:

$$\phi = \arctan\left(\frac{1}{\omega RC}\right). \tag{9.47}$$

If $\omega = 1/RC$, then $\phi = 45°$. If the *RC* network is designed to pass frequencies down to $f = 20\,\text{Hz}$, then typical values might be: $C = 1\,\mu\text{F}$, $R = 30\,\text{k}\Omega$. Then $RC = 0.3$ sec; and when $f = 5\,\text{Hz}$, $\omega = RC$ and we have a 45° phase shift. The phase drift approaches 90° as $\omega \to 0$, but the gain $\to 0$, so a *single RC* network can never produce a phase shift near 90°. Several networks, however, may produce a 180° phase shift and oscillation.

At higher frequencies, shunt capacitance becomes important in determining the gain and phase shift. Usually we have a shunt capacitance C_s between the signal lead and ground in parallel with a resistance R_s. The parallel impedance of C_s and R_s is:

$$Z = \frac{X_{C_s}R_s}{X_{C_s} + R_s} = \frac{R_s}{1 + j\omega R_s C_s}$$

$$|Z| = \frac{R_s}{\sqrt{1 + \omega^2 R_s^2 C_s^2}}. \tag{9.48}$$

Z falls to zero as ω approaches infinity. When $\omega = 1/R_sC_s$, the phase of Z changes by 45° compared to the low frequency phase. For $C_s = 10\,\text{pF}$, and $R_s = 10\,\text{k}\Omega$, $R_sC_s = 10^{-7}$ sec, and the 45° phase shift occurs at $f = 1.6\,\text{MHz}$. The phase shift approaches 90° as $\omega \to \infty$, but $Z \to 0$ as $\omega \to \infty$.

If a number of phase shifting networks exist in an amplifier, it is possible to have an accidental phase shift that produces positive feedback. If the amplifier gain at this frequency exceeds $1/\beta$, the amplifier will oscillate. It is customary to consider the amplifier gain and phase shift as a function of frequency by plotting the complex value of the loop gain $A_0\beta$ as a function of frequency in a polar plot. Such a plot is called a Nyquist diagram and is illustrated in Fig. 9.23. The magnitude of the loop gain $A_0\beta$ at a frequency ω is the distance from

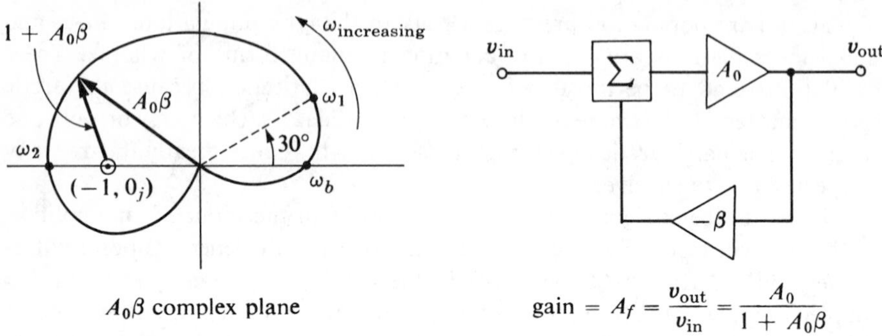

FIGURE 9.23 Nyquist diagram.

the origin out to the point on the $A_0\beta$ graph corresponding to the frequency ω. The desired phase of the amplifier for negative feedback and stable operation is $A_0\beta$ positive and purely real; this is termed the "midband" frequency ω_b and corresponds to the point of the $A_0\beta$ vector lying on the positive real axis. As the frequency differs from ω_b, both the magnitude and the phase of $A_0\beta$ change, and this is indicated by the tip of the $A_0\beta$ vector lying off the positive real axis. For example, at the frequency $\omega_1 > \omega_b$ in Fig. 9.23, the phase of $A_0\beta$ is 30° different from the phase at the midband.

If $A_0\beta = -1$, then the gain A_f with feedback $A_f = A_0/(1 + A_0\beta)$ diverges to infinity, and the system is unstable, i.e., oscillation results. In Fig. 9.23 this situation occurs at the frequency ω_2. Oscillation also results if $A_0\beta$ is negative (corresponding to positive feedback) and greater than one. It can be shown that if the complex vector drawn from the point $(-1, 0j)$ to the point $A_0\beta$, i.e., the vector $1 + A_0\beta$, does *not* enclose the point $(-1, 0j)$ then the circuit will be stable. This is the famous Nyquist criterion for stability.

Examples of stable and unstable Nyquist plots are shown in Fig. 9.24. The physical meaning of the vector $1 + A_0\beta$ enclosing the point $(-1, 0j)$ is simply that $A_0\beta$ has a magnitude greater than 1 ($A_0 > 1/\beta$), and the feedback is positive rather than negative ($\beta < 0$).

From Chapter 2, the phase shift of a network is often determined from the slope of the gain-versus-frequency plot. Thus, a clue that instability and oscillations may be encountered is often apparent from a rapidly changing gain-versus-frequency graph, especially if the gain is large at these frequencies. An example is shown in Fig. 9.25. The hump in the gain near ω_1 corresponds to the $A_0\beta$ loop gain approaching -1 and the $1 + A_0\beta$ vector almost encircling the point

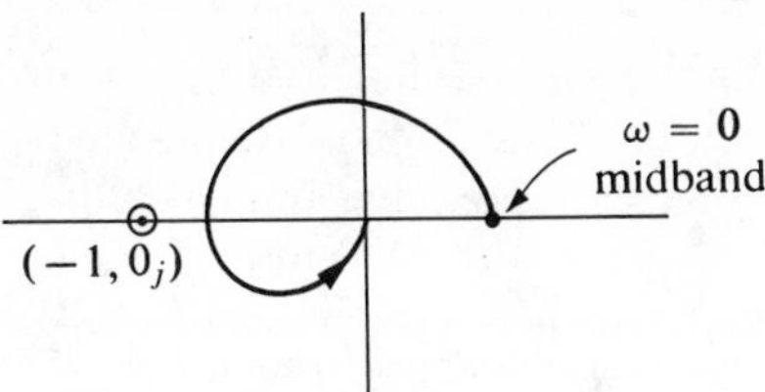

(a) *Stable dc coupled amplifier*

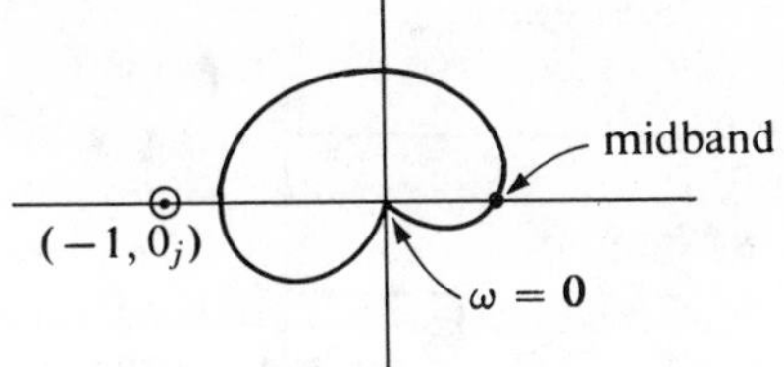
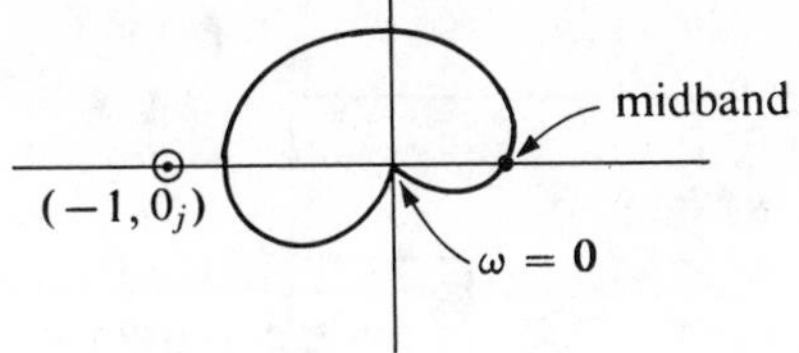

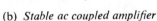

(b) *Stable ac coupled amplifier*

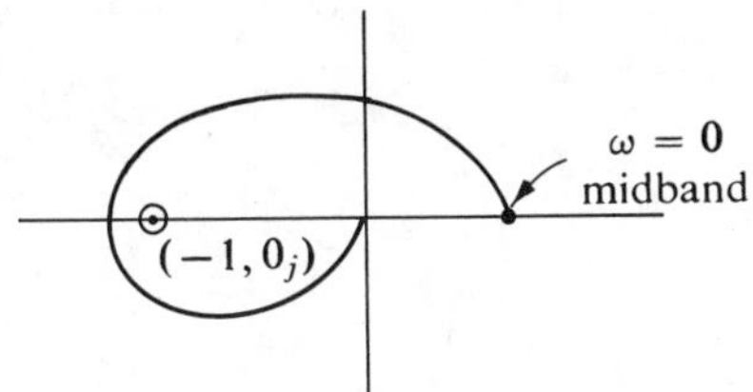

(c) *Unstable dc coupled amplifier*

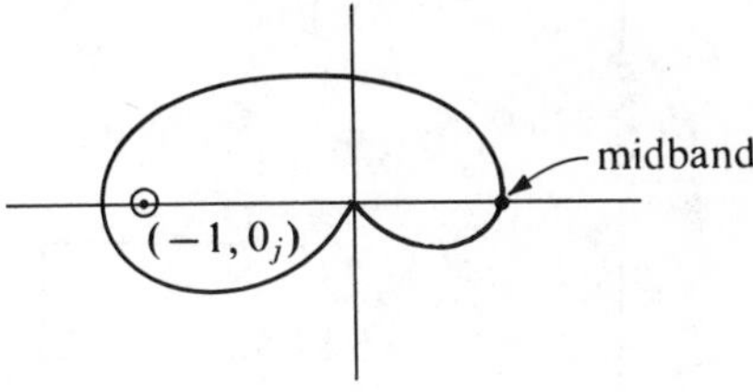

(d) *Unstable ac coupled amplifier*

FIGURE 9.24 Nyquist diagrams $A_0\beta$ complex plane.

$(-1, 0j)$. If the gain hump is high enough, the $A_0\beta$ curve will enclose the point $(-1, 0j)$, and the circuit will be unstable and will oscillate at ω_1. It is an unpleasant and common feature of such circuits that they will, in fact, always oscillate at ω_1 if $A_0\beta$ is negative and exceeds 1 at ω_1. They will not amplify signals benignly at frequencies different from ω_1. The reason is basically that there are always some Fourier components of noise voltages at the frequency ω_1; these voltages initiate the positive feedback and oscillations result because the gain is essentially infinite at ω_1.

One of the simplest electronic oscillators is one in which a transformer is used to couple the feedback from the output back to the input as is shown in Fig. 9.26. The output is a reasonably good sine wave of several volts peak-to-peak amplitude. R_1 and R_2 merely set the dc operating bias for the transistor. Both C_1 and C_E are bypass capacitors; their capacitive reactance should be

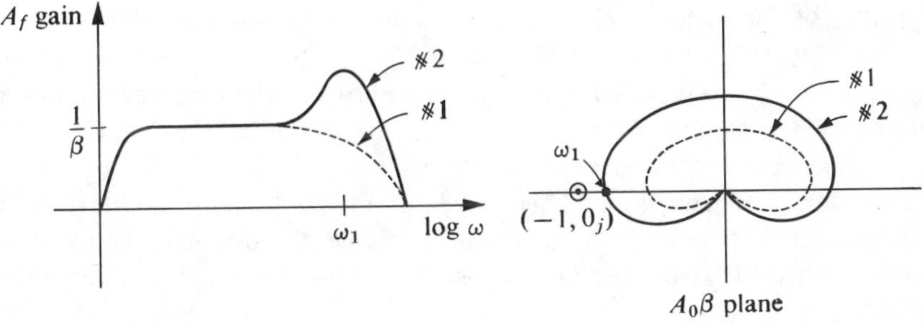

FIGURE 9.25 Almost unstable amplifier.

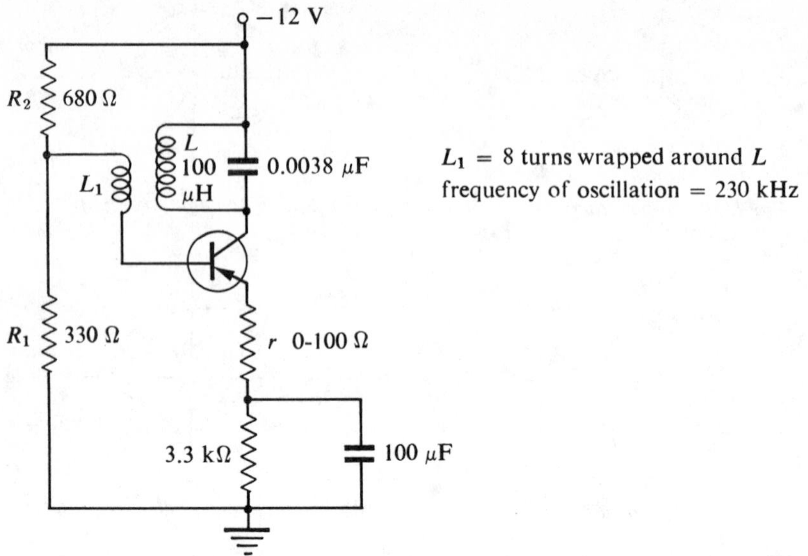

FIGURE 9.26 Transformer feedback oscillator.

small at the operating frequency. This circuit is sometimes called the tuned collector oscillator or the tuned plate oscillator when a vacuum tube is used.

The gain A_0 comes from the voltage gain inherent in a common emitter amplifier, and the feedback ratio β is obtained from the transformer coupling the base and collector. For an ideal transformer with no flux losses, β is given by the turns ratio:

$$\beta = \frac{v_{\text{feedback}}}{v_{\text{out}}} = \frac{N_1}{N}, \tag{9.49}$$

where N_1 is the number of turns in the base lead of inductance L_1 and N is the number of turns in the inductance L in the collector lead.

Notice that the parallel LC circuit in the collector means that there is a high collector impedance and consequently a high gain only at the resonant frequency of the LC tank: $\omega_0 = 1/\sqrt{LC}$. Thus, there is gain and feedback only at one frequency ω_0, and the oscillator output is therefore a sine wave of frequency $f_0 = \omega_0/2\pi = (1/2\pi)(1/\sqrt{LC})$. The purity of the sine wave output depends upon the Q of the LC tank; the higher the Q, the less the distortion of the output sine wave. A lower Q LC tank means there is gain and feedback over a wider range of frequencies, and hence the output is not a pure sine wave of only one frequency. If either L or C is made variable, the frequency of oscillation can be varied. It is usually best to use a variable capacitor C rather than a variable L.

The phase of the feedback voltage can be changed by 180° by reversing the two leads to either the primary or the secondary. So with one connection we will have positive feedback and an oscillator; reversing the leads produces negative feedback and an amplifier with minimum gain at ω_0.

The gain of the circuit can be adjusted by varying the unbypassed resistance r in the emitter lead. The larger r, the lower the gain A_0, and the smaller the output amplitude. If r is made too large so that $A_0\beta < 1$, the circuit will not oscillate.

Another oscillator circuit in which the 180° phase shift in the feedback from collector to base is easily seen is the RC phase shift oscillator shown in Fig. 9.27.

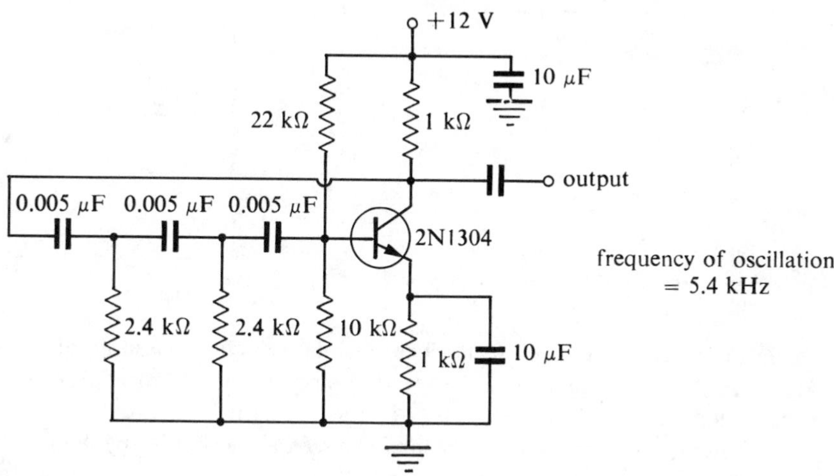

FIGURE 9.27 *RC* phase shift oscillator.

In this circuit, the 180° phase shift is obtained by three RC high-pass filters in series between the collector and the base. The resistance to ground in the third RC filter is essentially h_{ie} of the transistor, which is about 2 kΩ. Each filter contributes about 60° of phase shift and also attenuates the signal. For three identical RC filters, where the loading effects of one filter on the other are important, a voltage gain of at least 29 is required, and the frequency of oscillation is given by $f_0 = (1/2\pi)(1/RC)$. The RC phase shift oscillators are used mainly for low frequencies of the order of 10 kHz or less, because at higher frequencies appreciable phase shifts are introduced by stray capacitance and inductances.

A very useful oscillator circuit is the Colpitts circuit, shown in Fig. 9.28. Resistors R_1 and R_2 set the dc bias for the transistor along with R_E and C_E in the usual fashion. The dc emitter current is given by $I_E = V_E/R_E \cong (R_2V_{bb})/(R_1 + R_2) \times 1/R_E$. "$RFC$" stands for Radio Frequency Choke and is an inductance in the collector lead designed to have a large inductive reactance at the frequency of operation. For an oscillation frequency of 500 kHz, a 2.5 mH RFC would have an inductive reactance of $X_L = 79$ kΩ, which would provide adequate ac voltage gain. The ac voltage on the collector is coupled through

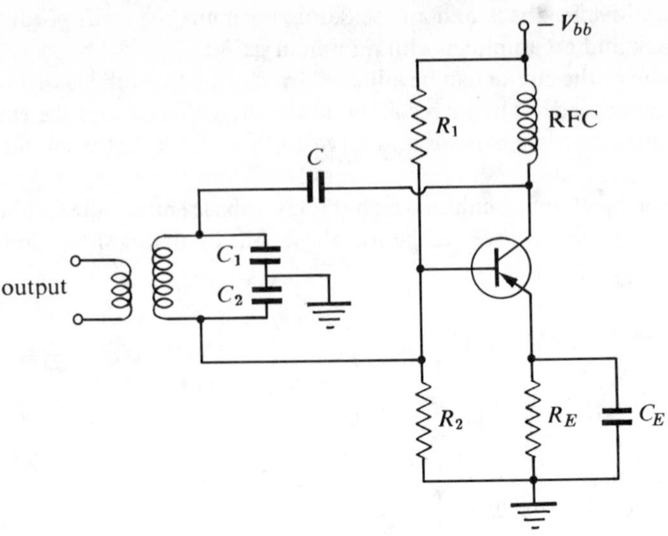

FIGURE 9.28 Colpitts oscillator.

the coupling capacitor C back to the base resonant circuit consisting of L and C_1 and C_2. Thus the reactance of C should be small at the operating frequency. For example, at $f = 500\,\text{kHz}$, $C = 0.1\,\mu\text{F}$, the capacitive reactance $X_C = 3\Omega$.

The feedback from collector to base can be seen to be positive by the following argument. The emitter is an ac ground, so we have a common emitter configuration. In this configuration, there is a 180° phase change from base to collector; that is, a positive going voltage on the base produces a negative going voltage on the collector. The base and collector are connected by the tank circuit consisting of L, $C_1 + C_2$, and there is a 180° phase change between the voltages on opposite ends of an LC tank. Hence a positive going voltage on the base produces an amplified negative going voltage on the collector, which is coupled back to the base as a positive going voltage by the L, $C_1 + C_2$ tank.

At high frequencies where air core transformers suffice, the output is usually transformer coupled from the inductance L. At lower, audio frequencies the output can be taken off the collector through a series coupling capacitor.

The frequency of oscillation is the resonant frequency of the L, $C_1 + C_2$ tank. As C_1 and C_2 are in series, their effective capacitance is $C_1C_2/(C_1 + C_2)$. Hence the resonant frequency is:

$$f_0 = \frac{1}{2\pi}\left[\sqrt{L\left(\frac{C_1C_2}{C_1 + C_2}\right)}\right]^{-1}. \tag{9.50}$$

The amount of feedback is determined by the ratio of C_1 to C_2. C_2 is connected between the base and emitter (the emitter is an ac ground because of C_E). Hence the smaller C_2 and the larger its capacitive reactance $1/\omega C_2$, the larger the feedback voltage to the base. Therefore, to change the oscillation frequency without changing the feedback ratio, the ratio C_1/C_2 should be kept constant while the effective series capacitance $C = C_1C_2/(C_1 + C_2)$ is changed.

The junction of the two capacitors, C_1 and C_2, can be thought of as a tap that takes a fraction of the ac voltage across the tank and applies it to the base as the feedback voltage. This fraction could equally well be obtained by tapping off the inductor L in the tank circuit. Such an oscillator is called a Hartley oscillator and is shown in Fig. 9.29(a).

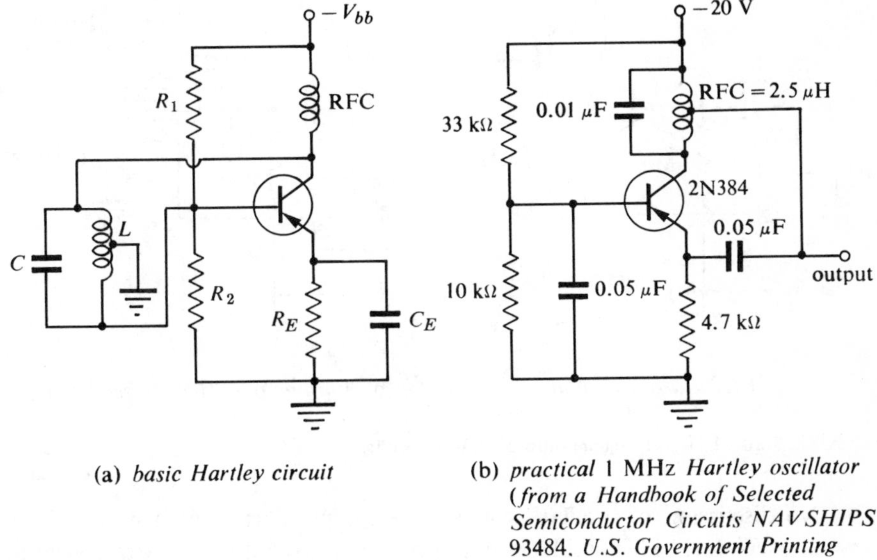

(a) *basic Hartley circuit*

(b) *practical* 1 MHz *Hartley oscillator (from a Handbook of Selected Semiconductor Circuits NAVSHIPS* 93484. *U.S. Government Printing Office*)

FIGURE 9.29 Hartley oscillator circuit.

Essentially the same analysis holds for the Hartley as for the Colpitts circuit. A practical 1 MHz Hartley circuit is shown in Fig. 9.29(b). The output is about 0.25 V and is taken off the unbypassed emitter resistor. The Colpitts tends to work better at higher frequencies where inadequate inductive coupling between the two ends of L in the Hartley circuit may be a problem. However in the Hartley circuit, it is easier to change the position of the tap to vary the feedback, and it is easier to build a variable frequency oscillator by making C a variable capacitor in the Hartley circuit.

Either transistors or vacuum tubes can be used in either the Colpitts or the Hartley oscillators. In the bipolar transistor circuits, the low base–emitter impedance (1–2 kΩ) tends to load the resonant LC tank circuit, because it is connected between the tap and one end of the tank. This problem can be eliminated by using either an FET or a vacuum tube. The FET gate–source impedance and the vacuum tank grid cathode impedance are each many orders of magnitude higher than the bipolar transistor base–emitter impedance. A single FET Hartley oscillator is shown in Fig. 9.30(a), and a vacuum tube pentode oscillator circuit is shown in Fig. 9.30(b). (Note the great similarity in the two circuits.) In Fig. 9.30(b), the screen grid acts as effectively as the plate.

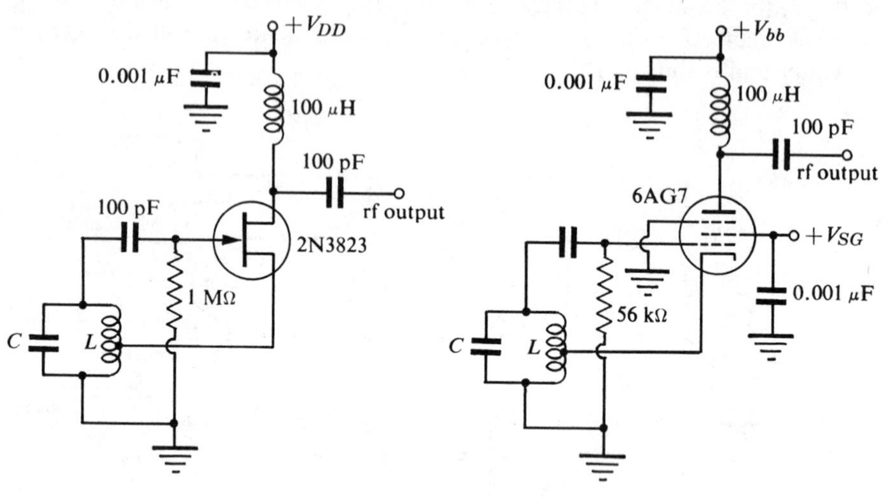

(a) FET *Hartley oscillator* (b) *Vacuum tube Hartley oscillator*

FIGURE 9.30 FET and vacuum tube oscillator circuits.

No discussion of oscillators would be complete without mention of the crystal oscillator. This oscillator uses a precisely ground thin quartz crystal to determine the oscillation frequency through the piezoelectric effect. The piezoelectric effect is the mechanical vibration of the crystal in response to an applied electric field, and the generation of an oscillating electric field as a result of its vibration. The quartz crystal is ground to the proper dimension so that one of its mechanical resonant frequencies is in the region of interest. The Q of the crystal's mechanical vibration is several orders of magnitude higher than that of a conventional LC tank whose Q is usually 100 at most. The crystal is in effect a very high Q LC resonant circuit. The crystal is mounted between two metal plates and has two terminals, one connected to each plate. Crystals can be purchased for \$5–\$10 each in frequencies from several kHz to nearly 100 MHz. Crystal oscillators find their principal application in low power, fixed-frequency circuits (tens of mW power) where frequency stability is the main consideration. Stability of the order of $\Delta f/f = 10^{-6}$ or 10^{-7} is easily obtainable with inexpensive crystals. The crystal can be used in almost any type of oscillator circuit, and greater frequency stability can be obtained by placing the crystal in a temperature-controlled oven, since temperature changes will produce slight changes in the crystal size which in turn results in small changes in the crystal frequency.

The crystal controlled oscillator of Fig. 9.31 uses the crystal in series with

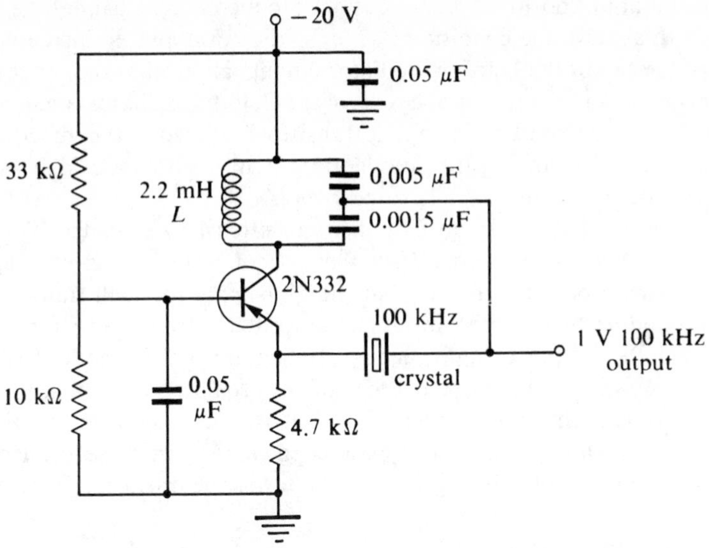

(*from a Handbook of Selected Semiconductor Circuits NAVSHIPS* 93484, *U.S. Government Printing Office*)

FIGURE 9.31 Crystal controlled 100 kHz oscillator.

the feedback to obtain feedback and oscillation at only the crystal frequency. The tuned circuit in the collector lead is merely to prevent oscillation at unwanted frequencies. The 33 kΩ and 10 kΩ resistors set the dc operating point in the usual way. The frequency stability claimed for this circuit is $\Delta f/f = 2 \times 10^{-7}$ for $\pm 10\%$ changes in the supply voltage, and $\Delta f/f = 1 \times 10^{-6}$ for a change in the output load from 100Ω to an open circuit.

Vacuum tube circuits are usually easier to use (and cheaper) when powers of several watts or more are required. However, when reliability, ruggedness, and light weight are extremely important (such as in rf communication for ambulance service, police, or fire departments, for example), solid state circuits may be used in spite of their higher cost and slightly increased circuit complexity. In military application, where weight and reliability are crucial and the cost is virtually immaterial, solid state circuits are invariably used.

9.6 PRACTICAL COMMENTS AND NEUTRALIZATION

As anyone who has ever tried to build an amplifier or oscillator knows, many things can go wrong in practice. Oscillators don't oscillate, and amplifiers tend to oscillate at strange frequencies, thereby obliterating the desired signal to be amplified. When an oscillator will not oscillate, the usual cure is to increase the amount of positive feedback and/or to increase the gain of the amplifier section to achieve the condition $A_0\beta \geqq 1$. For example, to increase β, the position of the tap in the Hartley oscillator circuit can be changed, or the ratio of the capacitances in the resonant circuit of the Colpitts oscillator can be varied. The gain can be increased by changing transistors, adjusting the impedance in the collector, or by changing the dc operating point to obtain a high value of h_{fe} or g_m for the transistor or FET, respectively.

Unwanted oscillations in an amplifier can often be eliminated by adding relatively large bypass capacitors from the V_{bb} connection to ground at each transistor. This procedure ensures that the V_{bb} supply for each transistor is a good ac ground and eliminates unwanted ac feedback from transistor to transistor through the V_{bb} power supply leads. In fact, it is good practice to include such bypass capacitors in the initial design of any oscillator or amplifier. Often, too, a small resistance, on the order of 1 kΩ, is put in series with the V_{bb} lead to each transistor along with the bypass capacitor, to increase the isolation between stages as is shown in Fig. 9.32. The lower the operating frequency of

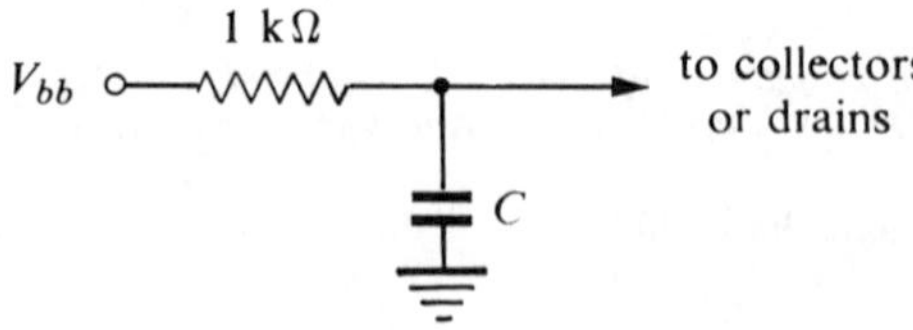

FIGURE 9.32 Power supply lead filtering to prevent unwanted coupling.

the circuit, the larger the bypass capacitance must be. $C = 100\,\text{pF}$ is adequate for frequencies over 10 MHz; whereas, $C = 1\,\mu\text{F}$ might be required for audio frequencies.

Filtering is often added to filament leads of high frequency vacuum tube circuits to prevent unwanted feedback from stage to stage via the filament supply. Radio frequency chokes and/or .001 μF bypass capacitors are often added in series with the filaments as shown in Fig. 9.33. The rf choke must

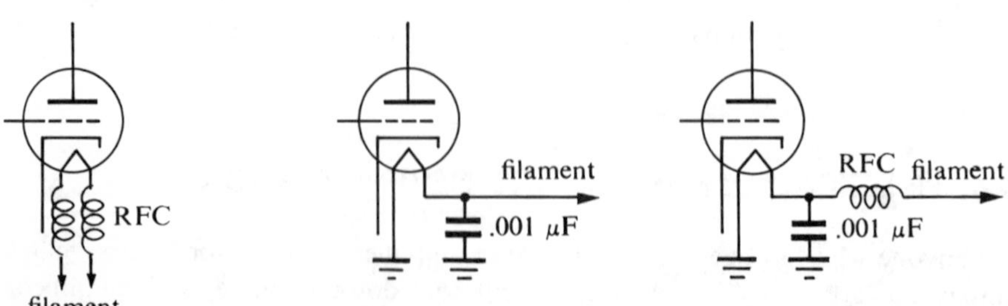

FIGURE 9.33 Vacuum tube filament filtering.

have sufficiently low dc resistance to avoid an excessive dc voltage drop across it. Often for circuits around 100 MHz, one can wind his own *RFC* with five to ten turns of number 22 wire. A one watt resistor of one megohm or greater resistance makes a convenient coil form; the two ends of the *RFC* are soldered to the two resistor leads.

Another source of unwanted oscillations in an amplifier is excessive lead length. The author as a teenager once laboriously and with great love constructed an eight tube superheterodyne AM receiver with three stages of i.f. amplification on what seemed like a chassis of reasonable size—approximately $8 \times 15 \times 3$ inches. The receiver oscillated at a large number of frequencies and sounded somewhat like a cage full of tomcats fighting. The cure is to build as compactly as possible using minimum lead lengths.

Particularly at frequencies above 1 MHz, it is desirable to use a tuned *LC* circuit in both the base and collector in transistor circuits, for the gate and drain in FET circuits, or for the grid and plate in vacuum tube circuits. Such a circuit is often called a "tuned base–tuned collector," "tuned gate–tuned drain," or "tuned grid–tuned plate" circuit, respectively, and is shown in Fig. 9.34 for an FET. This circuit will oscillate because of the inherent capacitive

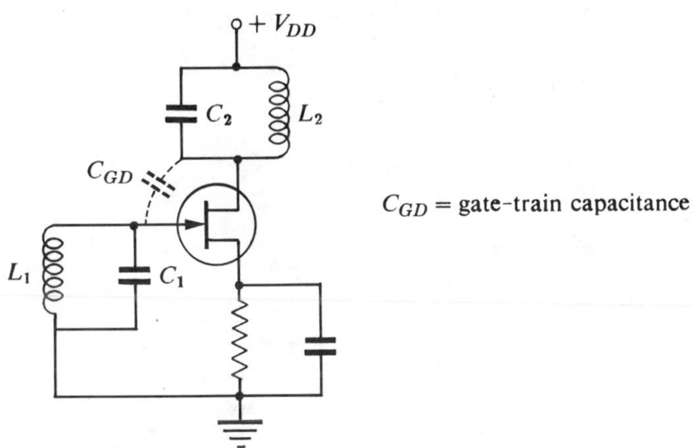

FIGURE 9.34 Tuned gate–tuned drain FET oscillator.

coupling between the input and output; that is, due to the base–collector, gate–drain, or grid–plate capacitance.

The gate–drain capacitance is usually *small*, of the order of several pF, and thus has a large impedance; e.g., if $C_{GD} = 2\,\text{pF}$, then $X_{CGD} = 1/\omega C_{GD} = 80\,\text{k}\Omega$ at 1 MHz. The C_{GD} and the gate–ground impedance R_G form a high-pass filter connecting the drain to the gate. This filter will introduce a phase shift of nearly $+90°$ if $X_{CGD} \gg R_G$. If the L_2C_2 tank circuit in the drain has a natural resonant frequency $\omega_D = 1/\sqrt{L_2C_2}$ higher than the operating frequency, then the impedance of the L_2C_2 tank is inductive and introduces a $+90°$ phase shift.

Thus the total phase shift introduced by L_2C_2 and C_{GD} is approximately 180°, which added to the 180° phase shift between gate and drain inherent in the common source configuration gives *positive* feedback: if the gate goes positive, the signal fed back to the gate through C_{GD} also goes positive. Such an oscillator is called a tuned gate–tuned drain oscillator.

An amplifier made from a bipolar transistor in the common emitter configuration, or an FET in the common source configuration, or a triode in the common cathode configuration with a tuned LC circuit in the input and output will tend to oscillate by this same argument. The common-base transistor amplifier, the common-gate FET amplifier, and the common-grid triode amplifier will not oscillate in this circuit configuration because of the excellent input-output isolation; i.e., the emitter–collector, the source–drain, and the cathode–plate capacitances are extremely low, well under 1 pF. In fact such circuits are mostly used for amplifiers at frequencies above 50 MHz for precisely this reason. A tuned grid–tuned plate pentode amplifier will not oscillate either unless the gain is very high, because of its extremely low value of control grid-plate capacitance (well under 1 pF) caused by the isolating effect of the screen grid between the control grid and the plate. To prevent such oscillation, one has to supply a negative-feedback signal to the input to cancel out the positive feedback as shown in Fig. 9.35. This process is called "neutralization." Either inductive

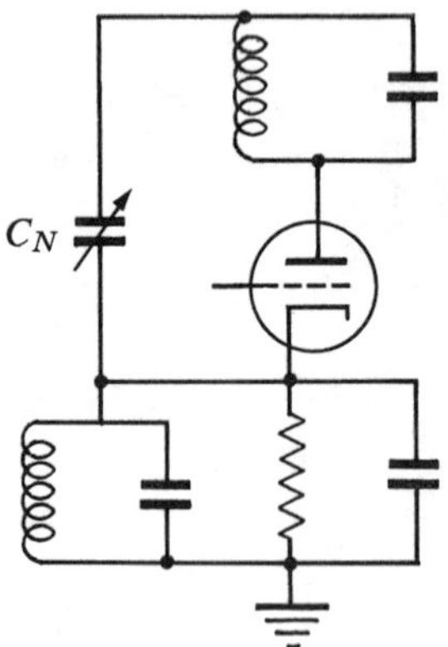

FIGURE 9.35 Neutralization.

or capacitive coupling can be used, but capacitive is more common. A certain amount of experimentation is usually necessary to eliminate the undesired oscillation.

Unwanted coupling or feedback between stages of a multi-stage amplifier can sometimes result from an overlap of rf grounding currents in the chassis, which serves as a common ground for the various amplifier stages. This overlap of currents and consequent coupling is greater when the various circuit grounds are connected to the chassis at different locations. Minimum overlap is thus obtained by connecting all the bypass capacitor grounds and other grounds to one point on the chassis, taking care to use as short ground leads as possible. A certain amount of experimentation is usually necessary. Figure 9.36 shows the proper and improper connection of an rf bypass capacitor.

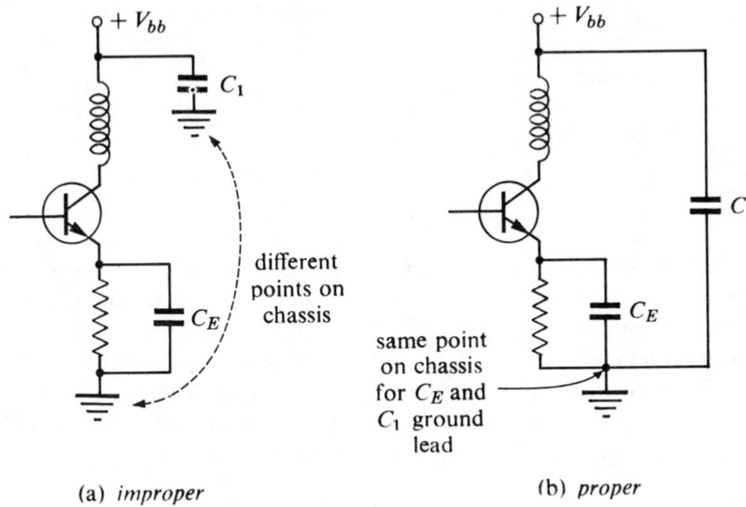

FIGURE 9.36 Ground connections.

problems

1. What is the principal disadvantage of negative feedback? What are the advantages?
2. (a) Derive the voltage gain expression $A_v = A_0/(1 + A_0\beta)$ for an amplifier with negative feedback factor β and gain A_0 without feedback. (b) Show by numerical substitution that if $A_0 = 10^6$ and $\beta = 10^{-2}$, the gain with feedback, A_v, varies by approximately 0.002% for a 20% change in A_0, the gain without feedback.
3. Prove that the percentage change ΔA_0 in the gain without feedback produces a change $\Delta A_f = \Delta A_0/A_0\beta$ in the gain with feedback.
4. Explain why an unbypassed emitter resistor in a common emitter amplifier results in decreased gain.
5. Give the amplifier characteristics if 1% of the output is fed back out of phase, i.e., $\beta = 0.1$.

	$\beta = 0$	$\beta = 0.1$
Voltage gain	1000	
Input impedance	20 kΩ	
Output impedance	5 kΩ	

6. Negative feedback affects distortion, gain, and impedance levels, but only for the circuit actually within the feedback loop. The bias divider chain, R_1R_2, is outside the feedback loop. The amplifier inside the dashed line *without* feedback has a gain

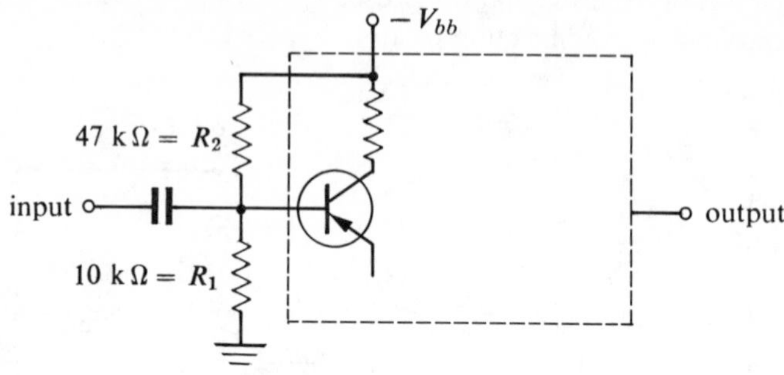

$A_0 = 10{,}000$ and input impedance 1 kΩ. Calculate (a) the input impedance of the *circuit* if there is no negative feedback and (b) the input impedance of the *circuit* if there is negative feedback with $\beta = 0.05$.

7. Explain (a) why negative feedback will increase rather than decrease the input impedance of an amplifier, and (b) why a large input impedance is usually desirable.

8. Calculate the voltage gain for the operational amplifier shown. Does the gain strongly depend upon A_0? Is this a good mode of operation?

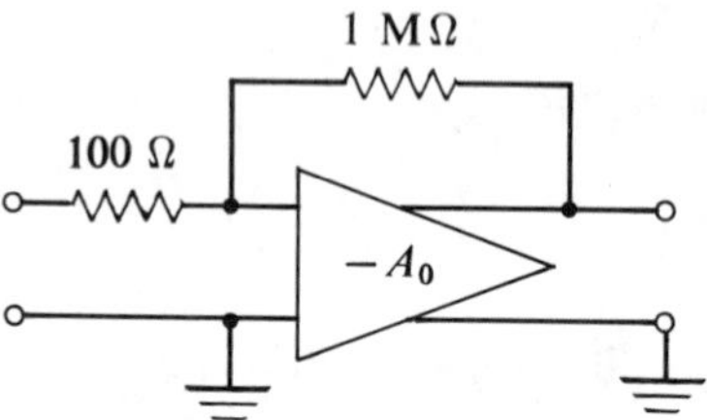

9. Prove that the voltage gain with feedback A_f for an operational amplifier must be substantially less than the open loop gain A_0 in order for A_f to depend only upon the ratio of R_f to R (use Fig. 9.12).

10. State the relationship between v_{out} and v_1 and v_2 for the operational amplifier shown.

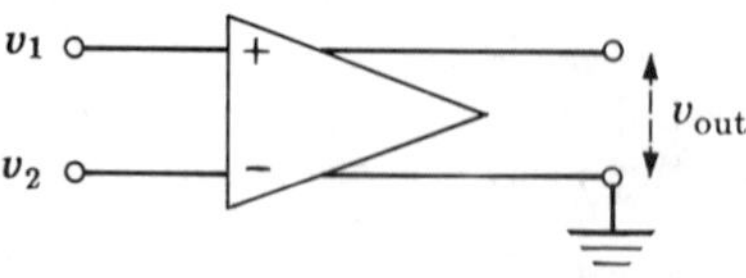

11. Sketch the expected output waveform for the circuits below.

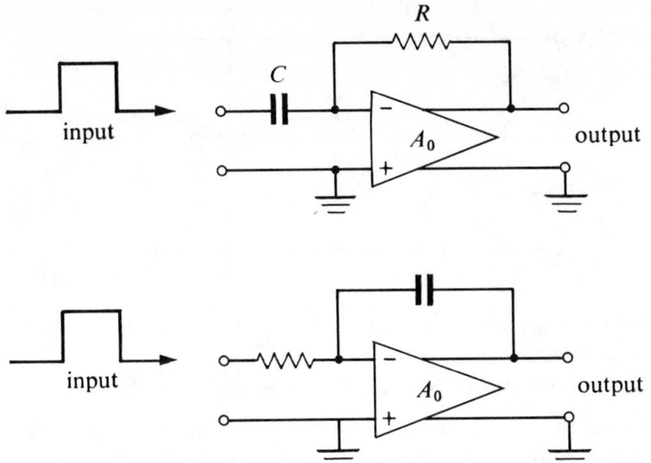

12. Prove that a voltage gain of at least 29 is necessary for the phase shift oscillator of Fig. 9.27 and that the frequency of oscillation is $f_0 = (1/2\pi\sqrt{6})(1/RC)$.
13. Calculate the feedback factor β for the Colpitts oscillator of Fig. 9.28.
14. A relaxation oscillator can be made with a neon tube as shown. The neon tube is filled with low pressure gaseous neon indicated by the solid dot inside the tube

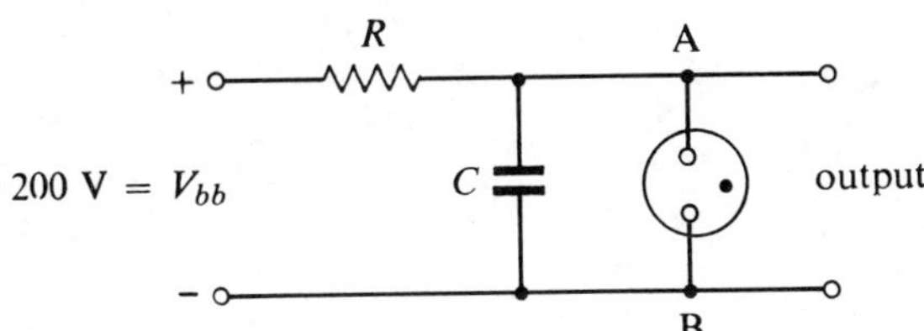

symbol. The tube will not conduct until the voltage across the tube V_{AB} exceeds a critical breakdown voltage of approximately 100 V. Then the tube suddenly conducts heavily and the voltage drop V_{AB} across the tube during conduction V_C is much less than the breakdown voltage. Typically $V_C \cong 10$ V. The effective tube resistance during conduction is very low, approximately 100 Ω.

(a) Sketch the output waveform versus time if $R = 500\,\text{k}\Omega$ and $C = 1\,\mu\text{F}$.
(b) How could you adjust the frequency of the waveform?
(c) How could you adjust the amplitude of the output?
(d) If the current flowing through the neon tube during the discharge of C must be limited, how would you change the circuit to accomplish this?

15. A relaxation oscillator can be made with a silicon-controlled rectifier (SCR) as shown.

(a) Sketch the output waveform versus time if the SCR breaks down when its anode-cathode voltage reaches 20 V.
(b) How could you adjust the frequency of the output?

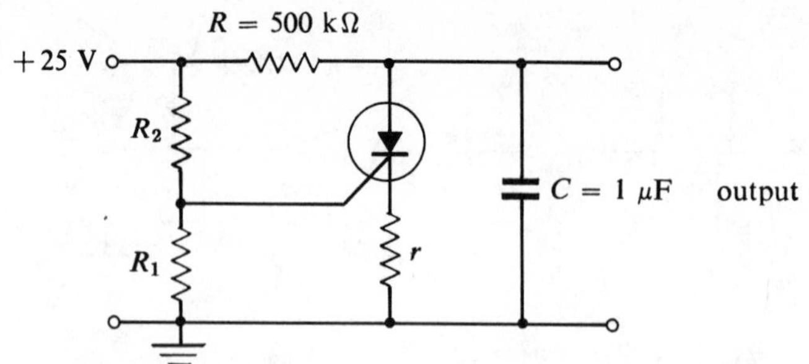

(c) If R_1 is increased so as to increase the gate current, how will the output be changed? Sketch.

(d) Explain how r affects the peak discharge current of the capacitor.

16. Briefly discuss the appropriate remedies for an oscillator that won't oscillate. You may assume that the components are all good and that the circuit has been properly wired.

10

HIGH FREQUENCIES

10.1 INTRODUCTION

Up to now we have tacitly been considering only circuits where the inevitable stray shunt capacitance between the signal carrying wire and ground, the inherent inductance in a straight piece of wire, and the radiated power may be neglected. In this chapter, we will show that such factors become crucial in determining circuit performance at higher frequencies.

High frequencies can mean different things to different people, but a tentative definition is those frequencies above approximately 1 MHz. "Shortwave radio" falls in the frequency range of 1–30 MHz. The term "very high frequency" (vhf) often is used to refer to frequencies from 30 MHz to around 200 MHz. Very high frequency TV channels 2–13 operate at frequencies from about 50 MHz to 200 MHz. Ultrahigh frequency (uhf) refers to frequencies from 200 MHz to around 700 MHz. For frequencies above 1000 MHz (= 1 GHz), the term "microwaves" is usually used. A classification may also be made in terms of the magnitude of the wavelength involved. For electromagnetic radiation $\lambda = c/f$, where c equals the speed of electromagnetic radiation 3×10^{10} cm/sec and f is the frequency in Hz. Thus, for high frequencies, $\lambda \cong 3 \times 10^4$ cm $= 300$ m; for vhf, $\lambda \cong 300$ cm $= 3$ m, for uhf, $\lambda \cong 60$ cm, and for microwaves, $\lambda < 30$ cm. As the frequency increases beyond about 3000 GHz, we enter the infrared region of the electromagnetic spectrum, and then the visible, ultraviolet, and X-ray regions as the frequency increases more. For frequencies higher than the infrared, quantum effects begin to predominate in most situations and the classical wave picture becomes increasingly unsatisfactory. If the wavelength λ is small compared to the linear dimension of the object with which the radiation interacts, then diffraction effects are small, i.e., the waves travel in straight lines and do not bend around obstacles. The electromagnetic spectrum is shown on a logarithmic scale in Fig. 10.1.

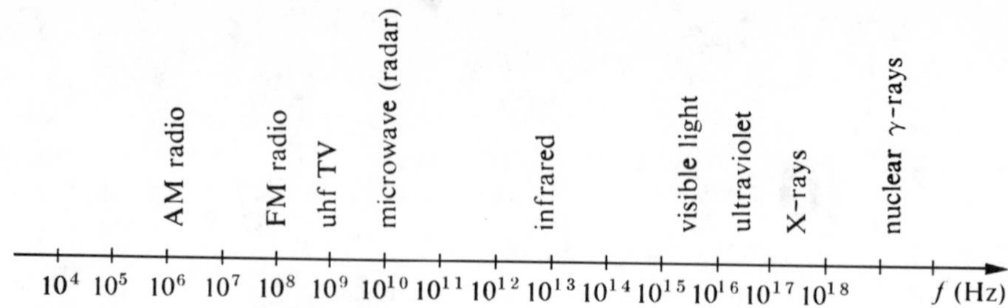

FIGURE 10.1 The electromagnetic spectrum.

According to the quantum theory, the energy of an electromagnetic wave of frequency f Hz comes in discrete bundles or quanta rather than being distributed uniformly through the wave. A quantum of electromagnetic radiation is called a "photon," and the energy E of one photon is given by $E = hf$, where h is Planck's constant: $h = 6.6 \times 10^{-27}$ erg sec $= 6.6 \times 10^{-34}$ joule sec. Quantum effects become significant usually only when $hf \geqq kT$; that is, when the energy of one photon is comparable to or larger than kT, which is the average amount of random thermal energy in a physical system per degree of freedom (e.g., the average translational energy of one molecule of any type is $\frac{3}{2}kT$). In the classical situation, $hf \ll kT$; the energy of a single photon is very difficult to detect because it is small compared to the random thermal energy. Thus classically, we usually are dealing with large numbers of photons, while quantum mechanically, a single photon has enough energy to stand out above the average thermal energy. Most of this chapter will deal with the classical situation in which individual photons can be neglected and the electromagnetic wave is treated as a continuous wave with the energy stored in the electric and magnetic fields.

10.2 ELECTROMAGNETIC RADIATION

The greatest triumph of nineteenth-century physics was the theoretical prediction by Maxwell and the experimental verification by Hertz that transverse electromagnetic waves could be produced by accelerating electric charges, and these waves would propagate through space at a speed of 3×10^{10} cm/sec, the speed of light, and in fact of all electromagnetic radiation. The spectrum of electromagnetic radiation is shown in Fig. 10.1. Such radiation carries linear momentum, angular momentum if it is circularly polarized, and energy. A simple sketch of the instantaneous picture of a plane polarized electromagnetic wave is shown in Fig. 10.2. The electric field vector **E** and the magnetic field **H** oscillate sinusoidally in time and space as given in (10.1) and (10.2).

$$\mathbf{E}_x = \mathbf{E}_0 \cos\left(\frac{2\pi z}{\lambda} - 2\pi\, ft\right) \tag{10.1}$$

$$\mathbf{H}_y = \mathbf{H}_0 \cos\left(\frac{2\pi z}{\lambda} - 2\pi\,\text{ft}\right) \tag{10.2}$$

Notice **E** and **H** are perpendicular to the direction of motion and to each other. The plane of polarization is taken by convention to be the plane containing the electric field vector **E**; the wave in Fig. 10.2 is polarized in the x direction. The

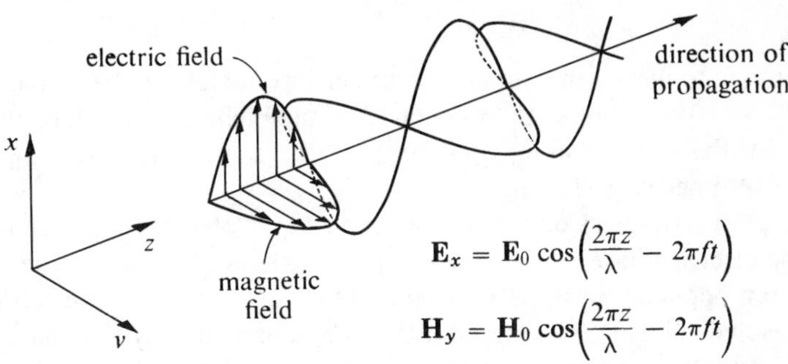

FIGURE 10.2 Classical electromagnetic wave picture.

linear momentum carried by the wave per unit volume is $p = W/c$ where W is the energy of the wave (stored in the electric and magnetic fields) per unit volume. The power P carried by the wave per unit area, that is, the energy per second passing per unit area, is $P = E_0H_0$. The direction of the power flow is the direction of the vector cross product $\mathbf{E} \times \mathbf{H}$ along the positive z axis in Fig. 10.2.

In electronic circuits, electric charge is in motion. A well-known result obtained directly from the fundamental Maxwell equations which describe all electromagnetic phenomena is that a point charge of q moving with an acceleration a radiates power outward in the form of electromagnetic radiation. The power radiated, P, is given by (cgs units):

$$P = \frac{2}{3}\frac{q^2}{c^3}a^2, \tag{10.3}$$

where q is the charge in statcoulombs, c is the speed of electromagnetic radiation (3×10^{10} cm/sec), and a is the magnitude of the acceleration in cm/sec^2. This power formula immediately tells us that the radiated electromagnetic power increases extremely rapidly as the frequency of the charge motion increases.

Consider a charge moving with simple sinusoidal motion; that is, the position of the charge is given by:

$$x = x_0 \cos \omega t. \tag{10.4}$$

The acceleration of the charge is simply the second time derivative of x with respect to time:

$$a = \frac{d^2x}{dt^2} = -\omega^2 x_0 \cos \omega t. \tag{10.5}$$

Notice that the acceleration varies as the square of the frequency of oscillation. Therefore the power radiated by a point charge q moving with simple harmonic motion of angular frequency ω varies as the *fourth* power of the frequency:

$$P \propto \omega^4 . \tag{10.6}$$

It may be safe to ignore the radiation from electrons sloshing back and forth in a wire at $f = 10\,\text{kHz}$, but at a frequency 100 times higher, $f = 1\,\text{MHz}$, the power radiated by the same charge, oscillating at the same amplitude X_0, will be $(100)^4 = 10^8$ times larger!

One of the simplest and most common types of radiating systems is an oscillating electric dipole. An electric dipole consists of a plus charge $+Q$ and an equal but opposite charge $-Q$ separated by a distance d. The electric dipole moment is defined as $P_e = Qd$, and if the dipole oscillates by the charges moving back and forth then the dipole radiates electromagnetic radiation. In practical units the total power radiated by an oscillating electric dipole is $P = 40\pi^2(S/\lambda)^2 I_0^2$ watts $= 80\pi^2(S/\lambda)^2 I_{\text{rms}}^2$; where I_0 is the peak oscillating current in amperes of the dipole, S is the length of the dipole, and $\lambda = c/f$ where f is the frequency of the oscillation in Hz.

From the expression $P = I_{\text{rms}}^2 R$, we see that the oscillating dipole acts like a resistance $R = 80\pi^2(S/\lambda)^2$ in terms of removing electrical energy from the circuit and radiating it into space. This is called the radiation resistance. For a dipole 3 cm long with a peak current of 10 mA, the average power radiated is 4×10^{-14} watts at 10 kHz, 4×10^{-10} W at 1 MHz, and 4×10^{-6} W at 100 MHz. It is quite reasonable that the power should vary as the oscillating current, because the larger the current the more oscillating charges there are to radiate. What is not quite so obvious is that the ratio of the dipole size S to the wavelength determines the radiated power. This dependence on $(S/\lambda)^2$ gives rise to the often quoted rule of thumb that if the size of the wire is comparable to the wavelength then the wire will radiate efficiently, i.e., will act as an antenna. In the previous example, $(S/\lambda) = 10^{-6}$ at 10 kHz, 10^{-4} at 1 MHz, and 10^{-2} at 100 MHz. If the dipole size equals the wavelength, then a 10 mA peak oscillating dipole current would radiate 4×10^{-2} watts or 40 milliwatts, which is enough power to run a short range radio transmitter. Practically speaking this means that any wire in a circuit operating at 1 MHz or higher has to be treated as a radiator of electromagnetic radiation, i.e., as an antenna. Other wires in the circuit will act as receiving antennas, and there is the possibility of strange and wondrous feedback loops introduced by this radiation. The net result is that if great care is not taken in wiring up circuits at frequencies above 1 MHz, they will invariably oscillate, and the higher the gain of the circuit the more prone they are to oscillate.

A transmitting antenna is a wire designed to radiate efficiently electromagnetic waves into space. As we have seen, for efficient radiation, the size of a dipole should be comparable to the wavelength of the radiation. One of the most common antenna shapes used is the half-wave antenna shown in Fig. 10.3.

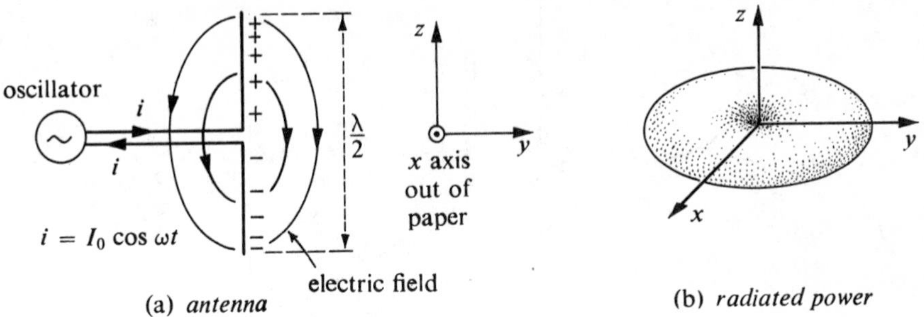

FIGURE 10.3 Half-wave antenna.

The two lead wires from the oscillator that drives the antenna are straight, parallel, and close together. They produce no net magnetic field because at any instant of time the currents in the two lead wires are equal and opposite. Hence they radiate no energy. The two $\lambda/4$ arms do radiate however. The current in these arms must fall to zero at the ends, and it can be shown that the current is given by

$$I = I_0 \cos \omega t \cos \frac{2\pi z}{\lambda}. \tag{10.7}$$

The power radiated by such a half-wave antenna is linearly polarized with the **E** vector parallel to the antenna direction, the z axis. The power is radiated outward in a cylindrically symmetric fashion with maximum power radiated in the equatorial (x–y) plane and zero power along the $\pm z$ axis. A sketch of the power radiation pattern is shown in Fig. 10.3(b). A calculation of the total power radiated is rather involved and yields the result:

$$P = 36.55 I_0^2 = 73.1 I_{\text{rms}}^2 .$$

Thus a half-wave antenna has a radiation resistance of 73.1 ohms. With a peak current of $I_0 = 10$ mA, it will radiate 3.655 milliwatts for any frequency.

Electromagnetic energy can be radiated predominantly in one direction by using several antennas spread the proper distance apart so that the radiated waves add constructively in the desired direction and interfere destructively in other directions. Such arrangements are called antenna "arrays." They are essential for communication from space vehicles, for example, where the power available is limited, and all the available power must be directed in a narrow beam toward the intended receiver. Two half-wave antennas, parallel to the z axis, spaced $\lambda/2$ apart, and driven in phase, will radiate predominantly along the

$\pm y$ axis and not along the x axis, for example. Such an array is shown in Fig. 10.4 with the radiated power pattern. More complicated antenna arrays con-

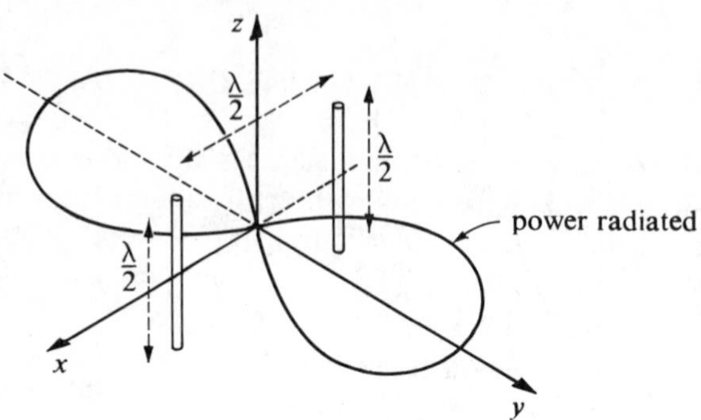

FIGURE 10.4 Two element, half-wave antenna array.

sisting of many antenna elements can be constructed to radiate power in extremely narrow "lobes."

Another basic type of antenna is the magnetic dipole antenna (Fig. 10.5). Such an antenna can be made from oscillating circular current whose magnetic field at large distances is the same as that from an oscillating magnetic dipole or bar magnet. The effective magnetic movement of such a current loop is iA in mks units where A is the area enclosed by the loop and i is the current. The

$$i = I_0 \cos \omega t$$

$$I_{\text{rms}} = \frac{I_0}{\sqrt{2}}.$$

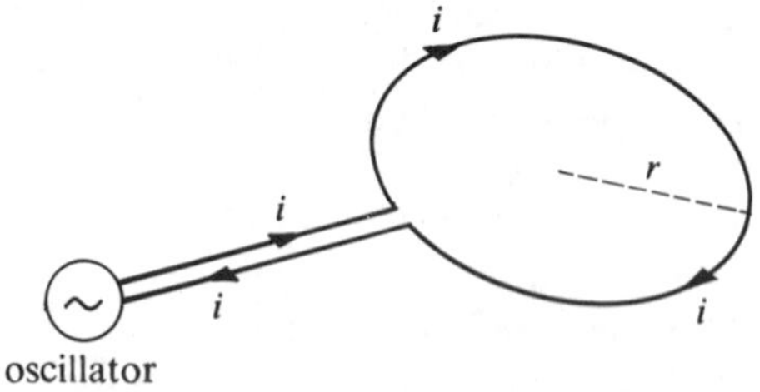

FIGURE 10.5 Magnetic dipole antenna.

total power radiated from such an antenna is given by:

$$P = 20\pi^2 \left(\frac{r}{\lambda\!\!\!^{-}}\right)^4 I_{\text{rms}}^2, \tag{10.8}$$

where $\lambda\!\!\!^{-} = \lambda/2\pi$.

Thus its radiation resistance is $20\pi^2(r/\lambda)^4$. If $r = \lambda$, the radiation resistance equals 197 ohms.

10.3 SERIES INDUCTANCE AND SHUNT CAPACITANCE

Any piece of wire has a nonzero self-inductance. This inductance depends upon the shape the wire is wound in, a coil with tightly spaced turns having more inductance than a loosely wound coil or a straight piece of wire. A straight piece of wire of finite cross-sectional area can be seen to have a self-inductance from Ampere's law, Faraday's law, and Lenz's law. Ampere's law implies there is a circular magnetic field inside the wire in a plane perpendicular to the length of the wire as shown in Fig. 10.6. Thus there is a magnetic flux inside the wire.

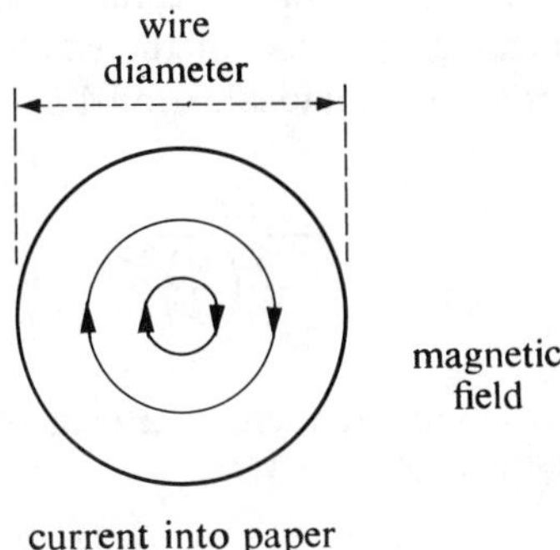

FIGURE 10.6 Magnetic fields and currents in current-carrying wire.

As the current changes the flux will change, and by Faraday's law an emf will be induced causing current loops to flow in planes perpendicular to the magnetic field lines. By Lenz's law, the induced currents will produce magnetic fields in such a direction as to oppose the *change* in the current. Thus there is an inductive reactance which opposes changes in current inherent in ac.

For example, a 4-inch long, perfectly straight piece of no. 18 wire has an inductance of $0.1\,\mu$H and therefore an impedance of $\omega L = 63\,\Omega$ at 100 MHz. Minimum inductance can be obtained when the cross-sectional area of the wire lead is a rectangle with a small height-to-width ratio, i.e., a flat strip of wire rather than a conventional wire with circular cross section. It is common practice in vhf and uhf circuits for leads to be made from strips of thin sheet metal, and the lead wires from transistors designed to operate at these frequencies are made from thin metal strips about 0.010 in. thick and 0.025 in. wide. Short flat ground wires are also critically important in high frequency circuits.

The distributed inductance of the leads and the body of a capacitor results in the capacitor acting like a series *LC* circuit. At frequencies below the self-resonant frequency, the capacitor acts like a capacitor, but above the self-resonant frequency, the capacitor acts like an inductor. At the self-resonant frequency, of course, the impedance of the capacitor is minimum. Self-resonant frequencies of capacitors range from 10 MHz for large capacitance values (0.01 μF) to several hundred megahertz for smaller capacitance values (50 pF).

Capacitance exists between any two conductors a finite distance apart. In particular there exists capacitance between a signal carrying wire and the chassis. This is called shunt capacitance because it literally shunts the ac signal to ground. Shunt capacitance is usually one of the most important reasons for signal attenuation at high frequencies. There is also capacitance between the elements of the transistor, e.g., capacitance between the base and emitter, base and collector, and collector and emitter. And, in vacuum tubes, there is capacitance between the various pairs of electrodes: grid–plate, grid–cathode, and plate–cathode.

The distributed capacitance between the turns of an inductor (e.g., an *RFC*) results in the inductor acting like a parallel *LC* tuned circuit whose self-resonant frequency is usually rather high because of the small value of the distributed capacitance *C*. The impedance of an ideal parallel *LC* circuit can be written as:

$$Z = \frac{j\omega L}{1 - \left(\frac{\omega}{\omega_0}\right)^2}, \tag{10.9}$$

where ω_0 is the angular resonant frequency. For a real inductor with impedance $(R + j\omega L)$ in parallel with C:

$$Z \cong \frac{j\omega L}{1 - \left(\frac{\omega}{\omega_0}\right)^2 + j\omega RC}. \tag{10.10}$$

To avoid the decrease in gain due to shunt capacitance, several steps can be taken: The output impedance driving the shunt capacitance can be made low, by use of a low output impedance circuit such as an emitter-follower as has already been discussed. Or, a small inductance can be placed in parallel with the shunt capacitance to form a parallel *LC* resonant circuit with a high impedance at high frequencies. Or, the shunt capacitance along with the inevitable series inductance of the leads can be incorporated into a transmission line structure which will be discussed later in this chapter. Or, an inductance can be inserted in such a way that its impedance, which increases with increasing frequency, tends to raise the gain at high frequencies to compensate the decrease in gain due to the shunt capacitance.

An example of this last technique is shown in Fig. 10.7 where an inductance has been inserted into the collector lead of an FET common source amplifier. This technique is sometimes called "shunt peaking." Such an amplifier normally has a gain of about 20–30, but with a bandwidth of only about 100 kHz. At

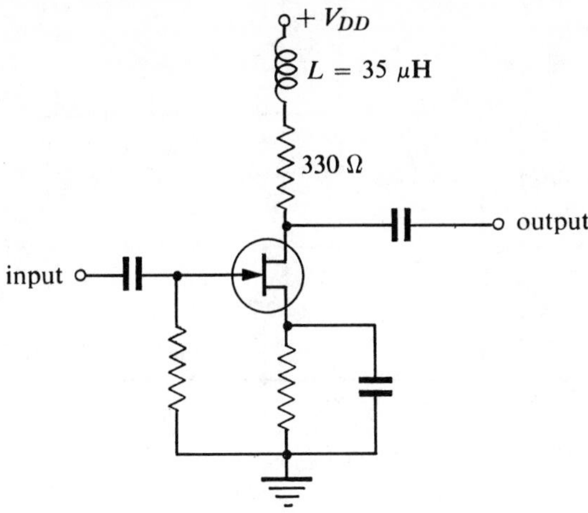

FIGURE 10.7 Shunt peaked FET amplifier.

frequencies of the order of MHz, the impedance in the collector circuit is essentially ωL, where L is the inductance of the coil. For example, at 4 MHz, $\omega L = 880\,\Omega$, which raises the gain above the low frequency gain that is determined by the $330\,\Omega$ collector resistor. Such amplifiers are commonly used in television sets and other high frequency circuitry and are often called "video" amplifiers. With pulse inputs, care must be taken in practice to avoid ringing due to damped oscillations of L with either the stray shunt capacitance between drain and ground or with the series coupling capacitance to the next stage. This ringing can be decreased by varying the value of the resistance in series with the choke; the higher the resistance the less ringing, but the lower the bandwidth.

10.4 HIGH FREQUENCY FET EQUIVALENT CIRCUIT AND THE MILLER EFFECT

Let us now consider how the inherent capacitances between the pairs of FET leads influence high frequency operation. The same type of analysis may be applied to bipolar transistor or vacuum tube circuits. A reasonably good high frequency FET equivalent circuit is shown in Fig. 10.8. The only difference between the low frequency equivalent circuit of Fig. 7.1 and the high frequency circuit of Fig. 10.8 is that three capacitances have been added. C_{GS} is the gate–source capacitance and is across the input. C_0 is the sum of the drain–source capacitance and any stray or shunt capacitance between the output lead and ground. C_{DG} represents the drain–gate capacitance and plays a major role in attenuating the high frequency gain, because it is effectively multiplied

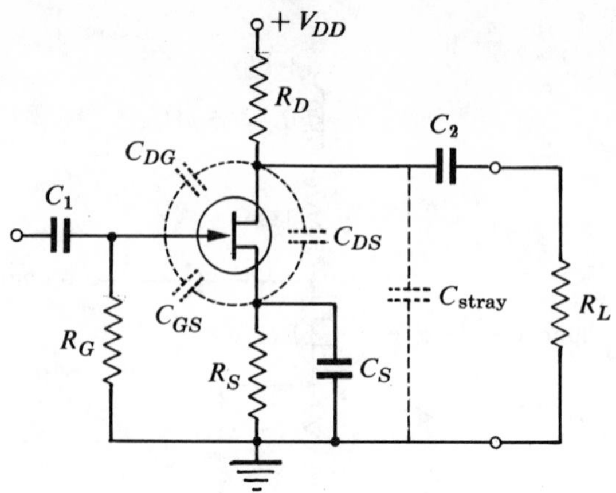

(a) *amplifier circuit*

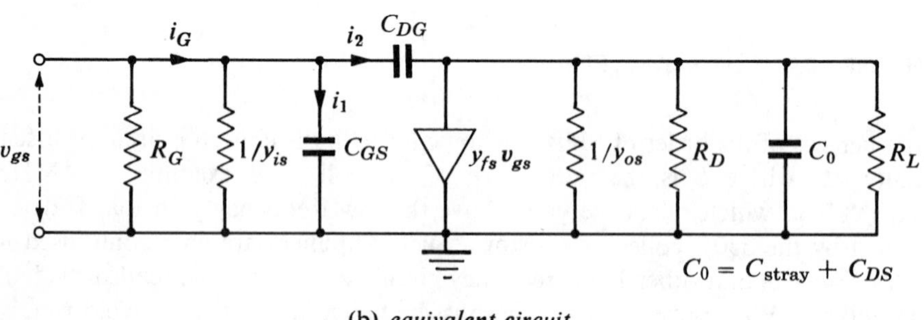

(b) *equivalent circuit*

FIGURE 10.8 High frequency FET circuit.

by the voltage gain of the amplifier and appears electrically to be connected across the input. The proof is as follows.

The input impedance is given by:

$$Z_{in} = \frac{v_{GS}}{i_G}, \tag{10.11}$$

and the ac gate current is given by

$$i_G \cong i_1 + i_2$$

or

$$i_G \cong \frac{v_{GS}}{\left(\dfrac{1}{j\omega C_{GS}}\right)} + \frac{v_{GS} - v_{DS}}{\left(\dfrac{1}{j\omega C_{DG}}\right)}. \tag{10.12}$$

We have assumed here that $1/j\omega C_{GS} \ll R_g$ or $1/y_{is}$. We recall from section 7.7 that the voltage gain A_V of the FET common source amplifier is given by:

$$A_V = \frac{v_{DS}}{v_{GS}} = \frac{-y_{fs}R_{DL}}{1 + y_{OS}R_{DL}} \cong -y_{fs}R_{DL}, \tag{10.13}$$

where $R_{DL} = R_D \parallel R_L$. Solving (10.13) for v_{DS} and substituting in (10.12), we find that:

$$i_G \cong \frac{v_{GS}}{\left(\dfrac{1}{j\omega C_{GS}}\right)} + \frac{v_{GS} + y_{fs}R_{DL}v_{GS}}{\left(\dfrac{1}{j\omega C_{DG}}\right)}$$

or

$$Z_{\text{in}} = \frac{v_{GS}}{i_G} = \{j\omega C_{GS} + j\omega C_{DG}(1 + y_{fs}R_{DL})\}^{-1}$$

$$Z_{\text{in}} \cong \{j\omega[C_{GS} + (1 + |A_V|)G_{DG}]\}^{-1}. \tag{10.14}$$

The coefficient of $j\omega$ in equation (10.14) is the effective capacitance ω across the input:

$$C_{\text{input}} \cong C_{GS} + (1 + |A_V|)C_{DG}. \tag{10.15}$$

If the voltage gain A_V is large, then the input capacitance can be many times larger than the gate–source capacitance C_{GS} of the FET. This enhancement of the input capacitance is called the "Miller effect" and holds for bipolar transistors and triode vacuum tubes in the common emitter and common cathode configuration, respectively. For bipolar transistors:

$$C_{\text{input}} \cong C_{BE} + (1 + |A_V|)C_{CB}, \tag{10.16}$$

where C_{BE} is the base–emitter capacitance and C_{CB} is the collector–base capacitance. For triode vacuum tubes:

$$C_{\text{input}} \cong C_{GK} + (1 + |A_V|)C_{PG}, \tag{10.17}$$

where C_{GK} is the grid–cathode capacitance and C_{PG} is the plate–grid capacitance.

It is important to emphasize that the Miller effect holds only for FET's, bipolar transistors, and triode vacuum tubes in the common source, common emitter, and common cathode configurations, respectively. The common–gate or common–drain FET amplifier, the common–base or common–collector bipolar transistor amplifier, the common–grid or common–plate triode amplifier, and the vacuum pentode in any configuration do not exhibit the Miller effect, and thus will amplify at high frequencies. The pentode does not exhibit a Miller effect even in the common cathode configuration, because the constant voltage screen grid between the control grid and the plate effectively reduces the grid–plate capacitance to an extremely low value, much less than one picofarad.

Capacitance values for an inexpensive FET (2N5459), an inexpensive bipolar transistor (2N4250), and a vacuum triode (12AX7) are given in Table 10-1.

TABLE 10-1

Device		Capacitance	
2N5459	FET	C_{GS} = 4.5pF	C_{DG} = 1.5pF
2N4250 (2N5087)	Bipolar Transistor	C_{BE} = 4.8pF	C_{CB} = 3.5pF
12AX7	Triode Vacuum Tube	C_{GK} = 1.6pF	C_{PG} = 1.7pF

Notice that the "inter-electrode" capacitance which is multiplied by the voltage gain in equation (10.14) is of the order of 2–4pF. If the voltage gain is on the order of 50, then a capacitance of 100–200pF will appear across the input. The impedance of a 100pF capacitor is only 167Ω at 10MHz, so it is easy to see how the Miller effect reduces the high frequency gain. Even at the relatively low frequency of 10kHz, the impedance of a 100pF capacitor is 167kΩ, which can have a drastic effect if it is in parallel with a high resistance. For example in an FET common-source amplifier, the gate-to-ground resistor may be 1 megohm. Thus at 10kHz, the gate-to-ground impedance would be 1 megohm in parallel with 167kΩ, which is only 143kΩ. Thus the input impedance can be significantly less than the value of the gate-to-ground resistor even in the audio frequency range.

It is also worth pointing out at this time that the "inter-electrode" or junction capacitance, which is multiplied by the gain in a bipolar transistor or an FET, is the capacitance inherent in the reverse-biased, p-n junction, and therefore is dependent on the voltage across the junction. The junction capacitance is given by $C = KV^{-1/2}$, where K is a constant depending upon the type of junction and V is the reverse bias voltage across the junction. For the 2N5087 bipolar transistor, for example, C_{CB} is 4.8pF when $V_{CB} = 1$V dc and falls to 1.7pF when $V_{CB} = 10$V dc. Thus, some improvement in high frequency operation can be obtained by increasing the reverse bias voltage between the base and collector for a bipolar transistor or between the gate and drain for an FET. This is impossible for a vacuum tube triode, of course; the triode grid-plate capacitance depends only upon the geometry of the tube electrodes and is thus constant for a given tube and independent of the tube voltages.

10.5 THE CASCODE AMPLIFIER

The Miller effect for an FET can be reduced by approximately a factor of 100 by connecting two FET's in a "cascode" arrangement shown in Fig. 10.9. Q_1 is an FET in a conventional common source configuration and Q_2 acts as the drain load of Q_1. Q_2 can also be thought of as a common gate amplifier. The relatively low input impedance of the common gate amplifier effectively reduces

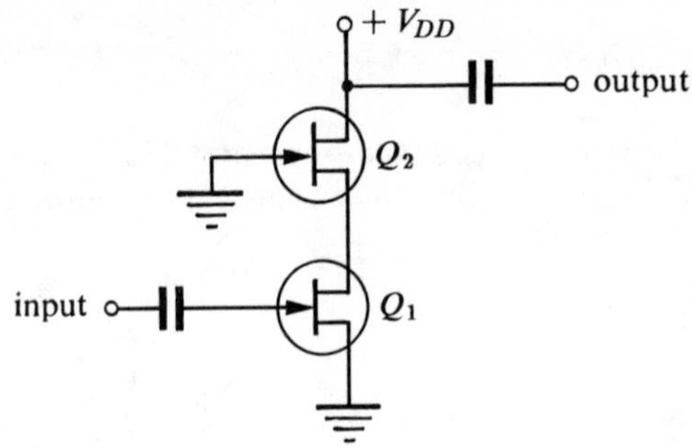

(a) *basic cascode circuit without biasing*

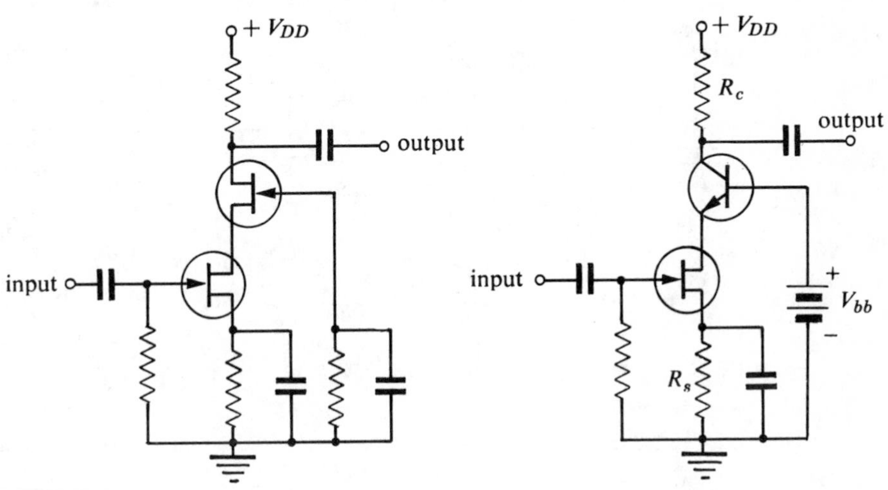

(b) *cascode circuit with biasing*

(c) *cascode circuit with bipolar transistor for Q_2*

FIGURE 10.9 FET cascode amplifier.

the voltage gain between the gate and collector of Q_1. Thus C_{DG} for Q_1 is not multiplied by a large factor. The voltage gain comes in Q_2, which is isolated from the input because the gate of Q_2 is an ac ground. Q_2 can be a bipolar transistor to achieve higher gain as is shown in Fig. 10.9(c). Notice also in Fig. 10.9(c) that the FET drain–source voltage is only approximately equal to $V_{bb} - I_s R_s$, while the transistor collector–emitter voltage (approximately equal to $V_{dd} - V_{bb}$) can be much higher. This allows one to use a large value of R_c to get a high voltage gain from Q while keeping the emitter current of Q_2 (which equals the drain current of Q_1) to a reasonably high value for a high transconductance from the FET Q_1. A cascode amplifier can also be made with two triodes.

10.6 THE GAIN-BANDWIDTH PRODUCT

Whenever a shunt capacitance is driven by an amplifier output as shown in Fig. 10.10, we have a low-pass RC filter in which R is the output resistance of the

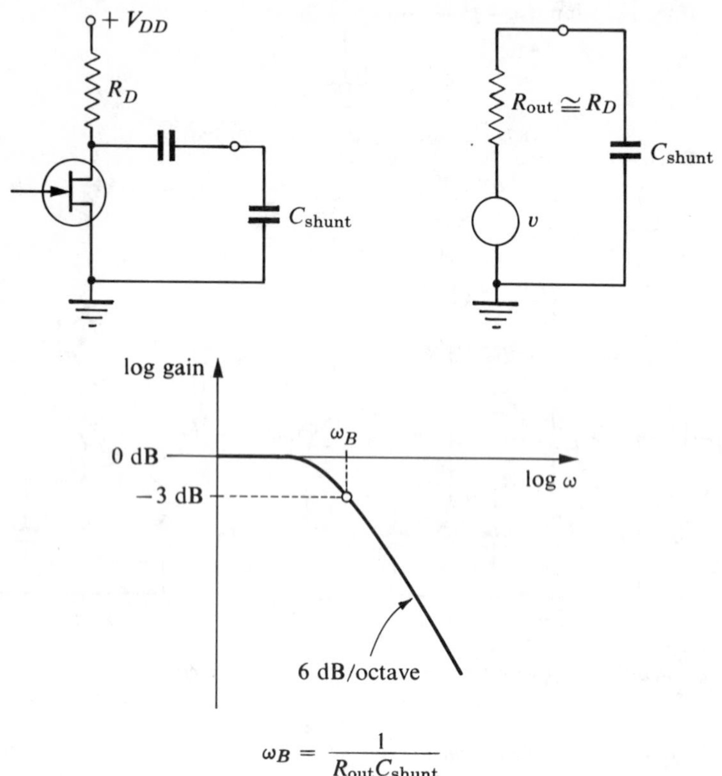

FIGURE 10.10 Amplifier driving capacitive load.

amplifier and C is the total shunt capacitance, i.e., the capacitance between the output terminal and ground. The upper break point where the voltage gain has fallen by 3 dB or a factor of $0.707 = 1/\sqrt{2}$ is given by $\omega_B = 1/R_{out}C_{shunt}$. The situation is somewhat more complicated but basically the same when the output impedance is complex (Z_0) so long as the real part of Z_0 is large compared to the imaginary part.

For an FET or pentode amplifier, from section 7.4, the mid-frequency gain is approximately given by $A_v \cong g_m R_D$, where g_m is the transconductance ($g_m = y_{fs}$) and R_D is the value of the drain or plate resistor. The output impedance of an FET or pentode amplifier is equal to the parallel combination of R_D and $1/y_{22}$

for an FET or R_D and r_p, the plate resistance, for a pentode. Usually $R_D \ll 1/y_{22}$ and $R_D \ll r_p$, so that the output impedance is approximately R_D. Thus we can write for the upper 3 dB angular frequency ω_B:

$$\omega_B = \frac{1}{R_{\text{out}}C_{\text{shunt}}} \simeq \frac{1}{R_D C_{\text{shunt}}}. \tag{10.18}$$

Eliminating R_D with $A_v = g_m R_D$ and rearranging, we obtain:

$$A_v \omega_B = \frac{g_m}{C_{\text{shunt}}}. \tag{10.19}$$

In words, the product of the voltage gain A_v and the upper 3 dB frequency equals the transconductance of the device divided by the total shunt capacitance between the output signal lead and ground. The bandwidth of the amplifier is conventionally taken to mean the frequency in hertz at which the gain has fallen 3 dB, or 0.707 times the mid-frequency gain. Thus, if the lower 3 dB frequency is small compared to ω_B, then we may say the bandwidth f is essentially equal to $\omega_B/2\pi$. Thus the voltage gain–bandwidth product is given by:

$$A_v \Delta f = \frac{g_m}{2\pi C_{\text{shunt}}}. \tag{10.20}$$

The significance of this gain-bandwidth product is that it depends only on characteristics of the *device*, g_m, and the *wiring* shunt capacitance, not upon the particular circuit used. The gain-bandwidth product is often denoted by f_T. For an FET with $g_m = 4000\mu\text{mhos} = 4 \times 10^{-3}$ mhos, and $C_{\text{shunt}} = 5$ pF, the gain–bandwidth product is:

$$A_v \Delta f = \frac{4 \times 10^{-3}\ \text{mhos}}{(2\pi)(5 \times 10^{-12}\text{F})} = 127\ \text{MHz}.$$

At a frequency of 127 MHz, the voltage gain will have fallen to unity. If $R_D = 2.2\ \text{k}\Omega$, the mid-frequency gain will be $A_v = g_m R_D = 4 \times 10^{-3} \times 2.2 \times 10^3 = 8.8$, and the upper 3 dB frequency is 14.5 MHz. Attempts to raise the gain at higher frequencies by increasing R_D while keeping the bandwidth are doomed to failure by equation (10.20). For example, if R_D is raised to 4.7 kΩ to increase the mid-frequency gain to 18.8, then the upper 3 dB frequency falls to 6.8 MHz. For a given device (g_m) and wiring (C_{shunt}), one can have a high gain over a small bandwidth or a low gain over a large bandwidth as illustrated in Fig. 10.11. However, amplifiers can be cascaded in series to achieve a much higher gain at a slightly reduced bandwidth so long as the output impedance of each stage is much less than the input impedance of the next stage. For example, two identical FET amplifiers in series, each with a gain-bandwidth product of 127 MHz, an upper 3 dB frequency of 14.5 MHz, and a mid-frequency gain of 8.8, would together have a gain of $(8.8) \times (8.8) = 77.3$ and an upper 3 dB

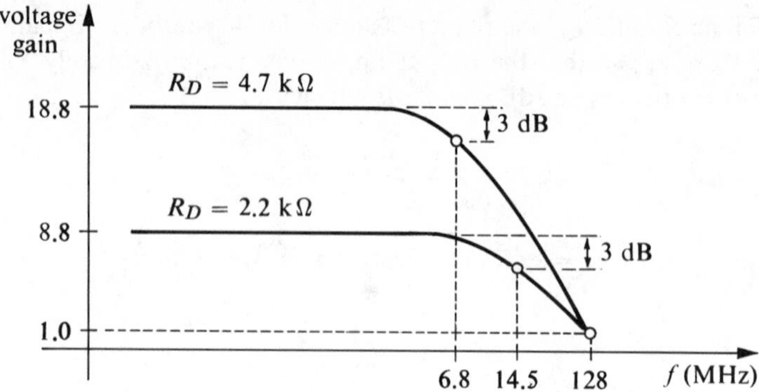

FIGURE 10.11 Constancy of gain–bandwidth product.

frequency or bandwidth of approximately 9.3 MHz. The new upper 3 dB frequency for two identical amplifiers (of any type) is given by:

$$\omega_{B2} = (\sqrt{2} - 1)^{1/2}\omega_B = 0.643\omega_B\,, \tag{10.21}$$

where ω_B is the upper 3 dB frequency of one amplifier.

10.7 DIFFUSION TIME AND TRANSIT TIME

An electric field exists only in the depletion region of a transistor, i.e., only at the reverse-biased base–collector junction if we neglect the small dc resistivity of the semiconductor material. Everywhere else in the transistor, charge carriers will move only by diffusion, which is the statistically probable tendency for charge carriers to move from regions of higher concentration toward regions of lower concentration. Hence, a fundamental limitation to the high frequency operation of bipolar junction transistors is the finite time required for the charge carriers to *diffuse* from the emitter across the base region to the beginning of the depletion region where they are accelerated by the electric field toward the collector. The transistor simply cannot react to a signal which changes appreciably in a time of the order of or shorter than this diffusion time. A typical diffusion time τ in a bipolar junction transistor is one microsecond, so frequencies of the order of or greater than $f = 1/\tau = 1$ MHz would not be amplified. Notice that this effect is entirely absent in an FET: an electric field exists along the entire channel from source to drain.

The diffusion time can be decreased by making the base very thin and pure, but only up to a certain point. One technique for enhancing the high frequency operation of a transistor is to deliberately vary the conductivity of the base material from the emitter side to the collector side by varying the base doping concentration. The doping is higher at the emitter side of the base and usually varies exponentially down to a low value near the collector. Thus the depletion region at the base–collector junction extends far into the base and results in a low junction capacitance which further aids high frequency operation. Such a transistor is called a "drift" transistor and effectively has a built-in electric field in the base to hurry the charge carriers along on their way from the emitter to the collector. The "diffused base" transistor works on essentially the same principle. Drift transistors can be made to operate with reasonable gains at several hundred MHz.

There are also complicated phase effects in the motion of charge carriers across the base which are beyond the scope of this book. The net effect of the diffusion time and phase effects is to result in α decreasing as the frequency increases according to:

$$\alpha_f = \frac{\alpha_0}{1 + j\dfrac{f}{f_\alpha}}, \tag{10.22}$$

where α_f is the value of α at the frequency f Hz. The α_0 is the low frequency value of α, and f_α is the "alpha cutoff frequency." This expression is similar in form to the gain versus frequency expression for a low-pass RC filter:

$$A_v = \frac{1}{1 + j\omega RC} = \frac{1}{1 + j\dfrac{\omega}{\omega_B}}, \tag{10.23}$$

where $\omega_B = 1/RC$ is the 3 dB or break point frequency. The α cutoff frequency is thus seen to be the frequency at which $\alpha_f = 0.707\,\alpha_0$, i.e., at which α is 3 dB down from its low frequency value. Typical values of the cutoff frequency are several MHz for diffusion transistors and up to 50 MHz for drift or diffused base transistors. However, one must remember that β or h_{fe} is a very sensitive function of α from the relation $\beta = \alpha/(1 - \alpha)$. Thus a very small change in α produces a very large change in β.

$$\Delta\beta = \frac{d\beta}{d\alpha}\Delta\alpha = \frac{1}{(1-\alpha)^2}\Delta\alpha = \frac{\beta^2}{\alpha^2}\Delta\alpha$$

or

$$\frac{\Delta\beta}{\beta} = \frac{\beta}{\alpha}\frac{\Delta\alpha}{\alpha} \cong \beta\frac{\Delta\alpha}{\alpha}, \tag{10.24}$$

where we have set $\beta/\alpha \cong \beta$. Expressed in words, the fractional change in β equals β times the fractional change in α; e.g., a 0.1% decrease in α produces a 10% decrease in β if $\beta = 100$. The net result is that the "beta cutoff frequency," f_β, where β is 0.707 of its low frequency value equals the alpha cutoff frequency divided by β:

$$f_\beta = \frac{f_\alpha}{\beta}. \tag{10.25}$$

For example if $f_\alpha = 20\,\text{MHz}$, and $\beta = 50$, then the β cutoff frequency, f_β, equals 400 kHz. It can be shown that the beta cutoff frequency f_β and the gain–bandwidth product f_T for a transistor are related by $f_T = \beta_0 f_\beta$, where β_0 is the low frequency value of β. Thus, if the gain–bandwidth product $f_T = 60\,\text{MHz}$, and $\beta_0 = 100$, then the transistor β would fall to 71 at 600 kHz.

A similar limitation exists in vacuum tube operation at high frequencies. An electron emitted from the cathode takes a certain finite time, the "transit time," to travel across to the plate. The larger the voltage difference between the cathode and plate the stronger the electric field, thus the shorter the transit time and the better the tube's high frequency response. The tube cannot respond to any signal which changes appreciably in a time of the order of or shorter than the transit time. An order of magnitude figure for the transit time τ can be obtained from the following argument. Assuming parallel plate geometry, we have $E = V/d$ for the electric field between electrodes a distance d apart across which a voltage difference V exists. An electron experiences a constant force $F = eE$, and therefore a constant acceleration. Thus:

$$d = \tfrac{1}{2}\,a\tau^2 = \tfrac{1}{2}\frac{F}{m}\,\tau^2 = \frac{eE}{2m}\,\tau^2$$

and

$$d = \frac{eV}{2md}\,\tau^2.$$

Solving for τ and putting in $d = 0.2\,\text{cm}$, $V \cong 100$ volts, $m = 9 \times 10^{-28}\,\text{gm}$, we find $\tau = \sqrt{2m/eV}\,d \cong 10^{-8}\,\text{sec}$. Thus we would expect tubes to be ineffective at frequencies on the order of $1/\tau = 10^8\,\text{Hz} = 100\,\text{MHz}$ or higher. Practically speaking, most vacuum tubes are ineffective at frequencies above several hundred MHz.

10.8 TRANSMISSION LINES

Nothing can travel faster than the speed of light $c \cong 3 \times 10^{10}\,\text{cm/sec}$. In particular, electrical signals, either voltages or currents, cannot propagate along a "line" faster than the speed of light. The term "line" is taken to mean any wire or collection of conductors designed to transmit electrical energy. This means that there is always a finite time required for any voltage or current change or signal to move down a line. If the voltage at one end of the line is changing relatively slowly, then the voltage at the other end will appear to change in exactly the same way; there will be no appreciable lag or phase difference between the voltages at the two ends. However, if the voltage at one end changes appreciably in a time comparable to (or less than) the time required for the voltage change to travel the length of the line, then there *will* be an appreciable phase difference between the voltages at the two ends. Quantitatively, if v is the speed of propagation down the line ($v < c$), and if L is the length of the line, then the time required to travel the length L is simply $\tau = L/v$. If the period of the voltage signal is T, then these phase differences will become important when $T \leqq \tau = L/v$. As $T = 1/f = \lambda/c$, we have $\lambda/c \leqq L/v$ where c is the speed of light and λ is the free space wavelength. Thus, $L \geqq v/c\lambda$ or $\lambda \leqq c/vL$ is the condition for appreciable phase difference between the voltages at the two ends of the line. Thus, for a given length wire L, these phase differences become more important at shorter wavelengths or higher frequencies. If, for example, the line is one-half wavelength long and if $v \cong c$, then the time for a voltage signal to travel down the line is just one-half a period, so the voltage at one end will differ in phase by 180° from the voltage at the other end. It will be shown later that the more series inductance L_s and shunt capacitance C_s there is in the line, the smaller the speed of propagation v. Hence, the larger L_s and C_s, the lower the frequency at which these phase differences become important.

We speak of a transmission "line" rather than a wire, because several wires or metal conductors of various shapes are usually used to transmit high frequency signals. One common transmission line is the conventional coaxial cable shown in Fig. 10.12(a), which consists of a center conductor surrounded

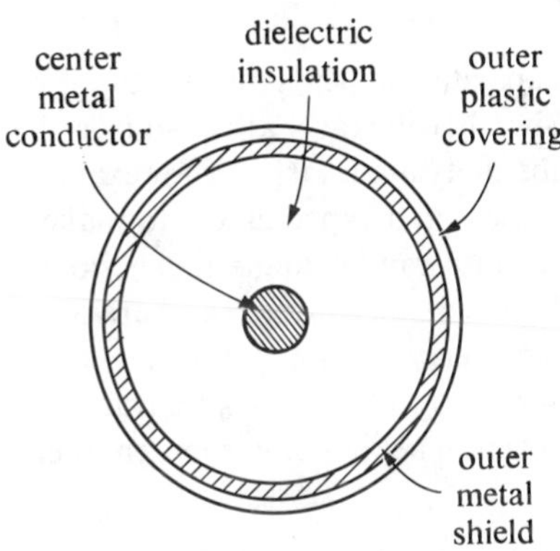

(a) *coaxial cable cross section*

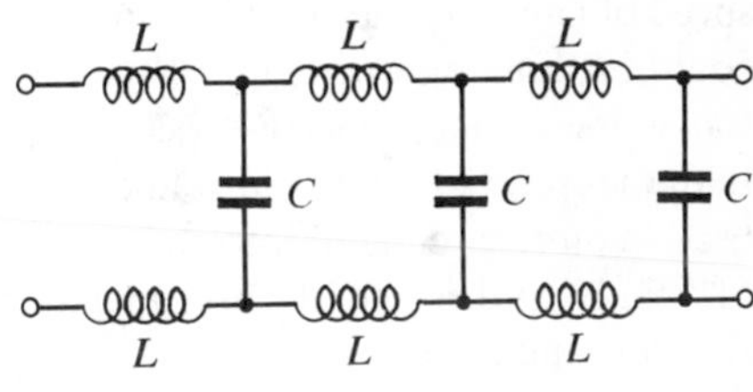

(b) *lumped circuit transmission line model*

FIGURE 10.12 Transmission lines.

by a cylindrical conductor called the shield, which is invariably grounded. The shield is usually braided copper covered with a plastic insulating material. The region around the center conductor is usually filled with a flexible dielectric, and the entire cable is flexible. There is a certain shunt capacitance per unit length associated with the cable. The exact value depends upon the geometry and the dielectric constant of the material between the inner conductor and the shield. A simple Gauss's law calculation yields for the capacitance per unit length:

$$C = \epsilon \ln \frac{R_2}{R_1}, \tag{10.26}$$

where R_2 is the outer radius, R_1 is the inner radius, and ϵ is the dielectric constant. For RG58A/U coaxial cable which is commonly used for frequencies up to several hundred MHz, $C = 28.5\,\text{pF}$ per foot or $0.93\,\text{pF/cm}$, and the series inductance per unit length is $0.0027\,\mu\text{H/cm}$.

A reasonably good electrical equivalent of a transmission line consists of a number of inductances and capacitances as shown in Fig. 10.12(b). The voltage across the line V and the current through the line each depend upon time t and the distance Z along the line. It can be shown that they obey the following equations:

$$\frac{\partial^2 V}{\partial Z^2} = LC\frac{\partial^2 V}{\partial t^2} \qquad \frac{\partial^2 I}{\partial Z^2} = LC\frac{\partial^2 I}{\partial t^2}, \tag{10.27}$$

which will be recognized as the standard wave equation by anyone who has studied partial differential equations. In any case, these equations imply that voltage or current signals propagate down the line with a speed:

$$v = \frac{1}{\sqrt{LC}}, \tag{10.28}$$

where L and C are the inductance and capacitance per unit length. For RG58A/U coaxial cable $v = 0.66c$; that is, signals will travel at two-thirds the speed of light. Notice that if L and C are deliberately made large, as for example could be done by wiring up a series of inductances and capacitances to make a transmission line, then the speed of propagation v can be made much slower than the speed of light. Such lines are usually called "delay lines" and are often used in pulse circuits to introduce a specific time delay. For example, if a series of $100\,\mu\text{H}$ chokes and $100\,\text{pF}$ capacitors are wired together to make a transmission line and if we have one choke and one capacitor per cm of line length, then:

$$v = \frac{1}{\sqrt{LC}} = \frac{1}{\sqrt{100 \times 10^{-6} \times 100 \times 10^{-12}}} = 10^7\,\text{cm/sec.}$$

Thus a signal will take $1\,\mu\text{sec}$ to travel along a 10 cm length of such a delay line.

A characteristic impedance can be defined for a transmission line. If the relationship between the voltage across the line and the current in the line is calculated, we obtain:

$$I = \frac{V}{\sqrt{L/C}}, \tag{10.29}$$

which implies that the line acts as an impedance $Z_0 = \sqrt{L/C}$. For RG58A/U cable $Z_0 = 53\,\Omega$. The standard TV twin lead has $Z_0 = 300\,\Omega$. Notice that Z_0 does not depend upon the length of the line, but rather only on the type of line.

The characteristic impedance of a transmission line is important, because if the load impedance attached to the output of a transmission line does not exactly equal the characteristic impedance of the line, then energy will be reflected from the load back up the line. This situation is called a mismatch and can be very serious at high powers and high frequencies. A line terminated with its characteristic impedance will have an input impedance equal to the characteristic line impedance regardless of the length of the line and the total shunt capacity of the line. This is a very desirable situation for transmitting high frequencies with minimum loss.

TABLE 10-2 TYPICAL CABLE DATA

Cable	*Characteristic Impedance*	*Velocity/c*	*Capacitance/foot*
RG58A/U	53Ω	0.66	28.5 pF
TV Twin Lead	300Ω	0.82	5.8 pF
RG59A/U	73Ω	0.66	21.0 pF

The amount of power reflected from the load is usually specified by giving the voltage-standing-wave-ratio or "VSWR." A mathematical solution of the transmission line equation shows that in general there are two waves traveling on the line, one in each direction:

$$V = V_+ e^{j(\omega t - 2\pi z/\lambda)} + V_- e^{j(\omega t + 2\pi z/\lambda)} . \tag{10.30}$$

The V_+ is the amplitude of the wave moving in the positive z direction, and V_- the amplitude of the wave moving in the negative z direction. These waves interfere to various degrees at different points along the line. At some points along the line the two waves will be exactly in phase and the voltage across the line will be:

$$V = (V_+ + V_-)e^{j\omega t} . \tag{10.31}$$

One-quarter wavelength ($\lambda/4$) farther down the line, each wave will have changed by 90° in phase, one by +90° the other by −90°; thus the two waves will be 180° out of phase and the voltage across the line will be:

$$V = (V_+ - V_-)e^{j\omega t} . \tag{10.32}$$

Thus, every $\lambda/4$, the voltage amplitude changes from a maximum to a minimum. The distance between two successive maxima or two successive minima is $\lambda/2$. The total voltage along the line due to the two waves traveling up and down the line is termed a "standing" wave; the positions of the maxima and minima remain the same. The voltage-standing-wave-ratio (VSWR) is defined as the quotient of the maximum voltage in the standing wave $(V_+ + V_-)$ and the minimum voltage in the standing wave $(V_+ - V_-)$:

$$\text{VSWR} \equiv \frac{V_+ + V_-}{V_+ - V_-}. \tag{10.33}$$

The transmitted power is proportional to V_+^2 and the reflected power to V_-^2. The VSWR can vary from 1.0 to ∞ as the reflected power varies from 0 to 100% of the transmitted power. If the power traveling back along the line from the load toward the driving source is zero, then the VSWR is equal to 1.0. If V_- approaches V_+, then the VSWR becomes infinite, and we have complete reflection.

The relationship between the VSWR and the percentage power reflected is given in Table 10-3.

TABLE 10-3 VSWR AND POWER REFLECTED

VSWR	*Percentage Power Reflected*
1.0	0%
1.5	4.0%
2.0	11%
∞	100%

In practical high-frequency circuits, impedance matching is usually used to eliminate power reflections from junctions between two lines of different impedance or between a line and a load. Impedance matching transformers are generally useful only below approximately 50 MHz. For higher frequencies where transmission lines are required, a quarter-wavelength line can be used to match the impedance of a line and a load or between two lines of different impedance. The condition for matching a line of impedance Z_1 to a load or other line of impedance Z_3 is that the impedance of the quarter-wavelength line Z_2 is: $Z_2 = \sqrt{Z_1 Z_3}$ (see Fig. 10.13).

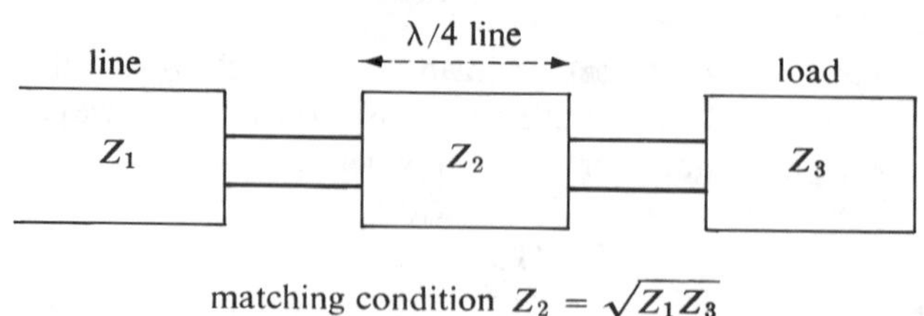

FIGURE 10.13 λ/4 line used as impedance matching device.

A quarter-wavelength line has another useful characteristic: When shorted at one end, it acts like a parallel LC resonant circuit. In general it is much easier at frequencies of hundreds of MHz to make a tuned resonant circuit with a quarter-wave shorted line than with discrete components, because the coil comprising the discrete inductance usually will have only one or one-half turns to resonate at such frequencies. The lead inductance is then comparable to the inductance of the turn of wire, and it is difficult to obtain the proper inductance because the wire layout in the circuit is so critical. Also, the loss of conventional capacitors tends to increase at frequencies above 100 MHz.

Consider a transmission line along the z axis and shorted at $z = 0$, and excited by a sinusoidal voltage at the other end as shown in Fig. 10.14. The

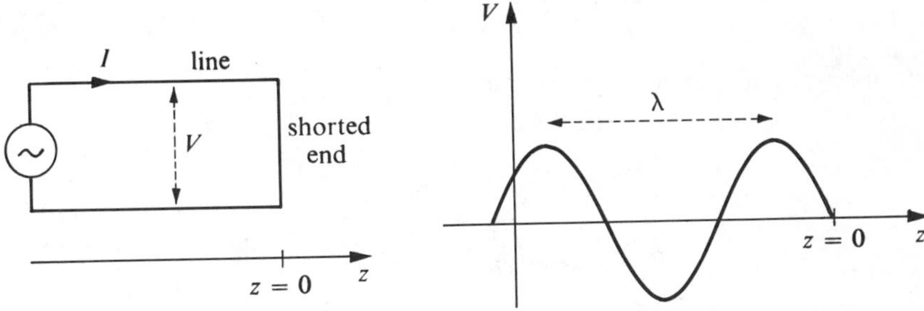

FIGURE 10.14 Shorted transmission line.

general solution of the transmission line equation consists of the sum of two waves traveling up and down the line, in the positive and negative z direction:

$$V = V(z, t) = V_+ e^{j(\omega t - 2\pi z/\lambda)} + V_- e^{j(\omega t + 2\pi z/\lambda)}. \tag{10.34}$$

At the shorted end, $z = 0$, V must equal zero—there cannot be any voltage developed across a short circuit. Thus,

$$V(0, t) = V_+ e^{j\omega t} + V_- e^{j\omega t} = 0,$$

which implies $V_+ + V_- = 0$ or $V_+ = -V_-$. The reflected voltage has the same magnitude as the transmitted voltage; the VSWR is infinite. Physically, this is reasonable because no energy can be dissipated in the shorted end, which can be regarded as a zero load; hence, the reflected power must equal the transmitted power. Putting $V_+ = -V_-$, we obtain for the voltage along the line:

$$V(z, t) = V_+ e^{j(\omega t - 2\pi z/\lambda)} - V_+ e^{j(\omega t + 2\pi z/\lambda)}$$

$$V(z, t) = V_+ e^{j\omega t}[e^{-j2\pi z/\lambda} - e^{j2\pi z/\lambda}]$$

or

$$V(z, t) = -2V_+ j e^{j\omega t} \sin 2\pi z/\lambda. \tag{10.35}$$

The voltage along the line varies sinusoidally with z, and equals zero every $\lambda/2$. The voltage is a maximum in absolute value every $\lambda/2$ along the line. This pattern is called a standing wave because the voltage wave appears to stand still with respect to z motion if one disregards the oscillation in time. That is, the z positions of the voltage maxima and zeroes do not move. For a fixed z, of course, the voltage oscillates from "+" to "−" with an angular frequency ω.

The quarter-wave line can be shown to act like an LC resonant circuit by considering the energy stored in the electric and magnetic fields associated with the voltage and current in the line. The current along the line can be shown to be shifted in z by $\lambda/4$ with respect to the voltage and is given by:

$$I = I(z, t) = 2\frac{V_+}{Z_0}\cos\frac{2\pi z}{\lambda}\,e^{j\omega t}. \tag{10.36}$$

We can rewrite the voltage and current as:

$$\begin{aligned} V(z, t) &= 2V_+ \sin\frac{2\pi z}{\lambda}\,e^{j(\omega t - \pi/2)} \\ I(z, t) &= 2\frac{V_+}{Z_0}\cos\frac{2\pi z}{\lambda}\,e^{j\omega t}. \end{aligned} \tag{10.37}$$

This tells us that the current and voltage are 90° = $\pi/2$ radians out of phase in time. Thus the voltage is zero everywhere along the line at $t = 0$, and the current is maximum. At this time all energy must be associated with the current; i.e., the energy is magnetic energy U_m and is given by:

$$U_m = \int_0^L \frac{1}{2}(L\,dz)I^2. \tag{10.38}$$

If the line is exactly $\lambda/4$ long, then:

$$U_m = \frac{V_+^2 L\lambda}{4Z_0^2}. \tag{10.39}$$

Remember L is the inductance per unit length, so $L\,dz$ is the inductance of the length dz of the line.

When $t = T/4$, the current is zero everywhere along the line, and all energy must be associated with the voltage; i.e., the energy must be electric field energy U_E in the charged capacitance of the line:

$$U_E = \int_0^L \frac{1}{2}(C\,dz)V^2. \tag{10.40}$$

If the line is exactly $\lambda/4$ long, then:

$$U_E = \frac{V_+^2 C\lambda}{4}. \tag{10.41}$$

Substituting in $Z_{\text{out}} = \sqrt{L/C}$, we find $U_m = U_E$. Thus we have the result that for a quarter-wavelength line the electric energy U_E equals the magnetic energy

U_m, and U_E and U_m are 90° out of phase in time. This is exactly the situation we have in a conventional parallel *LC* resonant circuit. The resonant frequency is determined by that frequency for which the line length equals $\lambda/4$.

In the quarter-wavelength line the voltage is maximum at the open, input end, and the current is maximum at the shorted end. Thus the electric field in the line will be maximum near the open, input end, and the magnetic field will be maximum near the shorted end. In an electron spin resonance experiment, for example, to stimulate the magnetic dipole transitions a strong oscillating *magnetic* field is required. Hence the sample should be placed near the shorted end of the line where the magnetic field is maximum. If a strong electric field were required, for example to stimulate *electric* dipole transitions, the sample should be placed near the open, input end.

10.9 DISTRIBUTED AMPLIFIERS

Nothing can be done to reduce the capacitance between ground and the various transistor or tube elements except to make better devices with lower capacitances, and there appears to be little room for improvement in this area. But, one can make a transmission line using these inherent capacitances by adding small inductances as shown in simplified form in Fig. 10.15. The inductances

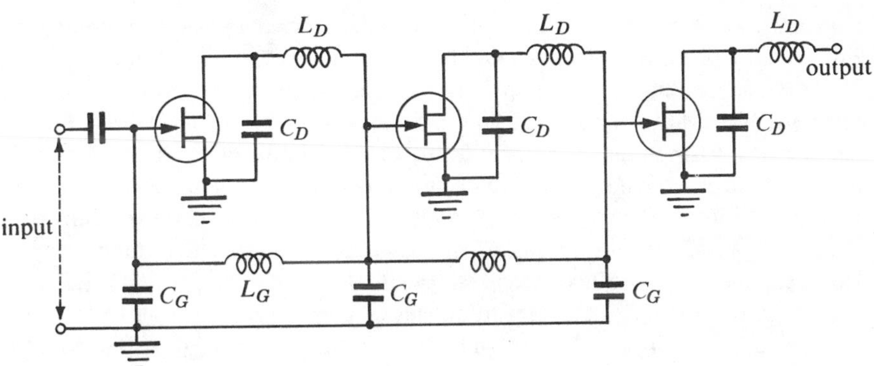

FIGURE 10.15 Distributed FET amplifier.

L_D and L_G are added to form two transmission lines, one in the gate circuit and one in the drain circuit. Both L_D and L_G are adjusted so that the time required for a signal to pass from one stage to the next is the same for each line. That is, the velocity of signal propagation must be the same for each line, $1/\sqrt{L_G C_G} = 1/\sqrt{L_D C_D}$. Thus the gate and drain voltages retain the proper phase and the signal is amplified. The net effect is to multiply effectively the FET transconductance by the number of stages, although this is difficult to realize in practice.

10.10 TUNED AMPLIFIERS

The simplest way to offset the high-frequency attenuation due to the shunt capacitance C_s to ground (of the transistors and the wiring) is to add an inductance in parallel with C_s as shown in Fig. 10.16(a). This forms a parallel LC

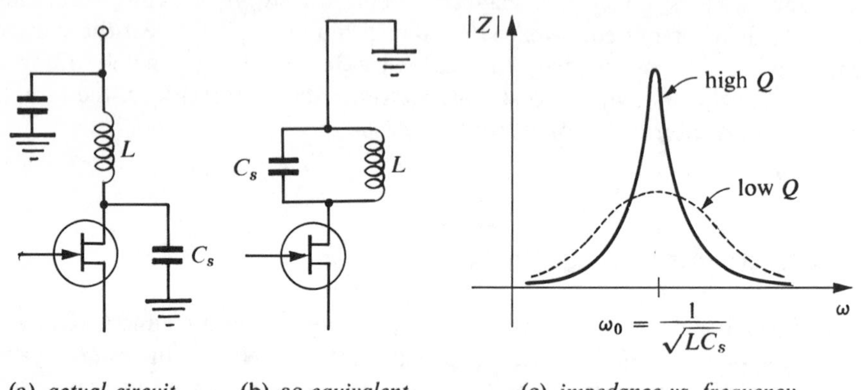

(a) *actual circuit* (b) ac *equivalent* (c) *impedance vs. frequency*

FIGURE 10.16 Tuned amplifier.

resonant circuit which will have a high impedance near its resonant frequency. For $C_s \approx 10\,\text{pF}$, a coil of 2 turns with a $\frac{1}{4}$-inch diameter will resonate at about 100 MHz depending upon the spacing between the turns, a tightly wound coil resonating at a lower frequency than a loosely wound coil. However, the impedance is maximum at the resonant frequency and falls off both above and below resonance at a rate depending upon the Q of the resonant circuit, as shown in Fig. 10.16(c). It is possible to have a bandwidth of several MHz over which the amplifier will operate if the resonant frequency f_0 is high enough. For example, if $f_0 = 50\,\text{MHz}$ and $Q = 20$, then $\Delta f = f_0/Q = 50\,\text{MHz}/20 = 2.5\,\text{MHz}$; whereas if the resonant frequency is lower a much smaller bandwidth f is obtained, e.g., if $f_0 = 1\,\text{MHz}$ and $Q = 20$, then $\Delta f = f_0/Q = 0.05\,\text{MHz} = 50\,\text{kHz}$. It is sometimes necessary to deliberately lower the Q to increase Δf for a given resonant frequency. If two tuned amplifier stages are cascaded in series with the output of one supplying the input to the next, as shown in Fig. 10.17(a), then the two tuned circuits can be deliberately tuned to slightly different resonant frequencies, which obtains a broader bandwidth. This is known as "stagger" tuning, and the gain versus frequency will be as shown in Fig. 10.17(b).

The output of stage 1 is RC coupled to the input of stage 2 in Fig. 10.17(a), through the coupling capacitor C_C and the gate-to-ground resistor R_G. Sometimes a transformer is used to couple the signal from one stage to the next as shown in Fig. 10.18.

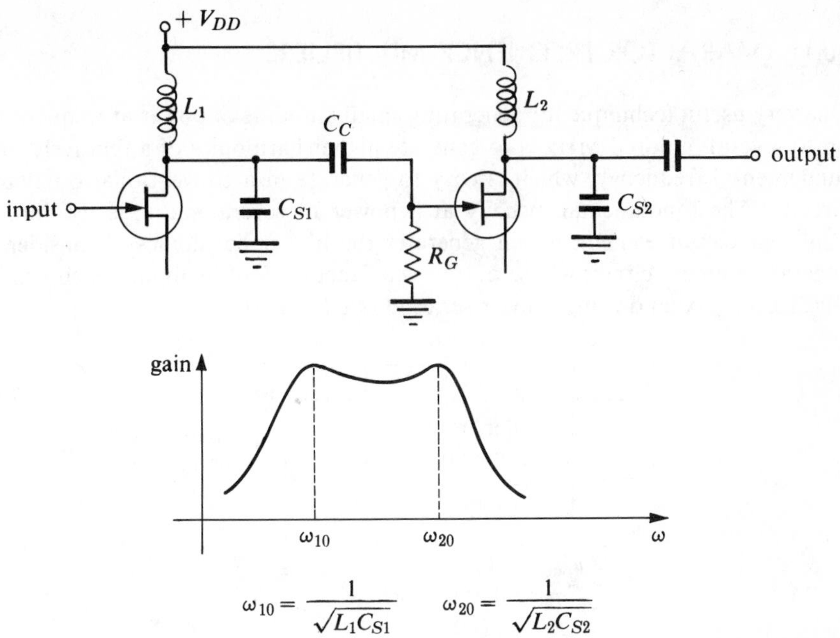

FIGURE 10.17 Two-stage tuned amplifier.

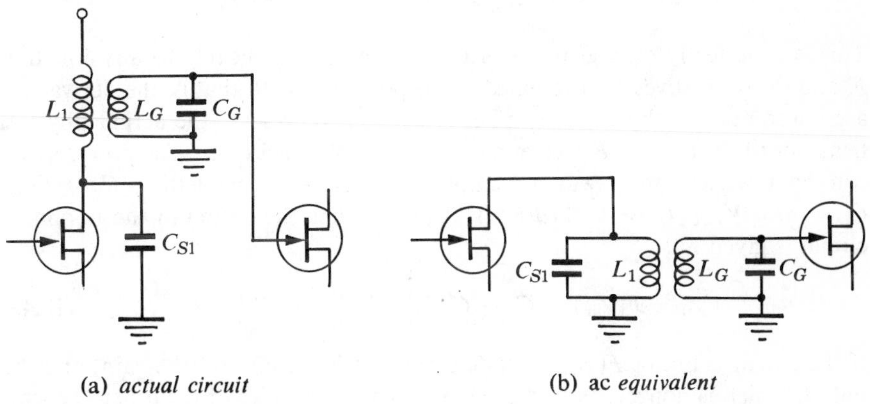

FIGURE 10.18 Tuned transformer coupling.

By Faraday's law of induction, $V = N\,d\phi/dt$, transformer coupling works very well with air core coils at frequencies from 1 MHz to 100 MHz because of the rapid flux change $d\phi/dt$ due to the high frequencies. The combination L_G and C_G forms a tuned resonant circuit in the gate of the second stage. The inductance L_G provides a dc path from the gate of the second FET to ground.

10.11 VARACTOR FREQUENCY MULTIPLIERS

One very useful technique for generating small amounts of power at frequencies up to several hundred MHz is to generate higher harmonics of a relatively low fundamental frequency, which is easy to generate in a conventional oscillator circuit. The fundamental, usually at a power of several watts, is fed into a nonlinear circuit element which generates the higher harmonics. Consider a general nonlinear circuit whose output is a function F of its input, as shown in Fig. 10.19. Expand F in a power series about $V_{in} = 0$:

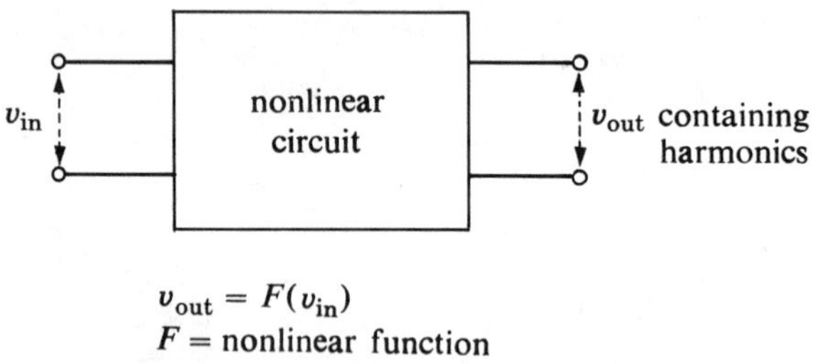

FIGURE 10.19 Harmonic generation by nonlinear circuit.

$$F = F(0) + \left(\frac{dF}{dV_{in}}\right)_0 V_{in} + \frac{1}{2!}\left(\frac{d^2F}{dV_{in}^2}\right)_0 V_{in}^2 + \frac{1}{3!}\left(\frac{d^3F}{dV_{in}^3}\right)_0 V_{in}^3 + \cdots \quad (10.42)$$

This is a perfectly general power series expansion, applicable to any function F, and the derivatives are evaluated at the point $V_{in} = 0$; that is, the derivatives are constants. It should be emphasized that F represents the actual physical behavior of the circuit; F is essentially the gain of the circuit. The power series can be rewritten in a way to emphasize this point by letting $C_0 = F(0)$, $C_1 = (dF/dV_{in})_0$, $C_2 = 1/2!\,(d^2F/dV_{in}^2)_0$, etc. Then the output of the nonlinear circuit is given by:

$$V_{out} = F(V_{in}) = C_0 + C_1V_{in} + C_2V_{in}^2 + C_3V_{in}^3 + \cdots \quad (10.43)$$

If the circuit is linear, $F(V_{in}) = AV_{in}$, where A is a constant (the gain) and the only C which is nonzero is C_1; $C_1 = A$. The magnitudes of the other C coefficients determine just how nonlinear the circuit is.

If the input is a perfectly pure sinusoidal wave, $V_{in} = V_0 \cos \omega t$, then the output in general is given by:

$$V_{out} = C_0 + C_1V_0 \cos \omega t + C_2V_0^2 \cos^2 \omega t + C_3V_0^3 \cos^3 \omega t + \cdots \quad (10.44)$$

The presence of higher harmonics can be seen by using the following trigonometric identities:

$$\cos^2 \omega t = \frac{1 + \cos 2\omega t}{2} \tag{10.45}$$

$$\cos^3 \omega t = \frac{3 \cos \omega t + \cos 3\omega t}{4}. \tag{10.46}$$

Thus the output can be written as:

$$V_{\text{out}} = C_0 + C_1 V_0 \cos \omega t + C_2 \frac{V_0^2}{2} + C_2 \frac{V_0^2}{2} \cos 2\omega t + \frac{3}{4} C_3 V_0^3 \cos \omega t + \frac{C_3}{4} V_0^3 \cos 3\omega t + \cdots \tag{10.47}$$

The output is seen to contain a dc term, a term in cos ωt, a term at twice the fundamental frequency (cos $2\omega t$), a term at three times the fundamental (cos $3\omega t$), and so forth. Notice that the amplitude of the higher harmonics goes down as the frequency goes up; that is, if the second harmonic at 2ω contains 20 mW of power then the third harmonic at 3ω might contain only 5 mW, depending on the particular type of nonlinear circuit used.

The most common nonlinear circuit element used to generate higher harmonics is a diode, although it should be emphasized that *any* nonlinear circuit will generate some harmonics. In particular, an overdriven amplifier whose output is a distorted version of its input will generate harmonics. However, a diode is inexpensive, extremely nonlinear because it clips off one polarity of the input, and generates considerable power in the harmonics.

Any diode will generate harmonics when driven by a sinusoidal signal, but the best diodes for frequencies above tens of MHz are termed "varactor" diodes, which are pn junction diodes specially made to have a low forward series resistance, a high back resistance, an appreciable junction capacitance, and a low stray capacitance. Another name for a diode especially designed to generate harmonics is "multiplier" diode. A varactor equivalent circuit is shown in Fig. 10.20(a) along with its circuit symbol. Often C_s and L can be considered negligibly small and R_J essentially infinite, so that R_s and C_J are the only important parameters.

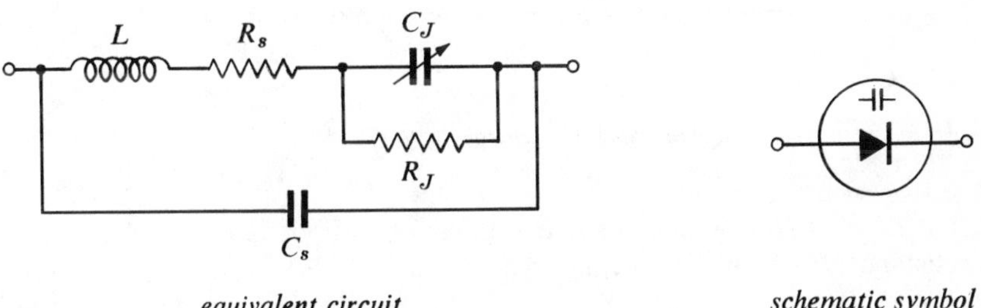

FIGURE 10.20 Varactor diode.

When the varactor diode is back-biased, there is a depletion region formed between the p- and the n-type materials as explained in Chapter 4. The depletion region contains no mobile charge carriers and so acts like a dielectric material. The p and n regions on either side of the depletion layer do contain mobile charge carriers and thus act like conductors. Hence the back-biased junction is similar to a capacitor. As the reverse bias voltage increases, the width of the depletion region increases, and thus the capacitance C_J between the p and n sides of the diode decreases. The capacitance varies inversely (approximately) as the square root of the reverse bias voltage V for most devices. The exact expression is:

$$C_J = \frac{\mathcal{K}}{\sqrt{W + V}}, \tag{10.48}$$

where W is the contact potential for the semiconductor junction (0.5 V for silicon) and $\mathcal{K}$ is a constant that depends upon the particular diode. Typical values are: for $V = 1\,\text{V}$, $C_J = 30\,\text{pF}$; for $V = 10\,\text{V}$, $C_J = 12\,\text{pF}$; for $V = 100\,\text{V}$, $C_J = 4.5\,\text{pF}$ for a 1N4937 diode.

It is clear from the fact that the capacitance depends upon the reverse bias voltage, that a varactor diode can be used as a tuning capacitor; instead of turning a knob to vary the capacitance, the reverse bias voltage is changed. Varactor diodes are available with capacitance ranges of up to ten-to-one as the reverse bias voltage changes, although a range of three-to-one is more common. Because of the high reverse resistance of a back-biased silicon junction, the varactor draws almost no current ($\approx 10^{-9}$ A), and therefore consumes almost no power. Hence the Q is very high and the varactor can be used at frequencies up to ten GHz or more. The basic circuit for a varactor-tuned circuit is shown in Fig. 10.21.

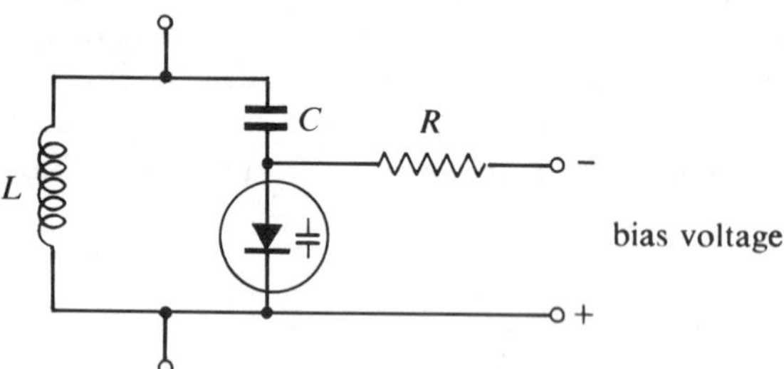

FIGURE 10.21 Varactor-tuned parallel *LC* circuit.

The dc bias voltage ensures that the varactor diode is reverse-biased, and the resistance R isolates the varactor from the battery. The resistance R can and should be rather large for maximum isolation. If the diode draws 10^{-9} A of current, then $R = 10\,\text{M}\Omega$ will mean that only a 0.01 V dc voltage drop appears across R. For high frequencies an rf choke is often used in place of R. The capacitance C should be much larger than the varactor capacitance C_J, so that C_J rather than C determines the resonant frequency. If $C_J = 100\,\text{pF}$, then C should be at least $0.01\,\mu\text{F}$. The circuit can be tuned either by varying the bias voltage or by applying an ac voltage across the varactor through a series capacitor. This technique is particularly useful in frequency-control circuits where a control voltage is required to adjust a frequency or tune an amplifier. There is an upper frequency limit for tuning the circuit by varying the dc bias voltage. Together R and C_J form a low-pass RC filter, so variations in the bias at frequencies above the break point, $f_B = \frac{1}{2\pi} RC_J$, will not be transferred to the varactor.

Varactor diodes are regularly used at this writing to double, triple, and quadruple input frequencies. A simple varactor tripler circuit is shown in Fig. 10.22. The series circuit L_1C_1 is tuned to the input or fundamental frequency f

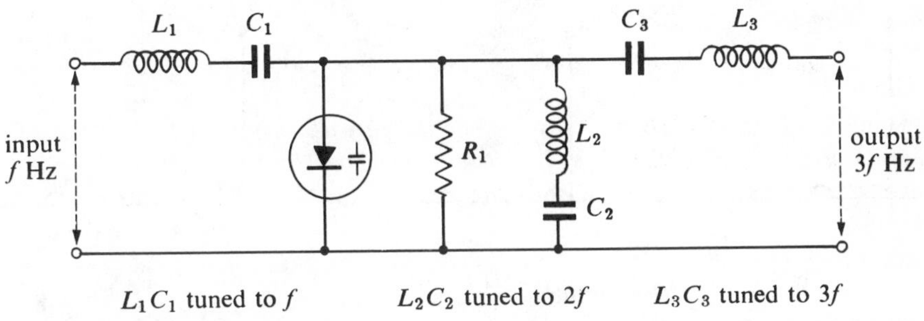

FIGURE 10.22 Varactor tripler circuit.

so that the varactor is driven strongly at f Hz. The resistance R_1 is usually of the order of $100\,\text{k}\Omega$ and serves to provide a dc return to ground, so that the varactor capacitance C_J does not charge up to a high dc negative voltage due to the negative swings of the input voltage. The varactor generates many higher frequency harmonics at $2f$, $3f$, $4f$, etc. The L_2C_2 circuit shorts out the second harmonic at $2f$, and the L_3C_3 circuit passes only the third harmonic through to the output. The fourth and higher harmonics are isolated from the output by L_3C_3. Efficiencies up to 70% to 80% can be obtained at output frequencies up to 500 MHz with such circuits, and varactor diodes are available that can handle up to 10 to 20 watts of rf power.

10.12 CLASSES OF AMPLIFICATION

If an amplifier is biased so that collector or drain or plate current flows during only part of the input signal cycle, then the output waveform is distorted and many harmonics are present in the output. Three general classes of amplifiers are usually defined: classes A, B, and C. In a class A amplifier, collector current flows during the entire input signal cycle. In class B, the bias is set so that collector current flows only during one-half the input signal cycle; that is, the input signal is essentially rectified as well as amplified. In class C, the bias is set so that collector current flows during less than one-half the input signal cycle; in other words, only the tips of the input signal cycle are amplified. Figure 10.23 shows the waveforms for the three classes.

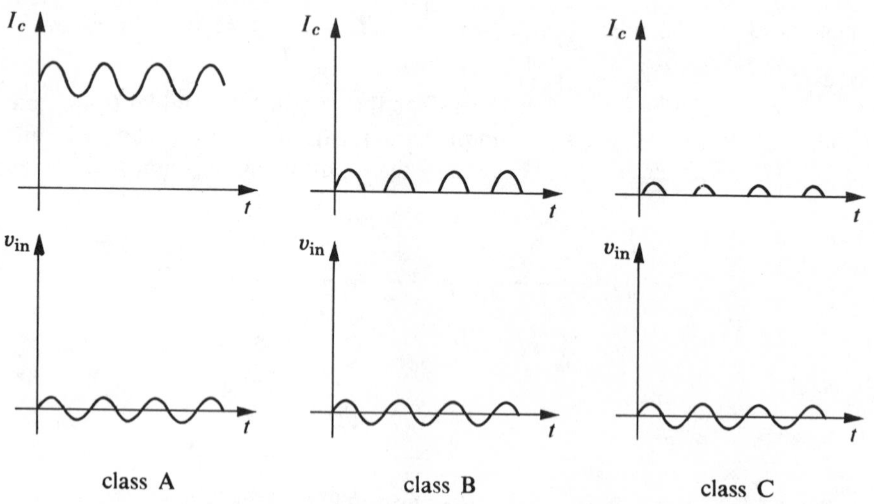

FIGURE 10.23 Classes of amplification.

Because of the distortion of the input waveform in class B and C amplifiers, the output contains many harmonic multiples of the input frequency. For example, if the input is at 10 MHz, the output will contain 10 MHz, 20 MHz, 30 MHz, 40 MHz, etc., harmonics.

All of the amplifiers we have considered up to now have been class A—the output waveform is a faithful, enlarged version of the input. All amplifiers that must pass a broad band of frequencies without appreciable harmonic distortion must be class A; for example, all audio high fidelity amplifiers must pass without distortion frequencies from 20 Hz to 20 kHz, a bandwidth of approximately 10 octaves. Running class B or C would result in a horribly distorted signal, for example, a pure input note at 400 Hz would produce output notes at 800 Hz, 1200 Hz, etc. In a high-frequency radio transmitter, however, with a carrier frequency at many megahertz, the frequencies passed in the signal sidebands will occupy only a very small fraction of an octave. Harmonic multiples of the signal will lie far beyond the width of the sidebands and thus can easily be rejected by a filter of modest Q in the receiver or transmitter.

Efficiency is a major consideration in the design of high power circuits, hence the use of class C amplifiers in high power radio transmitters. The high degree of harmonic distortion present in class C operation is of little consequence if the collector load is a tuned circuit with a reasonably high Q, because the higher harmonics are filtered out. This is typically the case in high power radio transmitters where there are usually tuned circuits coupling the various amplifier stages together, and the final output load, the antenna, is carefully tuned to the fundamental frequency of the input signal. If the rejection of the harmonics is not complete, some power will be broadcast at integral multiples of the fundamental. This is often the case with amateur radio transmitters, and because of this the allowed frequencies for amateur operation are mainly integral multiples of one another. For example, some amateur bands are, in terms of wavelength, 10, 20, 40, and 80 meters. A transmitter broadcasting on the 80 meter band can often be also heard on the 40, 20, and 10 meter bands.

The principal use of class B and especially C amplifiers is in high-frequency, high power radio transmitters. A little thought will show that class A is very inefficient, because collector current is continuously flowing even in the absence of a signal. Without going into an exact analysis of the efficiency of the various classes, class C is considerably more efficient than class A.

10.13 MICROSTRIP AND STRIP LINE TECHNIQUES

One of the most useful recent techniques for transmitting high-frequency signals is with special types of transmission lines called "microstrips" or "strip lines." In a microstrip line, there is one conducting strip resting on a large dielectric sheet in contact with a large grounded plane, as shown in Fig. 10.24(a). Electrically speaking, the grounded plane can be replaced by another conducting strip oppositely charged as shown in Fig. 10.24(b).

The second strip is called the "image" of the first, because its position is that of a mirror image of the actual strip. Thus, the actual microstrip construction of a conducting strip and a ground plane is electrically equivalent to a line of two conducting strips. The signal is applied to the conducting strip and most of the electric field lines go to the ground plane through the dielectric strip. An isolated signal-carrying conducting strip is shown in Fig. 10.24(c). The electric field lines all go from the strip out to infinity and tend to radiate more electric power than the microstrip configuration.

The characteristic impedance of a microstrip line with dielectric thickness h and conductivity strip width w is $Z_0 = \sqrt{\mu/\epsilon}\, h/w$, where μ is the permeability and ϵ is the dielectric constant of the dielectric. For most dielectric materials $\mu = \mu_0 = 4\pi \times 10^{-7}$ mks units, the permeability of free space; and $\epsilon = \kappa\epsilon_0$, where κ is the familiar dielectric constant ($\kappa = 1$ for vacuum, $\kappa > 1$ for all dielectrics) and $\epsilon_0 = 8.85 \times 10^{-12}$ mks units is the permittivity of free space. Thus the characteristic impedance is given by:

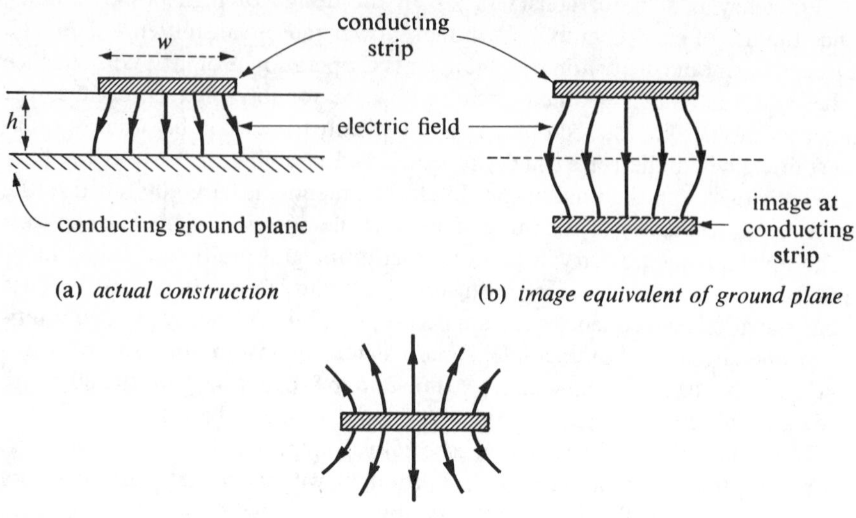

FIGURE 10.24 Microstrip construction and electric field lines.

$$Z_0 = \frac{1}{\sqrt{\kappa}} \sqrt{\frac{\mu_0}{\epsilon_0}} \frac{h}{w}. \tag{10.49}$$

The factor $\sqrt{\mu_0/\epsilon_0}$ has the numerical value of $377\,\Omega$. Neglecting fringing and leakage flux, we find that this expression for the characteristic impedance can be derived by elementary considerations, using only Gauss's law, Faraday's law, and Ampere's law, and is accurate to within 10% provided that h/w is 0.1 or less.

One can show also from elementary considerations that the propagation velocity v of electromagnetic waves down a microstrip line is given by $v = c/\sqrt{\kappa}$, where c equals the speed of light or 3×10^8 m/sec and κ is the familiar dielectric constant. For a Teflon-impregnated fiberglass dielectric, $\kappa = 2.65$, which gives $v = 1.84 \times 10^8$ m/sec. If $h/w = 0.1$, the characteristic impedance of such a line is $46\,\Omega$.

A more widely used type of transmission line is called a "strip line," and consists of a conducting strip and two ground planes as shown in Fig. 10.25.

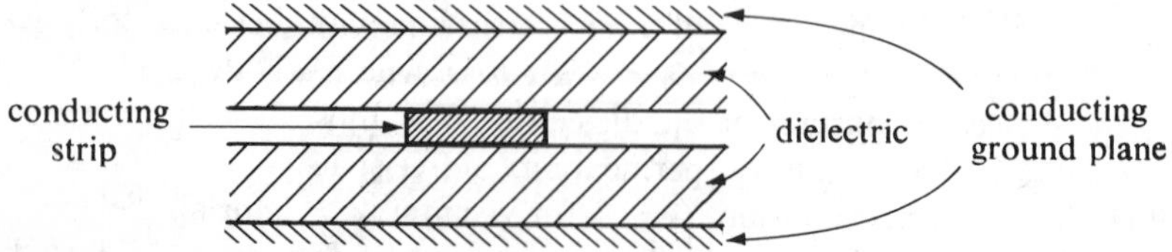

FIGURE 10.25 Strip line construction.

The advantages of strip line construction are that little power is lost by radiation even when the line dimensions are on the order of a wavelength, the characteristic impedance of the line can be fixed at any desired value by varying the dimensions of the conducting strip and the dielectric constant, and extremely compact, low loss circuits can be mass produced by modern manufacturing techniques. This last advantage is particularly advantageous for aircraft electronics. Strip line circuits can handle frequencies up to 10 GHz(= 10,000 MHz). Thus, a pulse with a rise time of the order of $(10\,\text{GHz})^{-1} = 10^{-10}$ sec = 0.1 nsec could be amplified with strip line technique. Present circuit techniques routinely deal with rise times of the order of 1–10 nsec.

X-band (10 GHz) receivers have been made using strip line techniques. The dielectric substrate is 5×10^{-2} cm thick sapphire with an effective dielectric constant of approximately 10. The conducting strip is 5×10^{-2} cm wide, and this produces a 50Ω characteristic line impedance which matches standard 50Ω connectors and components. The power loss of these lines is about 0.1 dB per cm at 10 GHz, which compares with 0.015 dB per cm for RG/58 coaxial cable at only 3 GHz and 0.0008 dB per cm for a $1'' \times 2''$ brass waveguide at 3 GHz.

With a certain amount of experience, conventional printed circuit techniques can be used to handle frequencies up to about 1GHz, or pulse rise time down to several nanoseconds. One side of the dielectric circuit board (usually fiberglass) is covered with copper and forms a ground plane. The conducting strips carrying the signals are on the other side of the circuit board, and so we have essentially a microstrip transmission line. Standard green fiberglass circuit board, copper coated on both sides, is used. The wiring is formed on one side by etching off the unwanted copper coating with acid. The remaining copper leads or strips approximately 0.1″ wide form transmission lines of approximately 50Ω characteristic line impedance, which will match 50Ω connectors and cables to minimize unwanted power reflection.

10.14 WAVEGUIDES

The lowest loss technique for transmitting power in the microwave frequency region uses waveguides as transmission lines. A waveguide is basically a closed metal pipe of high conductivity with no central conductor. Electromagnetic waves propagate down the inside of the waveguide in various modes. The waveguide may be rectangular or circular in cross section, as shown in Fig. 10.26. The exact mathematical treatment of the propagation of electromagnetic waves down a waveguide is beyond the scope of this book, but a brief summary will illustrate the electric and magnetic field configurations inside the guide.

All waveguide theory can be deduced from the four basic equations of electromagnetism, Maxwell's equations. In differential form in free space, they are:

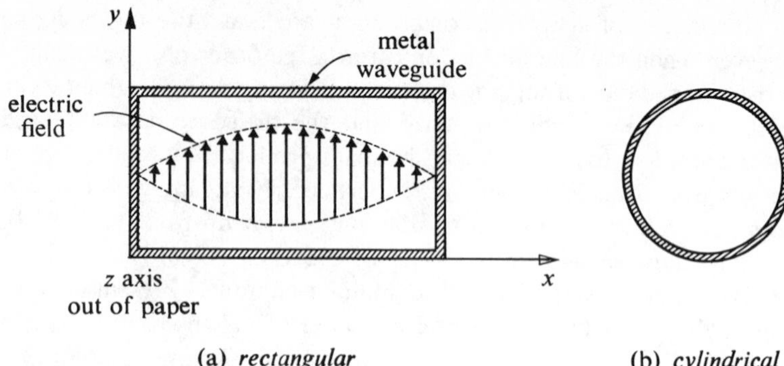

(a) *rectangular* (b) *cylindrical*

FIGURE 10.26 Waveguide cross sections.

$$
\begin{aligned}
\nabla \cdot \mathbf{E} &= 0 \\
\nabla \cdot \mathbf{H} &= 0 \\
\nabla \times \mathbf{E} &= -\mu_0 \frac{\partial \mathbf{H}}{\partial t} \\
\nabla \times \mathbf{H} &= \epsilon_0 \frac{\partial \mathbf{E}}{\partial t}.
\end{aligned}
\tag{10.50}
$$

The first equation says that the electric field **E** lines totally inside the waveguide are continuously closed loops; there are no charges on which **E** lines can start or stop. (**E** lines can start and stop on charges in the conducting walls of the waveguides.) The second equation says that the magnetic field **H** lines are continuous closed loops. This is always true in free space or any medium. The third equation says that a closed loop of **E** field lines exists only when the magnetic field inside the loop and perpendicular to the loop is changing in time. The fourth equation says that a closed loop of **H** field lines exists only when the electric field inside the loop and perpendicular to the loop is changing in time. These equations are covered in detail in any book on electricity and magnetism and will not be treated further here.

Basically waveguide propagation is treated by assuming the electric and magnetic fields inside the waveguide are sinusoidal in form and also obey Maxwell's equations. The assumed sinusoidal form is:

$$
\begin{aligned}
\mathbf{E} &= \mathbf{E}_0(x, y)e^{j2\pi\left(\frac{t}{T} - \frac{z}{\lambda_g}\right)} \\
\mathbf{H} &= \mathbf{H}_0(x, y)e^{j2\pi\left(\frac{t}{T} - \frac{z}{\lambda_g}\right)}.
\end{aligned}
\tag{10.51}
$$

That is, we have assumed the electric and magnetic fields each vary sinusoidally in time with a period T (and a frequency $f = 1/T$), and sinusoidally along the waveguide (which runs along the z axis) with a spatial periodicity λ_g, which is called the guide wavelength. These equations are substituted into the Maxwell equations, and after some algebraic manipulation the solutions fall naturally into two groups.

The first group consists of transverse electric (TE) waves in which the electric field is everywhere perpendicular to the long axis of the waveguide; that is, $E_z = 0$. The second group consists of transverse magnetic (TM) waves in which $H_z = 0$. In both cases the electric and magnetic field lines are everywhere perpendicular to one another. A third type of wave called the transverse electro magnetic waves (TEM) is possible in which $E_z = 0$ and $H_z = 0$, if the wavelength is much less than the interior dimension of the waveguide or if there is some conductor inside the waveguide.

The most common shape for a waveguide is one of rectangular cross section as shown in Fig. 10.27. We will now consider the TE waves, which are most

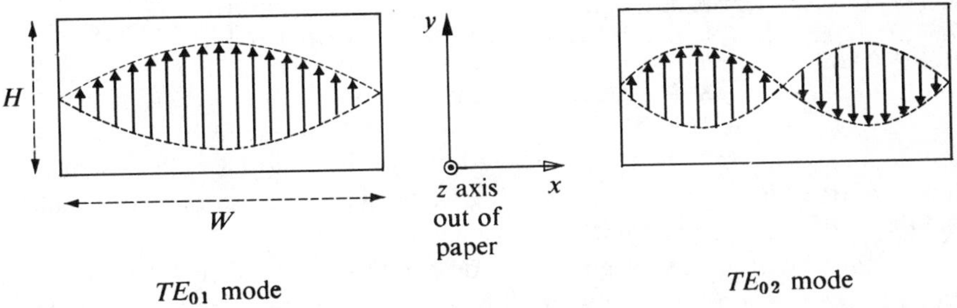

FIGURE 10.27 Electric fields in a rectangular waveguide.

often used. The most important boundary condition determining the configuration of the electric fields inside the waveguide is that the tangential component of the electric field must equal zero at the extreme edges of the waveguide. Here we are assuming zero resistance in the metal walls of the waveguide. The mathematical solutions require that the amplitude of the y component of the electric field vary sinusoidally with x; hence, we must have E_y varying something like that shown in Fig. 10.27. It can be shown that the maximum free-space wavelength that will propagate down a rectangular waveguide of height H and width W is $\lambda_{max} = 2W$. This corresponds to a minimum frequency for propagation, the "cutoff" frequency, so the waveguide essentially acts like a high-pass filter. Any frequency below cutoff is very rapidly attenuated. For example, a standard X-band rectangular waveguide measures $\frac{1}{2} \times 1$ in. on the outside, and $.4 \times .9$ in. on the inside. Thus $W = 0.9$ in. $= 2.28$ cm and the maximum free-space wavelength which will propagate down the guide is $\lambda_{max} = 4.56$ cm, corresponding to $f_{min} = c/\lambda_{max} = (3 \times 10^{10}\,\text{cm/sec})/4.56\,\text{cm} = 6.6\,\text{GHz}$. The smaller the waveguide, the higher the cutoff frequency. Table 10-4 gives outer waveguide dimensions and the frequency ranges.

One of the principal advantages of using waveguides is that the attenuation is extremely low. An uncovered wire several cm long carrying a 10 GHz signal would radiate most of the energy, thus producing essentially no transmission along the wire. The X-band brass waveguide, on the other hand, has a loss of only approximately 0.5 dB per foot. The disadvantages of waveguides

TABLE 10-4 WAVEGUIDE SIZES AND FREQUENCY

Military Type No.		*Waveguide Outer (inches)*	*Frequency Range*
RG 69/U		3.41 × 6.66	1.12–1.70 GHz
RG 104/U		2.31 × 4.46	1.7–2.66 GHz
RG 48/U	(S band)	1.5 × 3.0	2.6–3.95 GHz
RG 49/U Brass	(C band)	1 × 2	3.95–5.85 GHz
RG 50/U Brass	(C band)	0.75 × 1.5	5.85–8.2 GHz
RG 52/U Brass	(X band)	0.5 × 1	8.2–12.4 GHz
RG 91/U Brass	(Ku band)	0.391 × 0.702	12.4–18.0 GHz
RG 53/U Brass	(K band)	0.25 × 0.5	18.0–26.5 GHz
RG 96/U Silver	(Ka band)	0.22 × 0.36	26.5–40.0 GHz
WR19		0.174 × 0.268	40.0–60.0 GHz
RG 99/U Silver		0.141 × 0.202	60.0–90.0 GHz

are high expense, bulkiness, and heavy weight. For short lengths, microstrip lines or strip lines seem to be replacing waveguides because they are smaller, lighter, and easier to manipulate experimentally.

A short section of waveguide closed at both ends is called a "cavity" and can act like a resonant LC circuit with an extremely high Q, on the order of 10,000. The mathematical treatment is beyond the scope of this book, but a solution of Maxwell's equations inside the cavity satisfying the boundary conditions shows that oscillating electromagnetic fields can exist only at certain discrete frequencies that depend upon the cavity dimensions. For the cavity of rectangular cross section and length L as shown in Fig. 10.28(a), a standing electromagnetic wave which stores energy is possible only if the electromagnetic waves reflect back and forth along the z axis; so that $E_x = 0$ at the two ends where $z = 0$ and $z = L$. This requires that the cavity length be exactly $\lambda_g/2$, λ_g, $3\lambda_g/2$, etc., where λ_g is the wavelength in the guide. In the cylindrical cavity of Fig. 10.28(b), the electric field lines in the lowest mode are circular and must

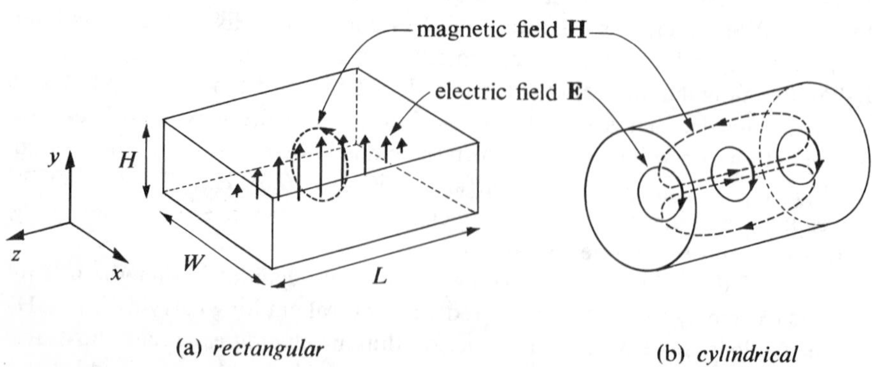

(a) *rectangular* (b) *cylindrical*

FIGURE 10.28 Microwave cavities.

go to zero at the edge of the cavity. Standing wave oscillations are set up only at certain discrete frequencies just as in the case of the rectangular cavity.

Cavities containing oscillating electromagnetic fields cannot be perfectly closed, of course; some way must be found to feed in enough energy to replace that lost inside the cavity during each cycle of oscillation. The principal loss of energy in a cavity is through joule heating I^2R in the cavity walls. Oscillating currents must flow in the walls as a result of the oscillating fields inside the cavity, and any resistance in the walls converts some electrical energy into heat energy. In actual practice the walls are usually machined as smooth as possible, then cleaned and plated with a thin layer of silver (which has the lowest resistance of any common metal), and then flashed with an extremely thin coating of gold to prevent the silver from corroding. Cavity Q's of up to 20,000 can be achieved with these techniques, with the cylindrical cavity being particularly useful.

Enough energy to compensate for the wall losses is fed in or "coupled" in by means of a short wire acting as an antenna or a coupling loop, or sometimes simply by a small hole or "iris." The wire antenna arrangement is used to transfer energy from an oscillator to the rectangular waveguide where it excites the dominant TE_{01} mode as shown in Fig. 10.29(a). Notice that the antenna

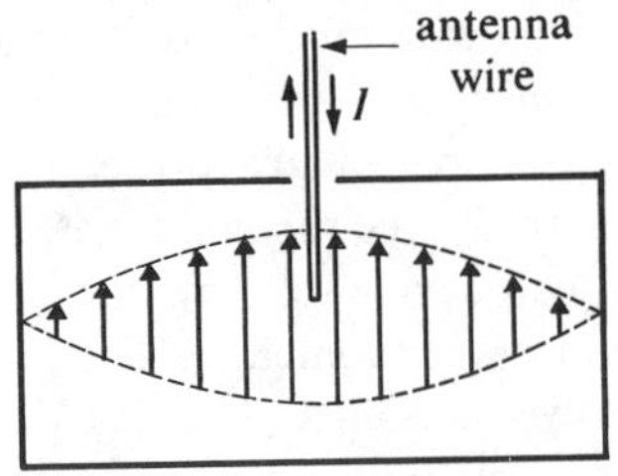

(a) *antenna wire coupling to rectangular waveguide*

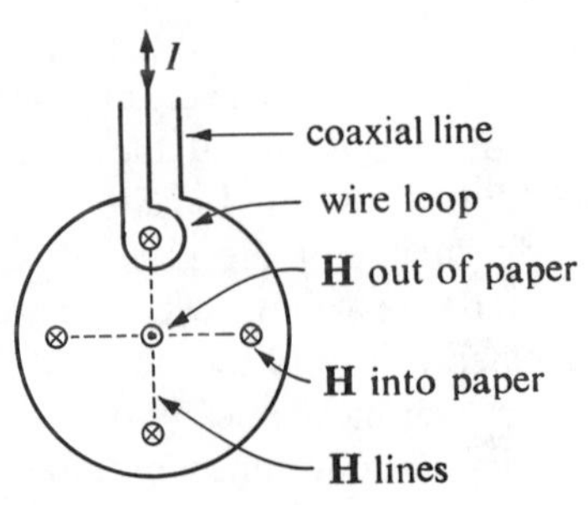

(b) *loop coupling to cylindrical cavity*

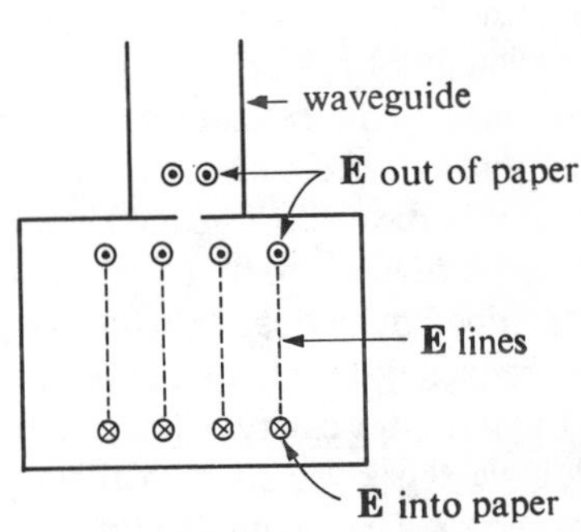

(c) *iris coupling to cylindrical cavity*

FIGURE 10.29 Coupling configurations.

wire is parallel to the *electric* field of the mode excited in the waveguide. Coupling via a loop is made so that the magnetic field of the loop (perpendicular to the plane of the loop) is parallel to the *magnetic* field of the mode excited in the cavity, as shown in Fig. 10.29(b). In iris coupling, shown in Fig. 10.29(c), the iris allows a small amount of the electric field from the dominant TE_{01} guide mode to "leak" into the cavity where it is in such a direction as to excite a cylindrical mode of oscillation. The fundamental point is that the excitation, either electric or magnetic, must be in such a direction as to reinforce the desired oscillations in the cavity; the exciting electric field coupled into the cavity must be parallel to the electric field of the desired oscillation, and similarly for the exciting magnetic field. This is analogous to the mechanical case where an impulse must be given in the proper direction in order to excite an oscillation.

10.15 MICROWAVE OSCILLATORS

At the present writing (1974) there are two devices in widespread use for generating low power microwaves, the reflex klystron and the various types of solid-state microwave diodes, such as the Gunn diode and the Impatt diode. The klystron is a special vacuum tube with all the disadvantages of vacuum tubes—mechanical fragility, filament power, and an expensive high voltage power supply requirement. It was the only source of low power microwave until approximately 1965, and is still preferred over the solid-state diode sources for some ultra-low noise applications such as a stable low noise microwave source for a sensitive electron spin resonance spectrometer. The exact theory of operation of the klystron is rather involved. Suffice it to say that electrons are emitted from a cathode and attracted toward a plate by a positive voltage on the order of several hundred volts. The plate is made in an open grid-like structure, and the electron beam passes through and is decelerated and reflected back toward the plate by a negatively charged electrode called the reflector, which is maintained at a voltage of several hundred volts negative with respect to the cathode. The plate is connected to a resonant cavity in the klystron. The electrons thus pass through the plate structure going in both directions, and if the reflector voltage is the proper value the electrons will "bunch up" and excite electromagnetic field oscillations in the cavity. The frequency of the oscillations is determined by two factors, the magnitude of the reflector voltage, which changes the frequency by about one or two megahertz per volt, and the physical size of the resonant cavity. The cavity size can be warped mechanically by a small knob on the klystron, and the klystron frequency thereby tuned over several gigahertz. The cavity oscillations are coupled to a waveguide by a small rod antenna protruding down into the waveguide and perpendicular to the broad face of the waveguide so as to excite the TE_{01} mode. The klystron is usually mounted in a tube socket directly on the broad face of the waveguide as shown in Fig. 10.30.

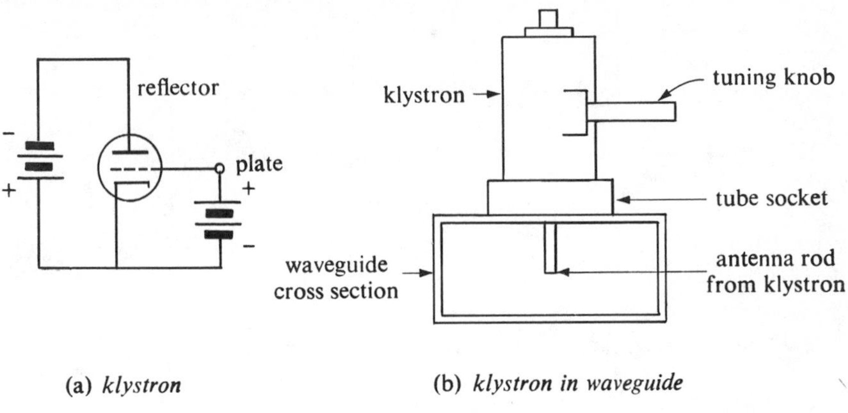

(a) *klystron* (b) *klystron in waveguide*

FIGURE 10.30 Reflex klystron.

Powers of the order of several hundred milliwatts can be generated by reflex klystrons. New low-noise reflex klystrons cost several hundred dollars, and the associated power supply also costs several hundred dollars. Higher power microwaves (10 watts or more) are usually generated by vacuum tubes called magnetrons, which incorporate a heavy permanent magnet to deflect the electron beam.

The recent development of low cost, solid-state microwave oscillators promises to revolutionize the microwave industry. The oscillators are basically nothing more than a relatively inexpensive solid-state diode mounted in a waveguide and supplied with 10–100 V dc at a current of tens of milliamperes. They are presently capable of generating up to several watts at frequencies up to tens of gigahertz, although powers of 10 to 50 mW are more common.

There are two general types of solid-state diode microwave oscillators: the Gunn diode and the Impatt diode. The Gunn diode is usually made from gallium arsenide (GaAs) and operates from a low dc voltage, 6–10 V dc. It is, however, considerably more expensive than the Impatt diode, which is made from silicon. For Gunn diodes there are two conduction bands, one containing electrons of high mobility and the other electrons of low mobility. As the voltage across the diode is increased, electrons are scattered from the high mobility band to the low mobility band. Thus, the average electron velocity and therefore the current *decreases* as the voltage increases. The resulting negative resistance results in oscillation. The Impatt diode operates off approximately 100 V dc and costs approximately ten dollars each in large quantities. The Impatt diode also appears to have less AM noise than the Gunn diode at the present time.

A typical waveguide mount for an Impatt diode microwave oscillator is shown in Fig. 10.31. The diode is mounted perpendicular to the broad face of the waveguide, one-half a guide wavelength from the end of the waveguide.

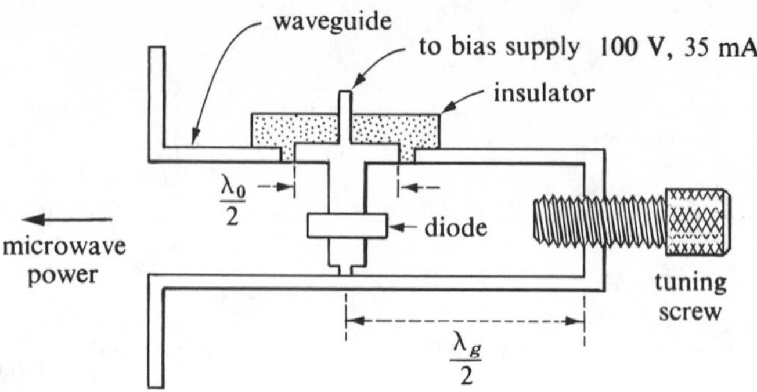

FIGURE 10.31 Impatt diode waveguide mount.

The frequency output changes by approximately 10 kHz per mA change in the bias current, and the tuning screw in the waveguide mount can change the frequency by approximately 160 MHz. Some of the newer Gunn diodes can be tuned over a 10% bandwidth, e.g., 1GHz out of 10 GHz.

The total cost of an Impatt diode oscillator is approximately $50, which opens up many new commercial markets such as Doppler radars for the measurement of speed, and burglar alarms. One interesting Doppler application is an automatic braking system for automobiles. A solid-state microwave oscillator–receiver is mounted in the front of the automobile with the microwave beam pointing forward. When the Doppler shift of the microwave beam reflected back to the receiver indicates an obstacle near the vehicle, the brakes are automatically applied. With this system it is possible to drive full speed at a brick wall and have the Doppler braking device automatically bring the vehicle to rest before striking the wall even though the driver continues to press down on the accelerator pedal.

references

For more information on electromagnetic waves and antennas, see Lorrain and Corson, *Electromagnetic Fields and Waves*, W. H. Freeman & Co., San Francisco, Calif., 1970. An excellent reference on experimental microwave techniques including waveguides and resonant cavities is *Electron Spin Resonance, A Comprehensive Treatise on Experimental Techniques* by Charles P. Poole, Jr., Interscience, 1967.

problems

1. Sketch the electric and magnetic fields for a 10 MHz electromagnetic wave moving from left to right. Label the space axis.

2. Calculate the power radiated by a half-wave antenna 10 cm long at 1 MHz and at 100 MHz if the **rms** antenna current is 20 mA in each case.
3. Sketch an antenna array which will radiate a lobe of power along the x axis if the antennas lie along the y axis direction.
4. Sketch the appropriate antenna orientation for a receiving *loop* antenna if the incident electromagnetic wave is plane polarized with the electric field vector along the x axis and is propagating along the positive z axis. *Hint:* Consider Faraday's law of induction.
5. Repeat Problem 4 for the case of a half-wave receiving antenna. *Hint:* The electrons in the receiving antenna must be accelerated by the *electric* field of the transmitted wave.
6. (a) Calculate the input impedance for the following FET amplifier at 1 kHz and

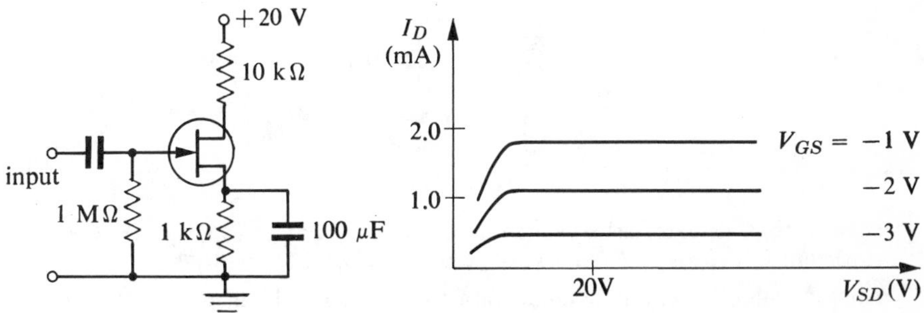

at 1 MHz. *Hint:* Consider the Miller effect. (b) How would increasing the source resistor from 1 kΩ to 3 kΩ change the answer to part (a)? By how much?
7. For an FET with a transconductance of 4500 μmhos and a 3.3 kΩ resistor in the drain, calculate (a) the gain–bandwidth product, (b) the midfrequency gain, and (c) the upper 3 dB frequency, if the total shunt capacitance between the drain and ground is 4.0 pF.
8. If the FET amplifier of Problem 7 is changed to have a drain resistance of 4.7 kΩ with an operating point of *twice* the dc drain current, calculate (a) the gain–bandwidth product, (b) the midfrequency gain, and (c) the upper 3 dB frequency. *Hint:* How does the transconductance depend upon the drain current?
9. Explain why the phase shift between the input and the output of a transmission line is larger at higher frequencies.
10. Calculate how long a length of RG58A/U cable would be required to delay a pulse by 0.1 μsec.
11. Calculate the phase difference between the input and output 10 MHz voltage for a 10 cm length of RG58A/U cable.
12. Show that for no reflected wave, the VSWR of a line is equal to unity. If half the transmitted power is reflected, calculate the VSWR. If 10% of the transmitted power is reflected, calculate the VSWR.
13. Show that the magnetic energy stored in a quarter-wave line is equal to $V_+^2 L\lambda/4Z_0^2$.
14. Show that the electric energy stored in a quarter-wave line is equal to $V_+^2 C\lambda/4$.
15. Calculate the resonant frequency of the tank circuit in the drain of the FET rf amplifier shown.

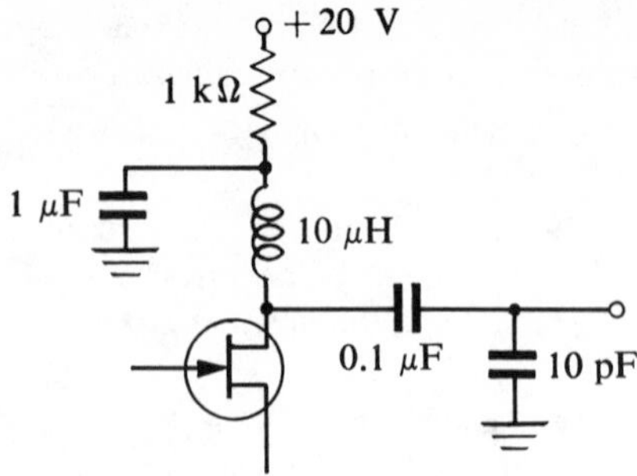

16. (a) Sketch an appropriate dc bias circuit for the varactor tuned, LC resonant circuit

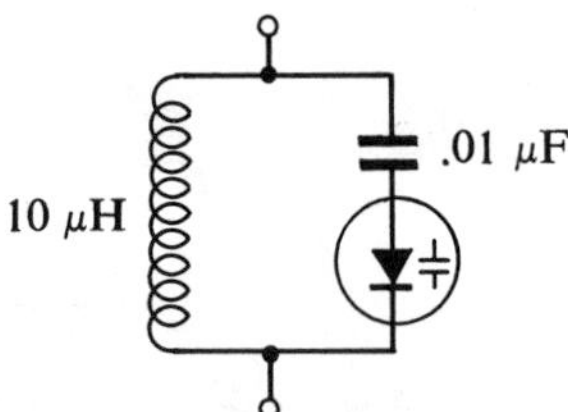

shown. (b) Calculate the bias voltage required to resonate at 10 MHz if the varactor capacitance is given by $C_J = KV^{-0.5}$, where $K = 45\,\text{pF} \times (\text{volt})^{1/2}$.

17. Approximately how rapidly can the varactor capacitance be changed in the following circuit if the dc bias is changed?

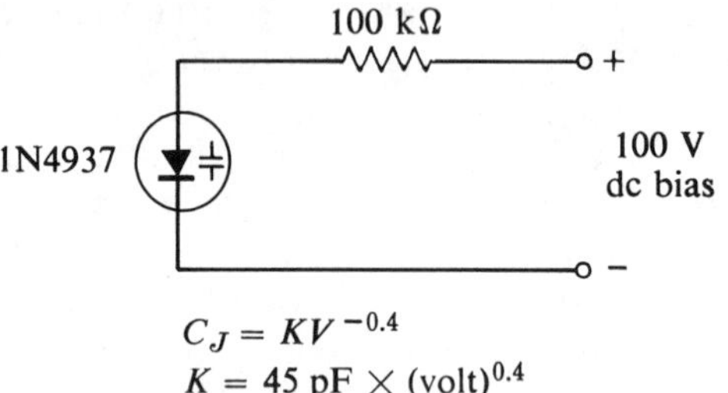

$$C_J = KV^{-0.4}$$
$$K = 45\text{ pF} \times (\text{volt})^{0.4}$$

18. Explain the purpose of L_2C_2 and L_3C_3 if the desired output frequency is twice the input frequency.

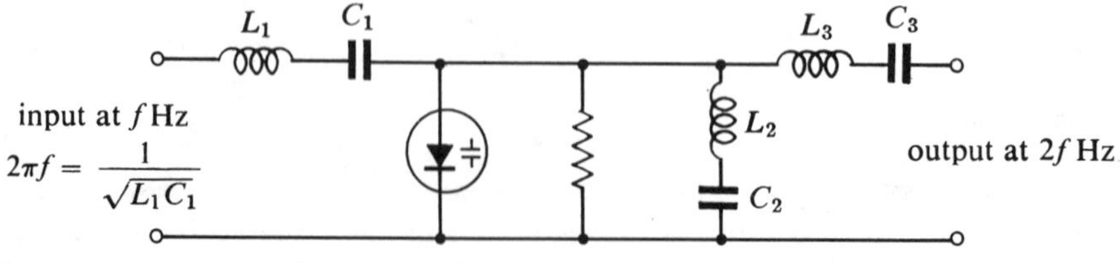

varactor frequency doubler

19. Calculate the characteristic impedance of a microstrip line if the dielectric thickness is 0.1 cm, the dielectric constant is 4, and the width of the conducting strip is 1 cm. Sketch the line.

20. Calculate the lowest frequency which can be propagated in the TE mode down a

rectangular waveguide whose exterior dimensions are 1×2 in. and whose walls are $\frac{1}{16}$ in. thick.

21. Calculate the Q of a rectangular microwave cavity if the peak electric field in the cavity is 1000 V/meter in the z direction, uniformly in the xy plane, and if the average power loss is 10 μW due to joule heating of the cavity walls. The frequency of the oscillation is 33 GHz. *Hint:* Consider the basic definition of Q in terms of energy stored and energy dissipated per cycle.

22. Calculate the maximum electric field intensity in a 50 mW beam of 10 GHz microwaves.

11

NOISE

11.1 INTRODUCTION

Noise, to the layman, is an unwanted or unpleasant acoustical disturbance such as the tubercular lady's cough during the pianissimo symphony passage or the garbage man's clanging the cans outside one's bedroom at 4 a.m. In this chapter, we shall take noise to mean any unwanted electrical disturbance that tends to obscure or otherwise hinder one's observation of the electrical signal of interest.

The first point, which must be made, is that it is never the absolute magnitude of the noise that is important, but only its magnitude *relative* to the magnitude of the signal. In other words the signal-to-noise ratio is important, not the size of the signal alone or the noise alone. The signal-to-noise ratio is usually defined as the ratio in dB of the signal power to the noise power and occasionally as the ratio of the rms signal voltage to the rms noise voltage. With a high signal-to-noise ratio and a small signal, a usable, large signal can be obtained by amplification, with a decent amplifier which itself does not introduce too much noise. However, with a low signal-to-noise ratio, amplification is useless. The amplifier cannot tell the difference between a signal voltage and a noise voltage, and so amplifies each impartially. Thus, even if the amplifier introduces a negligible amount of noise, the signal-to-noise ratio at the output will be the *same* as at the input. And, of course, all amplifiers introduce some noise of their own in addition to the noise present in the input signal. Hence, we come to the initially discouraging conclusion that the signal-to-noise ratio at the output of an amplifier is *always* less than the signal-to-noise ratio at the input.

Noise, as we have defined it, is any *unwanted* electrical disturbance, and so noise can be regular or random in nature. Noise in a circuit can be classified roughly into four categories: (1) interference, either regular or random, from sources outside the circuit, (2) inherent thermal noise generated within resistances in the circuit, (3) inherent shot noise which is the statistical fluctuation of current due to the discrete nature of the charge on the electron, and (4) transistor and tube flicker noise which is a random low-frequency noise increasing with decreasing frequency.

11.2 INTERFERENCE

Many electrical sources outside a circuit can introduce noise voltages into the circuit. Near and distant lightning flashes are a source of broad-band electromagnetic radiation ranging from a few kilohertz to hundreds of megahertz in frequency. Many man-made devices are also sources of electromagnetic radiation: automobile ignition systems, medical diathermy machines, and any machine that contains a spark gap, such as electric motors with sparking between the commutator and the brushes. Noisy electric power transformers and high voltage power lines are sources of 60 Hz and multiples of 60 Hz electromagnetic fields. The cure is to shield the circuit, particularly the input stage or the "front end," by metal plates or screens. If bolts or screws are used to fasten the metal shield to the chassis, the spacing of the bolts should be small compared to the wavelength of the interfering radiation. As a rule of thumb, one should be extremely careful in shielding against interference above several MHz; the shield should preferably be soldered rather than bolted in place. Double shielded cable might also be used at the input of high gain amplifiers to reduce the interference. If the source of the interference can be eliminated, fine, but if shielding is the only possibility the shielding should be at the source as well as at the circuit.

Interference at 60 Hz, the power line frequency, often is caused by "ground loops," which are closed loops of wire all points of which are supposed to be at ground potential. Because of the very low resistance of these ground loops, appreciable 60 Hz currents may be induced in them by the 60 Hz magnetic fields present around the power lines. These induced 60 Hz currents in the ground loops may then induce small 60 Hz voltages in the input leads of an amplifier or receiver. The cure is to avoid creating ground loops by bringing all the ground connections together at one point in the circuit. A certain amount of experimentation is usually necessary.

Mechanical vibration can often be converted into electrical noise. This noise is often called "microphonic" noise. Vibration may cover a broad range of frequencies in aircraft and rockets or may be relatively narrow-band. For example, the natural vibration frequency of most floors lies between 10 Hz and 20 Hz. The larger the object, of course, the lower the vibration frequency. In vacuum tube circuits vibration of the tube electrodes will change the interelectrode capacitance and the gain of the tube, thus changing the waveform of the signal. Transistors are, of course, immune to this type of noise because they are literally solid devices. However, in any circuit containing cables, vibration of the cable will result in noise voltages being generated in the braided outer shield, due to piezoelectric effects, flexing of the dielectric, and the flexing and intermittent contacts between the wires in the braid. This may be a serious problem at the input of a high gain amplifier where the signal itself may only be millivolts in amplitude. The cure for this cable noise is to eliminate the vibration by shock mounting the circuit, or to stop the source of the vibration, or to use a special low-noise cable. This cable has a cylindrical layer of *conducting* plastic extruded over the dielectric surrounding the central wire. A second concentric metal shield surrounds this conducting plastic, then a thin layer of polyethylene, and finally the standard outer braided metal shield. The conducting plastic and the inner shield serve to attenuate the noise generated by flexing of the braid.

If the interference frequencies differ substantially from the signal frequencies, then frequency selective filters can be used to eliminate the interference. These filters should, themselves, introduce little noise in the circuit and are often put in the circuit after an initial stage of amplification, i.e., after the "preamp," but not at the final output. If the interference consists of occasional spikes that are much larger than the signal, these spikes can be attenuated and the signal retained by using two diodes as shown in Fig. 11.1. Any voltage less than the

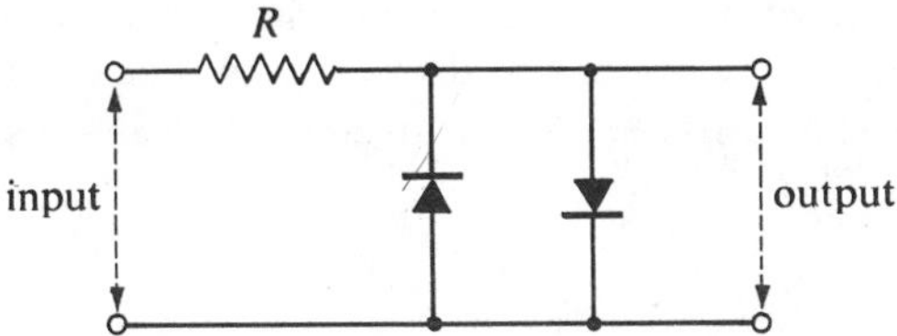

FIGURE 11.1 Diode limiter.

turn-on voltage of the diodes (0.2 V for Ge, 0.5 V for Si) will pass through unaffected, but a noise spike of either polarity larger than the turn-on voltage will cause one of the diodes to conduct and thus be shorted to ground. This circuit is particularly useful when a large noise spike will disable the following circuit for an appreciable time, e.g., by charging up a capacitor which may cut off the circuit. The capacitor then takes a finite time to discharge and during this time the circuit may be inoperative.

11.3 THERMAL NOISE OR JOHNSON NOISE

The thermal energy of matter is basically the random vibrational energy of the atoms. The higher the temperature, the more violent the motion and the larger the thermal energy per atom. The famous equipartition theorem of statistical mechanics says that for every mathematical degree of freedom of a physical system in equilibrium at $T°$ Kelvin or absolute (°C + 273° = °K), there is associated $\frac{1}{2}kT$ energy; where k is Boltzmann's constant, $k = 1.38 \times 10^{-16}$ erg/degree = 1.38×10^{-23} joule/degree, and T is the absolute temperature. Thus a point mass which has three degrees of freedom for motion in the x, y, and z directions will on the average have $\frac{3}{2}kT$ energy. Examples are free atoms and free electrons. A free electron at room temperature, $T = 20°\text{C} = 293\,°\text{K}$, will have $E = \frac{3}{2}kT = (\frac{3}{2})(1.38 \times 10^{-23}\ \text{J/°K})\ (293\,°\text{K}) = 6.1 \times 10^{-21}$ joule of energy. However, this is the *average* energy; some electrons will have more, some less. According to classical theory, this random thermal energy will completely die out as the temperature approaches absolute zero. Quantum theory, however, predicts a small residual or "zero point" vibrational energy even at absolute zero.

Because many of the electrons in a resistance are essentially free and are in constant random vibrational motion, the voltage difference between the two ends of any resistance will fluctuate randomly. J. B. Johnson, in 1928, showed that the power associated with such fluctuations varied linearly as the bandwidth B of the measuring instrument. For example, a sensitive rms power meter with a response from dc to 1000 Hz placed across a resistance might measure 0.01 μW power, but a power meter with a response from dc to 2000 Hz would measure 0.02 μW noise power. Johnson also found that the square of the noise voltage varied linearly with the resistance R. Noise power is proportional to the mean-square noise voltage, and the expression for the mean-square noise voltage generated by a resistance R at T degrees absolute in a bandwidth B is:

$$\overline{e_n^2} = 4kTRB, \tag{11.1}$$

where k is Boltzmann's constant. The bar over e_n^2 indicates an average. The rms noise voltage is:

$$(\overline{e_n^2})^{1/2} = (4kTRB)^{1/2}. \tag{11.2}$$

This noise is commonly referred to as Johnson noise, thermal noise, or resistor noise.

An approximate derivation follows from the equipartition theorem. The random statistical fluctuations of the electrons in the resistor produce a fluctuating difference in the electron density at the two ends of the resistor. This difference in electron density produces a voltage difference between the two ends of the resistor which, like all circuit elements, has an effective shunt capacitance C, as shown in Fig. 11.2(a). The energy associated with the voltage fluctuations is stored in the electric field of the shunt capacitance C. Thus, the energy of the voltage fluctuations is given by $E = \frac{1}{2}CV^2$, where V is the instantaneous voltage difference across the capacitor. By the equipartition theorem, if the resistor and capacitor are in thermal equilibrium at $T°$ Kelvin, the average energy stored in the capacitor must equal $\frac{1}{2}kT$. Thus:

R C

(a) *equivalent circuit for resistance and its inherent shunt capacitance*

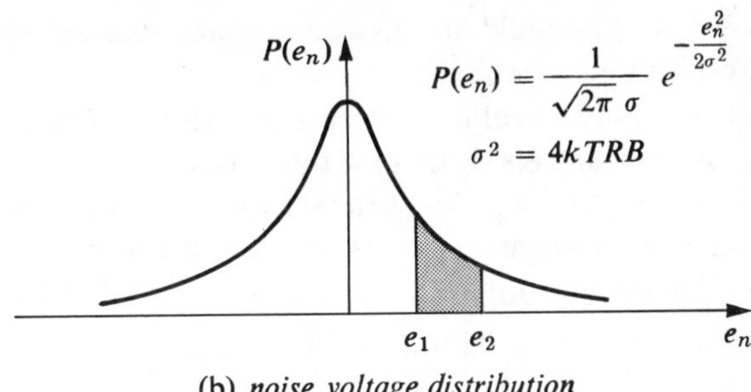

(b) *noise voltage distribution*

FIGURE 11.2 Johnson noise.

$$\tfrac{1}{2}kT = \tfrac{1}{2}C\overline{V^2}$$

or

$$\overline{V^2} = \frac{kT}{C}, \tag{11.3}$$

where $\overline{V^2}$ is the average-squared noise voltage. The bandwidth of the parallel RC circuit is from dc out to the frequency where the total parallel impedance is 3dB down from R.

$$B = \frac{1}{2\pi RC} \tag{11.4}$$

Solving equation (11.4) for C and substituting in (11.3), we obtain an expression for the average-squared noise voltage in terms of R:

$$\overline{V^2} = 2\pi kTRB, \tag{11.5}$$

which is close to the exact equation, (11.1).

The noise voltage developed across the resistance R can be of either polarity, and its magnitude varies statistically according to a Gaussian distribution:

$$P(e_n) = \frac{1}{\sqrt{2\pi}\,\sigma} e^{-e_n^2/2\sigma^2}, \tag{11.6}$$

where

$$\sigma^2 = 4kTRB.$$

The expression $P(e_n)\,de_n$ is the probability that the noise voltage is between e_n and $e_n + de_n$. The probability that the noise voltage is between e_{n1} and e_{n2} is $\int_{e_{1n}}^{e_{2n}} P(e_n)\,de_n$, shown as the shaded area in Fig. 11.2(b). Notice that small noise voltages are more probable than large ones. In statistical language, $\sigma = \sqrt{4kTRB}$ is the standard deviation of the random noise. The probability that the noise voltage is between $-\sqrt{4kTRB}$ and $+\sqrt{4kTRB}$ is 0.68 or 68%, and thus there is a 32% probability that the noise is greater in magnitude than $\sqrt{4kTRB}$. In noise calculations it is common practice to assume the noise voltage e_n exactly equals $\sqrt{4kTRB}$, but one should always keep in mind that the noise amplitude varies randomly with time.

An important mathematical consequence due to the "Gaussian" or "random" character of Johnson noise is that the total noise voltage from two or more dependent random noise sources adds as the square root of the sum of the squares of the individual noise voltages. For example, two resistances R_1 and R_2 in series generate independent Johnson voltages of $e_{n1} = \sqrt{4kTR_1B}$ and $e_{n2} = \sqrt{4kTR_2B}$, respectively. Thus the total noise voltage is given by:

$$e_{n\,\text{total}} = \sqrt{e_{n1}^2 + e_{n2}^2} = \sqrt{4kTR_1B + 4kTR_2B} = \sqrt{4kT(R_1 + R_2)B},$$

which is simply the noise voltage that would be generated by a single resistance $R_1 + R_2$. In other words, noise *powers* simply add, since power is proportional to the square of the voltages. This method of addition of independent noise sources is used repeatedly in analyzing the noise generated by different components in a circuit.

It also should be emphasized that it is only a *resistance* which generates noise; a pure reactance, either capacitive or inductive, generates no Johnson noise. For example, a parallel LC resonant circuit as shown in Fig. 11.3 will generate Johnson noise only due to the resistance R of the inductance. The noise voltage $e_n = \sqrt{4kTRB}$ appears across the resistance R, *not* across the tank terminals A and B. The noise voltage appearing between A and B, e_{nAB}, is con-

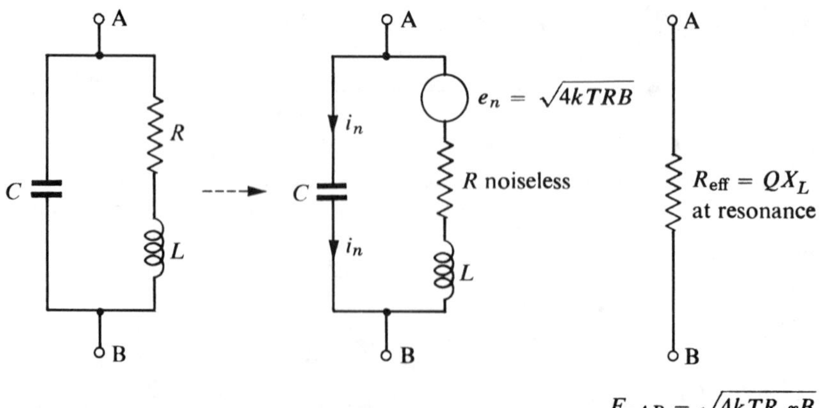

FIGURE 11.3 Noise generated in a parallel LC circuit.

siderably larger. $e_{n\text{AB}}$ equals the resulting noise current i_n times the capacitive reactance:

$$e_{n\text{AB}} = (i_n) \times \frac{1}{j\omega C} = \left(\frac{e_n}{j\omega L + R + \dfrac{1}{j\omega C}} \right) \times \frac{1}{j\omega C}.$$

At resonance, $\omega_0^2 = 1/LC$, and so $e_{n\text{AB}} = e_n/j\omega RC$. Eliminating ω using $\omega_0^2 = 1/LC$, we have:

$$e_{n\text{AB}} = \frac{e_n}{R\sqrt{C/L}} = \frac{\sqrt{4kTRB}}{R\sqrt{C/L}} = \sqrt{\frac{4kTBL}{RC}}. \quad \textbf{(11.7)}$$

But the Q of the circuit at resonance is given by $Q = \omega_0 L/R = (1/R)\sqrt{L/C}$, so the AB terminal noise voltage is $e_{n\text{AB}} = Q\sqrt{4kTRB}$. Thus, the noise voltage generated by the resistance R appears across the tank terminals Q times larger. The same result can be obtained by realizing the tank has a purely resistive impedance $R_{\text{eff}} = QX_L = \omega^2 L^2/R = Q^2 R$ at resonance. Thus, the AB terminal noise voltage at resonance is immediately calculated to be:

$$e_{n\text{AB}} = \sqrt{4kTR_{\text{eff}}B} = Q\sqrt{4kTRB}.$$

Johnson noise is often called "white noise" because its power density depends only upon the bandwidth B, not upon the frequency f. That is, *any* 1 kHz bandwidth, whether from 1 kHz to 2 kHz or from 1 MHz to 1.001 MHz, of Johnson noise will contain the same average noise power. The word "white" is used because white noise contains "all frequencies" just as white visible light contains a mixture of all of the colors of the spectrum.

A little thought will show that the formula $e_n^2 = 4kTRB$ cannot hold indefinitely as the frequency increases; for if it did, one resistance would develop an *infinite* noise power because $\int_0^\infty e_n^2\, df \to \infty$. It is clear that the noise power per unit bandwidth must decrease as the frequency increases. The Johnson noise formula holds up to a frequency on the order of $f = kT/h$, where h (Planck's constant) $= 6.67 \times 10^{-34}$ joule/sec, at which point the equipartition theorem of classical statistical mechanics breaks down and quantum statistical mechanics takes over. In classical theory the average energy per degree of freedom is kT ($\frac{1}{2}kT$ kinetic and $\frac{1}{2}kT$ potential); and as the frequency increases, the number of possible modes of oscillation increases, thus yielding an infinite power for high frequencies. In quantum theory, the average energy per degree of freedom is given by:

$$\overline{E} = \frac{hf}{e^{hf/kT} - 1}, \quad \textbf{(11.8)}$$

which approaches kT for low frequencies, $f \ll kT/h$, and which decreases monotonically as the frequency increases above $f = kT/h$. For room temperature

$T \cong 300°K, f = kT/h \cong 10^{13}$ Hz, which is in the infrared region of the electromagnetic spectrum; so we may safely use the Johnson noise formula up into the microwave region of tens of gigahertz. For a different perspective on the infinite power implied by the classical equipartition theorem, the interested reader should consult any modern physics book on the topic of the "ultraviolet catastrophe" and the Rayleigh–Jeans law of blackbody radiation.

Most practical resistors used in circuits exhibit *more* noise than the Johnson noise formula, $e_n^2 = 4kTRB$, so the Johnson noise expressions should be regarded as a lower limit on the noise generated with a resistor. Variations in noise power of up to a factor of 100 have been observed in resistors of the same resistance value made by the same company. Also, different type resistors exhibit different noise characteristics. In general, wire wound resistors exhibit the least noise, and standard composition carbon resistors the most, with metal film resistors occupying an intermediate position. It should be pointed out that the least noisy resistors are also the most expensive, and also that wire wound resistors have considerably more inductance than the other types, so they cannot be used in high-frequency circuits in general. Also, the higher the voltage across and the current through a resistor, the more excess noise is generated.

11.4 SHOT NOISE

In most electronic circuit problems, the basic quantization of the electronic charge can be neglected, and charge and current can be treated as continuous variables. However, all currents consist of the flow of electrons or ions (or holes) which are discrete charges, and random fluctuations in the number of charge carriers flowing past a given point in the circuit produce random fluctuations in the current. This occurs in transistors, vacuum tubes, and in all photoemission devices such as photomultiplier tubes and phototubes. Random fluctuations will occur in the collector or drain or plate current simply because the number of electrons arriving per second fluctuates. The resultant noise is termed "shot" noise, because the fluctuations are similar to those occurring when a hail of shot strikes a target. For a vacuum diode where the cathode emission current is limited only by the cathode temperature, that is, where each electron emitted from the cathode reaches the plate; the root-mean-square current noise is given by:

$$(\overline{i_n^2})^{1/2} = \sqrt{2eI_{dc}B}, \tag{11.9}$$

where e is the electronic charge, 1.6×10^{-19} coulomb, B is the bandwidth, and I_{dc} is the average or dc emission current. Another way of looking at this shot noise current is that $i_n = (\overline{i_n^2})^{1/2}$ is the random fluctuating ac current superimposed on the dc current. The exact expression for the shot noise for other devices will vary, but for most devices the general feature remains that the noise current is proportional to the *square root* of the dc current. Semi-empirical techniques are still used for determining the noise characteristics of actual circuits containing transistors and vacuum tubes. Also, like Johnson noise, shot noise is "white"; i.e., there is a constant power density per unit bandwidth, independent of frequency.

The shot noise expression can be derived from a basic statistical argument as follows. Consider a device emitting discrete charged particles from terminal A to B, which then flow through a resistance R as shown in Fig. 11.4. This

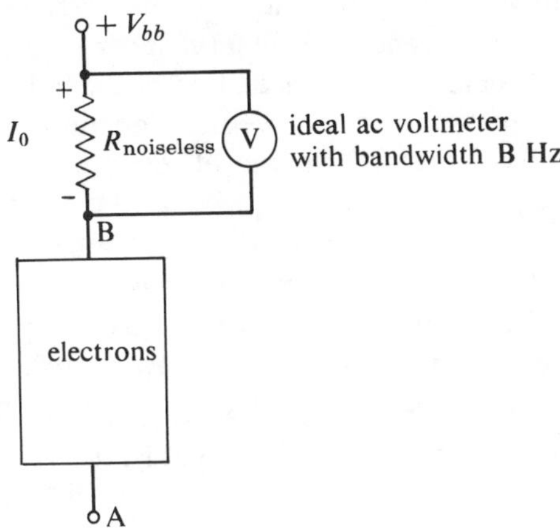

FIGURE 11.4 Shot noise.

device may be a temperature limited diode or a transistor or a vacuum tube. If the average dc current flowing is I_0, then the average number $\bar{r}$ of charge carriers flowing per second is given by $\bar{r} = I_0/e$, where e is the net charge on each carrier. The current may be different from I_0, however, due to random fluctuations in the number of charge carriers emitted from the device. If we let I be the instantaneous current flowing corresponding to $r = I_0/e$ charge carriers flowing per sec, then the fluctuating noise voltage developed across R will be given by $V_{\text{noise}} = (I - I_0)R = (r - \bar{r})eR$. A more meaningful way of looking at the fluctuations in the rate of charge carrier flow is to look at the absolute number n of charge carriers flowing in a time interval Δt: $n = r\Delta t$. In terms of n, the noise voltage is

$$v_n = (n - \bar{n})eR/\Delta t, \tag{11.10}$$

where $\bar{n} = \bar{r}\Delta t$ is the number of charge carriers flowing in Δt corresponding to the average dc current I_0. The mathematical theory of random processes can now be invoked to tell us a typical value for the fluctuation $(n - \bar{n})$ in the number of charge carriers flowing in a time Δt. If the average number of charge carriers flowing in Δt is $\bar{n}$, then the probability of n, the actual number flowing in Δt, is given by the standard Gaussian distribution:

$$P(n) = \frac{1}{\sqrt{2\pi}\,\sigma} e^{-(n-\bar{n})^2/2\sigma^2}. \tag{11.11}$$

Strictly speaking, the probability of having a number between n_1 and n_2 flow in Δt is given by $\int_{n_1}^{n_2} P(n)\,dn$. $P(n)$ is normalized to unity: $\int_0^\infty P(n)\,dn = 1$; that is, it is certain that some value of n occurs between 0 and ∞. Notice that it is most likely that $n = \bar{n}$, that the average number of charge carriers flows; i.e., $P(n)$ is maximum when $n = \bar{n}$. Notice also that if n is substantially different from $\bar{n}$, the probability of getting that value of n is extremely small.

The difference $n - \bar{n}$ between the actual number n of charge carriers and the *average* number $\bar{n}$ represents the fluctuation or the noise. To calculate the magnitude of the fluctuation or noise, we must obtain a representative value for $n - \bar{n}$. From the statistical nature of $P(n)$, we see that the larger $(n - \bar{n})^2/2\sigma^2$, the less probable that value of n. It is conventional to say that a representative or typical value of the variable in a Gaussian distribution is that value which makes the exponent equal to *minus one*. That is, $(n - \bar{n})^2/2\sigma^2 = 1$ or $(n - \bar{n}) = \pm\sqrt{2}\sigma$. This value of the fluctuation means that $P(n) = P_{\max}e^{-1} = 0.37P_{\max}$ and that there is a probability of $\int_{n'}^\infty P(n)\,dn = 0.08$ that the fluctuation is larger in value than $n' = \bar{n} + \sqrt{2}\sigma$, and a probability $\int_0^{n'} P(n)\,dn = 0.08$ that the fluctuation is less than $n' = \bar{n} - \sqrt{2}\sigma$. That is, there is a 16% chance that $|n - \bar{n}|$ exceeds $\sqrt{2}\sigma$.

It can be shown for most physical processes that $\sigma = \sqrt{\bar{n}}$; the standard deviation is approximately equal to the square root of the average number. Substituting $n - \bar{n} = \sqrt{2}\sigma = \sqrt{2\bar{n}}$ into equation (11.10) gives:

$$v_n = \frac{(n - \bar{n})eR}{\Delta t}$$

$$v_n = \frac{\sqrt{2\bar{n}}eR}{\Delta t}.$$

But $\bar{n}$ can be expressed in terms of the average dc current flowing, $\bar{n} = I_0\Delta t/e$. Thus the noise voltage is given by:

$$v_n = \sqrt{\frac{2I_0\Delta t}{e}}\,\frac{eR}{\Delta t} = \sqrt{\frac{2I_0e}{\Delta t}}\,R. \tag{11.12}$$

Notice the noise voltage increases as the time interval Δt during which we count the number of charge carriers *decreases*. In other words, if we average the noise voltage over a long time interval, Δt, the noise voltage will go to zero. If we identify this time interval with the inverse of the bandwidth of the voltmeter we are using to measure the noise voltage across R, $\Delta t = 1/B$; then $v_n = \sqrt{2I_0eB}R$. This is not unreasonable because $1/B$ is the fastest response time of the voltmeter; any fluctuations occurring in times less than $1/B$ simply do not affect the voltmeter. Thus the square of the noise voltage developed across R is:

$$v_n^2 = 2I_0eBR^2. \tag{11.13}$$

The noise power developed in R is:

$$P_n = \frac{v_n^2}{R} = 2I_0eBR, \tag{11.14}$$

and the noise current squared is

$$i_n^2 = \frac{v_n^2}{R^2} = 2I_0eB, \tag{11.15}$$

which is the shot noise expression. The two important factors are that the noise current, i_n, varies as the square root of the average or dc current I_0 and as the square root of the bandwidth B.

Some numerical values for the shot noise voltage developed across R are instructive. If $R = 1\,\text{k}\Omega$ and $I_0 = 1$ mA, then for various bandwidths B we have $v_n = 1.3 \times 10^{-7}\sqrt{B}$, or as shown below. Thus for an audio amplifier with a

B	v_n
1 Hz	0.013 μV
10 kHz	1.3 μV
1 MHz	13 μV
100 MHz	130 μV

bandwidth of 10 Hz–20 kHz, the noise voltage would be only several microvolts. The noise voltage becomes increasingly bothersome as the bandwidth increases. Again it should be pointed out that this noise source (as with any noise source) is most serious when it occurs at the front end of a high gain amplifier, where the noise will be amplified by all the following amplifier stages. Another way of looking at the seriousness of the problem is that at the front end of an amplifier the signal voltage is usually very small; thus the signal-to-noise ratio will be lowered appreciably by the presence of even a small amount of noise.

11.5 CALCULATION OF AMPLIFIER NOISE

An exact calculation of all the Johnson noise and shot noise generated within a practical amplifier would be impossibly difficult, but a few comments can be made. First, the "noise figure" is used to describe how much the amplifier degrades or decreases the signal-to-noise ratio. The noise figure, often abbreviated N.F., is defined in dB as ten times the logarithm of the signal-to-noise power ratio at the input divided by the signal-to-noise power ratio at the output:

$$\text{Noise Figure} = \text{N.F.} = 10 \log_{10} \frac{(\text{S/N})_{\text{input}}}{(\text{S/N})_{\text{output}}}.$$

The noise figure will always be greater than 0 dB because the output S/N ratio is always less than the input S/N ratio, due to the additional noise contributed by the amplifier. A perfect amplifier which adds no noise whatsoever thus has a N.F. of 0 dB. An amplifier which decreases the S/N power ratio by a factor of two would have a 3 dB N.F. In general most rf amplifiers in the 10 MHz to 500 MHz frequency region tend to have noise figures in the vicinity of 5 dB to 10 dB.

If the total effective noise contributed by an amplifier can be represented by a single voltage noise generator e_n at the input, or equivalently by a single noisy resistance R_i, where $e_n^2 = 4kTR_iB$, as is the case for most vacuum tube amplifiers, then a simple analysis will show that the noise figure of the amplifier–source combination depends only upon the ratio R_s/R_i, where R_s is the source resistance (see Fig. 11.5). Usually R_i is approximately equal to the input re-

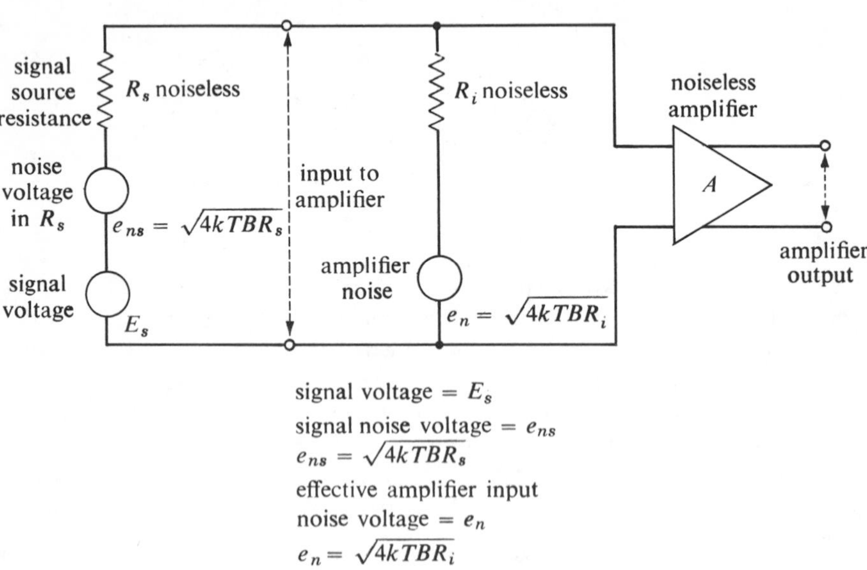

FIGURE 11.5 Noisy amplifier and signal source equivalent circuit.

sistance of the amplifier. The signal noise contribution is e_{ns}, and a fraction λe_{ns} appears across the amplifier input terminals where $\lambda = R_i/(R_i + R_s)$. A fraction $\epsilon = R_s/(R_i + R_s)$ of the noise voltage, e_n, also appears across the input, so the total input noise is:

$$\textit{total input noise to amplifier} = \sqrt{(\lambda e_{ns})^2 + (\epsilon e_n)^2}, \tag{11.16}$$

where we have used the fact that noise voltages from different (independent) sources add as the square root of the sum of the squares. The input signal voltage to the amplifier is λE_s, and the input voltage due to the noise in the signal is λe_{ns}. Thus the power noise figure is given by:

$$\text{N.F.} = 10 \log_{10} \frac{(\text{S/N})_{\text{input}}}{(\text{S/N})_{\text{output}}} = 10 \log_{10} \frac{\left(\dfrac{\lambda E_s}{\lambda e_{ns}}\right)^2}{\left[\dfrac{A\lambda E_s}{A\sqrt{(\lambda e_{ns})^2 + (\epsilon e_n)^2}}\right]^2}, \tag{11.17}$$

where A is the amplifier gain and we have used the fact that the S/N power ratio is the square of the S/N voltage ratio. Algebraic manipulation yields:

$$\text{N.F.} = 10 \log_{10} \left(\frac{\lambda^2 e_{ns}^2 + \epsilon^2 e_n^2}{\lambda^2 e_{ns}^2}\right).$$

Substituting for λ, ϵ, e_n, and e_{ns} yields after a little algebra:

$$\text{N.F.} = 10 \log_{10} \left[1 + \frac{\epsilon^2 e_n^2}{\lambda^2 e_{ns}^2}\right] = 10 \log_{10} \left[1 + \frac{\left(\dfrac{R_s}{R_i + R_s}\right)^2 4kTBR_i}{\left(\dfrac{R_i}{R_i + R_s}\right)^2 4kTBR_s}\right]$$

or

$$\text{N.F.} = 10 \log_{10} \left(1 + \frac{R_s}{R_i}\right). \tag{11.18}$$

It is worthwhile pointing out that when the amplifier is matched to a fixed source (i.e., when R_i is made equal to R_s) for maximum *power* transfer the noise figure is 3dB. For $R_i > R_s$, the noise figure decreases, but more signal power is lost due to the impedance mismatch. The same conclusion is obtained if we examine the output signal-to-noise power:

$$\left(\frac{\text{S}}{\text{N}}\right)_{\text{output}} = \frac{(A\lambda E_s)^2}{\left[A\sqrt{(\lambda e_{sn})^2 + (\epsilon e_n)^2}\right]^2}$$

or

$$\left(\frac{\text{S}}{\text{N}}\right)_{\text{output}} = \frac{E_s^2}{4kTBR_s\left(1 + \dfrac{R_s}{R_i}\right)}. \tag{11.19}$$

For a fixed source e_s and R_s, the output signal-to-noise power is maximum when $R_i \gg R_s$.

Modern noise theory has shown that the noise from a noisy four terminal network can be completely specified by *four* parameters, usually two noise generators and their correlations. In the case of vacuum tubes, only one noise parameter is necessary for frequencies below the microwave region. Usually, the effective input noise resistance R_T of the tube is used to specify the noise generated within the tube according to:

$$P_{ni} = 4kTBR_T \qquad P_{no} = GP_{ni}, \tag{11.20}$$

where P_{ni} is the effective noise at the input of the tube, P_{no} is the tube noise at the output, and G is the power gain of the tube. In the case of transistors, two noise parameters are necessary to represent the transistor noise. Usually a voltage and a current input noise generator are used with the correlations between the two neglected. All the noise in the entire transistor amplifier is represented by one equivalent noise voltage generator, e_n, and one equivalent current generator, i_n, at the input as shown in Fig. 11.6. e_n^2B is the mean

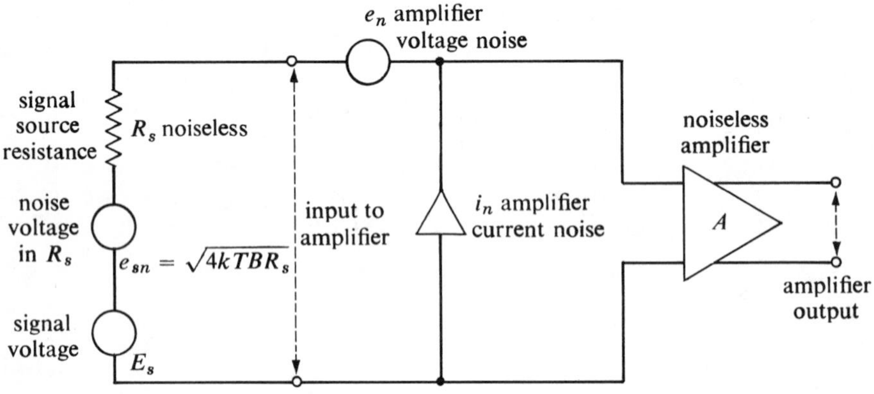

FIGURE 11.6 Transistor amplifier and source noise equivalent circuit.

square voltage generated in a bandwidth B, so e_n has units of volts/$\sqrt{\text{Hz}}$. Similarly, i_n has units of amperes/$\sqrt{\text{Hz}}$. The input signal source is represented as a voltage source e_s in series with a source resistance R_s. The Johnson noise generated by R_s is represented by a voltage generator $e_{sn} = \sqrt{4kTBR_s}$. B refers to the bandwidth of the amplifier, since we will be referring all noise calculations to the amplifier output. The amplifier voltage and current noise generators e_n and i_n are assumed to be completely independent of each other and to generate perfectly "white" noise, i.e., constant noise power per unit bandwidth.

We can now calculate the amplifier output signal-to-noise ratio from analyzing the equivalent circuit of Fig. 11.6. The signal input voltage to the amplifier is e_s, assuming negligible loading of the signal source, i.e., assuming the amplifier input impedance is large compared to R_s. Thus the signal output voltage from the amplifier is Ae_s. The noise voltage in the output of the amplifier is simply the amplifier gain A times the total effective input noise voltage to the amplifier. There are three noise sources at the input: the source noise e_{sn}, and the effective amplifier voltage and current noise sources, e_n and i_n. If we assume these three sources are random in nature and independent of one another, then the total noise is the square root of the sum of the squares. Hence the total amplifier output noise voltage is given by:

$$N_{\text{output}} = A\sqrt{e_{sn}^2 + e_n^2B + (i_nR_s)^2B}. \tag{11.21}$$

The expression i_nR_s is the noise voltage generated by the noise current flowing through the source resistance R_s. Putting the expressions for the three noise voltages in N_{output}, we obtain for the output voltage signal-to-noise ratio:

$$\left(\frac{\text{S}}{\text{N}}\right)_{\text{output}} = \frac{Ae_s}{A\sqrt{4kTBR_s + e_n^2B + i_n^2R_s^2B}}. \tag{11.22}$$

It is immediately obvious that the signal-to-noise ratio has been decreased by the amplifier, for the signal-to-noise ratio of the input signal source is:

$$\left(\frac{\text{S}}{\text{N}}\right)_{\text{input}} = \frac{e_s}{\sqrt{4kTBR_s}}. \tag{11.23}$$

The better the amplifier, the smaller e_n and i_n, and the less the degradation of the signal-to-noise ratio. The noise figure is a measure of how much the signal-to-noise ratio is reduced by the amplifier; the better the amplifier the smaller the noise figure. The noise figure can be expressed in terms of the noise generators; from (11.17), (11.22), and (11.23) we have:

$$\text{N.F.}_{\text{dB}} = 10\log_{10}\left[\frac{4kTR_s + e_n^2 + i_n^2R_s^2}{4kTR_s}\right] = 10\log_{10}\left[1 + \frac{e_n^2 + i_n^2R_s^2}{4kTR_s}\right]. \tag{11.24}$$

The "noise factor" F is defined as the ratio of the power signal-to-noise ratios:

$$F \equiv \frac{\text{(S/N) power input}}{\text{(S/N) power output}} = \left[1 + \frac{e_n^2 + i_n^2R_s^2}{4kTR_s}\right]. \tag{11.25}$$

A perfect amplifier which introduces no noise of its own ($e_n = i_n = 0$) has a noise figure of 0 dB and a noise factor of 1.0. An amplifier which decreases the signal-to-noise ratio by a factor of two has a noise figure of 3 dB and a noise factor of 2.0.

It should be pointed out that the noise figure of an amplifier by the preceding definition depends not only on the noise generated by the amplifier, e_n and i_n, but also on the source resistance R_s. One really should speak of the noise figure of the *amplifier–source combination* rather than of the amplifier alone. Two amplifiers cannot be compared on the basis of their noise figures alone; one must also take into consideration the source resistance. For example, two identical amplifiers with the same equivalent noise generators, e_n and i_n, will have different noise figures if used with different source resistances. The smaller the source resistance, the larger (and worse) the noise figure. This may seem puzzling, but a little thought will show that for a small R_s the source contributes very little noise and the amplifier most of the noise. Hence the signal-to-noise ratio is degraded substantially by the amplifier, and the noise figure is high. However, for a *fixed* value of source resistance, the lower the noise figure the better the output signal-to-noise ratio.

What really matters is the signal-to-noise ratio at the amplifier output, and to maximize this one obviously should use an amplifier which contributes as little noise, e_n and i_n, as possible. However, the output signal-to-noise ratio also depends upon R_s and T, the source temperature. The cooler the source, and the smaller the source resistance, the less the output noise and the greater the output signal-to-noise ratio. However, for a given source, with R_s and T fixed, what can we do? Even with a fixed amplifier noise contribution, e_n and i_n, we can improve the amplifier output signal-to-noise ratio by introducing an impedance matching device between the source and the amplifier. We recall that for a fixed source resistance R_s, the lower the amplifier noise figure, the higher the output signal-to-noise ratio. Hence we will try to minimize the noise figure by an impedance matching device that changes the effective source resistance which the amplifier sees. With an impedance matching device between the source and the amplifier input as shown in Fig. 11.7, the amplifier will see a new source resistance R_s'.

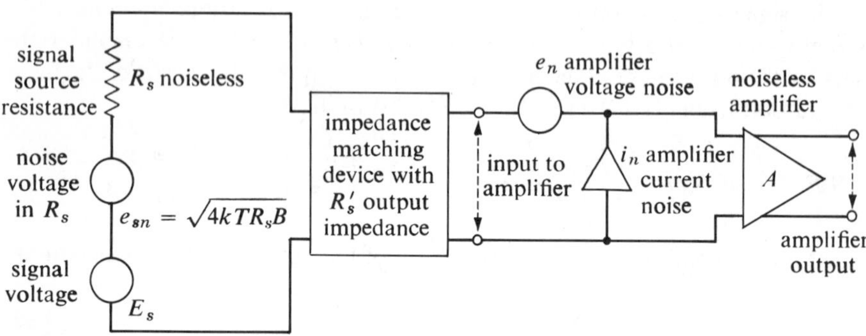

FIGURE 11.7 Amplifier with impedance matching device to transform source resistance.

Hence, the new amplifier noise figure is:

$$\text{N.F.}' = 10 \log_{10}\left(1 + \frac{e_n^2 + i_n^2 R_s'^2}{4kTR_s'}\right). \tag{11.26}$$

This new noise figure is now to be minimized. As R_s' approaches zero, the noise figure approaches infinity; and as R_s' approaches infinity, the noise figure also approaches infinity. Hence, there should be a finite value of R_s' for which N.F.′ is a minimum. We minimize N.F.′ by minimizing the noise factor, $F' = 1 + (e_n^2 + i_n^2 R_s'^2)/(4kTR_s')$, in the usual way by calculating dF'/dR_s', setting the derivative equal to zero, and solving for R_s'.

$$\frac{dF'}{dR'_s} = \frac{4kTR'_s 2i_n^2 R'_s - (e_n^2 + i_n^2 R'^2_s)4kT}{(4kTR'_s)^2} = 0$$

Solving for R'_s yields $R'_s = e_n/i_n$; the optimum value for R'_s is the ratio of the effective amplifier input noise voltage to the effective amplifier noise current. Substitution of $R_s = e_n/i_n$ yields the minimum possible noise factor.

$$F'_{\min} = 1 + \frac{e_n^2 + i_n^2(e_n/i_n)^2}{4kT(e_n/i_n)} = 1 + \frac{e_n i_n}{2kT} \quad \textbf{(11.27)}$$

or

$$\text{N.F.}_{\min} = 10 \log_{10}\left[1 + \frac{e_n i_n}{2kT}\right]. \quad \textbf{(11.28)}$$

Notice we still have the result that the smaller the amplifier noise contribution, e_n and i_n, the smaller the noise figure. And, the cooler the source, the smaller the noise figure.

The most commonly used impedance matching device is a transformer. It must be extremely well shielded and should have negligible primary winding resistance to avoid increasing the total source resistance. It should also have negligible power loss in the core material, which means a high quality low-loss laminated iron core for frequencies up to 30 kHz and a powdered iron, ferrite, or air core for higher frequencies. We recall that for a transformer with a turns ratio of $n = N_{\text{sec}}/N_{\text{pri}}$, the voltage is changed by a factor of n and the current by a factor of $1/n$, so the impedance is changed by a factor of n^2.

$$V_{\text{sec}} = nV_{\text{pri}}$$

$$I_{\text{sec}} = \frac{1}{n} I_{\text{pri}}$$

$$Z_{\text{sec}} = \frac{V_{\text{sec}}}{I_{\text{sec}}} = n^2 Z_{\text{pri}}. \quad \textbf{(11.29)}$$

For example, a step-up transformer with $n = 10$ will increase the primary impedance by a factor of 100. The transformer output terminals will appear to have a voltage generator of 10 times higher amplitude and an output impedance 100 times larger. This impedance transformation is illustrated in Fig. 11.8. The optimum source resistance is $R'_s = e_n/i_n$, so the optimum transformer turns ratio is:

$$R'_s = n^2 R_s = e_n/i_n$$

or

$$n = \sqrt{\frac{e_n/i_n}{R_s}}. \quad \textbf{(11.30)}$$

A common use of this technique is when the signal source has a low impedance, and the optimum source resistance $R_s = e_n/i_n$ is much higher. A

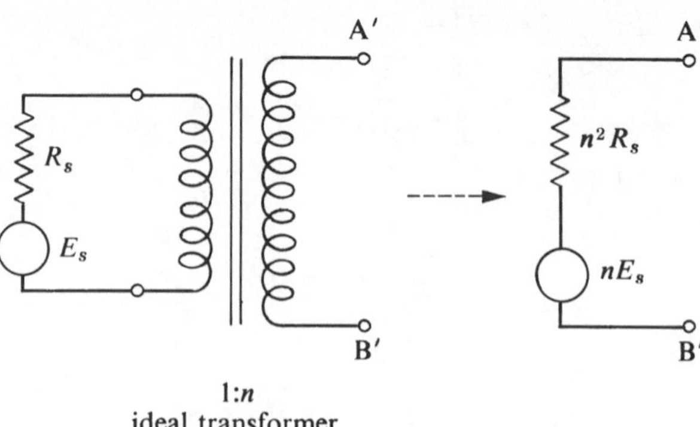

FIGURE 11.8 Transformer impedance transformation.

step-up transformer is then used to achieve a better signal-to-noise ratio. This situation occurs, for example, when a crystal diode is used to detect a microwave frequency; the diode produces a small dc output voltage proportional to the incident microwave power. The diode output impedance is typically about 50Ω, and e_n/i_n for the amplifier is usually on the order of 10^5 to 10^6 ohms. Hence a step-up transformer with turns ratio $n = (e_n/i_n)/R_s = \sqrt{10^6/50} = 141$ should be used. If, on the other hand, the source resistance is larger than the optimum value, e_n/i_n, a step down of the source resistance is called for. One example of such a source is a photomultiplier tube, which acts essentially as a constant current source, i.e., it has a very high output impedance. In this case the source noise is almost entirely shot noise and would be represented by a *current* noise generator i_{sn}. However, a step-down transformer is not a practical solution because its primary resistance is too high, thereby increasing the effective source noise. Thus the only alternative is to choose an amplifier with as large an e_n/i_n ratio as possible, which means a high input impedance FET amplifier usually.

An alternate technique for low noise counting of photons from weak sources uses modern amplitude discriminators and fast amplifiers to convert each photomultiplier voltage pulse resulting from one photon to a standardized voltage pulse. These pulses are then fed into a digital-to-analog converter, the output of which is proportional to the incident photon intensity. In this way low amplitude noise pulses and baseline drift are prevented from affecting the output.

The improvement in the signal-to-noise ratio from impedance matching is shown in Fig. 11.9 for a specific amplifier with $e_n = 2 \times 10^{-8}\,\mathrm{V}/\sqrt{\mathrm{Hz}}$ and $i_n = 2 \times 10^{-14}\,\mathrm{A}/\sqrt{\mathrm{Hz}}$. These amplifier noise parameters are typical of modern state-of-the-art transistorized amplifiers at frequencies 1 kHz–10 kHz. Notice the reduction in noise figure from 27 dB to 0.21 dB. It should also be pointed out that e_n and i_n do depend upon frequency; this topic will be treated later in the chapter.

The improvement in the output signal-to-noise ratio can be verified by considering the equivalent circuit of the amplifier and the signal source e_s volts with impedance R_s, as transformed by a transformer of turns ratio n, shown in Fig.

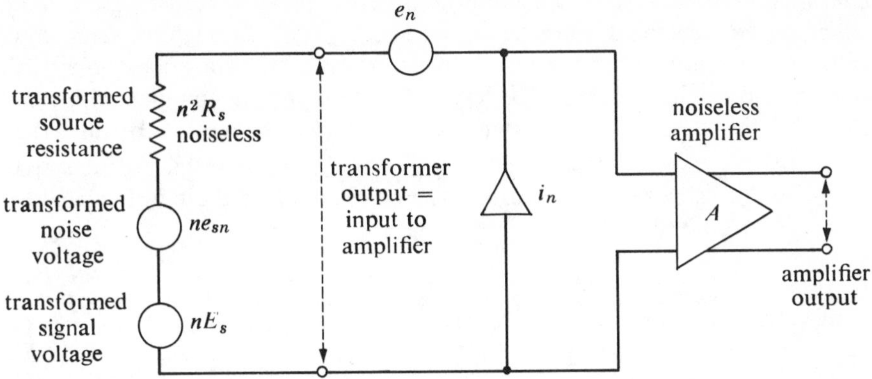

FIGURE 11.9 Equivalent circuit of source impedance matched to amplifier.

11.9. The voltage signal-to-noise ratio of the output without the impedance transformation is:

$$(S/N)_{output} = \frac{e_s}{\sqrt{(4kTR_s + e_n^2 + i_n^2R_s^2)B}}. \tag{11.31}$$

Without the impedance transformation, the noise input voltage to the amplifier is $E_{ni} = \sqrt{(e_{sn}^2 + e_n^2 + i_n^2R_s^2)B}$. With the impedance transformation, the input noise is $E_{ni} = \sqrt{(ne_{sn})^2 + e_n^2B + (i_nn^2R_s)^2B}$, where $e_{sn} = \sqrt{4kTR_sB}$ is the signal–noise voltage. The signal voltage output with the impedance transformation is simply $S_o = Ane_s$. Thus the voltage signal-to-noise ratio of the output with the impedance transformation is:

$$(S/N)_{output} \equiv \frac{Ane_s}{A\sqrt{(ne_{sn})^2 + e_n^2B + (i_nn^2R_s)^2B}},$$

$$= \frac{nE_s}{\sqrt{n^24kTR_sB + e_n^2B + i_n^2n^4R_s^2B}},$$

or

$$(S/N)_{output} = \frac{e_s}{\sqrt{\left(4kTR_s + \frac{e_n^2}{n^2} + i_n^2n^2R_s^2\right)B}}. \tag{11.32}$$

When n is very large, the $i_n^2n^2R_s^2$ term predominates in the denominator and the signal-to-noise ratio approaches zero. When n is very small the e_n^2/n^2 predominates and again the signal-to-noise ratio approaches zero. Thus, we expect an optimum value of n for which the signal-to-noise ratio is maximum. Physically, the reason an optimum signal-to-noise ratio can be realized with an impedance transformation is because the transformation affects the amplifier voltage noise source e_n and current noise source i_n differently. For a large value of n, the amplifier voltage noise e_n becomes negligible compared to the source noise, but the amplifier constant current noise becomes dominant because of the noise voltage $i_n n^2 R_s$ generated when the noise current i_n flows through the stepped-up value of the source resistance $n^2 R_s$. Thus, from (11.32), for an amplifier with large e_n and small i_n, n should be large. Similarly, for an amplifier with small e_n and large i_n, n should be small to minimize the noise. In all cases, $n = \sqrt{(e_n/i_n)/R_s}$ will give the best output signal-to-noise ratio. We calculate the optimum value of n in the usual way by maximizing the output signal-to-noise ratio $(\mathrm{S/N})_o$:

$$\frac{d}{dn}\left(\frac{\mathrm{S}}{\mathrm{N}}\right)_{\text{output}} = \frac{d}{dn}\frac{e_s}{\sqrt{B}}\left(4kTR_s + \frac{e_n^2}{n^2} + i_n^2 n^2 R_s^2\right)^{-1/2}$$

$$= \frac{e_s}{\sqrt{B}}\left(-\frac{1}{2}\right)\left(4kTR_s + \frac{e_n^2}{n^2} + i_n^2 n^2 R_s^2\right)^{-3/2}\left(\frac{-2e_n^2}{n^3} + 2i_n^2 n R_s^2\right).$$

Therefore,

$$\frac{d}{dn}\left(\frac{\mathrm{S}}{\mathrm{N}}\right)_{\text{output}} = 0 \quad \Rightarrow \quad \frac{2e_n^2}{n^3} = 2i_n^2 n R_s^2$$

or

$$n = \sqrt{\frac{e_n/i_n}{R_s}}, \tag{11.33}$$

which agrees with equation (11.30).

Substitution of $n = \sqrt{(e_n/i_n)/R_s}$ in the output signal-to-noise expression yields the maximum voltage signal-to-noise attainable at the amplifier output:

$$\text{Maximum}\left(\frac{\mathrm{S}}{\mathrm{N}}\right)_o = \frac{e_s}{\sqrt{4kTR_s + 2e_n i_n R_s}}. \tag{11.34}$$

Finally, it should be pointed out that a *transformation* of impedance is necessary to improve the signal-to-noise ratio. If the source resistance R_s is less than the optimum value of e_n/i_n, it is not possible to just add another resistance R_A in series with R_s to make $R_s + R_A = e_n/i_n$. This will *lower* the signal-to-noise ratio as can be seen by referring to the signal-to-noise expression above. More noise will be generated due to the increased thermal noise voltage generated in R_A, and also due to the amplifier current noise flowing through R_A, which produces an additional noise voltage in R_A.

11.6 FLICKER NOISE

In the preceding calculation of amplifier signal-to-noise ratio, we have assumed that the amplifier voltage and current equivalent noise generators e_n and i_n were independent of one another and also frequency independent. In general, however, e_n and i_n do depend upon frequency, because at low frequencies a new type of noise, called "flicker noise," is larger than either Johnson or shot noise.

Flicker noise empirically has a power dependence of f^{-1}, although the exponent can vary from about 0.9 to 1.4. That is, the flicker noise power is given by: $P_{fn} = kf^{-n}$, where $n \cong 1.0$. The f^{-1} frequency dependence is 3 dB per octave, or a factor of 10 increase in power for each decade of frequency decrease. Flicker noise usually is the major noise source below several hundred hertz in most electronic amplifiers. The frequency at which flicker noise equals the other noise in a device is often called the "corner frequency." A general graph of noise versus frequency is shown in Fig. 11.10. For semiconductor diodes (1N23,

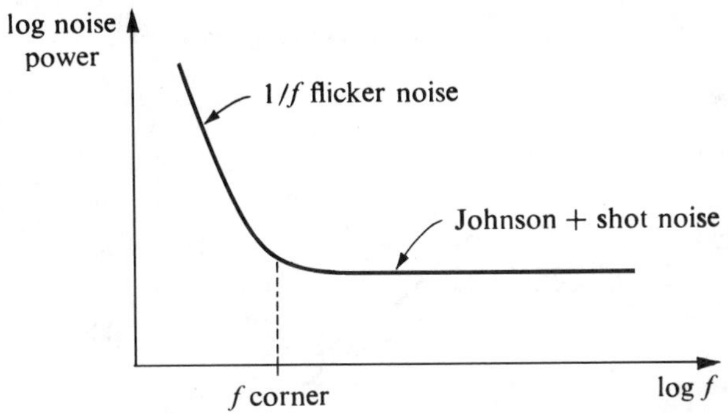

FIGURE 11.10 Noise spectrum.

etc.) used to detect microwave frequencies around 10 GHz, the corner frequency is on the order of tens of kHz. Newer Schottky barrier diodes have recently been developed with a corner frequency around 1 kHz.

Flicker noise is also occasionally referred to as "pink noise" to distinguish it from white noise. Because of the $1/f$ dependence, there is the same pink noise power in each octave or frequency decade, i.e., from 10 Hz to 20 Hz or from 50 Hz to 100 Hz. The exact cause of flicker noise is not well understood, but it appears to depend upon surface characteristics. Measurements which verify the $1/f$ frequency dependence have been made on some devices down to frequencies on the order of 10^{-4} Hz. It is clear that the $1/f$ dependence cannot extend down to dc because the total noise power would then be infinite. All transistors and vacuum tubes exhibit flicker noise, as do some resistors. Carbon composition resistors carrying substantial current exhibit $1/f$ noise, but metal film or wirewound resistors are much less noisy.

Examples of voltage and current effective input noise spectra are shown in Fig. 11.11 for three types of amplifiers. Notice that in all three amplifier types there is a characteristic 3 dB/octave increasing flicker voltage noise e_n as the frequency decreases below around 1 kHz. The current noise i_n tends to increase at higher frequencies because of the decreasing input capacitive reactance across which the constant voltage noise appears.

Hence there are several implications of the frequency dependence of e_n and i_n. The lowest noise region is usually somewhere between 100 Hz and 10 kHz. Thus it is best if the signal to be amplified falls in this region. Often this can be

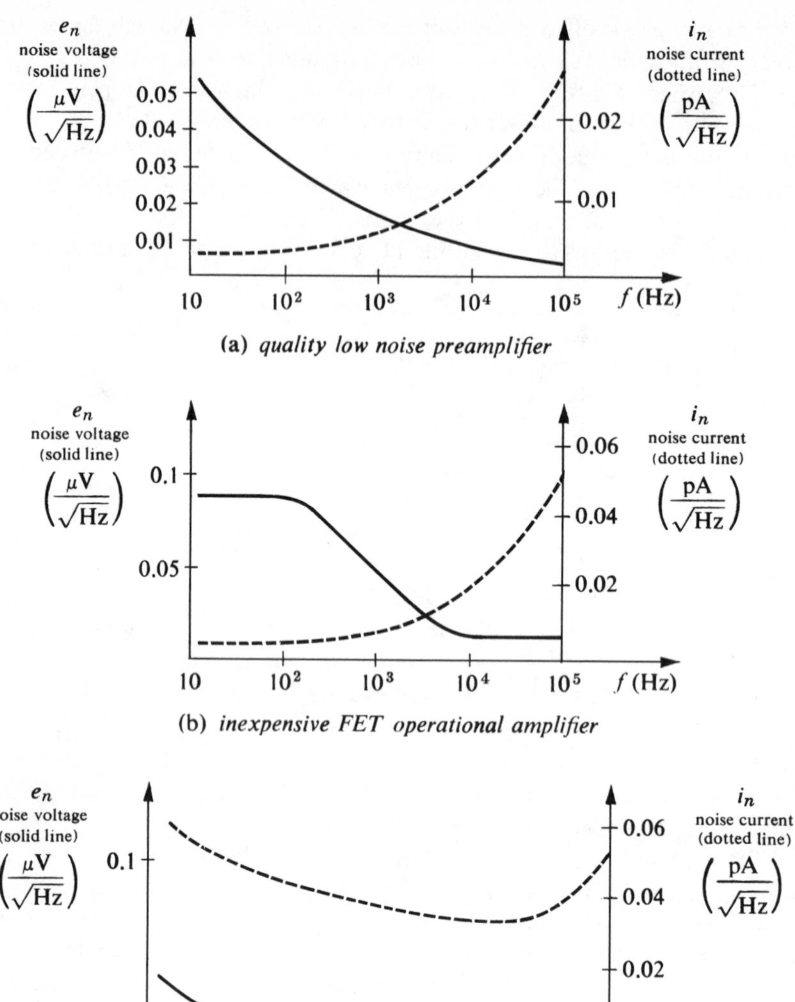

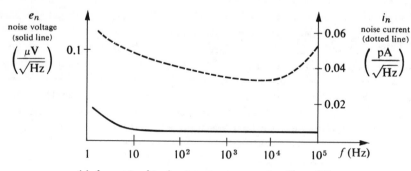

FIGURE 11.11 Typical amplifier effective input noise spectra.

done by modulating the signal in this frequency region. For example, a dc or slowly varying signal could be deliberately converted into a 1 kHz sinusoidal signal, the amplitude of which is proportional to the dc signal, amplified and then rectified and converted back into dc.

The amplifier bandwidth should also be the minimum necessary to amplify the Fourier frequency components of the signal. Any amplifier bandwidth in excess of this merely amplifies noise and thus degrades the signal-to-noise ratio.

The bandwidth is usually limited by putting in appropriate low-pass and high-pass filters.

The optimum value of the source resistance, $R_{\text{opt}} = e_n/i_n$, varies with frequency. Typically, e_n/i_n ranges from a high value of around 10 megohms at 100 Hz to a low value of around 100 kΩ at 100 kHz. Hence, great care must be taken in choosing the impedance matching ratio depending upon the frequency of the signal.

The flicker noise will also depend slightly upon the power supply voltage, so some experimentation in this respect is worthwhile. Most amplifier noise generators e_n and i_n are substantially independent of temperature, but in FET amplifiers the shot or current noise varies as the square root of the gate–channel leakage current, which in turn varies with temperature. The net effect is for the FET shot noise to double for approximately every 15 degrees of temperature rise. This effect is only noticeable, of course, above the corner frequency, i.e., where the shot noise is not masked by the low-frequency flicker noise.

11.7 NOISE TEMPERATURE

The noise figures of an amplifier is a somewhat ambiguous concept. Intuitively one would like the noise figure to depend only on the amplifier's characteristics, but as we saw in section 11.5, it depends upon the bandwidth, the source resistance, and the temperature of the source resistance. A different parameter for describing an amplifier's noise is the "noise temperature." Consider an amplifier driven by a source with resistance equal to the input impedance of the amplifier. If the source resistance were at 0 °K, it would generate no Johnson noise and all the noise in the amplifier output would be generated within the amplifier. If we now increase the source resistance temperature until the output noise power exactly doubles, the resistance is contributing an amount of noise power equal to that generated within the amplifier. Thus, the temperature *increase* of the resistance is a measure of the amplifier noise. Noise temperature is independent of the bandwidth of the amplifier and so can be used to compare amplifiers of different bandwidths, whereas one cannot make such a comparison on the basis of the amplifier noise figures.

The noise temperature is also linearly proportional to the noise power in the output, so the signal-to-noise ratio is inversely proportional to the noise temperature. An amplifier with a noise temperature of 100°K will have twice the output signal-to-noise power of an amplifier with a noise temperature of 200°K.

A noise temperature can be defined for circuits other than amplifiers, e.g., an antenna or a transmission line. An antenna driving a transmission line feeds a certain amount of noise into the line. The antenna impedance should be matched to the line impedance for optimum power transfer. Hence, if we replace the antenna by a resistance equal to the antenna impedance and heat the resistance up until the noise at the output of the line is the same as with the antenna connected, then the temperature of this resistance is the noise temperature of the antenna. The noise temperature is a measure of how much noise power the antenna produces, not of the actual physical temperature of the antenna material. Some of the noise in the antenna output is blackbody electromagnetic radiation received by the antenna from external sources. Hence, the antenna noise temperature often depends upon the direction the antenna points, i.e., on what the antenna "sees."

A noise temperature for a transmission line can be defined in a similar way. The line is replaced by a resistance equal to the characteristic impedance of the line. The temperature this resistance must have in order to generate a noise power equal to that generated by the line is the noise temperature of the line.

Because the noise temperature is always proportional to the actual noise power, the total noise power of a system is proportional to the sum of the noise temperatures of the various parts of the system. The same addition technique cannot be used to calculate the total system noise from the noise figures of the various parts.

11.8 LOCK-IN DETECTION

Perhaps the single most useful technique for increasing the signal-to-noise ratio is lock-in detection. In this technique, the desired signal is chopped or amplitude modulated at a frequency f_m, amplified, then synchronously detected or rectified to obtain a slowly varying signal, and finally fed through a low-pass filter to obtain the signal output. The block diagram and the waveforms are shown in Fig. 11.12. The lock-in amplifier or detector is also known as a "synchronous amplifier" or a "phase sensitive" amplifier or detector.

The signal plus noise is first chopped or amplitude modulated at a fre-

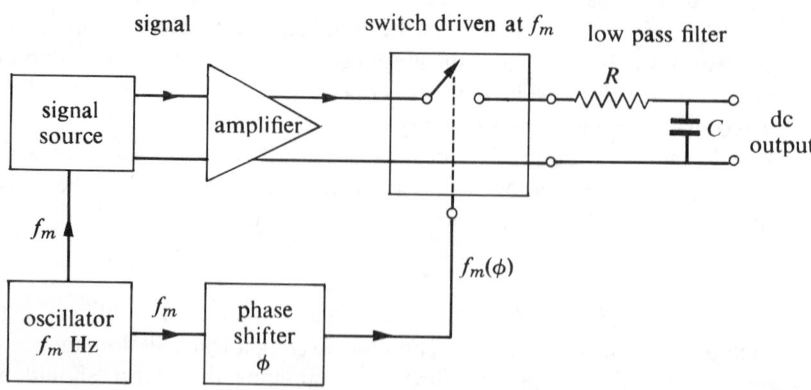

FIGURE 11.12 Lock-in detector block diagram.

quency, f_m Hz, supplied by a reference oscillator. The resulting chopped signal is amplified by an amplifier tuned to the frequency f_m, but with a sufficiently broad frequency response to pass the "shape" or envelope of the chopped signal, i.e., the sidebands of the signal on either side of f_m must also be passed. This is very similar to the problem of AM radio discussed at the end of Chapter 3. The bandwidth or equivalently the Q of this amplifier is usually adjustable; Q values usually range from 5 to 20. The slower the signal varies with time the closer the sidebands to the frequency f_m and the narrower the bandwidth (the higher the Q) required. One should always use as high a Q as possible to reject any noise at frequencies outside the signal sidebands.

The amplified chopped signal is then fed into a synchronous detector which is essentially a DPDT switch driven at the frequency f_m by the *same* reference oscillator which chopped the signal in the first place. A phase shifter is usually incorporated in the instrument so that the DPDT switch can be made to close on the left at the instant when the signal goes positive and close on the right when the signal goes negative. The output of the synchronous detector is proportional to the *cosine* of the phase angle between the signal and the switch frequency. The resulting signal output is unidirectional; it is dc, but its amplitude depends on the amplitude of the signal. This output is then fed through a buffer amplifier to a low-pass filter which smooths out or integrates the unidirectional signal pulses. The buffer amplifier serves to isolate the synchronous detector and the low-pass filter from one another. Any steady frequency which is not equal to f_m or an odd multiple of f_m, and not *phase coherent* with f_m, will be averaged to zero by the action of the DPDT switch. For example, a frequency $2f_m$ will contain a complete cycle of one positive and one negative pulse during the time interval $1/2f_m$, when the switch is closed to the left. A frequency $4f_m$ will contain no complete cycles in the time interval $1/2f_m$, etc. The switch action effectively multiplies the input signal by a square wave of frequency f_m, whose Fourier spectrum contains all the odd multiples of f_m (f_m, $3f_m$, $5f_m$, . . .). The filtering action of the output low-pass filter then averages all the even multiples of f_m and any noncoherent noise to zero, provided the time constant RC is large enough. The dc and slowly varying output will depend only upon f_m and its odd multiples. The amplifiers are usually tuned to reject $3f_m$ and higher frequencies, so in effect only f_m and very closely spaced sidebands contribute to the output. Notice that because the same oscillator chops the signal and also drives the DPDT switch the synchronous detector is "locked" to the frequency f_m. A small drift in the frequency f_m cannot detune the instrument, hence the name "lock-in" detector.

The improvement in the signal-to-noise ratio comes in several ways, but mainly from the narrowing of the effective bandwidth of the signal by making the signal vary slowly with time. Then all the noise outside the signal bandwidth can be rejected with the low-pass filter. The signal sidebands then lie very close to f_m, the chopping frequency, from $f_m - f_1$ to $f_m + f_1$, a bandwidth of $2f_1$ Hz. The synchronous detector converts the signal to dc; it effectively translates f_m to 0 Hz or dc. Thus, the signal sidebands in the output of the synchronous detector are now from 0 Hz (dc) out to f_1 Hz. Thus, the bandpass of the output low-pass filter need only be from dc to f_1, i.e., $1/(2\pi RC) \cong f_1$. By sweeping through the signal very slowly, f_1 is made very small (the sidebands are close to f_m), so that a large output filter time constant can be used. Thus, the effective bandwidth $f_1 \approx 1/RC$ is extremely small, and all the signal and very little noise is passed through this bandwidth. All the noise at frequencies above f_1 will be averaged to zero by the filtering action of the output low-pass filter. In most instruments, the time instant is adjustable over a wide range, from 0.001 to 10 seconds or more. One must always be careful to sweep through the signal slowly enough so that all the signal sidebands lie within $1/RC$ Hz of f_m. Practically speaking, this means the signal waveform should be spread out over $10RC$ or more seconds. The choice of the chopping frequency f_m can also be made in a region where the amplifiers, etc., produce very little noise.

Consider a specific experiment, electron spin resonance (ESR) or nuclear magnetic resonance (NMR). In ESR or NMR, the sample contains magnetic dipoles (electron spins or nuclear magnetic moments) which absorb energy only at one (or several discrete) values of an external magnetic field if a fixed-frequency electromagnetic field of the proper polarization is fed into the sample. For example, in ESR, free electrons bathed in 10,000 MHz (10 GHz) electromagnetic radiation in a microwave cavity will absorb 10 GHz energy only when the magnetic field equals 3751 gauss. In NMR, protons bathed in 42.6 MHz electromagnetic radiation inside a coil in a resonant LC tank will absorb 42.6 MHz energy only when the external magnetic field equals 10,000 gauss. The sample always absorbs energy over a very small but finite range of magnetic field known as the "line width," as shown in Fig. 11.13. The resonance signal is the energy absorption curve of Fig. 11.13(a). We could observe it by detecting and amplifying the absorbed energy as we increase or "sweep" the magnetic field from 3570 to 3572 gauss. However, if we sweep the field upward very slowly, the signal will vary slowly in time and contain only very low-frequency Fourier components. This would be difficult to amplify because of the low-frequency $1/f$ flicker noise present in all crystal diodes and amplifiers. If we sweep more rapidly, the signal Fourier components will be at higher frequencies, say from 100 Hz to 100 kHz; but this means an amplifier bandwidth of 100 Hz to 100 kHz, or approximately 100 kHz would be necessary to reproduce the signal with negligible distortion. (Leaving out any of the signal's frequency components by an inadequate amplifier bandwidth will distort the signal.) If the source resistance is 50 Ω and is at 300°K, then the source noise voltage is:

$$e_{sn} = \sqrt{4kTR_sB} = \sqrt{4 \times 1.38 \times 10^{-23} \times 300 \times 50 \times 10^5}$$
$$= 0.29\,\mu\text{V},$$

which practically speaking is enough to obscure many small resonance signals of great scientific interest.

What is done is to convert the resonance signal to essentially a single frequency by sinusoidally modulating the magnetic field at a frequency f_m, which

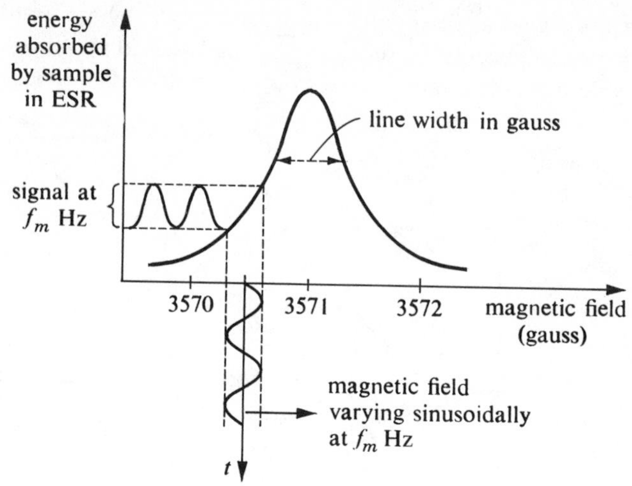

(a) *energy absorption curve*

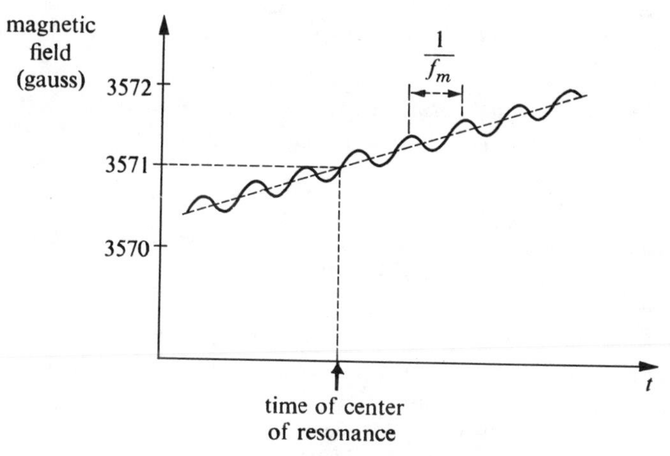

(b) *magnetic field sweep*

FIGURE 11.13 Electron spin resonance.

may range from 100 Hz to 100 kHz, while simultaneously sweeping the magnetic field slowly upward from 3570 to 3572 gauss. The amplitude of the magnetic field modulation is made much less than the line width. The energy absorbed by the sample is then amplitude modulated *at the modulation frequency* f_m, and the *amplitude* of the fluctuating energy absorbed is proportional to the *slope* of the energy absorption curve of Fig. 11.13. This amplitude is zero off to the left of the line, around 3570 gauss, increases and then decreases to zero at the center of the line at 3571 gauss, and then increases above 3571 gauss, but with a 180° phase shift relative to the signal on the left side of the line. The point is that we now have a signal at essentially one frequency f_m, whose amplitude traces out the slope of the energy absorption curve we wish to measure, as shown in Fig. 11.14. Because the signal frequency is essentially one frequency f_m, a very

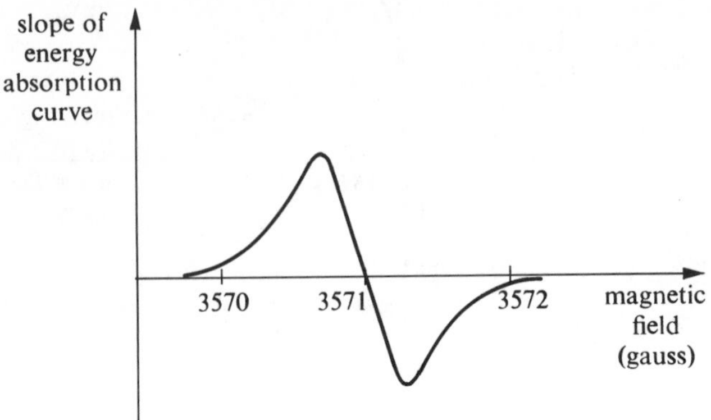

FIGURE 11.14 Slope of energy absorption curve.

narrowband amplifier which passes extremely little noise can be used. If the bandwidth is 1 Hz centered at f_m, then the rms signal noise voltage from the 50 ohm, 300°K source is:

$$e_{sn} = \sqrt{4kTR_sB} = \sqrt{4 \times 1.38 \times 10^{-23} \times 300 \times 50 \times 1}$$
$$= 0.00091\,\mu\text{V}.$$

We see that the rms noise voltage has decreased by a factor of $\sqrt{B} = \sqrt{10^5} = 316$, meaning the signal-to-noise ratio has been increased by a factor of 316.

The actual instrument works as follows. A moderately narrowband fixed-tuned amplifier, with a Q of approximately 10, amplifies the incoming signal at the frequency $f_m = 1/T_m$. This signal is then fed into a switch driven at the modulation frequency f_m. That is, the switch is closed for a period of time $T_m/2$, open for $T_m/2, \ldots$. The phase of the switch is adjusted by the phase adjustment ϕ, so that the switch is closed for precisely the entire positive half-cycle of the signal and open for the entire negative half-cycle. Thus, the capacitor C is charged up positively by the positive half-cycles of the signal, producing a positive dc output voltage proportional to the amplitude of the signal S. When the magnetic field sweeps past the center of the energy absorption curve, the phase of the signal abruptly shifts by 180°, so that the switch is now closed for the negative half-cycles of the signal and open for the positive half-cycles. The output voltage is now negative. The sign of the output voltage is thus dependent upon the phase of the input relative to the phase of the modulation frequency; hence the name "phase sensitive detector" is sometimes used instead of lock-in detector. The magnetic field must be swept slowly enough so that the signal amplitude does not change appreciably in the time RC, which is essentially the minimum time during which the dc output voltage can change. A typical value for RC is 1 second, so the energy absorption curve should be swept through no faster than once in 10 or 20 seconds. The effective frequency bandwidth of the switch and RC output filter is approximately $1/RC$, because it is a low-pass filter with a break-point frequency $\omega_B = 1/RC$.

Any noise voltage present along with the signal will be random in phase or "incoherent" with respect to the modulation frequency driving the switch. Therefore, short positive and negative noise spikes are equally likely during the time the switch is closed. And any low-frequency noise, such as $1/f$ flicker noise or 60 Hz or 120 Hz interference, will tend to give the capacitor a net charge of zero; because the positive and negative noise cycles will each be sampled equally by the switch if the output is averaged over many noise points, i.e., if the break-point frequency of the output filter is less than the noise frequency: $1/(2\pi RC) < f_{\text{noise}}$. This is equivalent to the earlier statement that the bandwidth of the switch-RC filter combination is essentially equal to $1/RC$.

In practical lock-in detectors, the switch is rarely a mechanical switch. It is usually an amplifier that is gated on and off by a square wave of frequency f_m. For example, a square wave of ten volts amplitude could be applied to the collector of a transistor amplifier.

In summary, only a signal frequency coherent or in phase with f_m, the modulation frequency driving the switch, will be passed through the lock-in detector. That is, a signal of f_m, $3f_m$, $5f_m$, etc., will produce a dc output, but the amplifiers will attenuate the frequency $3f_m$ and higher.

In microwave ESR experiments, the detector is usually a point contact silicon crystal diode, the 1N23 family for 10 GHz, for example. These diodes exhibit $1/f$ flicker noise as well as white noise, and their corner frequency is on the order of 50–100 kHz. The crystal noise is usually far larger than the amplifier noise, so a considerable increase in signal-to-noise is achieved by using a modulation frequency above the corner frequency. $f_m = 100$ kHz is a typical choice. Newer Schottky diodes have recently been made with the corner frequency down around 1 kHz, enabling one to use lower modulation frequencies, which are experimentally more convenient.

Lock-in detection is not confined to ESR and NMR experiments. It is particularly useful when the signal is at dc or a low frequency where $1/f$ amplifier flicker noise is large. For example, if the response of a complicated biological system to a light stimulus is to be measured, then the light can be chopped by a rotating slotted wheel at a frequency f_m. The desired signal is the electrical response of the system to the light and is taken off a transducer of some sort, for example, electrodes attached to the skull, etc.; see Fig. 11.15. The signal is usually buried in other electrical signals and noise, and therefore difficult to detect. But if the transducer output is fed into a lock-in detector with a reference frequency of f_m, then the lock-in output will be proportional to the system response to the light and not to any other stimulus which may produce electrical signals in the transducer. The essential point is to "chop" or amplitude modu-

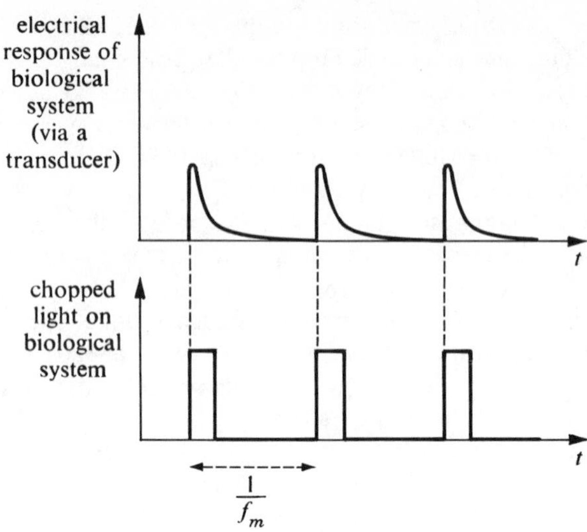

FIGURE 11.15 Light stimulus and electrical response.

late the stimulus at a frequency f_m and to use the same chopping frequency source to supply the reference input to the lock-in detector.

11.9 SIGNAL AVERAGING TECHNIQUES

Noise is random, not only in frequency but also in polarity. A noise voltage, averaged over a long time, will average to zero because there will be as many positive as negative noise pulses. A signal, on the other hand, has a definite, nonrandom polarity. Hence, if we can repeatedly average over a signal plus noise, then the noise contributions ought to get smaller and eventually average to zero, while each average should add to the signal amplitude.

There are several types of instruments available to perform such signal averaging, some digital and some analog. The signal must be repetitive, not necessarily periodic. The analog signal averager known as a "waveform eductor" works as follows. At the beginning of each sweep through the signal, a trigger pulse or command of some sort must start the sweep. The time during which the signal exists is divided up into many intervals called "channels," say 100; and the voltage in each interval is measured and stored on a separate capacitor, one for each interval. Thus on every sweep each capacitor receives a voltage proportional to the signal value plus the noise value during that interval for that particular sweep. The same process is repeated when the next trigger pulse comes along, and the capacitors accumulate more voltage each sweep. After N sweeps, each capacitor should have a voltage equal to N times the signal voltage v_s, plus the sum of N noise voltages, which add as the square root of the sum of the squares because the noise is random in time. Thus the voltage v stored after N sweeps is given by:

$$v = Nv_s + \sqrt{e_{n1}^2 + e_{n2}^2 + \cdots + e_{nN}^2}, \tag{11.35}$$

where v_s is the signal voltage present at each sweep and e_{n1} is the noise voltage present on the first sweep, etc. If the noise is constant in character for all the sweeps, i.e., if the experimental conditions do not change, then the square of the noise should be the same for each sweep:

$$e_{n1}^2 = e_{n2}^2 = \cdots = e_{nN}^2 \equiv e_n^2.$$

Thus,

$$v = Nv_s + \sqrt{Ne_n^2}$$

or

$$v = Nv_s + \sqrt{N}\sqrt{e_n^2} = Nv_s + \sqrt{N}e_{\text{rms}n}, \tag{11.36}$$

where $e_{\text{rms}n}$ is the root mean square noise. The signal-to-noise ratio after N sweeps is then:

$$\text{S/N} = \frac{Nv_s}{\sqrt{N}e_{\text{rms}n}} = \sqrt{N}\frac{v_s}{e_{\text{rms}n}}. \tag{11.37}$$

Hence, the signal-to-noise ratio increases as the square root of the number of sweeps through the signal. There is no theoretical upper limit on the improvement in the signal-to-noise ratio which can be obtained by this technique. The practical limits are: the stability and reproducibility of the trigger pulse—it must occur at precisely the same point in time relative to the signal; the ability of the capacitor storage to keep the stored signal and noise for the duration of the N sweeps; and, of course, the noise generated within the instrument.

Signal averagers are generally used in pulse rather than steady-state experiments, where the signal occurs once after some sort of stimulus. Lock-in techniques are usually used in steady-state experiments where the signal can be sinusoidally modulated. Pulse experiments are widely used in biology. The stimulus could be a light or sound pulse which produces a response that is usually buried in noise. The stimulus is usually also used to provide a trigger pulse for the signal averager, as shown in Fig. 11.16. The signal averaging technique

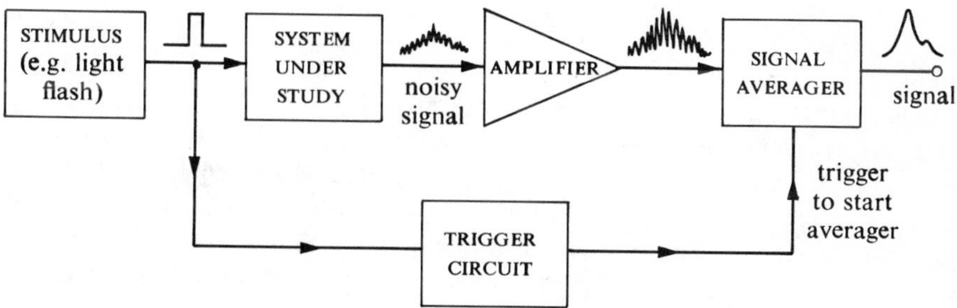

FIGURE 11.16 Signal averager in pulse experiment.

is not quite so good as the lock-in technique in extracting signals buried in a great deal of noise, but it does have the advantage of presenting the true signal shape as the output, rather than the approximate first derivative as in the lock-in technique.

The digital signal averager is sometimes called the "computer of average transients," or CAT, and in general is considerably more expensive than the analog waveform eductor. However, most general purpose computers with the proper program can be used as a CAT. In theory the CAT can be used for an infinite time to improve the signal-to-noise ratio without limit because of the permanent memory and the absence of drift in its digital circuitry, but there will always be some noise and drift in the input transducer and the input analog to digital converter. The CAT has the advantages that its memories in which the signal is stored are more permanent, and its output is already binary coded for further computer processing if desired. The intervals or channels in a waveform eductor can be made as short as one microsecond, while the shortest CAT interval is about 30 microseconds.

Signal averaging can also be used in pulsed resonance experiments such as NMR or ESR spin echo experiments. Modern instruments have time resolution on the order of five to ten nanoseconds and so can be used with extremely fast signals such as in photoluminescence and fluorescence.

11.10 CORRELATION TECHNIQUES

Information theory research has resulted in the development of correlation techniques for the recovery of signals from noise. There are basically two types of correlation: autocorrelation and crosscorrelation.

In autocorrelation, a repetitive signal is compared with a delayed version of itself. The product of the signal $v(t)$ and the same signal at an earlier time $v(t - \tau)$ is calculated for many different values of t, and the results added and averaged in time to form the autocorrelation function, C_{vv}, which depends upon τ and the shape of the signal v.

$$C_{vv}(\tau) \equiv \lim_{T\to\infty} \frac{1}{2T} \int_{-T}^{+T} v(t)v(t - \tau)\, dt. \qquad \textbf{(11.38)}$$

In practice, $v(t)$ usually contains a small repetitive signal plus a large amount of random noise. If τ equals the basic repetitive period of the signal $v(t)$, then $v(t)v(t - \tau)$ essentially equals v^2, and so the correlation coefficient will be large and positive for this value of τ. The same will hold when τ equals any integral number of signal periods. Thus the correlation function will have the same *period* as the signal. For random noise, however, which has no repetitive period, the correlation function C will never be periodic, and in general it can be shown that C will decrease with increasing τ. When $\tau = 0$, the correlation function equals the mean-square value of the signal:

$$C_{vv} = \lim_{T\to\infty} \frac{1}{2T} \int_{-T}^{T} v^2(t)\,dt. \tag{11.39}$$

Examples of several signals and their autocorrelation functions are shown in Fig. 11.17. The autocorrelation function of random white noise has no period

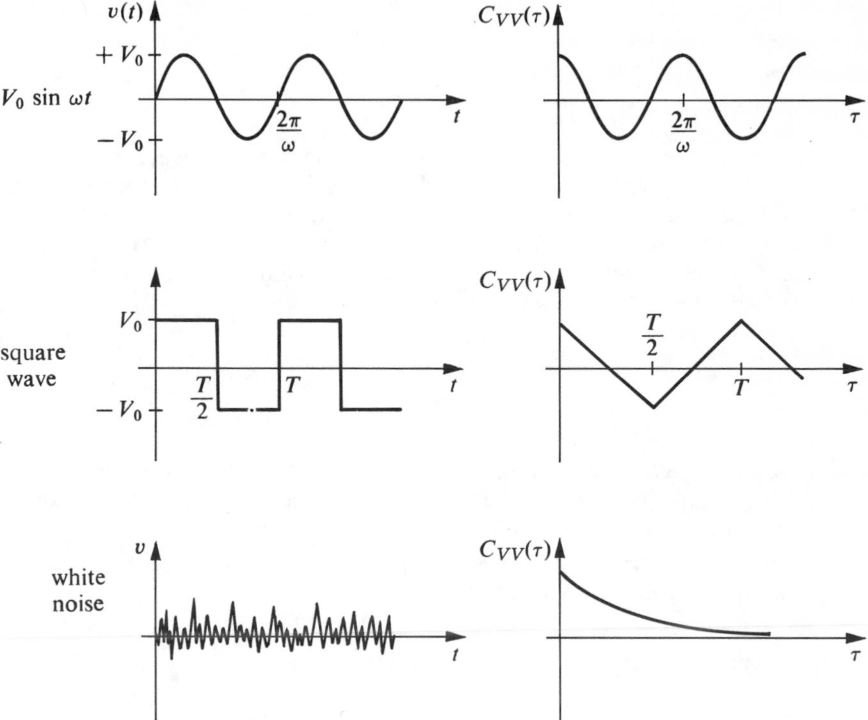

FIGURE 11.17 Examples of signals and their autocorrelation functions.

and has a correlation function which exponentially decays as a function of τ. Function B is the bandwidth of the white noise function.

In crosscorrelation, a repetitive signal is compared with another different signal for various time delays. If v and u are the two signals, their crosscorrelation function is defined as:

$$C_{vw}(\tau) \equiv \lim_{T\to\infty} \frac{1}{2T} \int_{-T}^{T} v(t)w(t-\tau)\,dt. \tag{11.40}$$

Signal averaging is really a special case of crosscorrelation; it is the crosscorrelation of the signal v with a series of delayed, narrow, and flat top pulses.

Each pulse looks at a narrow portion of the signal and the time averager adds up the signal and averages out the noise. The crosscorrelation function is then the signal v; the longer the integration time, $2T$, the higher the signal-to-noise ratio.

Crosscorrelation and autocorrelation instruments are now available to calculate one hundred points of the correlation function and smooth out the intervals between the points, thus producing an analog output which can drive a chart recorder or an oscilloscope to display the correlation function.

references

Further information on analog signal averaging and lock-in techniques can be found in Vol. 6, No. 1, of *Tek Talk* of the Princeton Applied Research Corporation, Box 565, Princeton, N.J., zip code 08540; in an article "Noise in Amplifiers" by S. Letzer and N. Webster in the August 1970 *IEEE Spectrum;* and in an article "Modern Signal-Processing Techniques for Overcoming Noise" by C. A. Nittrouer in the September and October 1968 *Electronic Instrument Digest.*

problems

1. Which would be a better source for a high fidelity music system: (a) a signal voltage of 10 mV along with a noise voltage of 1 μV or (b) a signal voltage of 1 V along with a noise voltage of 1 mV? Explain.
2. Calculate the Johnson noise voltage generated in a 1 megohm resistance at 300 °K as measured by a scope with a 10 MHz bandwidth.
3. For Johnson noise generated in a resistance R at T degrees, what is the percentage probability that the noise voltage is (a) greater than $+\sqrt{4kTBR}$, (b) less than (more negative than) $-\sqrt{4kTBR}$, (c) between 0 and $+\sqrt{4kTBR}$, and (d) between 0 and $-\sqrt{4kTBR}$?
4. How can one reduce the Johnson noise generated within a resistance?
5. Calculate the noise voltage appearing at resonance across the terminal AB of the following tank circuit, assuming negligible current is drawn from the tank.

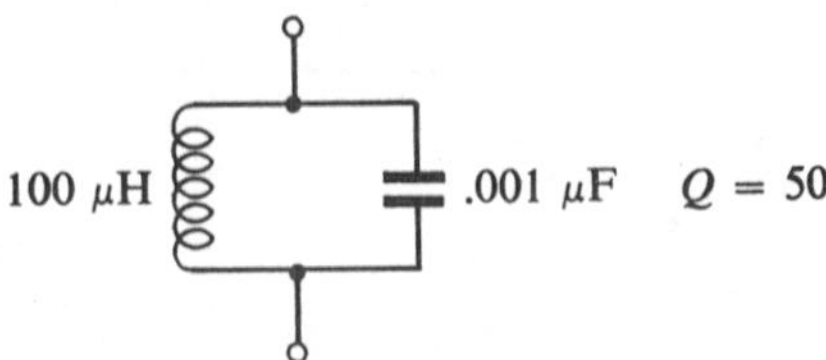

6. Sketch a rough but highly magnified graph of the current flowing through a diode carrying 112 mA current as seen by an oscilloscope with a bandwidth of 10 MHz.

7. Explain why the relative or percentage fluctuations of current due to shot noise are smaller at higher dc currents.
8. Calculate the noise figure in dB for a vacuum tube amplifier with an input impedance of 100kΩ driven by a source with a 10kΩ output impedance.
9. Show that the noise figure for a transistorized amplifier–source combination diverges to infinity for zero source resistance and also for infinite source resistance.
10. Calculate (a) the noise figure for a 50 ohm source driving a transistorized low-noise amplifier with an effective input voltage noise generator, $e_n = 5 \times 10^{-8}\text{V}/\sqrt{\text{Hz}}$, and an effective input current noise generator, $i_n = 4 \times 10^{-14}\text{A}/\sqrt{\text{Hz}}$, and (b) the input transformer turns ratio to maximize the output signal-to-noise ratio.
11. Explain why the turns ratio of the transformer matching the source impedance to the optimum low-noise impedance will depend upon frequency. Will the turns ratio be higher or lower at high frequencies? You may assume an ideal transformer; that is, the transformer loss is small and independent of frequency.
12. Briefly explain the meaning of the words "lock-in" in a lock-in detector.
13. Briefly explain how the lock-in detector output signal-to-noise ratio depends upon the output time constant.
14. Why does a larger output time constant in a lock-in detector require a slower passage through thè signal?
15. Explain why noise which is of random phase and polarity will not appear in the output of a lock-in detector, but noise whose frequency is an odd multiple of the reference frequency will contribute to the output.
16. Two equal resistances initially at different temperatures are suddenly connected in parallel. Discuss the noise voltages produced across the two resistors, the power flow, and the final equilibrium situation.
17. Calculate the approximate mean-square voltage expected at the output if the input is a 10MΩ resistance at 70°C. You may neglect the noise generated in the 10kΩ resistor.

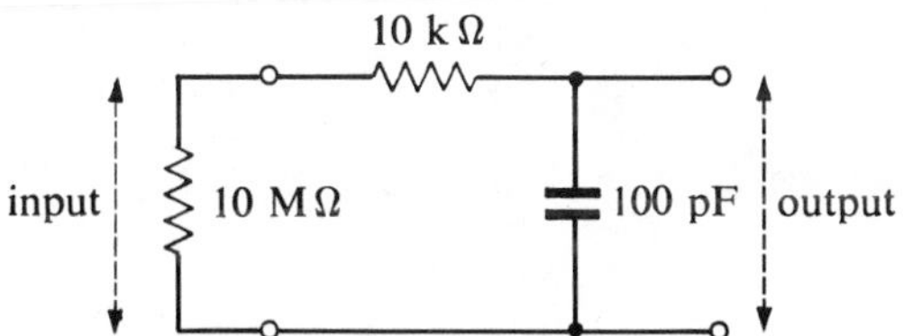

18. For the amplifier whose total effective noise can be represented by an effective noise generator at the input, $e_{na} = \sqrt{4kTBR_{\text{in}}}$, show explicitly that the output signal-to-noise power ratio is increased by a factor of:

$$\frac{R_1 + R_s}{R_1 + (R_s/n^2)},$$

if a stepdown transformer of turns ratio $n:1$ is used between the source and the

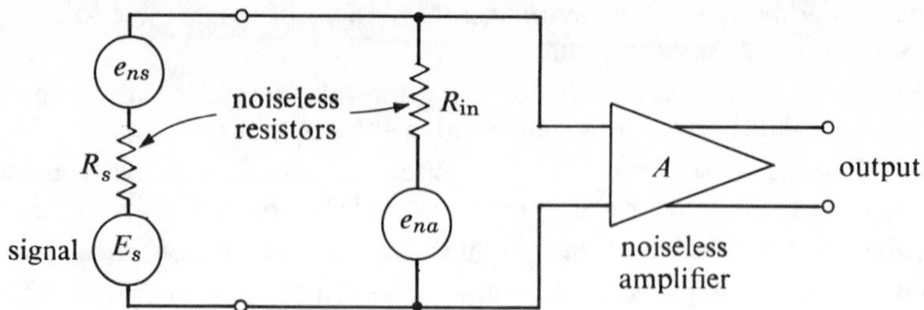

amplifier input. Thus, if R_s is already very small compared to R_1, the improvement will be negligible.

19. Show that the noise temperature of two resistances R_1 and R_2 in series is equal to their common temperature. *Hint:* Noise voltages add vectorially, i.e., as the square root of the sum of the squares.

20. Consider an amplifier with an effective input noise temperature T_{ai} driven by a source with a noise temperature T_s. Show that the total output noise power of the amplifier is given by:

$$N_{\text{out}} = GkB(T_{ai} + T_s),$$

where B is the bandwidth and G, the power gain of the amplifier. Assume the source is matched to the amplifier input, so the maximum source noise power is fed into the amplifier.

21. Explain how an antenna noise temperature can be higher than the actual temperature of the wire comprising the antenna.

22. If the signal-to-noise voltage average is 10 : 1 without signal averaging, what would be the best signal-to-noise voltage you would expect if the signal were swept through and averaged 1,000 times?

12

PULSE AND DIGITAL CIRCUITS

12.1 INTRODUCTION

Pulse and digital circuits are important for two basic reasons. First, pulses naturally occur in many areas of science, and they must be detected, amplified, and counted in various ways. The pulses may be light pulses, mechanical impulses, an electrical pulse from a nuclear radiation detector, or the pulses from a photomultiplier tube when individual light photons strike the photomultiplier photocathode. Secondly, any piece of information, whether a voltage level (e.g., 7.541 V), a counting rate, an age, height, criminal record, salary, or social security number can be converted into digital form using a digital language. The most commonly used digital language is binary; it has only two symbols or digits: "0" and "1." Each binary digit, whether a 0 or a 1, is termed a "bit." A series of bits grouped together to represent a single number is termed a "word." For example, 101 is a "three bit word," and 10110 is a "five bit word." Any number can be expressed in binary form as is illustrated in Table 12-1. It is clear that many more digits are required to represent a number in binary form than in decimal form. The advantage of the binary system is that the information can be rapidly processed by "digital" circuits in such a way as to minimize errors due to voltage drifts, component aging, distortion or noise, because the circuits need only distinguish between two voltage levels, say 0 V and +5 V, which represent the two binary digits "0" and "1." The two voltage levels representing "0" and "1" need not be precisely defined for modern circuitry; a typical allowable range for logical "0" may be from 0.1 V to 0.4 V, and for logical "1", from 2.4 to 4.5 V. This type of digital circuit should be contrasted with an "analog" circuit in which the circuit is required to respond to an infinite number of different voltages. For example, a slow drift of transistor characteristics or dc supply voltage will definitely affect the output of an analog circuit. The output will perhaps increase from 2.5 V to 2.6 V. The

TABLE 12-1 BINARY NUMBER SYSTEM

Decimal Number	*Binary Equivalent*
0	0000
1	0001
2	0010
3	0011
4	0100
5	0101
6	0110
7	0111
8	1000
9	1001
10	1010
11	1011
12	1100
13	1101
14	1110
15	1111
⋮	⋮

0.1 V increase is an error introduced by the drift. However, in digital circuits, errors introduced by slow drifts of voltage or device characteristics usually do not affect the digital form of the output. For, if the "1" voltage level drifts from 5.0 to 5.1 V, the operation of the circuit is usually unchanged and the output is unchanged, because all the digital circuit has to do is to distinguish *between* the "0" voltage level (0 V) and the "1" voltage level (5 V). It is relatively easy to design digital circuits so that normal drifts, aging, and noise do not blur the distinction between the "0" and the "1" voltage levels. In the previous case of 0 V and 5 V, a drift or noise pulse on the order of 2 or 3 volts would be necessary to confuse the binary circuit; i.e., to change a "0" into a "1", or vice versa.

The mathematical operations of addition, subtraction, multiplication, division, extracting roots, etc., can all be performed by digital circuits on numbers in binary form. A machine designed to do this at high speed is called a "digital computer." In fact a computer is really nothing more than a large, high-speed, adding machine connected to a typewriter or cathode-ray tube readout. A distinction is usually made between a "calculator" and a "computer." A calculator is a small machine which can only add, subtract, multiply, or divide. A computer contains a memory unit in which considerable information can be stored and retrieved at will. A computer can also store instructions or programs that tell it how to manipulate the input data which the operator supplies. A computer will do precisely what you tell it to, no more and no less. (There is a saying popular among computer programmers: "Garbage in, garbage out," usually abbreviated "GIGO.")

Computers are useful because they can perform literally millions of mathematical operations in an incredibly short time. Complicated integrals can be performed numerically, averages of large masses of data taken, etc. They are also useful because of the fact that with modern solid-state memory banks large quantities of information can be stored in a relatively small volume, and can be selectively retrieved almost instantaneously. Thus, for the first time in human history, it is technically feasible to store in a computer large quantities of information about every citizen in a country, and to have this information almost instantly available. The moral overtones of this "data bank" are just beginning to be explored. Obvious applications are in law enforcement, especially in tax collection, and the systematic study of national and international economies. It is not clear that a man *should* carry around with him a complete lifetime record of all his legal foibles including parking tickets, etc., although a good case might be made for storing records of convicted felons. What is perhaps the most serious flaw of such a computerized data bank is that expressed by "GIGO." If incorrect or outdated information spews forth from the computer "blessed" by modern man's worship of technology, then the individual concerned may be irreparably damaged by a blind faith in the computer output.

12.2 LOGIC CIRCUITS

In binary language there are only two digits: "0" and "1." It is conventional that the "1" state is thought of in an intuitively affirmative sense, something like a "yes," although this is not strictly necessary. That is, the "1" and "0" states are sometimes called "true" and "false" or "yes" and "no" states. In logical circuitry the two states are usually represented by any two different voltage levels, +5V and 0V, or +10V and +5V, or −5V and −10V, etc.; the only requirement is that the circuitry can reliably distinguish between the two voltage levels. From now on we will assume the more positive level represents the digit "1", and the more negative level, the digit "0." This system is called "positive logic." If the more negative level represents "1," the logic is called "negative logic." In most positive logic circuits, +5V and 0V levels represent "1" and "0," respectively.

It is necessary in computers and other useful circuits to make logical decisions and to perform mathematical operations. The larger of two numbers must be chosen, numbers must be added, subtracted, multiplied, divided, etc. The algebra of such statements and operations is called "Boolean" algebra. A few words on notation are in order: A and B each refer to one-bit words. Words with more than one bit can be treated "sequentially," i.e., the bits can be operated on by the same circuit at different times, or the bits can be treated "in parallel" at the same time, with many separate circuits required, one for each bit. In the next few sections, we will assume that all the A and B words contain only one bit for simplicity.

There are three basic logical operations: the "OR," the "AND," and the "NOT" operations. All mathematical operations can be performed with combinations of these three. Circuits that perform these operations are usually referred to as "gates." That is, an OR gate is a circuit which performs the operation Y = A OR B. The OR gate produces an output Y = 1 either when the A input equals 1 or when the B input equals 1, or when both inputs equal 1. The OR gate output is Y = 0 otherwise. There may be more than two inputs to a gate; e.g., a three-input OR gate would perform the operation Y = A OR B OR C. It is conventional (and confusing to the beginner) to write Y = A + B for the OR operation; this equation is read as "Y equals A or B." A "truth table" for any mathematical operation is a list of all the possible inputs and outputs presented in tabular form. The truth table for a two-input OR gate is shown in Fig. 12.1(a). Notice that the OR operation is commutative; that is, the order

$Y = A + B$

A	B	Y
0	0	0
0	1	1
1	0	1
1	1	1

(a) OR *truth table*

$Y = A \cdot B$

A	B	Y
0	0	0
0	1	0
1	0	0
1	1	1

(b) AND *truth table*

$Y = \overline{A}$

A	Y
0	1
1	0

(c) NOT *truth table*

FIGURE 12.1 Truth tables for the three basic logic operations.

in which we write the two inputs does not affect the output: A + B = B + A.

The AND operation for two inputs is written Y = A · B, or sometimes the dot is omitted, and we write Y = AB. The output Y of a two-input AND gate equals 1 only when both A and B each equal 1; otherwise Y equals 0. For more than two inputs, the AND output equals 1 only when each input equals 1. For three inputs A, B, and C, we would write Y = A · B · C. The AND operation is also commutative; A · B = B · A. The truth table for a two-input AND operation is shown in Fig. 12.1(b).

The third basic operation is the NOT operation, which simply changes a 1 input to a 0 output and a 0 input to a 1 output. It is written $Y = \overline{A}$ for a single input A. The horizontal bar over the letter A denotes "NOT A." The truth table for the NOT operation is shown in Fig. 12.1(c).

At this point it is worth pointing out that if we change from positive logic to negative logic in a circuit, we change an AND gate into an OR gate, and vice versa. That is, for the same physical circuit, merely changing the meaning of the two voltage levels at the input and the output will change the logical operation performed by the circuit. This is sometimes expressed by saying that an AND gate for positive pulses is an OR gate for negative pulses, keeping in mind that the two voltage levels must remain the same—the positive pulses must go from 0 V to +5 V, and the negative pulses from +5 V to 0 V. At the present writing, most transistor logic circuits use positive logic, while MOSFET logic circuits use negative logic.

Some of the most commonly used Boolean algebra expressions are shown below. Each expression can be proved by writing out a truth table.

$$\mathrm{A} + \overline{\mathrm{A}} = 1 \qquad (\mathrm{A} + \mathrm{B}) + \mathrm{C} = \mathrm{A} + (\mathrm{B} + \mathrm{C})$$

$$\overline{\overline{\mathrm{A}}} = \mathrm{A} \qquad \mathrm{A} \cdot \overline{\mathrm{A}} = 0$$

$$\mathrm{A} \cdot \mathrm{B} = \mathrm{B} \cdot \mathrm{A} \qquad \overline{\mathrm{A} \cdot \mathrm{B}} = \overline{\mathrm{A}} + \overline{\mathrm{B}}$$

$$\mathrm{A} + \mathrm{B} = \mathrm{B} + \mathrm{A} \qquad \overline{\mathrm{A}} \cdot \overline{\mathrm{B}} = \overline{\mathrm{A} + \mathrm{B}}$$

$$\overline{1} = 0 \qquad \mathrm{A} + \mathrm{A} \cdot \mathrm{B} = \mathrm{A}$$

$$\overline{0} = 1 \qquad \mathrm{A} + (\mathrm{B} \cdot \mathrm{C}) = (\mathrm{A} + \mathrm{B}) \cdot (\mathrm{A} + \mathrm{C})$$

$$(\mathrm{A} \cdot \mathrm{B}) \cdot \mathrm{C} = \mathrm{A} \cdot (\mathrm{B} \cdot \mathrm{C}) \qquad \mathrm{A} \cdot (\mathrm{B} + \mathrm{C}) = (\mathrm{A} \cdot \mathrm{B}) + (\mathrm{A} \cdot \mathrm{C})$$

12.3 THE OR CIRCUIT

There are four important practical logic circuits: the OR, the AND, the NOT, and the FLIPFLOP circuit. The OR circuit is designed to tell whether one or more "1" levels are present at one or more inputs; if a "1" is present at any one (or more) of the inputs, then a "1" level is produced at the output of the OR circuit. The truth table for an OR circuit with two inputs is shown in Fig. 12.2(a). Remember that A + B means A "OR" B; that is, either we have A or

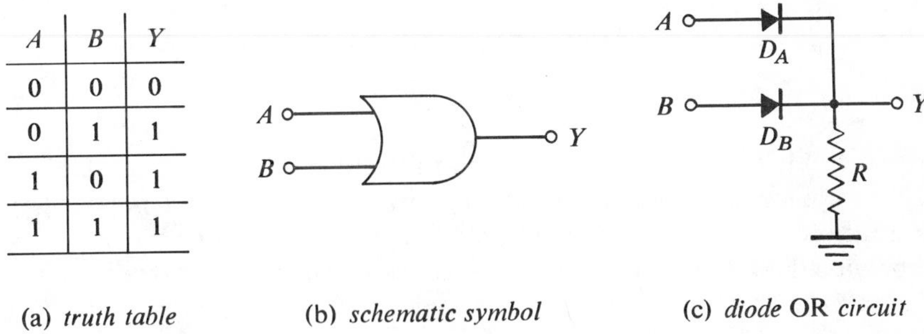

A	*B*	*Y*
0	0	0
0	1	1
1	0	1
1	1	1

(a) *truth table* (b) *schematic symbol* (c) *diode* OR *circuit*

FIGURE 12.2 OR logic circuits.

we have B or both. The A "AND" B = A · B means we have both A and B and is covered in section 12.4.

In other words, in order for a logical "1" level voltage to appear at the output of an OR circuit, a logical "1" level voltage must be present at either input A or B, or at both inputs. It should be pointed out that in the latter case, the output is of the same amplitude as if only one input "1" were present. That is, the output voltage level is the same for a "1" at A or at B, or for a "1" at both A and B. The standard schematic symbol for a diode OR circuit is shown in Fig. 12.2(b), and a specific diode OR circuit built to work on positive logic is shown in Fig. 12.2(c). Logic circuits employing only diodes and resistances are often termed "DRL," for diode-resistor-logic. Those employing only transistors are termed "TTL," often said "T-squared-L," for transistor-transistor logic. TTL usually refers to circuits in which the transistors are either saturated (fully on) or almost completely off. Emitter-coupled-logic, termed "ECL," generally uses transistors not driven to saturation and, as the name implies, coupling is taken from the emitter for high speed and low output impedance.

In the simple diode OR circuit of Fig. 12.2(c), a positive voltage level of E volts (a "1") at input A will forward bias diode D_A and reverse bias diode D_B, so that $V_{out} = E - V_{D_A}$, where V_{D_A} is the turn-on voltage of diode D_A (0.6 V for a silicon diode). Usually the "1" voltage level is made large compared to 0.6 V, so we have $V_{out} \cong E_A$. A similar argument shows that for a positive pulse of amplitude E_B at input B, $V_{out} \cong E_B$. If identical "1" voltage levels E are presented at *both* inputs A and B, then we also have $V_{out} = E$, which follows from the following argument:

$$E = V_{D_A} + V_{out}$$

$$E = V_{D_B} + V_{out}.$$

Adding yields:

$$2E = V_{D_A} + V_{D_B} + 2V_{out}$$

or

$$E = \frac{1}{2}(V_{D_A} + V_{D_B}) + V_{out}.$$

$$V_{out} \gg \frac{1}{2}(V_{D_A} + V_{D_B}) \cong 0.6\text{V}$$

implies

$$E \cong V_{out}, \tag{12.1}$$

provided only that E is much larger than the diode turn-on voltage, 0.6 V. If "1" voltage levels are present at both inputs A and B, but if one pulse is larger than the other, then the output will equal the larger of the two inputs. This is so because if $E_A > E_B$, for example, then diode D_B will be *reverse*-biased and $V_{out} \cong E_A$.

A more precise analysis takes into account the source resistance R_S of the inputs and the forward resistance of the conducting diodes. The resistance of a reverse-biased silicon diode can be taken to be infinite, because it is of the order of 10^8 ohms, which is almost always many orders of magnitude larger than any other resistance in the circuit.

The simple diode OR circuit of Fig. 12.2(c), including source resistances R_{SA} and R_{SB} and forward diode resistance R_f, is shown in Fig. 12.3. If there is a "1" input at A, we have $E_A > 0$ and $E_B = 0$; and $E_A = i_A(R_{SA} + R_f + R)$. Thus:

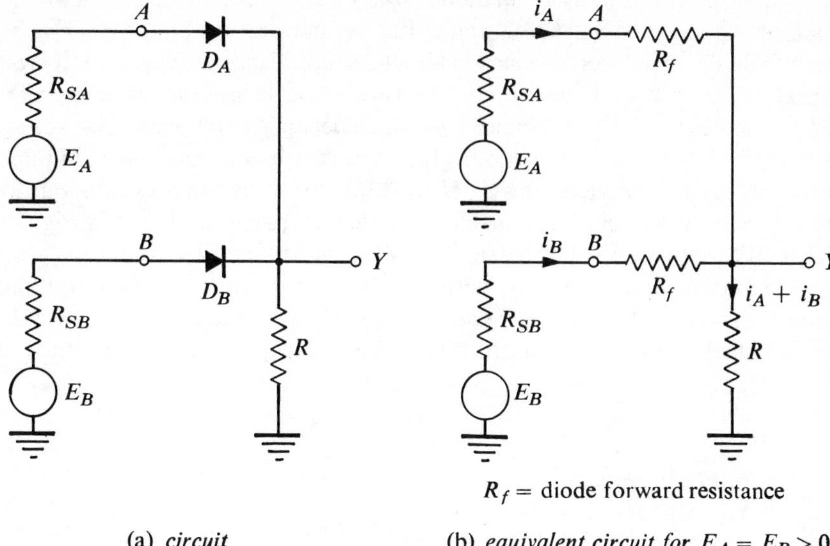

(a) *circuit* (b) *equivalent circuit for* $E_A = E_B > 0$

FIGURE 12.3 Diode OR logic circuit including source resistances.

$$V_{out} = i_A R = \frac{E_A}{1 + \dfrac{R_{SA} + R_f}{R}}.$$

If $E_A = E_B$, and the diodes are identical, then by symmetry $i_A = i_B$. The two Kirchhoff loop equations are:

$$E_A = i_A R_{SA} + i_A R_f + (i_A + i_B)R$$
$$E_B = i_B R_{SB} + i_B R_f + (i_A + i_B)R.$$

Adding the two equations:

$$2E = (i_A + i_B)(R_S + R_f) + 2(i_A + i_B)R.$$

Using $V_{out} = (i_A + i_B)R$, we eliminate $(i_A + i_B)$ and obtain for the output:

$$V_{out} = \frac{E}{1 + \dfrac{R_S + R_f}{2R}}. \qquad \textbf{(12.2)}$$

Typically, $R_f = 50\Omega$, $R_S \cong 50\Omega$, and $R = 1\,\text{k}\Omega$–$10\,\text{k}\Omega$, so:

$$V_{out} \cong E.$$

Let us now consider the effects of shunt capacitance on the rise and fall times of the pulses in a diode OR circuit. A forward-biased diode is essentially a pure resistance, from about 10 to 100 ohms depending upon the current flowing through the diode. For a silicon diode a 1 mA current corresponds to a forward diode resistance of about 600Ω, and a 100 mA current to about 6Ω. For germanium diodes, 1 mA corresponds to about 250 Ω, 100 mA to about 2.5 Ω forward resistance. A reverse-biased diode, however, is a large resistance R_r (about $10^8 \Omega$) in parallel with a capacitance C_D, which depends upon the reverse voltage. $C_D \cong KV^{-1/2}$, where V is the reverse bias and K is a constant depending upon the diode. A typical value for C_D is 10–20 pF for a reverse bias of about 5 V. Thus, the equivalent circuit of a simple diode OR gate is as shown in Fig. 12.5 with a "1" input at A. Diode D_A is forward-biased and D_B is reverse-biased.

We have included any stray wiring capacitance C_S between the output and ground. Thus, neglecting R_{SB} because $R_{SB} \ll R_r$ and setting $R_r = \infty$, we have the simplified equivalent circuit of Fig. 12.4(a). The point here is that the total

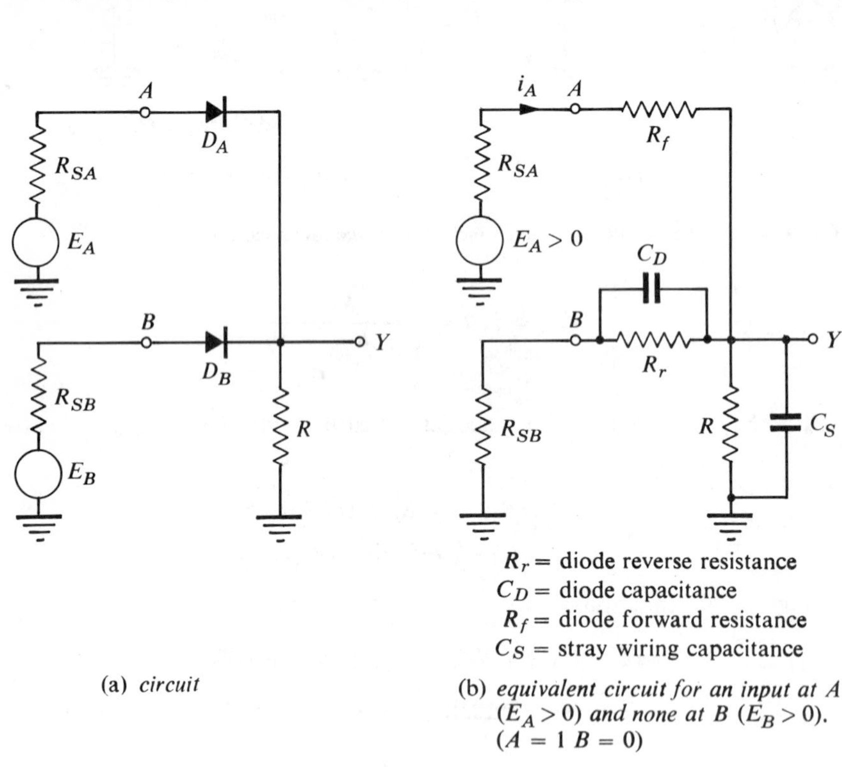

(a) *circuit*

(b) *equivalent circuit for an input at A ($E_A > 0$) and none at B ($E_B > 0$). ($A = 1$ $B = 0$)*

FIGURE 12.4 Diode OR circuit.

capacitance across the output is $C_S + C_D$. If there are more than two inputs, each input terminal at 0 V has a back-biased diode, and thus contributes a capacitance C_D across the output. Then, the total capacitance across the output is $C_{out} = C_S + nC_D$, where n is the number of reverse-biased diodes.

The waveform of the output pulse is shown in Fig. 12.5(b). The rise time, T_1, of the leading edge of the output pulse is given by $T_1 = 2.2(R_{SA} + R_f)C_{out}$,

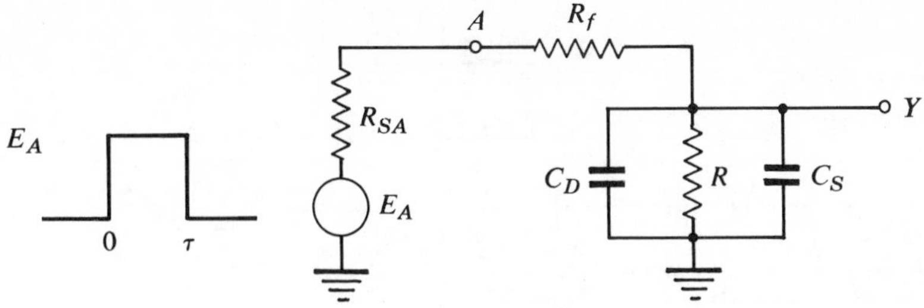

(a) *a simplified equivalent circuit*

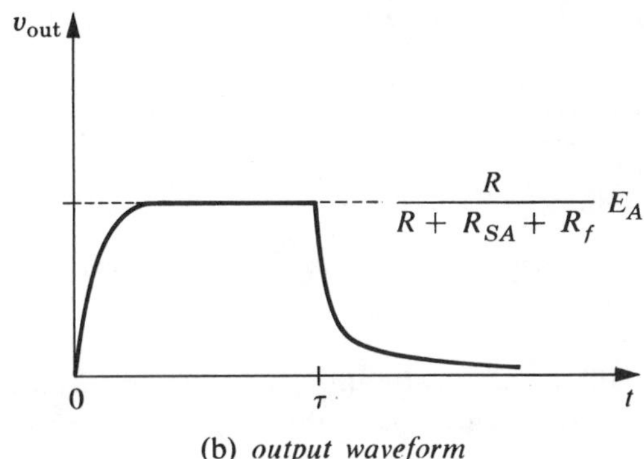

(b) *output waveform*

FIGURE 12.5 Diode OR circuit showing capacitances.

because C_{out} is charged through R_{SA} and R_f in series. However, the fall time is considerably longer; because when the input voltage level at A drops to zero diode D_A is now also reverse-biased. The output voltage falls to zero by capacitor C_{out} discharging *through* R, which is usually much larger than $R_{SA} + R_f$. The fall time for the output pulse is given by $T_2 = 2.2RC_{out}$. If $R_S + R_f = 50\Omega$, $R = 10\text{k}\Omega$, $C_S \cong 10\text{pF}$, and $C_D = 10\text{pF}$, then $T_1 = 2.2\text{nsec}$ and $T_2 = 0.44\,\mu\text{sec}$.

It is possible to construct an OR logic circuit using transistors instead of diodes. Such logic is often abbreviated "TTL" for "transistor-transistor-logic." A recent development in computer circuitry along this line is "emitter-coupled logic," often abbreviated "ECL." An ECL three-input OR logic circuit is shown in Fig. 12.6. T_1 is connected in the common-base configuration. If a "1" input is present at any one or more of the inputs A, B, and C, then that transistor conducts, which raises the voltage at the emitter of T_1. This in turn raises the voltage at the collector of T_1, which turns T_2 on, thus producing a "1" output. The advantages of ECL are extremely high speed and a small propagation delay, of less than one nanosecond between the input and output. Notice that the output is taken off the emitter; thus the name and the physical advantage that the output impedance is low, thus aiding in high-speed operation. The biasing of the transistors is such that the voltage level swings are small and the tran-

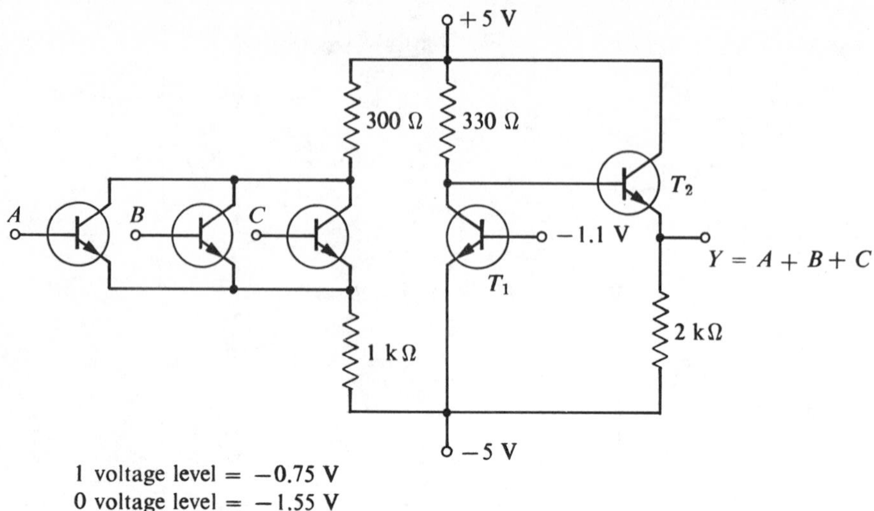

FIGURE 12.6 Emitter-coupled logic OR gate (TTL). (Courtesy of Motorola Semiconductor Products Inc.)

sistors are never driven into saturation, i.e., for npn transistors the collector voltage is always positive with respect to the base voltage.

12.4 THE AND CIRCUIT

The AND circuit produces a "1" output only when each input is "1" simultaneously. It is sometimes called a coincidence circuit, because it produces an output only when input signals are present in coincidence, i.e., at the same time. The truth table for the logical AND operation for two inputs A and B and the schematic symbol for the AND circuit are shown in Fig. 12.7. Notice that A "AND" B is written A · B; not A + B, which means A "OR" B.

$Y = A \cdot B$

A	B	Y
0	0	0
0	1	0
1	0	0
1	1	1

(a) *truth table*

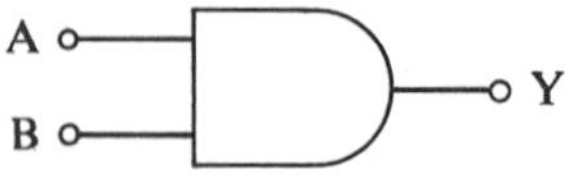

A
B
Y

(b) *schematic symbol*

FIGURE 12.7 AND circuit.

A simple AND diode circuit designed to work on positive pulses is shown in Fig. 12.8. If no input pulse is present at either A or B, then both diodes are

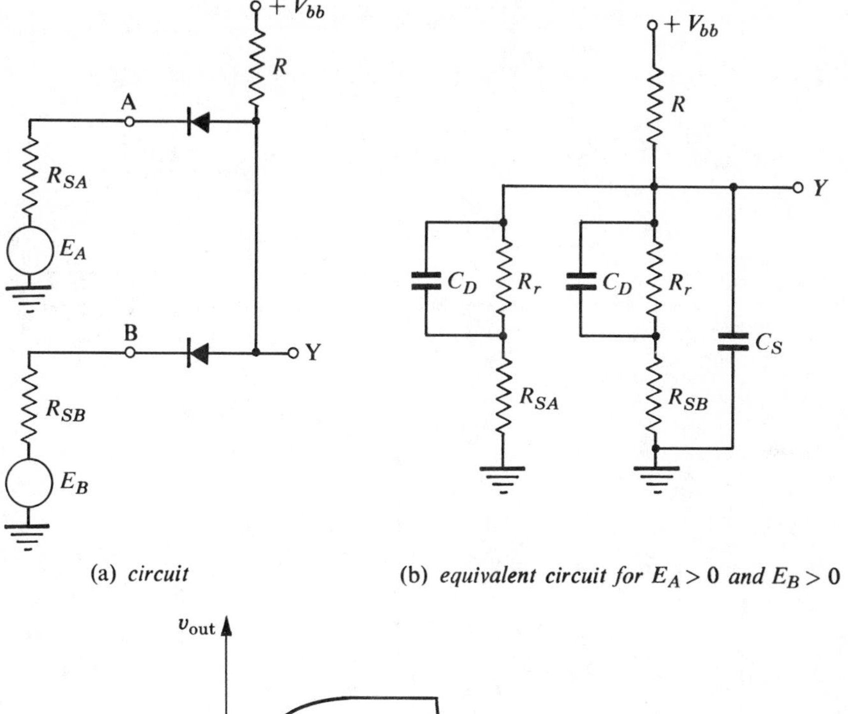

(a) *circuit*

(b) *equivalent circuit for* $E_A > 0$ *and* $E_B > 0$

(c) *output waveform for ideal positive pulses at A and B*

FIGURE 12.8 Diode AND circuit.

conducting; and the output voltage at Y is very low because $R_{SA} \ll R$, $R_{SB} \ll R$, and $R_f \ll R$. The output is:

$$V_{out} = \frac{\dfrac{R_f + R_S}{2}}{\dfrac{R_f + R_S}{2} + R} V_{bb}. \tag{12.3}$$

Typically the forward resistance of each diode is $R_f = 100\Omega$; if $R_S = 50\Omega$ and $R = 10\,\text{k}\Omega$, then $V_{out} = 0.00745 V_{bb}$. If $V_{bb} = 5\text{V}$, $V_{out} = 0.037\text{V}$. If there is one input (at A), so that the input A terminal voltage is more positive than the output voltage, then the diode D_A is reverse biased and the output voltage is still low:

$$V_{\text{out}} = \frac{R_f + R_{SB}}{R_f + R_{SB} + R} V_{bb} = 0.0148 V_{bb}. \tag{12.4}$$

If $V_{bb} = 5\text{V}$, then $V_{\text{out}} = 0.074\text{V}$. We have assumed here that negligible cur-

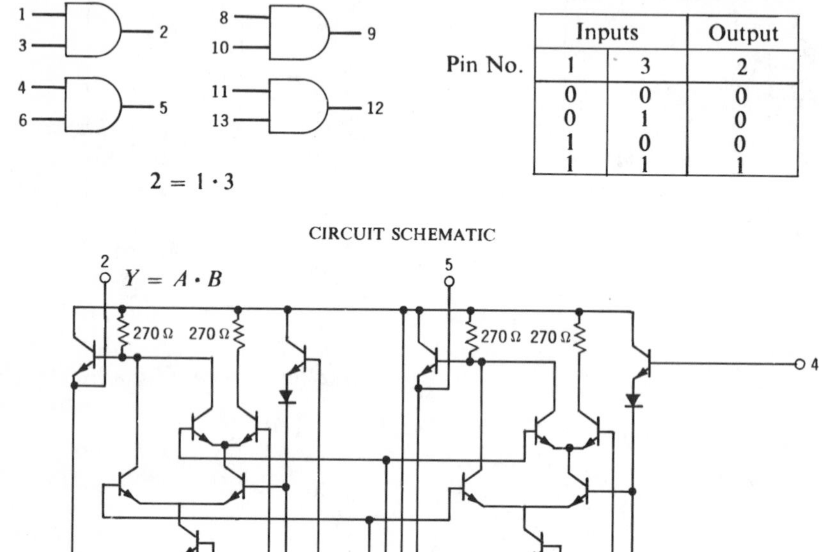

AND GATE
SAMPLE TRUTH TABLE

	Inputs		Output
Pin No.	1	3	2
	0	0	0
	0	1	0
	1	0	0
	1	1	1

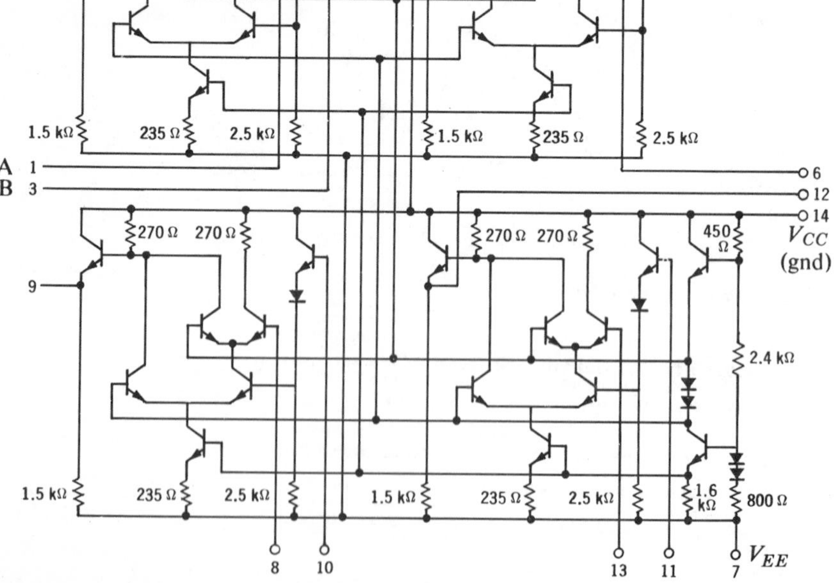

FIGURE 12.9 Emitter-coupled AND gate for positive logic. (Courtesy of Motorola Semiconductor Products Inc.)

rent flows through the reverse-biased diode D_A. If, however, there is a positive input pulse present at both the A and B inputs, then *both* diodes will be turned off, and the output voltage will be given by:

$$V_{out} = \frac{R_r + R_S}{\dfrac{R_r + R_S}{2} + R} V_{bb} = 0.998 V_{bb}. \qquad (12.5)$$

Typically, the reverse resistance of each reverse-biased diode R_r is $10^8 \Omega$, so the output is essentially equal to V_{bb}. If $V_{bb} = 5$V, then $V_{out} = 4.99$V. A little thought will show that to get a large output for two inputs and a small output for one or no inputs requires that:

$$R_S + R_f \ll R \ll R_r, \qquad (12.6)$$

which is rather easy to satisfy.

The rise time of the output pulse may be calculated by noting that there is an output only when there is an input pulse present simultaneously at both input terminals. Thus, both diodes D_A and D_B are reverse biased, and the total capacitance, $C_{out} = 2C_D + C_S$, between the output terminal and ground must charge up through the resistance R. Thus, the time constant for the rise of the output pulse is RC_{out}, and the rise time equals $2.2RC_{out}$. The fall time of the output pulse is shorter, because when no input pulses are present at inputs A and B, the diodes are forward-biased and each input places a resistance of $R_S + R_f$ across the output. Thus, the total resistance between the output and ac ground is the parallel combination of R and $(R_S + R_f)/2 \equiv R'$. The time constant for the fall of the output pulse is $[R'R/(R' + R)]C_{out}$; the fall time is $[2.2\, R'R/(R' + R)]C_{out}$. For $R = 10\,\text{k}\Omega$, $C_{out} = 10$pF, $R_f = 100\Omega$, $R_s = 50\Omega$, the rise time $= 0.22\,\mu$sec, and the fall time $= 0.0016\,\mu$sec.

An emitter-coupled AND gate for positive pulses is shown in Fig. 12.9. A positive pulse must be present at each input 1 and 3 in order to allow the output voltage at 2 to rise to the logical "1" level. If either or both inputs are "0," then the output is at logical "0" level.

12.5 THE NOT CIRCUIT

A NOT circuit or gate or inverter has the simple property of inverting or changing the binary input. That is, a "1" input produces a "0" output, and vice versa. The truth table for a NOT circuit is shown in Fig. 12.10(a). A bar written over the input A often is used to denote NOT A. The schematic symbol for a NOT circuit or gate is a triangle, either with or without an open circle at the point of the triangle. A common emitter amplifier is an obvious choice for a NOT circuit, as shown in Fig. 12.10(c), because of the 180° phase inversion between the input and output. For positive logic, the npn transistor is normally

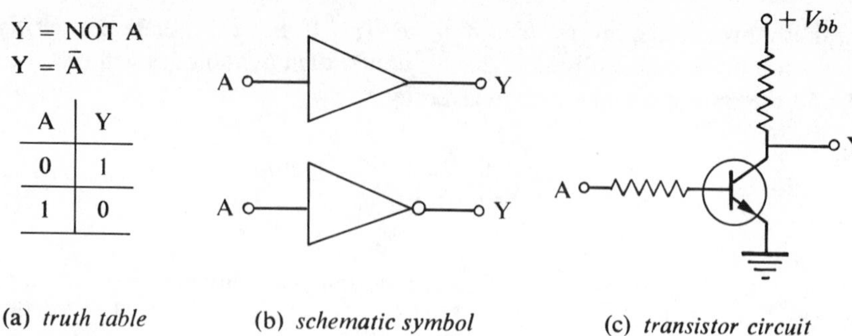

A	Y
0	1
1	0

(a) *truth table* (b) *schematic symbol* (c) *transistor circuit*

FIGURE 12.10 NOT circuit.

biased so that it is cut off. Thus for a "0" input, the output at the collector is at a high positive voltage, V_{bb}, corresponding to the "1" state. With a positive input, the transistor is turned on and the output voltage changes to near 0 volts, the "0" state.

The "OR," the "AND," and the "NOT" circuits or gates form the basis of all modern digital computers. It can be shown that any mathematical operation can be performed in a computer by the appropriate combination of OR, AND, and NOT gates.

12.6 THE NOR CIRCUIT

The NOR circuit or gate is a combination of an OR circuit followed by a NOT or inverter circuit. The truth table and schematic symbol are shown in Fig. 12.11 for two inputs A and B. The NOR circuit can be thought of as a circuit which produces a "1" output only when *none* of the inputs is "1." Notice that

Y = NOT (A OR B)

$Y = \overline{(A + B)}$

A	B	A + B	Y
0	0	0	1
0	1	1	0
1	0	1	0
1	1	1	0

(a) *truth table*

$Y = \overline{(A + B)}$

A, B → Y

(b) *schematic symbol*

FIGURE 12.11 The NOR operation.

the small open circle at the output distinguishes the "NOR" gate schematic symbol from the OR gate schematic symbol. Notice that changing from positive logic to negative logic changes the NOR operation to the NAND operation (see section 12.7).

12.7 THE NAND CIRCUIT

The NAND circuit or gate is a combination of an AND circuit followed by a NOT or inverter circuit. The truth table and schematic symbol are shown in Fig. 12.12 for two inputs A and B. Notice that the small open circle at the

Y = NOT (A AND B)

$Y = \overline{A \cdot B}$

A	B	A · B	Y
0	0	0	1
0	1	0	1
1	0	0	1
1	1	1	0

(a) *truth table*

$Y = \overline{A \cdot B}$

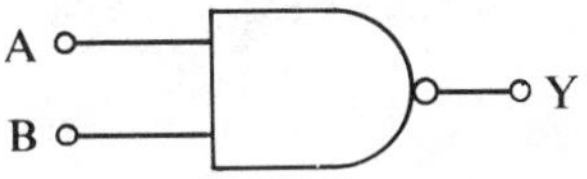

A

B

Y

(b) *schematic symbol*

FIGURE 12.12 The NAND operation.

output distinguishes the NAND gate schematic symbol from the AND gate schematic symbol. Also notice that changing from positive to negative logic changes the NAND operation to the NOR operation.

12.8 THE EXCLUSIVE OR CIRCUIT

The exclusive OR circuit or gate produces a "1" output if and only if *one* of the inputs produces "1", and a "0" output otherwise. The truth table and schematic symbol are shown in Fig. 12.13 for two inputs A and B. Inspection of the truth table will show that the Boolean algebra expression for the exclusive OR operation can be written as the product of (A OR B) and NOT (A AND B): $Y = (A + B) \cdot (\overline{A \cdot B})$. Or we can write Y = (A and NOT B) or (NOT A and B): $Y = (A \cdot \overline{B}) + (\overline{A} \cdot B)$.

Block diagrams of two exclusive OR circuits made from basic AND and OR gates are shown in Fig. 12.14. Notice that an inverting amplifier with unit gain

A	B	Y	A + B	A · B	$\overline{A \cdot B}$
0	0	0	0	0	1
0	1	1	1	0	1
1	0	1	1	0	1
1	1	0	1	1	0

(a) *truth table*

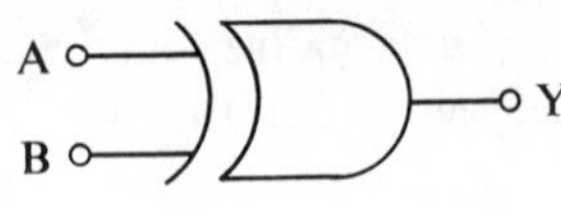

(b) *schematic diagram*

FIGURE 12.13 The exclusive OR operation.

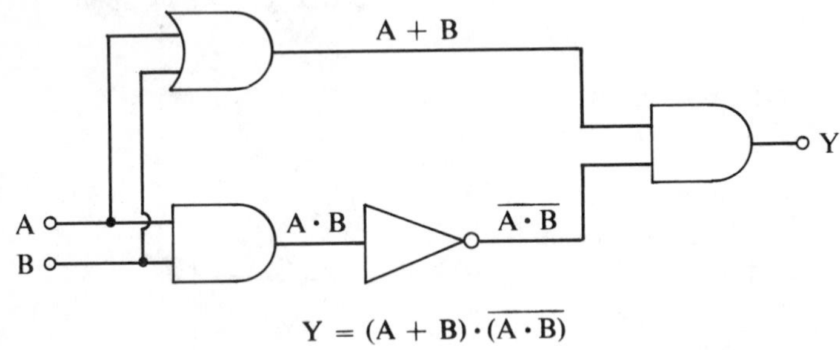

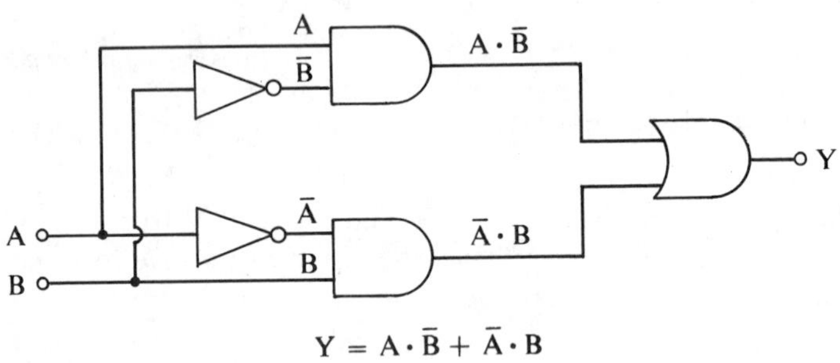

FIGURE 12.14 Exclusive OR circuits.

is used to convert A to $\overline{A}$ and B to $\overline{B}$; that is, a "1" input produces a "0" output, and vice versa.

12.9 THE EXCLUSIVE NOR CIRCUIT

The exclusive NOR circuit or gate produces a "1" output if and only if the two inputs A and B are the same, both "0" or both "1." The truth table and schematic symbol are shown in Fig. 12.15 for two inputs A and B. The exclusive

$Y = \bar{A} \cdot \bar{B} + A \cdot B$

A	B	Y
0	0	1
0	1	0
1	0	0
1	1	1

(a) *truth table*

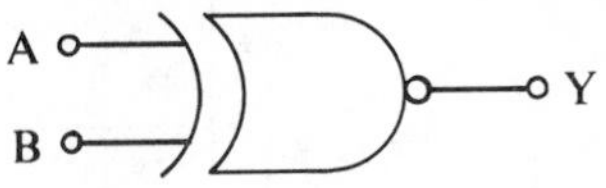

(b) *schematic symbol*

FIGURE 12.15 The exclusive NOR circuit.

NOR gate can be made from an exclusive OR gate of Fig. 12.14 by merely adding a NOT circuit at the output. Also notice that changing from positive to negative logic changes the exclusive NOR operation to the exclusive OR operation.

12.10 BINARY ADDITION

All modern digital computers are essentially very high-speed adding machines and memory circuits designed to handle numbers in binary form. They also can perform the operations of subtraction, multiplication, and division, but these are basically special cases of addition. Two binary numbers can be added together by a computer in two ways, in parallel or serially. In parallel addition, each column of the numbers is added simultaneously, while in serial addition, the columns are added one at a time, starting with the first or least significant bit.

The fundamental arithmetic of binary addition is contained in four rules:

1. $0 + 0 = 0$
2. $0 + 1 = 1$
3. $1 + 0 = 1$
4. $1 + 1 = 0$ but 1 must be carried over to the next higher column.

Thus, $00 + 01 = 01$ using rule (2) for the first column and (1) for the second column. (The columns are numbered from right to left.) $01 + 10 = 11$ using (3) for the first column and (2) for the second column. $01 + 01 = 10$ using (4) for the first column and (1) and (2) for the second column. These four rules of binary addition can be expressed as the truth table of Fig. 12.16.

From inspection of the truth table, we see that the sum is precisely the exclusive OR operation in each of the four possible cases, and that the digit to be carried is the AND operation applied to the two digits to be added, A and B. Thus, we see that to add two bits we need a circuit with two inputs, one for A

A	B	SUM BIT S	CARRY BIT C
0	0	0	0
0	1	1	0
1	0	1	0
1	1	0	1

FIGURE 12.16 Binary addition truth table.

and one for B, and *two* outputs: one for the sum S and one for the carry bit C. Such a circuit is shown in Fig. 12.17.

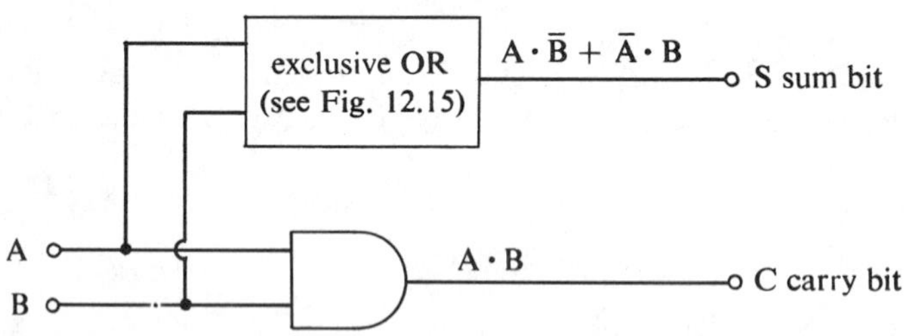

FIGURE 12.17 Binary adder ("half adder").

The adder circuit of Fig. 12.17 is often called a "half adder" since two such circuits are necessary to complete an addition—one circuit to add the two digits in one column, and another circuit to add in the carry bit from the next smaller column. A circuit to add two two-bit binary numbers is shown in Fig. 12.18. The addition of two two-bit binary numbers can be written:

$$S_2S_1 = A_2A_1 + B_2B_1. \tag{12.7}$$

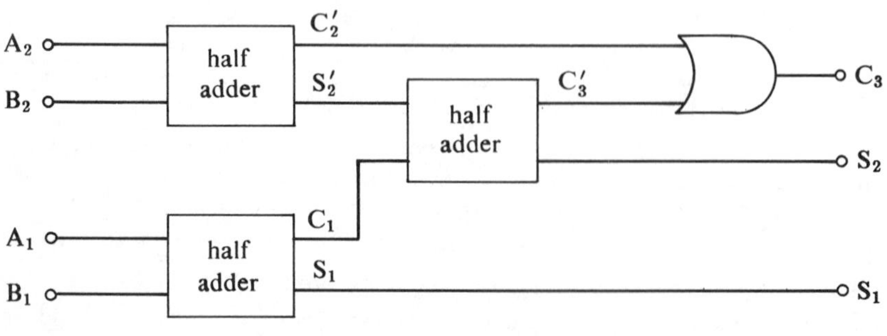

FIGURE 12.18 Complete parallel adder for adding two two-bit words.

A_2A_1 is a symbolic way of writing a two-bit binary number; the subscript refers to the column, e.g., if the two-bit number is 10, then $A_2 = 1$, $A_1 = 0$.

Let A_1 = the bit in the first column for the number A.
A_2 = the bit in the second column for the number A.
B_1 = the bit in the first column for the number B.
B_2 = the bit in the second column for the number B.
S_1 = the bit in the first column for the number Sum.
S_2 = the bit in the second column for the number Sum.
C_1 = the carry bit resulting from the addition of A_1 and B_1.
C_2' = the carry bit resulting from the addition of A_2 and B_2.

A word about notation is in order. The bits in a word are numbered. In most logic terminology, the least significant bit on the extreme right is numbered with a subscript 1, and the subscripts increase from right to left. Thus a three-bit word would be symbolically represented by $A_3A_2A_1$. However, IBM numbers the bits from *left* to *right* starting with zero; in IBM notation a three-bit word would be represented $A_0A_1A_2$. The circuit enclosed by the dotted line is termed a "full adder." The full adder has three inputs: two for the two bits from the two words which are being added and a carry input for the carry bit resulting from the addition of the next lower column. One full adder is required for each bit of a word to be added except for the least significant bit, which needs only a half adder. For example, the addition of two five-bit words would require four full adders and one half adder. All the possible states for the parallel adder are shown in Table 12-2. Notice that the inputs A_1 and B_1 must be

TABLE 12-2 LOGIC STATES FOR PARALLEL ADDER

Addition	A_2	A_1	B_2	B_1	C_1	S_1	C_2'	S_2'	C_3'	S_2	C_3
00 + 00 = 00	0	0	0	0	0	0	0	0	0	0	0
00 + 01 = 01	0	0	0	1	0	1	0	0	0	0	0
00 + 10 = 10	0	0	1	0	0	0	0	1	0	1	0
00 + 11 = 11	0	0	1	1	0	1	0	1	0	1	0
01 + 00 = 01	0	1	0	0	0	1	0	0	0	0	0
01 + 01 = 10	0	1	0	1	1	0	0	0	0	1	0
01 + 10 = 11	0	1	1	0	0	1	0	1	0	1	0
01 + 11 = 00 + Carry 1	0	1	1	1	1	0	0	1	1	0	1
10 + 00 = 10	1	0	0	0	0	0	0	1	0	1	0
10 + 01 = 11	1	0	0	1	0	1	0	1	0	1	0
10 + 10 = 00 + Carry 1	1	0	1	0	0	0	1	0	0	0	1
10 + 11 = 01 + Carry 1	1	0	1	1	0	1	1	0	0	0	1
11 + 00 = 11	1	1	0	0	0	1	0	1	0	1	0
11 + 01 = 00 + Carry 1	1	1	0	1	1	0	0	1	1	0	1
11 + 10 = 01 + Carry 1	1	1	1	0	0	1	1	0	0	0	1
11 + 11 = 10 + Carry 1	1	1	1	1	1	0	1	0	0	1	1

present simultaneously at the two inputs to half adder number 1, and that A_2 and B_2 must be present simultaneously at the inputs to number 2. That is, the two columns are treated simultaneously or in parallel.

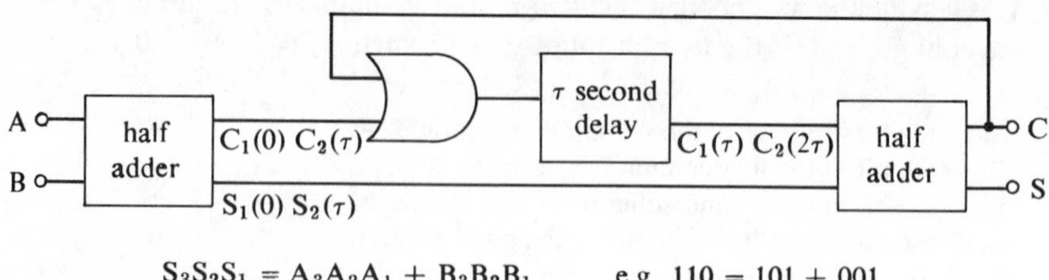

$S_3S_2S_1 = A_3A_2A_1 + B_3B_2B_1$ e.g. $110 = 101 + 001$

FIGURE 12.19 Serial adder.

The other method of addition of binary numbers, serial addition, adds A_1 and B_1 first, and then at a slightly later time adds A_2 and B_2 plus any carry digit that might have resulted from the earlier addition of A_1 and B_1. A serial adder circuit is shown in Fig. 12.19. Notice that the two binary numbers A and B must be in the form of a train of pulses such that the pulse for A_1 occurs at precisely the same time as the pulse for B_1, and so on.

The time delay circuit delays the carry pulse by exactly τ seconds, the time between the pulses. This is necessary so that the carry pulse is added to the

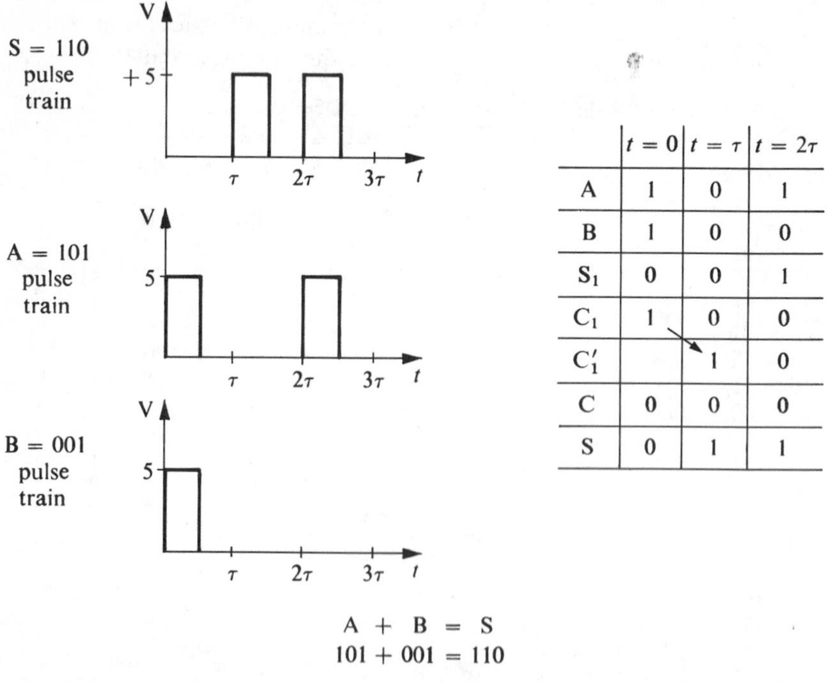

	$t = 0$	$t = \tau$	$t = 2\tau$
A	1	0	1
B	1	0	0
S_1	0	0	1
C_1	1	0	0
C_1'		1	0
C	0	0	0
S	0	1	1

(a) *pulses vs. time* (b) *table of logic values at various times*

FIGURE 12.20 Pulses in serial addition of 101 + 001 = 110.

next higher digit. The train of pulses for the sum 101 + 001 = 110 is shown in Fig. 12.20(a). Notice that we read a binary number 001 from right to left, while the train of pulses representing 001 is read from left to right if time increases from left to right. Figure 12.20(b) shows the digits present in the serial adder at the different times 0, τ, and 2τ.

Parallel addition is seen to require a full adder for each bit in addition to the least significant bit which requires only a half adder. However, the time required to perform the addition is merely the time required for the pulses to pass through the adder circuits, which is usually on the order of ten nanoseconds. In serial addition, by contrast, far fewer components are needed, but the time required to perform the addition is $N\tau$, where N is the number of bits in the binary numbers being added. In summary, parallel addition is fast but expensive; serial addition is slow and cheap.

12.11 THE PULSE-HEIGHT ANALYZER

One of the most useful instruments in the modern physics laboratory is the pulse-height analyzer. Its function is to take a large number of electrical input pulses of varying amplitudes and sort them according to their amplitudes. The analyzer usually displays the results as a graph of number of pulses N versus pulse amplitude V, on a cathode-ray tube or printed out on an electric typewriter

A typical application of a multichannel, pulse-height analyzer is shown in Fig. 12.21. In this setup, it is desired to count the number of gamma rays γ_1 of

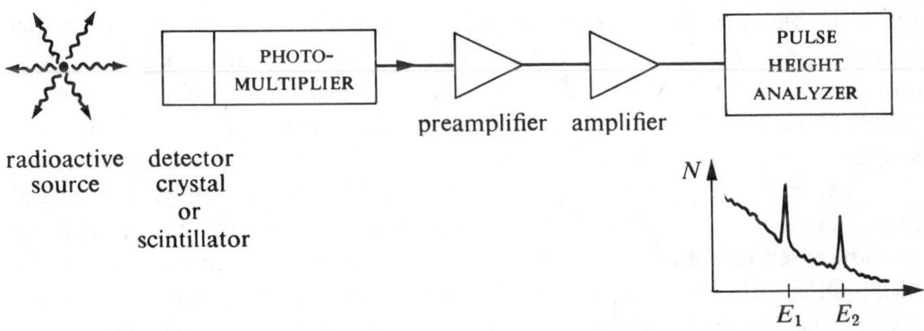

FIGURE 12.21 Typical use of multichannel pulse-height analyzer in nuclear physics counting experiment.

energy E_1 and gamma rays γ_2 of energy E_2 emitted from the radioactive Co^{60} source. The gamma rays that strike the detector are converted into small-amplitude electrical pulses of short duration. The scintillator crystal produces a light pulse if the gamma ray interacts with the crystal via the photoelectric effect, Compton scattering, or pair production. The photomultiplier tube converts these very small light pulses into electrical pulses. The preamplifier has a low output impedance to drive the cable. The detector may be any number of types, but in all modern detectors the quantity of electric charge produced by each charged particle in the detector is linearly proportional to the energy *loss* of the particle in passing through the detector. If the gamma ray or particle is totally absorbed in the detector, then the charge produced is linearly proportional to the total energy of the gamma ray or particle. See Appendix H for further details on detectors and photomultiplier tubes.

The graph of number of counts, N, versus energy, E, in Fig. 12.21 is really a bar graph. That is, the pulse-height analyzer divides up the energy range (pulse amplitude) into a finite number of intervals, or "channels." For example, if the range of pulse amplitudes is from 0 V to 5 V, each channel might be 0.05 V wide. Thus there would be 100 channels. All pulses with amplitudes between 2.10 V and 2.15 V would be counted as going in the same channel.

The sharp peaks in the γ spectrum at E_1 and E_2 are photopeaks, because each photopeak results from a γ-ray interacting with the detector crystal via the photoelectric effect in which most of the γ-ray energy is given to the photoelectron produced. There will be one photopeak for each γ-ray of a different energy, and the better the detector, the larger the photopeak for a given γ radiation intensity.

A "single-channel analyzer" is an instrument which counts the number of pulses between V_0 volts and $V_0 + \Delta V$ volts amplitude. V_0 is sometimes called the "base line" and ΔV is called the "channel width" or the "window width." A complete energy spectrum of a radioactive source can be obtained using a single-channel analyzer and an appropriate radiation detector and amplifier. One sets the base line V_0 and the window ΔV, and then counts. Then one resets the base line ΔV volts higher and counts again. This process is repeated until the desired spectrum is covered. Clearly, this procedure is extremely slow and laborious for slow counting rates and when a high resolution spectrum (small ΔV) is required. When a single-channel analyzer is used in this way, it is said to be operated in the "differential mode," yielding a differential spectrum.

The "integral mode" of operation of a single-channel analyzer counts all those pulses with an amplitude greater than a certain set level V_0. That is, the window extends from V_0 up to the maximum amplitude the analyzer is capable of handling. If, for example, one wishes to count a single gamma ray, one can set the base line V_0 in the valley just below the photopeak of the gamma ray, as shown in Fig. 12.22, provided there are no other gamma rays of higher energy in the source. Then only those pulses corresponding to photoelectric effect electrons in the detector will be present in the output of the analyzer. The many pulses of lesser amplitude due to Compton electrons and scattered or other low energy radiation will be rejected by the analyzer.

A single-channel analyzer determines whether a pulse is between V_0 and $V_0 + \Delta V$ volts in amplitude. A block diagram for such an instrument is shown in Fig. 12.23. We will assume positive logic; a pulse present means a positive voltage level representing logical "1." No pulse present means a low voltage

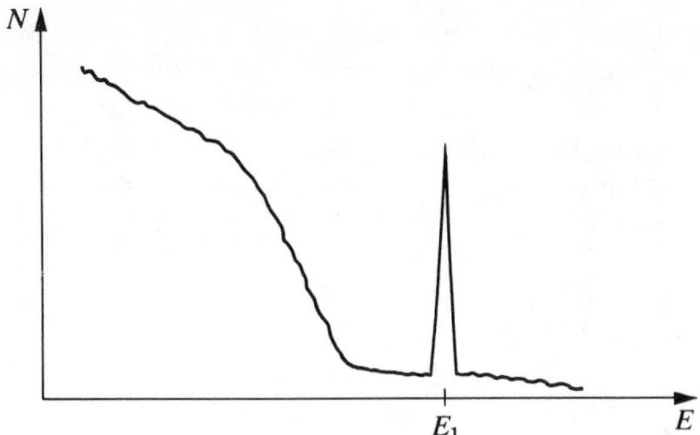

FIGURE 12.22 Gamma ray pulse–height spectrum for one gamma ray of energy E_1.

level representing logical "0." The V_0 level discriminator (sometimes called the "lower-level discriminator" or LLD) produces an output pulse A only if the input pulse is greater than V_0, and the $V_0 + \Delta V$ discriminator (sometimes called the "upper-level discriminator" or ULD) produces an output B only if the input pulse is greater than $V_0 + \Delta V$. The inverter circuit at the output of the ULD thus produces a pulse $\overline{B}$ only when the input pulse is less than $V_0 + \Delta V$. Thus if the input pulse is in the channel of width ΔV with base line V_0, there will be an output pulse at A and not at B, and A = 1, B = 0, and $\overline{B}$ = 1. Only then will the AND gate produce an output pulse, $A \cdot \overline{B} = 1$. If the input pulse is larger than $V_0 + \Delta V$, then there will be a pulse at *both* A and B, A = 1, B = 1, $\overline{B}$ = 0, and no analyzer output, $A \cdot \overline{B} = 0$. If the input pulse is less than V_0, then there will be no pulse at either A or B, A = 0, B = 0, $\overline{B}$ = 1, and no analyzer output, $A \cdot \overline{B} = 0$. The level discriminators are basically Schmitt trigger circuits set to "flip" when the dc level of the input exceeds a certain voltage.

A multichannel analyzer simultaneously measures the pulses in all the channels with base lines from 0 V to a maximum pulse amplitude. The first channel covers amplitudes from 0 V to ΔV V, the second from ΔVV to $2\Delta V$V, etc. Modern multichannel analyzers usually come with 256, 512, 1024, 2048, or 4096

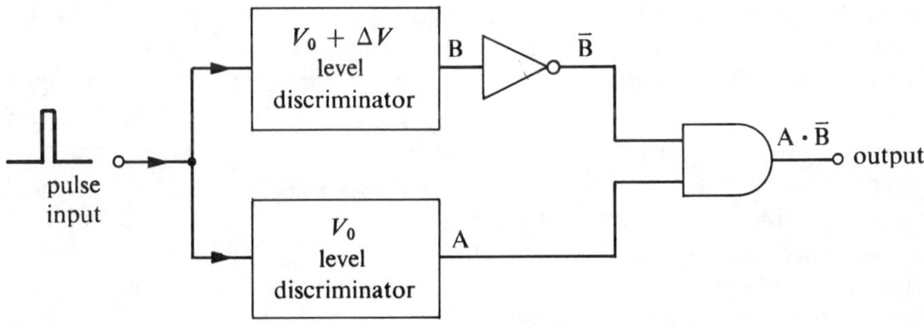

FIGURE 12.23 Single-channel analyzer block diagram.

channels. The output pulse spectrum is usually displayed on an oscilloscope screen as a bar graph of number of pulses on the vertical axis versus pulse amplitude (energy) on the horizontal axis. The number of pulses in each channel is represented by the vertical position of a dot on the oscilloscope screen. Thus, for a 512-channel analyzer, there will be 512 dots on the scope face. A sketch of a pulse-height spectrum is shown in Fig. 12.24 for a Cs^{137} source which

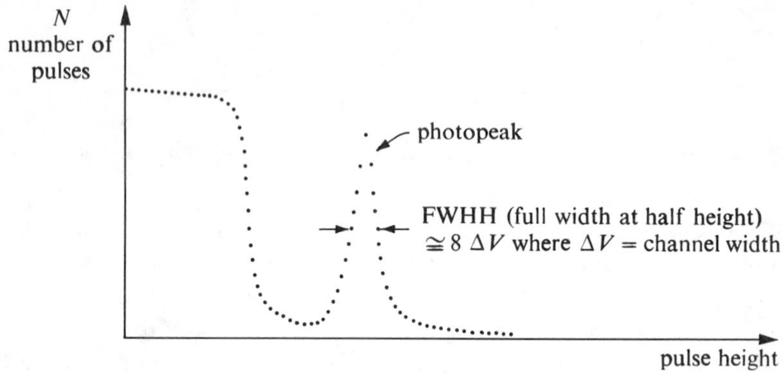

FIGURE 12.24 Multichannel analyzer pulse-height spectrum.

emits only one gamma ray of 662 KeV energy. Notice that the width of the photopeak can be determined by counting the number of dots in the peak between the points of half-maximum height.

In the radiation detector and linear amplifier of Fig. 12.21, pulses coming into the analyzer will be linearly proportional to the energy lost by the radiation in the detector. Thus the problem is to measure the height or amplitude of the pulses and to keep a record of how many pulses fall into each channel. Most modern multichannel analyzers work according to the block diagram of Fig. 12.25. The first step in the analyzer is to convert the input pulse amplitude from analog form (e.g., 2.34 V) to digital form by an "analog-to-digital" converter, sometimes abbreviated ADC.

The small capacitor C is initially uncharged. The input pulse, whose amplitude V_0 is proportional to the energy lost in the detector, charges up C to a certain voltage. The switch S_1 is then opened, and C is discharged to zero volts with a constant current discharge circuit. Thus the time τ for C to discharge is given by $\tau = Q/i$, where Q is the charge on C and i is the constant discharge current. An oscillator or clock continuously generates a very stable frequency in the range 5 MHz to 50 MHz in most analyzers. A gate connects this constant frequency clock to a counter with the gate normally closed so that no clock pulses reach the counter. The gate is opened when the capacitor C begins to discharge and is closed when C is fully discharged to zero volts, which time is sensed by the "zero crossing detector." Thus a burst of clock sine waves or pulses lasting for τ seconds is fed to the counter. The number of clock pulses is given by $N = f\tau = FQ/i$, where f is the constant clock frequency. Thus the

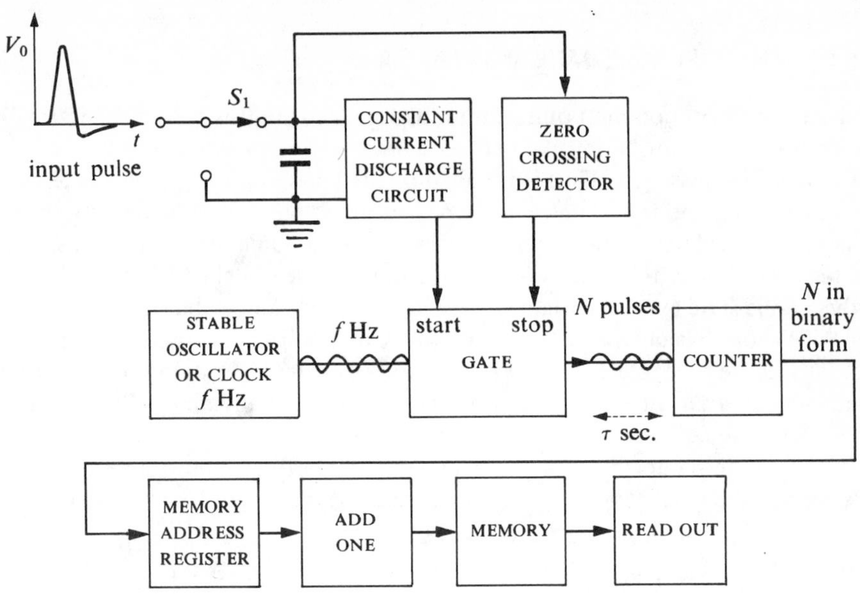

FIGURE 12.25 Block diagram of multichannel analyzer.

number of pulses read by the counter is linearly proportional to the charge Q on the capacitor C, which is in turn proportional to the input pulse amplitude.

The problem is now to store the input pulse amplitude in digital form in the appropriate place, called the "memory." In a 1024-channel analyzer there will be 1024 storage "bins" in the memory, each one corresponding to a certain channel or range of pulse heights. The memory "address" refers to the location of a storage bin in the memory, *not* to the number stored at that location. The "memory address register" of the analyzer is a temporary storage device to hold the pulse amplitude N while looking up the correct address in the memory. The "add one" circuit adds one to the number stored in the memory bin corresponding to the counter reading. For example, each time the counter reads a number of clock pulses corresponding to a pulse amplitude of 1.79 V, the number of 1.79 V pulses stored in the memory is increased by one.

The graph of number of pulses versus pulse amplitude is obtained by converting the information stored in the memory to analog form with a digital-to-analog converter and displaying the analog voltages on a cathode-ray tube; or the information can be typed out in a table. With the cathode-ray tube display, the horizontal deflection voltage is proportional to the memory address, low addresses corresponding to small pulse amplitudes, and high addresses corresponding to large pulse amplitudes. The vertical deflection voltage is proportional to the number stored in the memory at the various addresses. The cathode ray tube display will appear like Fig. 12.24.

12.12 THE BISTABLE MULTIVIBRATOR

Perhaps the most common pulse circuit in a memory unit is a "multivibrator." A multivibrator can be loosely defined as a circuit with essentially two states or conditions possible. Thus it is suited to the binary language; one state can represent "0", the other state "1." A "bistable" multivibrator or "flip-flop" has *two* stable states. The circuit will very rapidly assume one of the two stable states regardless of the initial condition, and it can be switched from one to the other stable state by appropriate voltage inputs. A "monostable" multivibrator or "one shot" has only *one* stable state to which it automatically returns, if it is initially in the other state. An "astable" or "free-running" multivibrator automatically alternates between the two states; it is essentially an oscillator which generates a square wave.

An emitter-coupled transistor bistable multivibrator is shown in Fig. 12.26. Because of the exact symmetry between T_1 and T_2, one might conclude that each

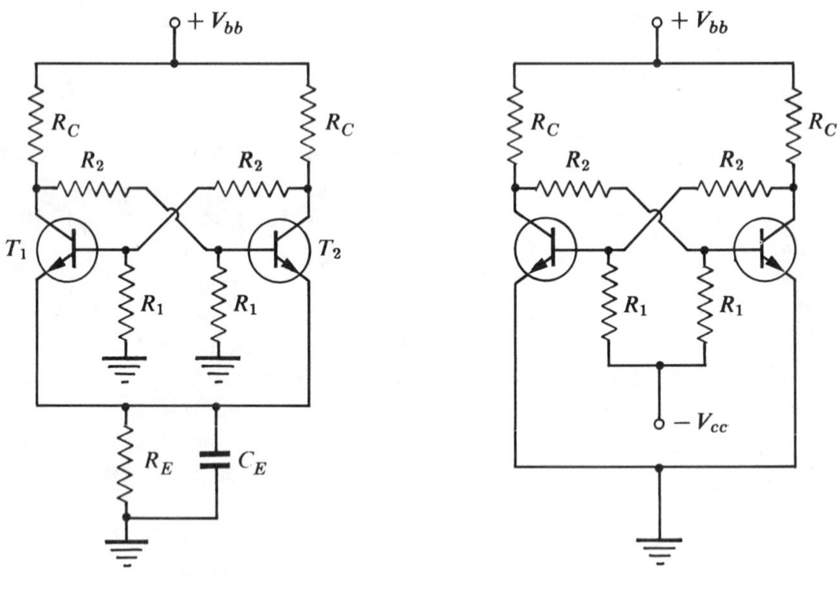

FIGURE 12.26 Bistable multivibrator.

transistor will conduct equally, i.e., have the same collector current. We will now show because of the R_1 and R_2 resistor voltage dividers, which couple the collector of one transistor to the base of the other, that one transistor will always be on and the other off. Suppose that each transistor is conducting. There will be some voltage fluctuation due to noise or a turn-on transient at all points in the circuit, in particular at the bases. Suppose the base of T_2 goes slightly more positive. Then T_2 will draw more collector current, and the voltage at the collector of T_2 will decrease, i.e., become less positive. The fraction $R_1/(R_1 + R_2)$ of this *drop* in the collector voltage is applied to the base of T_1, thus turning T_1 off more. T_1 draws less collector current, and its collector voltage rises, i.e., becomes more positive. A fraction $R_1/(R_1 + R_2)$ of this positive-going voltage is applied to the base of T_2, reinforcing the original positive-going fluctuation. The cycle starts over again, and the net result is to turn T_2 on strongly and to cut T_1 off. The *on* and the *off* transistor can be distinguished because the collector voltage of the on transistor, T_2, will be low and that of the off transistor, T_1, will be high, essentially equal to $+V_{bb}$.

The values of R_1 and R_2 must be chosen so that there are two stable states, that is (for the npn transistors of Fig. 12.26), the base of the off transistor is held negative with respect to the emitter, and the base of the on transistor is held 0.6 V at least (for silicon transistors) positive with respect to the emitter. That is, if $V_{C_{on}}$ is the collector voltage of the on transistor and $V_{off} = V_{bb}$ is the collector voltage of the off transistor, then we must have two conditions satisfied:

$$\frac{R_1}{R_1 + R_2} V_{bb} \geqq V_E + 0.6\,\text{V}$$

$$(\textit{on transistor held on}) \tag{12.8}$$

$$V_{B_{off}} = \frac{R_1}{R_1 + R_2} V_{C_{on}} < V_E$$

$$(\textit{off transistor held off}). \tag{12.9}$$

The value of V_{bb} is chosen less than the transistor breakdown voltage and large enough to operate the transistors in a region where they have reasonable gain. Typically, V_{cc} is from 3 V to 18 V. $V_E = I_E R_E$, where I_E is the current drawn by the on transistor; I_E is usually several milliamperes and V_E must be less than V_{bb}. Usually V_E is on the order of several volts. Thus R_E is determined from $R_E = V_E/I_E$. If we now assume that the on transistor is fully turned on or "saturated," then its collector–emitter voltage will be very small, of the order of 0.1 volt. Then $V_{C_{on}} \cong V_E$, and equation (12.9) becomes $R_1/(R_1 + R_2) < 1$, which is clearly true. The off transistor is kept off, because its base–emitter junction is reverse-biased. It is usually desirable to have the base–emitter junction of the off transistor at least one volt reverse-biased. If the reverse bias were too large, the off transistor would be too hard to turn on, or possibly the base–emitter junction might break down. If the reverse bias were too small, a noise pulse, voltage transient, or pickup voltage might turn the off transistor on accidentally. Thus (12.9) can be written:

$$V_E - V_{B_{off}} = V_E - \frac{R_1}{R_1 + R_2} V_E \geqq 1\,\text{V} \quad \text{or} \quad \frac{R_2}{R_1 + R_2} V_E \geqq 1\,\text{V}. \tag{12.10}$$

We now have two equations in the two unknowns, R_1 and R_2:

$$\frac{R_1}{R_1 + R_2} V_{bb} \geqq V_E + 0.6\,\text{V} \qquad \frac{R_2}{R_1 + R_2} V_E \geqq 1\,\text{V}, \tag{12.11}$$

but they are homogeneous equations; that is, there is no constant term. Each equation, in other words, specifies only the ratio R_1/R_2. Putting in $V_{bb} = 12\,\text{V}$, $V_E = 4\,\text{V}$ (for $R_E = 1\,\text{k}\Omega$, $I_E = 4\,\text{mA}$), we have:

$$\frac{12R_1}{R_1 + R_2} \geqq 4.6 \qquad \text{and} \qquad \frac{4R_2}{R_1 + R_2} \geqq 1,$$

which imply

$$R_1 > 0.6R_2 \qquad \text{and} \qquad R_1 < 3R_2. \tag{12.12}$$

So, we must have $0.6R_2 < R_1 < 3R_2$, which is easy to satisfy; $R_1 = R_2$ would do nicely. The absolute magnitude of R_1 and R_2 can be determined by requiring that the divider current I_D flowing through R_1 and R_2 be large compared to the base current of the on transistor, $I_{B_{\text{on}}} = I_E/\beta = 4\,\text{mA}/100 = 0.04$ mA. But (see Fig. 12.27) $R_1 + R_2$ cannot be too small; otherwise too much current will be drawn through R_C and the voltage at the collector of the off transistor will drop appreciably below V_{bb}. So, we desire $I_D R_C \ll V_{bb}$. R_C is determined from:

$$V_E = 4\,\text{V} \qquad \text{and} \qquad I_c = 4\,\text{mA},$$

$$V_{CE_{\text{on}}} \cong 0.1\,\text{V},$$

$$R_C = \frac{V_{bb} - (V_E + 0.1)}{I_c} = \frac{12\,\text{V} - 4.1\,\text{V}}{4\,\text{mA}}$$

$$R_C = \frac{7.9\,\text{V}}{4\,\text{mA}} \cong 2\,\text{k}\Omega.$$

If we choose $I_D = 10I_{B_{\text{on}}} = 0.4\,\text{mA}$, then the voltage drop across $R_C = 2\,\text{k}\Omega$ due to I_D is $I_D R_C = 0.4\,\text{mA} \times 2\,\text{k}\Omega = 0.8\,\text{V}$, which is small compared to V_{bb}. Thus $R_1 + R_2 = V_{C_{\text{off}}}/I_D = 11.2\,\text{V}/0.4\,\text{mA} = 28\,\text{k}\Omega$. Thus $R_1 = R_2 = 14\,\text{k}\Omega$. The dc voltages and resistances for the bistable multivibrator are shown in Fig. 12.27(b).

In order to take advantage of the existence of the two stable states of a bistable multivibrator, we must be able to switch from one state to the other. With T_1 off and T_2 on, if we can somehow turn the off transistor T_1 on a bit, then the coupling action of the R_1R_2 networks will cause the circuit to "flip" over into the other state with T_1 on and T_2 off. This can be accomplished in several ways, usually by feeding in voltage pulses. The pulses which cause the circuit to flip are called "trigger" pulses or "clock" pulses.

In emitter triggering, shown in Fig. 12.28(a), negative pulses are applied to the emitter to flip the circuit. The base of the off transistor T_1 will be at $+2.05\,\text{V}$ if $R_1 = R_2$. Thus, the reverse bias on the base–emitter junction of T_1 is $4\,\text{V} - 2.05\,\text{V} = 1.95\,\text{V}$. Thus, a negative pulse of $1.95\,\text{V}$ or larger on the emitter will make the n-type emitter of T_1 negative with respect to its p-type base, thus

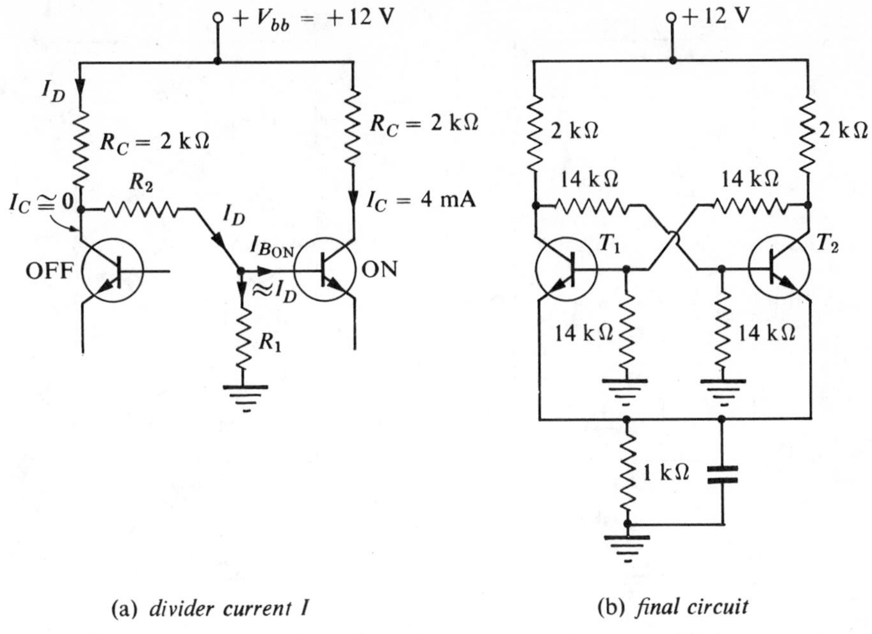

(a) *divider current I* (b) *final circuit*

FIGURE 12.27 Bistable multivibrator.

turning T_1 on. The collector voltage of T_1 will fall, thus turning T_2 off, which in turn turns T_1 on more and the circuit will flip.

The multivibrator can also be triggered by applying negative trigger pulses through two "steering" diodes D_1 and D_2, as shown in Fig. 12.28(b). The "cathodes" (negative sides) of the two diodes are connected together at point p and are tied to the $V_{bb} = +12$V supply through R_D. The emitter resistor is bypassed to hold the dc emitter voltage constant.

Notice that with no trigger pulse both diodes are reverse-biased, so no dc current flows through R_D, and point P is at $V_{cc} = +12$V. Thus, diode D_2 has a reverse bias voltage of 7.9 V, while diode D_1 has a reverse bias of only 0.8 V. To make a diode conduct, one must apply a voltage equal and opposite to the reverse bias plus the diode turn-on voltage. Thus a negative trigger pulse applied through C_1 greater than $0.8\text{V} + 0.6\text{V} = 1.4\text{V}$ in amplitude will turn on diode D_1, which will then "steer" the negative pulse to the collector of the off transistor T_1 and through R_1 and R_2 to the base of the on transistor T_2, thereby tending to turn the on transistor off. So long as the trigger pulse is less than $7.9 + 0.6 = 8.5\text{V}$ in amplitude, diode D_2 remains nonconducting. Thus, the on transistor's collector current decreases, and its collector voltage rises. This positive-going voltage is coupled to the base of T_1, which is thereby turned on and the circuit flips.

Each negative trigger pulse flips the multivibrator from one state to the other. Thus the frequency of the rectangular wave on either collector will

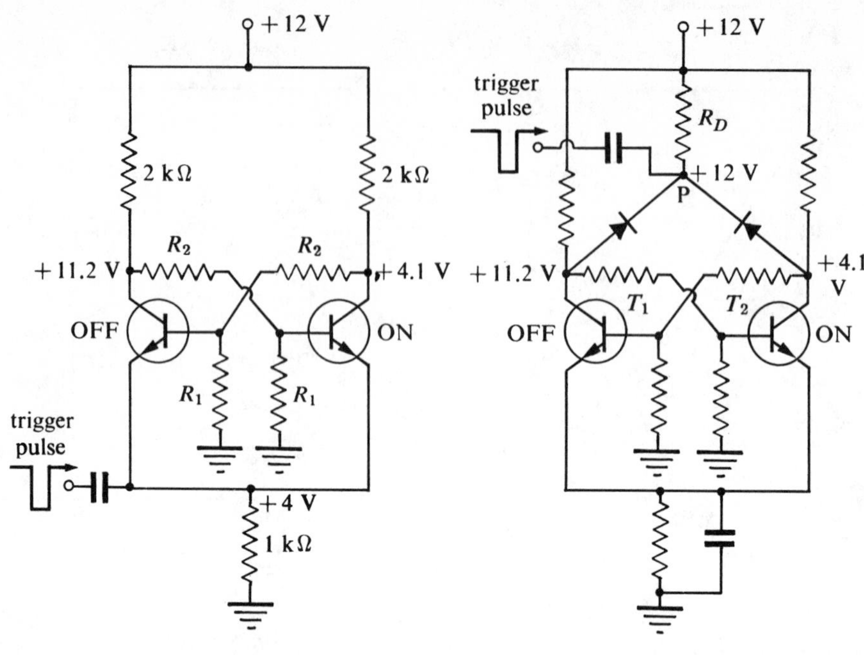

(a) *emitter triggering* (b) *collector triggering with steering diodes*

FIGURE 12.28 Bistable multivibrator triggering techniques.

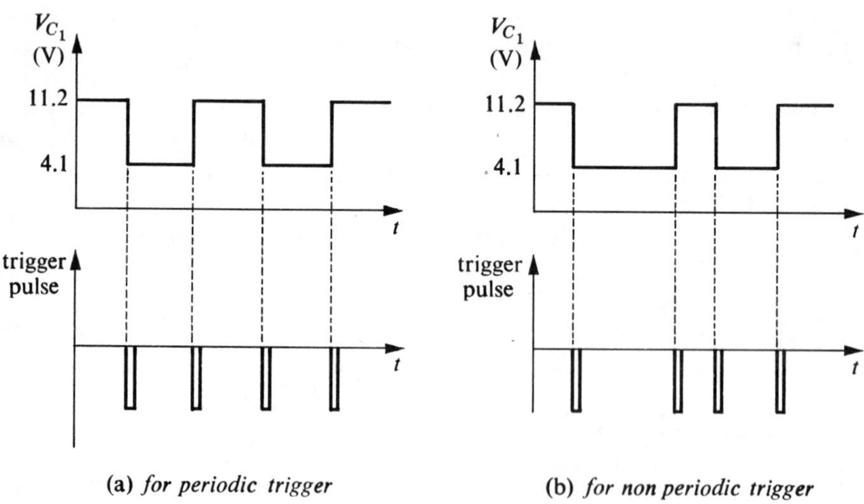

(a) *for periodic trigger* (b) *for non periodic trigger*

FIGURE 12.29 Waveforms in bistable multivibrator.

be precisely one-half of the trigger pulse frequency, as shown in Fig. 12.29(a). If the trigger pulses are not periodic, then the circuit will still flip once for each trigger pulse as shown in Fig. 12.29(b). A series of bistable multivibrators can be used to divide a frequency by precisely 2^n, where n is the number of multivibrators. All that must be done is to convert the square wave output on the collector of one multivibrator into a suitable trigger pulse for the next stage. One common technique is to differentiate the collector square-wave with a simple RC differentiator, and use the short spikes produced as trigger pulses for the next stage. With steering diode triggering, only the negative spikes will trigger the next multivibrator, so the frequency will be divided by two. The input to the two steering diodes is usually called the "clock," "trigger," or "toggle" input and is sometimes denoted by the symbol CP or T.

Two other inputs are usually given special names in flip-flops; they are called the "reset" or "clear," and "set" or "preset" inputs, or the "R" and "S" inputs, respectively (shown in Fig. 12.30). The voltage level on the collector

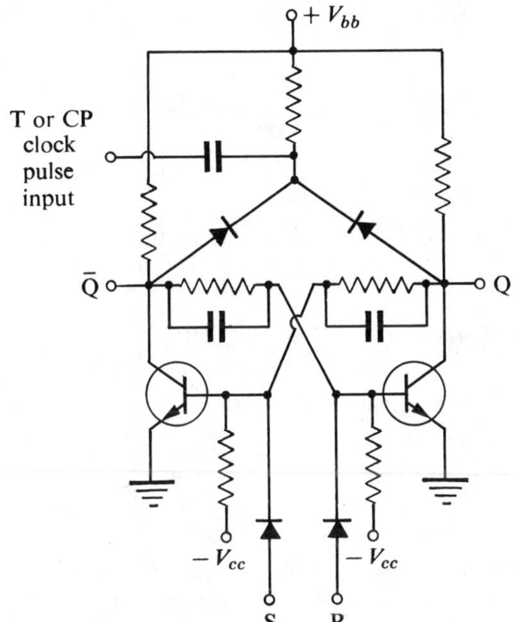

R	S	Q
0	0	no change
0	1	1
1	0	0
1	1	indeterminate

FIGURE 12.30 RS clocked flip-flop.

of one transistor is either high (+11.2 V) or low (+4.1 V), and thus respectively represents "1" or "0" in a positive logic system. (In practical integrated circuit systems, the voltage levels are usually nearer +4 V and 0 V for "1" and "0," respectively.) The standard or cleared state of a flip-flop is with T_1 off and T_2 on; thus the voltage at the collector of T_1 is high or at logic "1" and the collector of T_2 is low or at logic "0." The output of T_2 is usually denoted by the symbol Q, and of T_1 by $\overline{Q}$ ("NOT Q"). Thus the standard or cleared state has $Q = 0$ and $\overline{Q} = 1$. (The only other possible stable state is $Q = 1$ and $\overline{Q} = 0$.)

A positive pulse input to the R, or clear, input will turn T_2 on and make Q = 0 and $\overline{Q}$ = 1. In other words the Q output has been "reset" or "cleared" to 0. A positive pulse input to the S, or set, input will set Q = 1 and $\overline{Q}$ = 0. If it is desired to start with a series of flip-flops in a counter all in the cleared state with Q = 0, then a positive clearing pulse would be applied simultaneously to the R inputs. If both the R and the S inputs are 0, then the state of the flip-flop is unchanged. If both R and S equal 1, then the R input tends to turn T_2 on and make Q = 0, but the S input is simultaneously trying to turn T_1 on and make $\overline{Q}$ = 0. In other words the flip-flop is being forced in two opposite directions. The resulting condition of the flip-flop is indeterminate; it depends on the inherent asymmetries in the circuit. Clearly one should never have both R and S simultaneously equal to 1.

A type of RS flip-flop can be made from two cross-coupled, positive logic NOR gates as shown in Fig. 12.31. Because a NOR gate has a 1 output only

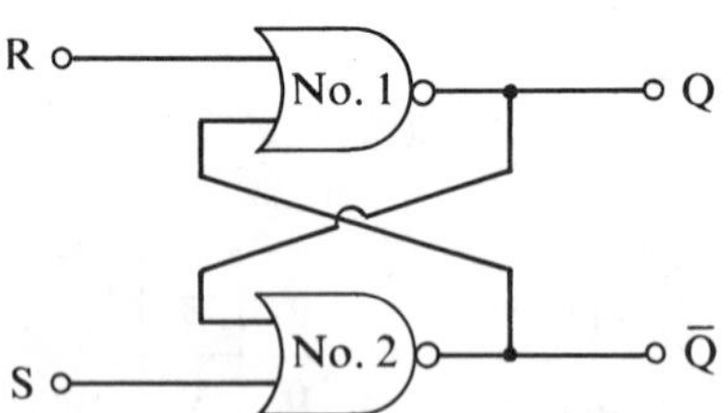

R	S	Q	$\overline{Q}$
0	0	no change	
0	1	1	0
1	0	0	1
1	1	0	0

FIGURE 12.31 Flip-flop made from NOR gates.

when *neither* input is 1, we immediately see that if Q = 1 there is a 1 input to NOR gate #2, thus making $\overline{Q}$ = 0. Similarly, if $\overline{Q}$ = 1, Q = 0. Thus there are only two possible outputs: Q = 0 and $\overline{Q}$ = 1, or Q = 1 and $\overline{Q}$ = 0. The operation of the set and clear inputs can be understood as follows. Suppose Q = 1 and $\overline{Q}$ = 0. If the R input is 1 and the S input is 0, then Q is immediately changed to Q = 0 because of the R = 1 input to NOR gate #1. NOR gate #2 now has two 0 inputs which makes $\overline{Q}$ = 1. Thus R = 1 and S = 0 has cleared or reset the outputs to Q = 0 and $\overline{Q}$ = 1. Similar arguments show that R = 0 and S = 1 will set the outputs to Q = 1 and $\overline{Q}$ = 0. If both S and R are equal to 1, then both Q and $\overline{Q}$ must equal 0, which is impossible for a conventional transistor flip-flop (Fig. 12.31), but is possible for this type of circuit. If both R and S are 0, then the output can be either Q = 1 and $\overline{Q}$ = 0, or Q = 0 and $\overline{Q}$ = 1; in other words the outputs will not be changed by R = S = 0 inputs. The R and S inputs are effective in determining the state of the flip-flop regardless of the clock input. In this sense the R and S inputs are termed "asynchronous" inputs; they override the clock or are independent of the clock. There is no fixed phase relationship between the clock pulse and the flip-flop output Q; the flip-flop output is set or cleared immediately after the R or S input.

There is another type of input, termed "synchronous," in which the inputs affect the condition of the flip-flop only *after* the clock pulse "commands" or "allows" it. These synchronous inputs are usually denoted by the letters J and K. The flip-flop output is determined by the particular JK inputs, but the flip-flop assumes the state only *after* the clock pulse. In other words, there is a definite phase relationship between the clock pulse and the flip-flop output Q. In the JK flip-flop, the outputs are fed back into the inputs in such a way as to eliminate the indeterminate state of the flip-flop when both J and K equal 1. A JK flip-flop changes its state when J and K each equal 1. The JK flip-flop will be discussed in detail in Chapter 13.

A bistable multivibrator constructed from discrete components (rather than using an integrated circuit as will be discussed in the next chapter) will flip up to many MHz with some care in construction and choice of transistors. There are two limitations to the maximum frequency at which the multivibrator can flip, shunt capacitance and storage effects. In order to turn the on (npn) transistor off, its base must be made more negative, and any shunt capacitance between base and ground must be charged up negatively. This shunt capacitance is due to the inherent base–emitter capacitance of the transistor and to any wiring capacitance between the base wire and ground. A proper choice of transistor and careful wiring practice will minimize this. However, the remaining shunt capacitance can be charged up much more quickly by the trigger pulse if we connect a relatively small capacitance C_2 across R_2, as shown in Fig. 12.32. The capacitor C_2 is often called a "speed-up" capacitor and is usually several hundred picofarads. C_2 effectively shorts out R_2 as far as the rapidly

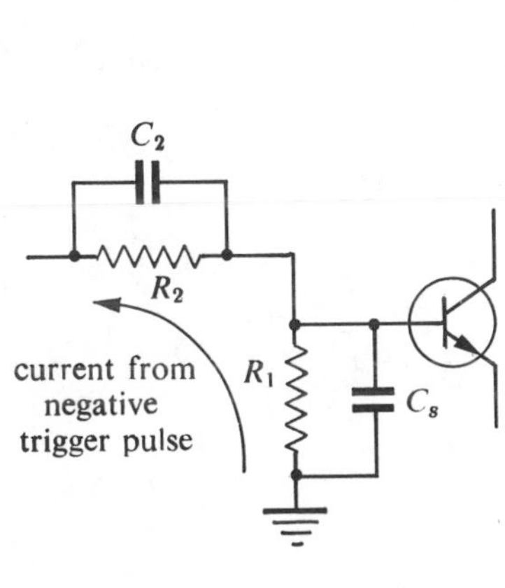

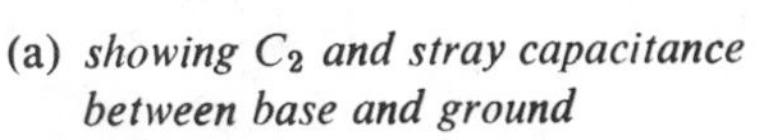
(a) *showing* C_2 *and stray capacitance between base and ground*

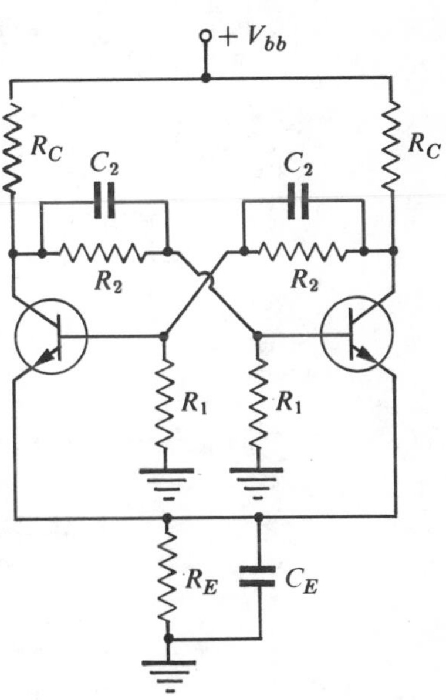

(b) *complete bistable multivibrator*

FIGURE 12.32 Speed-up capacitor C_2.

changing trigger pulse is concerned, so the shunt capacitance can be charged more quickly.

The second factor which limits the high-frequency operation is that of charge storage in the saturated on transistor. For an npn transistor, electrons are injected from the emitter into the base where they are minority carriers in the p-type base region. A fraction $\alpha \cong 0.98$ to 0.99 of them *diffuse* across the base to the collector, because there is almost no electric field in the base material. The electrons eventually are swept into the collector by the electric field present in the depletion region at the base–collector junction. When the transistor is saturated, the electric field at the base–collector junction is very small, because the voltage across the transistor is very small. Thus the depletion field electric field does not penetrate as deeply into the base region and the electrons "see" a thicker base region across which they must diffuse. There is a considerable amount of charge effectively "stored" in the base when the transistor is saturated. If the transistor is not saturated, the charge carrier density in the base falls to zero at the base-collector junction and is large at the emitter-base junction where the carriers are injected. In saturation, the charge carrier density is greater than zero at the base–collector junction; and if the base is suddenly made to go negative in an attempt to cut the transistor off, the collector current will continue to flow for a finite time until all the charge carriers stored in the base region have been removed. This is illustrated in Fig. 12.33.

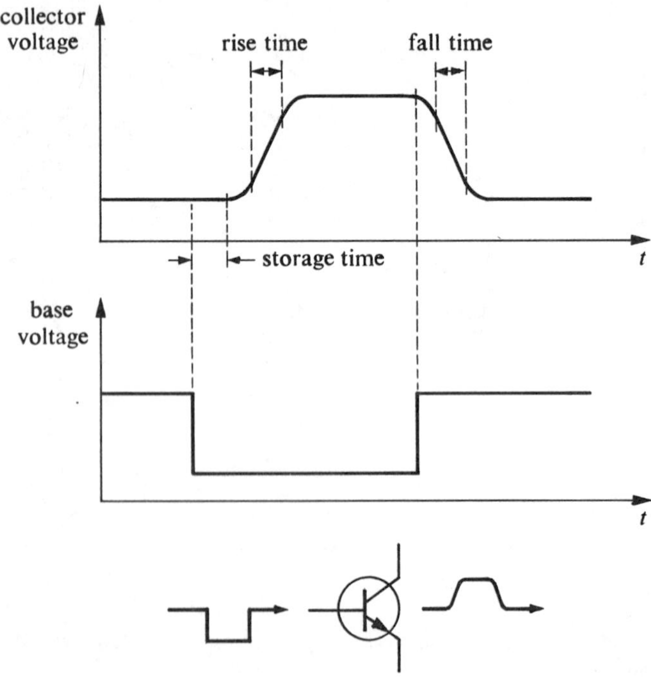

FIGURE 12.33 Storage time T_s in saturated npn transistor.

If the on transistor in a multivibrator is not driven into saturation, the multivibrator is said to be "nonsaturated." Clearly, for a given transistor, a nonsaturated multivibrator will operate at higher frequencies than a saturated one. Modern nonsaturated flip-flops have switching times on the order of one nanosecond. However, saturated multivibrators do have several advantages over nonsaturated ones. The output voltage waveform on the collector is essentially independent of the transistor characteristics; the maximum and minimum voltages are V_{bb} and $V_E + V_{CE\text{Sat}} \cong V_E$, because the transistors alternate between a completely on and a completely off state. The circuit design is also easier; the choice of the divider resistances R_1 and R_2 is not critical. Modern transistors are being made with storage times on the order of 20 nanoseconds. In practice, one usually uses an integrated circuit multivibrator for high-frequency operation, which will be discussed in the next chapter.

12.13 THE MONOSTABLE MULTIVIBRATOR

A simple change in the base bias network for one transistor converts the bistable multivibrator into a monostable multivibrator with only one stable state: T_1 off and T_2 on. The T_2 is maintained on by its base being tied to V_{bb} through R_3; T_1 is held off by R_1 and R_2 just as in the case of the bistable circuit. Notice that there is no dc coupling from the collector of T_1 to the base of T_2, but only ac coupling through C_3. The capacitor C_3 is charged with one plate at $11.2\text{V} \cong V_{bb}$ and the other at $V_E + 0.6\text{V} = 4.6\text{V}$, as shown in Fig. 12.34(a), when the circuit is in the stable state.

The monostable multivibrator can be flipped to the state with T_1 on and T_2 off by applying a negative trigger pulse to the base of T_2, which will decrease I_{C2}, thus making V_{C2} go more positive. This positive voltage swing is coupled to the base of T_1 by the $R_1R_2C_2$ network. Thus T_1 is turned on slightly, and its collector voltage drop is coupled to the base of T_2 through C_3. T_2 is turned off more, and the circuit quickly assumes the state with T_1 on and T_2 off. The circuit switches states very quickly, in times on the order of $1\,\mu\text{sec}$ or less. However, this state is not stable; it will last only as long as the base of T_2 is more negative than the emitter. Because the base of T_2 is connected to V_{bb} through R_3, eventually it will be raised up in voltage to the point where T_2 starts to conduct, and the circuit will flip back to the original stable state. The time the circuit is held in the unstable state with T_1 on and T_2 off is determined by how fast C_3 is charged through R_3. In the unstable state, the voltages are shown in Fig. 12.34(b). When the circuit quickly flips from the stable to the unstable state, the voltage on the collector of T_1 drops from $+11.2\text{V}$ to $+4.1\text{V}$, a drop of 7V. The voltage on each plate of C_S must also drop by 7V; thus just after the circuit has flipped into the unstable state the voltages are as shown in Fig. 12.34(b). This configuration is clearly unstable, and a charging current I will start to flow through R_3 to charge C_3. As C_3 charges up, the voltage on its right-

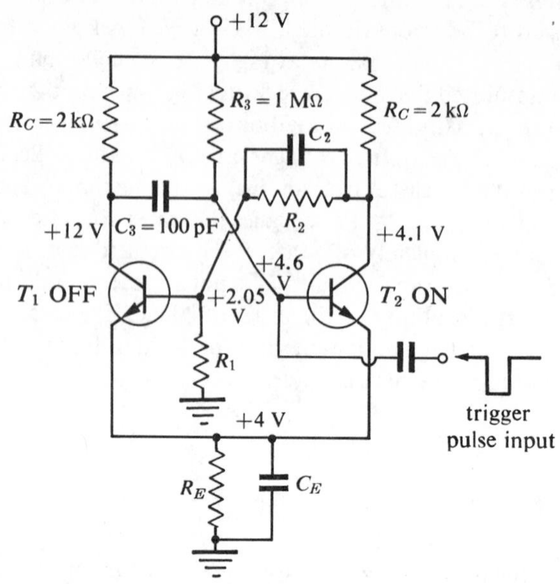

(a) *in stable state*

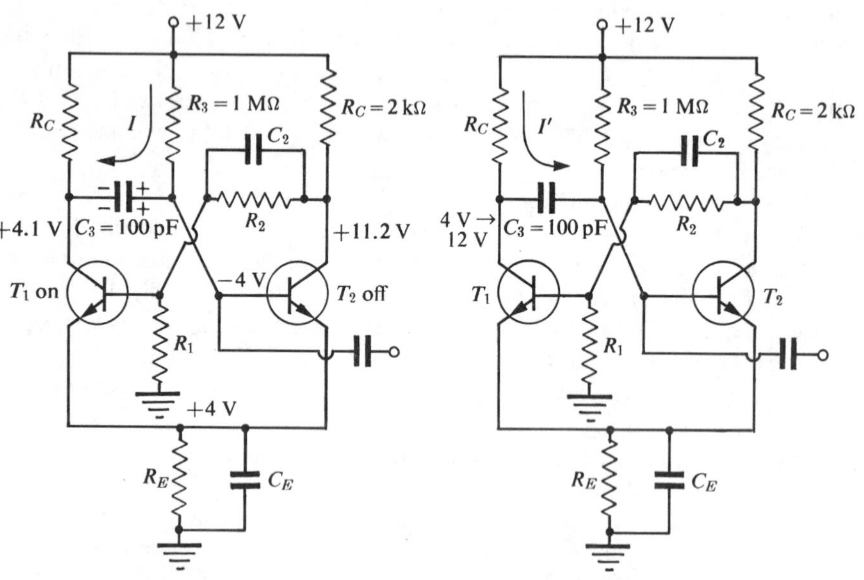

(b) *in unstable state*

(c) *in process of flipping from unstable to stable state*

FIGURE 12.34 Monostable multivibrator.

hand plate, which is connected to the base of T_2 (now off), will gradually become more positive. When the base of T_2 becomes slightly positive with respect to its emitter, T_2 will start to conduct, its collector voltage will fall, and the circuit will flip back to the stable state with T_2 held on and T_1 held off. The stable state, with T_1 off, has T_1's collector voltage at +11.2V. Hence, when T_1 turns off, its collector voltage must rise from 4V to 11.2V, and this means C_3 must charge up so that its left-hand plate goes from +4V to +11.2V. Thus, a charging current I' must flow through R_C to charge up C_3. The time constant for this charging will be $T = R_C C_3 \cong (2\,\text{k}\Omega)(100\,\text{pF}) = 20\,\mu\text{sec}$, which means the collector voltage on T_1 does not immediately rise from 4V to 11.2V but rather takes the shape shown in Fig. 12.35. The collector voltage on T_2 is also not perfectly rectangular because the base-to-ground capacity must be charged and discharged as T_1 turns on and off. However, this capacitance is on the order of only 10pF and is charged through $R_C = 14\,\text{k}\Omega$, so it charges and discharges in a time of the order of $14 \times 10^3 \cdot 10^{-11} = 14 \times 10^{-8}$ sec. The time τ the circuit remains in the unstable state depends upon how rapidly C_3 charges up through R_3; thus τ depends upon R_3C_3: the larger R_3C_3, the larger τ. τ may be taken to be on the order of R_3C_3; thus for $R_3 = 1$ megohm and $C_3 = 0.001\,\mu\text{F}$, $\tau = (10^6\Omega)(10^{-9}\text{F}) = 10^{-3}$ sec = 1 msec. The waveforms will be as shown in Fig. 12.35. The shape of the output on the collector of T_2 will be essentially independent of the exact shape or magnitude of the trigger pulse. All that is necessary is that the trigger pulse be sufficiently large to decrease the collector current of T_2 when it is in the saturated *on* state. The coupling from T_2 to T_1, and vice versa, through R_1R_2 and C_3 will then serve to flip the circuit.

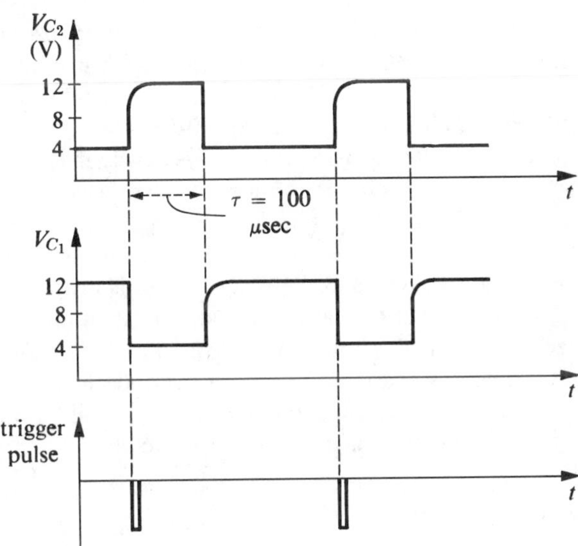

FIGURE 12.35 Monostable multivibrator waveforms.

A monostable multivibrator is often used to provide a known time delay for a pulse. For example, if it is desired to delay a pulse by 100μsec in a circuit, the pulse can be used to trigger a monostable multivibrator which is designed to put out a 100μsec wide pulse. The pulse on the collector of T_2 (see Fig. 12.35) is then differentiated and the negative derivative from the trailing edge of the collector pulse will be exactly 100μsec delayed with respect to the leading edge of the original trigger pulse. Another example of the use of a monostable multivibrator is when a more energetic pulse is desired. The available pulse may be too small in amplitude or of too short duration to start or stop a counter, for example. In this case, the initial pulse is used as the trigger pulse, and the leading edge of the monostable multivibrator output pulse will occur at the same time as the trigger pulse, and will be larger and more energetic than the trigger pulse.

12.14 THE ASTABLE MULTIVIBRATOR

If we change the base biasing so that each base is coupled to the other collector through a capacitor and is also tied to $+V_{bb}$ through a resistance, then we have a symmetrical circuit in which each transistor is biased like the normally *on* transistor of the monostable circuit. This circuit is called an *astable* or free-running multivibrator and is shown in Fig. 12.36(a). The name comes from the fact that the circuit will automatically flip or switch back and forth between the T_1 on, T_2 off state to the T_1 off, T_2 on state. In other words, the circuit is a square-wave oscillator. By an argument similar to that used in section 12.12 for the bistable multivibrator, it can be shown that the circuit will quickly assume a state with T_1 strongly on and T_2 off, or vice versa. This conclusion follows from the collector-to-base coupling through the capacitors C_1 and C_2. However, the voltage of the base of the off transistor will always slowly increase, until the off transistor is turned on and the circuit switches to the other state. That is, if T_2 is off, the current I_2 gradually charges up C_2 through R_2 until T_2 turns on; if T_1 is off, the current I_1 gradually charges up C_1 until T_1 is turned on. Thus, the time T_1 is off is proportional to R_1C_1, and the time T_2 is off is proportional to R_2C_2. If $R_1C_1 = R_2C_2$ and the transistors are identical, an essentially square wave is produced on each collector, as shown in Fig. 12.36(b). If $R_1C_1 > R_2C_2$, then an asymmetrical "square" wave is produced on the collector, as shown in Fig. 12.36(c). Occasionally such an astable multivibrator will fail to start oscillating, but usually only if the supply voltage is turned on slowly.

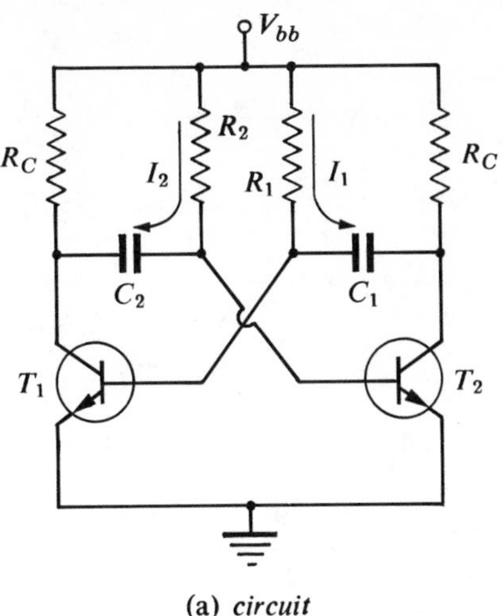

(a) *circuit*

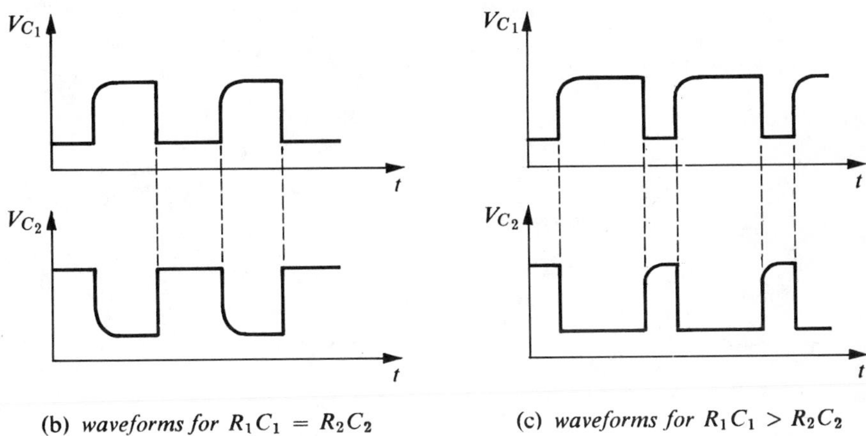

(b) *waveforms for* $R_1C_1 = R_2C_2$

(c) *waveforms for* $R_1C_1 > R_2C_2$

FIGURE 12.36 Astable multivibrator.

12.15 THE SCHMITT TRIGGER CIRCUIT

It is often useful to discriminate between pulses or waveforms on the basis of their absolute magnitude, for example, to reject all pulses smaller than 2.0 volts and accept those larger than 2.0 volts. One circuit which will accomplish this is the Schmitt trigger circuit shown in Fig. 12.37.

The circuit is seen to be asymmetrical. The coupling from the collector of T_1 to the base of T_2 is like that in the bistable multivibrator, but the base of T_1 is not tied to the collector of T_2. An analysis shows that there are two stable

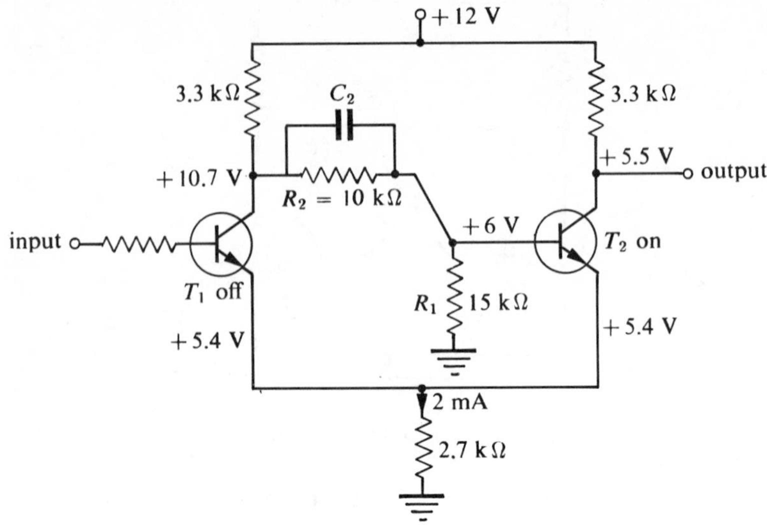

(a) *stable state with* $V_{in} < V_E + 0.6\text{ V} = 6.0\text{ V}$

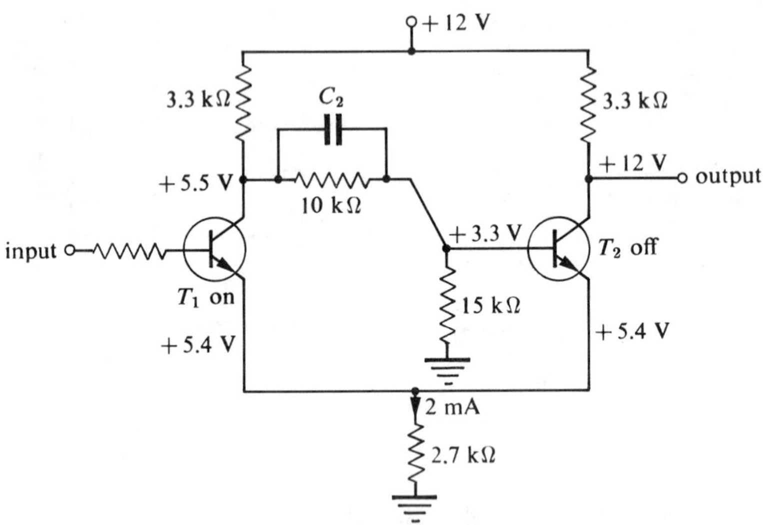

(b) *stable state with* $V_{in} > 6.0\text{ V}$

FIGURE 12.37 Schmitt trigger.

states. If T_1 is off and T_2 is on, the dc voltages will be as shown in Fig. 12.37(a). Assume T_2 draws 2 mA; thus both emitters are at +5.4 V. T_1 will remain off unless its base voltage rises above 5.4 V plus its base–emitter turn-on voltage, about 0.6 V for a silicon or 0.25 V for a germanium transistor. For T_1 silicon,

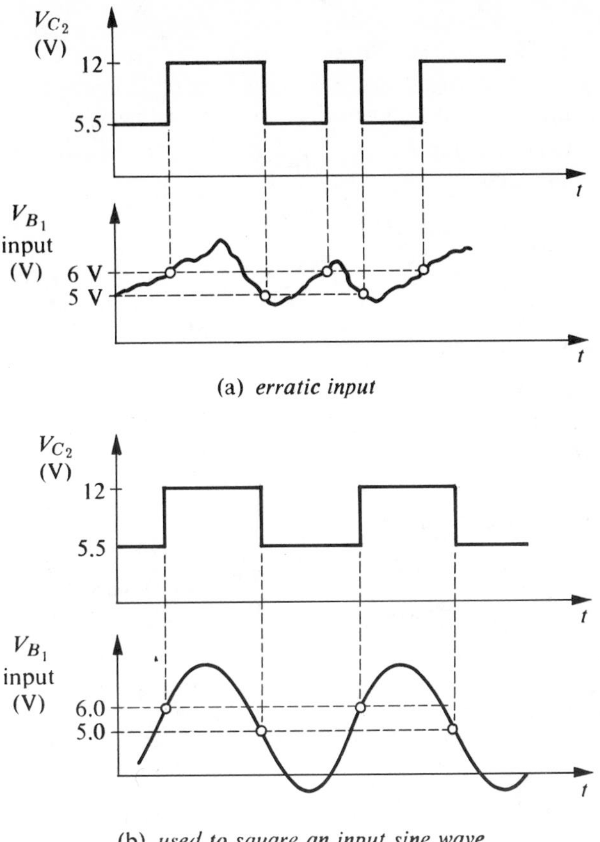

FIGURE 12.38 Schmitt trigger waveforms.

if the input voltage at the base of T_1 does rise above 6.0 V, T_1 will start to draw current, its collector voltage will fall, and the base of T_2 will be driven negative with respect to its emitter through the $R_1R_2C_2$ coupling network. With T_1 on and T_2 off, the circuit is stable in the state shown in Fig. 12.37(b), so long as the input voltage remains above +6.0 V (to keep T_1 on). If the input voltage falls well below 6.0 V, T_1 will be turned off and T_2 on again. This does not happen until the input voltage falls approximately 0.5 to 1.0 volt below 6.0 V, because in the T_1 on, T_2 off state, the base–emitter junction of T_2 is strongly reverse biased by about 2.7 volts. Thus the circuit waveforms are as shown in Fig. 12.38(a).

Notice that the output taken from the collector of T_2 is constant at +12 V and is completely independent of the input wave shape so long as the input is above 6.0 V. With a sine wave input, the output will be a rectangular wave as shown in Fig. 12.38(b); this is sometimes called a "squaring" circuit.

problems

1. Give the following numbers in binary form: 1, 4, 9, 16, and 20.
2. Briefly explain why a digital circuit is more immune to temperature fluctuations, supply voltage drifts, and component aging than analog circuits.
3. Write truth tables for: (a) $Y = A \cdot B \cdot C$ and (b) $Y = A + B + C$.
4. Write the truth tables for:

   (a) $Y = A + \overline{B}$
   (b) $Y = \overline{A} \cdot B$
   (c) $Y = A \cdot B \cdot \overline{C}$
   (d) $Y = \overline{A} + B + C$.

5. Write the truth tables for:

   (a) $Y = (A \cdot B) + C$
   (b) $Y = A \cdot (B + C)$
   (c) $Y = (A \cdot B) + \overline{C}$
   (d) $Y = A \cdot (B + \overline{C})$.

6. Write the truth table for the following.

7. Sketch the output Y for the following inputs: (a) 2.0 V positive pulse at A and B and (b) 0 V at A, 2.0 V positive "pulse" at B. The diodes are silicon.

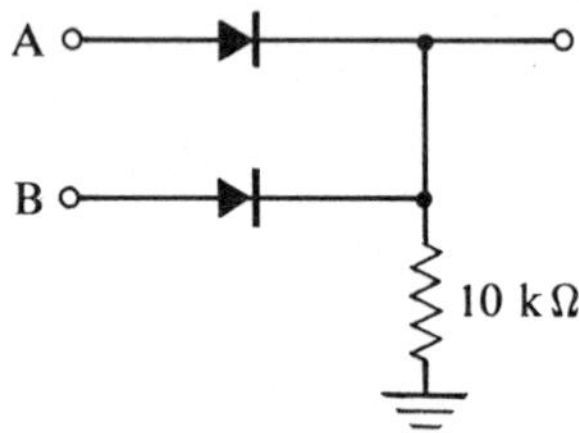

8. Sketch the output Y for the following inputs: (a) 2.0 V positive pulse at A and at B and (b) 2.0 V positive pulse at A and 0 V at B.

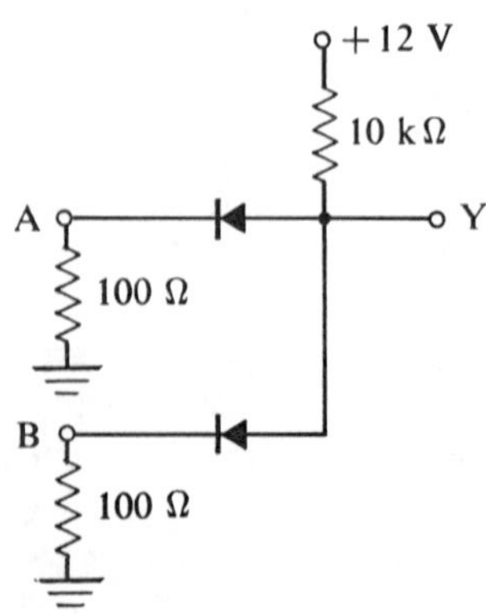

9. Write the truth table for the following.

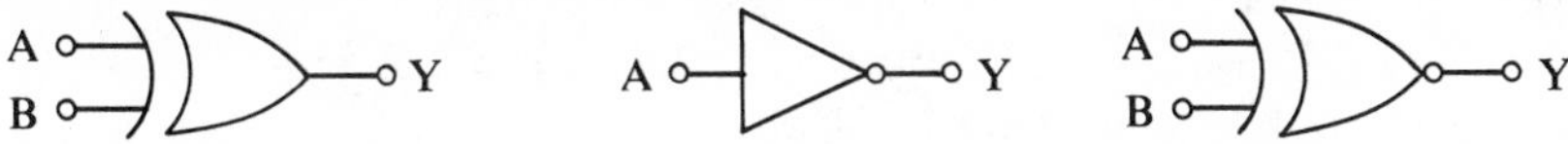

10. Sketch a block diagram using basic AND, OR, and NOT circuits to add two two-bit numbers in parallel fashion.
11. Explain "channel width" and "base line" with respect to a pulse-height analyzer.
12. Explain the difference between a single-channel and a multichannel analyzer.
13. Sketch a graph of a multichannel analyzer output for a gamma source with a single 160keV gamma ray. Label the axes. You may assume that an appropriate gamma ray detector and pulse amplifier supply the input to the analyzer.
14. Calculate the energy of the unknown gamma ray and the width of its peak in keV from the following spectrum, if the standard 662keV γ-ray peak from Cs^{137} occurs at channel number 330 with the same analyzer and amplifier settings.

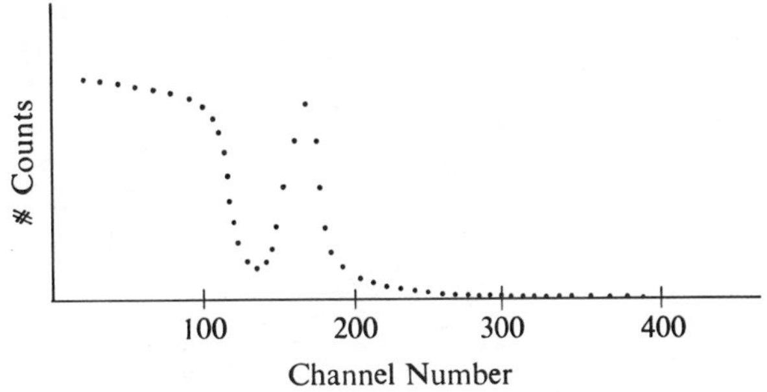

15. Calculate R_1 and R_2 for the emitter-coupled bistable multivibrator shown.

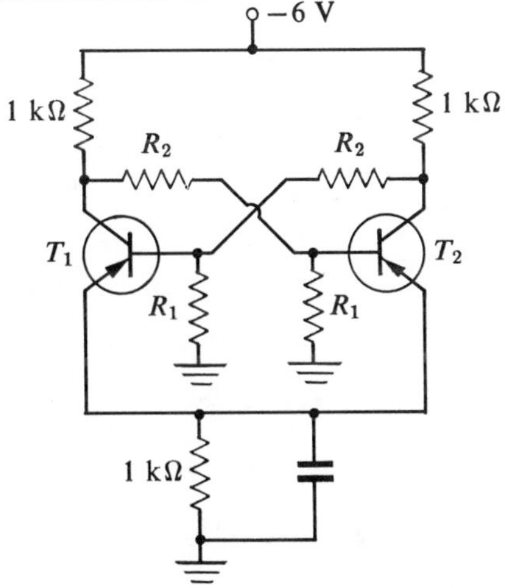

16. Sketch an appropriate diode–resistor network to make a clock input for the multivibrator of Problem 15. The clock pulses are positive. Sketch the collector voltage of T_2 for several clock pulses.
17. Modify the bistable multivibrator of Problem 15 to work on a separate bias supply, so that no emitter resistor is used.
18. Sketch a bistable multivibrator or flip-flop made from two NOR *gates* and carefully show that it acts as a flip-flop by an argument using the NOR gate truth tables.
19. Using a block diagram, show how a monostable multivibrator and a differentiating and diode clipping network can be used to delay a positive voltage pulse by 100 μsec. That is, the output of your circuit should be a positive pulse occurring exactly 100 μsec after the input pulse.
20. Sketch (to scale) the output of a Schmitt trigger circuit set to trigger at 2.0 V for an increasing input, and at 1.5 V for a decreasing input. The input waveform is

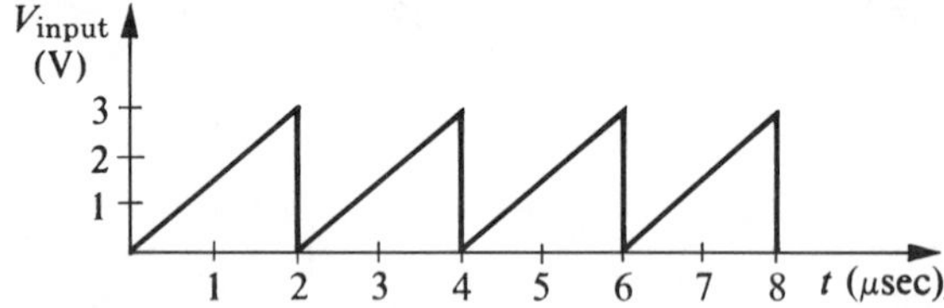

shown. Assume the collector resistor is 4.7 kΩ in the Schmitt trigger and the collector current for the on state is 2 mA and $V_{bb} = 18$ V.

13

INTEGRATED CIRCUITS

13.1 INTRODUCTION

An integrated circuit (I.C.) is a collection of interconnected transistors, diodes, resistors, and capacitors mounted in one package or case with as many as fourteen leads.

The word "integrated" does not refer to the mathematical process of adding together an infinite number of infinitesimally small terms, but rather to the fact that all the transistors, diodes, and resistors are formed from a single small piece of semiconductor material called a "chip" or a "die." If only one chip is present in the case, the I.C. is called "monolithic"; if several chips are mounted inside the case the I.C. is called "hybrid" (see Fig. 13.1). From now on we will refer only to monolithic I.C.'s. The case is usually no bigger than that of an ordinary transistor. Some integrated circuits contain several thousand transistors and resistors, and so extreme miniaturization is possible.

Because of their extremely small size, integrated circuits tend to be restricted to low power applications. They simply are not physically large enough to dissipate more than several watts of power, and many I.C.'s dissipate only tens of millwatts. Their small size, however, does enable them to operate at high frequencies because of the small shunt capacitance and series inductance—from the short "leads" between the components comprising the I.C. The cost of an I.C. is considerably less than the total cost of the separate components and labor which would be necessary to make the same circuit from discrete components. The small size also means that the different transistors are usually in intimate thermal contact with one another. It is relatively easy to produce matched pairs of transistors and resistors in integrated circuits, because the transistors are all made from the same piece of semiconductor and so will tend to have the same thermal drift characteristics as well as being at the same temperature.

FIGURE 13.1 Hybrid integrated circuit. (Courtesy of Motorola Semiconudctor Products Inc.)

One ingenious example of modern I.C. technology uses both n channel and p channel enhancement type MOSFETs in pairs. Power is drawn only when switching states; no power at all is consumed in the steady state.

Today it is impossible to make inductances or capacitors with large values in monolithic integrated circuits. Capacitances larger than several tens of picofarads are rare. Thus, one often finds dc coupled rather than *RC* coupled integrated circuit amplifiers. Special circuits are often used to avoid the necessity of large bypass capacitors.

Tuned circuits are also difficult to produce in integrated circuits, and thus any tuned *LC* tanks, for example, are usually added external to the I.C. using discrete components. It is also modern design philosophy to minimize the number of interconnections between the various I.C.'s comprising the circuit, even to the extent of using more complicated circuits.

13.2 INTEGRATED CIRCUIT CONSTRUCTION

At this writing most I.C.'s are made from silicon. A small, carefully selected, ultra pure (<0.01 parts per million impurities typically) silicon single crystal is used as a seed crystal to grow a larger ultra pure silicon single crystal. The seed is dipped into p-type molten silicon and slowly raised up into a cooler region. A large p-type silicon crystal about one inch diameter and six inches long can be grown in this way. This large crystal is sawed into "wafers" approximately 0.010 inches thick, which are polished and chemically etched to form a very smooth, chemically clean silicon surface. The p-type wafers form the "substrate" and are typically 0.005–0.010 inches thick. A thin layer (0.001 inches) of n-type silicon is grown on top of the p-type substrate; this n-type layer is "epitaxial," which means it continues the crystal structure of the substrate. The epitaxial n-type layer eventually forms the collectors of the npn transistors in the I.C. The wafer is then oxidized at very high temperatures (about 1000 °C) in a furnace to form a thin insulating layer of SiO_2 covering the entire epitaxial n-type surface of the wafer. In some processes, one starts with an n-type substrate and grows an epitaxial p-type layer on it.

A complicated series of photoetching and diffusing processes follows to build up the desired n–p–n transistor structure, the p–n diode structure, and the appropriate resistances. If it is desired to remove certain areas of the SiO_2 surface layer and to leave other areas, the entire wafer is covered with a uniformly thin film of photosensitive emulsion, called "photo resist." The wafer is illuminated with ultraviolet light through a precisely made mask or stencil, which exposes only certain areas. The exposed areas of the photo-resist layer become polymerized while the unexposed areas remain unpolymerized. A chemical solvent is then used which dissolves away only the unpolymerized photo resist. Thus we are left with only certain areas of the SiO_2 layer covered with polymerized photo resist. The entire wafer is then etched in hydrofluoric acid to etch away the exposed SiO_2, exposing the substrate, but not attacking the SiO_2 covered by the polymerized photo resist. The SiO_2 remaining acts as a mask or stencil for diffusing n-type impurities part way into the exposed p-type substrate, the impurities diffusing much more readily into the exposed silicon than into the SiO_2.

A similar series of photoetching processes is used to build up the desired circuit, which may consist of dozens of transistors, diodes, resistors, and perhaps one or two capacitors. The resistances are formed from strips of doped silicon etched to the proper width and thickness to yield the desired resistance. It is difficult to manufacture high resistances by such techniques, and few I.C.'s contain resistances greater than 100 kΩ. Small capacitances on the order of 1–10 pF can be made with SiO_2 as the dielectric and a vapor deposited layer of aluminum as one of the plates. The final network of "wires" connecting the components with one another is formed by vacuum depositing a thin layer of aluminum over the entire wafer, and etching away the undesired areas.

The completed I.C. is cut away from the wafer and mounted in some sort of case with wire leads. Three cases are common (see Fig. 13.2): a hermetically

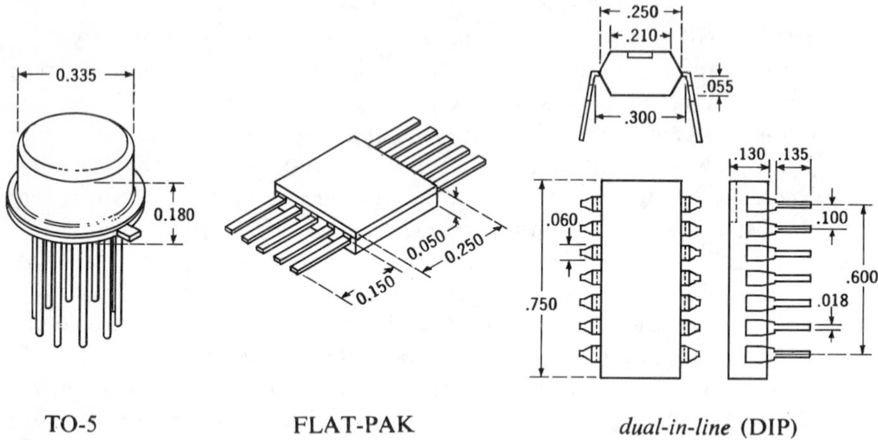

FIGURE 13.2 Three common integrated circuit cases.

sealed metal case (T0-5, etc.) which looks somewhat like an individual transistor except for the greater number of leads, the "flat-pak" which looks like a small rectangular solid with leads protruding from two sides, and the dual-in-line package, which is perhaps the most common. Great care is taken to form a good electrical and mechanical connection between the small I.C. aluminum conductors and the wire leads which are typically 0.019-inch-diameter gold plated kovar, which is an alloy of iron, nickel and cobalt having a low coefficient of thermal expansion.

13.3 SPECIAL DESIGN FEATURES OF INTEGRATED CIRCUITS

It should be emphasized that an I.C. is really nothing new; it is merely a collection of transistors, resistors, and diodes, etc., in a miniature form. However, the manufacturing techniques necessary to make the I.C. impose some constraints on circuit design which are not present when one is working with discrete components. Thus the circuit diagram of an I.C. may look, at first glance, rather different from a circuit made from discrete components. For example, the absolute value of resistance in an I.C. cannot be controlled to much better than 20%, large values ($>100\,\text{k}\Omega$) of resistances require special techniques and are rare, and large values of capacitance (>30 pF) are impractical.

The absence of large capacitors in an I.C. makes it impossible to stabilize a common emitter bipolar transistor amplifier stage with a large bypass capacitor across the emitter resistor. In an I.C. it is much easier to make several additional transistors and diodes to stabilize a common emitter amplifier stage than to make one large capacitor. Hence a basically simple common emitter amplifier may look rather complicated in an I.C. Advantage is usually taken in the I.C. design of the tight thermal coupling between the various components on the chip; that is, all the transistors and resistors will be locked at essentially the same temperature by virtue of their being constructed from the same semiconductor chip. This property is used in the stabilization scheme of Fig. 13.3(a).

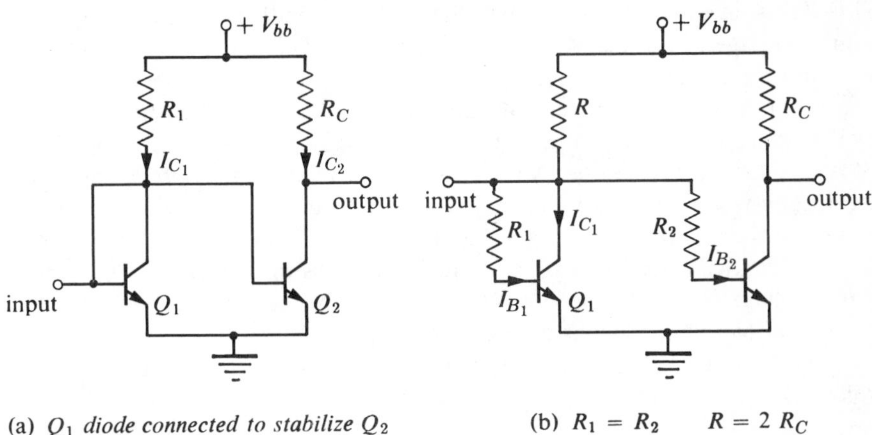

(a) Q_1 *diode connected to stabilize* Q_2 (b) $R_1 = R_2 \quad R = 2\,R_C$

FIGURE 13.3 I.C. bias stabilization circuits.

Q_1 and Q_2 are two npn transistors on the same chip, but Q_1 is connected as a diode, its base being connected to its emitter. Q_2 is the amplifier transistor in the common–emitter configuration. Both Q_1 and Q_2 are made from the same chip and are at the same temperature, their base-emitter junctions connected in parallel. Thus, Q_1 and Q_2 will have the same base–emitter voltage and consequently the same collector current, which can be set to the desired value by choice of R_1, since $I_{C_1} = (V_{bb} - V_{BE})/R_1$. If the temperature of the chip increases, thereby tending to increase I_{C_2}, V_{BE_1} decreases and thus decreases the forward-bias V_{BE_2}, and thereby stabilizes I_{C_2}.

Another bias stabilization technique is shown in Fig. 13.3(b). Resistance R_1 equals R_2, and thus the base of Q_1 is at the same voltage as the base of Q_2. A Kirchhoff voltage equation yields:

$$V_{bb} = (I_{C_1} + I_{B_1} + I_{B_2})R + I_{B_1}R_1 + V_{BE_1} \tag{13.1}$$

or

$$I_{C_1} = \frac{V_{bb} - V_{BE_1}}{R} - \left(1 + \frac{R_1}{R}\right) I_{B_1} - I_{B_2}.$$

If the transistors are identical, $I_{B_1} = I_{B_2}$ and thus $I_{C_1} = (V_{bb} - V_{BE_1})/R - (2 + R_1/R)I_B$. Usually $V_{bb} \gg V_{BE_1}$ and the I_B term can be neglected if R_1/R is not too large. Thus the collector current of Q_1, or Q_2, is given by:

$$I_{C_1} = I_{C_2} \cong V_{bb}/R. \tag{13.2}$$

Thus the operating point of both transistors can be set by adjusting V_{bb} and R. The dc collector voltage of Q_2 can be varied by adjusting R, and is given by:

$$V_{C_2} = V_{bb} - I_C R_C = V_{bb}(1 - R_C/R). \tag{13.3}$$

R_C is usually chosen to achieve the desired gain and output impedance for Q_2. Thus if R is set equal to $2R_C$, we have the desirable condition that the collector-emitter voltage of Q_2 equals one-half of the supply voltage: $V_{C_2} = V_{bb}/2$.

Another difficulty encountered in designing I.C.'s is that constant current sources to deliver small currents of the order of microamperes cannot be made by the usual technique of using a large series resistance of the order of megohms, because such high resistance values cannot be made easily by modern I.C. fabrication techniques. The technique used is to generate a very small stable voltage of the order of millivolts using the difference in the base–emitter voltage for two identical transistors carrying different collector currents. This small, stable voltage is then applied across a resistance of the order of 1 kΩ–10 kΩ to produce a microampere constant current.

The relationship between the collector current I_C and the base–emitter voltage V_{BE} is given by:

$$I_C = I_0 e^{qV_{BE}/kT} \qquad \text{or} \qquad V_{BE} = \frac{kT}{q} \ln \frac{I_C}{I_0}, \tag{13.4}$$

where q is the electronic charge 1.6×10^{-19} coulombs, k is Boltzmann's constant 1.38×10^{-23} joules/degree, and T is the temperature of the transistor semiconductor material in degrees absolute or Kelvin. I_0 is the collector current which flows when the base is shorted to the emitter, i.e., when $V_{BE} = 0\text{V}$. In integrated circuits, I_0 is essentially constant for different transistors on the same chip. Thus, for two transistors operating at different collector currents I_{C_1} and I_{C_2}, the difference in their base–emitter voltages will be given by:

$$\Delta V_{BE} = V_{BE_2} - V_{BE_1} = \frac{kT}{q} \ln \frac{I_{C_2}}{I_{C_1}}. \tag{13.5}$$

The disadvantage of this technique is that the small voltage generated is linearly dependent upon the absolute temperature T. Thus a change in temperature from $20°\text{C} = 293°\text{K}$ to $30°\text{C} = 303°\text{K}$ would change V_{BE} by about $(303 - 293)/293 = 3.3\%$.

There are many other circuit configurations which are more appropriate to I.C.'s than to circuits made from discrete components. For example, shifting the dc level of a signal from one stage to another cannot be done with large series coupling capacitors as is done in circuits using discrete components, because in I.C.'s, such large capacitors are impractical.

13.4 THE μA709 OPERATIONAL AMPLIFIER

As we have discussed in Chapter 9, operational amplifiers are dc coupled high gain amplifiers which can be made to perform a number of mathematical operations: multiplication, division, differentiation, integration, addition, and subtraction. One of the most widely used integrated circuit operational amplifiers is the high gain, general purpose μA709 made by Fairchild and costing approximately a dollar each in small quantities. The circuit diagram and lead configuration for the TO-99 case and the flat pak are shown in Fig. 13.4, and several things are immediately apparent.

First, there are fourteen transistors and fifteen resistors in one 709! A quick back-of-the-envelope calculation shows that a comparable amplifier made from discrete components would cost of the order of \$20 or more for parts alone. Second, there are *no* capacitors in the circuit, so the amplifier is dc coupled. Third, there are two inputs, one inverting and one non-inverting. For the inverting input, the output is 180° out of phase with respect to the input; for the non-inverting input, the output is in phase with respect to the input. If a single input is used, the other input is grounded. Fourth, normally two supply voltages are needed, one positive and one negative with respect to ground. Normally ± 15V supplies are used, but other voltages can be used—from ± 6V to ± 18V. The two supply voltages need not be equal so long as the total voltage drop from $+V$ to $-V$ is from 18V to 30V. If the 709 is used as a differential amplifier with both the inverting and non-inverting inputs used, there must be a dc connection from the input to ground.

Perhaps the most important practical feature of the 709 operational amplifier is that external frequency compensation in the form of *RC* networks must be added to prevent unwanted high-frequency oscillations. The general frequency compensation schematic is shown in Fig. 13.5. High-gain amplifiers tend to oscillate as anyone who has ever built one "from scratch" will testify. The basic reason is that as the amplifier gain falls off at high frequencies the phase of its output is shifted with respect to its input. If the output shifts 180° in phase, then the negative feedback which stabilizes the amplifier at low frequencies will have changed to *positive* feedback, and voila: oscillations will occur if the amplifier loop gain $A_0\beta$ is greater than 1.0 at this frequency. The cure for this unwanted oscillation is to make sure that the loop gain is less than 1.0 when the output has shifted 180° in phase. In amplifiers made with discrete components, fairly large capacitors can be added so that the voltage gain falls off at a rate near 6dB/octave = 20dB/decade, which corresponds to a maximum phase shift of 90° as the frequency approaches infinity. Such a circuit,

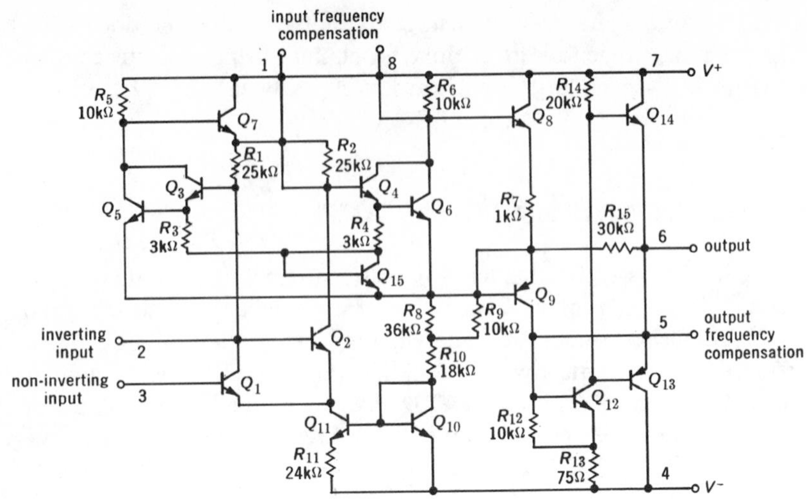

(a) *schematic diagram*

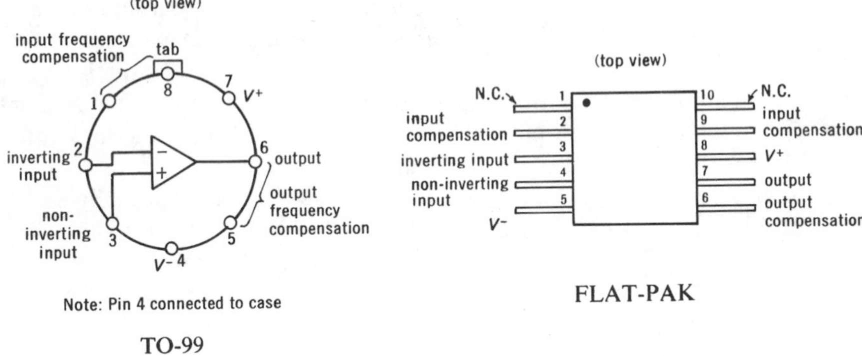

(b) *connection diagram*

FIGURE 13.4 μA709 I.C. operational amplifier. (Courtesy of Fairchild Semiconductor Components Group, Fairchild Camera and Instrument Corp.)

therefore, cannot oscillate, because the phase shift can never reach 180°. The appropriate compensation for the μA709 is shown in Fig. 13.5. Notice that for more negative feedback (higher β), i.e., for lower overall amplifier gain $1/\beta$, larger capacitances are required. There are literally dozens of other graphs available for the μA709 in the manufacturer's data sheets: gain versus supply voltage and temperature, etc.

However, in integrated circuits, such large capacitors cannot be incorporated into the chip, and the gain falls off at a rate determined by the internal capacitances. Above 2 MHz, the μA709 gain falls off at approximately 18 dB/

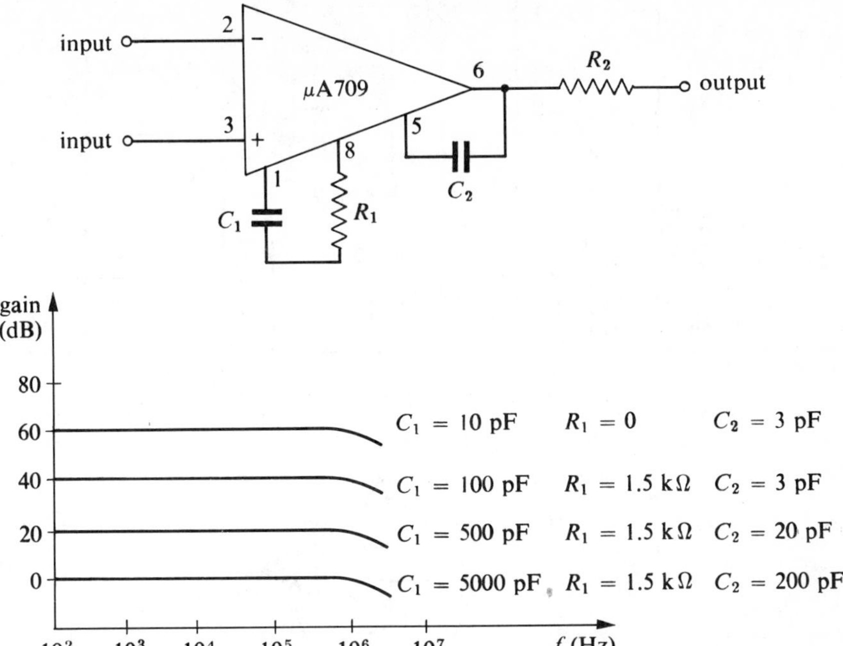

FIGURE 13.5 External frequency compensation for a μA709 I.C.

octave. One must add external components (usually R and C) to make the loop gain fall off at a modest 6dB/octave. As the loop gain falls off at high frequencies, the operational amplifier becomes less "ideal"; its input impedance falls, its output impedance rises, and its stability fails.

An I.C. operational amplifier which has an inherent 6dB/octave gain roll-off and which therefore does not require external frequency compensation is the Fairchild μA741. Its schematic diagram is shown in Fig. 13.6. It has the same pin configuration as the μA709, but it has a lower bandwidth than the μA709. Notice that it has one 30pF capacitor in the circuit, whereas the 709 has none. This 30pF capacitor serves to roll off the gain at 6dB/octave and thereby stabilize the amplifier. A photograph of the 741 chip shows that this single 30pF capacitor occupies approximately 25% of the chip area, which is dramatic evidence of why I.C.'s so rarely contain capacitors. It is often cheaper to use a circuit containing several dozen more transistors, diodes, and resistors just to avoid one capacitor.

A simple μA709 single-ended, operational amplifier together with measured characteristics is shown in Fig. 13.7. A simple differentiator made with a μA709 is shown in Fig. 13.8 along with a sketch of the waveform for a step function input. The capacitor used must be high quality.

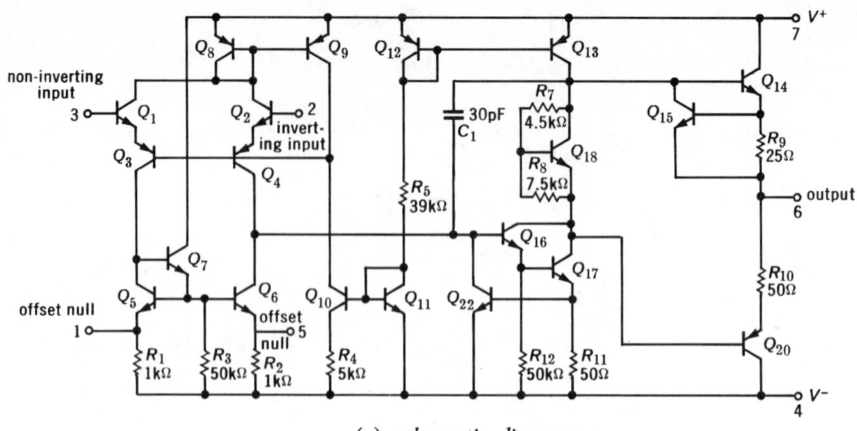

(a) *schematic diagram*

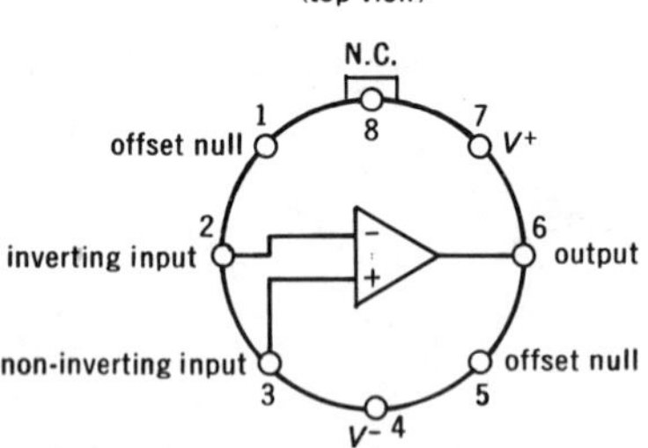

(b) *connection diagram*

FIGURE 13.6 μA741 I.C. operational amplifier. (Courtesy of Fairchild Semiconductor Components Group, Fairchild Camera and Instrument Corp.)

13.5 I.C. RF AMPLIFIER

The Fairchild μA703 I.C. amplifier can be used at frequencies up to 200 MHz with the addition of external tuned circuits. Figure 13.9 shows the μA703 schematic diagram and its pin diagram. Notice that Q_3 and Q_4 form a simple differential amplifier pair. Transistor Q_5 forms a constant current source for Q_3 and Q_4, and Q_2 stabilizes the collector current of Q_5. The power gain available at 30 MHz from the amplifier of Fig. 13.10(a) is about 35 dB with a noise figure of 6 dB. The bandwidth depends upon the Q of the external tuned circuitry but is of the order of 1 MHz.

The 100 MHz amplifier circuit of Fig. 13.10(b) differs obviously from the 30 MHz circuit in the smaller capacitances and inductances used in the external LC tuned circuits, but also in the presence of a 0.001 μF bypass capacitor from pin 5 to ground. This capacitor is necessary to ensure that the diode-connected

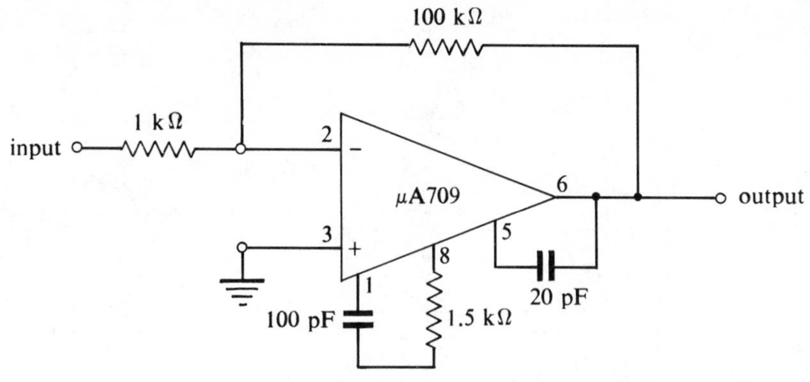

Measured Circuit Performance

1 kHz voltage gain	100
1 kHz input impedance	700 kΩ
1 kHz output impedance	150 Ω
upper 3 dB frequency	500 kHz

FIGURE 13.7 μA709 amplifier.

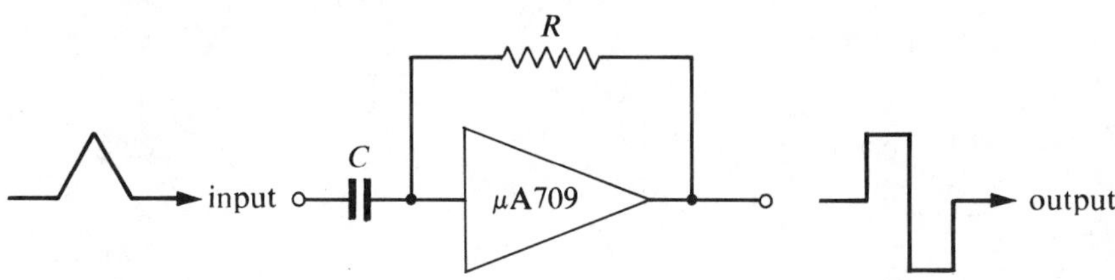

FIGURE 13.8 μA709 differentiator.

Q_1 and Q_2 are a true ac short at 100 MHz, in order to make the base of Q_4 a true 100 MHz ground. The power gain at 100 MHz is about 20 dB with a 6 dB noise figure, and a bandwidth of the order of 5 MHz. In both circuits, pin 1 must be connected to ground with a bypass capacitor to decouple the power supply, i.e., to make sure the power supply is a good ac ground at the frequency of operation.

With care, the μA703 can even be used at 200 MHz, yielding a power gain of about 14 dB. Note that the power gain goes down substantially as the frequency of operation is increased. This behavior is common to most high-frequency amplifiers, because of the increased bypassing of high frequencies by the fixed shunt capacitances inherent in the circuit.

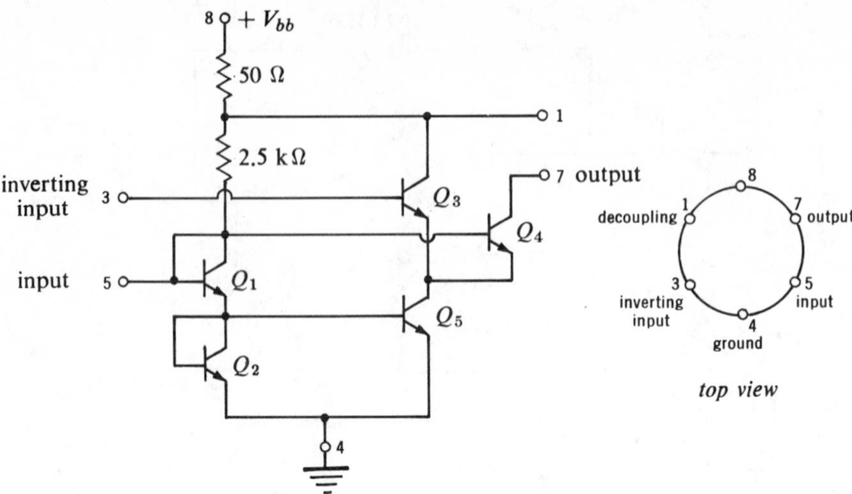

FIGURE 13.9 μA703 I.C. rf amplifier. (Courtesy of Fairchild Semiconductor Components Group, Fairchild Camera and Instrument Corp.)

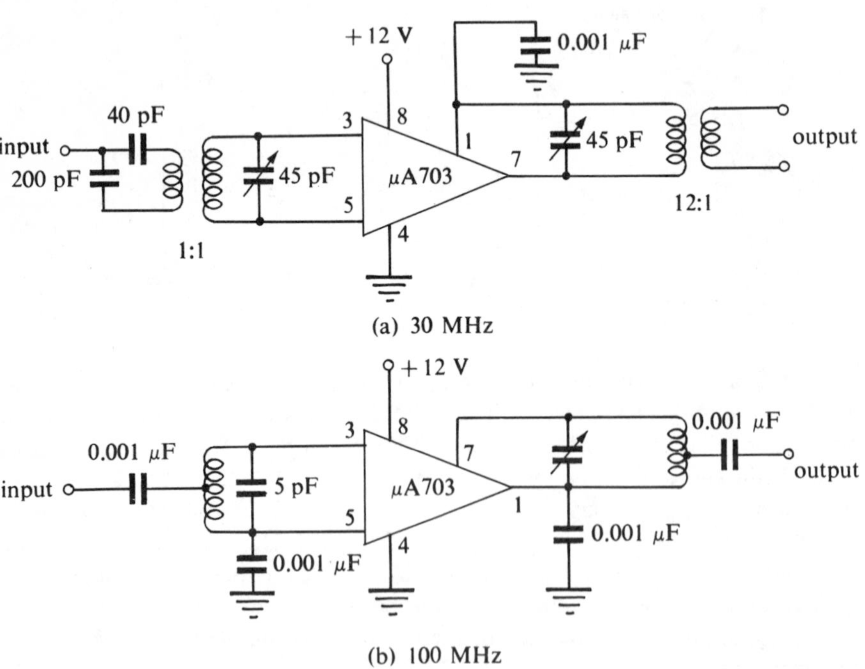

FIGURE 13.10 μA703 amplifier circuits. (Courtesy of Fairchild Semiconductor Components Group, Fairchild Camera and Instrument Corp.)

13.6 AN I.C. TTL NAND GATE

One of the simplest logic circuits is the NAND gate as discussed in Chapter 12 and shown in Fig. 13.11(a). A modern I.C. TTL ("transistor-transistor logic") NAND gate is the Texas Instruments SN7400, which contains four separate two-input positive logic NAND gates. A few general comments are in order to

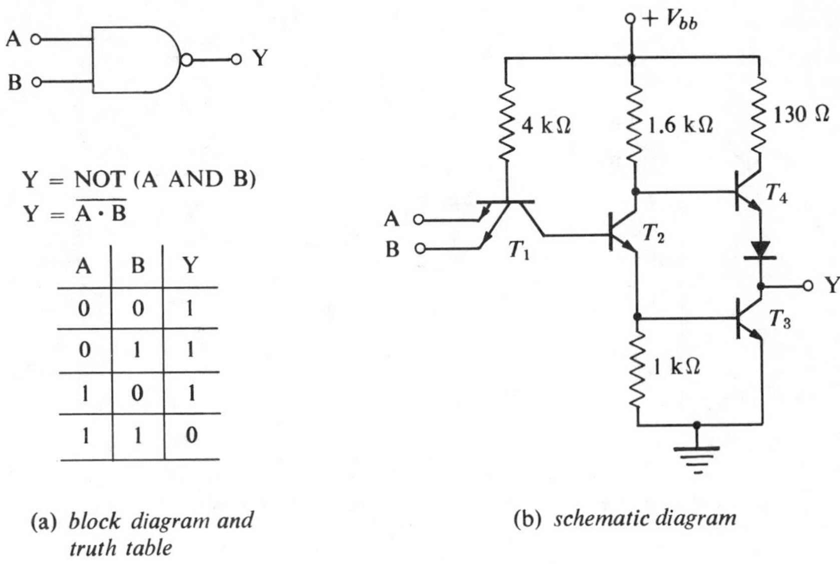

A	B	Y
0	0	1
0	1	1
1	0	1
1	1	0

(a) *block diagram and truth table*

(b) *schematic diagram*

FIGURE 13.11 SN7400 NAND gate. (Courtesy of Texas Instruments Inc.)

clarify the I.C. nomenclature. The first word "dual" or "triple" or "quadruple" or "hex" merely refers to the number of separate gates or circuits in the package. The word "positive" or "negative" refers to the type of logic used. Thus, a "dual two-input positive NAND gate" contains two positive logic NAND gates each with two inputs. A "quadruple two-input positive NAND gate" contains *four* positive logic NAND gates in the package. The schematic for one two-input NAND gate is shown in Fig. 13.11(b), and has one noteworthy feature: the input transistor T_1 has two emitters. Each emitter must be driven positive in order to raise the base voltage of T_2 and turn T_2 on, which turns T_3 on and T_4 off and causes the output voltage at Y to drop to the logical 0 level of approximately 0.1V. If one or both of the inputs are at logical 0, then T_2 is biased off and T_3 is off and T_4 on, holding Y up at the logical 1 level of approximately +4V.

13.7 AN I.C. TTL NOR GATE

Another simple logic circuit is the NOR gate as discussed in Chapter 12 and shown in Fig. 13.12(a). A modern I.C. TTL NOR gate is the Texas Instruments SN7402 which contains four separate two-input NOR gates designed to work on positive logic. The schematic for one two-input NOR gate is shown

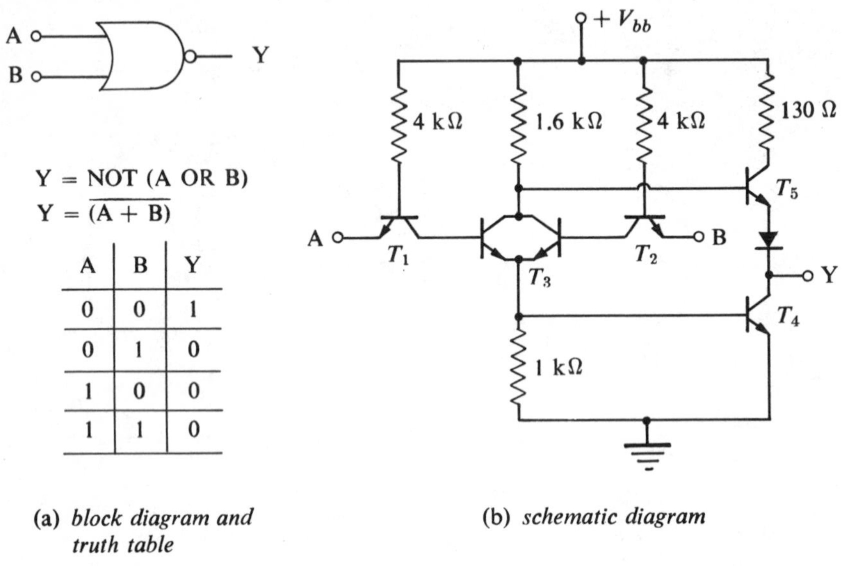

A	B	Y
0	0	1
0	1	0
1	0	0
1	1	0

(a) *block diagram and truth table* (b) *schematic diagram*

FIGURE 13.12 SN7402 NOR gate. (Courtesy of Texas Instruments Inc.)

in Fig. 13.11(b). Notice that the transistor T_3 has two bases. If either input A or B is at logical 1 level, then T_3 is turned on, which turns T_4 on and T_5 off, thus causing the output voltage at Y to drop to the logical 0 level. If both inputs are at logical 0 level, then T_3 is normally biased off, which turns T_4 off and T_5 on and which raises the output voltage at Y to the logical 1 level.

13.8 AN I.C. TTL INVERTER

It is often desired to change a logical 1 to a logical 0 and vice versa, i.e., to "invert" the input. One such inverter is the TI SN7404 shown in Fig. 13.13. The SN7404 contains six separate inverters and is thus called a "hex inverter." With 0 input, T_2 is biased off, so that T_4 is conducting and thus Y is near V_{cc}, i.e., at the logical 1 level. With a 1 input at A, T_2 is turned on, T_4 off, and T_3 on, thus creating a logical 0 at the output.

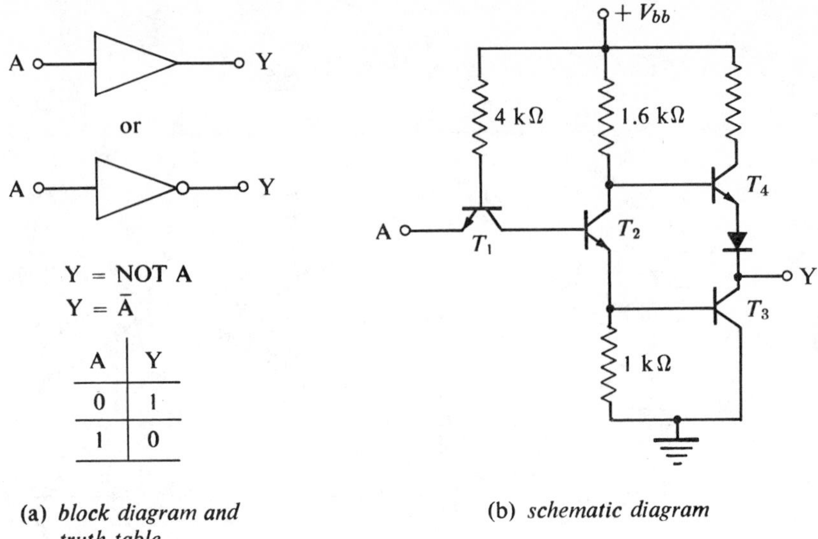

(a) *block diagram and truth table*

(b) *schematic diagram*

FIGURE 13.13 Inverter. (Courtesy of Texas Instruments Inc.)

13.9 I.C. TTL FLIP-FLOPS

There are two general types of flip-flops commonly used, the "RS" and the "JK" flip-flop. Both types have set and reset input, a clock or toggle input, and two inputs called R and S, or J and K. The symbol Q denotes the logical level of the collector of one transistor in the flip-flop; the other collector must therefore be $\bar{Q}$. In the RS flip-flop, the R and S inputs (logic levels) determine whether Q = 0 or Q = 1, i.e., determine which transistor is turned on by the clock pulse. In the commonly used "edge-triggered" flip-flops, the output level Q does not change until the trailing edge of the clock pulse. In other words, the output changes in Q are "clocked" by the pulse fed into the clock input, or the R and S inputs are "synchronous." The set or preset and the clear or reset inputs are asynchronous and work essentially instantaneously; a 1 set input immediately makes Q = 1 and a 1 reset or clear input immediately makes Q = 0 regardless of the clock input. A base, edge-triggered clock RS flip-flop is shown in Fig. 13.14 along with the truth table. The set and clear inputs and the cross coupling networks have been omitted for clarity.

"Edge" triggering is achieved by differentiating the input clock pulse with the C_3R_3 network. The diodes D_1 and D_2 pass only the negative derivative spikes, but only if the dc voltages of points P_1 and P_2 are low enough. The diodes D_S and D_R clip off the positive derivative spikes. Thus, a logical 1 input (+5 V) at S will raise the dc voltage at P_1 and reverse bias D_1 so that the negative spike cannot pass through diode D_1 to the base of T_1. A logical 0 input at R

R	S	Q
0	0	indeterminate
0	1	1
1	0	0
1	1	no change

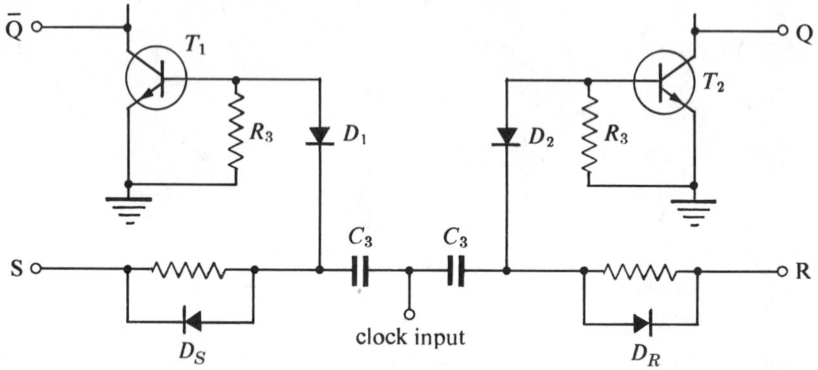

FIGURE 13.14 Clocked RS flip-flop.

will allow the negative spike to pass through diode D_2 and turn off T_2, thus making Q = 1. Similarly a 0 input at S and a 1 input at R will make Q = 0 at the time of the trailing edge of the clock pulse. If both R and S inputs are at logic 0, then the negative spikes from the differentiated edge of the clock pulse are applied to the bases of *both* transistors T_1 and T_2, and the resulting state depends upon the asymmetry of the circuit. Either transistor may be on. If both R and S inputs are at logic 1, the negative spikes are prevented from reaching the base of *either* transistor, because both diodes D_1 and D_2 are reverse biased, and the state of the flip-flop does not change.

The R and S inputs are often multiple inputs to an AND gate as shown in Fig. 13.15. In order for an S input to be fed to the flip-flop, there must be inputs S_1, S_2 and S_3 simultaneously.

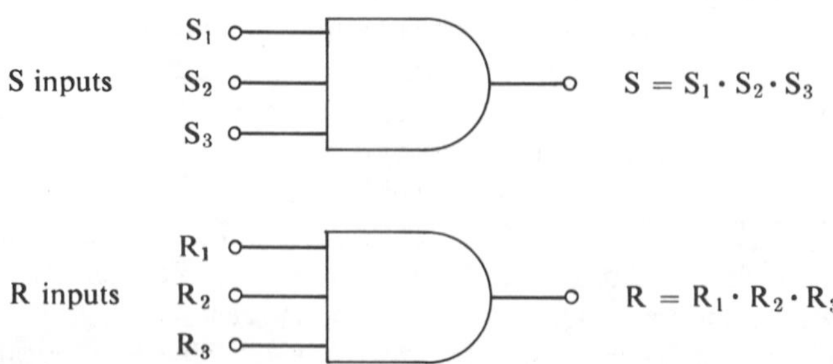

FIGURE 13.15 Multiple R and S inputs.

The JK flip-flop is slightly more versatile than the RS flip-flop and differs in having the Q and $\overline{Q}$ outputs connected to the inputs as shown in Fig. 13.16.

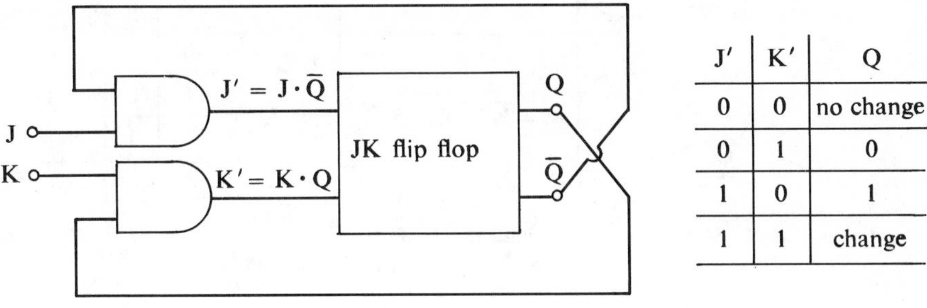

J′	K′	Q
0	0	no change
0	1	0
1	0	1
1	1	change

J	K	Q before clock pulse	J′	K′	Q after clock pulse
1	1	1	0	1	0
1	1	0	1	0	1

FIGURE 13.16 JK flip-flop and truth table.

The truth table for the JK flip-flop in Fig. 13.16 shows that for J = 0 and K = 1, Q = 0, and that for J = 1 and K = 0, Q = 1. Thus J is analogous to R and K is analogous to S in the RS flip-flop of Fig. 13.14. However, there is no ambiguous outcome for the JK flip-flop as there is for the RS flip-flop. If J and K are both at logical 0, then the fact that Q is one of the K inputs and $\overline{Q}$ one of the J inputs requires that no J′ or K′ input signal is fed to the flip-flop (J′ = 0 and K′ = 0); hence there is no change in the condition of the flip-flop. Recall that in the case of the RS flip-flop, when $R = 0$ and $S = 0$, the state of the flip-flop was ambiguous and determined solely by the asymmetry of the circuit. For J = 1 and K = 1, the flip-flop must change its state, e.g., if Q = 0 and $\overline{Q}$ = 1, then J′ = 1, K′ = 0, which makes Q = 1. If Q = 1 and $\overline{Q}$ = 0, then J′ = 0, K′ = 1, which makes Q = 0.

The Texas Instruments SN7470 JK flip-flop is a modern monolithic integrated circuit, bistable multivibrator, or flip-flop, capable of flipping at frequencies up to 35 MHz. It costs less than a dollar in small quantities. Such high speed operation is almost impossible to obtain and much more expensive with a multivibrator made from discrete components. A block diagram is shown in Fig. 13.17 along with a pin diagram for the dual-in-line package. The schematic is considerably different and more complicated than that for the two-transistor, bistable multivibrator in Chapter 12. There is one clock input, one clear input, and one preset input. The 7470 works on positive logic. A logical 0 or low input to "clear" sets Q to logic 0; a low input to "preset" sets Q to logic 1. The 7470 is an "edge-triggered" flip-flop; that is, it switches

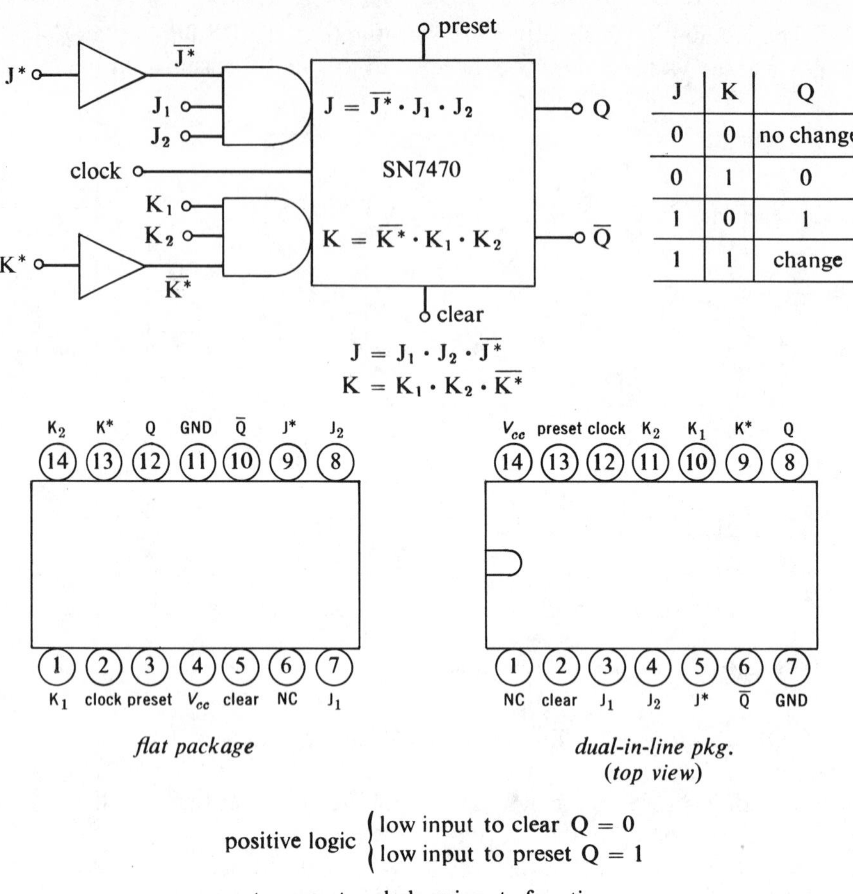

FIGURE 13.17 SN7470 flip-flop. (Courtesy of Texas Instruments Inc.)

from one state to the other on the negative-going *edge* of the clock pulse as is shown in Fig. 13.18. Notice that the $J^\star$ and $K^\star$ inputs are followed by inverting amplifiers. Thus a $J^\star = 1$ input produces a $\overline{J^\star} = 0$ input to the J AND gate. In order to get $J = 1$, we must have $J_1 = 1$ and $J_2 = 1$ and $J^\star = 0$, and similarly for the K input.

The use of an SN7470 JK flip-flop as a divide-by-two counter is shown in Fig. 13.19. To divide the clock frequency by 2, we desire the output of the flip-flop, Q, to change for each clock input pulse. Thus, we want $J = 1$ and $K = 1$ for all clock input pulses. Because $J = J_1 \cdot J_2 \cdot \overline{J^\star}$, we must have $J_1 = 1$ and $J_2 = 1$ and $J^\star = 0$. Thus we must keep J_1 and J_2 at the logical 1 voltage level of +4V, and $J^\star$ must be grounded. Similarly, to make $K = 1$, we must keep K_1 and K_2 at the logical 1 level and $K^\star$ at ground. The flip-flop

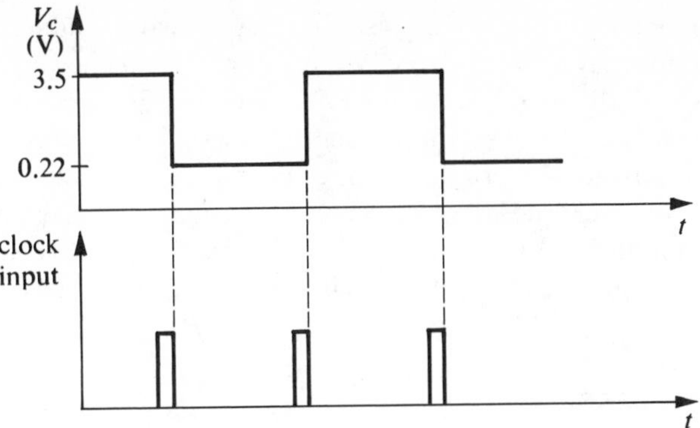

FIGURE 13.18 Clock and output waveform for the JK flip-flop.

may be cleared, i.e., Q = 0, by momentarily grounding the clear input by momentarily closing switch *S*.

Often flip-flops are used in pairs; one is called the "master" and the other the "slave." The usual master–slave flip-flop operation is fourfold: (1) Decouple slave flip-flop from master flip-flop, (2) enter information into master, (3) disconnect the inputs from the master, and (4) transfer information from master to slave. The advantage of a master–slave flip-flop is that no coupling capacitors are required.

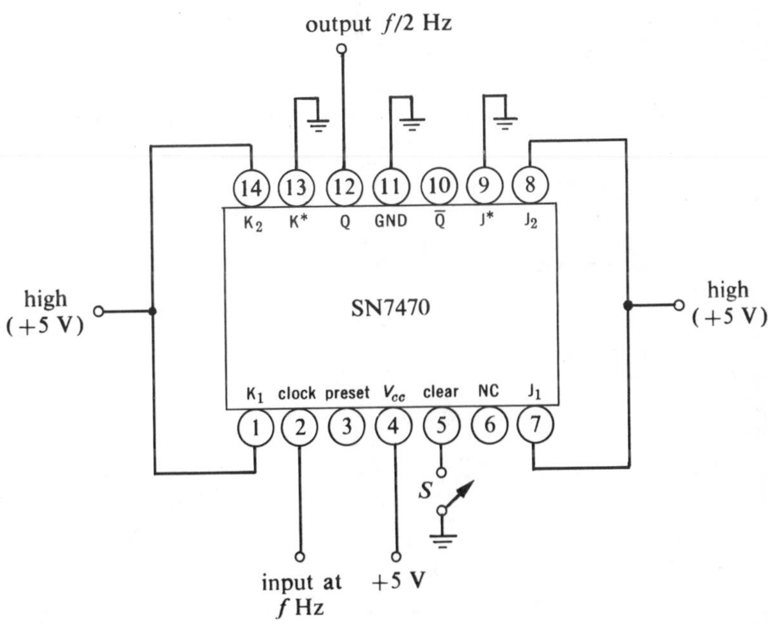

FIGURE 13.19 SN7470 connected as a divide-by-two counter. (Courtesy of Texas Instruments Inc.)

One common use for flip-flops is to make a decade counter in which an input clock frequency of f Hz produces an output frequency of $f/10$ Hz. A single flip-flop will divide the input clock frequency by 2, two flip-flops in series will divide by 4, etc. To obtain a frequency division by 10, we must introduce logical feedback around several flip-flops to produce a $\div 2$ section driving a $\div 5$ section. A commercial decade counter using four master–slave flip-flops and capable of counting up to 18 MHz input clock frequency is the TI SN7490 whose block diagram is shown in Fig. 13.20. The reset connections have been omitted for simplicity.

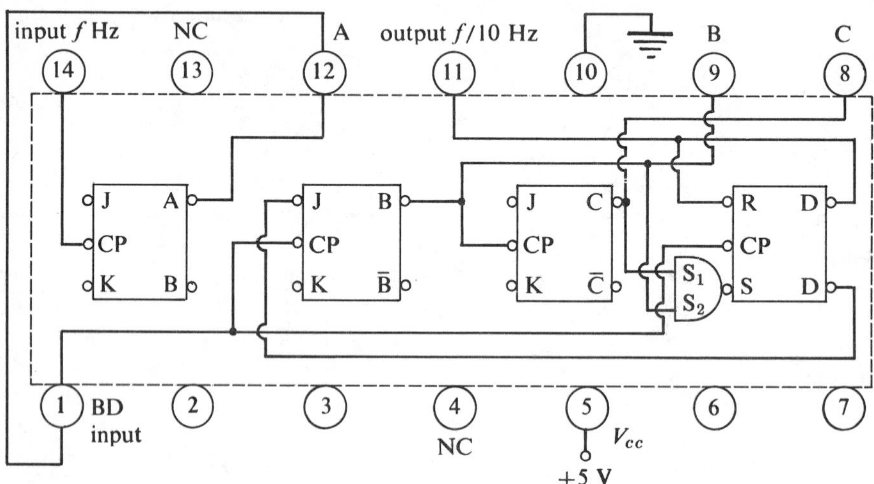

FIGURE 13.20 I.C. SN7490 decade counter. (Courtesy of Texas Instruments Inc.)

The first flip-flop has outputs A, $\overline{A}$, the second B, $\overline{B}$, the third, C, $\overline{C}$, and the fourth D, $\overline{D}$. In the SN7490 the divide-by-two stage flip-flop A is not

TABLE 13-1 SN7470 DECADE COUNTER TRUTH TABLES

Count	D	C	B	A
		Count Sequence		
0	0	0	0	0
1	0	0	0	1
2	0	0	1	0
3	0	0	1	1
4	0	1	0	0
5	0	1	0	1
6	0	1	1	0
7	0	1	1	1
8	1	0	0	0
9	1	0	0	1

internally connected to the divide-by-five stage flip-flop B, C, and D. Hence to use as a decade counter, the output of the first flip-flop, A, at pin 12 must be externally connected to the clock pulse input of the second flip-flop, CP, at pin 1. The count sequence is shown in Table 13-1.

13.10 AN I.C. TTL MONOSTABLE MULTIVIBRATOR

Integrated circuit monostable multivibrators are available in which one supplies an external timing capacitor to determine the output pulse width. One such I.C. is the Texas Instruments SN74121 monostable multivibrator, which will produce an output pulse of from 30 nsec to 40 sec width and 3.3 V amplitude triggered by a particular voltage level at the input. The hysteresis or backlash (Chapter 12) is 0.2 V or less.

A monostable multivibrator can also be made from a linear integrated circuit, the "differential voltage comparator," which is a high gain amplifier with differential input and single-ended output. One such circuit is the Fairchild μA710 high speed comparator whose circuit is shown in Fig. 13.21(a). A monostable multivibrator can be made with the μA710 using the circuit of Fig. 13.21(b). The circuit will trigger whenever the input pulse is more negative than V_{ref}.

13.11 DESIGN EXAMPLE USING I.C.'s

The availability of many different types of linear and logic integrated circuits has revolutionized electronic circuit design in the last five years. Almost no one builds a complicated circuit out of discrete components anymore. Not only is apparatus made from integrated circuits easier to construct and cheaper, but it usually works better than that made from discrete components. At the present time, discrete component circuits are still to be preferred in some cases; for example, when an exceptionally low noise figure is desired or when a very large bandwidth is required.

To illustrate the modern way of thinking about circuit design, we will consider the problem of designing a single-channel pulse height analyzer. We recall from Chapter 12 that a single-channel analyzer with a base line set at V_0 volts and a window width of ΔV volts is essentially a gate which will pass only pulses with amplitudes between V_0 and $V_0 + \Delta V$. Thinking in terms of I.C. amplifiers, level detectors or Schmitt trigger circuits, and gates, one is led rather quickly to the block diagram of Fig. 13.22 for a single-channel analyzer.

The first stage of the low noise pulse amplifier may perhaps be a discrete component FET amplifier to obtain a low noise figure. The LLD and the ULD are the "low-level discriminator" and "upper-level discriminator," respectively.

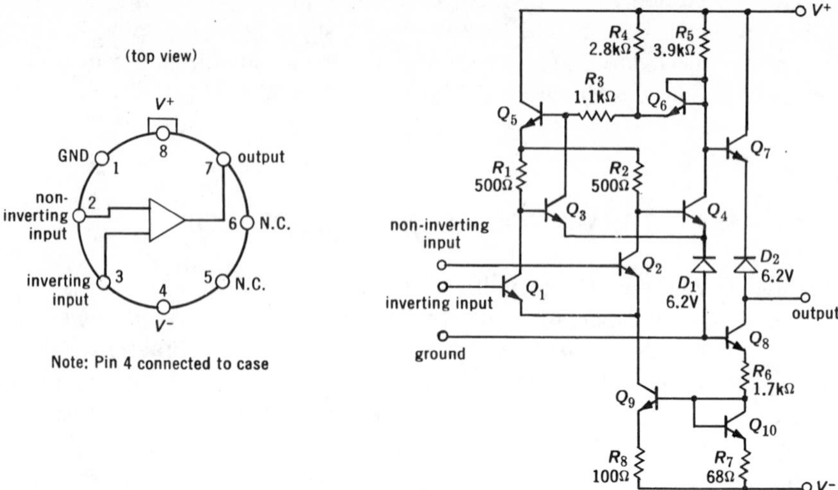

(a) *schematic diagram*

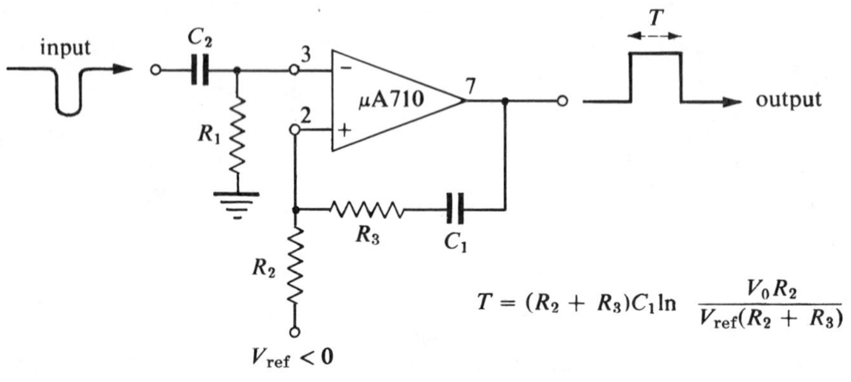

$$T = (R_2 + R_3)C_1 \ln \frac{V_0 R_2}{V_{\text{ref}}(R_2 + R_3)}$$

(b) *monostable multivibrator*

FIGURE 13.21 μA710 high speed comparator. (Courtesy of Fairchild Semiconductor Components Group, Fairchild Camera and Instrument Corp.)

They produce output pulses only when their inputs are greater than V_0 and $(V_0 + \Delta V)$, respectively. The LLD and ULD could each be an integrated circuit level discriminator, or a Schmitt trigger circuit such as the Texas Instruments SN74121 mentioned earlier. The truth table for the single-channel analyzer is shown in Fig. 13.22(b). An input pulse below V_0 will trigger neither the LLD nor the ULD, thus A = 0 and B = 0. Thus $\overline{\text{B}} = 1$ and the final AND gate has a 0 and a 1 input, yielding $\text{Y} = \text{A} \cdot \overline{\text{B}} = 0 \cdot 1 = 0$. An input pulse between V_0 and $V_0 + \Delta V$ will trigger the LLD making A = 1, but *not* the ULD making

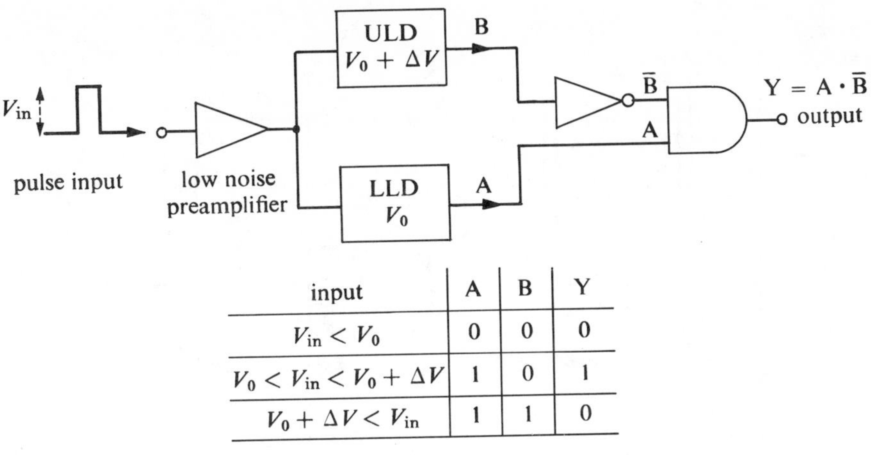

input	A	B	Y
$V_{in} < V_0$	0	0	0
$V_0 < V_{in} < V_0 + \Delta V$	1	0	1
$V_0 + \Delta V < V_{in}$	1	1	0

FIGURE 13.22 Single-channel pulse height analyzer.

B = 0. Thus $\overline{B}$ = 1, and the analyzer output is Y = A · $\overline{B}$ = 1 · 1 = 1 signifying a pulse in the channel. An input pulse above $V_0 + \Delta V$ in amplitude will trigger both the LLD and the ULD, making A = 1 and B = 1. Thus, $\overline{B}$ = 0, and the output Y = A · $\overline{B}$ = 1 · 0 = 0. Thus, the circuit of Fig. 13.22 will pass only pulses with amplitudes between V_0 and $V_0 + \Delta V$, which is precisely what a single-channel analyzer must do.

There is, however, one complicating feature which must be considered. If a pulse is larger than $V_0 + \Delta V$V in amplitude, then the pulse will spend some time in the window between V_0 and $V_0 + \Delta V$V, on the leading edge from t_1 to t_2, and on the trailing edge from t_3 to t_4 as shown in Fig. 13.23. The output of the A · $\overline{B}$ gate will therefore contain two narrow pulses, of width $(t_2 - t_1)$ and $(t_4 - t_3)$, the widths depending upon the shape of the input pulse. These are "false" pulses because they do not correspond to the peak input pulse amplitude lying in the channel. They must be discriminated against in some way. One way is to change the type of discriminator so that it responds only to the peak voltage of the input pulse, or to shape the input pulse so that the rise and fall times are extremely small, thus producing exceedingly narrow outputs from the A · $\overline{B}$ gate when the input pulse amplitude is greater than $V_0 + \Delta V$. Both of these solutions are rather sophisticated, but there is a third practical solution: the use of a "strobe" pulse. Notice that the desired output pulse from the A · $\overline{B}$ gate occurs at the same *time* as the peak of the input pulse as shown in Fig. 13.23(b). Thus, if we can generate a narrow pulse called the "strobe" pulse, occurring at precisely the same time as the *peak* of the input pulse, then the correct output pulses from the A · $\overline{B}$ gate will be only those in coincidence with the strobe pulse; i.e., only those occurring at the same time as the strobe pulses. A block diagram of this arrangement is shown in Fig. 13.24. The

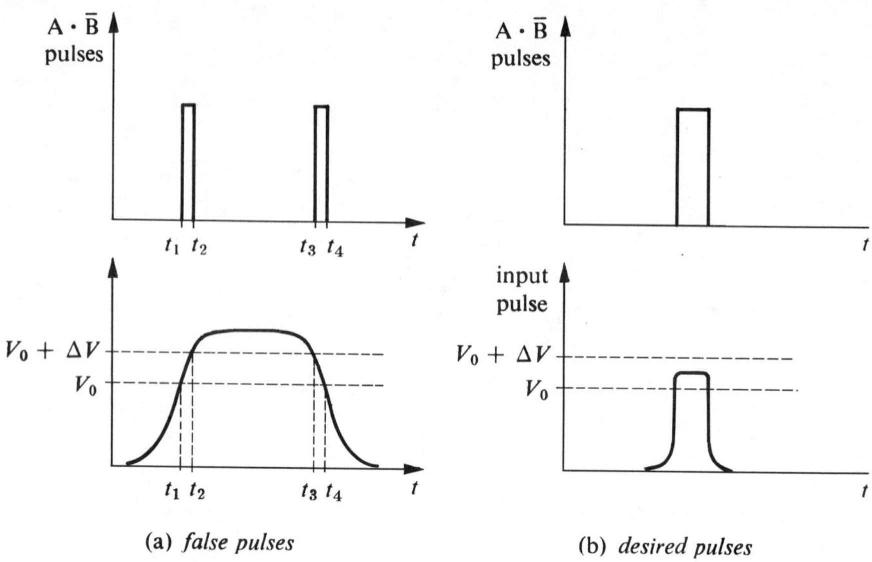

FIGURE 13.23 Pulses in pulse height analyzer.

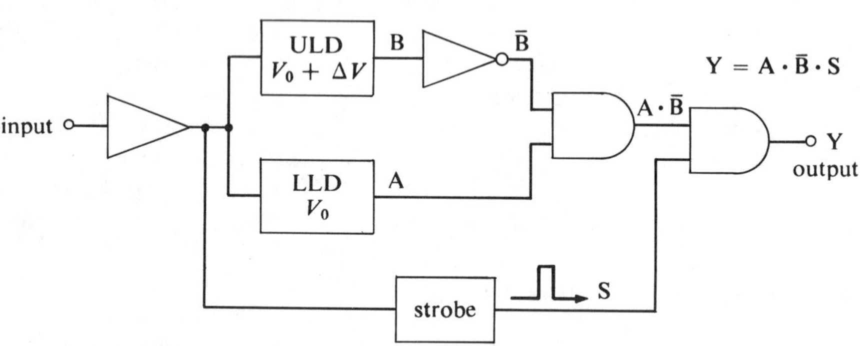

FIGURE 13.24 Single-channel analyzer with strobe.

strobe pulse and the A · B̄ pulse are inputs to an AND gate or coincidence circuit. The strobe circuit should be a monostable multivibrator set to produce a narrow output pulse at the same time as the peak of the input pulse.

Let us now consider the problem of the two level discriminators. The basic decision which must be made in a pulse height analyzer is whether an input pulse is larger or smaller than a certain reference voltage. In other words the input pulse amplitude must be compared with a stable reference voltage. Integrated circuits are available to perform this comparison and are called "comparators." The Fairchild μA710 high speed differential comparator and the μA711 dual comparator are two such integrated circuits. Comparators are essentially differential amplifiers with a very narrow linear range of operation.

The two inputs of a differential amplifier or a comparator are labeled + and − for the noninverting and the inverting inputs, respectively. A positive pulse into the noninverting input produces a positive pulse at the output, while a positive pulse into the inverting input produces a negative pulse at the output. The output depends only upon the difference between the two inputs and can be written as $V_{out} = A(V_{ni} - V_i)$; where V_{ni} is the input to the noninverting input, V_i is the input to the inverting input, and A is the gain.

A comparator designed to work on positive logic is constructed so that the output voltage is either about +3V at logical 1 or 0V at logical 0. The comparator action is shown in Fig. 13.25. Notice that if V_{ni} exceeds V_i by

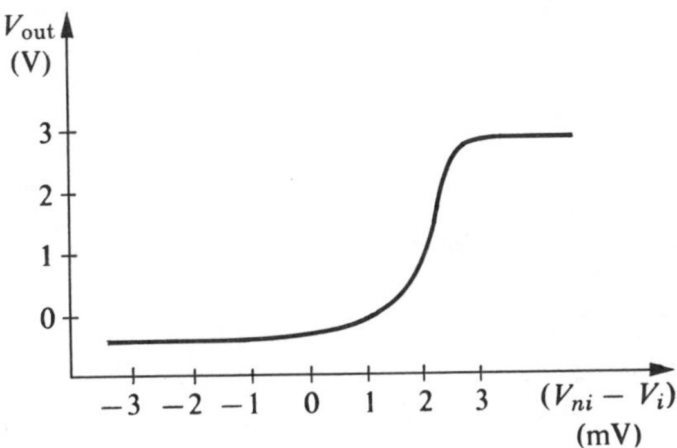

FIGURE 13.25 Comparator action.

more than only several *millivolts*, the output voltage is essentially constant at + 3.2V, and if V_{ni} is less than V_i by 1 millivolt or more, the output is at − 0.5V. Thus the comparator output essentially is either +3 V (logical 1) or 0 V (logical 0) depending only on whether $V_{ni} > V_i$ or $V_{ni} < V_i$. Many comparisons are made possible by setting one of the inputs (V_{ni} or V_i) at a constant reference voltage which may be either positive or negative.

The single-channel analyzer of Fig. 13.24 can be constructed from a μA711 dual comparator whose simplified schematic is shown in Fig. 13.26. Notice there are two identical sides to the circuit, one for each comparator. The outputs from the two comparators are fed into an OR circuit consisting of T_1 and T_2. If *either* T_1 or T_2 is turned on by a positive pulse at their base, then point E goes positive; and if E goes positive enough to make the zener diode, D_3, conduct, then the output goes positive yielding a logical 1 output. If D_3 is not conducting, then the output is at logical 0. Notice that the base of T_3 is grounded, thus preventing the output from ever going more negative than about 0.6V, with the exact maximum negative output voltage depending on the temperature.

The two discriminators of the single-channel analyzer can now be made from two comparator sections of the μA711. We desire an output only when the LLD fires, i.e., when $V_{in} > V_o$, *and* when the ULD does not fire; $B = 0$ or

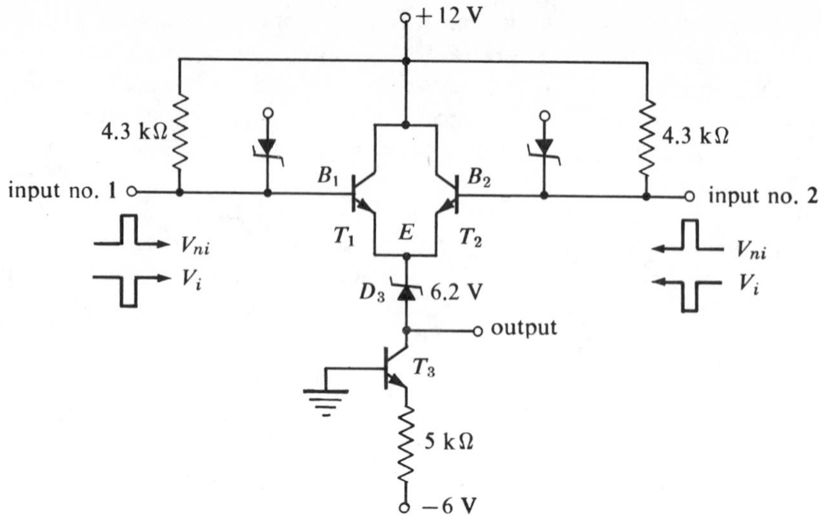

FIGURE 13.26 μA711 dual comparator simplified schematic. (Courtesy of Fairchild Semiconductor Components Group, Fairchild Camera and Instrument Corp.)

$\bar{B} = 1$, i.e., when $V_{in} < V_o + \Delta V$. However, care must be taken because of the OR circuit wired into the μA711. We cannot use positive pulses to represent a discriminator firing, because the output comes from an OR circuit: there would then be an output when the LLD fires and the ULD does not fire ($V_o < V_{in} < V_o + \Delta V$) and also "false" outputs when both the LLD and the ULD fire. The solution is to use a zero voltage to represent the LLD firing and a positive voltage to represent the ULD firing as shown in Fig. 13.27. The lower edge of the window V_o volts is used as a reference input to the noninverting input of comparator no. 1, and the upper edge of the window $V_0 + \Delta V$ volts is used for the reference input to the inverting input of comparator no. 2. Thus if $V_{in} < V_0$, the output of comparator no. 1 at B is nearly equal to +4 V; and the output of comparator no. 2 at A is 0 V. If $V_0 < V_{in} < V_0 + \Delta V$, the output of comparator no. 1 is 0 V, and of comparator no. 2 is 0 V. If $V_0 + \Delta V < V_{in}$, then the output of comparator no. 1 is 0 V and of comparator no. 2 it is +4 V, as shown in Table 13-2.

TABLE 13-2 COMPARATOR VOLTAGES FOR VARIOUS INPUT VOLTAGES

V_{in}	*Comparator Output* No. *1* (*volts*) B	*Comparator Output* No. *2* (*volts*) A	Y = A + B *Output* (*volts*)
$V_{in} < V_o$	+4	0	+4
$V_0 < V_{in} < V_0 + \Delta V$	0	0	0
$V_0 + \Delta V < V_{in}$	0	+4	+4

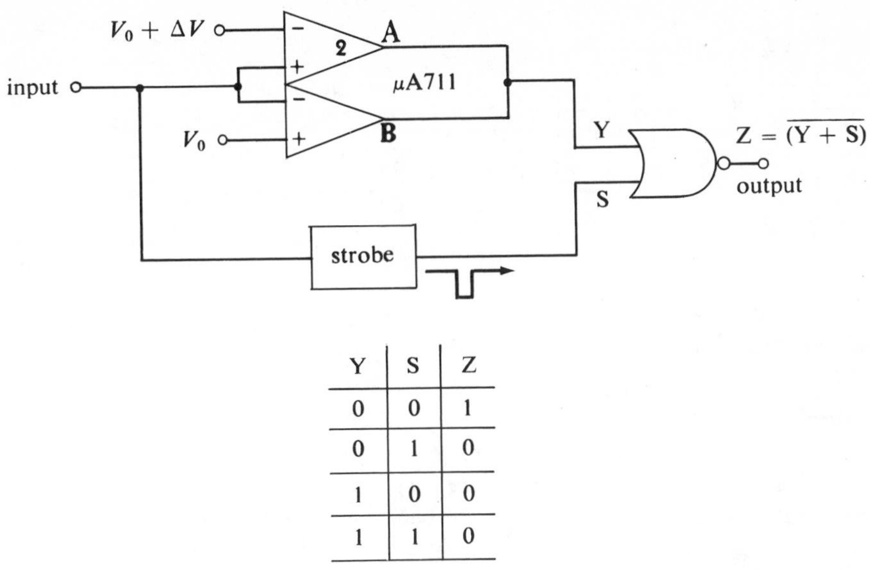

Y	S	Z
0	0	1
0	1	0
1	0	0
1	1	0

FIGURE 13.27 Single-channel analyzer with strobe.

The net result is that the output of the μA711 is low at logical 0 only when the input pulse lies in the channel or window between V_0 and $V_0 + \Delta V$. This output is in the form of a negative pulse, and must be counted only when it is in coincidence with the strobe pulse. An AND gate designed to work on negative pulses would perform this function, but NOR gates are more commonly available so we use a NOR gate designed to work on positive logic such as the Texas Instruments SN7402, shown in Fig. 13.27. Notice that the logical 1 and 0 in the NOR gate truth table refer to positive logic voltage levels; i.e., logical 1 represents a voltage of +4V, logical 0 represents a voltage of 0V. It is also worthwhile to point out that the "negative" pulses from the strobe monostable multivibrator and the μA711 comparator go from +4V down to 0V. The SN7402 NOR gate is an AND gate for those negative pulses as can be seen from the truth table of Fig. 13.27. This is an example of the general rule mentioned earlier that a NOR gate for positive logic is an AND gate for negative logic.

13.12 I.C. REGULATED POWER SUPPLIES

An ideal dc power supply (see Fig. 13.28) will deliver a constant output voltage over a wide range of currents due to a varying load and in spite of temperature or line voltage fluctuations. A battery with zero internal resistance would be such an ideal supply. If the internal resistance is greater than zero, the output

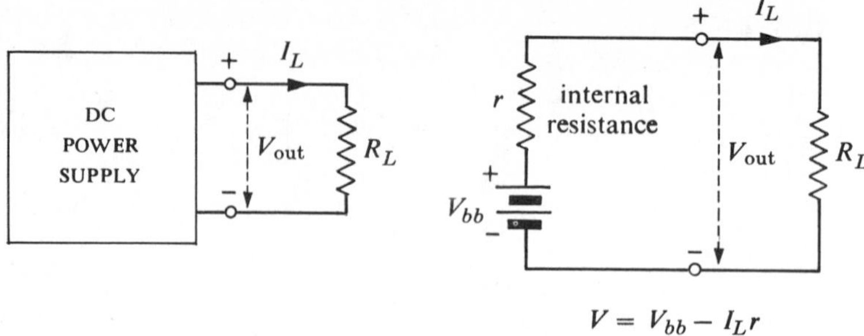

FIGURE 13.28 Power supply.

or terminal voltage will decrease as the load current I_L drain from the supply increases. Thus, an ideal power supply would have a zero output impedance, which leads us to the thought of using a dc coupled emitter follower circuit with the output taken off the emitter. The basic circuit is shown in Fig. 13.29(a).

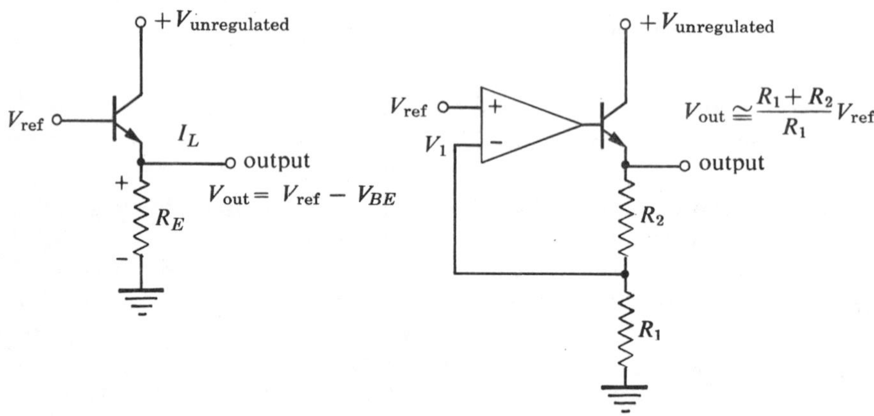

(a) *simple emitter follower* (b) *with differential amplifier in feedback loop*

FIGURE 13.29 Regulated power supply.

Notice that a constant voltage, the "reference voltage," must be supplied to the base in order to produce an output that is approximately constant: $V_{out} = V_{ref} - V_{BE}$. The output voltage can therefore be no more stable than the reference. The output can, however, supply much more current than the reference supply, which need only supply enough current to drive the base. The simple emitter follower of Fig. 13.29(a) can be improved greatly by amplifying the feedback signal from output to the base. Recall that the unbypassed emitter resistance R_E provides the negative feedback. If the emitter current increases due to the unregulated voltage on the collector increasing, then the emitter becomes more positive, which decreases the base–emitter forward bias and thereby decreases the emitter current. A dc coupled differential amplifier with gain A_0 can be added to the feedback to improve the constancy of the output as is shown in Fig. 13.29(b). If the error amplifier gain A_0 is large enough, the dc output voltage wili be given by $V_{out} \cong [(R_1 + R_2)/R_1]V_{ref}$. The reference voltage is usually supplied from a very stable low current zener diode. Let the output voltage of the error amplifier be V_B since it is the base voltage of the power pass transistor. We have $V_B = V_{BE} + V_{out}$, $V_B = A_0(V_{ref} - V_1)$, and $V_1 = [R_1/(R_1 + R_2)]V_{out}$. Eliminating V_1 yields:

$$V_{out} = \frac{A_0 V_{ref} - V_{BE}}{1 + A_0 \dfrac{R_1}{R_1 + R_2}}. \tag{13.6}$$

$V_{BE} \cong 0.6\text{V}$, and usually $V_{ref} > 1\text{V}$ and $A_0 \geq 1000$, so $A_0 V_{ref} \gg V_{BE}$. Thus:

$$V_{out} \cong \frac{A_0 V_{ref}}{1 + A_0 \dfrac{R_1}{R_1 + R_2}}. \tag{13.7}$$

And for a high gain A_0, $A_0 R_1/(R_1 + R_2) \gg 1$ usually, so that a good approximation for the output is:

$$V_{out} \cong V_{ref} \frac{R_1 + R_2}{R_1}. \tag{13.8}$$

This is, as the reader will probably already have recognized, a special case of the negative feedback amplifier gain expression $A_f = A_0/(1 + A_0\beta)$. Here $\beta = R_1/(R_1 + R_2)$ is the fraction of the output voltage fed back to the input. Notice that the output voltage is independent of the unregulated voltage and the error amplifier gain. The stability of the output voltage depends only upon the stability of the reference V_{ref} and of the resistors R_1 and R_2.

A practical circuit using the μA709 I.C. operational amplifier is shown in Fig. 13.30. A conventional full-wave diode rectifier is used to supply the unregulated voltage. A zener diode, D_1, is used to supply the reference voltage, and two output pass transistors are used in place of the single power pass transistor of Fig. 13.29(b). There are also available complete monolithic integrated circuit voltage regulators containing their own zener diode voltage reference, error amplifier, and low power pass transistor. They require only a few external discrete components and will regulate load currents up to several hundred mA. With the addition of a separate power pass transistor they can supply regulated output voltages at currents of several amperes.

One such I.C. voltage regulator is the Fairchild μA723, whose circuit diagram is shown in Fig. 13.31; two practical regulator circuits are shown in Fig. 13.32.

The ratio of R_1 and R_2 determines the output voltage as we have just seen. $V_{out} = [(R_1 + R_2)/R_1]V_{ref}$ with $V_{ref} = 6.2\text{V}$ from the zener diode built into the

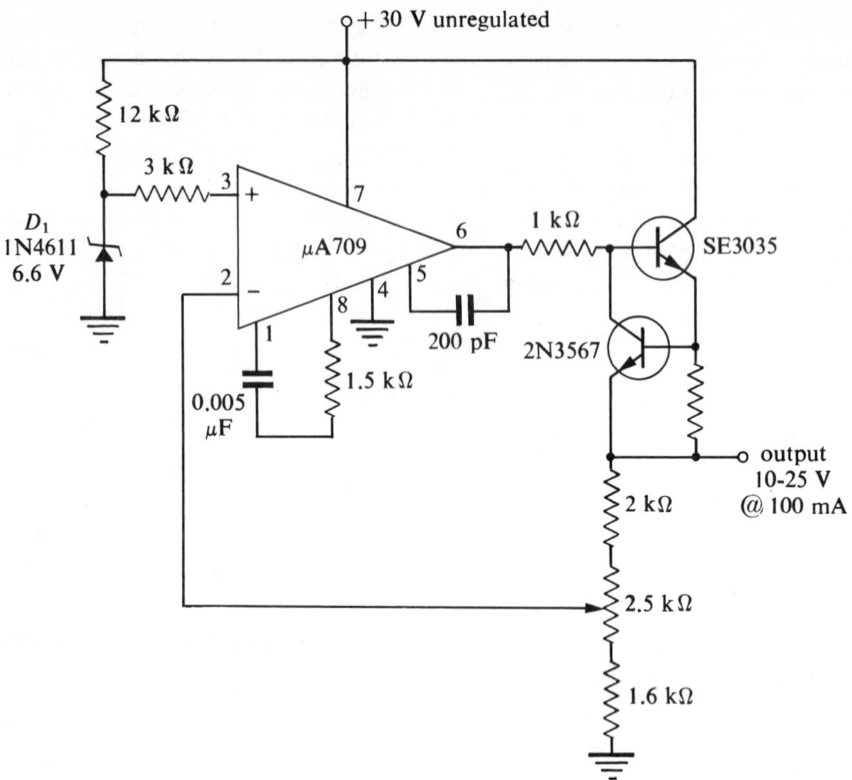

FIGURE 13.30 Regulated power supply using μA709 as error amplifier. (Courtesy of Fairchild Semiconductor Components Group, Fairchild Camera and Instrument Corp.)

μA723 I.C. The circuit is short circuit protected by the combination of the external resistance R_{SC} and transistor Q_{16} which drives the Darlington output pair Q_{14} and Q_{15}. The collector–emitter voltage of Q_{16} is given by $V_{CE_{16}} = V_{BE_{14}} + V_{BE_{15}} + I_L R_{SC} = 1.2\text{V} + I_L R_{SC}$, and its base–emitter voltage $V_{BE_{16}} = I_L R_{SC}$. If $I_L R_{SC}$ is less than 0.6 V, the turn-on voltage for Q_{16}, then Q_{16} is off and the regulator is functioning normally. However, if $I_L R_{SC}$ rises above 0.6 V due to an increase in the load current, then Q_{16} is turned on and pulls the base of Q_{14} negative; thus turning Q_{14} off, which in turn turns Q_{15} off. If $R_{SC} = 5$ ohms, this protective shut-off action will commence when $I_L R_{SC} = 0.6\text{V}$, or when $I_L = 0.6\text{V}/R_{SC} = 0.6\text{V}/5\Omega = 120\,\text{mA}$. This protective action is almost instantaneous, as it merely involves turning the transistor Q_{16} on. Different values of R_{SC} will result in the protective action or current limiting taking place at different load currents. This action is temperature dependent because the turn-on voltage of Q_{16} (as with any transistor) is temperature dependent. The higher the temperature, the lower the turn-on voltage, typically 0.6 V at 25°C and 0.5 V at 125°C. Thus with $R_{SC} = 5$ ohms, and the device at 125°C (A tempera-

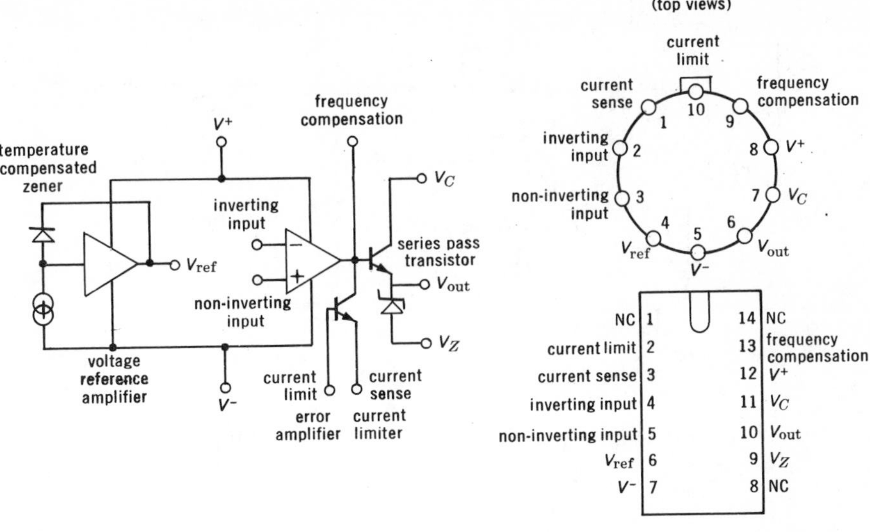

Note: On metal can, pin 5 is connected to case

equivalent circuit *connection diagrams*

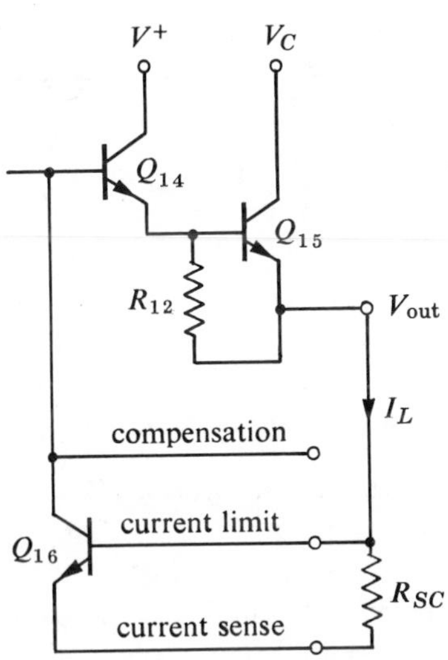

short circuit protection

FIGURE 13.31 μA723 schematic. (Courtesy of Fairchild Semiconductor Components Group, Fairchild Camera and Instrument Corp.)

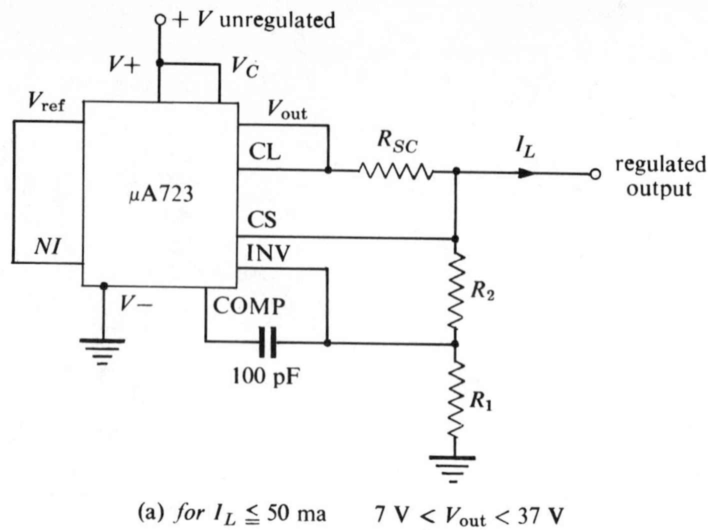

(a) *for* $I_L \leqq 50$ ma $7\text{ V} < V_{out} < 37\text{ V}$

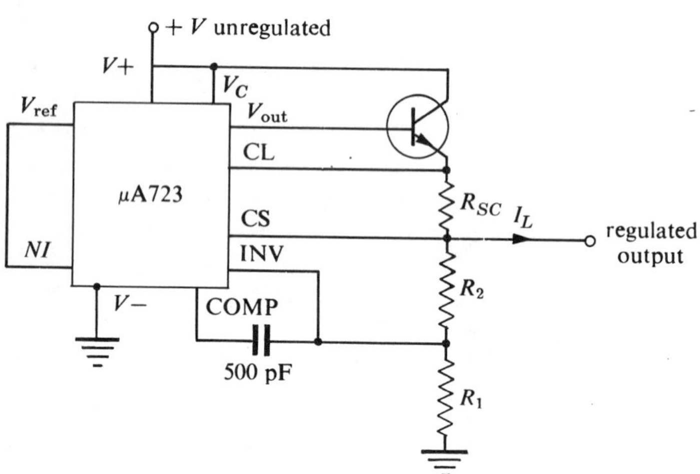

(b) *with external pass transistor* $I_L \leqq 10$ A

FIGURE 13.32 μA723 regulated power supplies. (Courtesy of Fairchild Semiconductor Components Group, Fairchild Camera and Instrument Corp.)

ture to be avoided!), the current limiting would take place at approximately $I_L = 0.5\text{V}/5\Omega = 100\text{mA}$. With $R_{SC} = 0$ ohms, the output impedance of the μA723 is approximately 0.05 ohms up to 1 kHz, rising to 1 ohm at several hundred kilohertz.

The output voltage regulation for this type of I.C. regulator in Fig. 13.32(a) is typically of the order of 0.01% for a 3 volt change in the 110V line voltage, and 0.03% for a 50 mA change in the load current. For the high current regulated supply of Fig. 13.32(b), the same voltage regulation may be expected, and approximately a 0.1% change in the output voltage for a 1 A change in the load current.

13.13 COMPLEMENTARY MOSFET CIRCUITS

One ingenious example of modern I.C. technology uses both n channel and p channel enhancement type MOSFET's in complementary pairs to make various types of logic circuits. One unusual advantage of such circuits is that extremely low power consumption is achieved because power is consumed only while the circuit is switching states and not while the circuit is in the steady state. Other advantages include higher noise immunity, relative immunity to temperature fluctuations, operation from a single polarity voltage supply, and a large fan-out capability.

The basic operation of such complementary symmetry MOSFET circuits can be seen by considering the inverter circuit of Fig. 13.33, which operates on positive logic. Notice that the output is taken from the two drains which are connected together, and that the input is applied to the two gates which are connected together. Because the two enhancement MOSFET's are of different types, one n channel and the other p channel, it is impossible to forward bias *both* simultaneously so as to draw current from the power supply. Thus no steady state power is consumed regardless of the logic state of the input. If the input is at logic zero, then the upper p channel MOSFET is conducting, the lower n channel MOSFET is not conducting, and the output voltage is approximately equal to the supply voltage V_{DD}, representing logic 1. If, on the other hand, the input is at logic 1, the lower p channel MOSFET is conducting and

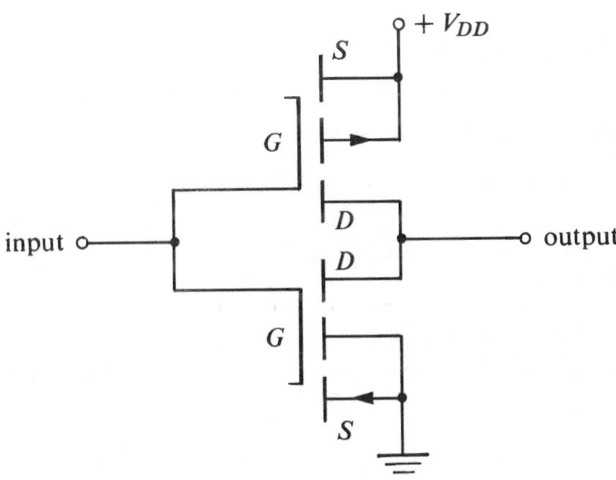

FIGURE 13.33 Complementary MOSFET inverter.

the upper n channel MOSFET is not conducting, thus producing an output voltage of almost zero, representing logic 0.

Only when the circuit is switching from one state to another is a surge of current drawn from the power supply as the circuit capacitances are charged and discharged. For moderate switching frequencies, this represents an extremely low average power consumption, of the order of microwatts. Many other logic circuits can be made with similar circuitry—flipflops, counters, NAND and NOR gates, etc. In all such circuits the MOSFET's always occur in complementary pairs, one n channel and one p channel so that there is no steady state power drawn from the power supply.

problems

1. Briefly state the similarities and differences between integrated circuits and circuits made from discrete components.
2. Sketch a block diagram for a μA709 amplifier with a voltage gain of 1000, including explicit values for the external compensation networks.
3. Explain the purpose of the 0.001 μF capacitor between pin 1 and ground of the μA703 rf amplifier. Why isn't this 0.001 μF capacitor incorporated inside the I.C.?
4. Show by means of truth tables that the two-input NOR gate Texas Instrument SN7402 logic equation can be written $Y = \overline{(A + B)}$ or $Y = \overline{A} \cdot \overline{B}$.
5. Explain the difference between "synchronous" and "asynchronous" flip-flop inputs.
6. Explain the difference between RS and JK flip-flops.
7. Show, by means of a block diagram, how you could extend the range of a 100kHz frequency counter with a decade counter used as a "prescaler," i.e., used to reduce the frequency of the input to the 100kHz frequency counter. By how much is the effective range of the frequency counter increased? What will the precision of the counter reading be with and without the prescaler? Illustrate with an input of 1,732,590 Hz and 325,708 Hz.
8. Explain the use of a strobe pulse in a single-channel analyzer.
9. (a) Calculate the regulated output voltage.

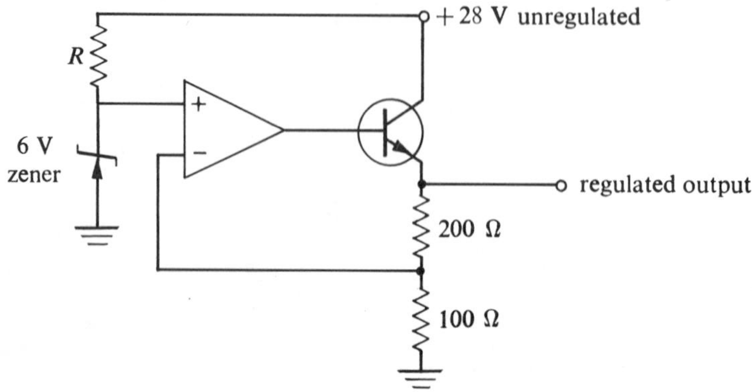

(b) Calculate a reasonable value for R if the zener diode can safely dissipate a maximum power of 500 mW.

(c) Calculate the power dissipated in T_1 if the load current is 120 mA.

10. Using an upper-level discriminator (ULD), a lower-level discriminator (LLD), and appropriate AND, OR, and NOR gates, sketch a block diagram for a pulse analyzer circuit which will pass only pulses less than V_1 and also pass pulses greater than V_2. ($V_2 > V_1$.)

references

1. "*μA723 Precision Voltage Regulator Application Notes.*" 46pp. Copyright, Fairchild Semiconductor, 1968. 20-BR-0064-78. Very clearly written survey of many different types of regulators that can be made from the μA723: low voltage, high voltage, positive or/and negative output voltage, floating or nonfloating, series or shunt regulators, etc.
2. "*The Microelectronics Data Book.*" approx. 800pp. Motorola Semiconductor Products Inc. 2nd ed. Dec. 1969. A vast collection of data sheets—all the Motorola integrated circuits: gates, flip-flop, adders, regulators, amplifiers, microwave devices, comparators, etc. Includes a table of Motorola replacement for I.C.'s made by other companies. An updating service is available at $2/yr. from Motorola Inc., Technical Information Center, P.O. Box 20901, Phoenix, Arizona, Zip Code 85036.
3. "*Fairchild Semiconductor Integrated Circuit Data Catalog 1970.*" 570pp. BR-BR-0015-29 100M. Data sheets on the Fairchild integrated circuits. List of application notes and technical papers available upon request. Fairchild Semiconductor, 313 Fairchild Drive, Mountain View, California, Zip Code 94040.
4. "*Integrated Circuits: Design Principles and Fabrication.*" ed. by R. M. Warner Jr. and J. N. Fordernualt. McGraw-Hill, N.Y., 1965. Basic semiconductor theory, I.C. design philosophy, but mainly devoted to the fabrication techniques for integrated circuits.
5. "*TTL Integrated Circuits Catalog from Texas Instruments.*" 420pp. Data sheets on the Texas Instruments Transistor–Transistor–Logic (TTL) integrated circuits. Includes list of application notes available upon request from Texas Instruments Incorporated, Marketing and Information Services, P.O. Box 5012, M.S. 308 Dallas, Texas, Zip Code 75222.

APPENDIX A

COMPONENTS

Resistors, Capacitors, Inductors, and Transformers

A.1 RESISTORS

The principal types of resistors, their characteristics and applications are given below.

TYPE	*CHARACTERISTICS*
Carbon	Inexpensive, common. Useful up to hundreds of MHz. Wide range of resistance values. 1 Ω–22 MΩ 1/8 W–2 W.
Metal Film	Stable, high voltage rating, available up to 10 kV. Low noise, low temperature coefficient: 25–100 ppm/°C. Very small inductance due to helix shape of film. Low distributed capacitance causes drop off of impedance at higher frequencies, e.g., 10% drop at 300 kHz for 10 MΩ at 3 MHz for 1 MΩ at 30 MHz for 100 kΩ at 100 MHz for 10 kΩ at 300 MHz for 1 kΩ
Wirewound	High power capabilities available—up to 1000 W. Low temperature coefficient: 20–100 ppm/°C. Maximum voltage 500–1000 V. Relatively high inductance, thus useful only at frequencies below several hundred Hz.
Potentiometers (variable resistors)	*Wirewound*—For low frequency; ac impedance tends to rise well above dc resistance at tens of MHz (high precision). *Non-Wirewound*—Better high-frequency performance (less inductance), lower precision. *Metal Film*—Useful up to 50 MHz.

A.2 CAPACITORS

The principal types of capacitors and their characteristics are given below.

TYPE	*CHARACTERISTICS*
Paper dielectric	Primarily for frequencies less than 10MHz. Inexpensive. 0.001 μF–10 μF, 100 V–2000 V max.WVDC (working voltage dc).
Polyester dielectric	0.001 μF–0.5 μF Compact, inexpensive. Resistance greater than $10^{10}\,\Omega$. Less than 1% power factor at 1 kHz. Temperature coefficient: 150 ± 50 ppm/°C, −40 °C to +75 °C, 500 WVDC. Constant capacitance up to 20 kHz. (1% low at ~20 kHz).
Polystyrene film dielectric	Resistance greater than $10^{14}\,\Omega$, 0.001 μF–0.5 μF. −40 °C to +85 °C. Less than .005% power factor. Temperature coefficient: 150 ± 50 ppm/°C. 500 WVDC. Long term stability ±0.2%.
Ceramic	1 pF–1 μF. High capacitance per unit volume. Many different temperature coefficients available.

Temperature Compensated Type	*Temp. Coeff.*
P100	+100 ppm/°C
N750	−750 ppm/°C
NPO	0
N030	−30 ppm/°C

General purpose "high K." Inexpensive, wide capacitance tolerance, large temperature coefficient.

Special types available to carry up to 10 amperes rf current at frequencies up to several hundred MHz.

Disc ceramic bypass capacitors form a series resonant circuit due to lead inductance. The following table gives the self-resonant frequency for various values assuming ½-inch lead length.

Capacitance Value	*Self-Resonant Frequency**
0.01 μF	15 MHz
0.001 μF	55 MHz
0.0001 μF	165 MHz

* (From *Radio Amateur's Handbook*, courtesy of American Radio Relay League.)

TYPE	*CHARACTERISTICS*
	Such bypass capacitors act properly only at frequencies approximately equal to or less than their self-resonant frequency.
Electrolytic (Tantalum)	High capacitance per unit volume. Polarized. 0.01–5000 μF. Capacitance falls off and power loss increases appreciably at 10 kHz and higher. Special types available for rf bypass applications up to 10 MHz. Approximately 300 maximum WVDC.
Electrolytic (Aluminum)	1 μF–10,000 μF. Extremely high capacitance values per unit volume. Polarized. Generally good only at frequencies less than 50 kHz. Commonly used as filters in power supplies. 450 V max. WVDC.
Mica	5 pF–0.005 μF. Stable, high Q, close capacitance tolerance. Good for tuned circuits. 500 V max. WVDC.

C in pF from colors in (2), (3), and (4) positions per the resistor color code; e.g., (2) = yellow, (3) = violet, and (4) = brown means C = 470 pF. (1) = white for commercial—black for mil. spec. (5) = tolerance black ±20%, silver ±10%, gold ±5%, and brown ±1%. (6) = brown ±500 ppm/°C and ±3% drift; yellow −20 to +100 ppm/°C and ±0.1% drift; and green 0 to 70 ppm/°C and ±0.05% drift.

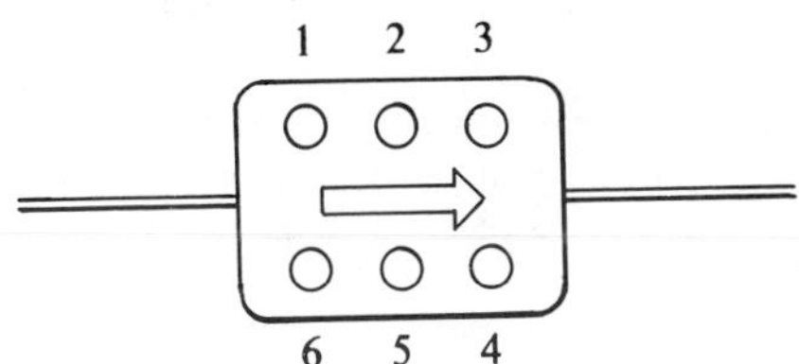

FIGURE A1.1 Molded mica capacitor color code.

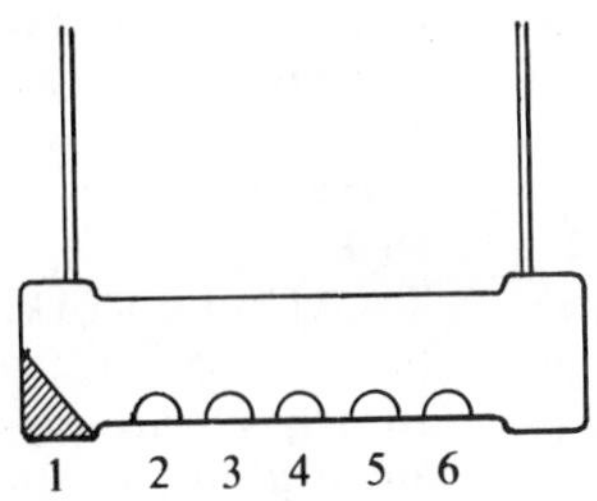

FIGURE A1.2 Ceramic tubular capacitor color code.

(1) and (2) determine temperature coefficient.

(1) Black "NPO" zero	(1) Silver "Y5D" ±3.3%
(2) missing temp. coeff.	(2) Brown −30 °C to +85 °C

(1) Brown "Z5U" +22%–56% capacitance variation +10 °C to +85 °C
(2) Gray

C in pF from colors in (3), (4), and (5) positions by the resistor color code; e.g., (3) = orange, (4) = orange, and (5) = red means C = 3300 pF. (6) = black ±20%, white ±10%, and green ±5%.

A.3 INDUCTORS (RF CHOKES)

Type	*Self-Resonant Freq.*	*Max.* dc *Current*	*Q*	dc *Resistance*
25 mH ferrite core 3 section (Miller 6308)	470 kHz	65 mA	102	82 Ω
2.5 mH iron core 3 section (Miller 4666)	1.7 MHz	160 mA	80	15 Ω
1 mH phenolic core 3 section (Miller 4652)	3.7 MHz	160 mA	59	19 Ω
100 μH iron core single layer (Miller 4632)	12 MHz	400 mA	107	3 Ω
10 μH iron core single layer (Miller 4622)	40 MHz	1.5 A	69	0.11 Ω
1 μH phenolic core single layer (Miller 4602)	190 MHz	2.0 A	60	0.05 Ω
0.1 μH phenolic core single layer (Miller 4580)	500 MHz	3.0 A	68	0.017 Ω

rf chokes should be used at frequencies less than their self-resonant frequency.

(Courtesy of J. W. Miller Inc.)

A.4 TRANSFORMERS

Type	*Characteristics*
Laminated iron core	Used up to several hundred hertz. Higher frequencies require smaller core. Special audio transformers available for 20 Hz–20 kHz.
Powdered iron core	Many different types available for use from 60 Hz–250 MHz.
Ferrite iron core (ceramic ferromagnetic)	High core resistivity, stable, low eddy current loss. Used from 1 kHz–1 GHz. Constant permeability available up to 30 MHz.
Air core	10 MHz–450 MHz typical. At 450 MHz only approximately one turn needed per winding. For higher frequencies microwave cavity and waveguide techniques are used.

APPENDIX B

BATTERIES

Any two different metals (called "electrodes") placed in a solution containing ions (called the "electrolyte") will generate a voltage difference between the two metals. For example a carbon and a zinc electrode in a solution of ammonium chloride will generate a voltage difference of 1.5 volts. This is the familiar "flashlight battery." Such an arrangement of two electrodes in an electrolyte is called a "cell." Several cells mounted in one package are called a "battery"; e.g., four 1.5V carbon–zinc cells connected in series form a 6V battery. Strictly speaking, a 1.5V carbon–zinc "flashlight battery" is really a "cell." In everyday usage, however, the word "battery" refers to either a cell or a battery. There are six basic types of batteries whose main characteristics and applications are listed below.

Battery Type	*Volts per Cell*	*Characteristics*	*Application*
Carbon–zinc ("Flashlight Battery")	1.5	Inexpensive, widely available in many sizes. Voltage falls off gradually with use. Poor shelf life especially at temperatures above 90°F. Poor performance below 32°F.	Inexpensive equipment, toys.
Alkaline–manganese	1.5	Good for relatively high current applications, but no better than carbon–zinc for low current applications. Good low temperature performance. Up to twice as much energy as carbon–zinc. Voltage falls off gradually with use. Good shelf life.	Radios, toys, movie cameras, electronic flash.
Mercury	1.35	Good at high temperatures (up to 130°F). Poor below 40°F. Voltage remains con-	Portable scientific equipment, TVs, radios, not exposed to low temperatures.

Battery Type	*Volts per Cell*	*Characteristics*	*Application*
		stant with use, until a sudden fall off when "used up." Excellent shelf life. Constant ampere–hour capacity for different load currents.	
Silver oxide	1.5	Good at low temperatures. Voltage remains constant with use. Excellent shelf life. Mainly for low current applications. Good shelf life at 10°F or cooler.	Portable instruments, hearing aids, electric watches.
Nickel–cadmium	1.25	Good at both high and low temperatures, −4°F to +113°F. Voltage remains constant with use—typically from 1.25–1.1 volts per cell. May be recharged separately or in series. For parallel recharging each cell or battery must have its own current limiting resistor. Charging current should equal ampere–hour capacity ÷ 10.	Critical portable equipment—radiation detectors, radios, satellites, alarms.
Lead–acid ("automobile battery")	2.0	Capable of supplying high currents (up to 100 amperes). Very heavy. Rechargeable. Exudes flammable hydrogen gas when recharged. Corrosive sulfuric acid electrolyte.	Automobiles, portable high power equipment.

(Courtesy of Union Carbide Corporation.)

A no-load or "open circuit" measurement of the battery terminal voltage may or may not identify a bad battery. If the battery terminal voltage is low, then the battery is certainly bad; if high then the battery may be good or bad. As a general rule batteries should be tested by measuring the terminal voltage with a load which draws approximately one half the recommended current; e.g., if the recommended current is 100 mA or less the battery terminal voltage should be measured with a load drawing 50 mA from the battery. A low terminal voltage indicates a bad battery.

A final comment: Always use a voltmeter *not an ammeter* to test a battery. Connecting an ammeter across a battery will usually destroy the ammeter and damage the battery. For more details see "*Eveready*" *Battery Applications and Engineering Data*, © 1968 by Union Carbide Corp., Unitech Division, Associated Educational Services Corp. A subsidiary of Simon & Schuster Inc., 1 West 39th St., N.Y., N.Y., 10018.

APPENDIX C

MEASURING INSTRUMENTS

Most electrical meters (voltmeters, ammeters, ohmmeters, etc.) display their readings on a moving coil or D'Arsonval meter which is shown in Fig. C1.1. A coil of N turns and area A of fine wire is wound on a very lightweight frame and pivoted on jeweled bearings in the gap of a small permanent magnet with

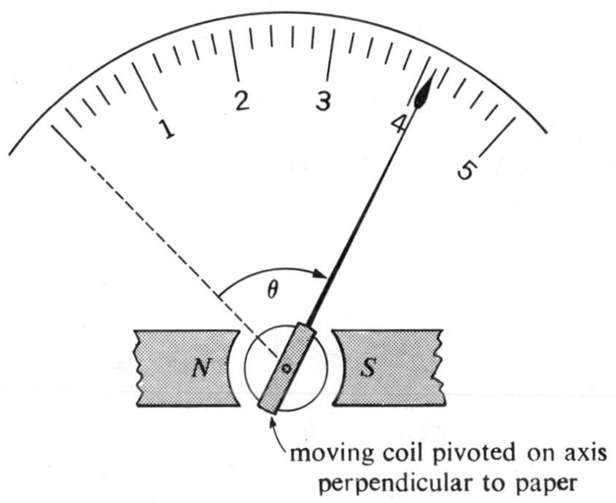

FIGURE C1.1 D'Arsonval meter movement.

magnetic induction B (the magnetic "field"). When a current I flows through the coil, the magnetic field exerts a torque $NIBA$ on the coil, and the coil rotates through an angle θ until the opposing torque of the coil suspension $K\theta$ equals the magnetic torque. Thus, $K\theta = NIBA$. A thin pointer attached to the coil indicates the deflection θ.

C.1 VOLTMETERS

An ideal voltmeter has an infinite resistance; when connected across a circuit element to measure the voltage drop across that element it will draw zero current and therefore will not disturb the circuit. An ideal ammeter has zero resistance;

when connected in series with a wire whose current is being measured, it will not impede the current flow at all.

Hence, if a D'Arsonval meter is being used for a voltmeter, the smaller the current drawn from the circuit to actuate the meter, the better. Since the meter deflection is given by $\theta = (NIBA)/K$, the meter coil should have a large number of turns, a large area, and should move in a strong magnetic field. Most meters are rated either in terms of the current required for a full scale deflection (e.g., $50\,\mu$A), a number fixed for any particular meter, or in terms of "ohms per volt," which means the resistance of the meter divided by the *full scale* voltage rating. The two ratings are equivalent; the meter ohms per volt equals the reciprocal of the full scale meter current.

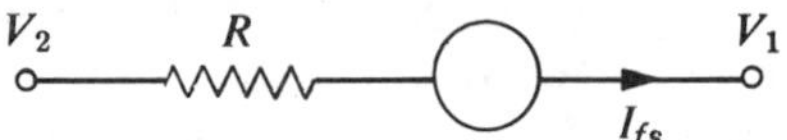

FIGURE C1.2

$$V_2 - V_1 = V\text{ (full scale)} = I_{fs}R$$

where,

R = meter resistance.
I_{fs} = current for full scale meter deflection.
V_{fs} = full scale meter voltage.

$$\frac{1}{I_{fs}} = \frac{R}{V_{fs}}$$

Most meters are $50\,\mu$A, or equivalently 20,000 ohms per volt, and have a resistance of about 2000 ohms in the meter coil. Thus a $50\,\mu$A, 2000-ohm meter would yield a full scale deflection with only $V = I_{fs}R = 50 \cdot 10^{-6} \cdot 2 \cdot 10^3 = 0.1\,$V. For higher voltages to correspond to a full scale deflection, additional resistance must be added in series to the meter. It is important to realize that the total resistance of the meter depends upon the *full scale* voltage. The resistance R is adjusted to change the full scale voltage, as shown in Fig. C1.3. If R = 2 megohms, V_{fs} = 100V; if $R = 200\,\mathrm{k}\Omega$, V_{fs} = 10V for a $50\,\mu$A, 20,000-

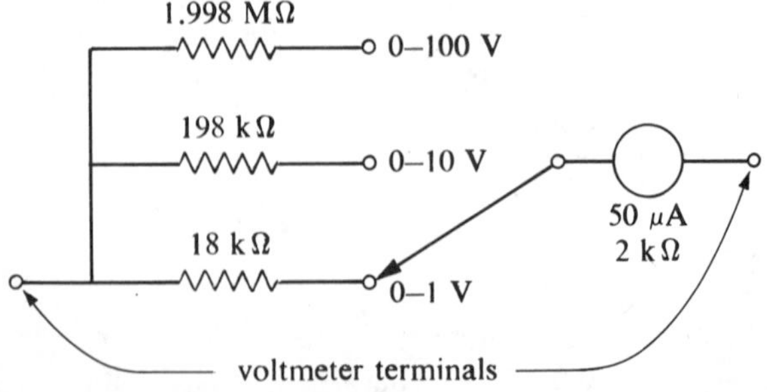

FIGURE C1.3 Multirange voltmeter schematic.

ohms-per-voltmeter. Thus, the effective meter resistance on any scale equals the ohms per volt multiplied by the *full scale* voltage readings for that scale. Inexpensive meters require 1 mA for a full scale deflection and thus have 1000 ohms per volt. Some new meters have 10 μA or 100,000-ohms-per-volt movements.

C.2 AMMETERS

An ammeter can be made from a sensitive 50 μA meter by shunting part of the current around the meter through a shunt resistance R_s as shown in Fig. C1.4.

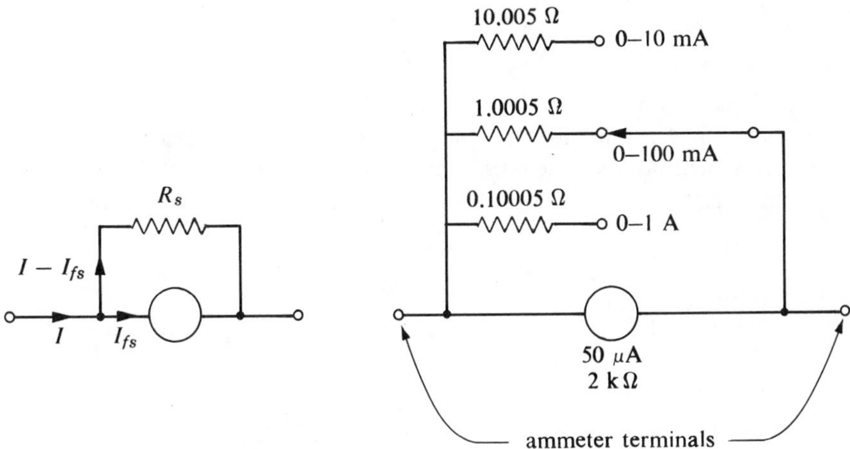

FIGURE C1.4 Multirange ammeter schematic.

If the desired full scale meter reading is I, the meter resistance $R_m = 2\,\text{k}\Omega$, and the meter full scale deflection $I_{fs} = 50\,\mu\text{A}$ through the meter, then:

$$(I - I_{fs})R_s = I_{fs}R_m$$

$$R_s = \frac{I_{fs}}{(I - I_{fs})}R_m.$$

Thus for a 2000 Ω, 50 μA meter, a full scale deflection for 100 mA will occur if $R_s = 1.0005\,\Omega$. An ammeter with several current ranges is shown in Fig. C1.4.

C.3 OHMMETERS

An ohmmeter can be made from a microammeter, a resistance R_1, and a battery by the general circuit of Fig. C1.5. The meter reading depends upon the resistance being measured R_x and therefore the meter can be calibrated in terms of ohms $I = V_{bb}/(R_x + R_1)$. A little thought will show that this circuit cannot be adapted to produce an ohmmeter with different ranges. For, when

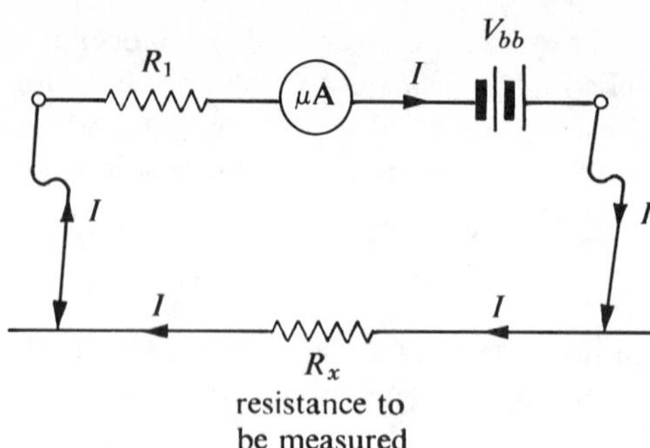

FIGURE C1.5 Basic ohmmeter circuit.

R_x is zero, the meter should read its full scale deflection, say 50 μA corresponding to zero ohms for R_x. This fixes the value of R_1 ($R_1 = V_{bb}/I_{fs}$) provided V_{bb} is constant, and therefore fixes the sensitivity of the ohmmeter. A practical circuit for a multirange ohmmeter is shown in Fig. C1.6. The zero adjustment, R_0,

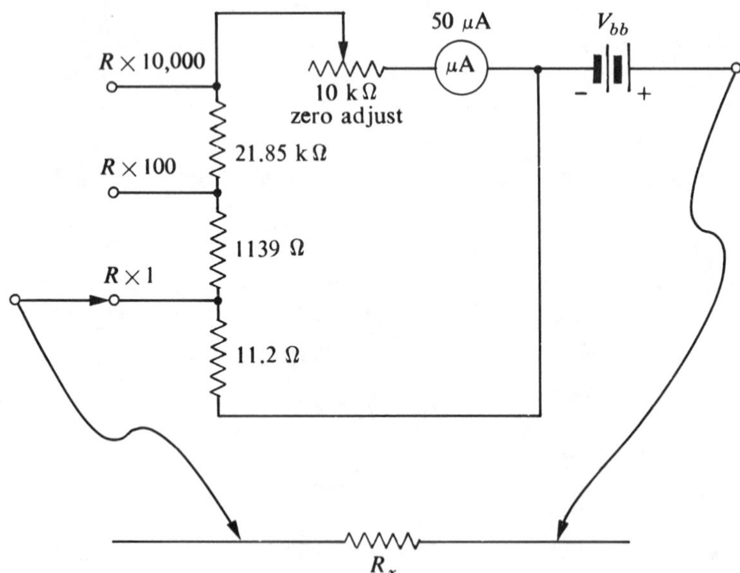

FIGURE C1.6 Multirange ohmmeter schematic.

should be made frequently and must be made each time the scale is changed. When the battery voltage V_{bb} falls, the first symptom is usually that the meter can't be zeroed on the $R \times 1$ scale. The battery voltage is 1.5 V for the $R \times 1$ and $R \times 100$ scales and 7.5 V for the $R \times 10{,}000$ scale.

C.4 COMMERCIAL METERS

Most modern test meters have ranges to measure voltage, current, and resistance in the same instrument and are often referred to as "multimeters" or "VOM's", standing for "volt–ohm–milliammeter." One such modern instrument is the Simpson 260 shown in Fig. C1.7. The meter is 50 μA full scale or

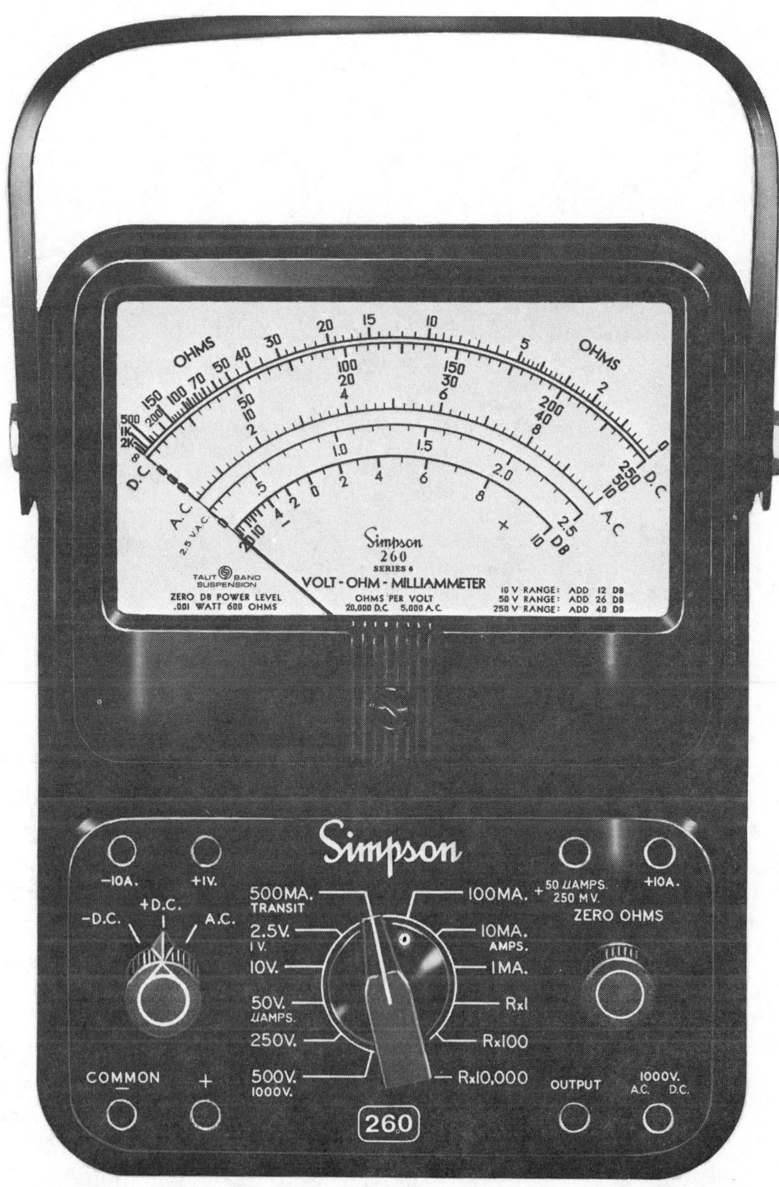

FIGURE C1.7 Modern VOM (Volt–Ohm–Milliammeter). (Courtesy of Simpson Electric Co.)

equivalently 20,000 ohms per volt and can also measure ac as well as dc voltages. On the ac range, the sensitivity is only 5,000 ohms per volt. The basic ac voltage circuit is shown in Fig. C1.8.

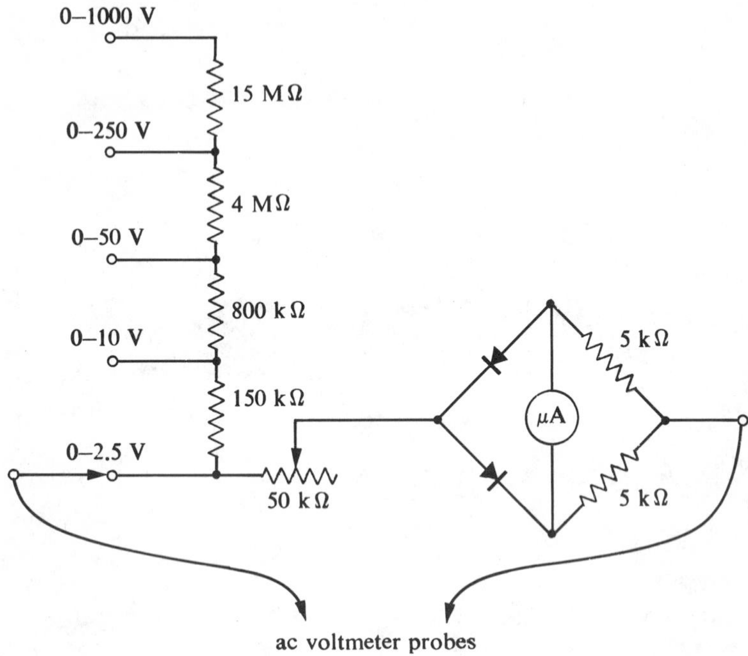

FIGURE C1.8 AC voltmeter schematic.

A word of caution is now in order. Such meters can be burned out very easily if a high voltage is accidentally present at the probes while the meter is set on a resistance scale, or a low voltage scale, or a current scale. A severe overload may burn out the meter movement without any noticeable meter deflection, puff of smoke, or any other sign. To prevent this sort of damage, one must be careful to: (1) turn off the voltage in the circuit when measuring resistance, (2) set the meter range *before* poking around in the circuit to find the appropriate terminals, (3) identify the circuit terminals properly, or (4) use a "goof-proof" meter.

A "goof-proof" meter or burnout proof meter is one which is protected against overloads by an electronic circuit inside the meter. In the Triplett model 630-PLK, an overload causes a yellow button to pop out and the meter is inoperative until the button is manually pressed in. The one thing to be careful of in such a meter is to be sure the small 30 volt battery which powers the electronic protective circuit is operative. A bad 30V battery means the meter is not protected at all and can be damaged by an overload, e.g., putting a 110V voltage across the meter probes when it is set on a resistance scale, and so forth.

A second general class of multimeter or VOM has a much higher input impedance, usually 10 megohms for *all* of the voltage scales. These meters used to be called "VTVM's" standing for vacuum tube voltmeters, but now are often made with FET's instead of vacuum tubes. One such meter is the Triplett 310 shown in Fig. C1.9.

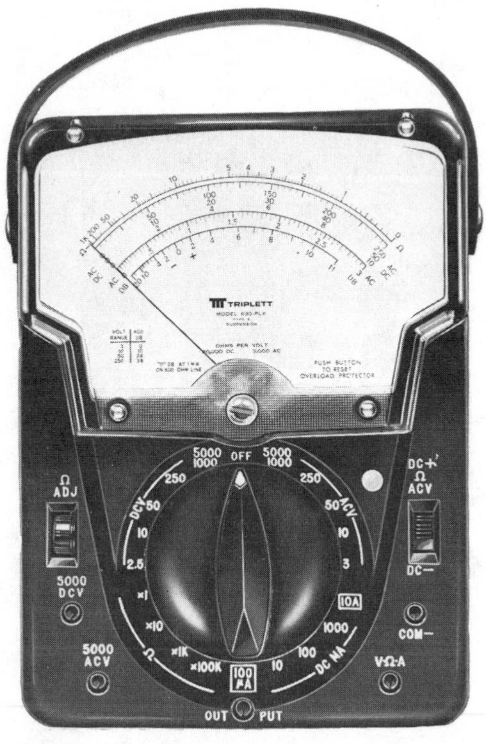

a) burnout proof meter *b) modern high input impedance meter* ("VTVM")

FIGURE C1.9 Laboratory meters. (Courtesy of Triplett Corp.)

C.5 OSCILLOSCOPES

The cathode ray oscilloscope or "scope" is probably the single most useful electrical instrument of them all. It can measure dc or ac voltages directly, currents indirectly, and most commonly is used to display the actual waveform of a voltage; that is, to plot a graph of voltage versus time on the face of the cathode ray tube.

A simplified block diagram of the oscilloscope is given in Fig. C1.10. The basis of the oscilloscope is the cathode ray tube, which is essentially a TV picture tube. The interior is highly evacuated and the inside of the broad face is coated with a phosphor which emits light when struck by electrons. A hot filament emits electrons and an accelerating voltage and series of focusing electrodes

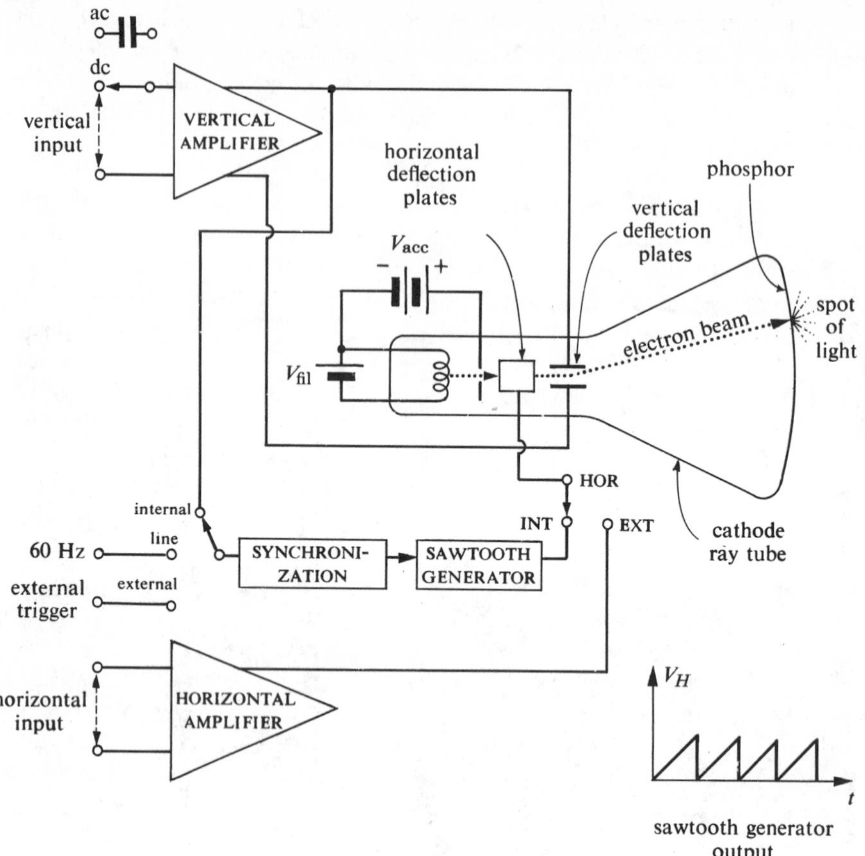

FIGURE C1.10 Cathode ray oscilloscope block diagram.

produce a thin beam of electrons aimed from the neck of the tube toward the broad face. Thus, a small light spot would be seen in the center of the screen provided nothing deflects the electron beam. The signal voltage to be observed is fed into the vertical input where it is amplified and applied to the vertical deflection plates. If the signal voltage is positive, the upper deflection plate is made positive with respect to the lower plate, and the electron beam is deflected upward. On the face of the cathode ray tube, the bright spot moves upward. If the signal voltage is negative, then the lower deflection plate is made positive with respect to the upper and the spot moves downward.

At the same time the electron beam is moving up and down in response to the signal voltage, it is also being swept horizontally by a sawtooth voltage generated within the oscilloscope and applied to the horizontal deflection plates. As the sawtooth increases linearly, the electron beam is swept horizontally across the face of the cathode ray tube from left to right (as seen by the observer) at a constant speed. This speed is read off the knob labeled "sweep speed" or "time/div.," etc., depending on the particular scope used. The speed is usually expressed in time for the beam to move one horizontal division on the CRT, e.g., 50 μsec/div. rather than in cm/sec. The steeper the sawtooth voltage (the shorter the period T_s) the faster the sweep speed and the smaller the time/div. When the sawtooth horizontal sweep drops rapidly to zero again, the electron beam is quickly deflected back to the left. During this short time, the tube is usually blanked out, i.e., the electron beam is actually stopped during this time, so that the observer cannot see the rapid right-to-left traverse of the beam. It is instructive for the beginner to set the horizontal sweep speed to a very low value, say 0.5 sec/div. with no vertical signal input, and watch the beam steadily move across the face of the cathode ray tube at a constant speed of 2 div./sec from left to right. Be sure to set the intensity or brightness control to a low enough value so that the phosphor is not burned by the beam spot. For higher sweep speeds, say 10 msec/div. or 50 μsec/div., the beam spot moves so rapidly that the observer sees a smooth horizontal line because the phosphor glows for approximately 0.1–0.5 sec after the electron beam strikes it. Thus a new sweep comes along to excite the phosphor before the light from the preceding sweep has completely disappeared.

With a sinusoidal input of frequency f and period $T = 1/f$, if the sawtooth period T_s equals T, then one sine wave will be displayed; if T_s equals $2T$ (a slower sweep) then two sine waves will be displayed, etc. The purpose of the "synchronization" or "sync" or "trigger" controls is to ensure that the horizontal sweep starts (is triggered) at the *same point* on the input signal voltage for each sweep. Thus, successive sweeps will lie on top of one another, yielding one stationary display on the face of the cathode ray tube. If the horizontal sweep starts at different points on the input signal voltage for different sweeps, then the waveforms displayed will not lie on top of one another and a blurred picture will be produced. The scope is then said to be "out of sync" or not triggering properly.

The simplest type of synchronization or triggering is the "internal" or "automatic" sweep mode in which the horizontal sawtooth is automatically started when the vertical signal input reaches a certain low voltage value. No external adjustments are required, and this usually results in a stable display unless the signal is extremely low in amplitude and/or high in frequency. Usually there are two internal triggering modes, + and −, meaning that the sweep starts when the vertical signal input goes positive or negative, respectively. The "external" triggering mode means that the horizontal sweep sawtooth does not start unless a signal voltage from outside the scope is fed into the "ext. sync." jack. On many scopes there is a "line" triggering mode which means that the horizontal sweep sawtooth is started in phase with the line voltage (usually 110V 60Hz). This triggering is useful in displaying waveforms which are phase coherent with the 110V line voltage. For example, if a high-frequency signal appears blurred when the conventional internal triggering is used due to the presence of a 60Hz amplitude modulation, then switching to line triggering will "lock in" the 60Hz envelope of the signal, or in fact any envelope which is phase coherent with the line, e.g., 120Hz, and so forth. Often this is the case when the power supply voltage is insufficiently filtered, and a strong 60 or 120Hz ripple voltage appears superimposed on the dc supply voltage.

C.6 BRIDGES

One general class of measuring instruments which do not require a calibrated meter is the bridge. Bridges can be used to measure resistance, capacitance, and inductance as well as other parameters. In a bridge, the meter is used as a null indicator; that is, when the bridge is properly adjusted the meter reads zero and the value of the parameter being measure (e.g., resistance) is determined by the values of the bridge components.

The simplest bridge is perhaps the Wheatstone bridge, shown in Fig. C1.11.

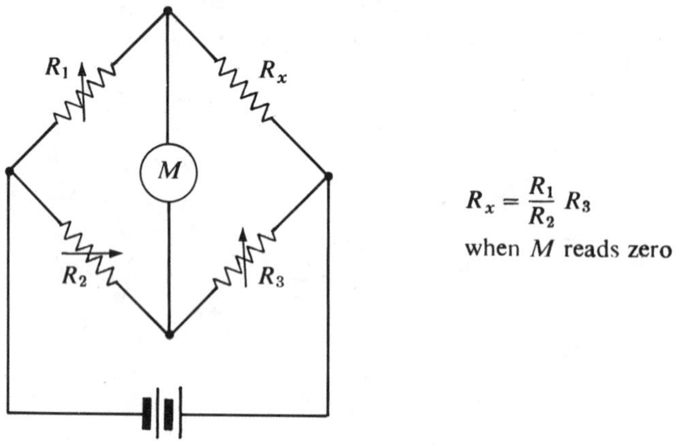

FIGURE C1.11 The Wheatstone bridge.

It can be used to measure the resistance R_x of an unknown resistor by adjusting R_1, R_2, R_3 until the meter (a very sensitive microammeter) reads zero. Then the bridge is said to be in balance and $V_A = V_B$. Thus, $R_x/R_3 = R_1/R_2$ and $R_x = (R_1/R_2)R_3$.

If an ac signal source is used instead of a battery, then we have an ac bridge that can be used to measure L or C. An ac capacitance bridge is shown in Fig. C1.12 for relatively high Q capacitors C_x, $Q \geq 1$. Value R_x is the series resistance of the capacitor C_x (R_x is zero for an ideal capacitor). The null indicator may be an ac meter or a sensitive oscilloscope. The ac source can be at any frequency, hence the variation of capacitance with frequency can be studied. When the bridge is balanced so that the meter or scope reads zero, then $V_A = V_B$ and the unknown capacitance is given by $C_x = (R_2/R_1)C_3$.

An ac inductance bridge is shown in Fig. C1.13 for high Q inductors, $Q > 1$. Notice that R_x is the effective *parallel* resistance of L_x. Thus the higher R_x, the

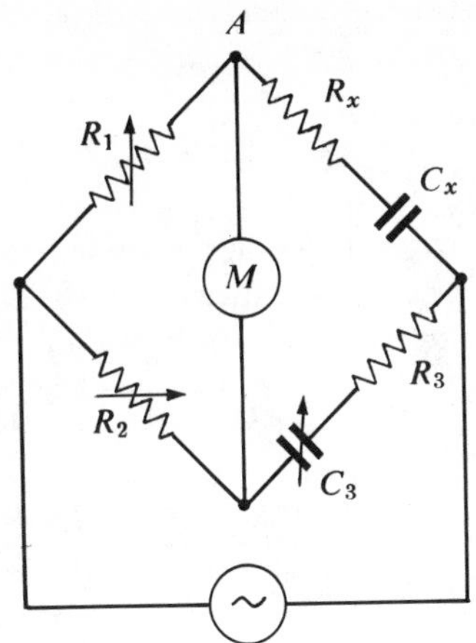

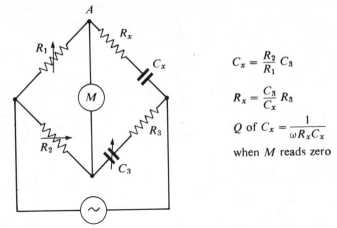

$$C_x = \frac{R_2}{R_1} C_3$$

$$R_x = \frac{C_3}{C_x} R_3$$

$$Q \text{ of } C_x = \frac{1}{\omega R_x C_x}$$

when M reads zero

FIGURE C1.12 An ac capacitance bridge.

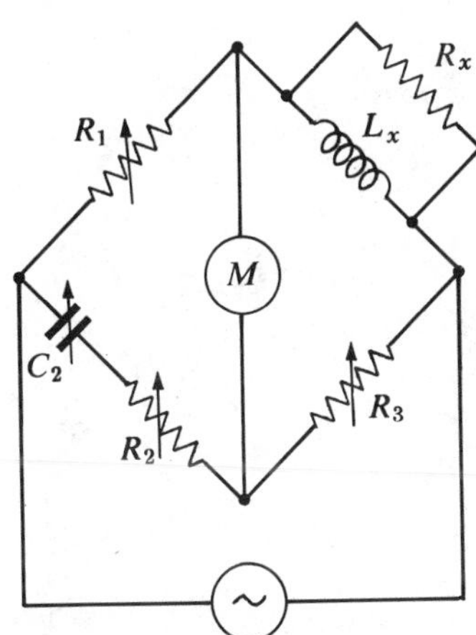

$$L_x = R_1 R_3 C_2$$

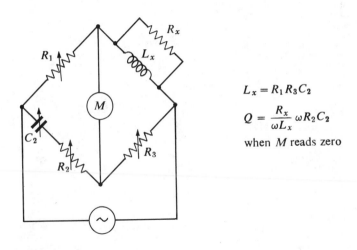

$$Q = \frac{R_x}{\omega L_x} \omega R_2 C_2$$

when M reads zero

FIGURE C1.13 An ac inductance bridge.

higher the Q: $Q = R_x/\omega L_x$. In making capacitance and inductance measurements, it is well to remember that often the impedance of the component being measured is not purely capacitive or purely inductive. However, the impedance of *any* real (lossy) component can always be written as:

$$z = a + jb,$$

where a and b are real numbers. And, the equivalent circuit for any real (lossy) component can be written as *either* a series or a parallel combination of a resistance and a reactance.

Component	*a*	*b*	*z*
pure resistance	positive	zero	R
pure capacitance	zero	negative	$j\left(-\frac{1}{\omega C}\right)$
pure inductance	positive	zero	$j(\omega L)$
lossy capacitance	positive	negative	$R_s + j\left(-\frac{1}{\omega C_s}\right)$ or $\frac{R_p\left(\frac{1}{j\omega C}\right)}{R_p + \left(\frac{1}{j\omega C}\right)} = \frac{R_p + j(-\omega R_p^2 C)}{1 + \omega^2 R_p^2 C^2}$

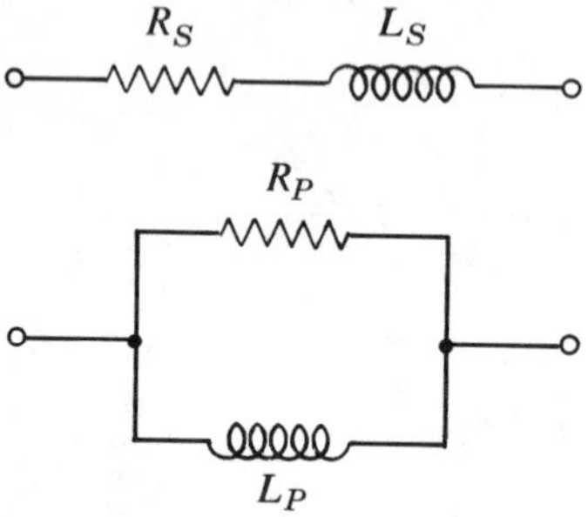

R_S L_S

R_P

L_P

(b) *lossy inductance*

lossy inductance	positive	positive	$R_s + j(\omega L_s)$ $\frac{R_p(j\omega L_p)}{R_p + j\omega L_p} = \frac{R_p\omega^2 L_p^2 + j\omega L R_p^2}{R_p^2 + \omega^2 L_p^2}$

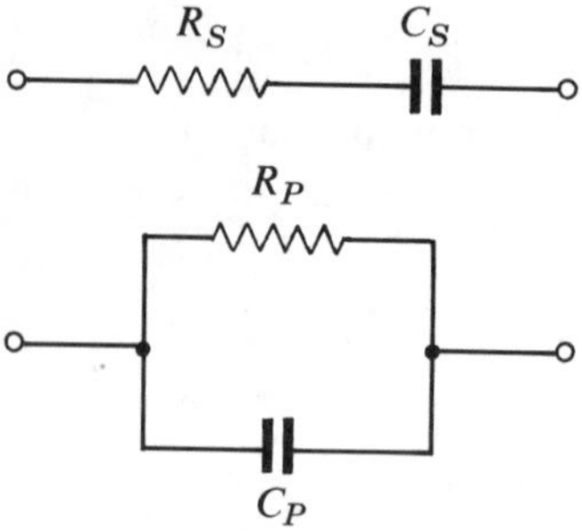

R_S C_S

R_P

C_P

(a) *lossy capacitance*

Commercial impedance bridges are available that incorporate dc and ac resistance bridges and ac inductance capacitance bridges in one instrument. The bridge circuits in this section were taken from the General Radio Type 1650 Impedance Bridge and were used by the kind permission of the General Radio Company.

APPENDIX D

CABLES AND CONNECTORS

Almost all cables used in scientific apparatus and quality electrical equipment are of the coaxial type; an outer grounded conducting sheath completely surrounds one or more inner conductors that carry the desired electrical signals. The principal advantages of such a configuration are shielding and safety. The outer shielding is usually a grounded single sheath of braided copper but may consist of two sheaths for extra shielding for use in noisy environments or where almost no extraneous pick-up can be tolerated. Different types of insulating jackets are available, the most common being black polyethylene (type 111a) which gives good long term stability, and moisture and abrasion resistance from $-55°C$ to $+85°C$. Polyvinylchloride is, in general, inferior. Teflon jackets are available where high temperatures and/or corrosive chemicals may be encountered. The following table lists eight of the more commonly used single-center conductor, coaxial cables along with their mechanical and electrical characteristics.

A variety of coaxial connectors is available to allow one to connect and disconnect cables to other cables and equipment. Connectors come in various sizes to go with the different diameter cables and also with various characteristic impedances to match the cable impedance. Connectors can also be classified according to the type of coupling, whether threaded or bayonet (push-and-twist), and the maximum voltage rating. Threaded connectors are much more resistant to mechanical vibration than the bayonet type, whereas the bayonet type is most convenient for laboratory use. Usually the connector is chosen to go with the cable. Most of the power loss usually occurs in the cable rather than the connector.

Many connectors including BNC are available in several types with drastically different electrical and mechanical properties. For example, the regular BNC connector line is useful only up to 150 MHz, whereas the improved BNC is useful up to 10 GHz. The standard solder connection between the connector and the cable braid will withstand approximately a 30 lb pull, whereas the BNC crimp connection will withstand a 90 lb pull. Some of the most useful connectors are described in the following table and are illustrated in Fig. D1.1.

For frequencies below 150 MHz, BNC connectors and RG58/U cable are a good general purpose choice. For higher frequencies, great care should be

Cable Type	*Inner Conductor*	*Outer Diameter*	*Dielectric*	*Charac-teristic Impedance*	*Maximum Voltage*	dB/ft *Attenuation*			pF/ft *Capaci-tance*	*Use*
						10 MHz	100	1000		
RG58/U	1 copper	0.200″	PE (polyethylene)	53.5Ω	1900 V	1.4	5.3	20	28.5	small g. p. (general purpose)
RG55B/U	1 silver-plated copper	0.206″	PE	53.5Ω	1400 V	1.3	4.8	17	28.5	double-braided shield, g. p.
RG59/U	1 copper weld	0.191″	PE	75Ω	2300 V	1.1	4.0	14	20.5	small g. p.
RG8A/U	7 copper	0.415″	PE	52Ω	4000 V	0.56	2.1	8.8	29.5	med. diam., g. p.
RG9B/U	7 silver-plated copper	0.430″	PE	50Ω	4000 V	0.45	2.3	9.0	30.0	microwave double-braided shield
RG174/U	7 copper weld	0.105″	PE	50Ω			20 (at 400 MHz)		30.0	very small diam. low att.
RG62A/U	1 copper weld	0.242″	Air	93Ω	750 V	0.82	2.7	9.0	13.5	small diam. low cap.
RG63B/U	1 copper weld	0.415″	Air	125Ω	1000 V	0.60	2.0	6.5	10.0	med. diam. low cap.

(a) *coaxial cable*

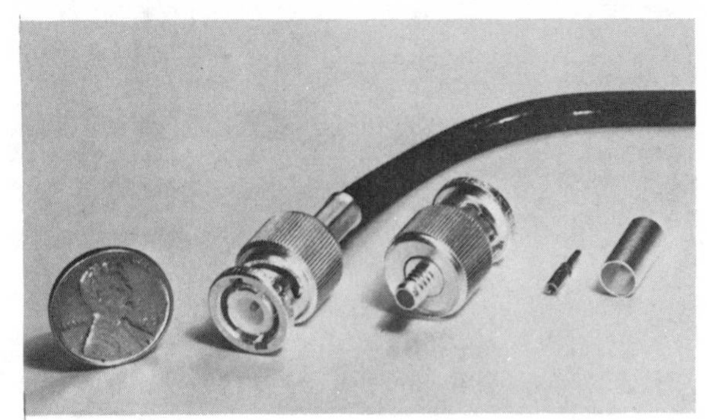

(b) BNC *male connector* (*MIL crimp style*)

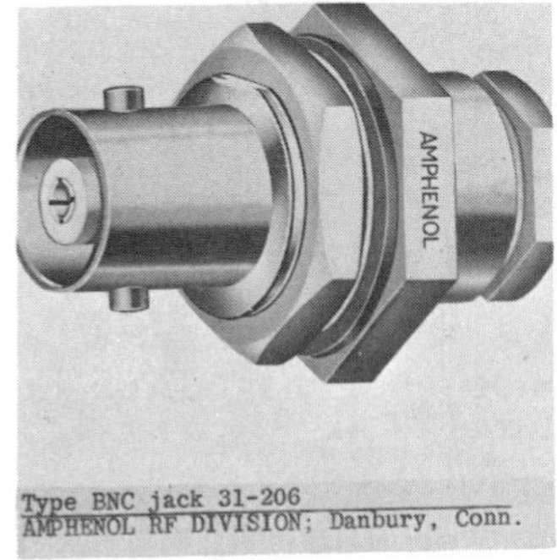

(c) BNC *female jack*

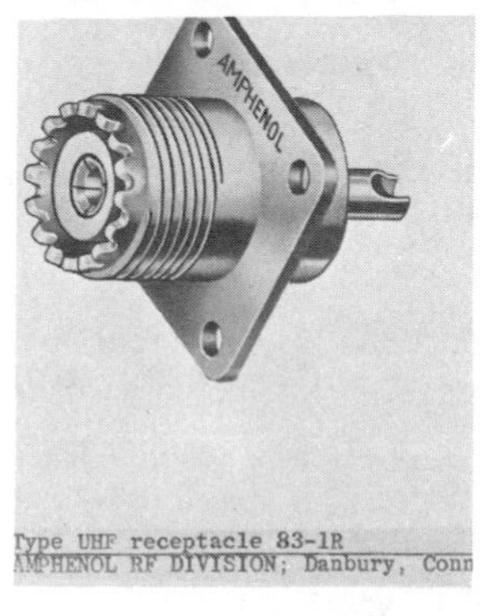

(d) uhf *female panel receptacle*

(e) *type* N *female panel receptacle*

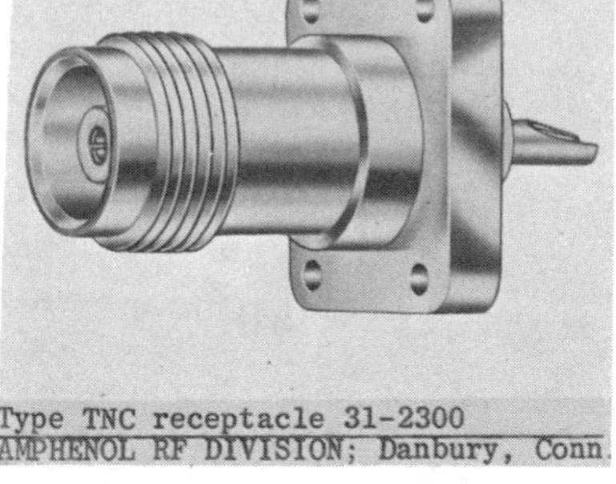

(f) *type* TNC *connector*

FIGURE D1.1 Coaxial connectors and cable. (Courtesy of Amphenol rf Division.)

CONNECTORS

Type	*Coupling*	*Impedance*	*Peak Voltage*	*Maximum Frequency*	*Typical Cable*
BNC (regular)	Bayonet	non-constant	500 V	300 MHz	RG–55,58,59,162
BNC (improved)	Bayonet	50 Ω	500 V	10 GHz*	RG–55,58,59,162
TNC	Threaded	50 Ω	500 V	10 GHz*	RG–55,58,59,162
N	Threaded	50,70 Ω	500 V	10 GHz*	RG–8,9
UHF	Threaded	—	500 V	400 MHz	RG–8,9

* The maximum frequency of operation really depends upon the connector *and* cable. If an appreciable impedance discontinuity exists at the cable-connector interface, an appreciable percentage of power will be reflected and a high VSWR will exist. As a general rule of thumb, if the electrical length of the connector is $\lambda/50$ or less, an impedance mismatch is not serious. For a BNC connector pair (male and female), this corresponds to frequencies of less than approximately 150 MHz. Above this frequency the cable and connector impedance should be carefully marked (e.g., don't use a 50 Ω connector with a 75 Ω cable), and extreme care should be taken in the attachment of the cable to the connector: no gap should exist between the cable core and the connector insulator, all shoulders should be square, and so forth. Careful attention should be paid to the actual operation of cutting and soldering the cable to the connector.

paid to the cable–connector interface and improved BNC connectors should be used. Type UHF connectors should not be used much above 300 MHz. There are many other types of connectors for high voltage applications, extreme mechanical strength, etc. For example, the type APC connector is a precision threaded 7 mm 50 ohm type for use up to 18 GHz with rigid air dielectric lines. The type SMA connector is a subminiature threaded 3 mm 50 ohm type for use up to 18 GHz.

APPENDIX E

COMPLEX NUMBERS

E.1 DEFINITION

Any complex number z means simply $z = a + jb$ where $j^2 = -1$ $(j = \sqrt{-1})$, and a and b are real numbers (positive or negative). a is called the real part of z, and b is called the imaginary part of z. Both a and b must have the same dimensions, e.g., both ohms or both volts and so forth. The $a + jb$ form for a complex number is often called the "Cartesian" form. Examples: $z = 2 + 3j$, $z = -4 + 2j$, $z = R + j\omega L$, $z = R - j/\omega C$.

E.2 COMPLEX CONJUGATE

The complex conjugate of z, written as z^*, means $z^* = a - jb$. To obtain the complex conjugate of any complex number, no matter how complicated, merely replace j by $-j$ and $-j$ by j. Examples: If $z = 2 + 3j$, $z^* = 2 - 3j$. If $z = re^{j\theta}$, $z^* = re^{-j\theta}$.

E.3 GEOMETRIC REPRESENTATION

We can represent any complex number by a single point in two-dimensional space by plotting the imaginary point along the vertical axis and the real part

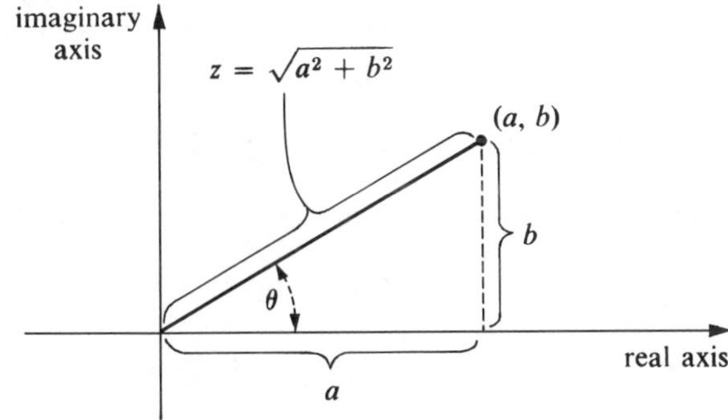

FIGURE E1.1

along the horizontal axis. The "absolute value of z" or the "magnitude" of z, written as $|z|$, means the distance from the origin to the point (a, b): $|z| = +\sqrt{a^2 + b^2}$. Notice $|z|$ is always real and positive; and z can represent an impedance, a voltage, or a current.

E.4 POLAR FORM

Any complex number z can also be written in polar form $z = |z|e^{j\theta}$. Using the trigonometric identity $e^{j\theta} = \cos\theta + j\sin\theta$, we see that $z = |z|\cos\theta + j|z|\sin\theta$. Thus, the real part of z is given by $a = |z|\cos\theta$ and the imaginary part by $b = |z|\sin\theta$. Also $|z| = +\sqrt{a^2 + b^2}$ and $\theta = \arctan b/a$. In polar form, the complex conjugate of z is written: $z^* = |z|e^{-j\theta}$.

E.5 EQUALITY

Two complex numbers $z_1 = a_1 + jb_1$ and $z_2 = a_2 + jb_2$ are equal if and only if $a_1 = a_2$ and $b_1 = b_2$. In words, their real *and* imaginary parts must be equal separately. If z_1 and z_2 are written in polar form, then $z_1 = z_2$ if and only if $|z_1|e^{j\theta_1} = |z_2|e^{j\theta_2}$: $|z_1|\cos\theta_1 = |z_2|\cos\theta_2$ *and* also $|z_1|\sin\theta_1 = |z_2|\sin\theta_2$.

E.6 ADDITION

Two complex numbers $z_1 = a_1 + jb_1$ and $z_2 = a_2 + jb_2$ can be added or subtracted by the following rule:

$$(z_1 \pm z_2) = (a_1 \pm a_2) + j(b_1 \pm b_2).$$

It is very inconvenient to add two complex numbers in polar forms, and very easy in the Cartesian form $a + jb$. Addition commute: $z_1 + z_2 = z_2 + z_1$. Example: $(2 + 3j) + (4 + 5j) = 6 + 8j$.

E.7 MULTIPLICATION

Two complex numbers $z_1 = a_1 + jb_1$ and $z_2 = a_2 + jb_2$ can be multiplied by the following rule:

$$z_1z_2 = (a_1 + jb_1)(a_2 + jb_2) = (a_1a_2 - b_1b_2) + j(a_1b_2 + a_2b_1).$$

If z_1 and z_2 are expressed in polar form, multiplication is very easy:

$$z_1z_2 = |z_1|e^{j\theta_1}|z_2|e^{j\theta_2} = |z_1||z_2|e^{j(\theta_1+\theta_2)}.$$

Multiplication of complex numbers commutes: $z_1z_2 = z_2z_1$.

E.8 DIVISION

Two complex numbers can be divided by the following rule:

$$\frac{z_1}{z_2} = \frac{a_1 + jb_1}{a_2 + jb_2} = \frac{a_1 + jb_1}{a_2 + jb_2} \times \frac{a_2 - jb_2}{a_2 - jb_2} = \frac{a_1a_2 + b_1b_2}{a_2^2 + b_2^2} + j\frac{a_2b_1 - a_1b_2}{a_2^2 + b_2^2},$$

where we have multiplied and divided z_1/z_2 by $(a_2 - jb_2)$ to get z_1/z_2 in the Cartesian form $a + jb$. Division is much easier if the number z_1 and z_2 are in polar form:

$$\frac{z_1}{z_2} = \frac{|z_1|e^{j\theta_1}}{|z_2|e^{j\theta_2}} = (|z_1|/|z_2|)e^{j(\theta_1-\theta_2)}.$$

E.9 MISCELLANEOUS

It is often desired to calculate quickly the magnitude of a complex number. Perhaps the most useful technique is to multiply the number by its complex conjugate and take the square root of the product:

$$zz^* = (a + jb)(a - jb) = a^2 + b^2 = |z|^2.$$

Therefore $|z| = \sqrt{zz^*}$.

Some common ac *RLC* circuits and their complex impedances are given below.

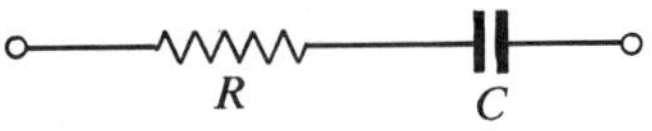

R C

$$z = R + \frac{1}{j\omega C} = R - \frac{j}{\omega C}$$

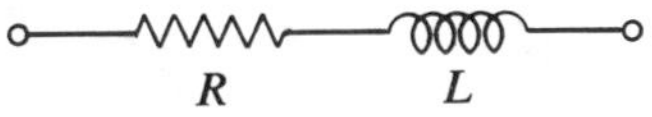

R L

$$z = R + j\omega L$$

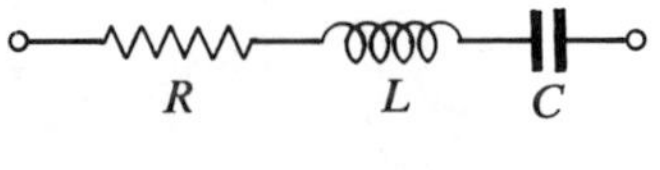

R L C

$$z = R + j\omega L + \frac{1}{j\omega C} = R + j\left(\omega L - \frac{1}{\omega C}\right)$$

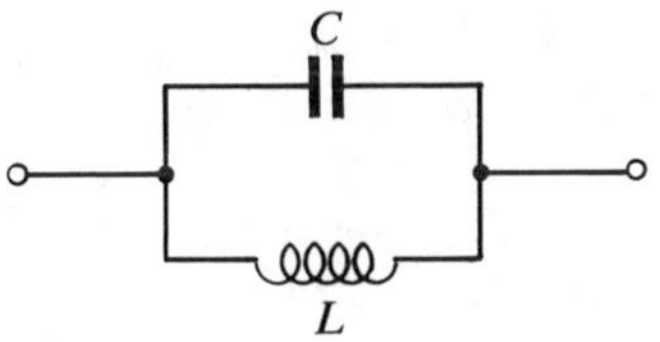

C L

$$z = \frac{(j\omega L)(1/j\omega C)}{j\omega L + 1/j\omega C} = \frac{L/C}{j(\omega L - 1/\omega C)}$$

$$= \frac{-j/\omega C}{1 - 1/\omega^2 LC}$$

APPENDIX F

DETERMINANTS

A determinant is a single number obtained from a square array of numbers by the following rule. (A pair of vertical lines enclosing the square array of numbers denotes the determinant of that array.) For a 2×2 array, the determinant of A is given by:

$$|A| = \begin{vmatrix} a_{11} & a_{12} \\ a_{21} & a_{22} \end{vmatrix} = a_{11}a_{22} - a_{21}a_{12},$$

e.g.,

$$\begin{vmatrix} 3 & 2 \\ 4 & 7 \end{vmatrix} = 3 \times 7 - 4 \times 2 = 21 - 8 = 13$$

For a 3×3 array:

$$|A| = \begin{vmatrix} a_{11} & a_{12} & a_{13} \\ a_{21} & a_{22} & a_{23} \\ a_{31} & a_{32} & a_{33} \end{vmatrix} = a_{11}\begin{vmatrix} a_{22} & a_{23} \\ a_{32} & a_{33} \end{vmatrix} - a_{12}\begin{vmatrix} a_{21} & a_{23} \\ a_{31} & a_{33} \end{vmatrix} + a_{13}\begin{vmatrix} a_{21} & a_{22} \\ a_{31} & a_{32} \end{vmatrix}.$$

The rule for expansion is to go across any row or down any column in the determinant and multiply with alternating signs each element a_{ij} by its "minor," which is the determinant obtained by crossing out the *ith* row and the *jth* column. This technique of expansion in minors is usually too cumbersome for determinants larger than 3×3. For larger determinants a computer is usually used. For a 3×3 determinant, the minor of a_{22} is:

$$\text{minor of } a_{22} = \begin{vmatrix} a_{11} & a_{13} \\ a_{31} & a_{33} \end{vmatrix} = a_{11}a_{33} - a_{13}a_{31}.$$

The signs of the multipliers of the minors is determined by alternating signs starting with + in the upper left corner; e.g., expanding in terms of the second column would give $A = -a_{12} \times$ its minor $+ a_{22} \times$ its minor $- a_{32} \times$ its minor.

In solving circuit problems, one usually writes down the Kirchhoff voltage equations that are linear in the currents and voltages. That is, they are of the form:

$$\begin{aligned}
a_{11}i_1 + a_{12}i_2 + \cdots + a_{1n}i_n &= v_1\\
a_{21}i_1 + a_{22}i_2 + \cdots + a_{2n}i_n &= v_2\\
a_{31}i_1 + a_{32}i_2 + \cdots + a_{3n}i_n &= v_3\\
\vdots \qquad\qquad\qquad\qquad &\quad \vdots\\
a_{n1}i_1 + a_{n2}i_2 + \cdots + a_{nn}i_n &= v_n .
\end{aligned}$$

The a coefficients are resistances or impedances in a circuit problem. One can always obtain the same number (n) of equations as currents so the a coefficients of the currents form a *square* array. Usually the voltages $v_1 \ldots v_n$ are known (some of them may be zero), and we are solving for the currents. The theory of determinants says that the currents are given by:

$$i_1 = \frac{\begin{vmatrix} v_1 & a_{12} & a_{13} & \cdots & a_{1n} \\ v_2 & a_{22} & a_{23} & \cdots & a_{2n} \\ \vdots & & & & \vdots \\ v_n & a_{n2} & a_{n3} & \cdots & a_{nn} \end{vmatrix}}{\begin{vmatrix} a_{11} & a_{12} & a_{13} & \cdots & a_{1n} \\ a_{21} & a_{22} & a_{23} & \cdots & a_{2n} \\ \vdots & & & & \vdots \\ a_{n1} & a_{n2} & a_{n3} & \cdots & a_{nn} \end{vmatrix}} \qquad i_2 = \frac{\begin{vmatrix} a_{11} & v_1 & a_{13} & \cdots & a_{1n} \\ a_{21} & v_2 & a_{23} & \cdots & a_{2n} \\ \vdots & & & & \vdots \\ a_{n1} & v_n & a_{n3} & \cdots & a_{nn} \end{vmatrix}}{\begin{vmatrix} a_{11} & a_{12} & a_{13} & \cdots & a_{1n} \\ a_{21} & a_{22} & a_{23} & \cdots & a_{2n} \\ \vdots & & & & \vdots \\ a_{n1} & a_{n2} & a_{n3} & \cdots & a_{nn} \end{vmatrix}}, \text{ etc.}$$

The determinant in the denominator is called the determinant of the coefficients and occurs in the denominator for all the current expressions. The determinant in the numerator for the *kth* current is the determinant of the coefficients except that the *kth* column has been replaced by the (known) voltages, $v_1 \ldots v_n$. Some useful properties of determinants are:

1. The determinant equals zero if any row or any column contains all zeros.
2. If each term in any row or column is multiplied by the same number, then the value of the determinant is multiplied by that number.
3. If any row (or column) is added to or subtracted from any other row (or column), term by term, the value of the determinant is unchanged. For example, adding the first column to the second column:

$$\begin{vmatrix} 3 & 2 \\ 4 & 7 \end{vmatrix} = 3 \times 7 - 4 \times 2 = 21 - 8 = 13$$

$$\begin{vmatrix} 3 & 2 \\ 4 & 7 \end{vmatrix} = \begin{vmatrix} 3 & 5 \\ 4 & 11 \end{vmatrix} = 3 \cdot 11 - 5 \cdot 4 = 33 - 20 = 13.$$

This technique is often useful in creating more zeros as elements of the determinant before expanding the determinant in minors. The more zeros present, the easier the arithmetic:

$$|A| = \begin{vmatrix} 2 & 1 & -2 \\ 1 & 2 & -1 \\ -1 & 1 & -1 \end{vmatrix}.$$

Add the first column to the third column:

$$|A| = \begin{vmatrix} 2 & 1 & 0 \\ 1 & 2 & 0 \\ -1 & 1 & -2 \end{vmatrix}.$$

Now expand along the third column:

$$|A| = -2\begin{vmatrix} 2 & 1 \\ 1 & 2 \end{vmatrix} = -2(2\cdot 2 - 1\cdot 1) = -2(4 - 1) = -6.$$

Notice that at least one of the constant terms $v_1 \ldots v_2$ must be non-zero in order for non-zero current solutions to exist. In other words, there must be some non-zero voltage source v in order to ensure that some current flows. A circuit example might generate three Kirchhoff equations (the coefficients of the currents are in ohms):

$$i_1 + 3i_2 + 0 = 0$$

$$i_1 + 2i_2 + 3i_3 = 3(\mathrm{V})$$

$$0 + i_2 + i_3 = 0$$

The currents are given by:

$$i_1 = \frac{\begin{vmatrix} 0 & 3 & 0 \\ 3 & 2 & 3 \\ 0 & 1 & 1 \end{vmatrix}}{\begin{vmatrix} 1 & 3 & 0 \\ 1 & 2 & 3 \\ 0 & 1 & 1 \end{vmatrix}} = \frac{-3\begin{vmatrix} 3 & 3 \\ 0 & 1 \end{vmatrix}}{\begin{vmatrix} 2 & 3 \\ 1 & 1 \end{vmatrix} - \begin{vmatrix} 3 & 0 \\ 1 & 1 \end{vmatrix}} = \frac{-9}{-1-3} = +\frac{9}{4} = +2.25\mathrm{A}.$$

$$i_2 = \frac{\begin{vmatrix} 1 & 0 & 0 \\ 1 & 3 & 3 \\ 0 & 0 & 1 \end{vmatrix}}{\begin{vmatrix} 1 & 3 & 0 \\ 1 & 2 & 3 \\ 0 & 1 & 1 \end{vmatrix}} = \frac{\begin{vmatrix} 3 & 3 \\ 0 & 1 \end{vmatrix}}{-4} = \frac{3}{-4} = -0.75\ \mathrm{A}.$$

$$i_3 = \frac{\begin{vmatrix} 1 & 3 & 0 \\ 1 & 2 & 3 \\ 0 & 1 & 0 \end{vmatrix}}{\begin{vmatrix} 1 & 3 & 0 \\ 1 & 2 & 3 \\ 0 & 1 & 1 \end{vmatrix}} = \frac{-3\begin{vmatrix} 1 & 3 \\ 0 & 1 \end{vmatrix}}{-4} = \frac{-3}{-4} = +0.75\ \mathrm{A}.$$

The negative sign for i_2 merely means that i_2 flows opposite to the direction assumed in writing down the three Kirchhoff equations.

APPENDIX G

CHARACTERISTIC CURVES AND PARAMETERS FOR SELECTED SEMICONDUCTOR DEVICES

Characteristic curves are given here for several diodes and transistors which will probably serve for 90% of most discrete component circuits. The curves were taken with a Tektronix type 576 curve tracer. The diode curves plot forward current passing through the diode vertically versus the forward voltage drop across the diode horizontally. The germanium and silicon transistor curves plot the collector current vertically versus the collector–emitter voltage horizontally for various constant values of the base current. The lowest line is for zero base current. The curves have been reflected so as to look the same for both pnp and npn transistors. The field effect transistor curves plot drain current vertically versus drain–source voltage horizontally for various constant values of the gate–source voltage. The highest line is for zero gate–source voltage.

For other devices not listed here, refer to the various manufacturers' handbooks. The single most useful (and the largest) is the "big brown book" of Motorola: *The Semiconductor Data Book*, Motorola Inc., Technical Information Center, P.O. Box 20912, Phoenix, Arizona, Zip Code 85036.

Another useful handbook is: *Preferred Semiconductors and Components from Texas Instruments* Catalog CC202, Texas Instruments Inc., P.O. Box 5012, Dallas, Texas, Zip Code 75222.

For integrated circuit data see: *TTL Integrated Circuits Catalog from Texas Instruments* Catalog CC201–R, Texas Instruments Inc., P.O. Box 5012, Dallas, Texas, Zip Code 75222, or *The Microelectronics Data Book*, Motorola Inc., Technical Information Center, P.O. Box 20912, Phoenix, Arizona, Zip Code 85036.

G.1 DIODES

PRV = peak reverse voltage, V_F = forward voltage drop, $@I_F$ forward current, and t_{rr} = reverse recovery time.

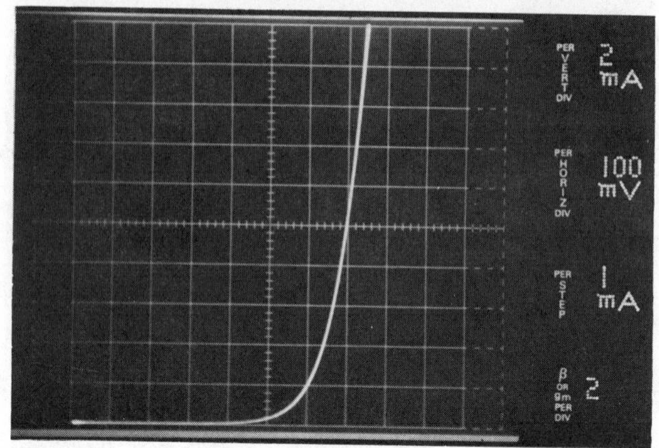

IN4154	($.20 ea.)*	*PRV*	V_F	$@I_F$	I_R	t_{rr}
	Silicon signal diode	25 V	1.0 V	@ 30 mA	0.1 μA	4.0 μsec

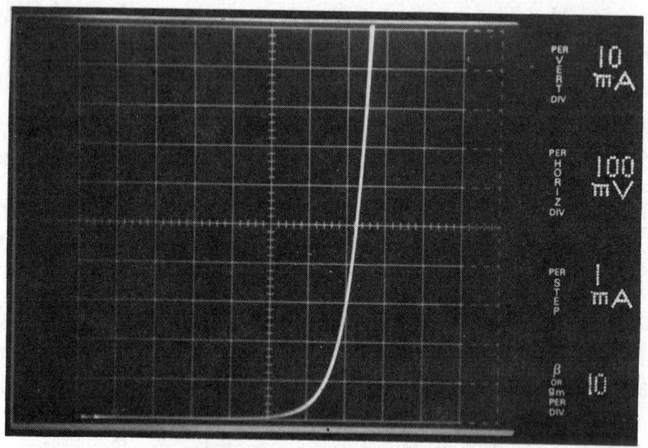

IN4004	($.10 ea.)*	*PVR*	V_F	$@I_F$	I_R	t_{rr}
	Silicon rectifier diode	400 V	1.6 V	@ 1.0 A	0.03 mA	30 μsec

* Prices shown are approximate and for small quantities.

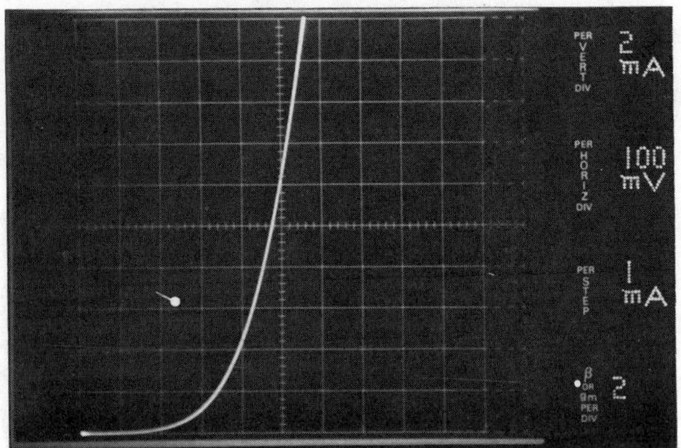

IN34A Germanium point contact signal diode ($.25 ea.)

PRV	V_F	I_F	I_R
60 V	1.0 V	@ 5 mA	500 μA @ 50 V

G.2 GERMANIUM TRANSISTORS

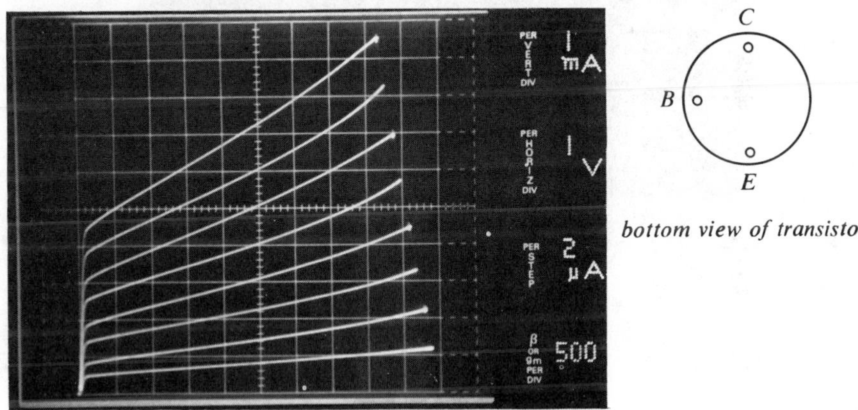

bottom view of transistor

2N1308 npn switching or general purpose, metal case ($1 ea.)

P_{max}	$V_{CB\,max}$	$V_{CE\,max}$	h_{FE}	$V_{CE\,sat}$	f_{hfb}
150 mW	25 V	—	> 80 @ 10 mA	0.2 V @ 10 mA	15 MHz

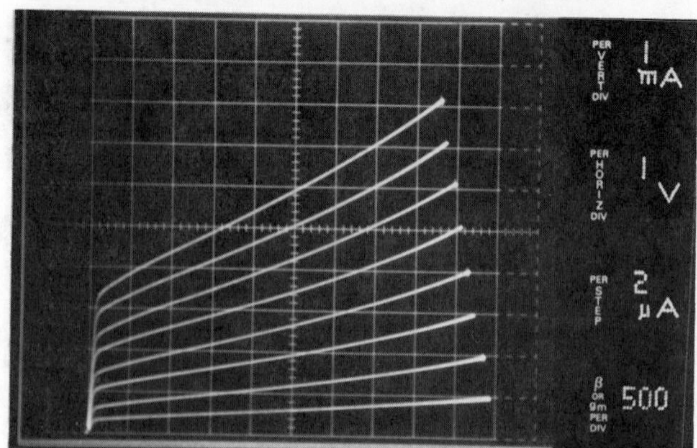

2N1309 pnp switching or general purpose, metal case ($1 ea.)

P_{max}	$V_{CB\,max}$	$V_{CE\,max}$	h_{FE}	$V_{CE\,sat}$	f_{hfb}
150 mW	30 V	—	>80 @ 10 mA	0.2 V @ 10 mA	15 MHz

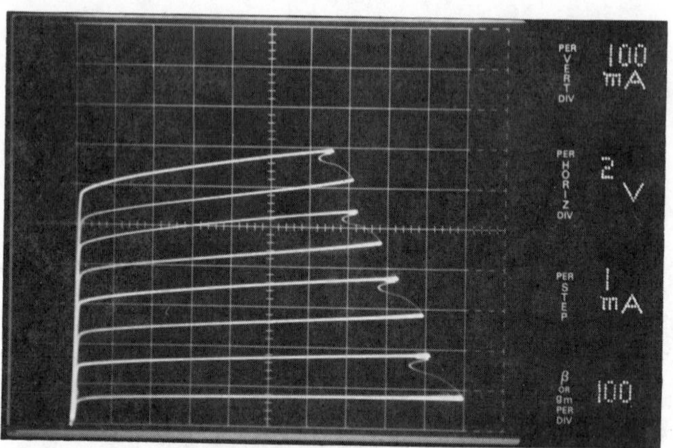

2N376A pnp audio power ($2 ea.)

P_{max}	$V_{CB\,max}$	$V_{CE\,max}$	h_{FE}	$V_{CE\,sat}$	f_{hfe}
90 W	50 V	40 V	35–120 @ 700 mA	1.75 V @ 5 A	5 kHz

G.3 SILICON TRANSISTORS

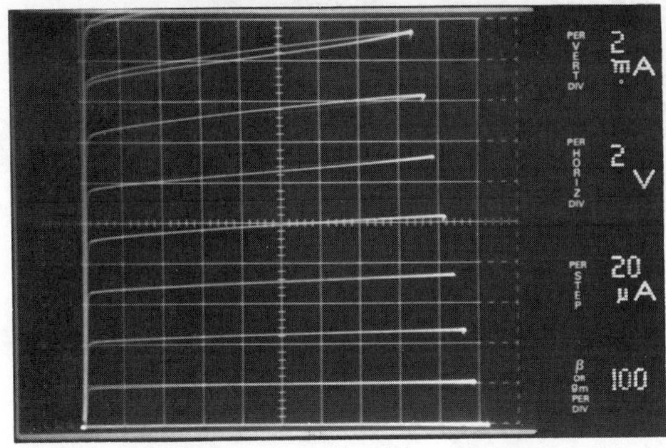

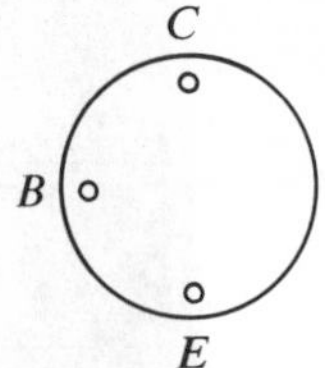

bottom view of transistor

2N2222 npn general purpose, metal case, switching ($.25 ea.)

P_{max}	$V_{CB\,max}$	$V_{CE\,max}$	h_{FE}	$V_{CE\,sat}$	f_T*
500 mW	60 V	30 V	100–300 @ 150 mA	0.4 V @ 150 mA	250 MHz

2N3565 npn audio general purpose, plastic case ($.25 ea.)

P_{max}	$V_{CB\,max}$	$V_{CE\,max}$	h_{FE}	$V_{CE\,sat}$	f_T*
200 mW	30 V	25 V	150–600 @ 1 mA	—	40 MHz

* f_T = gain–bandwidth product.

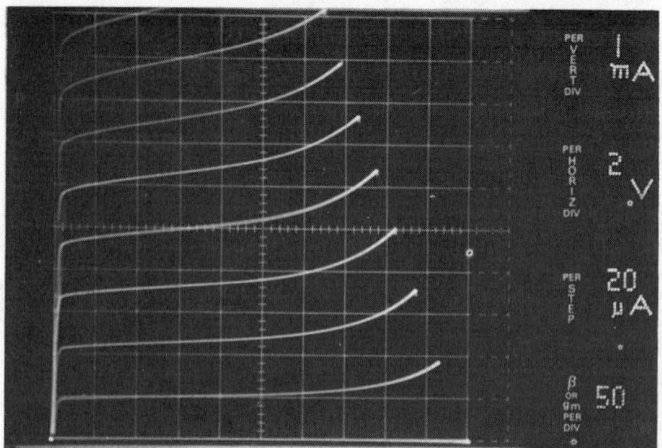

2N2369 npn switching,
hi-speed,
metal case ($.40 ea.)

P_{max}	$V_{CB\text{max}}$	$V_{CE\text{max}}$	h_{FE}	$V_{CE\text{sat}}$	f_T
360 mW	40 V	40 V	40–120 @ 10 mA	0.25 V @ 10 mA	500 MHz

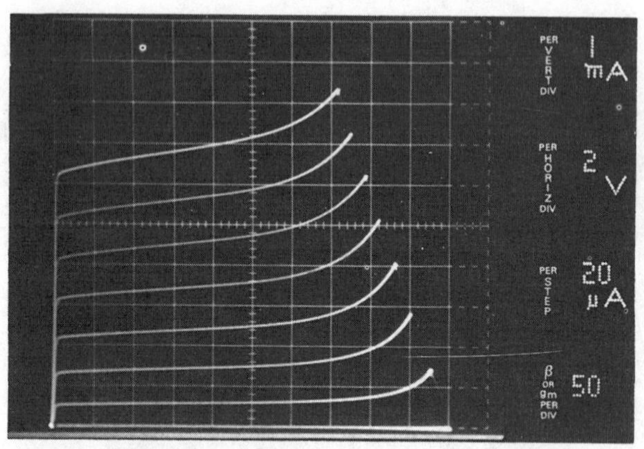

2N3646 npn switching,
hi-speed,
plastic case ($.30 ea.)

P_{max}	$V_{CB\text{ max}}$	$V_{CE\text{ max}}$	h_{FE}	$V_{EC\text{ sat}}$	f_T
200 mW	40 V	15 V	30–120 @ 30 mA	0.2 V @ 30 mA	350 MHz

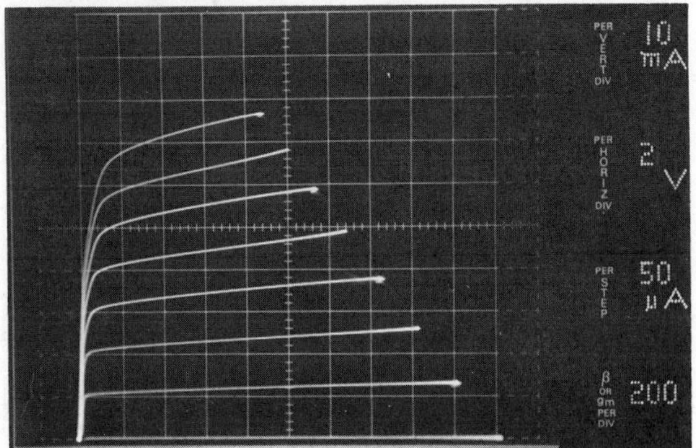

2N2907A pnp switching,
metal case,
hi-speed ($.25 ea.)

P_{max}	$V_{CB\max}$	$V_{CE\max}$	h_{FE}	$V_{CE\text{sat}}$	f_T
1.8 W	60 V	60 V	100–300 @ 150 mA	0.4 V @ 150 mA	200 MHz

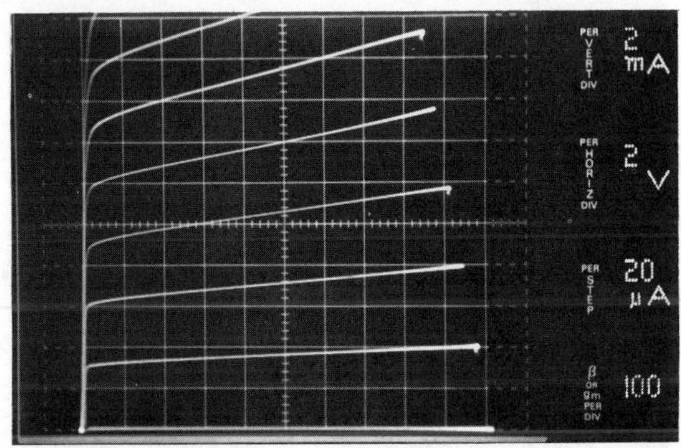

2N3251 pnp switching,
metal case,
hi-speed ($.25 ea.)

P_{max}	$V_{CB\max}$	$V_{CE\max}$	h_{FE}	$V_{CE\text{sat}}$	f_T
360 mW	50 V	40 V	100–300 @ 10 mA	0.25 V @ 10 mA	300 MHz

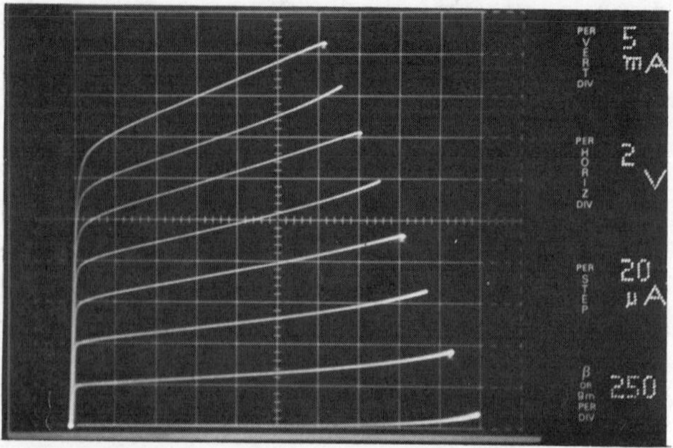

2N3638 pnp switching
($.12 ea.)
plastic case

P_{max}	$V_{CB\max}$	$V_{CE\max}$	h_{FE}	$V_{CE\text{sat}}$	f_T
300 mW	25 V	25 V	>30 @ 50 mA	0.25 V @ 50 mA	100 MHz

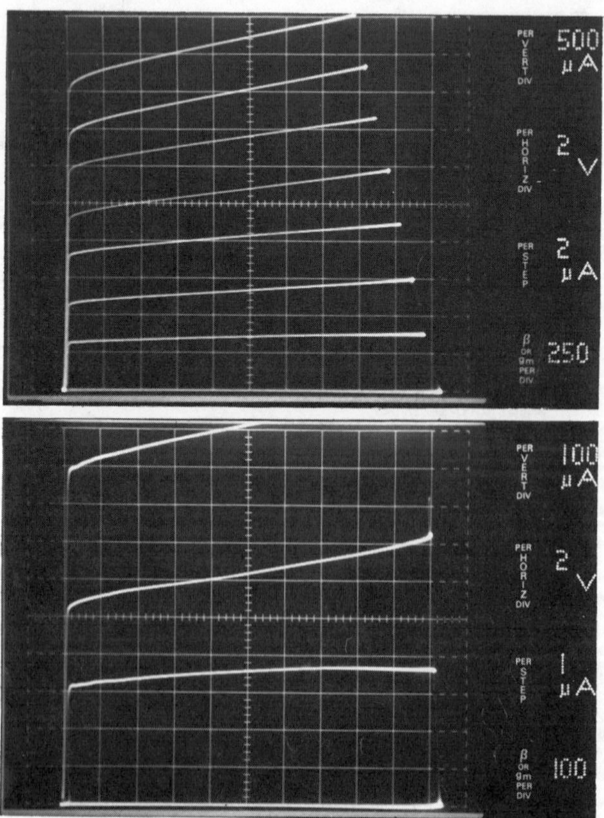

2N4250 pnp audio
(2N5086, low noise
2N5087) low I_C (\$.80 ea.)

(expanded scale)

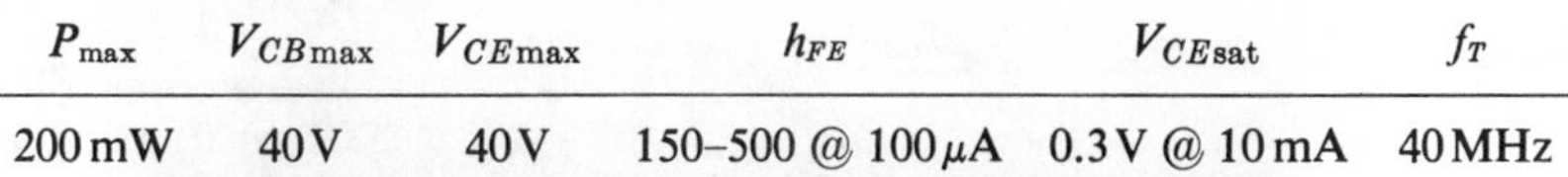

$P_{\max}$	$V_{CB\max}$	$V_{CE\max}$	h_{FE}	$V_{CE\text{sat}}$	f_T
200 mW	40 V	40 V	150–500 @ 100 μA	0.3 V @ 10 mA	40 MHz

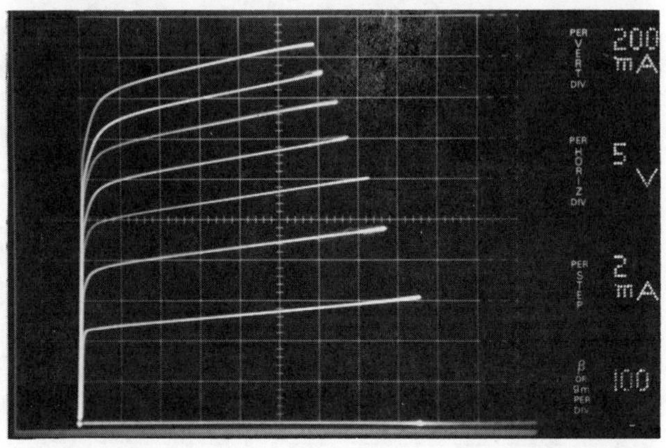

2N3055 npn audio
power (\$2 ea.)

$P_{\max}$	$V_{CB\max}$	$V_{CE\max}$	h_{FE}	$V_{CE\text{sat}}$	f_{hfe}
115 W	100 V	70 V	20–70 @ 4 A	1.1 V @ 4 A	20 kHz

G.4 FIELD EFFECT TRANSISTORS

I_{DSS} = drain–source current with gate connected to source. (max. drain current)

I_{GSS} = gate–source current with drain connected to source. (max. gate leakage current)

C_{ISS} = maximum input capacitance

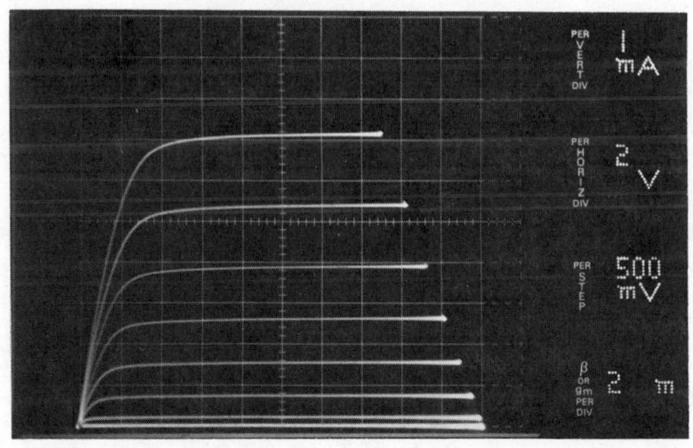

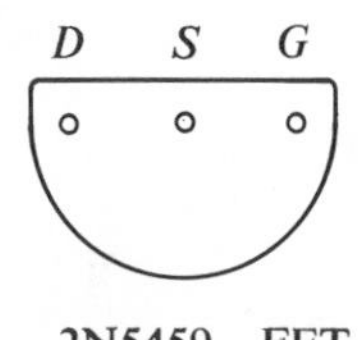

2N5459 FET

2N5459 n-channel junction, general purpose up to 20 MHz ($1 ea.)

I_{DSS}	I_{GSS}	$V_{GSS\,\max}$	y_{fs}	C_{ISS}
4–16 mA	1 nA	25 V	2000–6000 μmhos @ I_{DSS}	7 pF

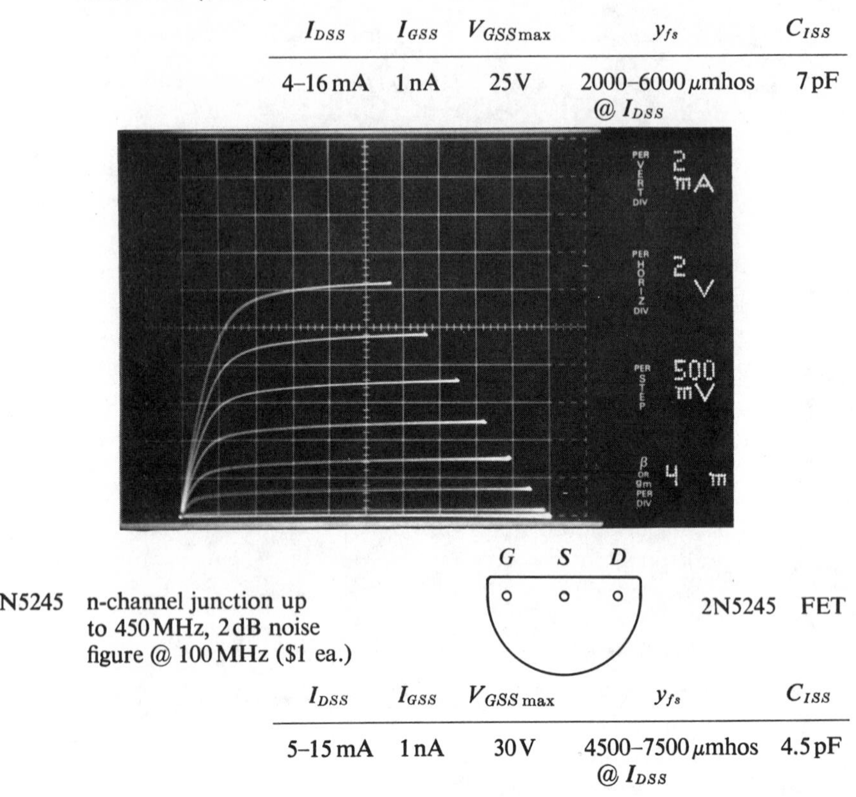

G S D

2N5245 FET

2N5245 n-channel junction up to 450 MHz, 2 dB noise figure @ 100 MHz ($1 ea.)

I_{DSS}	I_{GSS}	$V_{GSS\,\max}$	y_{fs}	C_{ISS}
5–15 mA	1 nA	30 V	4500–7500 μmhos @ I_{DSS}	4.5 pF

APPENDIX H

NUCLEAR RADIATION DETECTORS

The function of a pulse height analyzer in nuclear physics is to sort out the various types of radiation according to the size of the voltage pulses they produce. The energy of the radiation lost in the detector material is proportional to the total quantity of electric charge Q produced in the detector, because the energy required to produce one ion pair is approximately constant. Thus, the radiation energy loss is proportional to the integral of the current i from the detector: $Q = \int i\,dt$. Various special preamplifiers are used to convert this current pulse to a voltage pulse such that the *amplitude* of the voltage pulse is proportional to the radiation energy loss in the detector. Thus, the pulse height analyzer will yield a distribution function of the number of different radiations with different energies.

Charged particles of energy E produce ionization and have a finite range R in any material. $R = \int_0^E dE/(dE/dx)$, where dE/dx is the energy loss per unit path length, which depends upon the particle energy and the material. However, gamma rays do not have a finite range. The intensity of gamma rays falls off exponentially with the distance x into the absorber: $I = I_0 e^{-Kx}$. The constant K depends upon the absorber material and the energy of the gamma ray. Heavier elements such as lead have larger values of K and hence "absorb" gamma rays better than light elements.

A gamma ray of energy E_γ interacts with the detector in one of three ways: by the photoelectric effect, the Compton effect, or by pair production. In the photoelectric effect, the γ-ray is absorbed by an atom and an orbital electron is ejected (the "photoelectron"). The energy E_{pe} of the photoelectron is given by $E_{pe} = E_\gamma - \phi$, where ϕ is the binding energy of the electron in the atom. Usually ϕ is on the order of 1–10eV, and E_γ is greater than 10keV; so for all practical purposes $E_{pe} \cong E_\gamma$. Thus, a 1.16MeV γ-ray from Co^{60} will produce a 1.16MeV photoelectron, which will give up its energy by producing ion pairs in the detector material. The photoelectron may, however, leave the detector *before* it comes to rest, in which case the energy released in the detector will be less than E_γ.

In the Compton effect, the γ-ray scatters off a quasi-free electron in the detector. A lower energy scattered gamma ray γ' and an energetic Compton electron are produced. The Compton electron clearly has an energy E_{ce} less than the incident γ-ray energy: $E_{ce} = E_\gamma - E_{\gamma'}$. Thus, even if the Compton electron is completely absorbed in the detector, the energy produced in the detector will be less than E_γ.

In pair production, the γ-ray interacts with a nucleus of the detector, and a positron–electron pair is produced. The total kinetic energy of the pair $E_p + E_e$ equals E_γ minus $2m_0c^2$, the rest energy of the pair, where m_0 = electron rest mass = 0.51 MeV.

$$E_p + E_e = E_\gamma - 2m_0c^2 = E_\gamma - 1.02\,\text{MeV}.$$

Clearly this process cannot occur unless the incident γ-ray energy is greater than 1.02 MeV.

The net result is that for γ-rays striking a detector, a series of various size pulses of electric charge is produced. Anywhere from 100 to 10^9 ion pairs may be produced for each γ-ray interacting with the crystal, depending upon the interaction process and the γ-ray energy. The maximum amplitude pulses result from photoelectrons produced in and completely absorbed by the detector. Those pulses are called the "photopeak." Many smaller amplitude pulses are produced from photoelectrons escaping the detector and from Compton electrons and pair production.

Most modern γ-ray detectors are either silicon or germanium p–n junctions across which a strong electric field is applied to collect the positive and negative ions produced by the radiation. Such a detector is shown in Fig. H1.1. The

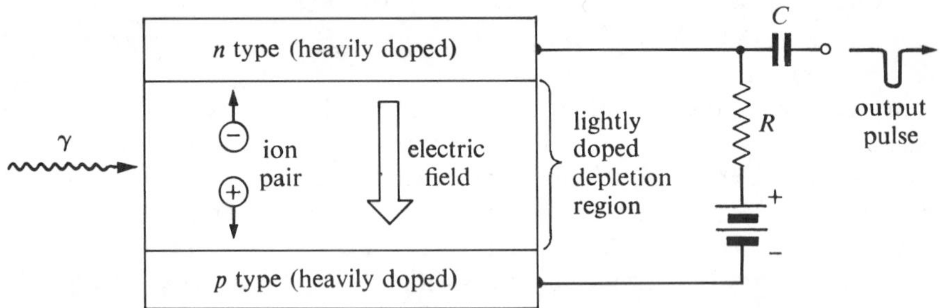

FIGURE H1.1 Solid-state radiation detector.

applied external voltage reverse biases the junction, and an electric field is produced in the central lightly doped depletion region. Thus, each time a pulse of free charge is produced in the central region by radiation, a voltage pulse is produced at the output from the charge flowing through the resistance R. Many modern detectors are "lithium drift" detectors, in which a small concentration of donor lithium ions has been "drifted" into the central region during manufacture by the application of heat and an applied electric field. The presence of the donor lithium ions tends to cancel out the effect of the acceptor impurities in the central region.

The output pulses from solid-state detectors tend to be from 1 to 10μsec in duration, and hence extremely fast amplifiers must be used to amplify them without serious attenuation. The amplifiers must also be linear, to retain the linear relation between radiation energy and pulse amplitude or pulse height. Other radiation detectors are often used, in particular scintillation detectors which produce a small flash of light when struck by a charged particle. Gamma rays can be detected by such scintillators only if they produce secondary charged particles: photoelectrons, Compton electrons, or electron–positron pairs. The scintillator is usually either plastic, a sodium iodide crystal doped with thallium, NaI(Tl); or a cesium iodide crystal doped with thallium, CsI(Tl). The scintillator is usually fastened directly to the face of a photomultiplier tube which converts the light pulses to electrical pulses. The pulses from a plastic scintillator are several nanoseconds long, while those from an alkali halide crystal are about one microsecond long.

The photomultiplier tube is extremely sensitive to small flashes of light. It consists of a photocathode electrode at negative high voltage followed by a series of 10 or 12 other electrodes (called dynodes), each at a slightly less negative voltage as shown in Fig. H1.2. The incident light strikes the photocathode and causes the emission of several photoelectrons. (The photocathode is designed to have a very low work function.) These photoelectrons are then accelerated so as to strike the first dynode where electron multiplication occurs. Each electron striking the dynode causes anywhere from 3 to 6 "secondary" electrons

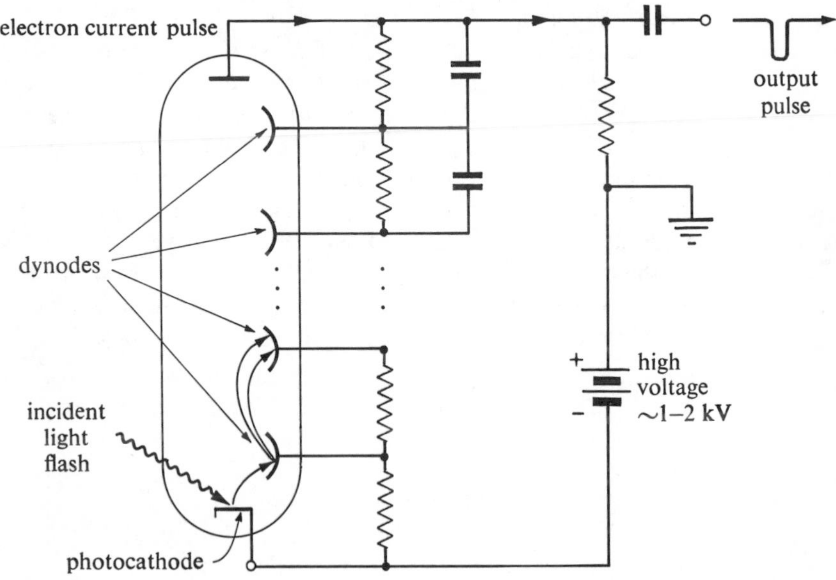

FIGURE H1.2 Photomultiplier tube.

to be emitted from the dynode. These secondary electrons are then accelerated to the second dynode, and so on. Thus, a considerable multiplication of electrons occurs, and a large output current pulse is present at the last dynode. Photomultiplier tubes are so sensitive that individual light photons can be detected.

Gas-filled detectors were very commonly used in the past; the Geiger tube, shown in Fig. H1.3, was perhaps the most common. It consists of a metal

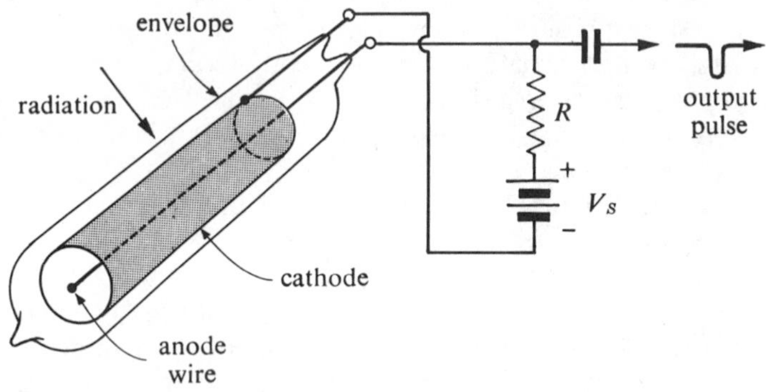

FIGURE H1.3

cylinder, usually grounded, with a positively charged wire at the center. Radiation passing through the gas produces ion pairs that are accelerated by the electric field between the anode cylinder and the wire cathode. For a Geiger tube, this electric field is strong enough ($V_s \cong 1500$ V) so that ion multiplication occurs; that is, the electron initially produced by the radiation is accelerated by the electric field and picks up sufficient energy to ionize more gas molecules and so on. The result is that a large voltage pulse about 1–5 μsec long and about one volt in amplitude is produced at the output, and the amplitude of the pulse is independent of the energy of the incident radiation. For lower supply voltages, less ion multiplication occurs, and the amplitude of the output pulses is proportional to the energy of the incident radiation. Such an instrument is called a proportional counter. Very low supply voltages and a slightly different type of tube are used in ionization chambers.

APPENDIX I

SUGGESTED LABORATORY EXPERIMENTS

The following experiments take approximately two laboratory periods each unless otherwise noted. The experiments gradually become more sophisticated as the semester progresses, and the material covered in the lectures becomes more advanced. For every experiment, however, the student is deliberately given the minimum possible instruction in writing. I feel it is essential to require the student to calculate the values of the various resistors and capacitors in the circuits. Otherwise, the experiments quickly degenerate into "cookbook" exercises that require little or no original thinking. In this way the student also quickly gains a feeling for the approximate magnitudes of various components and voltages.

It is essential that the instructor be available for most of the laboratory period to answer questions. It is also essential that the instructor already have done the experiment himself rather than just have looked at the apparatus and instruction sheet. Numerous small practical questions arise in the students' minds, particularly in the first few weeks, which can be answered quickly only by someone who has actually performed the experiment himself. For these reasons, if a student is used to supervise part of the laboratory period, he should have a detailed practical background in the subject matter. Indeed, he should have taken the course himself if at all possible. A good senior physics major, for example, could probably perform well as a laboratory instructor if he did well in the course the previous year. However, a faculty member still should give the lectures and be available for the first hour of each laboratory period.

Students typically work in groups of two at a laboratory bench which contains: one commercial oscilloscope with probe (a 1x probe is sufficient); one homemade, variable voltage, short circuit protected, dc power supply; one commercial broadband sine wave generator (a square wave or pulse generator is necessary for several experiments); and a homemade circuit breadboard vertically mounted on a table top size relay rack. A photograph of a bench setup with two breadboards is shown in Fig. I1.1. There is a common supply of resistors, transistors, capacitors, and other small parts centrally located in the laboratory. Each bench also has its own set of two BNC–BNC cables, banana plugs, clip leads, and a dozen or so spring clip solderless connectors which fit snugly into the holes in the breadboard.

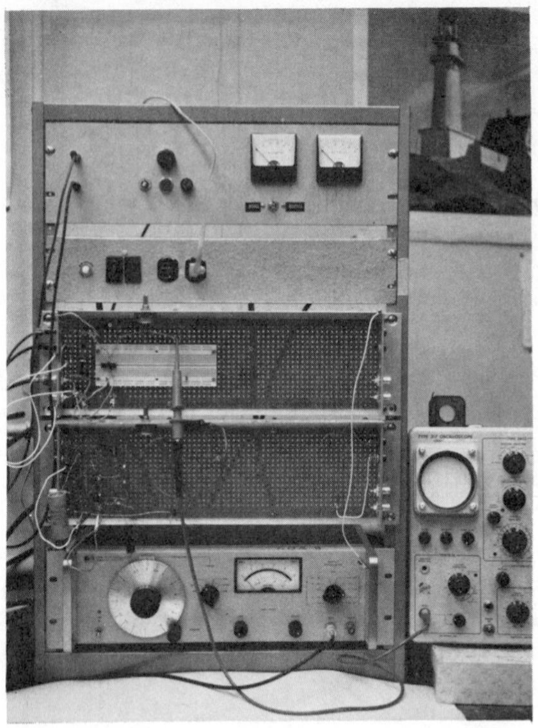

FIGURE I1.1 Laboratory bench for two students.

The breadboard consists of a 4.8 × 16.99 × $\frac{1}{16}$ inch piece of phenolic punched terminal board (Vectorboard No. 64A18 @ $1.50) mounted in the vertical plane on the front of the table model relay rack. This board contains 1,152 0.091 in. diameter holes on 0.265 in. centers in a rectangular grid. The spring clip push-in terminals (Vector No. T30N–2 @ $7.00/100) fit snugly in the holes and provide solderless tie points for circuit wiring. The phenolic board is securely screwed to an aluminum U-channel on the top and bottom, which are in turn screwed to the relay rack. A vertical $\frac{1}{8}$ in. thick aluminum angle is at each end of the phenolic board on which are mounted female banana plugs for power supply leads and several female BNC chassis connectors for feeding signals in and out of the breadboard. A horizontal $\frac{1}{16}''$ thick aluminum plate 2″ wide is mounted just above the breadboard resting on the tops of the $\frac{1}{8}''$ angle. Several $\frac{3}{8}''$ diameter holes in this plate provide convenient mounting for potentiometers. Hundreds of different electronic experiments can be easily performed at this bench.

For integrated circuit experiments, a small ($2\frac{1}{4}''$ × $6\frac{1}{2}''$) plug-in terminal board (SK–10 "Universal Component Socket") is used, @ $18. (El Instruments Inc., 61 First St., Derby, Conn.) Dual-in-line integrated circuits plug directly into the SK–10 socket along with various resistors and capacitors. Number 20 or 22 hook-up wire also fits into the socket.

The short-circuit protected, variable dc voltage power supply shown in Fig. I1.2 has been used for most of the experiments. The only critical thing about the circuit is that the power transistor T_1 be germanium and the short circuit protection diode D_5 be silicon. Normally D_5 does not conduct unless a large enough current is drawn through transistor T_1 so that $V_{EB} + I_E R_2$ exceeds the turn-on voltage of D_5 (0.55 V). When D_5 conducts, the output current from the supply is limited. With $R_2 = 1$ ohm, the output current will be limited to a maximum of about 300 mA. A smaller value of R_2 would result in a higher limiting value of the output current; e.g., if $R_2 = 0.5$ ohm, then the maximum output current would be about 600 mA.

EXPERIMENT 1—KIRCHHOFF'S LAWS

1. Design and construct two circuits using the variable voltage dc power supply and several resistors (at least three) to demonstrate (a) Kirchhoff's voltage law and (b) Kirchhoff's current law.
   Measure voltages with the oscilloscope to verify the two Kirchhoff laws. Check the circuits you propose to use with the instructor *before* you construct them.
2. Using resistor(s) and capacitor(s), design and construct a series circuit to demonstrate Kirchhoff's voltage law for ac voltages. Use $f = 1$ kHz. Measure the voltages across the components with the scope. Carefully sketch the waveforms.
   How does the fact that the *phase* of the voltage and current through a capacitor differ by 90° affect the measurements? How must you trigger the scope to measure the phase of V_R and V_c?

EXPERIMENT 2—*RLC* CIRCUITS

1. Design and wire up a low-pass *RC* filter circuit with a break point somewhere between 1 kHz and 50 kHz.
   Measure the voltage gain $|A_v|$ as a function of frequency with a sinusoidal input. Graph on log-log paper.
2. Repeat the above for a high-pass *RC* filter circuit.
3. Design and construct a parallel *LC* resonant circuit and measure the voltage gain A_v as a function of frequency. Graph. Choose a resonant frequency somewhere in the range 50 kHz–200 kHz. Do for $r = 0$ and for $r \cong R_L$ where R_L = resistance of L. Calculate the Q for each value of r and compare with the experimental value.

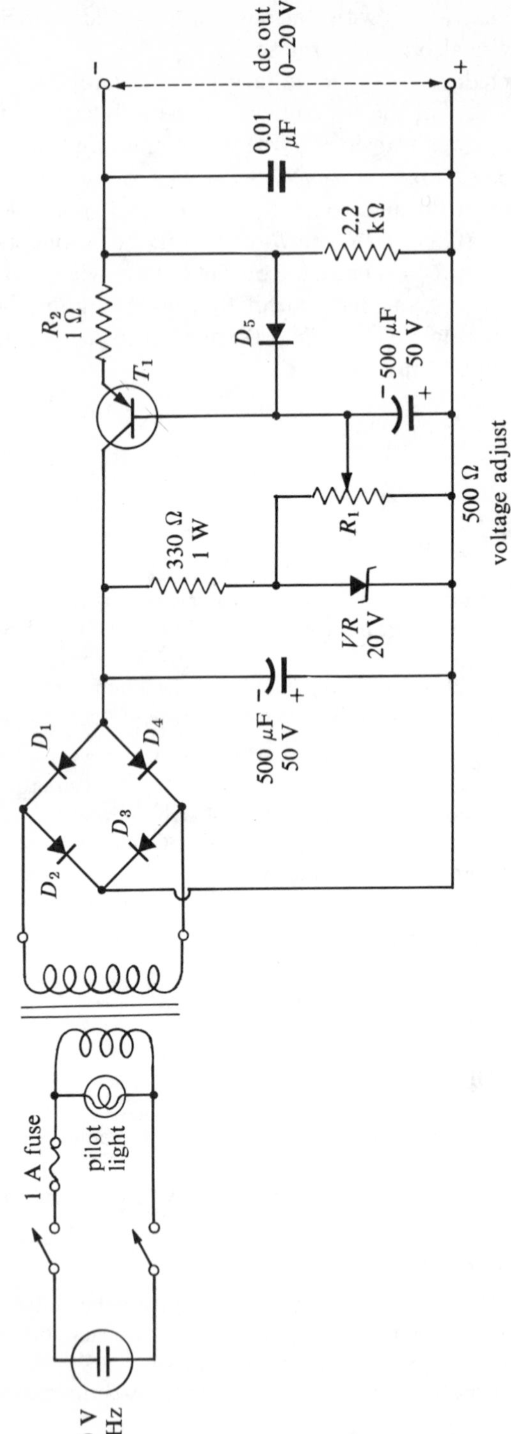

D_1–D_5 = 1N2484 or 1N4004 silicon rectifier diode
VR = 20 V zener diode 1 watt
T_1 = 2N376 or equivalent Germanium power transistor

FIGURE I1.2 Variable voltage dc short-circuit protected power supply.

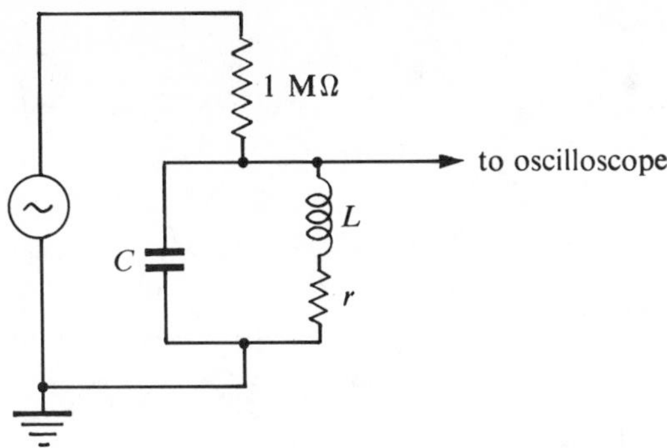

parallel LC circuit

4. Repeat for a series *LC* circuit.
For (3) and (4) use an rf choke with $L = 100\mu\text{H}$ or $10\mu\text{H}$. ($1\mu\text{H} \equiv 10^{-6}\,\text{H}$)

5. If time permits (and your interest is sufficient), repeat (3) briefly (with $r = 0$) using an *air core* coil of approximately 10 to 20 turns of wire for L. Is the Q larger or smaller?

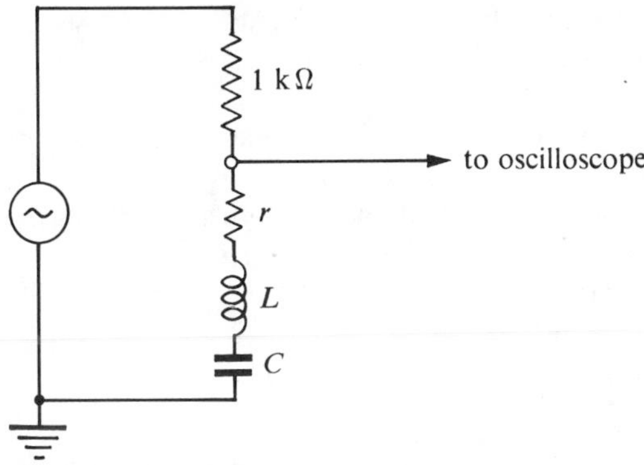

series LC circuit

Hint: To facilitate data taking, first run through the available range of frequencies quickly, taking a mental note of $|A_v|$ for $f = 10\,\text{Hz}$, $1\,\text{kHz}$, $100\,\text{kHz}$, $1\,\text{MHz}$. If the break point (parts 1 and 2) or the resonant peak or dip (parts 3 and 4) falls within the frequency range of your instrument, then go back and record the data, taking closely spaced data points only in regions where A_v is a rapidly varying function of frequency. In general, a $|A_v|$ versus f plot need contain only from 10 to 20 points for this experiment.

EXPERIMENT 3—DIFFERENTIATING AND INTEGRATING CIRCUITS

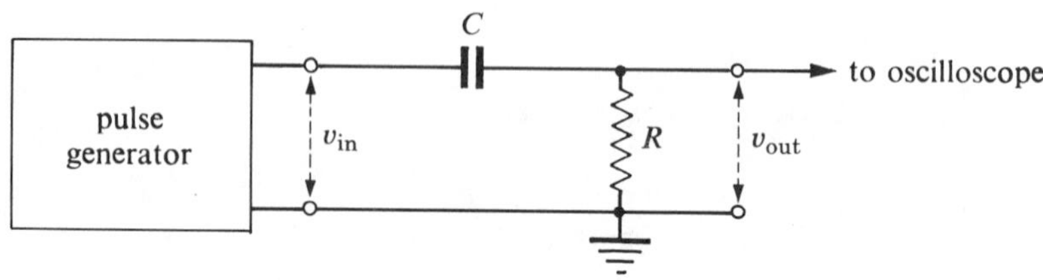

differentiating circuit

1. *Differentiating Circuit.* Hook up the pulse generator (GR 1217) to the scope alone and become familiar with the various controls on the pulse generator. Take the output of the pulse generator between the red terminal marked "+" and the metal terminal directly underneath it. Use a very low output pulse amplitude (almost fully counterclockwise). Set the pulse width at several hundred microseconds and the pulse repetition frequency ("PRF") at about 300 or 1,000 per second. Remember that the pulse width must be less than $(\text{PRF})^{-1}$.
   Hook up the following circuit. Do *not* use an electrolytic condenser for C; use a mica or paper condenser. Use $R \cong 1\,\text{k}\Omega$ or $10\,\text{k}\Omega$.
   Carefully sketch the input and output waveforms *to scale* for the three cases:
   (a) $RC \ll \tau$ $\quad RC = \tau/100 \quad$ Try $R = 10\,\text{k}\Omega$, $C = 100\,\text{pF}$
   (b) $RC = \tau$ $\quad$ Try $R \cong 10\,\text{k}\Omega$, $C = 0.01\,\mu\text{F}$
   (c) $RC \gg \tau$ $\quad RC \cong 100\tau \quad$ Try $R = 100\,\text{k}\Omega$, $C = 0.1\,\mu\text{F}$
   Measure the "sag."

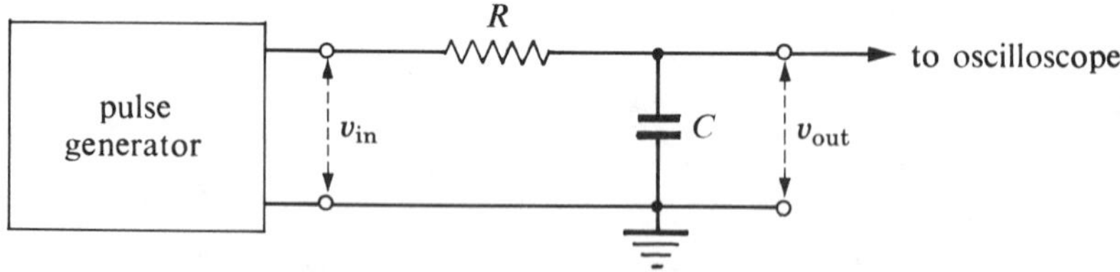

integrating circuit

2. *Integrating Circuit.* Hook up the following circuit. Use $R \cong 1\,\text{k}\Omega$ or $10\,\text{k}\Omega$. Carefully sketch the input and output waveforms to scale for the three cases (a) $RC \ll \tau$, (b) $RC = \tau$, and (c) $RC \gg \tau$.
   Results: (1) Explain the data in terms of a qualitative argument involving the Fourier components of the input pulse. (2) Derive mathematically the shape of the output pulse for the two circuits assuming a perfectly square (zero rise time) input pulse. Hand in separately as a problem.

EXPERIMENT 4—Q AND THE FOURIER ANALYSIS OF A SQUARE PULSE

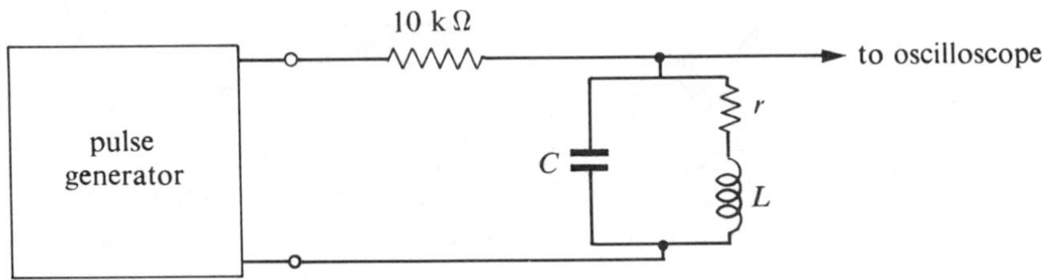

1. Construct a parallel LC circuit with a resonant frequency of several hundred kilohertz. Hook it up to a sine wave generator and measure its Q by measuring the resonant frequency, f_0, and the bandwidth, Δf, the full frequency width at half the maximum. $Q = f_0/\Delta f$
   Approximately 10 points will be sufficient to measure the Q.
2. Measure the Q of the LC circuit from $Q = \omega L/R_L$ where R_L is the dc resistance of the inductance.

3. Hook up the square wave generator and observe the damped sinusoidal oscillations (called "ringing") across the LC circuit for two or three values of r. Start with $r = 0$. Measure the Q by measuring the damping of the oscillations. (You will have to solve the differential equation for a damped oscillator to do this.)

EXPERIMENT 5—DIODE CURVE

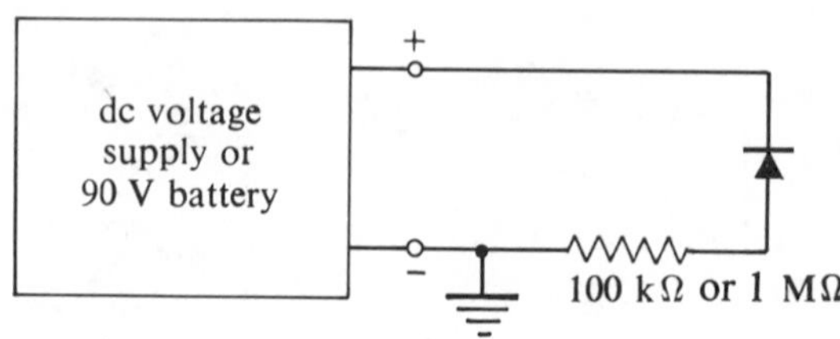

reverse current

1. (*Reverse current direction.*) Hook up the following circuit using a silicon diode (IN2484, IN4004, IN4154, IN4148, etc.)

   Measure the reverse current for several voltages by measuring the voltage drop across $R = 100\text{k}\Omega$. (For best sensitivity, use a 5 mV/cm scope.) Do not exceed the peak inverse voltage ("PIV" or "PRV") of the diode. Measure the reverse current using a 90 V battery in place of the variable voltage source. Try placing the scope in series with the diode in place of R to measure the current.

   Questions: (a) Why not measure the reverse current with an ammeter in series with the diode? (b) Why not measure the voltage drop across R with a VOM?

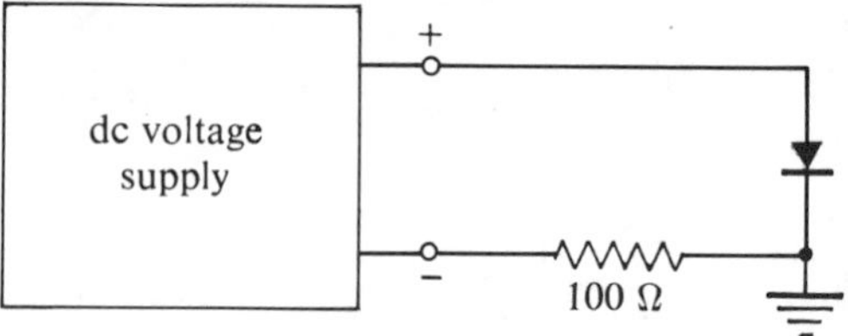

2. (*Forward current direction.*) Hook up the following circuit. Note the change in the position of the ground.

   Measure the forward current by measuring the voltage drop across R. Measure the voltage drop across the diode. You may need to change the value of R to obtain a reasonable voltage drop across R.

   Results: (a) Plot I through the diode versus V across the diode. Use an expanded scale for the reverse current.

   Repeat the above for an IN34A or other germanium diode. (b) Compare the reverse current of the silicon diode with that of the germanium diode. (c) Compare the turn-on voltage of the silicon diode with the germanium diode.

EXPERIMENT 6—POWER SUPPLY

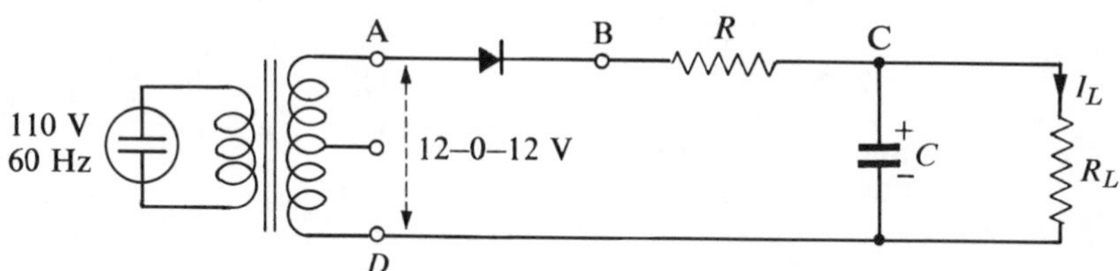

half-wave circuit

1. *Half-Wave Circuit.* Measure V_{AD} with the scope. Compare with the labeled transformer secondary voltage. Wire up the above circuit. Measure I_L, V_{AD}, V_{AB}, V_{BC}, V_{CD}, for $R_L \cong 300, 200, 100, 50$ ohms. Measure the ripple amplitude and frequency with the scope on ac coupling. Sketch carefully to scale for each value of R_L.

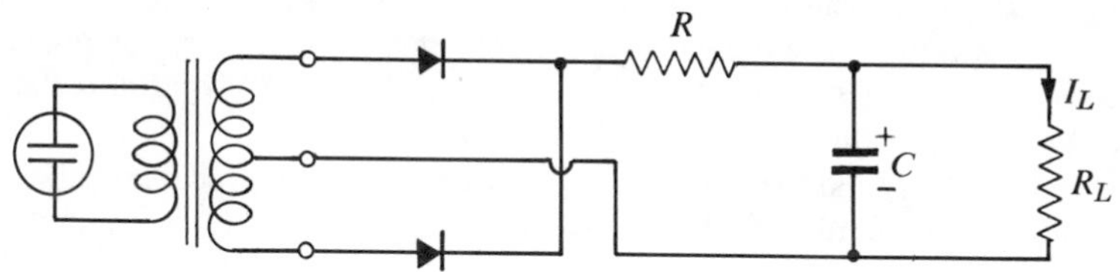

full-wave circuit

2. *Full-Wave Circuit.* Repeat the above measurements.

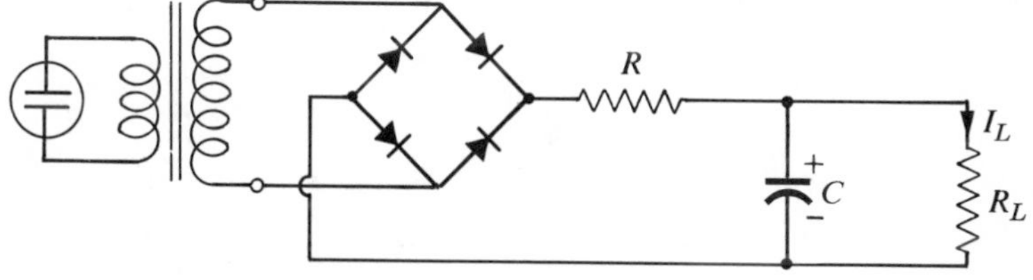

full-wave bridge circuit

3. *Bridge Full-Wave Circuit.* Repeat the above measurements.
4. Try two *RC* filter circuits in series. Measure V_{out} dc, ripple, and I_L.

   Questions: (a) Explain why *R* can't be increased indefinitely. (b) Explain why the ripple amplitude depends upon the load current. (c) Estimate the forward resistance of the diodes used. (d) Summarize your data in a table giving V_{out} dc, I_L, ripple. (e) Graph V_{out} dc versus I_L.

EXPERIMENT 7—ZENER DIODE

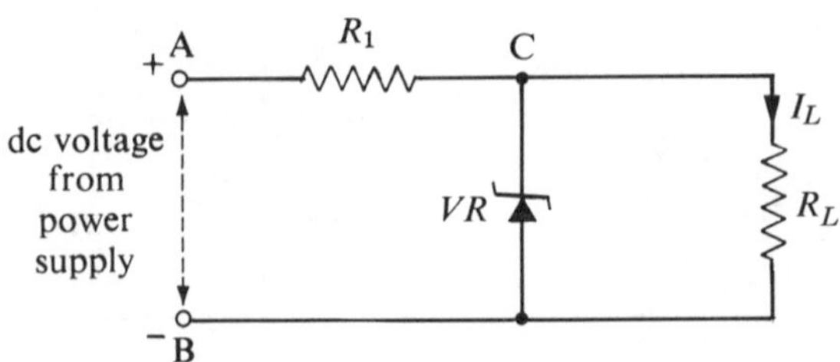

1. Measure the current–voltage curve of your zener diode. Be sure to increase the reverse voltage enough so that the diode breaks down. Also, you must be sure to limit the current through the zener diode so as not to exceed the maximum power dissipation for your diode. The power dissipated in the zener diode when it has fired is given by $P = IV_z$, where I is the current through the diode and V_z is the zener breakdown voltage.
2. Hook up your zener diode and a resistor R_1 to the output of your power supply from the last experiment (with one RC filter section). The value of R_1 must be carefully chosen to protect the zener diode. *Question:* When is maximum power dissipated in the zener? When $R_L \to 0$ or when $R_L \to \infty$? Note also that we must have $V_{AB} > V_z$ in order to "fire" the zener.

   Measure V_{AB}, V_{CB}, and the ripple voltage across the load R_L for reasonable load currents I_L : 1 mA, 10 mA, and 50 mA, with and without the zener–R_1 combination.
3. Determine the ac dynamic resistance of your zener diode. *Hint:* Consider the small ripple voltage.

EXPERIMENT 8—DC TRANSISTOR CURVES

Look up the maximum voltage, current, and power ratings of the transistor and do not exceed these ratings at any time during the experiment. Transistors can be destroyed in several milliseconds if too high a voltage is applied. Hook up one of the following circuits:

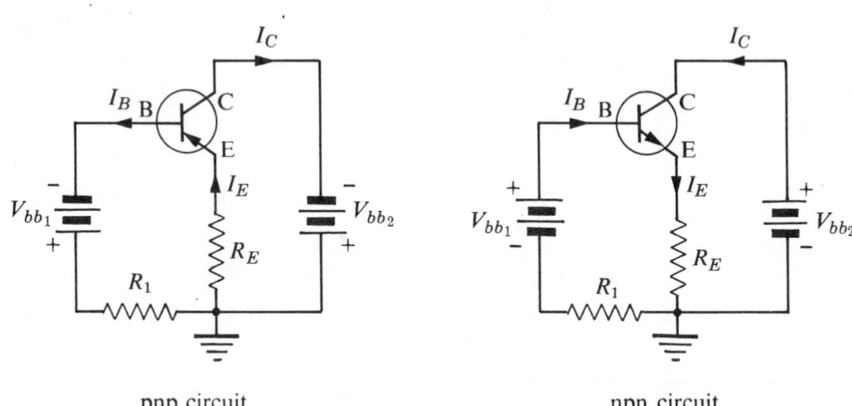

pnp circuit npn circuit

1. Plot a graph of I_C versus V_{CE} for various *fixed* values of I_B. Before you take any data, draw the $I_C V_{CE} = P_{\text{max}}$ curve on your graph, and shade in the region corresponding to $P > P_{\text{max}}$. *Never* run the transistor in this shaded region. I_B can be determined by measuring the voltage drop V_1 across R_1 and using $I_B = V_1/R_1$. Be sure to keep I_B constant by varying V_{bb_1} as V_{CE} is changed. V_{CE} can be measured directly with the oscilloscope, I_C by measuring the voltage drop V_E across R_E and using $I_E = V_E/R_E$. Use a 0–25 mA I_C scale, 0–20 V V_{CE} scale. Try $R_1 = 100\,\text{k}\Omega$. Approximately 4 or 5 points are sufficient for each fixed value of I_B. $I_C = I_E - I_B \cong I_E$.
2. Plot a graph of V_{BE} versus I_B for various fixed values of V_{CE}. V_{BE} can be measured directly with the scope, and as in part 1, $I_B = V_1/R_1$. Use $V_{CE} = 5\text{V}$, 10V, and 15V. Use a 0–1.0V V_{BE} scale, 0–60μA I_B scale.
3. Place a resistor of several thousand ohms in series with the collector. With the transistor collector current $I_C \cong 10\text{mA}$, observe the collector currents as a function of time after you short out R_E. Do not exceed $I_{C\text{max}}$. (a) Explain your observations qualitatively in terms of the temperature dependence of the number of majority charge carriers. (b) Why does R_E prevent this behavior?
4. From an inspection of your graph of I_C versus V_{CE} for various values of I_B, estimate α. Is α a constant?

EXPERIMENT 9—COMMON EMITTER AMPLIFIER, PART I

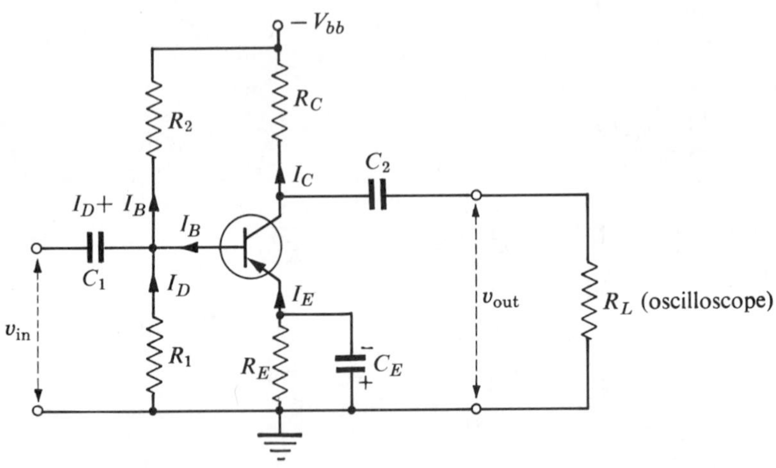

pnp circuit

1. Sketch the I_C versus V_{CE} curves for your transistor in your lab book.
2. Draw the load line for the desired application. This fixes V_{bb} and $(R_C + R_E)$.
3. Choose the operating point ("quiescent point"). This fixes I_B, I_C, V_{CE}. Choose $I_D \gg I_B$.
4. Pick R_E to be several hundred ohms. Calculate $V_E = I_E R_E$.
5. $V_B = V_E + 0.6\text{V}$ for a Si transistor.
   $V_B = V_E + 0.3\text{V}$ for a Ge transistor.
6. Calculate R_1 and R_2 from V_{bb}, V_B, I_B, I_D.
7. Calculate R_C knowing $(R_C + R_E)$ and R_E.
8. Calculate C_1 to make $1/\omega C_1 \ll (R_1 \| R_2)$ at lowest frequency.
9. Calculate C_2 to make $1/\omega C_2 \ll R_L$ at lowest frequency.
10. Check $R_E/(R_1 \| R_2)$ ratio. (More on this later.) $(R_1 \| R_2)/R_E$ should be between 5 and 10 for good thermal stability.
11. Measure the actual dc voltage values at various points of your amplifier and record on the schematic diagram in your lab book. Compare with the calculated voltages. Remember that $I_C \gtrsim 1\text{ mA}$ and $V_{CE} > 1$ or 2V in order for a transistor to operate properly. If either of these conditions is not satisfied, you must change the operating point to achieve them—by changing R_1, R_2, and R_E, and perhaps R_C.
12. Using a small sinusoidal input to give negligible output distortion, measure the voltage gain as a function of frequency. Plot on semilog paper. Ten points should be enough for a graph.

EXPERIMENT 10—COMMON EMITTER AMPLIFIER, PART II

1. Using a load line different from that of the previous experiment, measure the voltage gain versus frequency and graph.

2. (a) Increase the input amplitude slowly and observe the waveform of the output for distortion. Sketch the input and the output for several input amplitudes. Use $f = 1\,\text{kHz}$. Record the maximum input and output voltage for negligible output distortion.

(b) Change the operating point (by changing R_1 and R_2) to minimize the output distortion, i.e., symmetrize the clipping of the positive and negative peaks of the output.

Hint: As the input voltage goes positive the output voltage goes negative, and vice versa for either a pnp or a npn transistor. Record the maximum input and output voltage for negligible output distortion.

3. Measure the input and output impedances at 1 kHz and 1 MHz. *Hint:* It is easiest to measure the output signal voltage in measuring the input impedance.

EXPERIMENT 11—COMMON COLLECTOR AMPLIFIER (EMITTER FOLLOWER)

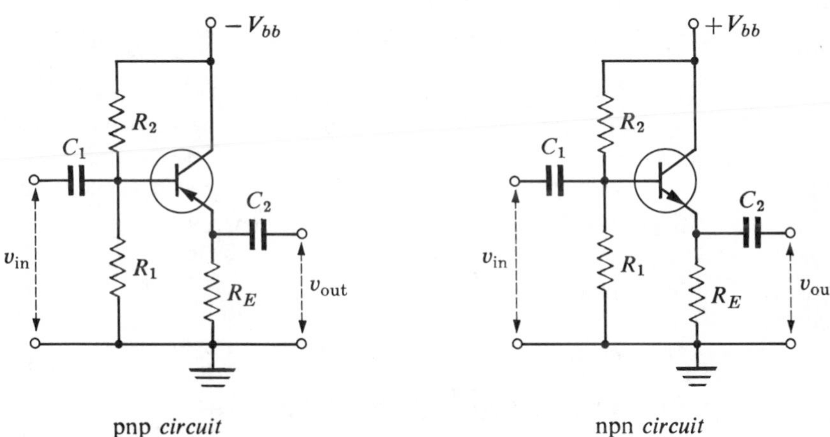

pnp *circuit* npn *circuit*

Construct the above circuit using calculated, reasonable values for the components. (Show the brief calculations in your lab notebook.) Record the actual dc voltages at various circuit points on the schematic in your lab book. Remember: $I_C \gtrsim 1\,\text{mA}$ and $V_{CE} > 1\text{–}2\,\text{V}$.

Measure the following with the hp oscillator 50 ohms output as the input to your circuit:

1. Voltage gain as a function of frequency. Graph on semilog paper. Approximately 5 to 10 points are sufficient for the graph.
2. Input impedance.
3. Output impedance at 100 Hz, 10kHz, 1 MHz.

Repeat (1) and (3) for the 600 ohm output of the hp oscillator.

Repeat (3) with a 5kΩ resistor (approximately) in series with C_1:

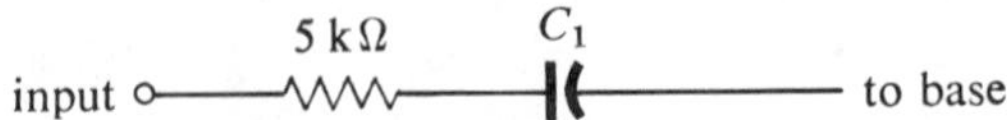

EXPERIMENT 12—MULTISTAGE AMPLIFIER EXPERIMENT (4 LAB PERIODS)

Design, construct, and debug a multistage amplifier to give a voltage gain of at least 20 with negative voltage feedback. There is no single, unique solution to this design problem; various compromises will have to be made. Try to achieve: a large bandwidth, a low output impedance, and a high input impedance. Measure $|A_v|$ as a function of frequency and graph on semilog paper. Measure Z_{in} and Z_{out} at several frequencies.

EXPERIMENT 13—COMMON BASE AMPLIFIER

1. Design, construct, and debug a common base amplifier. Calculate the dc voltages expected at various points in the circuit. Measure and record the actual dc voltages on your schematic. Remember: $I_C \gtrsim 1$ mA and $V_{CE} > 1$–2V.
2. Measure the voltage gain as a function of frequency. Graph on semilog paper.
3. Measure the output impedance. What is the limiting factor in the high-frequency response? (Consider the load presented to the amplifier by the scope.)
4. Measure the input impedance.

EXPERIMENT 14—FET CHARACTERISTIC CURVE

1. Measure the "output" characteristic curves for your FET, i.e., plot the drain current, I_D, as a function of the source–drain voltage, V_{SD}, for various constant gate–source voltages, V_{GS}. The circuit is shown below. Notice that the gate–source junction must be *reverse* biased.

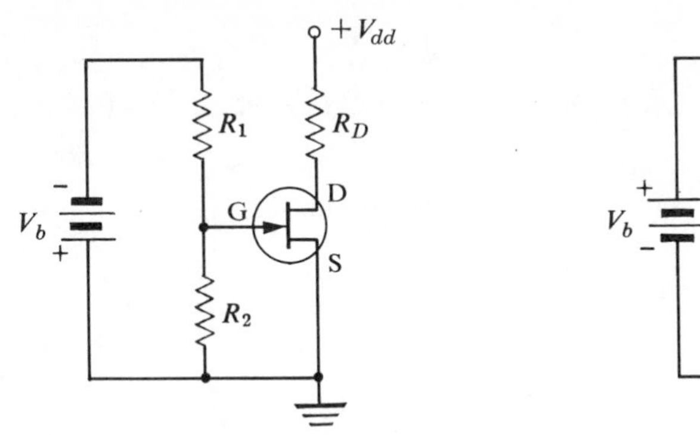

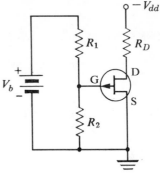

n *channel circuit* p *channel circuit*

Notice that the battery V_{bb} need not supply a large current, because the gate–source impedance of the FET is very high (on the order of tens of megohms); so R_2 can be large—100kΩ or more to draw little battery current. The ratio of R_1 and R_2 can be varied to change V_{GS}; a potentiometer can be used conveniently here as shown.

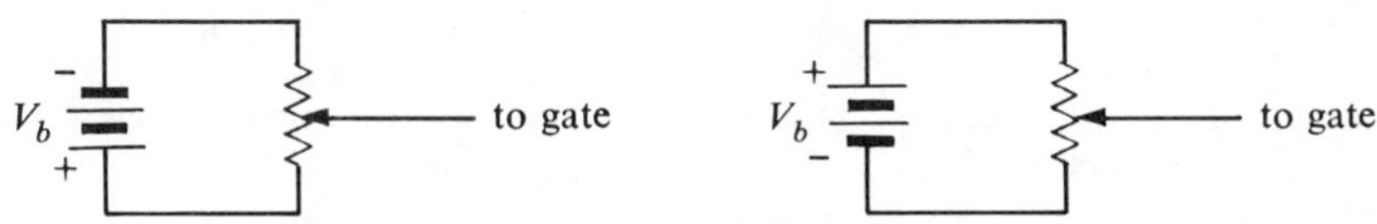

potentiometer

V_{dd} is a variable voltage power supply. V_{SD} and V_{dd} can be measured with the oscilloscope on dc coupling in the usual fashion. I_D can be calculated from Ohm's law:

$$I_D = (V_{dd} - V_{SD})/R_D.$$

Question: Does R_2 have to be readjusted to keep V_{GS} constant as V_{SD} is varied? Each graph for a fixed value of V_{GS} (say 1.0 V) should contain only approximately 5 or 6 points. Be sure to vary V_{GS} by small amounts; i.e., do graphs for $V_{GS} = 0$ V, 0.5 V, 1.0 V, and so forth until pinch-off is achieved.

Question: What is the pinch-off voltage for your FET?

2. From your graph, determine y_{22} and y_{21} for your FET.
3. What can you conclude from the curves about V_{SD} if your FET is to have a reasonably high transconductance?

EXPERIMENT 15—FET COMMON SOURCE, RC COUPLED AMPLIFIER

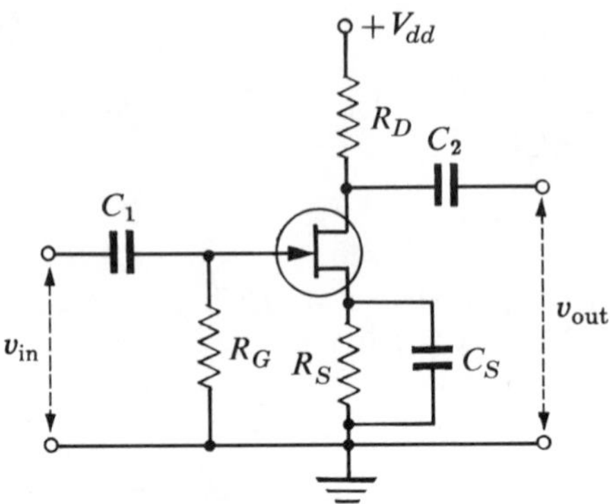

Design, construct, and debug the amplifier. You may take the minimum signal frequency to be amplified as 1000 Hz.

1. Draw the y parameter ac equivalent circuit and derive the voltage gain expression. From your experimentally determined values of y_{21} and y_{22}, calculate an approximate value of R_D to achieve a voltage gain of 10 or 15.
2. Assuming $R_D \gg R_S$, draw the dc load line on your I_D versus V_{SD} graph. Choose an operating point and calculate R_S.
3. Pick a reasonable R_G and calculate C_1, C_2, and C_S.
4. Construct the amplifier and measure the dc voltages on the gate and drain. Compare with the voltages predicted from your operating point.
5. Measure the voltage gain as a function of frequency and plot on semilog paper. Compare with the predicted voltage gain.
6. Measure Z_{in} and Z_{out}. Compare your experimental value of Z_{out} with the value calculated from your equivalent circuit.

EXPERIMENT 16—FET COMMON DRAIN AMPLIFIER (SOURCE FOLLOWER)

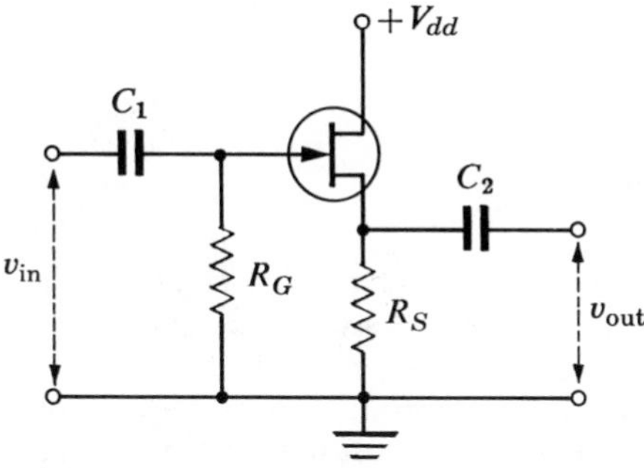

1. Design, construct, and debug the amplifier. You may take the minimum signal to be amplified as 1000 Hz.
   Choose an operating point to minimize the output impedance and maximize the voltage gain. Remember that the transconductance depends upon the drain current. Record the dc voltage values and compare with the predicted values.
2. Measure the voltage gain as a function of frequency. Graph on semilog paper.
3. Measure the output impedance at several frequencies. Change the dc drain current by approximately a factor of two and measure the output impedance.
4. Draw the ac *y* parameter equivalent circuit and derive: (a) voltage gain (b) output impedance. Compare with your data obtained in (2) and (3).

EXPERIMENT 17—VACUUM TUBE AMPLIFIER

WARNING: The +300 V dc supply is *EXTREMELY DANGEROUS*. *Use only one hand* in poking around the circuit. The other hand should be *in your pocket*.

Choose a vacuum tube triode or pentode and look up the base diagram in a tube manual. The diagram will look like this.

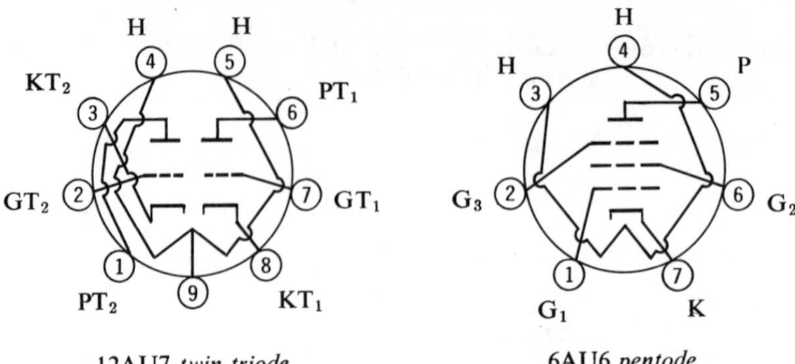

12AU7 *twin triode* 6AU6 *pentode*

The basic *RC* coupled self-bias amplifier circuit is similar to the common source FET circuit. A cathode resistor R_K provides the necessary grid-to-cathode voltage bias. Both C_K and C_S are bypass capacitors, R_P is the resistor in the plate circuit across which the ac output voltage is developed, C_1 and C_2 are coupling capacitors, and R_S is the screen resistance to supply the proper screen voltage (usually +100 V to +200 V) from the V_{bb} supply.

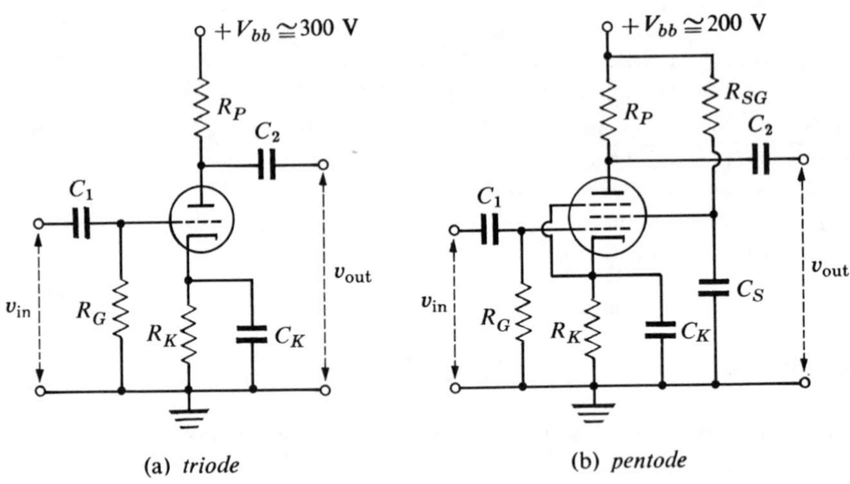

(a) *triode* (b) *pentode*

The operating point is chosen by looking at the load line drawn on the tube characteristic curves of plate current I_P versus plate–cathode voltage. The characteristic curves for the particular tube used will be found in the tube manual.

ANOTHER WARNING: The +300 V dc is *EXTREMELY DANGEROUS.* Keep one hand in your pocket when poking around the circuit.

1. Design the amplifier circuit for the tube type chosen.
2. Construct and debug.
3. Measure and record:
   (a) the dc voltages at various points in the circuit.

(b) the voltage gain as a function of frequency. Graph.
(c) the output and input impedances at an intermediate frequency.

EXPERIMENT 18—TRANSFORMER FEEDBACK OSCILLATOR

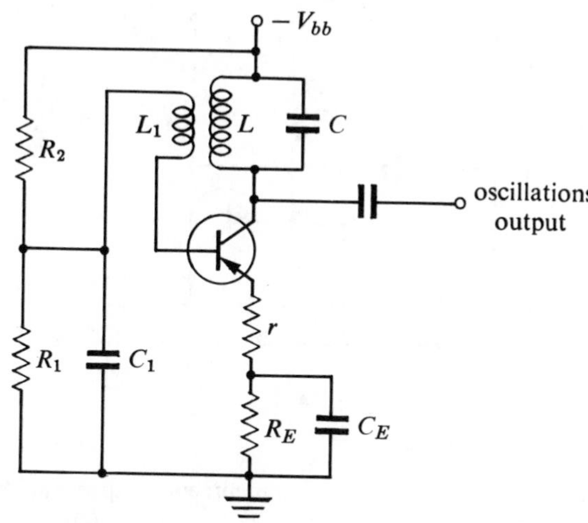

1. Calculate the values of the components so that $I_C \simeq 0.5\,\text{mA}$. Both C_1 and C_E are bypass capacitors at the frequency of oscillation determined by LC. The L and C should resonate in the range 50–200kHz. Use $r = 0$. Construct the circuit with L_1 shorted out. Measure the dc voltage values and compare with the predicted ones. A small rf choke can be used for L with several turns of insulated wire wrapped around it for L_1.
2. Feed in a small sinusoidal signal through a coupling capacitor to the base and measure the voltage gain as frequency over the range 50–200kHz. Graph $L_1 = 0$ $C_1 = 0$.
3. Disconnect the sinusoidal input and hook up L_1. The amount of feedback is determined by the coupling between L and L_1 and by the number of turns in L_1. The circuit should oscillate if the gain is sufficient and the phase of the feedback is positive. The feedback phase can be changed by reversing the leads to L_1. If the circuit fails to oscillate, the value of R_2 may be too low and R_2 may consequently be loading the L_1-L transformer, thus lowering the gain. The remedy is to increase R_1 and R_2 while keeping the same operating point.
4. Once the circuit is oscillating, reverse the leads to L_1 and verify that the oscillations stop. Explain in terms of phase.
5. With the circuit oscillating, insert various values of r in the emitter lead and observe the oscillations. Explain your observations.
6. What starts the oscillations?
7. How could this circuit be made into a variable frequency oscillator?

EXPERIMENT 19—COLPITTS OSCILLATOR

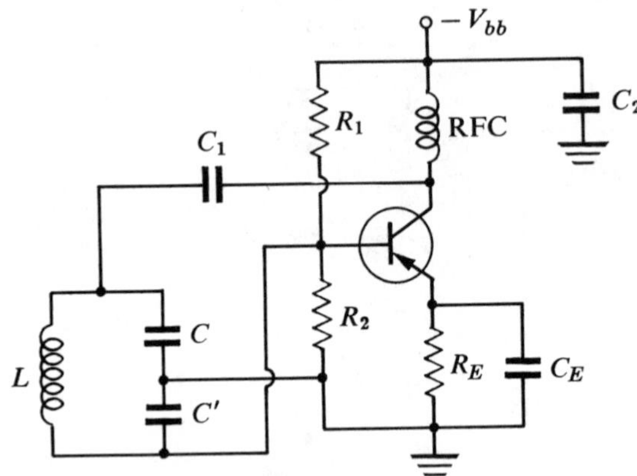

1. Calculate R_1, R_2, and R_E for a collection current of approximately 0.5 mA. The value R_2 should not be too small (it might be necessary to let $R_2 \to \infty$) to avoid lowering the Q of the LC resonant circuit. $1/\sqrt{LC_{\text{eff}}}$ determines the oscillation frequency. Capacitor C_1 is a dc blocking capacitor; caclulate its approximate capacitance. Why is it necessary? Why is the RFC necessary? Both C_2 and C_E are bypass capacitors.
2. Vary the ratio C/C' and observe the circuit performance.
3. What determines the feedback ratio?
4. Would this circuit be useful for a variable frequency oscillator?

EXPERIMENT 20—PHASE SHIFT OSCILLATOR

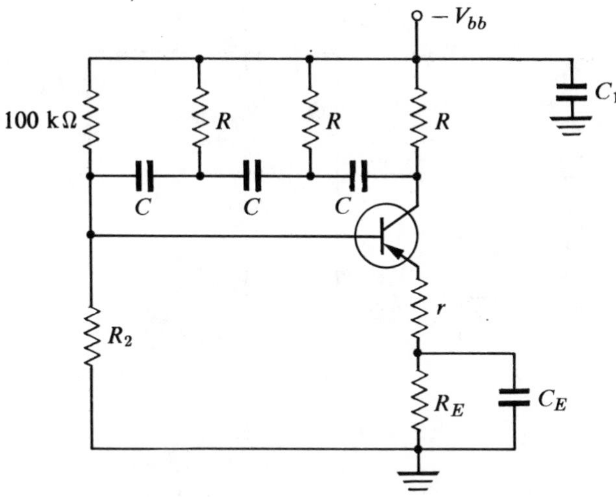

1. Calculate R and C for an oscillation frequency in the range 10kHz–100kHz. Derive the expression for RC in terms of the desired oscillation frequency.
2. Calculate R_2 and R_E so that the transistor draws about 0.5 mA. Use $R \cong 10\text{k}\Omega$ in the phase shifting network. C_1 is a bypass capacitor.
3. Sketch the output waveform for various values of r. Start with $r = 0$.
4. Would this circuit be useful for a variable frequency oscillator?

EXPERIMENT 21—HARTLEY OSCILLATOR

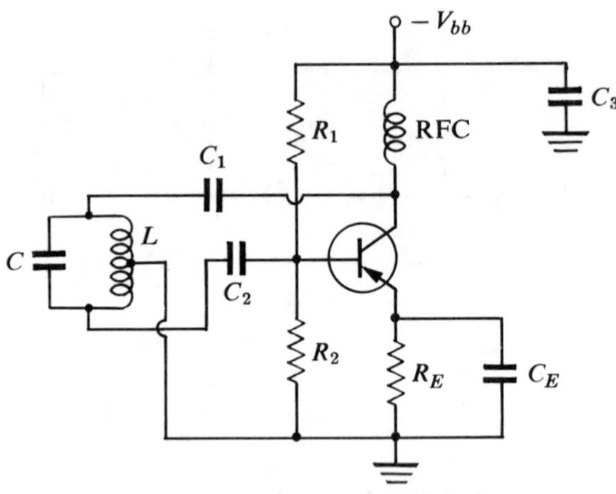

1. Calculate R_1 and R_2 and R_E for a collector current of approximately 0.5 mA. The resistance R_2 should not be too small to avoid lowering the Q of the LC resonant circuit. You may have to let $R_2 \to \infty$.
2. Vary the position of the tap and observe the circuit performance. LC determines the oscillation frequency. C_1 and C_2 are dc blocking capacitors; *calculate* their approximate capacitance. Why are they necessary? Why is the RFC necessary? Both C_3 and C_E are bypass capacitors.
3. What determines the feedback ratio β?
4. How could this circuit be made into a variable frequency oscillator?

EXPERIMENT 22—THE "BINARY" OR BISTABLE MULTIVIBRATOR OR "FLIP-FLOP"

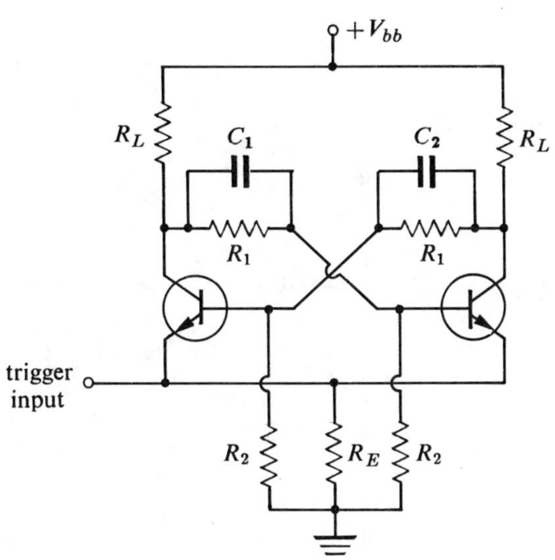

1. Construct and debug. Measure the dc collector voltages with no trigger input. The output may be taken from either (or both) collectors.

$$R_1 \geqq R_2 \geqq R_L \gg R_E .$$

   C_1 is small (approx. 100 pF–1,000 pF). Determine V_{BE} to be certain the reverse bias is sufficient to drive the "off" transistor to cutoff. Determine I_B to be certain the "on" transistor will be in saturation.

2. Sketch the input and output waveforms. Label the time axis quantitatively. What is the maximum frequency at which the circuit will flip?
3. Increase the capacitance of the C_1's and note the effect upon time of switching. Sketch the output waveform for several values of C_1.

EXPERIMENT 23—THE BINARY OR DIVIDE-BY-TWO COUNTER (FOUR LAB PERIODS)

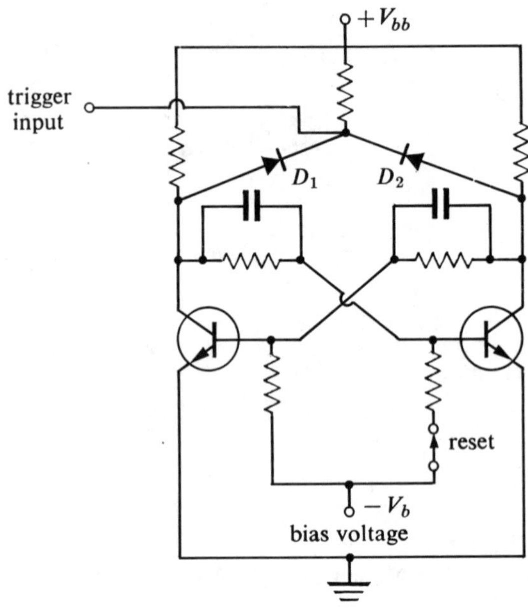

1. Construct and debug. Explain the function of the two steering diodes D_1 and D_2.
2. Construct a scale of 16 counter.
3. Using appropriate feedback, convert your scale of 16 counter to a decade counter.

EXPERIMENT 24—THE ASTABLE OR FREE RUNNING MULTIVIBRATOR

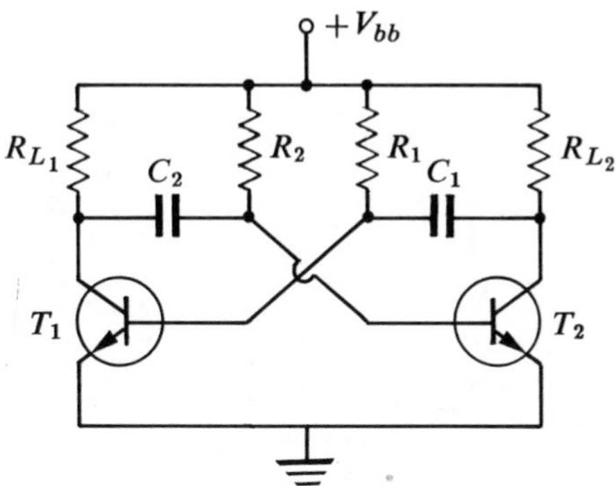

1. Construct and debug. Let $R_{L_1} = R_{L_2}$ and $R_1 = 2R_2$ with $R_{L_1} \ll R_1$. Let $C_1 = 2C_2$ and determine C_1 so that the pulse width is on the order of 1 millisec. Sketch the output waveform.
2. After establishing good outputs with the above criteria, let $R_1 = R_2$ and sketch the output waveform.
3. Let $C_1 = C_2$ and sketch the output waveform.
4. Would it be possible to make this a variable frequency oscillator? A variable pulse width oscillator? How rapidly can your circuit oscillate?
5. Redesign the circuit so that one transistor is normally "on" and the other "off". Feed in a short trigger pulse to turn the "on" transistor "off", and a square pulse will be generated at the collectors. This circuit is called a "monostable" multivibrator, because it has only one stable state.

EXPERIMENT 25—THE SCHMITT TRIGGER

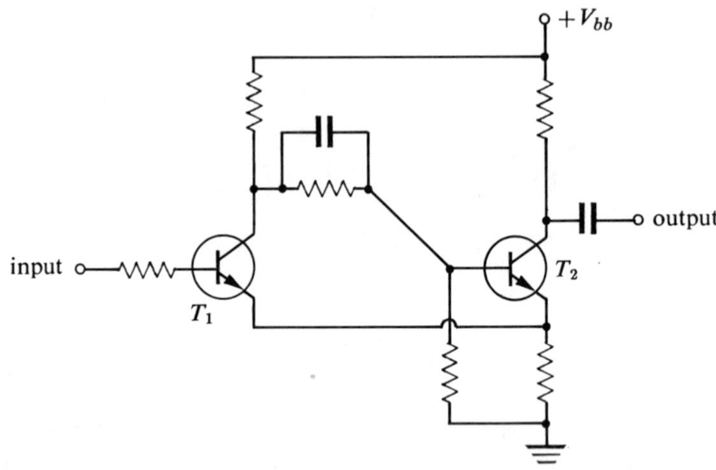

1. Construct and debug. Design so that T_1 is cutoff, T_2 conducting. Measure the dc voltages.
2. Adjust the circuit elements so that the circuit triggers on 1 V, 2 V, 3 V, 4 V, and 5 V input pulses.
3. With the circuit at the 4 V triggering configuration, put a sine wave on the input and sketch the output.

EXPERIMENT 26—INTEGRATED CIRCUIT OPERATIONAL AMPLIFIER

In this experiment, we will construct an operational amplifier and a differentiator from one of the most widely used integrated circuit operational amplifiers, the Fairchild "μA709".

Each 709 comes in a TO–99 cylindrical metal case with eight leads or in a plastic dual-in-line package (DIP) with 14 leads. The base diagrams are as follows. The μA709 is constructed on a single silicon chip and contains 14 transistors and 15 resistors which form a high gain amplifier. The basic circuit in this experiment is as follows. The R_1C_1 and R_2C_2 compensation networks are necessary to avoid oscillation at high frequencies, and their values for various amplifier gains can be read off the μA709 data sheet. Notice the higher the amplifier gain, the smaller R_1, C_1, R_2, and C_2 are.

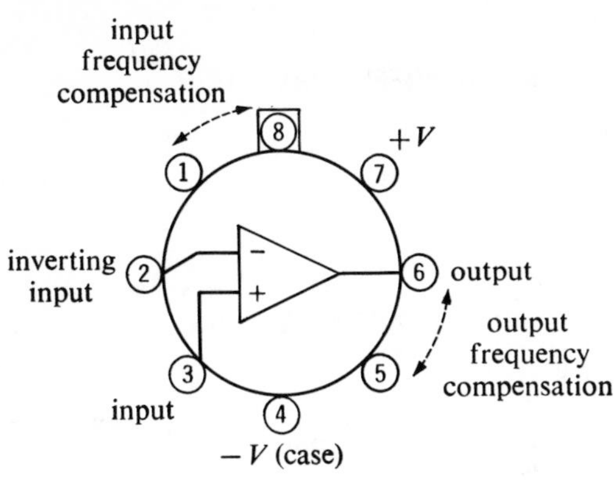

TO-99 *base diagram*

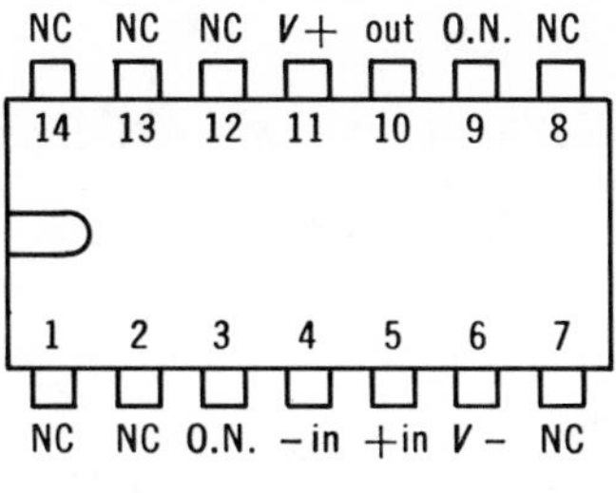

NC = no connection
−in = inverting input
O.N. = offset null
+in = non inverting input

DIP *pin diagram*

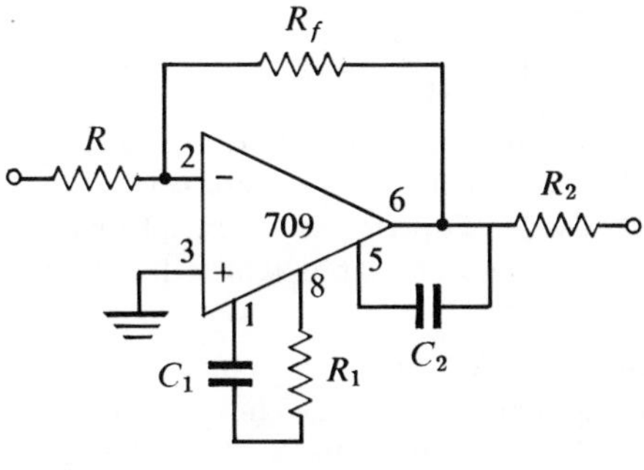

gain $A \cong R_f/R$

1. Construct the operational amplifier with a voltage gain of 10. Use $R = 1\,\text{k}\Omega$. If you have only a single power supply with floating output (neither terminal grounded) to use, you will have to use the following circuit to obtain the V^+ and V^- voltages for the μA709.

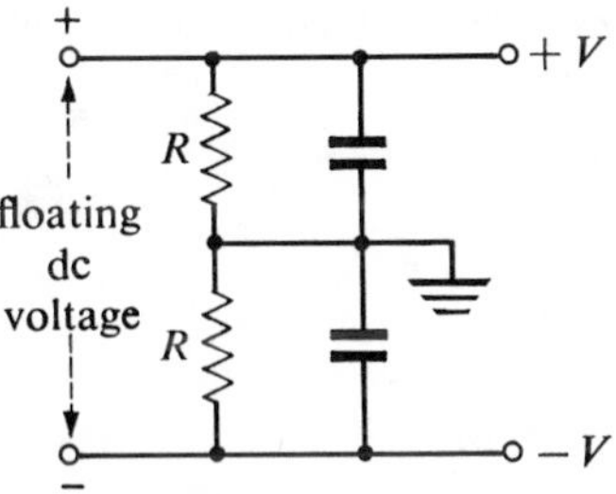

2. Measure A_v versus frequency. Graph.
3. Measure the input and output impedances.
4. Repeat for a voltage gain of 100. Use $R = 1\,\text{k}\Omega$.

EXPERIMENT 27—INTEGRATED CIRCUIT DIFFERENTIATOR

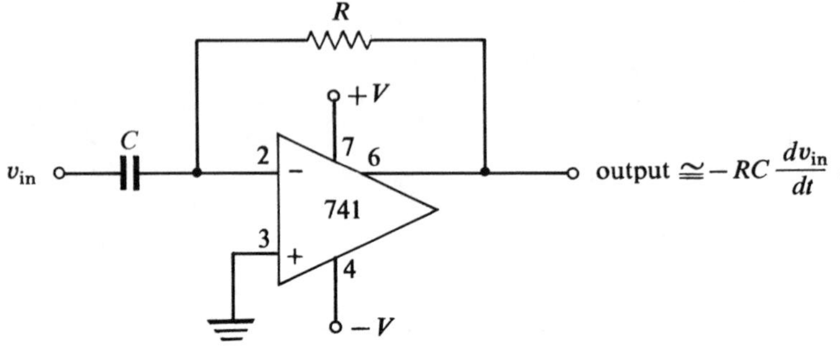

An operational amplifier can be used to differentiate an input voltage as shown in the following circuit. The integrated circuit we are using is a μA741, which is almost identical to the μA709 used in Experiment 26, except that the μA741 requires no external compensation networks. The pin diagrams for the μA709 and the μA741 are identical.

1. Construct and debug the differentiator. You may have to make a V^+, V^- supply from a single floating supply as described in Experiment 26. The value of R and C must be chosen with some care. Start with $C \cong 0.2\,\mu\text{F}$ and $R = 1\,\text{k}\Omega$. With a square wave input of approximately 0.5 V amplitude, sketch the input and output waveforms for various frequencies. Start with approximately a 250 Hz square wave.
2. With $R = 1\,\text{k}\Omega$, vary C and observe the effect on the output. Sketch what happens if C is made too large ($C > 1\,\mu\text{F}$). What happens if C is made too small ($C < 0.01\,\mu\text{F}$)? With $C \cong 0.005\,\mu\text{F}$?
3. With $C = 0.2\,\mu\text{F}$, vary R and observe the effect on the output. Sketch.
4. Add a small capacitor (100pF–1000pF) in parallel with R. Does this improve things? Why?
5. If time permits interchange R and C to make an integrator.

EXPERIMENT 28—INTEGRATED CIRCUIT NAND GATE FLIP-FLOP

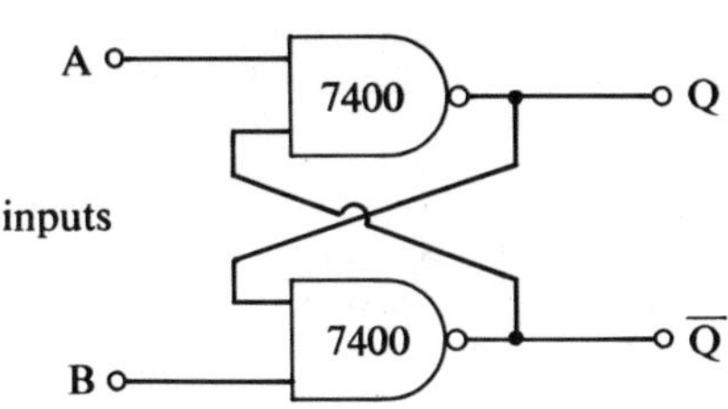

A	B	Q
0	0	X
0	1	1
1	0	0
1	1	NC

X = indeterminant
NC = no change

A flip-flop or bistable multivibrator can be constructed from two NAND gates connected as shown below because there are only two stable states for the logic levels. That is, if pulses of logic level 1 are alternately fed into the inputs A and B, the output (either Q or $\overline{Q}$) level will alternate between the logic 1 and the logic 0 levels.

The NAND gate used in this experiment is a 7400 Quad Two Input NAND gate for positive logic. The pin diagram for a dual-in-line package and a schematic for one of the four gates is shown below.

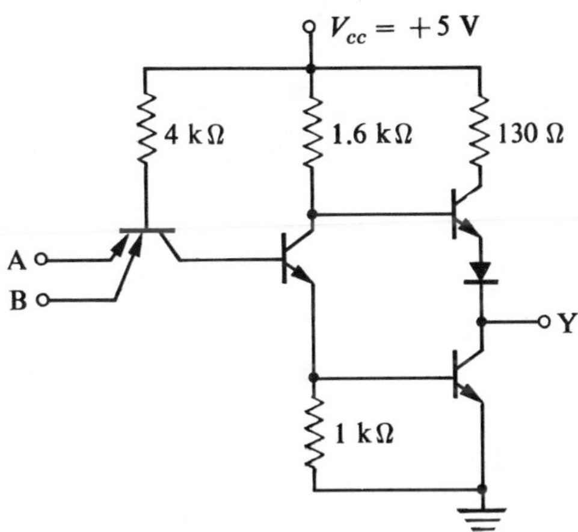

7400 *schematic*

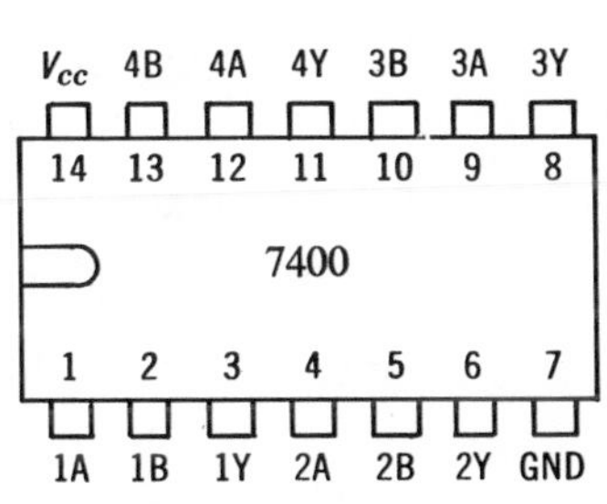

DIP *pin diagram*

	Voltage	*Current*
Logic 1 input	$\geqq 2$V	$40\,\mu$A @ 2.4V, 1 mA @ 5.5V
Logic 0 input	<0.8V	-1.6 mA
Logic 1 output	3.3V	
Logic 0 output	0.22V	

1. Hook up a flip-flop using two of the four NAND gates in the 7400 I.C. Measure the dc output voltage with a high impedance dc voltmeter or the scope. Flip the circuit and measure the output voltage. A clip lead and a 2–5 V battery can be used to provide a steady logic 1 input. Grounding an input provides a logic 0 input.
2. Feed in a series of positive pulses of 3–4 V amplitude from a pulse generator through an appropriate steering network to operate the flip-flop as a divide-by-two counter and sketch the input and output waveforms. How fast will your circuit flip?
3. Make a flip-flop using two cross coupled NOR gates. The 7402 Quad Two Input NOR gate schematic and pin diagram are given below.

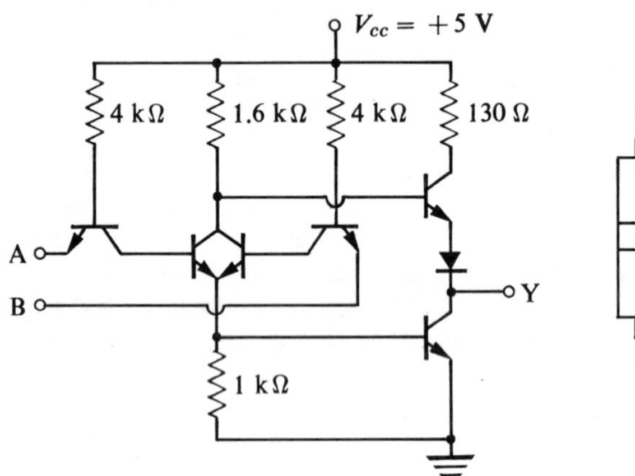

7402 *schematic*

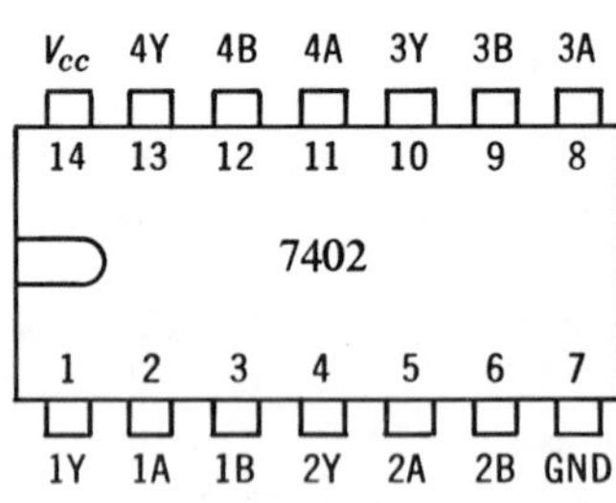

DIP *pin diagram*

EXPERIMENT 29—INTEGRATED CIRCUIT VOLTAGE FOLLOWER (ONE LAB PERIOD)

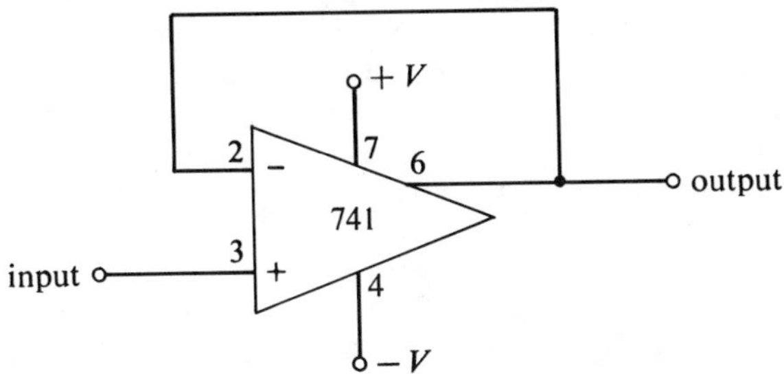

1. Connect a 741 operational amplifier as a voltage follower.
2. Measure the voltage gain for several input pulses. What is the phase of the output relative to the input?
3. Measure the input and output impedance.

EXPERIMENT 30—INTEGRATED CIRCUIT MONOSTABLE MULTIVIBRATOR

The 74121 integrated circuit monostable multivibrator can be dc triggered or pulse triggered with an output pulse whose duration can be varied from 40 μsec to 40 sec with the addition of an external timing capacitor. The dual-in-line pin diagram is given below along with a simplified block diagram.

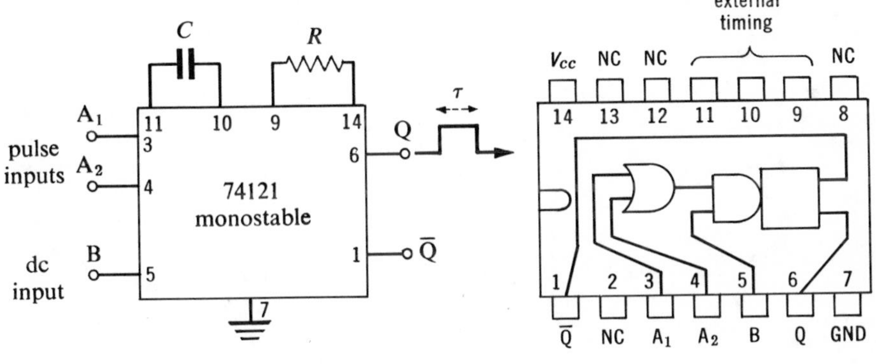

DIP *pin diagram*

(Courtesy of Texas Instruments.)

The input logic 1 level $\geqq$ 2V, input logic 0 level $\leqq$ 0.8V. The length or duration τ of the output pulse is determined by the external C and R connected to terminals 9, 10 and 11. Minimum pulse width of about 30–40nsec is obtained when pins 10 and 11 are open ($C = 0$) and pin 9 is connected to pin 14. The value of C may be from 0 to 1000μF, and R from 2kΩ to 40kΩ. The output pulse width is given by:

$$T = RC \ln 2.$$

There are two types of inputs: dc or Schmitt triggering (B, pin 5) or pulse triggering (pins 3 and 4). Input B is a positive Schmitt trigger input or level detector. The 74121 will produce an output pulse when B passes above the logic 1 voltage level of 2V provided only that either A_1 or A_2 is at logic 0 voltage. The only requirement is that the B input change less rapidly than one volt per second. In other words, the B input is used if an output pulse is desired when the dc B input rises above approximately 2V.

The pulse trigger inputs A_1 and A_2 are negative edge triggered. The 74121 will produce an output pulse when either A_1 or A_2 goes to logic 0 level provided that B is a logic 1 level. The pulse input at A_1 or A_2 should change more slowly than one volt per microsecond. In other words the A_1 or A_2 inputs are used when an output pulse is desired with its leading edge coincident with the negative going edge of the A_1 or A_2 input.

1. Hook up the 74121 and trigger with pulses at A_1 or A_2. Use R and C to obtain an output pulse width of approximately one millisecond. Sketch the input and output.
2. Hook up the 74121 with the Schmitt input at B and vary the input voltage slowly and observe the output.

EXPERIMENT 31—INTEGRATED CIRCUIT FLIP-FLOP

$$K = K_1 \cdot K_2 \cdot \overline{K^*}$$
$$J = J_1 \cdot J_2 \cdot \overline{J^*}$$

t_n		t_{n+1}
J	K	Q
0	0	Q_n
0	1	0
1	0	1
1	1	$\overline{Q}_n$

t_n = before clock pulse
t_{n+1} = after clock pulse

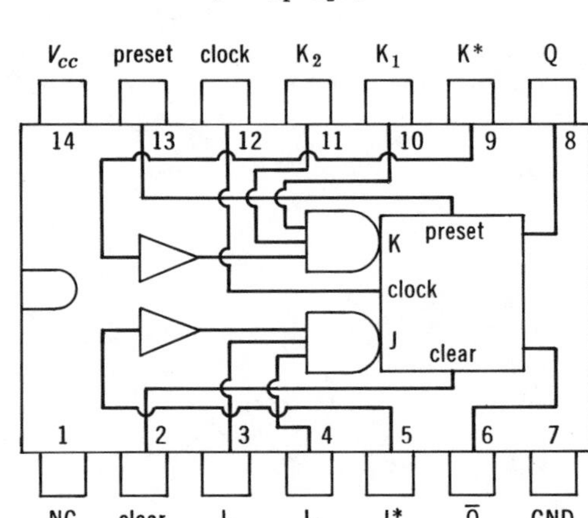

7470 DIP *pin diagram*

The 7470 integrated circuit flip-flop is a high speed edge-triggered JK flip-flop, which can flip at frequencies up to approximately 30 MHz with a variety of trigger or clock inputs. The logic block diagram and pin diagram are given above.

The preset and clear inputs work only when the clock input is low ($\leqq 0.8$ V). A low preset makes Q = 1; a low set makes Q = 0.

1. Hook up the 7470 as a simple divide-by-two counter. The input pulse train (~4 V pulse amplitude > 20 μsec width) should be fed in the clock input, and inspection of the truth table shows that both J and K must be high (at logic 1) in order for the output Q to change. $Q_{n+1} = \overline{Q}n$. Thus we desire $J_1 = J_2 = 1$ and $\overline{J}\star = 1$ which means $J\star = 0$. Similarly, we desire $K_1 = K_2 = 1$ and $\overline{K}\star = 1$ or $K\star = 0$. Thus J★ and K★ inputs should be *grounded.* Sketch the input (clock) pulse train and the output (Q) pulse train to the same time scale.
2. Devise a two-transistor driver circuit to drive two light emitting diodes so that one diode is lit when Q = 1 and the other is lit when Q = 0.
3. Connect two 7470's in series so that you have a divide-by-four counter. Sketch the input and output pulse trains to the same time scale.
4. Devise a feedback scheme to convert four 7470s in series into a decade counter, i.e., a divide-by-ten counter.

EXPERIMENT 32—INTEGRATED CIRCUIT VOLTAGE-CONTROLLED OSCILLATOR

The Signetics NE566T is an inexpensive integrated circuit voltage-controlled oscillator which can oscillate over a frequency range of 50kHz to 500kHz with a frequency stability of 100ppm/°C. The approximate frequency f of oscillation is determined by an external RC network: $f = 1/(6RC)$, and the frequency can be controlled by the dc voltage applied at pin 5. Either a square wave or a sawtooth output waveform is available on pins 3 and 4, respectively. The pin diagram is given below.

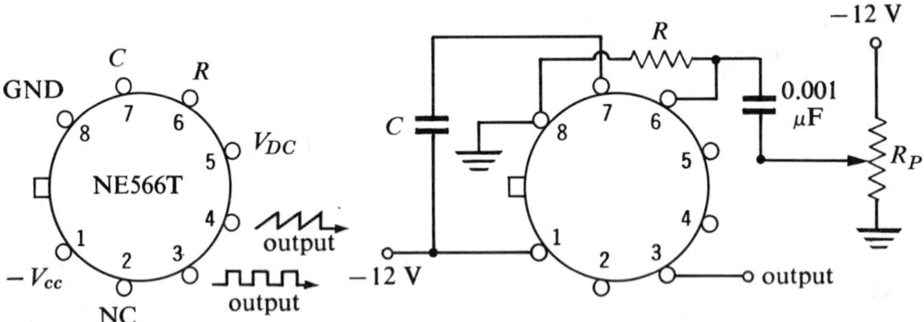

(Courtesy of Signetics Corp.)

1. Hook up the NE566T with a variable dc voltage from 0 to -12V at pin 5. Use $R = 10\text{k}\Omega$, $C = 50\text{pF}$. The oscillation frequency should then be approximately 100kHz. Sketch the output waveform at pins 3 and 4. Vary the dc voltage at point 5 by means of a potentiometer R_p (100kΩ) and measure the output frequency for several voltages.
2. The oscillator can be used to measure temperature by using a thermistor as part of the dc voltage determining network for pin 5. Using an inexpensive thermistor with a resistance of the order of 10kΩ at room temperature, measure the output frequency for various temperatures of the thermistor.

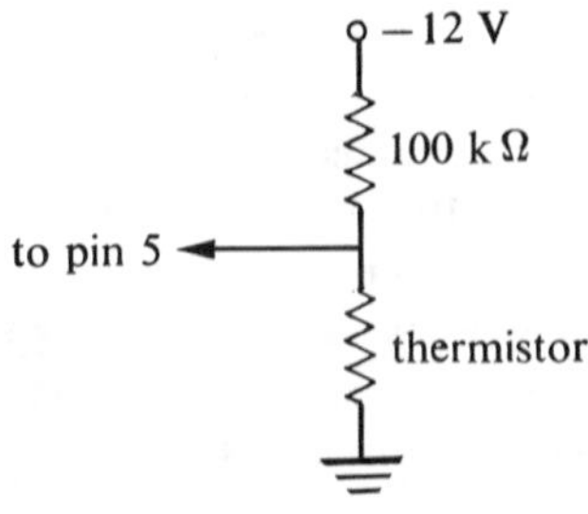

INDEX

A

Accelerated charge, radiation from, 327
Acceptor atom, 130
Alpha (α), 162, 163
Alpha cutoff frequency, 341
AM (amplitude modulation) radio, 92
Ammeter, 495
Ampere, 2
Amplification, 163
Amplification factor, 269
Amplitude, 36
Analog computer example, 304
Analyzer, pulse height, 427
AND circuit, 416
  DRL AND gate, 417
  TTL AND gate, 418
AND truth table, 410
Anode, 262
Antenna, 328
  array, 329
  half-wave, 329
  magnetic dipole, 330
Astable multivibrator, 444
  circuit, 445
  waveforms, 445
Avalanche diode, 142

B

Bandwidth, 90, 101, 339

Base, 157, 158
Base line restoration, 150, 152, 153
Battery, 11
  internal resistance, 12
  types and characteristics, 489
  testing, 490
Beat, 97
Beat frequency, 97
Beta (β), 162, 163
Beta cutoff frequency, 341
Bias, 159
Bias, cathode, 267
Binary addition, 423
Binary language, 407
Binary number system, 408
Binding energy, 260
Bistable multivibrator, 431
  circuit, 435
  triggering, 436
  waveforms, 436
Bit, 407
  carry, 424
  sum, 424
Boolean algebra, in computers, 409
Bridges, 502
  capacitance, 502
  inductance, 502
  Wheatstone, 502
Burnout proof meter, 498

C

Cable, 505
  capacitance, 201
  coaxial, 343
  equivalent circuit, 202
  low noise, 373
  types, 506
Capacitance:
  cable, 201
  definition, 41
  junction, 135, 336
  shunt, 332
  stray, 45
Capacitive reactance, 46, 47
Capacitor:
  blocking 181
  bypass, 268
  coupling, 181
  types and characteristics, 486
Capacitors, 41
  energy stored in, 46
  parallel connection, 44
  polarized, 41
  schematic symbol, 41
  series connection, 44
  types, 485
  unpolarized, 41
  variable, 43
  water pipe analogy, 43
Carrier, 92
Carry bit, 424
Cascode amplifier, 337
CAT (Computer of Average Transients), 402
Cathode:
  bias, 267, 272
  follower, 274, 275
Cathode ray oscilloscope, 499
  automatic sweep mode, 501
  external sync, 501
  sweep speed, 500
  sync, 501
  trigger, 501
Cavity, microwave, 362
Cavity coupling, microwave, 363
Characteristic impedance, 344
Charge, 1
Charge storage, in transistors, 440
Class A amplification, 356
Class B amplification, 356
Class C amplification, 356
Clear, flip-flop input, 467
Clock pulse, 465
CMOS integrated circuits, 483
Coaxial cable, 343
Color code, resistor, 5, 6
Collector, 157, 158
Colpitts oscillator, 313
Common base:
  amplifier circuit, 220

Common base *(cont.)*:
configuration, 214
current gain, 219
experiment, 544
input impedance, 219
output impedance, 219
voltage gain, 219
Common cathode configuration, 274
Common collector:
amplifier circuit, 212
configuration, 206
current gain, 208
emitter follower, 209
experiment, 543
input impedance, 208
output impedance, 208
voltage gain, 208
Common drain:
amplifier, 247
amplifier circuit, 248
configuration, 235
current gain, 249
experiment, 547
input impedance, 248
output impedance, 247
voltage gain, 248
Common emitter:
amplifier, 177–183
amplifier circuit, 206
configuration, 165, 192
current gain, 197
experiment, 542
input impedance, 198
output impedance, 200
voltage gain, 196
Common gate:
amplifier circuit, 250
configuration, 235
Common grid configuration, 274
Common plate configuration (cathode follower), 274, 275
Common source:
amplifier, 243
configuration, 235
current gain, 246
experiment, 546
input impedance, 247
output impedance, 247
Common source *(cont.)*:
power gain, 246
voltage gain, 245
Comparator, differential voltage, 471
Compensated voltage divider, 108–111
Complementary symmetry integrated circuits, 483
Complex numbers, 509
Complex voltage plane, 56
Compton effect, 428, 527
Computer, digital, 408
Conduction band, 118
Conduction in semiconductors, 126
Connector, 505
types, 506
Contact potential, 133
Corner frequency, 391
Correlation techniques, 402
Coulomb, 1
Coupling:
RC, 350
transformer, 351
Crystal lattice, 117
Crystal oscillator, 316
Current, 1
alternating (ac), 2, 35
direct, 2
sign convention, 1

D

Darlington circuit, 213
D'Arsonval meter, 492
Decibel, definition, 59
Decoupling network, 293
Delay lines, 344
Density of states, 121
Depletion region, 133
Determinants, 512
Differential amplifier:
inverting input, 457
non-inverting input, 457
Differential input, 299
Differential voltage comparator, 471
Differentiating circuit (*RC*), 101–108

Differentiator, 301, 461
Diffused base transistor, 341
Diffusion time, 340
Digital:
  circuits, 407
  computer, 408
Diode, 137
  avalanche, 142
  characteristic curves, 518
  clipping circuit, 150, 151
  equation, 142
  Gunn, 365
  Impatt, 366
  limiter, 373
  multiplier, 353
  reference, 142
  temperature stabilization, 176
  tunnel, 142-144
  varactor, 353
  zener, 9, 10
Diode, vacuum, 262
  cathode, 262
  filament, 262
  plate (anode), 262
DIP (dual-in-line) I.C. case, 454
Dipole, electric, 328
Distributed amplifiers, 349
Divider, voltage, 22
  Norton equivalent of, 28
  Thevenin equivalent of, 27
Donor atom, 128
Doping, 128
Doppler radar, 366
Down converter, 97
Drift transistor, 341
Drain, 226
DRL logic, 412

E

ECL logic, 415
Edge, flip-flop triggering, 465
Eductor, waveform, 400
Electric dipole, 328
  transitions, 349
Electromagnetic radiation, 326
Electron spin resonance, 349, 396, 397
Electron volt, 115
Emitter, 157, 158
Emitter follower, 209, 213
Energy:
  band, 118
  gap, 118
Epitaxial layer, 453
Equipartition theorem, 373
Equivalent circuit, 187
  FET, 238
  high frequency FET, 334
  pentode, 273
  transistor, 191
  triode, 269
Excess resistor noise, 378
Experiments, 531-567
EXCLUSIVE NOR circuit, 422
EXCLUSIVE OR circuit, 421

F

Fall time, 83
Farad, 41
Feedback, 281
  positive, 306-317
Fermi-Dirac statistics, 120
Fermi energy, 122, 126, 131
Fermi function, 121, 122
Field effect transistor (FET), 225
  construction, 225
  curves, 527
  high frequency equivalent circuit, 334
  pinch-off voltage, 228
  schematic symbol, 231
  temperature effects, 238
  Y parameters, 234, 239
Filament, 260
Filter:
  high-pass *RC*, 57
  low-pass *RC*, 61
Filtering power supply, 318
Firing angle, 146
Flat-pak I.C. case, 454
Flicker (1/*f*) noise, 390

Flip coil, 303
Flip-flop (*see also bistable multivibrator*), 432
  clear input, 467
  clock pulse, 465
  edge triggering, 465
  JK, 438, 467–470
  master-slave, 469
  present input, 467
  RS, 437
  RS clocked, 465
  synchronous inputs, 465
Forward:
  bias, 136
  current transfer ratio, 190
  transconductance, 236
Fourier:
  analysis, 84–97
  coefficients, 85
  of single pulse, 89
  of square wave, 87
  theorem, 85
  transform, 89
Freerunning multivibrator, 444
Frequency, 36
  compensation, 457
Full-wave:
  bridge rectifier, 150
  rectifier, 148, 149

G

Gain, 163, 164
  bandwidth product, 295, 338, 339
Gate, 226, 410
Gaussian noise distribution, 375
Geiger counter, 530
Geiger Mueller tube, 83
Giga (prefix), 2
Grid, 264
  leak bias, 276
Ground, 3
  loops, 372
  rf connections, 321
Guide wavelength, 361
Gunn diode, 365

H

Half adder, 424
Half-wave:
  antenna, 329
  rectifier, 147
Harmonic generation, 95, 353
Hartley oscillator, 315
Henry, 49
High-pass filter (*RC*), 57
  breakpoint, 59
  gain, 58
  gain vs. frequency, 59
  phase shift in, 61
h parameter:
  conversions, 218
  current gain expression, 217
  equivalent circuit, 188–191
  input impedance expression, 217
  output impedance expression, 217
  voltage gain expression, 217

I

Ideal current source, 24
Ideal voltage source 24
I_{DSS}, 241
IGFET (Insulated Gate Field Effect Transistor), 251
Impatt diode, 365, 366
Impedance, 65
  characteristic, 344
  matching, 346
  matching to obtain noise reduction, 386
  matching transformer, 387
  mismatch, 291
Induced voltage, 49, 53
Inductance, 48
  of coil, 51
  of straight wire, 331
Inductive reactance, 54
Inductors, 50
  energy stored in, 52
  parallel connection, 51

Inductors *(cont.)*:
schematic symbol, 49
series connection, 51
Insulated gate field effect transistor, 251
Integrated circuit, 451
chip, 451
hybrid, 451
kovar lead, 454
monolithic, 451
monostable multivibrator (SN74121), 471
rf amplifier (μA703), 460, 461, 462
substrate, 452
TTL flip-flop, 465
TTL inverter, 464
TTL NAND gate, 463
TTL NOR gate, 464
Integrating circuit (*RC*), 97-101
Integrator, 301
Interference, 372
Inverting input, 457
Ionization chamber, 530
Ion pair, 262
Iris, microwave, 363

J

JK flip-flop, 438, 465
decade counter, 470
divide-by-two counter, 469
SN7470, 467, 468
truth table for, 467
waveforms, 469
Johnson noise, 373
Junction, abrupt, 132
Junction capacitance, 135, 336

K

Kilo (prefix), 2
Kirchhoff's laws, 17
Klystron, 365

L

Laboratory:
bench, 532
experiments, 531
Limiter, diode, 373
Line width, in magnetic resonance, 396
Lithium drift detector, 528
Load line, 167, 267, 268
equation, 21
transistor, 21
Lobes, 330
Local oscillator, 97
Lock-in detection, 394, 395
Logic:
DRL, 412
ECL, 415
negative in computers, 409
positive in computers, 409
TTL, 412, 415
Loop currents, 19
Low noise cable, 373
Low-pass filter (*RC*), 61
break point, 62
gain, 62
gain vs. frequency, 62
phase shift in, 63

M

Magnetic dipole:
antenna, 330
transitions, 349
Magnetic saturation, 50
Majority carrier, 129
Master slave flip-flop, 469
Matching, power, 15
Maxwell's equations, 360
Mechanical oscillator, 308
Mega (prefix), 2
Memory address register, 431
Metal oxide semiconductor field effect transistor (MOSFET), 251
Micro (prefix), 2

Microphonic noise, 264, 372
Microstrip, 357, 358
Microwave:
cavities, 362
oscillators, 364
Microwaves, 325
Midband frequency, 310
Miller effect, 245, 335
Milli (prefix), 2
Minority carrier, 129
Mixer, 97
Mobility, 127
Modulation, 92
Monostable multivibrator, 432, 441, 442
integrated circuit, 471
waveforms, 443
MOSFET, 251
base, 253
bulk gate, 253
depletion type, 252
enhancement type, 253
substrate, 253
Multimeter, 497
Multiplier diode, 353
Multivibrator:
astable (free running), 444
asynchronous input, 438
bistable (flip-flop), 431
clear input, 437
clock input, 437
monostable (one shot), 441
nonsaturated, 441
preset input, 437
reset input, 437
saturated, 441
set input, 437
speed-up capacitors, 439
synchronous input, 438
toggle input, 437
Mutual inductance, 52

N

NAND circuit, 421
Nano (prefix), 2
Negative feedback, 281, 282
amplifier circuits, 290–295
input impedance, effect on, 284
output impedance, effect on, 286
voltage gain, 283
Negative logic, 409
Neutralization, 320
Noise, 371
excess resistor, 378
factor, 385
figure, 382, 383, 385
flicker or $1/f$, 390
gaussian distribution, 375
impedance matching to reduce, 386
microphonic, 372
phase sensitive detection to reduce, 394–398
pink, 391
shot, 378
signal averaging to reduce, 400
signal-to-noise ratio, 381
temperature, 393
thermal or Johnson or resistor, 373–374
transistor amplifier, 384
vacuum tube amplifier, 383
white, 377, 379, 384
Non-inverting input, 457
NOR circuit, 420
Norton's theorem, 28
NOT:
circuit, 419
truth table, 410
npn transistor, 157, 158
n-type semiconductor, 129, 131
Nyqnist diagram, 310

O

Offset voltage, 298
Ohmmeter, 495
Ohm's law, 6
Ohms-per-volt meter characteristic, 493

Open loop gain, 296
Operational amplifier:
  addition, 298
  amplifier differentiator, 461
  amplifier μA709, 457, 458, 461
  amplifier μA741, 459, 460
  amplifiers, 296
  analog computer, 304
  differentiator, 301
  integrator, 301
  offset voltage, 298
  summing point, 296
  virtual ground, 298
  voltage follower, 300
  voltage gain, 297
OR circuit, 411
  DRL OR gate, 412
  TTL, 415
  TTL OR gate 416
OR truth table, 410
Oscillation, in snowshoe hare population, 308
Oscillator, 306
  Colpitts, 313
  crystal, 316
  Hartley, 315
  mechanical, 308
  microwave, 364
  phase shift, 313
  relaxation, 323
  tuned gate–tuned drain, 320
Oscilloscope (*See also cathode ray oscilloscope*), 499

P

Pair production, 527
Parallel, addition, 423, 425
Parallel, energy storage, 74
Parallel, *RLC* resonance, 72
Passive circuit element, 4
Pauli exclusion principle, 119
Pentode:
  amplifier circuit, 273
  cathode follower circuit, 277
Pentode *(cont.)*:
  equivalent circuit, 273
  schematic symbol, 271
  screen grid, 271
  suppressor grid, 270
  vacuum, 270
  video amplifier, 276
  voltage gain, 274
Phase, 36, 37
  in *RC* circuit, 46
  shift circuits, 65, 66
  shift minimum, 64
  shift oscillator, 313
Phase sensitive detector, 394, 398
Photoelectric effect, 428, 527
Photomultiplier:
  dynode, 529
  photocathode, 529
  tube, 529
Photopeak, 430, 528
Photo resist, 453
Pico (prefix), 2
Piezoelectric effect, 316
Pinch off, 227
Pinch-off voltage, 228
Pink noise, 391
Plate, 262
  resistance, 269
p-n junction, 132
pnp transistor, 157, 158
Polarization, 326
Positive:
  feedback, 281, 306–317
  logic, 409
Potential, electric, 3
Potentiometer, 9
  linear taper, 9
  logarithmetic taper, 9
Power, 13
  ac, 39
  average, 39, 40
  matching theorem, 15
  supply, integrated circuit regulated, 477
  supply, practical laboratory, 533
  supply, short circuit protection, 480
Powers-of-ten prefixes, 2
Preset, flip-flop input, 467

Probe, oscilloscope divider, 111
Proportional counter, 530
p-type semiconductor, 130, 131
Pulse, 83
  amplitude, 84
  frequency, 84
Pulse height analyzer, 427,
  base line, 428
  channel, 428
  channel width, 428
  differential mode, 428
  integral mode, 428
  lower level discriminator (LLD), 429, 471
  memory, 431
  single channel, 471
  strobe pulse, 473
  upper level discriminator (ULD), 429, 471
  window width, 428
Pulse width, 84
Pulses, 407

Q

Q (quality factor), 75
  definition, 76
  frequency response, in terms of, 78
  of parallel *RLC* circuit, 76
  of series *RLC* circuit, 76
Quarter-wavelength line, 346

R

Radian, 36
Radiation resistance, 328
Rayleigh–Jeans law, 378
Reactance:
  capacitive, 46, 47
  inductive, 54
Recombination, 133
Rectification, 137
Reference diode, 142
Register, memory address, 431
Relaxation oscillator, 323
Repetition rate, 84
Resistance, 4
  dynamic, 5
  linear, 5
  negative, 5
  nonlinear, 5
  radiation, 328
Resistor:
  characteristics and types, 485
  noise, 374
  power rating, 14
  schematic symbol, 7
Resistors:
  parallel connection, 8
  series connection, 7
Resonant frequency, 70
Reverse bias, 134
Reverse current, 134
Reverse voltage transfer ratio, 190
Ripple, 148
Rise time, 83
RLC series circuit, 65
  complex impedance, 68
  differential equation, 67
  impedance, 71
  phase of current, 69
rms amplitude, 36
RS flip-flop, 437, 465
  Clocked, 466

S

Sag, 102
Saturation, of transistor, 440
Schmitt trigger, 445
  circuit, 429, 446
  waveforms, 447
Scintillator, 83
Screen grid, 270, 271
Secondary electrons, 271
Self:
  bias, 267, 272

Self *(cont.)*:
resonant frequency, 332
Series:
addition, 423, 426
RLC resonance, 70
Shot noise, 378
Shunt capacitance, 332
In transistors, 439
Shunt peaking, 332
Sidebands, 94
Signal averaging, 400
Signal-to-noise ratio, 381
Silicon controlled rectifier, 144
Single channel analyzer, 428
Source, 226
Source follower, 247
Space charge effect, 271
Speed-up capacitors, 439
Squaring circuit, 447
Stability factor, 173
Stagger tuning, 350
Standing wave, 345, 347
Storage time, 440
Strip line, 357, 358
Sum bit, 424
Summing point, 296
Suppressor grid, 270
Synchronous:
detector, 394, 398
flip-flop imputs, 465

T

Temperature stabilization, diode, 176
Thevenin's theorem, 26
Thermal noise, 373
Thermal runaway, 167
Thermionic emission, 260
Time constant, 105
Transconductance, 233, 236
Transformer:
coupling, 351
impedance matching, 387
Transformers, 53
schematic symbols, 54
Transformers *(cont.)*:
types, 485
Transistor:
amplifier noise, 384
characteristic curves, 519–525
construction, 157
npn, 158
pnp, 158
Transit time, 137, 342
Transmission:
impedance, 344
line, 342
mismatch, 344
quarter-wavelength, 346
Transverse:
electric waves (TE), 361
electromagnetic waves (TEM), 361
magnetic waves (TM), 361
Triode:
amplification factor, 269
amplifier circuit, 267
characteristic curves, 266
equivalent circuit, 269
grid, 264
plate, 262, 264
plate resistance, 269
schematic symbol, 264
twin, 276
Tripler circuit, varactor, 355
Truth table, 411
TTL:
gate, 416
logic, 412, 415
Tubes:
anode or plate, 262
cathode, 259, 260, 261
envelope, 261
filament, 259, 260, 261
grid, 264
vacuum, 259
Tuned:
amplifiers, 350
base–tuned collector circuit, 319
gate–tuned drain oscillator, 320
plate–tuned grid circuit, 319
Tunnel diode, 142–144
Turn-on voltage, 136
Twin triode, 276

U

Ultraviolet catastrophe, 378
Unit cell, 117

V

Vacuum tube amplifier noise, 383
Valence band, 118
Valve, 259
Varactor:
  diode, 353
  frequency multiplier, 352
  tripler circuit, 355
  tuning, 354
Volt, 3
Voltage, 3
  doubler, 149, 152
  follower, 300
Voltmeter, 492
  ohms-per-volt, 493
VOM (volt-ohm-milliammeter), 497
VSWR (voltage-standing-wave-ratio), 345, 346
VTVM meter, 499

W

Watt, 13
Waveform eductor, 400
Waveguides, 359
White noise, 377, 379, 384
Word, in computer language, 407

Y

y parameter equivalent circuit, 233

Z

Zener diode, 9, 10, 142
Zero point energy, 374